Silicon Nanomaterials Sourcebook

VOLUME TWO

Series in Materials Science and Engineering

Recent books in the series:

Advanced Thermoelectrics: Materials, Contacts, Devices, and Systems
Zhifeng Ren, Yucheng Lan, Qinyong Zhang (Eds)

Silicon Nanomaterials Sourcebook: Two-Volume Set
Klaus D. Sattler (Ed)

Silicon Nanomaterials Sourcebook: Low-Dimensional Structures, Quantum Dots, and Nanowires, Volume One
Klaus D. Sattler (Ed)

Silicon Nanomaterials Sourcebook: Hybrid Materials, Arrays, Networks, and Devices, Volume Two
Klaus D. Sattler (Ed)

Conductive Polymers: Electrical Interactions in Cell Biology and Medicine
Ze Zhang, Mahmoud Rouabhia, Simon E. Moulton (Eds)

Physical Methods for Materials Characterisation, Third Edition
Peter E J Flewitt, Robert K Wild

Multiferroic Materials: Properties, Techniques, and Applications
J Wang (Ed)

Computational Modeling of Inorganic Nanomaterials
S T Bromley, M A Zwijnenburg (Eds)

Automotive Engineering: Lightweight, Functional, and Novel Materials
B Cantor, P Grant, C Johnston

Strained-Si Heterostructure Field Effect Devices
C K Maiti, S Chattopadhyay, L K Bera

Spintronic Materials and Technology
Y B Xu, S M Thompson (Eds)

Fundamentals of Fibre Reinforced Composite Materials
A R Bunsell, J Renard

Novel Nanocrystalline Alloys and Magnetic Nanomaterials
B Cantor (Ed)

3-D Nanoelectronic Computer Architecture and Implementation
D Crawley, K Nikolic, M Forshaw (Eds)

Computer Modelling of Heat and Fluid Flow in Materials Processing
C P Hong

High-K Gate Dielectrics
M Houssa (Ed)

Metal and Ceramic Matrix Composites
B Cantor, F P E Dunne, I C Stone (Eds)

High Pressure Surface Science and Engineering
Y Gogotsi, V Domnich (Eds)

SILICON NANOMATERIALS
SOURCEBOOK VOLUME II
Hybrid Materials, Arrays, Networks, and Devices

Editor: Klaus D. Sattler

CRC Press
Taylor & Francis Group
Boca Raton London New York

CRC Press is an imprint of the
Taylor & Francis Group, an **informa** business

CRC Press
Taylor & Francis Group
6000 Broken Sound Parkway NW, Suite 300
Boca Raton, FL 33487-2742

First issued in paperback 2020

CRC Press is an imprint of Taylor & Francis Group, an Informa business
No claim to original U.S. Government works

ISBN-13: 978-1-4987-6378-3 (hbk)
ISBN-13: 978-0-367-78207-8 (pbk)

Library of Congress Cataloging-in-Publication Data

Names: Sattler, Klaus D., editor.
Title: Silicon nanomaterials sourcebook / edited by Klaus D. Sattler.
Other titles: Series in materials science and engineering.
Description: Boca Raton, FL : CRC Press, Taylor & Francis Group, [2017] |
Series: Series in materials science and engineering | Includes
bibliographical references and index. Contents: volume 1. Low-dimensional
structures, quantum dots, and nanowires
Identifiers: LCCN 2016059471| ISBN 9781498763776 (v. 1 ; hardback ; alk.
paper) | ISBN 1498763774 (v. 1 ; hardback ; alk. paper)
Subjects: LCSH: Nanosilicon. | Nanostructured materials.
Classification: LCC TA418.9.N35 S5556 2017 | DDC 620.1/15--dc23
LC record available at https://lccn.loc.gov/2016059471

Visit the Taylor & Francis Web site at
http://www.taylorandfrancis.com

and the CRC Press Web site at
http://www.crcpress.com

Contents

Series Preface

This international series covers all aspects of theoretical and applied optics and optoelectronics. Active since 1986, eminent authors have long been choosing to publish with this series, and it is now established as a premier forum for high-impact monographs and textbooks. The editors are proud of the breadth and depth showcased by published works, with levels ranging from advanced undergraduate and graduate student texts to professional references. Topics addressed are both cutting edge and fundamental, basic science and applications-oriented, on subject matter that includes: lasers, photonic devices, nonlinear optics, interferometry, waves, crystals, optical materials, biomedical optics, optical tweezers, optical metrology, solid-state lighting, nanophotonics, and silicon photonics. Readers of the series are students, scientists, and engineers working in optics, optoelectronics, and related fields in the industry.

Proposals for new volumes in the series may be directed to Lu Han, senior publishing editor at CRC Press, Taylor & Francis Group (lu.han@taylorandfrancis.com).

Preface

Silicon is one of the most technologically important materials today owing to its omnipresent significance in microelectronics. Its nanoscale forms, such as nanocrystals, porous silicon, quantum wells, or nanowires, have stimulated great interest among scientists because of their special physical properties, such as light emission, field emission, and quantum confinement effects. The progress made in the synthesis of silicon nanostructures in recent years has attracted considerable attention. Today, large quantities of silicon nanomaterials can be produced, and they are investigated with the most advanced analytical instruments available.

While silicon is the essential semiconductor material for modern microelectronic devices, this sourcebook shows a much wider range of applications, which are possible for silicon on the nanometer scale. Mostly inspired by the discovery of new carbon allotropes, methods have been developed in the last two decades for silicon to obtain similar low-dimensional and nanoscale morphologies and structures. When the size of silicon is reduced to the 1–100 nm range, quantum confinement can significantly affect the properties and performance of the material. Another inspiration came from the discovery of visible light emission from porous silicon, which inspired many scientists to start research on nanoscale silicon. Electronic and photonic studies of these materials have revealed peculiar effects and have subsequently been extended toward biomedicine with applications in tissue engineering, drug-delivery, biosensing, radiotherapy, and sonodynamic therapy. This is possible because of the good biocompatibility and biodegradability of nanoscale silicon and tunable surface derivatization. Fabrication of sub-wavelength nanostructures has allowed for the development of antireflection materials as well as other photon management structures such as materials with light-trapping properties, with applications in optoelectronic devices, photodetectors, and phototransistors.

Silicon Nanomaterials Sourcebook provides an introduction to synthetic methods used for the production of various silicon nanoscale morphologies and structures. Among these methods are solution synthesis and microwave-assisted synthesis, pulsed laser ablation, electrodeposition and plasma synthesis, metal-assisted chemical edging, interface functionalization, and nanoscale interface manipulations.

Volume One of the sourcebook covers low-dimensional silicon nanostructures such as nanosheets, clusters, nanoparticles, nanocrystals, nanowires, and nanotubes. Structural, electronic, and photonic properties of these materials may differ significantly from the silicon bulk properties.

Volume Two focuses on functional and industrial nanosilicon, describing materials such as nanowire, nanopencil and nanopore arrays, core–shell nanostructures, or porous silicon templates. These nanostructures have interesting antireflection, photonic, and thermoelectric properties. They have a wide range of applications as sonosensors and solar cells, for Li-ion batteries and energy storage, and in biomedicine, solar energy conversion, chemical and biological sensing techniques, DNA sequencing, and quantum information.

The sourcebook comprehensively covers the many aspects of silicon nanomaterials. It reflects the interdisciplinary nature of this field bringing together physics, chemistry, materials science, molecular biology, engineering, and medicine. Its contents include growth mechanisms and fundamental properties as well as electronic device, energy storage, and biomedical and environmental applications. It is a unique reference for industrial professionals and university students, offering deep insight into a wide range of areas from science to engineering. While addressing the current knowledge and the latest advances, it also includes basic mathematical equations, tables, and graphs. This provides the reader with the tools necessary to understand the current status of the field as well as future technology development of nanoscale silicon materials and structures.

Editor

Klaus D. Sattler pursued his undergraduate and master's courses at the University of Karlsruhe in Germany. He received his PhD under the guidance of Professors G. Busch and H.C. Siegmann at the Swiss Federal Institute of Technology (ETH) in Zurich, where he was among the first to study spin-polarized photoelectron emission. In 1976, he began a group for atomic cluster research at the University of Konstanz in Germany, where he built the first source for atomic clusters and led his team to pioneering discoveries such as "magic numbers" and "Coulomb explosion." He was at the University of California, Berkeley, for 3 years as a Heisenberg fellow, where he initiated the first studies of atomic clusters on surfaces with a scanning tunneling microscope.

Dr. Sattler accepted a position as professor of physics at the University of Hawaii, Honolulu, in 1988. There, he initiated a research group for nanophysics, which, using scanning probe microscopy, obtained the first atomic-scale images of carbon nanotubes directly confirming the graphene network. In 1994, his group produced the first carbon nanocones. He has also studied the formation of polycyclic aromatic hydrocarbons and nanoparticles in hydrocarbon flames in collaboration with ETH Zurich. Other research has involved the nanopatterning of nanoparticle films, charge density waves on rotated graphene sheets, band gap studies of quantum dots, and graphene folds. His current work focuses on novel nanomaterials and solar photocatalysis with nanoparticles for the purification of water.

He is the editor of the sister reference, *Carbon Nanomaterials Sourcebook* (CRC Press, 2016), *Fundamentals of Picoscience* (CRC Press, 2014), and the seven-volume *Handbook of Nanophysics* (CRC Press, 2011). Among his many other accomplishments, Dr. Sattler was awarded the prestigious Walter Schottky Prize from the German Physical Society in 1983. At the University of Hawaii, he teaches courses in general physics, solid state physics, and quantum mechanics.

Contributors

Olena I. Aksimentyeva
Faculty of Chemistry
Ivan Franko National University of Lviv
Lviv, Ukraine

Simona Boninelli
MATIS IMM-CNR
Catania, Italy

Maher Boulos
Tekna Plasma Systems Inc.
Sherbrooke, Canada

Boyuan Cai
Nanophotonic Research Center
Key Laboratory of Optoelectronic Devices
    and Systems of Ministry of Education and
    Guangdong Province
College of Optoelectronic Engineering
Shenzhen University
Shenzhen, People's Republic of China

Hyung Hee Cho
Department of Mechanical Engineering
Yonsei University
Seoul, South Korea

Chia-Yun Chou
Department of Chemical Engineering
The University of Texas at Austin
Austin, Texas

María L. Dell'Arciprete
Instituto de Investigaciones Fisicoquímicas Teóricas
    y Aplicadas (INIFTA)
CONICET - Faculty of Science
University of La Plata
La Plata, Argentina

Elisabetta Dimaggio
Dipartimento di Ingegneria dell' Informazione
Università di Pisa
Pisa, Italy

Richard Dolbec
Tekna Plasma Systems Inc.
Sherbrooke, Canada

Evelina P. Domashevskaya
Faculty of Physics
Voronezh State University
Voronezh, Russia

Aleksandra I. Efimova
Physics Department
M. V. Lomonosov Moscow State University
Moscow, Russia

Frank Endres
Clausthal University of Technology
Clausthal-Zellerfeld, Germany

Xing Fang
School of Mechanical Engineering
Shanghai Jiao Tong University
Shanghai, People's Republic of China

Barbara Fazio
IPCF-CNR
Messina, Italy

Giorgia Franzò
MATIS IMM-CNR
Catania, Italy

Bora Garipcan
Institute of Biomedical Engineering
Bogazici University
Istanbul, Turkey

Leonid A. Golovan
Physics Department
M. V. Lomonosov Moscow State University
Moscow, Russia

Kirill A. Gonchar
Physics Department
M. V. Lomonosov Moscow State University
Moscow, Russia

Mónica C. Gonzalez
Instituto de Investigaciones Fisicoquímicas Teóricas
    y Aplicadas (INIFTA)
CONICET - Faculty of Science
University of La Plata
La Plata, Argentina

Roxana M. Gorojod
CONICET - Universidad de Buenos Aires
Instituto de Química Biológica Ciencias Exactas y
    Naturales (IQUIBICEN)
Facultad de Ciencias Exactas y Naturales
Departamento de Química Biológica
Laboratorio de Disfunción Celular en
    Enfermedades Neurodegenerativas y
    Nanomedicina
Buenos Aires, Argentina

Petra Granitzer
Institute of Physics
University of Graz
Graz, Austria

Abdelbast Guerfi
Hydro-Quebec Research Center
Varennes, Canada.

Jiayin Guo
Tekna Plasma Systems Inc.
Sherbrooke, Canada

Sayak Dutta Gupta
School of Materials Science and Engineering
Indian Institute of Engineering Science and Technology
Shibpur, India

Atrayee Hazra
School of Materials Science and Engineering
Indian Institute of Engineering Science and Technology
Shibpur, India

Johnny C. Ho
Department of Physics and Materials Science
State Key Laboratory of Millimeter Waves and
    Centre for Functional Photonics (CFP)
City University of Hong Kong
Hong Kong, People's Republic of China

Pierre Hovington
Hydro-Quebec Research Center
Varennes, Canada.

Gyeong S. Hwang
Department of Chemical Engineering
The University of Texas at Austin
Austin, Texas

Fabio Iacona
MATIS IMM-CNR
Catania, Italy

Alessia Irrera
IPCF-CNR
Messina, Italy

Baohua Jia
Centre for Micro-Photonics
Faculty of Science Engineering and Technology
Swinburne University of Technology
Melbourne, Australia

Anna S. Kalyuzhnaya
Department of Chemistry
M. V. Lomonosov Moscow State University
Moscow, Russia

Dong-Hee Kang
School of Materials Science and Engineering
Gwangju Institute of Science and Technology
Gwangju, South Korea

Eun K. Kang
School of Electrical Engineering and
    Computer Science
Gwangju Institute of Science and Technology
Gwangju, South Korea

Beom Seok Kim
IFW Dresden
Dresden, Germany

Seong-Min Kim
School of Materials Science and Engineering
Gwangju Institute of Science and Technology
Gwangju, South Korea

Mónica L. Kotler
CONICET - Universidad de Buenos Aires
Instituto de Química Biológica Ciencias Exactas y
    Naturales (IQUIBICEN)
Facultad de Ciencias Exactas y Naturales
Departamento de Química Biológica
Laboratorio de Disfunción Celular en
    Enfermedades Neurodegenerativas y
    Nanomedicina
Buenos Aires, Argentina

Ken Kurosaki
Graduate School of Engineering
Osaka University
Osaka, Japan

and

JST, PRESTO
Saitama, Japan

Dominic Leblanc
Hydro-Quebec Research Center
Varennes, Canada

Seyeong Lee
School of Materials Science and Engineering
Gwangju Institute of Science and Technology
Gwangju, South Korea

Yong T. Lee
School of Electrical Engineering and Computer
    Science
Gwangju Institute of Science and Technology
Gwangju, South Korea

Wendong Liu
State Key Laboratory of Supramolecular Structure
    and Materials
International Joint Research Laboratory of Nano-
    Micro Architecture Chemistry
College of Chemistry
Jilin University
Changchun, People's Republic of China

Xianglei Liu
School of Energy and Power Engineering
Nanjing University of Aeronautics and Astronautics
Nanjing, People's Republic of China

Xueyao Liu
State Key Laboratory of Supramolecular Structure
    and Materials
International Joint Research Laboratory of Nano-
    Micro Architecture Chemistry
College of Chemistry
Jilin University
Changchun, People's Republic of China

Zewen Liu
Institute of Microelectronics
Tsinghua University
Beijing, People's Republic of China

Salvatore Mirabella
MATIS IMM-CNR
Catania, Italy

and

Dipartimento di Fisica e Astronomia
Università di Catania
Catania, Italy

Maria Miritello
MATIS IMM-CNR
Catania, Italy

Liubomyr S. Monastyrskii
Faculty of Electronics and Computer Technologies
Ivan Franko National University of Lviv
Lviv, Ukraine

Dario Narducci
Department of Materials Science
University of Milano Bicocca
Milan, Italy

Igor B. Olenych
Faculty of Electronics and Computer Technologies
Ivan Franko National University of Lviv
Lviv, Ukraine

Liubov A. Osminkina
Department of Physics
M. V. Lomonosov Moscow State University
Moscow, Russia

and

National Research Nuclear University "MEPhI"
Moscow, Russia

Alp Özgün
Institute of Biomedical Engineering
Bogazici University
Istanbul, Turkey

Zingway Pei
Graduate Institute of Optoelectronic Engineering
Department of Electrical Engineering
National Chung Hsing University
Taichung, Taiwan

Giovanni Pennelli
Dipartimento di Ingegneria dell' Informazione
Università di Pisa
Pisa, Italy

Enrico Prati
Istituto di Fotonica e Nanotecnologie
Consiglio Nazionale delle Ricerche
Milano, Italy

Francesco Priolo
MATIS IMM-CNR
Catania, Italy
and
Dipartimento di Fisica e Astronomia
Università di Catania
Catania, Italy
and
Scuola Superiore di Catania
Università di Catania
Catania, Italy

Oliver Puffky
Institute of Applied Physics
Abbe Center of Photonics
Friedrich Schiller University Jena
Jena, Germany

Mallar Ray
School of Materials Science and Engineering
Indian Institute of Engineering Science and
    Technology
Shibpur, India

Davide Rotta
INPHOTEC - TeCIP Institute
Scuola Superiore Sant'Anna
Pisa, Italy
and
Laboratorio Nazionale di Reti Fotoniche
Consorzio Nazionale Interuniversitario per le
    Telecomunicazioni
Pisa, Italy

Klemens Rumpf
Institute of Physics
University of Graz
Graz, Austria

Seiji Samukawa
Advanced Institute for Materials Research and
    Institute of Fluid Science,
Tohoku University
Sendai, Japan

Xuefeng Song
State Key Lab of Metal Matrix Composites
School of Materials Science and Engineering
Shanghai Jiao Tong University
Shanghai, People's Republic of China

Young M. Song
School of Electrical Engineering and Computer
    Science
Gwangju Institute of Science and Technology
Gwangju, South Korea

Martin Steglich
Institute of Applied Physics
Abbe Center of Photonics
Friedrich Schiller University Jena
Jena, Germany

Zhuang Sun
State Key Lab of Metal Matrix Composites
School of Materials Science and Engineering
Shanghai Jiao Tong University
Shanghai, People's Republic of China

Andrew P. Sviridov
Department of Physics
M. V. Lomonosov Moscow State University
Moscow, Russia

Vladimir A. Terekhov
Faculty of Physics
Voronezh State University
Voronezh, Russia

Antonio Terrasi
MATIS IMM-CNR
Catania, Italy
and
Dipartimento di Fisica e Astronomia
Università di Catania
Catania, Italy

Subramani Thiyagu
Nanostructured Semiconducting Materials Group
International Center for Materials
    Nanoarchitectonics
National Institute for Materials Science
Tsukuba, Ibaraki, Japan

Victor Y. Timoshenko
Physics Department
M. V. Lomonosov Moscow State University
Moscow, Russia

Sergey Y. Turishchev
Faculty of Physics
Voronezh State University
Voronezh, Russia

Yifan Wang
Institute of Microelectronics
Tsinghua University
Beijing, People's Republic of China

Lisong Xiao
Institute for Combustion and Gas Dynamics –
    Reactive Fluids (IVG)
University of Duisburg-Essen
Duisburg, Germany

Fei Xiu
Key Laboratory of Flexible Electronics (KLOFE) and
Institute of Advanced Materials (IAM)
Jiangsu National Synergetic Innovation Center for
    Advanced Materials (SICAM)
Nanjing Tech University
Nanjing, People's Republic of China

Bai Yang
State Key Laboratory of Supramolecular Structure
    and Materials
International Joint Research Laboratory of Nano-
    Micro Architecture Chemistry
College of Chemistry
Jilin University
Changchun, People's Republic of China

Cheng Yang
State Key Lab of Metal Matrix Composites
School of Materials Science and Engineering
Shanghai Jiao Tong University
Shanghai, People's Republic of China

Young J. Yoo
School of Electrical Engineering and Computer
    Science
Gwangju Institute of Science and Technology
Gwangju, South Korea

Myung-Han Yoon
School of Materials Science and Engineering
Gwangju Institute of Science and Technology
Gwangju, South Korea

Aikebaier Yusufu
Graduate School of Engineering
Osaka University
Osaka, Japan
and
Research Institute of Nuclear Engineering
University of Fukui
Fukui, Japan

Karim Zaghib
Hydro-Quebec Research Center
Varennes, Canada

Zhuomin M. Zhang
George W. Woodruff School of Mechanical
    Engineering
Georgia Institute of Technology
Atlanta, Georgia

Changying Zhao
School of Mechanical Engineering
Shanghai Jiao Tong University
Shanghai, People's Republic of China

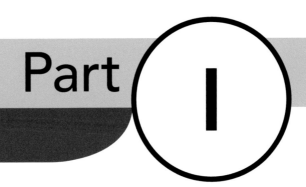

Part I

Arrays, hybrids, and core–shell

1 Formation and optical properties of silicon nanowire arrays

Anna S. Kalyuzhnaya, Aleksandra I. Efimova, Leonid A. Golovan, Kirill A. Gonchar, and Victor Y. Timoshenko

Contents

1.1 INTRODUCTION

The history of silicon nanowire (SiNW) arrays started almost 60 years ago when Treuting and Arnold (1957) reported on the formation of "silicon whiskers" in the vapor deposition process. Further progress was achieved in 1964 when Wagner and Ellis employed the vapor–liquid–solid (VLS) technique to produce silicon nanowires of about 100 nm in diameter. However, the SiNW arrays have not become merely a brilliant manifestation of technological abilities but have undergone further developments. New approaches for their formation have been developed and are widely used. Nowadays, SiNW arrays attract more and more researchers' interest in various fields of study because of the possible practical applications of the SiNW-based devices. Indeed, silicon is one of the fundamental materials of modern microelectronics, and silicon technologies are well developed, which makes integration of the SiNW arrays with the present electronics easy. It is worth noting the broad spectrum of possible practical applications. For example, SiNW arrays allow the control of wetting from superhydrophobicity to complete wetting of the nanostructure by the applied voltage (Krupenkin et al. 2004). They look to be promising as anodes in batteries, in particular due to incorporation of huge amount of lithium into silicon (Chan et al. 2008; Baek et al. 2016), low-threshold field emitters (Huang et al. 2007; Kumar et al. 2014). In comparison with crystalline silicon (c-Si), SiNW arrays demonstrate lower thermal conductivity and practically the same electric conductivity and Seebeck coefficient, which makes them good thermal-isolating and thermoelectric materials (Boukai et al. 2008; Hochbaum et al. 2008; Calaza et al. 2015). Single SiNWs seem to be very promising for their employment in sensorics, for example, as a basis for molecular or, with additional functionalization, biological sensitive resistors or field-effect transistors (Cao et al. 2014; Williams et al. 2014). They use the generation of the carriers and, as a result, variation of the conductivity in the SiNWs covered with molecules (e.g., gas or lipids). Employment of the SiNW arrays allows researchers to consider more sophisticated sensors. For example, an array of the vertical SiNWs with palladium tops between two electrodes can serve as a hydrogen detector because of palladium expansion in hydrogen atmosphere resulting in their connection and a current flowing through the device. Low gas ionization voltage in SiNW arrays could be useful for gas detection (Sadeghian and Islam 2011). Adsorption of molecules at SiNWs will result in variation of the mass and, as a result, change in mechanical resonance frequency.

However, special interest should be paid to optical properties and optical applications of the SiNW arrays. First, single SiNWs are very promising for photonics (Brönstrup et al. 2010), for example, as waveguides (Khorasaninejad and Saini 2010) and nonlinear optical devices (Singh et al. 2015; Kuyken et al. 2016). SiNW arrays could enrich their optical properties. Indeed, in the array of the nanowires of high refractive index and 100 nm in diameter, strong scattering is inevitable. In turn, the strong scattering causes significant light trapping, which results in an increase of efficiencies of both linear and nonlinear optical interactions. For example, total reflection (i.e., sum of specular and diffuse ones) of SiNW arrays exhibits enhancement in the transparency spectral region and huge (above 95%) absorption of visible light. The latter effect, which gave the SiNW arrays the name of *black silicon* (Liu et al. 2014), promises wide applications of the formed structures in sensing and photovoltaics as solar cells or emitting elements. The light trapping could be also responsible for the efficient nonlinear optical interactions in SiNW arrays.

Formally, SiNW arrays could be considered as macroporous silicon (according to IUPAC classification; see Rouquerol et al. 1994). In contrast to micro- or mesoporous silicon (with pores less than 2 nm and 50 nm, respectively), optical properties of SiNW arrays in visible and near-infrared (IR) regions can hardly be described in terms of so-called effective refractive index (Theiß 1997), and new approaches should be developed. Besides, the SiNW arrays are in close connection with the highly discussed metasurfaces (Yu and Capasso 2014; Glybovsky et al. 2016).

Thus, it would be very instructive to discuss structural properties of the SiNW arrays formed by different techniques and the influence on their optical properties, including light absorption and reflection, Raman scattering, photoluminescence, and nonlinear optical effects with respect to their possible applications, which seem to be very promising.

1.2 FORMATION AND STRUCTURE OF SILICON NANOWIRE ARRAYS

The SiNW arrays can be produced by means of various techniques, with their structural properties being determined by the way they were formed. Next, we review several of the most popular techniques.

1.2.1 VAPOR–SOLID–LIQUID TECHNIQUE

The oldest but still very popular technique of the SiNW formation is their growth with the help of VLS method firstly suggested by Wagner and Ellis in 1964 (Figure 1.1). The ingenious idea of the VLS technique is growing the silicon nanowires from silicon vapor at a substrate in the presence of metal (often gold) particles. Si vapor is formed either in the reactions of either hydrogen $SiCl_4$ reduction or in decomposition of SiI_2 or SiH_4 into simple substances. Crystalline or amorphous silicon as well as SiO_2 or Si_4N_3 wafers can be employed as the substrate.

Apart from the chemical vapor deposition (CVD) mechanism of the VLS nanowire growth described earlier, the silicon atoms participating in the VLS process can be obtained in the electron beam evaporation (EBE) process (Becker et al. 2008). Another variation of the VLS technique is synthesis of the SiNWs with the help of laser ablation of a target consisting of silicon and a catalyst, for example, iron. The ablation products form liquid nanodroplets during their cooling in inert gas. Deposition of the droplets on the substrate results in growth of the SiNWs according to the VLS mechanism (Morales and Lieber 1998).

The pivotal role of metal particles is determined by formation of a liquid Si–metal alloy at relatively low temperature. A silicon vapor excess oversaturates the gold–silicon alloy (eutectic) with Si, and the latter continuously freezes out on the liquid–solid interface (droplet–solid silicon), thus forming nanowires.

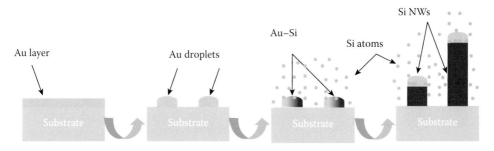

Figure 1.1 Schematic drawing of the VLS process as it takes place in a plasma or chemical vapor deposition experiment. The first two steps yield the formation of gold droplets by heating a continuous gold layer on a silicon substrate; the third step shows Si atoms to reach the substrate and to be incorporated into the liquid Au–Si droplet above the eutectic temperature (Au–Si: 373°C); and in the last step, supersaturation of the Au–Si droplet with Si leads to the growth of the SiNW at a higher growth velocity than the continuous silicon layer in between the droplets takes. (Reprinted from Sivakov, V., et al., *Nanowires – Fundamental Research*, InTech, Rijeka, Croatia, pp. 45–80, 2011. With permission.)

In this way, the Si "pillars" covered with a metal "cap" are formed. The nanowire grows when silicon vapor and the metal, often called the catalyst, are present.

The majority of the publications concerning the VLS growth of the SiNWs are focused on gold-catalyzed processes, but other metals (Ag, Al, Cd, Co, Cu, Dy, Fe, Ga, In, Mg, Mn, Ni, Os) can also be used as catalysts in this technique. The VLS growth of the SiNWs with the help of different metals and their influence on the electronic properties of the nanoparticles were systematized by Schmidt et al. (2009). The metals used as catalysts in the VLS method are divided into three types (A, B, and C) depending on the features of the metal–silicon phase diagram (Figure 1.2). The type A catalysts, for example, gold (Wagner and Ellis 1965) and silver (Nebol'sin et al. 2005), are characterized by a simple phase diagram with the only eutectic point and without compound phases—silicides, with the eutectic point being located at silicon concentration above 10 atomic % (Figure 1.2a). Although the Al–Si phase diagram is similar to the Au–Si one, the synthesis of the nanowires was reported at the temperatures higher (Osada et al. 1979) and lower (Whang et al. 2007) than the eutectic point. Whereas the former case corresponds to the VLS mechanism, in the latter one the vapor–solid–solid (VSS) mechanism was suggested. The VSS mechanism is based on the existence of a Si–Al solid solution with a very small amount of Si (about 1% at 500°C) (pocket at the left side in the phase diagram, Figure 1.2b). To the right of the boundary marked as VSS, two phases (the Si–Al solid solution and pure silicon) coexist. Thus, excess of Si due to the atoms from vapor results in precipitation of silicon in the form of wires (Wang et al. 2006).

Arrays, hybrids, and core-shell

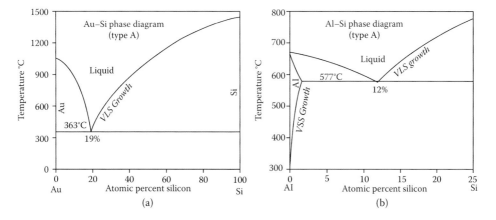

Figure 1.2 Metal–Si phase diagrams: (a) Au–Si, (b) Al–Si.

(Continued)

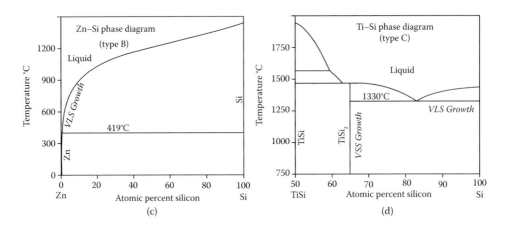

Figure 1.2 (Continued) Metal–Si phase diagrams: (c) Zn–Si, (d) Ti–Si. (Reprinted from Schmidt, V., et al.: Silicon nanowires: a review on aspects of their growth and their electrical properties. *Adv. Mater.*, 2009. 21. 2681–2702. Copyright Wiley-VCH Verlag GmbH & Co. KGaA. With permission.)

The type B of the silicon–catalyst phase diagrams also contains an eutectic point, but, in contrast to the first type, it lies at very low silicon concentrations (less than 1%); it is realized for Zn, Cd, Ga, and In. The phase diagrams of the type C catalysts (e.g., Ti, Fe, Dy, Pd, Pt, Cu, Ni) contain silicide phases and the eutectic temperatures are high, so the VLS growth requires a temperature higher than 800°C. In some cases, at lower temperatures, the growth via the VVS mechanism takes place, which, however, results in much worse crystallinity of the SiNW. Thus, for the B and C types of catalysts, one should meet more stringent requirements than for the type A catalysts (Ag, Au, Al); this is why the latter ones are used more often.

The advantages of gold as a catalyst material are as follows: low eutectic point (373°C); wide two-phase region of temperatures and compositions (19%–100%) between the eutectic line and the liquidus, in which Si and the liquid containing Si and gold coexist and silicon can be precipitated; and the absence of compounds in the Si–Au system. It is very important that gold is intoxic and stable in the air at high temperatures (Schmidt et al. 2009). However, the gold admixture negatively influences the electronic properties of the nanowires (see later discussion); this is one of the reasons for searching other catalysts or other SiNW formation techniques.

The formed nanowires are crystalline. The growth directions are reported to be (111) (Wagner and Ellis 1964), (110), and (112) (Wu et al. 2004). The first direction is explained by the fact that a single solid–liquid interface parallel to (111) plane has the lowest free energy, whereas the growth in (110) direction is characterized by lower energy of termination at the solid–vacuum interface. It is worth noting that the latter case takes place preferably for the thinnest SiNWs. In this case, the catalyst–SiNW interface has a V-shaped form (Figure 1.3; Wu et al. 2004). Simulation indicates a possibility of the growth direction control by means of change of neutral gas with plasma in the VLS process, which results in (110) oriented SiNWs (Mehdipour and Ostrikov 2013).

The SiNW arrays produced by means of the VLS method can be described as separated nanowires whose rare location corresponds to the location of metal catalyst on the silicon wafer. The diameter of the nanowire is determined by the size of the metal particles and ranges from hundreds of micrometers to tens of nanometers (Schmidt et al. 2009). For example, gold is deposited on the silicon wafer using the ion beam method. To achieve a uniform Au-cluster diameter distribution thermal annealing, plasma treatment, or the combination of both is used (Weber et al. 2006). It is worth noting that both the diameter of the SiNWs and the forms of the gold tops depend on the formation process. In particular, SiNWs formed with the help of EBE have semispherical gold top and larger diameter, whereas the ones formed with the help of CVD are characterized by smaller diameter and almost spherical tops (Figure 1.4) (Becker et al. 2008).

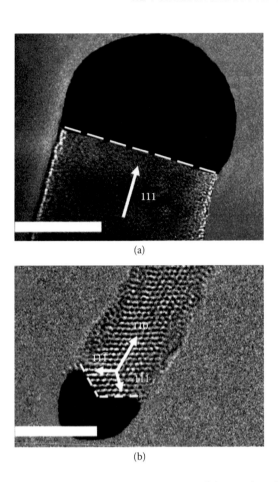

(a)

(b)

Figure 1.3 High-resolution transmission electron microscopy images of the catalyst alloy/SiNW interface: (a) for ⟨111⟩ growth direction (scale bar 20 nm), (b) for ⟨110⟩ growth direction (scale bar 5 nm). (Reprinted with permission from Wu, Y., et al., Controlled growth and structures of molecular-scale silicon nanowires, *Nano Lett.*, 4, 433–436, 2004. Copyright 2004 American Chemical Society.)

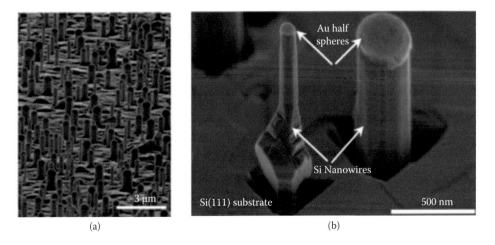

(a)   (b)

Figure 1.4 Scanning electron microscopy images of the nanowires grown by EBE technique at 650°C, 80 mA evaporation current (a, b)   *(Continued)*

Arrays, hybrids, and core–shell

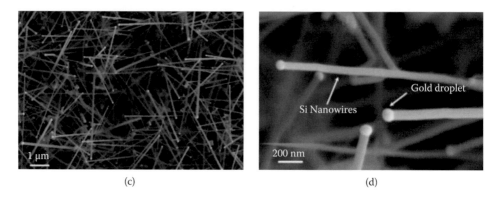

(c)   (d)

Figure 1.4 (Continued) Scanning electron microscopy images of the nanowires grown by CVD technique (c, d). (Reprinted from Becker, M., et al.: Nanowires enabling signal-enhanced nanoscale Raman spectroscopy. *Small.*, 2008. 4. 398–404. Copyright Wiley-VCH Verlag GmbH & Co. KGaA. With permission.)

1.2.2 METAL-ASSISTED WET CHEMICAL ETCHING

A very popular technique of the SiNW array formation is wet etching with the help of a metal acting as a catalyst of the process. In contrast to VLS, metal-assisted chemical etching (MACE) does not require sophisticated equipment and is free of contamination of the nanowires with the metal catalysts.

One- and two-step MACE processes are known. The one-step process (MACE-I) includes simultaneous reduction of metal ions and the oxidation and dissolution of silicon. For example, MACE process with silver reduced from the aqueous solution with the help of the Si valence band (VB) electron can be described by the following half-reactions:

$$Ag^+ + e^-_{Si,VB} \rightarrow Ag^0(solid) \quad (E_0 = 0.80 \text{ V}) \tag{1.1}$$

$$Si^0 + 6F^- \rightarrow SiF_6^{2-} + 4e^- \quad (E_0 = 1.24 \text{ V}) \tag{1.2}$$

where E_0 are standard electrode potentials. Reaction (1.2) is considered as a composite one proceeding through silicon oxidation to silica:

$$Si(solid) + 2H_2O \rightarrow SiO_2 + 4H^+ + 4\ e^-_{Si,VB} \tag{1.3}$$

$$SiO_2(solid) + 6HF \rightarrow H_2SiF_6 + 2H_2O \tag{1.4}$$

The Ag^+ ion is reduced in the vicinity of the silicon wafer accepting an electron from Si, forming an Ag^0 particle at the surface. Simultaneously, silicon under the silver particle is dissolved, forming a pit in the wafer and the silver nucleus falls into the pit (Figure 1.5). The excess oxidation causes accumulation of electrons on the surface of the nucleus, which attracts the Ag^+ ions to the vicinity of the silver particle, leading to the growth of Ag particles immobilized in the pits of the wafer surface. Due to the presence of Ag^+ in the solution and silicon at the wafer, these reactions go continuously. Thus, simultaneous deposition of Ag and etching of Si results in the formation of vertically aligned SiNWs, which can reach up to a length of 50 μm at ambient conditions (Hochbaum et al. 2008).

A two-step metal-assisted chemical etching (MACE-II) process, which separates silver nucleation and the electroless etching procedure in different aqueous solution (Figure 1.6), is used more often because it allows the silver nanoparticle size to be controlled and aligned columns to be formed. The first step of the process is metal (often noble metals as Ag, Au, Pt, and Rh; Yae et al. 2012) nanoparticle formation and deposition on the Si surface, which is practically the same as in the case of MACE-I. The second step of the MACE-II process employs other oxidizers in an HF-containing solution, such as Fe^{3+} or H_2O_2. During this stage, metal nanoparticles act as catalysts and cause Si dissolution underneath; thus, they sink into the Si wafer, forming the nanowires. If necessary, the metal particles can be removed after the process completion by rinsing the samples in HNO_3.

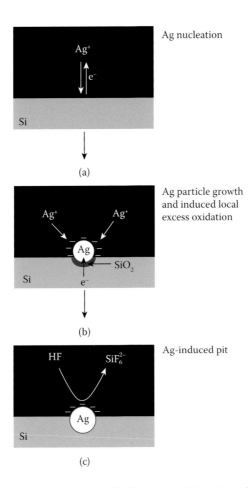

Figure 1.5 (a–c) Mechanism of Ag deposition onto a Si substrate in HF/AgNO₃ solution. (Reprinted from Peng, K.Q., et al.: Fabrication of single-crystalline silicon nanowires by scratching a silicon surface with catalytic metal particles. *Adv. Funct. Mater.,* 2006. 16. 387–394. Copyright Wiley-VCH Verlag GmbH & Co. KGaA. With permission.)

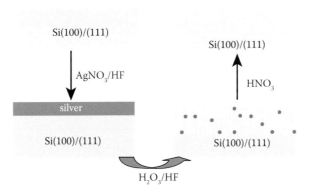

Figure 1.6 Schematic illustration of the etching process of silicon wafers using a sequence of two solutions; Solution I is based on AgNO₃/HF, and Solution II is based on H₂O₂/HF. After Solution I treatment, a quasicontinuous silver layer forms (schematically indicated by a layer named silver) on the silicon surface consisting of densely aligning Ag nanoparticles. (Reprinted with permission from Sivakov, V.A., et al., 2010, Realization of vertical and zigzag single crystalline silicon nanowire architectures, *J. Phys. Chem. C.,* 114, 3798–3803. Copyright 2010 American Chemical Society.)

Arrays, hybrids, and core–shell

Let us consider in more detail the reactions that proceed at the second step. First, the cathode reaction taking place at the metal surface is:

$$H_2O_2 + 2H^+ + 2e^- \rightarrow 2H_2O \quad (E_0 = 1.763\ V). \tag{1.5}$$

To describe anode reactions, which are Si oxidation and dissolution, three scenarios are considered (Huang et al. 2011):

(a) direct dissolution of Si in tetravalent state

$$Si + 6HF \rightarrow H_2SiF_6 + 4H^+ + 4e^- \tag{1.6}$$

(b) direct dissolution of Si in divalent state

$$Si + 4HF_2^- \rightarrow SiF_6^{2-} + 2HF + H_2\uparrow + 2e^- \tag{1.7}$$

(c) Si oxide formation followed by its dissolution, coincides with the case of MACE-I

$$Si + 2H_2O \rightarrow SiO_2 + 4H^+ + 4e^- \tag{1.3}$$

$$SiO_2 + 6HF \rightarrow H_2SiF_6 + 2H_2O. \tag{1.4}$$

Reduction of H_2O_2 at the metal surface (reaction 4) results in hole (h^+) generation. Hole injection from metal into Si in its turn is crucial for the Si dissolution (reaction 7). As it follows from half-reactions (1.2) and (1.5), the holes are injected into the Si valence band (Figure 1.7a). Metal–Si contact is a Schottky barrirer (Lai et al. 2016), which is negative for n-type Si and positive for p-type Si (Figure 1.7b and 1.7c). Thus, the holes injected into n-type Si are trapped near the Si–metal interface, whereas for p-type Si, drift of the holes and their mobility

Arrays, hybrids, and core–shell

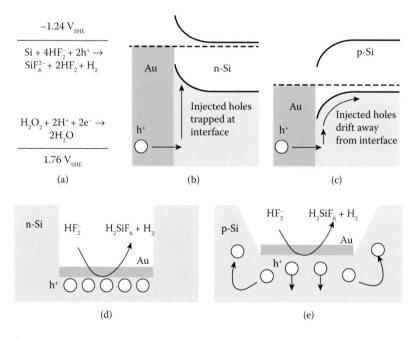

Figure 1.7 (a) Schematic of reduction potentials of the two MACE half-reactions relative to the standard hydrogen electrode, VSHE. (b, c) Band-diagram of Au–Si interface for n- and p-type Si, respectively. (d, e) Schematic of the resulting morphology of MACE for n- and p-type Si, respectively. (Reprinted with permission from Lai, R.A., et al., 2016, Schottky barrier catalysis mechanism in metal-assisted chemical etching of silicon, *Appl. Mater. Interfaces*, 8, 8875–8879. Copyright 2016 American Chemical Society.)

should be taken into account. As a result, the etched pores for *n*-type Si are anisotropic with strict vertical walls, whereas MACE in *p*-type Si causes broadening of the etched pores to the surface (Lai et al. 2016).

Instead of HF, a solution of NH_4HF_2 or NH_4F can be used with similar results (Brahiti et al. 2012; Gonchar et al. 2016).

Thus, all the aforementioned facts indicate importance of such parameters as Si wafer doping, oxidant concentration, amount of surface defects, and so on. The detailed discussion of the factors that influence the structural properties of the SiNW arrays is given in the reviews by Huang et al. (2011) and Chiappini (2014).

The etching is anisotropic and goes predominantly in two crystallographic directions: ⟨100⟩ and ⟨111⟩. The former one is a well-known pore growth direction in anodic or alkaline etching caused by rather weak bonds in this direction. Typically the ⟨100⟩-oriented SiNWs occur if the metal particles are small and isolated. At non-(100) surfaces, the etching direction depends on oxidant concentration: ⟨100⟩ and ⟨111⟩ for low and high concentration, respectively. The lack of the oxidant deep inside the wafer as a result of long etching time can result in a change of the etching direction. Whereas isolated metal nanoparticles can freely move in the direction of preferable etching, interconnected nanoparticles forming the metal mesh at the wafer surface do not possess this freedom. As a result, at low-symmetry surface, for example, (110), vertically aligned SiNWs occur (Huang et al. 2009; Han et al. 2014) (Figure 1.8).

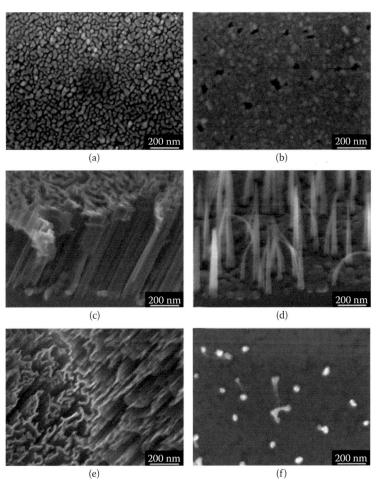

Figure 1.8 SEM images of isolated silver particles (a) and silver layer with pores (d) deposited on the (110) substrate. Bird's-eye view (b) and plan-view (c) SEM images of the substrate etched with isolated silver particles. Bird's-eye view (e) and plan-view (f) SEM images of the substrate etched with the help of silver layer with pores. (Reprinted with permission from Huang, Z., et al., 2009, Ordered arrays of vertically aligned (110) silicon nanowires by suppressing the crystallographically preferred <100> etching directions, *Nano Lett.*, 9, 2519–2525. Copyright 2009 American Chemical Society.)

Doping of the initial silicon wafer is of great importance for the etched pores orientation. For example, highly *n*-doped (resistivity of the order of 10 mΩ·cm) wafers are characterized by the lack of the vertical etching, with the predominant etching being parallel to the wafer surface (Figure 1.9; Qu et al. 2009). The explanation is that when the silver particles are nucleated by the defective sites, such as the dopants, a great amount of the defects leads to a continuous layer of silver particles at the surface. As the etching proceeds below the metal particles, the vertical etching is not possible. In this case, optimization of the processing conditions is required, for example, reduction of $AgNO_3$ concentration, which results in lower density of the silver particles at the surface.

Both in MACE-I and MACE-II processes, silver and silicon (cathode and anode) constitute electrochemical cell with electromotive force, or reduction potential, of 2.04 V. Electrochemical etching is well known to result in the formation of porous Si (por-Si). Indeed, SiNW sidewalls are found to be covered with por-Si film with nanocrystal size ranging from 2 to 50 nm (Hochbaum et al. 2009; Sivakov et al. 2010b). Transmission electron microscopy (TEM) reveals the por-Si cover of the SiNWs as a roughness of the sidewalls (Figure 1.10). It is worth noting that arrays of all-por-Si nanowires can be obtained by application of the MACE process to por-Si film, preliminarily prepared with the help of electrochemical etching of c-Si wafer (Jung et al. 2016).

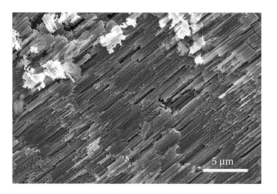

Figure 1.9 SEM image of *n*-Si (100) wafer of 0.008–0.02 Ω·cm resistivity after etching in a solution containing 5 M HF and 0.02 M $AgNO_3$ for 2 hours. (Reprinted with permission from Qu, Y., et al., 2009, Electrically conductive and optically active porous silicon nanowires, *Nano Lett.*, 9, 4539–4543. Copyright 2009 American Chemical Society.)

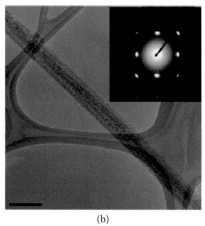

(a)   (b)

Figure 1.10 (a) A cross-sectional SEM of the porous nanowire array. Scale bar is 10 μm. (b) TEM image of the porous nanowire from which the selected area electron diffraction (panel b inset) pattern was obtained. The diffraction pattern indicates that the nanowire is single crystalline. Scale bar is 200 nm. (Reprinted with permission from Hochbaum, A.I., et al., 2009, Single crystalline mesoporous silicon nanowires, *Nano Lett.*, 9, 3550–3554, Copyright 2009 American Chemical Society.) *(Continued)*

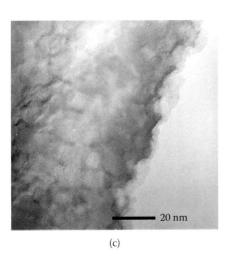

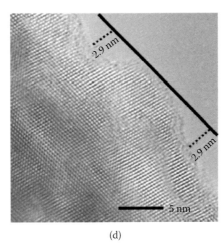

(c)   (d)

Figure 1.10 (Continued) High-resolution TEM images (c) and (d) of the SiNW. A peak to valley height of the sidewall roughness lies typically between 2.5 and 3.5 nm. (Reprinted with permission from Sivakov, V.A., et al., 2010, Roughness of silicon nanowire sidewalls and room temperature photoluminescence, *Phys. Rev. B*, 82, 125446. Copyright 2010 by the American Physical Society.)

1.2.3 PERIODIC SiNW ARRAYS

Periodic SiNW arrays, that is, photonic-crystal structures, are of special interest due to their unique optical properties.

These arrays can be fabricated with the help of reacting ion etching (RIE). The pattern at the wafer surface for the further etching can be made of photoresist by means of photolithography (Dhindsa and Saini 2016; Ko et al. 2016). Then, the RIE forms the SiNWs, sometimes with periodic diameter modulation. In the former case, the typical SiNW size is of about micrometers. To reduce the diameter of the nanowires, they can be additionally oxidized with later silicon oxide removal. As a result, structures with SiNW diameter modulated in the range from 0.3 to 0.5 μm were produced (Figure 1.11). Another approach to the periodic SiNW array formation is employing silica nanospheres spin-coated at the silicon wafer, where they form hexagonal packed structure (Figure 1.12a) (Zhu et al. 2009; Choi et al. 2015). Further, RIE in CHF$_3$/Ar gas combination reduced the silica nanosphere sizes (Figure 1.12b). Thus, the silica nanoparticle

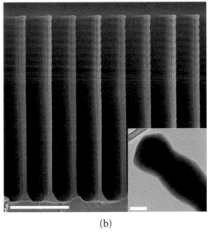

(a)   (b)

Figure 1.11 SEM images of a silicon microwire array after the RIE process (a) and oxidation and oxide removal (b). The scale bars are 5 μm. Inset at panel (b) is TEM image of a single SiNW with scale bar of 200 nm. (Reprinted from Ko, M., et al.: Periodically diameter-modulated semiconductor nanowires for enhanced optical absorption. *Adv. Mater.*, 2016. 28. 2504–2510. Copyright Wiley-VCH Verlag GmbH & Co. KGaA. With permission.)

Arrays, hybrids, and core–shell

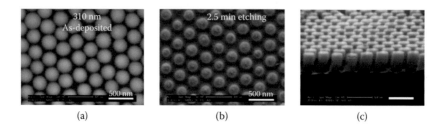

(a)   (b)   (c)

Figure 1.12 Silica nanospheres as-deposited at Si wafer (a) and RIE etched for 2.5 min (b). (c) SiNWs formed by means of MACE of c-Si wafer with metal deposited through the silica nanoparticle mask. Scale bar is 500 nm. (Reprinted with permission from Choi, J.Y., et al., 2015, Fabrication of periodic silicon nanopillars in two-dimensional hexagonal array with enhanced control on structural dimension and period, *Langmuir*, 31, 4018–4023. Copyright 2015 American Chemical Society.)

mask was formed at the Si surface; thereafter, Au with an interfacial layer of Ni was deposited through this mask. Then, the silica nanospheres were removed with the help of sonification, and the MACE process resulted in the formation of the SiNW arrays (Figure 1.12c).

1.3 ELECTRONIC PROPERTIES OF SiNW ARRAYS

During the SiNW growth with the help of the VLS technique, the catalyst metal slowly contaminates the nanowires. This doping with metal atoms strongly influences the electronic properties of the SiNW arrays. The character of this influence depends on the position of energy states of the impurity in the Si energy band gap. For example, the impurity levels close to the conduction band result in the occurrence of *n*-type semiconductor, as in the cases of Bi and Te (Miyamoto et al. 1976). If the impurity levels are close to the valence band energy, the *p*-type doping takes place (Al, In; Nebol'sin et al. 2003). Such metals as Au, Zn (Miyamoto et al. 1976; Nebol'sin et al. 2003), Cu (Wagner and Ellis 1965), Fe (Morales and Lieber 1998), Pb (Miyamoto et al. 1976), and Co (Wang et al. 1998) produce energy levels close to the band-gap middle. Employing Ag, Al, Pt, Pd, and Ni allows one to avoid both high doping of the SiNWs and recombination in them. The positions of the impurity energy levels are shown in Figure 1.13. It is worth noting that more controllable doping of the SiNWs could be done by supplying the dopant during growth, for example, by adding gaseous dopant precursors such as phosphine (PH_3) or diborane (B_2H_6) during the CVD process (Schmidt et al. 2009). Detailed discussion of doping and electronic states in SiNWs can be found in the review by Schmidt et al. (2009).

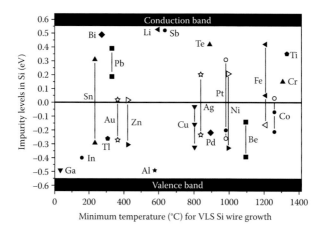

Figure 1.13 Energy levels of various impurities in Si with respect to the middle of the band gap as a function of the minimum temperature necessary for the VLS growth. Levels above the band-gap middle that are marked with solid symbols are donor levels, whereas open symbols indicate acceptor levels. Analogously, full symbols below the band-gap middle are acceptor levels, whereas open symbols are donor levels. (Reprinted from Schmidt, V., et al.: Silicon nanowires: a review on aspects of their growth and their electrical properties. *Adv. Mater.*, 2009. 21. 2681–2702. Copyright Wiley-VCH Verlag GmbH & Co. KGaA. With permission.)

In the SiNW formed by the MACE technique, diffusion of silver atoms results in the formation of the deep-level states with activation energy of 0.53 eV (Venturi et al. 2015).

1.4 LIGHT REFLECTION AND ABSORPTION IN SiNW ARRAYS

Optical properties of SiNW arrays strongly depend upon their morphology, which can be described by d (diameter), P (pitch), and L (length) of individual nanowires (that is, the thickness of the array), individual shape and surface quality of nanowires, and roughness of the interfaces. Most of the reflectance measurements were carried out using integrating spheres, which could provide one with data on total, diffuse, or specular type of reflection.

Dense SiNW arrays fabricated via various modifications of MACE or RIE techniques on c-Si substrates can drastically suppress reflection over a wide spectral bandwidth ranging from 300 to 1000 nm that corresponds to the c-Si above the band-gap region. Such arrays look matte black and are often referred to as black silicon (Figure 1.14). For example, low total reflectance down to a few percent was registered from thin (L ~ 250 nm) arrays of SiNWs obtained by MACE treatment (Figure 1.15) (Koynov et al. 2006).

Very low total reflectance was reported for the arrays of pyramidal SiNWs. For example, such structure of $L \sim 1$ μm formed by RIE showed as low as 1.21% reflectance (Bett et al. 2014). A similar result of less than 1% average reflectance in a 300–800 nm spectral range could be obtained from thicker arrays using

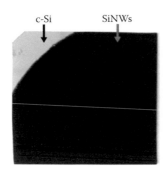

Figure 1.14 A photo of the SiNW array sample. (Reprinted from Osminkina, L.A., et al.: Optical properties of silicon nanowire arrays formed by metal-assisted chemical etching: evidences for light localization effect. *Nanoscale Res. Lett.*, 2012, 7, 524. Copyright Wiley-VCH Verlag GmbH & Co. KGaA. With permission.)

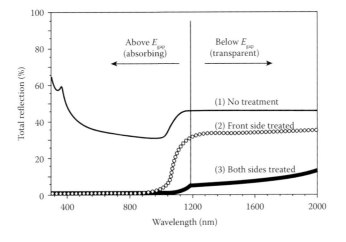

Figure 1.15 Total reflectance spectra of (1) double-side polished c-Si wafer; (2) one-side MACE treated and (3) double-side MACE treated c-Si wafer. Vertical dot line denotes the wavelength corresponding to indirect band gap of c-Si E_{gap}. (Reprinted with permission from Koynov, S., et al., 2006, Black nonreflecting silicon surfaces for solar cells, *Appl. Phys. Lett.*, 88, 203107–203107. Copyright 2006, American Institute of Physics.)

Arrays, hybrids, and core-shell

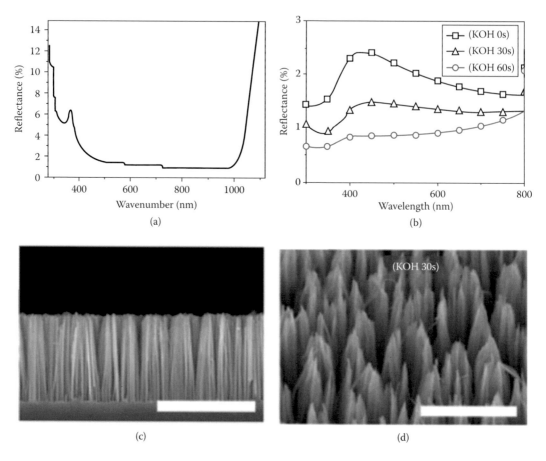

Figure 1.16 (a) Total reflectance spectrum of the SiNW arrays formed by RIE. (Reprinted from Bett, A.J., et al., *The 29th European PV Solar Energy Conference and Exhibition*, 22–26 September 2014, pp. 987–991. Amsterdam, 2014. With permission.) (b) Reflectance spectra of the tapered SiNW arrays formed by MACE and etching in KOH. (c) Cross-sectional SEM image of the SiNW array after MACE. Scale bar is 10 μm. (d) 30°-tilted SEM image that represents the morphological change of the SiNWs due to the postetching in KOH, etching time is 30 s. Scale bar is 5 μm. (Reprinted from Jung, J.Y., et al., 2010, A strong antireflective solar cell prepared by tapering silicon nanowires, *Opt. Expr.*, 18, A286–A292. With permission of Optical Society of America.)

additional tapering of bunched SiNWs by post-MACE dipping in KOH (Jung et al. 2010; pre-dipped separate SiNWs of d = 50–200 nm, Figure 1.16).

Low reflectance in the region under consideration is also characteristic of sufficiently thin columnar nonbunched or slightly bunched arrays fabricated via MACE. For example, Peng et al. (2005) communicated 1.4% or much less reflectance; Chang et al. (2011) demonstrated the average specular and total reflectance as low as 0.03% and 1.49%, respectively, for L = 4.52 μm, d < 50 nm, P < 300 nm columnar arrays (Figure 1.17).

Disordered silicon nanowires, which are fabricated by bottom-up approaches, also exhibit antireflective properties, though not so strongly pronounced (e.g., about 10% total reflectance of the structures made by metal organic vapor phase epitaxy [MOVPE] technique) (Muskens et al. 2008).

The remarkably low reflectance of the structures with different d and P was accounted for (1) individual shapes of separate SiNWs, for example, nanocones, bunching effect, and/or rough SiNW/substrate transition layer (as demonstrated in Figure 1.17b), which results in effective light concentration in the SiNW array (Li et al. 2015); (2) multiple scattering, which increases total light path (the transport mean free path of light) inside the array, resulting in strong enhancement of the probability for interactions between light and wires, that is, light trapping (Peng et al. 2005; Jung et al. 2010; Chang et al. 2011; Lajvardi et al. 2015).

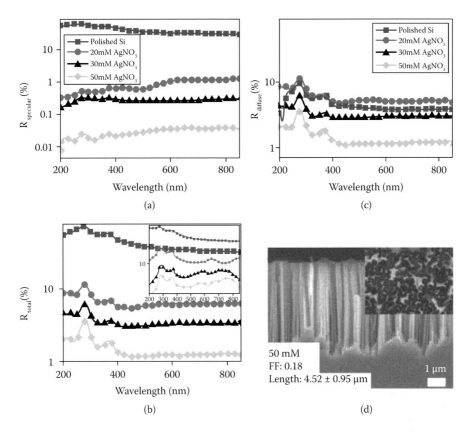

Figure 1.17 (a) Specular, (b) total, and (c) diffuse reflectance of the SiNWs obtained while using different AgNO₃ concentrations in MACE processing. (d) The cross-sectional and top-view SEM images of the SiNW arrays fabricated with 50 mM AgNO₃. The FF is a filling factor, that is, the area ratio of the SiNWs to the total substrate surface at the air/SiNW array interface. (Reprinted from Chang, H.C.H., et al., 2011, Nanowire arrays with controlled structure profiles for maximizing optical collection efficiency, *Energy Environ. Sci.*, 4, 2863–2869. By permission of The Royal Society of Chemistry.)

Absorptance of strongly dense SiNW arrays, calculated as $A = 1 - T - R$, where T is transmittance and R is reflectance, evidences that the amount of the absorbed light is significantly increased compared to untreated c-Si surface (Figure 1.18a). Even rather thin SiNW arrays (Koynov et al., 2006) demonstrate nearly 100% absorptance in the c-Si opacity region, that is, if the photon energy is above band gap. Relatively thicker arrays of black silicon (Bett et al. 2014) also showed nearly 100% absorptance in the c-Si opacity/above band-gap region. As one can see from Figure 1.18, the absorptance came close to the Yablonovitch limit of a perfect scatterer (Yablonovitch et al., 1982).

Up to 100% absorptance was also demonstrated by thick ($L = 2$–$40\ \mu m$) SiNW arrays with individual nanowires of minimum $d = 5$–$10\ nm$, typically larger up to $d = 20$–$100\ nm$ or bundled together (Figure 1.19a) (Tsakalakos et al. 2007; Sivakov et al. 2009). It should be pointed out that these results were obtained for SiNWs fabricated from silicon deposited on a transparent substrate. Hence, a thin (200 nm–1 μm) intermediate Si layer remained under the SiNW array (Figure 1.19b). High absorption was also manifested by SiNW arrays formed on amorphous Si substrates (Koynov et al. 2006; Zhu et al. 2009). In the latter paper, highly ordered middle-dense SiNW arrays of $d \sim 300\ nm$, $L \sim 600\ nm$ were made via a chlorine-based RIE process from a thin a-Si:H film grown on an indium tin oxide (ITO)-coated glass substrate. Deliberately assembled silica particles were used as an etch-mask during the RIE for perfect SiNW arrangement. In this single paper, absorbance was not calculated but directly measured by mounting samples at the center of the integrating sphere. It would be instructive to compare influence of the SiNW form (cones or pillars) on the efficiency of the light absorption. The contradictory data are

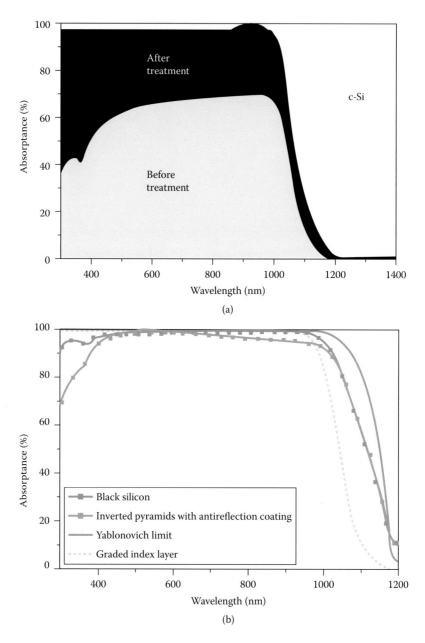

Figure 1.18 (a) Absorptance spectra of the Au-MACE processed c-Si and the untreated Si sample. (Reprinted from Koynov, S., et al., 2006, Black nonreflecting silicon surfaces for solar cells, *Appl. Phys. Lett.*, 88, 203107–203107. Copyright 2006, American Institute of Physics.) (b) Absorptance of the samples with black silicon arrays and inverted pyramids with antireflection coating, respectively, on the front side and a planar rear side; simulated absorptance of a graded index layer and Yablonovich limit of a perfect scatterer. (Reprinted from Bett, A.J., et al., *The 29th European PV Solar Energy Conference and Exhibition*, 22–26 September 2014, pp. 987–991, Amsterdam, 2014. With permission.)

known so far. According to Zhu et al. (2009), Si nanocone array is more effective in the light absorption than the arrays of nanopillars (Figure 1.20). In contrast, Li et al. (2015) reported on better light concentration by the nanocone array, which, however, gave no advantages in absorption in comparison with the pillar array.

Another situation is realized in the transparency region below the c-Si band gap (sub-band-gap region). Although Koynov et al. (2006) demonstrated that there was no additional absorption by SiNW arrays

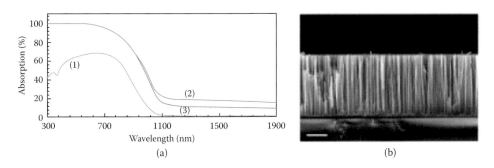

Figure 1.19 (a) Absorptance data for (1) solid Si film of L = 11 µm, (2) SiNW array on the glass substrate, (3) the same SiNW array on the glass after annealing at 400°C. (b) Cross-sectional SEM image of a L = 10 µm SiNW array on the glass substrate with a ~1 µm thick solid Si intermediate layer. Scale bar is 4 µm. (Reprinted from Tsakalakos, L., et al., *J. Nanophoton.*, 1, 013552, 2007. With permission.)

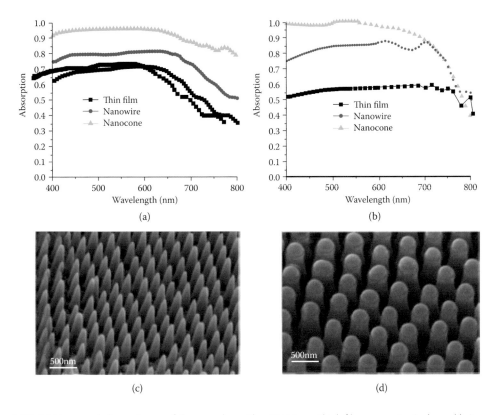

Figure 1.20 (a) Measured absorptance of the samples with a-Si:H 1 µm thick film, nanowire (column-like) arrays, and nanocone arrays as the top layer on ITO coated glass substrate and (b) corresponding calculated absorptance spectra. SEM images of the (c) nanocone and (d) columnar nanowire arrays, respectively. (Reprinted with permission from Zhu, J., et al., 2009, Optical absorption enhancement in amorphous silicon nanowire and nanocone arrays, *Nano Lett.*, 9, 279–282. Copyright 2009 American Chemical Society.)

within an experimental error of 2%, most of the authors (Tsakalakos et al. 2007; Sivakov et al. 2009; Bett et al. 2014) communicated about 20% absorptance of MACE- and RIE-formed SiNW arrays. This absorption of the SiNW arrays in the c-Si sub-band-gap region was partially attributed to strong infrared light trapping coupled with the presence of surface states on the nanowires, which absorb the below band-gap light. The states and related defects could arise from the aforementioned roughness of the nanowires typically of 1–5 nm. Thereby, the sub-band-gap absorption in the highly arranged (Wu et al. 2001;

Arrays, hybrids, and core–shell

Sivakov et al. 2009) and disordered (Tsakalakos et al. 2007) SiNW arrays was attributed to defect states and nanostructures on the surface of individual wires.

Another explanation of the sub-band-gap absorption was discussed by Bett et al. (2014). The passivation of the samples by Al_2O_3 revealed no effect on the absorption value; thus, surface defects were excluded by the authors from the potential reasons. The contribution of the free-carrier absorption (FCA) in the near-infrared region was simulated, yet about 4% absorption remained unaccounted for (Figure 1.21).

Numerous efforts were made over the last few years to explain the data on low reflection and high absorption of the SiNW arrays. Adachi et al. (2013) reviewed the results of optical modeling for ordered arrays of cylindrical SiNWs of different diameter (d = 25–200 nm), pitch (P = 50–600 nm), and length (L = 1–5 µm) via numerical solution of Maxwell's equations using transfer matrix method, full wave finite elements, and finite difference time domain (FDTD) for isolated layers in the air or layers on top of a reflecting or nonreflecting film. The rigorous coupled-wave analysis (RCWA) method was used for well-separated SiNWs (Zhu et al. 2009). The calculations demonstrated (1) high absorptance of the structures; (2) a shift of the absorption end to the low-energy region with the pitch increase that was ascribed to the transfer from the effective medium regime to light scattering in the structure; and (3) the emergence of additional resonant absorption bands under certain array parameters.

Resonant adsorption with one or several local extremums in the above Si band-gap region was predicted and experimentally manifested for sparse SiNW arrays formed by inductively coupled-plasma RIE preceded by electron beam lithography (Seo et al. 2011). The arrays of distant cylindrical SiNWs (d = 90–140 nm, $P \sim 1$ µm, L = 1 µm) exhibited multicolor reflection (Figure 1.22). The reflectance spectra showed a dip whose position shifted to longer wavelengths as the radius of the nanowires increased. The dip was attributed to the nanowire fundamental waveguide mode. The FDTD method simulations confirmed the hypothesis (Seo et al. 2011).

Ordered (photonic-crystal) structures based on SiNWs have their features determined by the array geometry. For example, thin L = 150–500 nm arrays of well-separated conical SiNWs with $P \sim 150$–350 nm and d/P = 0.3–0.7 that were fabricated by KBr dry etch (Bodena et al. 2008) showed wavelength-dependent

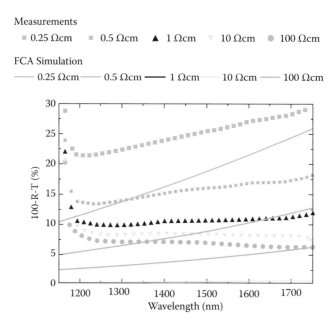

Figure 1.21 Measured absorptance (symbols) and FCA simulation (lines) for SiNW arrays formed on differently doped Si substrates. (Reprinted from Bett, A.J., et al., *The 29th European PV Solar Energy Conference and Exhibition*, 22–26 September 2014, pp. 987–991, Amsterdam, 2014. With permission.)

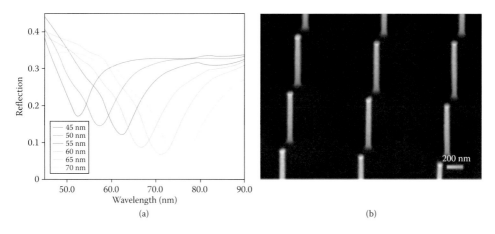

Figure 1.22 (a) Measured reflectance spectra of SiNW arrays. The colors of the curves correspond to the colors of the arrays. (b) 30°-tilted SEM image of an array with the nanowire pitch P = 1 µm, diameter d = 90 nm, and length L = 1 µm. (Reprinted with permission from Seo, K., et al., 2011, Multicolored vertical silicon nanowires, *Nano Lett.*, 11, 1851–1856. Copyright 2011 American Chemical Society.)

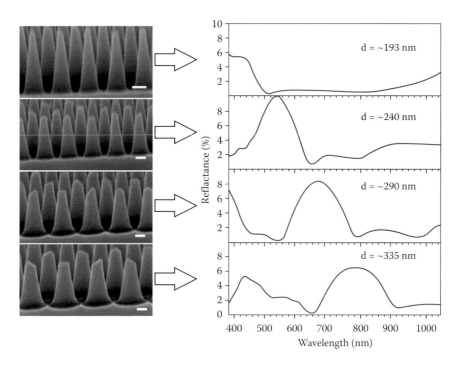

Figure 1.23 SEM images (a) and measured reflectance spectra (b) of the SiNW arrays with different pitches. The scale bars are 100 nm. (Reprinted with permission from Bodena, S.A., & Bagnall, D.M., 2008, Tunable reflection minima of nanostructured antireflective surfaces, *Appl. Phys. Lett.*, 93, 133108. Copyright 2008, American Institute of Physics.)

reflectance with several local minima in the region under consideration (Figure 1.23). Optical modeling using RCWA for cosine-based form of separate SiNWs showed similar spectral features. At the same time, simulations revealed that the stacks of higher tapered pillars (L up to 1800 nm) should result in lower R and exhibit the same properties as graded refractive index coatings for $d \ll \lambda$.

It would be very instructive to compare contributions of specular and diffuse reflections into the total reflection of the SiNW array. In particular, this comparison could elucidate peculiarities of the light

Arrays, hybrids, and core–shell

propagation in such medium as arrays consisting of aligned SiNWs that demonstrate strong scattering and are not periodically ordered. For this purpose, the dependence of the reflectance on the SiNW array thickness as well as measurement of the photon lifetime in the SiNW array can be employed.

The comparison of total and diffuse reflectance spectra of similar structures carried out for a set of different wavelengths (1250, 1126, 1064, and 417 nm) revealed that starting from about 2 μm, the thick SiNW arrays exhibited entirely diffuse reflection at all the wavelengths (compare Figure 1.24a and 1.24b). At the same time, substantial contribution of specular reflection for thin arrays was observed (Efimova et al. 2016). It is the simultaneous competitive action of specular and diffuse contributions to the total reflectance, decreasing and increasing with the SiNW array thickness, respectively, that is responsible for its nonmonotonic dependence of total reflectance on L.

To analyze the reasons for the high reflection of the SiNW, the photon lifetime in them was found (Efimova et al. 2016) by means of optical femtosecond heterodyning. It is a favorable method of scattering media study (Johnson et al. 2003; Bestem'yanov et al. 2004) based on the measurement of cross-correlation function $C(\tau)$ for incident, A, and scattered by the sample, S, waves with the help of a Michelson interferometer and a femtosecond laser:

$$C(\tau) = \int_{-T/2}^{T/2} A(t - \tau) S(\tau)\, dt \tag{1.8}$$

Here, T is the vibration period of the movable interferometer mirror used to introduce time delay. Fourier transformation provides a means of extracting the signal S at the frequency of the movable mirror vibration $\nu = 2v_m/\lambda$, where v_m is the mirror velocity and λ is the laser wavelength. The power of the cross-correlation function at this frequency can evidence the type of radiative transport in the sample, including diffusion approximation (Johnson et al. 2003).

The study revealed that for the SiNWs of the thickness above 4 μm, the cross-correlation function power had an expressed exponential tail, indicating diffusion-like radiative transport in the SiNW arrays (Figure 1.25). The typical photon lifetimes were found to be 0.5 and 1.4 ps for the SiNW arrays with thicknesses of 4.5 and 16 μm, respectively.

Obviously, the photon lifetime in the SiNW array strongly depends on the structural parameters such as diameter of the SiNW, their alignment, and so on. Thus, for the set of SiNW arrays obtained at different substrates (Table 1.1), the multifold photon lifetime increase was found (compare the photon lifetime

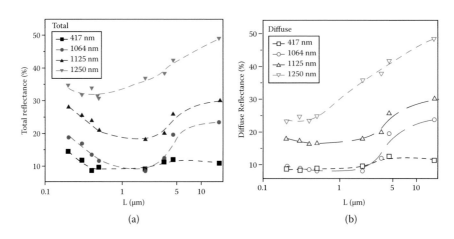

Figure 1.24 Total (a) and diffuse (b) reflectance of the SiNW arrays versus the SiNW array thickness L for the wavelengths of 417 nm, 1064 nm, 1125 nm and 1250 nm; the lines are eye-guides. (Reprinted with kind permission from Springer Science+Business Media: *Opt. Quant. Electron.*, Enhanced photon lifetime in silicon nanowire arrays and increased efficiency of optical processes in them, 48, 2016, 232–240, Efimova, A., et al., Copyright 2016.)

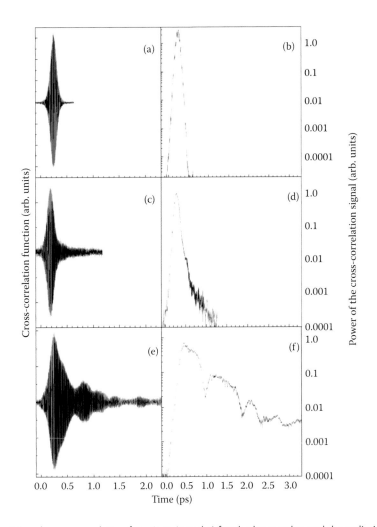

Figure 1.25 Auto- (a) and cross-correlation functions (c and e) for the laser pulse and the radiation scattered by the SiNW arrays with thicknesses of 4.5 (c) and 16 μm (e), respectively, and powers of the corresponding signal (b, d, and f) versus the time. (Reprinted with kind permission from Springer Science+Business Media: *Opt. Quant. Electron.*, Enhanced photon lifetime in silicon nanowire arrays and increased efficiency of optical processes in them, 48, 2016, 232–240, Efimova, A., et al., Copyright 2016.)

Table 1.1 Characteristics of SiNW array samples

SAMPLE	DOPING	SPECIFIC RESISTANCE (OHM·CM)	SURFACE ORIENTATION	NANOWIRE DIAMETER (NM)
A	As (n)	0.001–0.005	(111)	300–500
B	B (p)	0.7–1.5	(100)	50–200
C	B (p)	1–20	(111)	100–350

of 650 fs and laser pulse duration of 80 fs for a wavelength of 1250 nm) (Figure 1.26). As one can see, the highest photon lifetime is realized in the sample B with well-aligned SiNWs, whereas bunching SiNWs in the sample C results in less photon lifetime and, at last, less aligned flake-like SiNWs (sample A) do not demonstrate any sufficient light trapping. Despite stronger scattering at a wavelength of 625 nm, light absorption results in a shorter photon lifetime.

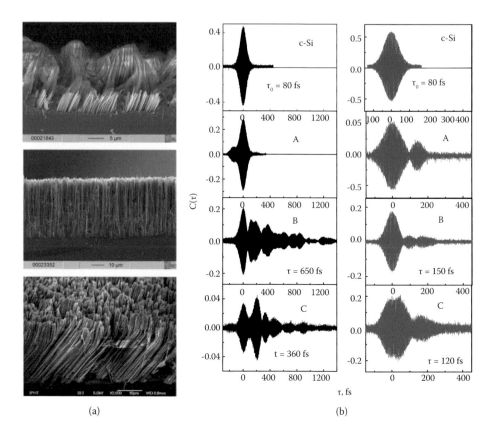

Figure 1.26 (a) Top to down: SEM cross-sectional images of SiNW arrays for the samples A, B, and C (Table 1.1). (b) Cross-correlation function of the 80-fs laser pulse and the pulse backscattered by the SiNW array for wavelengths of 1250 nm (left panel) and 625 nm (right panel). (Gonchar 2015.)

1.5 PHOTOLUMINESCENCE

Typical photoluminescence (PL) spectra of the SiNW arrays formed by means of MACE technique are broad band (Figure 1.27a) or two bands (Figure 1.27b), ranging from near-IR to visible (PL photon energy from 0.8 eV to 2.5 eV).

The near-IR band of the PL spectra in SiNW array is present both in initial c-Si and in SiNW arrays, although the latter signal is several times higher (Figure 1.26). The PL signal of the sample A (see Table 1.1) wafer is an order of magnitude weaker than one for low-doped p-Si (100) or p-Si (111) wafers, which could be caused by extra PL quenching in heavily doped Si due to Auger recombination. The most probable reason of the NIR PL efficiency enhancement under excitation with the light at 1064 nm is the light trapping. It is worth noting that although sample B demonstrates the highest photon lifetime (Figure 1.26), the maximal signal of the three samples is for sample C, which possesses the lower doping level and, as a result, less optical losses and irradiative recombination. The PL efficiency depends on quality of the initial c-Si substrate. In particular, a lower number of defects (longer charge-carrier lifetime) results in higher PL signal (Figure 1.28c). Dependence of the PL signal on the SiNW layer thickness is nonmonotonic and reaches its maximum at a SiNW layer thickness of 2–3 μm. The PL intensity decrease at larger layer thicknesses can be attributed to the fact that, as a result of long-duration chemical etching, the SiNWs get thinner and more porous, which intensifies nonradiative recombination at their surfaces. Nevertheless, the high interband PL intensities of the SiNWs in comparison with the case of the c-Si substrates indicates relatively low nonradiative recombination rates in the SiNWs as well as higher generation rate of photoexcited charge carriers in the SiNWs because of the efficient scattering of light compared to scattering in the cSi substrates (Gonchar 2014). IR PL in SiNW arrays is often considered as a phonon-assisted radiative recombination of the carriers inside the SiNW core. Oxide at the SiNW walls as well as Si nanocrystals with a broader band gap can enhance this process by suppressing surface recombination (Demichel et al. 2008, 2011; Ghosh et al. 2014). Larger size and higher areal density of these Si nanocrystals result in several times stronger PL signal. It is worth noting that IR PL was reported in SiNWs formed by VLS with Au or Cu technique at low temperature only (Demichel et al. 2008, 2011).

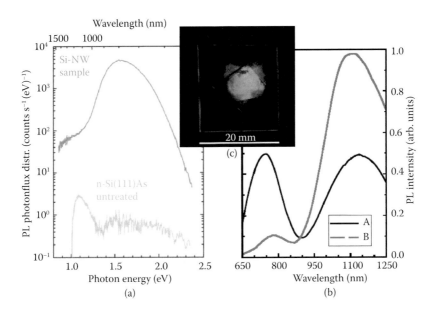

Figure 1.27 (a) Room temperature visible PL of SiNW array and heavily doped *n*-type Si (111): As, excitation at a wavelength of 488 nm. (Reprinted with permission from Sivakov, V.A., et al., 2010, Roughness of silicon nanowire sidewalls and room temperature photoluminescence, *Phys. Rev. B*, 82, 125446. Copyright 2010 by the American Physical Society.) (b) PL spectra of SiNW arrays formed by MACE of *n*-type Si (111):As (A) and *p*-type Si (100): (B), excitation at a wavelength of 532 nm (Gonchar, 2015). (c) Sample irradiated by a nitrogen laser at a wavelength of 337 nm. Visible blue emission next to the orange emission from the SiNW sample is due to fluorescence of the glass substrate. (Reprinted with permission from Sivakov, V.A., et al., 2010, Roughness of silicon nanowire sidewalls and room temperature photoluminescence, *Phys. Rev. B*, 82, 125446. Copyright 2010 by the American Physical Society.)

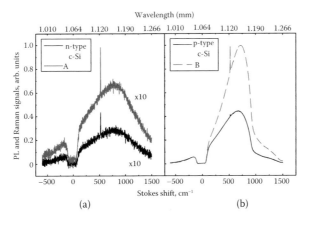

Figure 1.28 Photoluminescence and Raman spectra of the substrates and SiNWs samples A (a) and B (b) under excitation at 1064 nm (Gonchar 2015). The peak at 520 cm^{-1} is the Raman scattering signal.   *(Continued)*

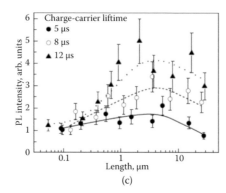

(c)

Figure 1.28 (Continued) (c) PL signal intensities normalized to the signal intensity of the corresponding initial c-Si substrates versus the SiNW layer thickness (SiNW length) for the samples obtained by means of MACE of low-doped *p*-Si (100) with different charge-carrier lifetimes. (Reprinted with kind permission from Springer Science+Business Media: *Semiconductors*, Optical properties of nanowire structures produced by the metal-assisted chemical etching of lightly doped silicon crystal wafers, 48, 2014, 1613–1618, Gonchar, K.A., et al., Copyright 2014.)

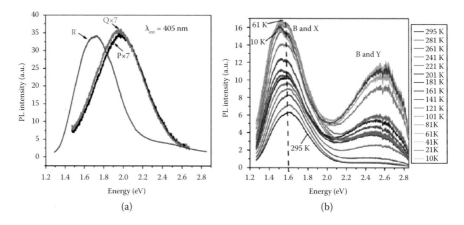

(a) (b)

Figure 1.29 (a) Comparison of visible PL spectra for SiNW samples with different nanocrystal sizes. Samples P, Q, and R have Si nanocrystal average cross-section of 37.9, 39.4, and 56.5 nm², respectively. (b) Temperature dependent visible PL spectra of SiNW array formed by MACE-I technique. The vertical dashed line shows a red-shift of the center of the low-energy visible PL band with lowering temperature. (The source of the material Ghosh, R., et al., Origin of visible and near-infrared photoluminescence from chemically etched Si nanowires decorated with arbitrary shaped Si nanocrystals, *Nanotechnology*, 25, 2014, 045703 is acknowledged.)

Visible PL in MACE-formed SiNW arrays (Sivakov et al. 2010; Zhang et al. 2013; Oda et al. 2014; Ghosh et al. 2014) cause a lot of interest because SiNWs are of too large diameter for quantum-size effect. The visible PL is often connected with formation of Si nanocrystals (por-Si) covering the SiNW walls. A smaller size of the Si nanocrystals results in higher PL photon energy (Figure 1.29a). Apart from the visible PL band centered about 1.6 eV detected at room temperature, another PL band of higher photon energy (2.6 eV) is revealed with temperature decrease (Figure 1.29b).

1.6 RAMAN SCATTERING

Raman scattering is certainly a powerful and useful method of research. That is why the study of features of the process in the SiNW array seems very interesting both for studying the SiNW arrays themselves and for developing new tools for sensitive diagnostics. Again, we have to distinguish between separate SiNWs fabricated by VLS technique or its derivatives and strongly scattering arrays of the SiNWs formed by MACE. In the former ones, the effect of the SiNW diameter on the Raman spectra and the possibility of enhancement of the Raman scattering efficiency due to plasmon effects are of great interest, whereas the latter ones are a very good subject for studying influence of the collective effects such as multiple light scattering on the Raman scattering efficiency.

The Raman spectra and their peak positions can depend on the SiNW diameter as it was found in the experiments with the SiNW arrays formed by means of pulsed laser vaporization and further VLS process. (Adu et al. 2005). The obtained results are in good agreement with the calculations carried out in the frames of the phonon confinement model proposed by Richter et al. (1981) for spherical particles and extended by Campbell and Fauchet (1986) to cylindrical particles (Figure 1.30).

The gold droplets of 20–500 nm diameter and semispherical shape placed at the top of the SiNWs grown by the VLS mechanism are well suited to exploit the surface-enhanced Raman scattering (SERS) effect and could be used for tip-enhanced Raman spectroscopy (TERS). A combination of a nanowire-based TERS probe and a SiNW-based SERS substrate promises optimized signal enhancement up to detection of single molecules (Becker et al. 2008). The Si-based devices for TERS can be also done with the help of the MACE method (Kazemi-Zanjani et al. 2013).

The SiNW arrays formed by means of MACE demonstrate significant enhancement of the Raman scattering efficiency in comparison with the initial c-Si substrate (Timoshenko et al. 2011) (Figure 1.28a and 1.28b). Although silver nanoparticles are employed for the SiNW formation, plasmonic effects in them are not responsible for the Raman signal rise; moreover, the Raman scattering efficiency increases after rinsing the sample in HNO_3 and Ag nanoparticle removal (Golovan et al. 2012b). The Raman scattering increase was also reported in vertically ordered SiNW arrays with diameters ranging from 30 to 60 nm formed by means of RIE in comparison with original silicon on insulator and c-Si wafers. The Raman enhancement per unit volume of the first-order phonon peak increases with increasing nanowire diameter. The results are understood using a model based on the confinement of light (Khorasaninejad et al. 2012a).

The Raman scattering efficiency rise is accompanied with less expressed orientation dependence of the Raman signal, which is an obvious result of strong light scattering. Enhancement of the Raman signal in SiNWs takes place even at the SiNW layer thickness less than 1 μm and increases with the SiNW layer length increase, tending to saturation for the SiNW layer thickness above 5 μm (Figure 1.31). The Raman signal from thin layers is formed partially in c-Si and partially in the SiNW array. However, if the SiNW array thickness exceeds the transport mean free path, the Raman signal tends to saturate.

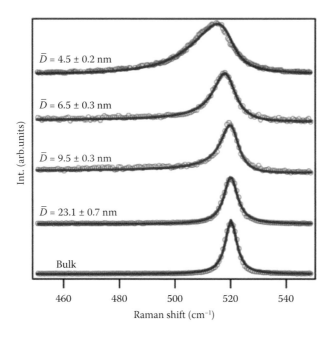

Figure 1.30 Raman spectra for c-Si and SiNWs of different average diameter $\bar{D}$. Open circles represent the experimental data and solid lines represent the fitting. (Reprinted with permission from Adu, K.W., et al., Confined phonons in Si nanowires, *Nano Lett.*, 5, 2005, 409–414. Copyright 2005 American Chemical Society.)

Arrays, hybrids, and core–shell

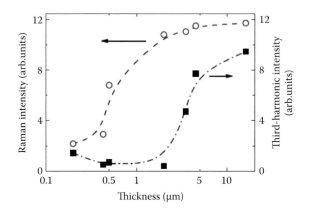

Figure 1.31 Raman and third-harmonic intensities versus SiNW layer thickness; the lines are eye-guides. Excitation at a wavelength of 1064 nm. The results are normalized to the corresponding signals from the c-Si substrate. (Reprinted with kind permission from Springer Science+Business Media: *Opt. Quant. Electron.*, Enhanced photon lifetime in silicon nanowire arrays and increased efficiency of optical processes in them, 48, 2016, 232–240, Efimova, A., et al., Copyright 2016.)

Since the efficiency of the processes in such scattering media as SiNW arrays strongly depends on the wavelength, it is worth paying special attention to the variation of the Raman scattering efficiency with the wavelength. Comparison of the Raman scattering efficiency for three different samples (see Table 1.1) of SiNWs for various wavelengths was carried out. Since Raman scattering efficiency depends on the free-carrier concentration, which differs for different samples, we should take ratio of Raman signal for SiNW I_{SiNW} and corresponding crystal substrate I_{wafer} (Figure 1.32). For sample B, which is characterized by the longest photon lifetime of all three samples (see Figure 1.26), the ratio I_{SiNW}/I_{wafer} increases with the decrease of excitation wavelength; for sample C with less photon lifetime, it is almost constant, whereas for sample A it falls. Discussing the variation of the Raman scattering efficiency with the excitation wavelength, we should keep in mind the interplay of two important

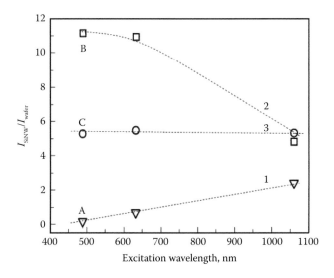

Figure 1.32 Ratio of Raman signals of SiNW arrays and c-Si wafer versus excitation wavelength for samples A (down triangles), B (squares), and C (circles). Dashed lines are eye-guides. (Reprinted with kind permission from Springer Science+Business Media: *Semiconductors*, Dependence of the efficiency of Raman scattering in silicon nanowire arrays on the excitation wavelength, 47, 2013, 354–357, Bunkov, K.V., et al., Copyright 2013.)

factors: the light scattering and the light absorption. Both the factors increase when the wavelength decreases. However, their actions are opposite: rise of the scattering efficiency results in increase of the photon lifetime in the medium, whereas the absorption reduces it.

All the samples demonstrate $I_{SiNW}/I_{wafer} > 1$ for excitation wavelength of 1064 nm, which is characterized by very low absorption. However, in the sample A with SiNWs bunching to clusters of rather large size (several micrometers), the light scattering is not enough to increase the photon lifetime in case of absorbing media. In contrast, aligned and separate SiNWs in sample B ensure longer photon lifetime, which increases with the wavelength decrease despite absorption. At last, the sample C is in intermediate position, since its less ordered SiNWs of larger diameter allow the light scattering to compensate for absorptions. Thus, structural parameters of the SiNW array determine the photon lifetime, which, in its turn, controls Raman scattering efficiency (Timoshenko V Yu, et al. 2011).

Investigation of the SiNWs fabricated by MACE of low-doped c-Si wafers by micro-Raman spectroscopy revealed the strong heating of SiNWs under visible laser irradiation with relatively low intensity. The Raman spectrum of SiNWs changes with excitation laser intensity and length of SiNWs. The continuous downshift of the Raman spectrum from 520.5 to 517 cm^{-1} under excitation with intensity up to 1 kW/cm^2 is well explained by heating SiNW arrays; average temperature increase can reach 150 K. The photo-induced heating is explained by low thermal conductivity coefficient of SiNW; its value was estimated as 0.1 W/(m K), which is three orders of magnitude lower than the corresponding value for c-Si. The low thermal conductivity of the investigated SiNWs prepared by the MACE method can be related to an effect of microporous shells of SiNWs, which increases the phonon scattering (Rodichkina et al. 2015). Thermal effects on Raman scattering under tight focusing were also studied for SiNWs grown by the VLS method; significant heating (up to 1000 K) by radiation of the common lasers with power less than 1 mW was reported (Anaya et al. 2013, 2014).

1.7 NONLINEAR OPTICAL EFFECTS

Nonlinear optics is a division of optics, which deals with propagation of such intense light that the medium polarization loses its linearity with respect to the wave amplitude E, and nonlinear terms should be taken into account. It is nonlinear optics that is responsible for wave mixing and frequency conversion, including optical harmonic and parametric generations, ultrashort laser-pulse generation, optical switching, and various other effects and applications. Connections of these nonlinear (quadratic and cubic with respect to E) polarization terms P_{NL} and E are given by the following equations:

$$P_{NL,i}(\omega) = \chi_{ijk}^{(2)}(\omega;\omega_1,\omega_2)E_{1,j}E_{2,k},$$

$$P_{NL,i}(\omega) = \chi_{ijkl}^{(3)}(\omega;\omega_1,\omega_2,\omega_3)E_{1,j}E_{2,k}E_{3,l},$$

(1.9)

where $\chi^{(2)}$ and $\chi^{(3)}$ are nonlinear susceptibility tensors of the second and third orders, correspondingly, ω, ω_1, ω_2, ω_3 are frequencies of interacting waves, with ω being their combination, and subscripts i, j, k, l denote Cartesian coordinates.

Among its various virtues, Si possesses rather high cubic nonlinear susceptibility $\chi_{ijkl}^{(3)}$, which is responsible for four-wave mixing and light self-action. In contrast, due to central symmetry of the c-Si crystallographic cell $\chi_{ijk}^{(2)} = 0$, which, in particular, results in extremely weak second-harmonic generation. Nevertheless, for the SiNWs of 37 nm in diameter and 650 nm in length fabricated using electron beam lithography of (100) Si wafer followed by dry etching, Khorasaninejad et al. (2012b) reported an SHG signal increase by at least a factor of 80 as compared to bulk silicon. The effect can be caused by both light trapping in the SiNW layer and reduction of the surface high symmetry.

Analogous to Raman scattering, efficiency of the third-harmonic (TH) generation was found to strongly depend on the photon lifetime in the SiNW layer (Figure 1.33). The same samples (see Table 1.1) that demonstrate the rise or fall of the Raman signal exhibit increased or decreased TH generation efficiency, correspondingly. The aforementioned rise of the TH signal for crossed polarizer

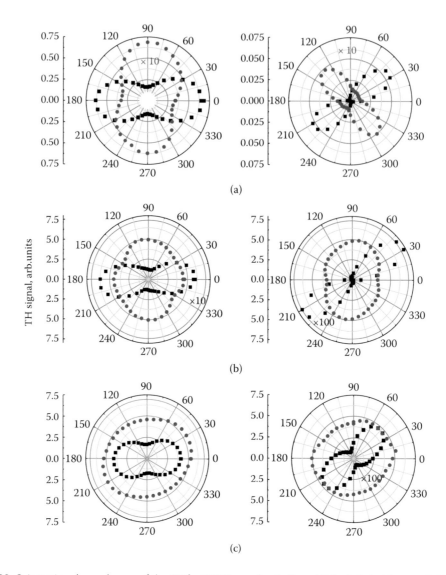

Figure 1.33 Orientation dependences of the TH for SiNW samples A, B, and C (red circles) for parallel (left column) and perpendicular (right column) polarizations of the TH and fundamental radiation. The black squares are related to the corresponding signals for c-Si substrates. The laser radiation was focused on the sample at the angle of incidence of 45°. Angle of 0° corresponds to p-polarized fundamental radiation (Adapted from Golovan LA, et al. (2012a). *ALT Proceedings 1*. doi: 10.12684/alt.1.81.)

and analyzer was caused by the light depolarization as a result of strong light scattering in the SiNW array. Variation of the orientation dependences for the TH in SiNW arrays in comparison with c-Si could indicate that the dependence is imposed by the geometry of SiNWs. For example, tilted SiNWs in sample A determine the TH orientation dependence, cone bunches of SiNWs in sample C result in practically isotropic TH orientation dependence, whereas sample B with nonbunched SiNWs demonstrates anisotropy of the TH. Thus, depending on the structure of the SiNW ensemble, both fall and one- or two-orders of magnitude rise of the Raman and TH signals in SiNWs in comparison with the corresponding c-Si substrates have been observed.

The dependence of the TH signal on the thickness of the SiNW layer is nonmonotonic (Figure 1.30): with the SiNW length increase the TH signal decreases for thinner layers (0.2–2 µm) and increases for thicker layers. This effect partially connected with reflection properties of the SiNW film (Figure 1.24).

While the fundamental wavelength lies in the region of transparency of the structures and the TH is generated along the whole thickness of the SiNW layers, the TH is efficiently absorbed (e.g., $\alpha^{-1} \sim$ 125 nm for c-Si at 417 nm). The presence of specular component in the reflectance spectra implies that for thin SiNW arrays the registered TH signal is generated both in the SiNWs and at the c-Si surface and passes through a scattering SiNW medium to get outside. Besides, low coherence length for the TH generation also results in a decrease of the TH signal. With the SiNW layer thickness growth, the region where the collected TH signal is generated moves up from the array/c-Si interface and for thick layers it is entirely located in the SiNW array. At these thicknesses we observe a significant increase in the TH intensity, which corresponds to the growth of the diffuse reflectance at 1250 nm (Figure 1.24). This fact correlates with the enhanced fundamental photon lifetime in thick SiNW arrays (Figure 1.25).

SiNWs grown by the MACE method and ultrasonicated in toluene demonstrated a four-order of magnitude rise of cubic susceptibility in comparison with c-Si (Huang et al. 2012), which opens up broad possibilities to employ SiNWs in photonics.

Coherent anti-Stokes Raman scattering (CARS) is a widely used technique of nonlinear optical spectroscopy. This is a nonlinear optical process of parametric interaction of three waves with frequencies ω_1, ω_2, and ω_3 resulting in generating a wave at the frequency $\omega_{CARS} = \omega_3 + (\omega_1 - \omega_2)$. For the sake of simplicity of experimental setup, a degenerative optical interaction, in which the same laser is used for ω_1 and ω_3, is often employed. CARS signal occurs at any combination of frequencies ω_1 and ω_2, leading to the so-called nonresonant background. When the difference frequency, $\omega_1 - \omega_2$, is at or near the resonance with some vibrational or electronic transition, it results in a significant rise of the signal at ω_{CARS}. Being a coherent nonlinear optical process, the CARS efficiency is sensitive to the phase matching of interacting waves. CARS is recognized to be an invaluable tool for studying microobjects and their molecular environment, chemical and biological objects, and for frequency conversion in various media, including Si-based ones.

The CARS spectrum collected from the c-Si sample consists of the background and a peak at 1008 nm, corresponding to there sonance with the phonon frequency of c-Si (Figure 1.34a). The most significant feature of the CARS signal from the SiNW arrays is a substantially higher CARS signal than that from the c-Si. SiNW arrays containing Ag nanoparticles exhibit a band of a relatively high CARS signal for the wavelengths less than 850 nm; for the SiNW arrays without Ag nanoparticles, the band of the high CARS signal is in the spectral interval from 950 to 1000 nm (Golovan et al. 2012). The resonant signal at 1008 nm for the SiNWs sample containing Ag nanoparticles is as high as for the c-Si, whereas it is four times higher for the SiNW sample without Ag nanoparticles (cf. insets in Figure 1.34).

These results can be explained by taking into account both the effects of light scattering in the SiNWr ensembles and the electron and hole states, whose transition frequencies are resonant to the CARS frequencies.

1.8 DEVICE APPLICATIONS

SiNWs are extensively explored as active components of chemical and biosensors whose responses are mainly related to changes of the electronic and optical properties of SiNWs due to specific and non-specific adsorption of molecules. Most of the proposed sensor concepts are dealing with an effect of the molecular adsorption on electrical conductivity of SiNWs (Peng et al. 2009). Another part of the sensor demonstration is based on applications of the surface-enhanced Raman scattering (SERS) in SiNWs covered with plasmon nanoparticles (Qiu et al. 2006; Zhang B et al. 2008; Sun et al. 2009; Wang et al. 2010; Yi et al. 2010; Huang et al. 2013). Modification of the photoluminescence properties of rough SiNWs formed by MACE has been also proposed to determine the partial oxygen pressure in air (Georgobiani et al. 2015).

Porous n-type SiNW arrays fabricated by MACE were investigated for gas sensing (Peng et al. 2009). Two gold electrodes were deposited on the top surface of SiNW arrays (see Figure 1.35a), and the resulting sample was exposed to gaseous NO with various concentrations ranging from 500 ppb to 100 ppm with recording the resistance change, ΔR (Figure 1.35b). It was suggested that NO

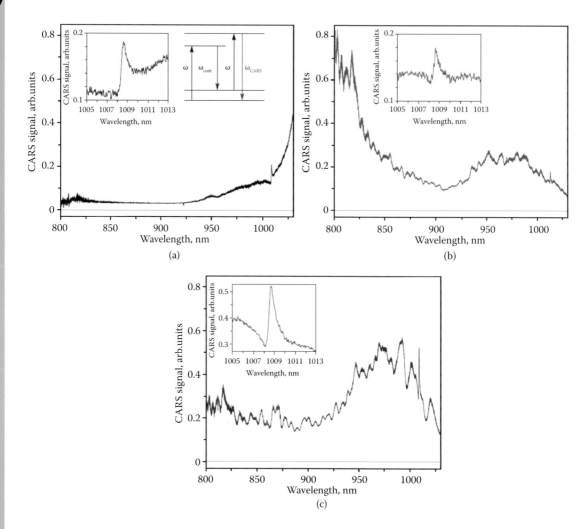

Figure 1.34 CARS spectra for c-Si (a), SiNWs with Ag nanoparticles (b), and SiNWs without Ag nanoparticles (c). Insets exhibit resonance peak corresponding to the phonon frequency of Si at 520 cm⁻¹ for the corresponding samples. Schematic energy diagram of the CARS process is shown in the right inset of (a). (The source of the material Golovan, L.A., et al., Coherent anti-Stokes Raman scattering in silicon nanowire ensembles, *Laser Phys. Lett.*, 9, 2012, 145–150 is acknowledged.)

molecules donated electrons to *n*-type SiNWs; thus, the electrical conductivity increased during the NO adsorption.

In et al. (2011) fabricated porous top electrodes deposited on the top surface of vertical SiNW arrays by employing two separate nanosphere lithography steps, that is, the first step for fabricating SiNWs and the second step for forming a porous top electrode. The resulting periodically porous gold top electrode was found to contact the tips of SiNWs and allowed gas molecules to access the surface of SiNWs easily; thus, sensing performance considerably increased compared to the counterpart with a continuous top electrode (see Figure 1.36).

As a novel nanostructured sensing device for detecting H_2, Noh et al. (2011) fabricated Pd-coated rough SiNW arrays. The Pd deposition was done on as-prepared SiNWs with surface morphology that was locally clustered and even slanted. Thus, it was realized that an inversely tapered shape where the very top regions of SiNWs are thickly covered with Pd can be locally connected with each other. Upon H_2 exposure, volume expansion of Pd occurs, and local bridge for electrical current flow is formed through the contact of Pd particles on tips of neighboring SiNWs (Figure 1.37a). The detection limit was 5 ppm with 6% sensitivity. The sensitivity increased with increasing H_2 concentration (Figure 1.37b).

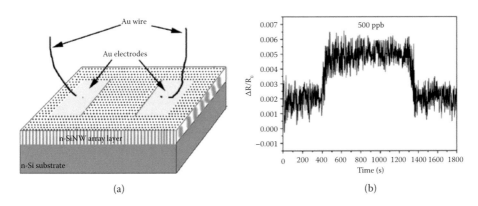

(a)   (b)

Figure 1.35 (a) Schematic representation of SiNWs sensor on NO; (b) detection of 500 ppb of NO by using the SiNWs sensor. (Reprinted with permission from Peng, K.Q., et al., Gas sensing properties of single crystalline porous silicon nanowires, *Appl. Phys. Lett.*, 95, 2009, 243112. Copyright 2009, American Institute of Physics.)

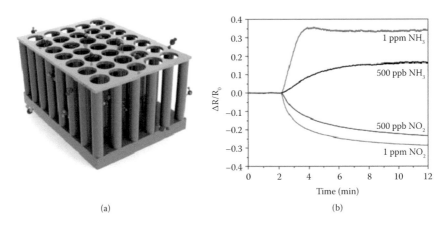

(a)   (b)

Figure 1.36 (a) Schematic illustration of the periodically porous top electrode SiNW sensor concept; (b) Sensor response to various concentrations of NO_2 and NH_3 following 2 min of clean air. (The source of the material In HJ, et al., Periodically porous top electrodes on vertical nanowire arrays for highly sensitive gas detection, *Nanotechnology*, 22, 2011, 355501 is acknowledged.)

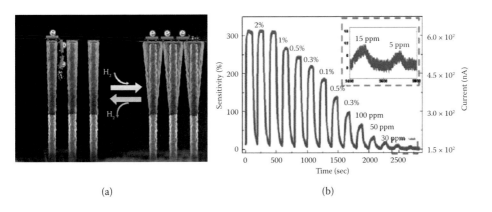

(a)   (b)

Figure 1.37 (a) Schematic illustration of SiNWs/Pd sensor; (b) time-dependent detection of H_2 by using the SiNWs/Pd sensor. (Reprinted from Noh, J., et al., High-performance vertical hydrogen sensors using Pd-coated rough Si nanowires, *J. Mater. Chem.*, 21, 2011, 15935–15939. By permission of The Royal Society of Chemistry.)

Arrays, hybrids, and core–shell

SiNWs were investigated as a SERS substrate prepared in a simple and reproducible way. Typical SERS-active materials are silver or gold nanoparticles. Ag-capped SiNWs were prepared and tested for the SERS analyses by using dye molecules of rhodamine 6G (Qiu et al. 2006). They found that the length of SiNWs influenced the SERS intensity. The higher SERS intensity was detected for shorter SiNWs, and it was explained by the rigid nature of individual nanowire.

A highly sensitive label-free immunoassay based on SERS with Ag-coated SiNWs was proposed (Zhang ML et al. 2008). The SERS spectra demonstrated a low detection limit for immune reactions. Immunocomplex formed with 4 ng of mouse immunoglobulin G (mIgG) and goat-antimouse immunoglobulin G (gamIg) on the substrate gave distinguishable Raman bands with shifted positions and changed intensities (Figure 1.38).

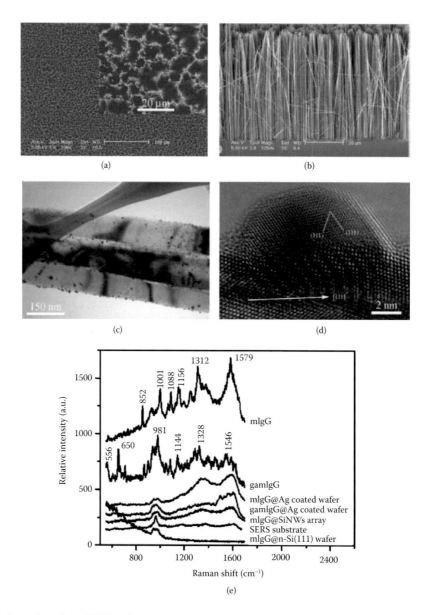

Figure 1.38 Examples of the SERS application using Ag-coated SiNWs: (a) and (b) images of as-prepared SiNW arrays; (c) and (d) are TEM and HRTEM of the SERS-active substrate, respectively; (e) SERS spectra of mIgG, gamIgG by using Ag-coated SiNWs and their corresponding controls. (Reprinted with permission from Zhang, M.L., et al., A surface-enhanced Raman spectroscopy substrate for highly sensitive label-free immunoassay, *Appl. Phys. Lett.*, 92, 2008, 043116. Copyright 2008, American Institute of Physics.)

The enhancement of the Raman signal was related to the dipolar resonance on metal surface and chemical effect between molecules and metal on the surface of SiNWs.

Huang et al. (2013) demonstrated ultrasensitive single-molecule detection with high selectivity by utilizing nanogap-free SERS system, in which Ag coating was realized on the surface of hexagonally packed SiNWs. The prepared nanogap-free SERS system detected long double-strand DNA with 14% accuracy. The observed high reproducibility was related to long inter distance between the nanowires (~150 nm) and continuous Ag coating on SiNW surfaces. The large inter wire space allowed large biomolecules access to the surface. Surface plasmon can propagate along the continuous Ag layer, and it results in the enhanced wide-range electric field (Figure 1.39).

PL properties of porous and nonporous SiNWs were found to be sensitive to molecular oxygen in ambient atmosphere (Georgobiani et al. 2015). Porous SiNWs grown on heavily doped c-Si substrates possess exciton PL in the range of 1.3–2.0 eV, whereas nonporous SiNWs prepared from low-doped c-Si exhibit an intensive line of the interband PL at 1.1 eV. The exciton PL intensity related to small-sized Si nanocrystals was found to depend on partial pressure of molecular oxygen, a quencher of the PL intensity, in the surrounding atmosphere. The interband PL intensity exhibits a superlinear dependence on the laser excitation intensity that indicates a large number of equilibrium charge carriers in the porous sample. It is established that porous SiNWs partially retain equilibrium charge carriers, which influence the sensitivity of PL properties to the molecular surroundings and improve reversibility of the sensor response (Figure 1.40).

SiNW arrays were also explored for improving efficiency of Si-based solar cells. A simple method has been developed for creating rectifying contacts that yield photovoltaic behavior from single SiNW prepared by an Au-catalyzed VLS method (Kelzenberg et al. 2008). The use of an aluminum-catalyzed VLS method for solar cells fabrication is reviewed by Hainey et al. (2016). Al-catalyzed and Au-catalyzed SiNWs as p-type cores were fabricated for both $p–n$ and $p–i–n$ radial junction solar cells (Ke et al., 2011) (Figure 1.41). However, open-circuit voltage V_{oc} were found to be greatly reduced for devices fabricated with Al-catalyzed cores compared to those with Au-catalyzed cores (Kempa et al. 2012). $P–n$ junction formation using ultra-high-vacuum-grown Al-catalyzed SiNWs fabricated on n-type Si (111) substrate was also demonstrated (Moutanabbir et al. 2011).

SiNW-based solar cells on glass substrates were fabricated by chemical etching with silver nitrate and hydrofluoric acid of 2.7 μm multicrystalline p^+nn^+ doped silicon layers (Figure 1.42) (Sivakov et al. 2009). Total reflectance less than 10% in the spectral region 300–800 nm and a strong broadband optical absorption more than 90% in the spectral region about 500 nm were registered. The highest V_{oc} and short-circuit current density (J_sc) were 450 mV and 40 mA/cm^2, respectively, at a maximum power conversion efficiency of 4.4%.

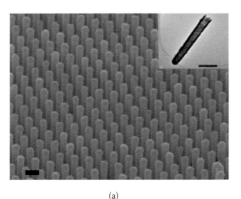

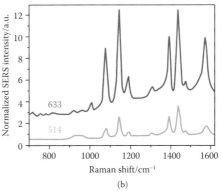

(a)   (b)

Figure 1.39 (a) SEM image of Ag-coated SiNWs, inset shows a TEM image. Scale bars are 200 nm. (b) SERS spectra of 4-aminothiophenol under excitation with 633 and 514 nm lasers of the same power, respectively. (Reprinted with permission from Huang, J.A., et al., Ordered Ag/Si nanowires array: wide-range surface-enhanced Raman spectroscopy for reproducible biomolecule detection, *Nano Lett.*, 13, 2013, 5039–5045. Copyright 2013 American Chemical Society.)

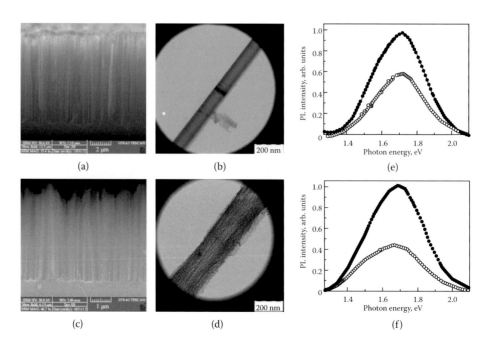

Figure 1.40 (a) and (b) SEM and TEM images of nonporous SiNWs, respectively; (c) and (d) SEM and TEM images of porous SiNWs; (e) and (f) PL spectra of nonporous and porous SiNWs, respectively, in gaseous nitrogen (black circles) and in molecular oxygen (open circles) atmosphere at 0.5 bar. (Reprinted with kind permission from Springer Science+Business Media: *Semiconductors*, Structural and photoluminescent properties of nanowires formed by the metal-assisted chemical etching of monocrystalline silicon with different doping level, 49, 2015, 1025–1029, Georgobiani, V.A., et al., Copyright 2015.)

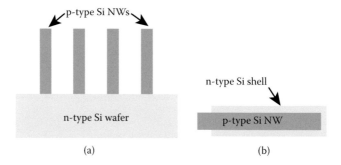

Figure 1.41 Schematics of (a) axial junction, (b) radial junction. (Reprinted with permission from Hainey, M.F., & Redwing, J.M., Aluminum-catalyzed silicon nanowires: growth methods, properties, and applications, *Appl. Phys. Rev.*, 3, 2016, 040806. Copyright 2016, American Institute of Physics.)

Another promising candidate for photovoltaics is hybrid solar cells. Power conversion efficiency of 13.01% has been reported by Yu et al. (2013) for the hybrid solar cells based on SiNWs and PEDOT:PSS conductive polymers. Wang et al. (2016) demonstrated the potential of thin film Si/PEDOT:PSS hybrid cells incorporated with SiNWs for light trapping. The surface treatment process, which consisted of oxygen plasma exposure, was found to improve the surface quality of SiNWs. The optimized cell SiNWs achieved the power conversion efficiency of 7.83%. The surface treatment process was found to remove surface defects and passivate the SiNWs and substantially improve the average open circuit voltage from 0.461 to 0.562 V for the optimized cell (Figure 1.43).

Thus, SiNW arrays formed by various techniques are novel media with an extremely broad spectrum of their structural, electronic, and optical properties, which can differ significantly from the ones for c-Si.

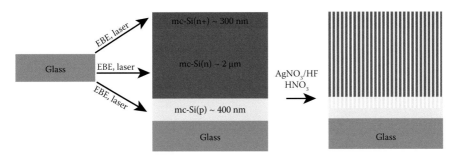

Figure 1.42 Schematic cross-sectional view of the mc-Si *p–n* junction layer stack on a glass substrate and SiNWs after chemical etching in AgNO$_3$/HF/HNO$_3$ solutions. (Reprinted with permission from Sivakov, V., et al., Silicon nanowire-based solar cells on glass: synthesis, optical properties, and cell parameters, *Nano Lett.*, 9, 2009, 1549–1554. Copyright 2009 American Chemical Society.)

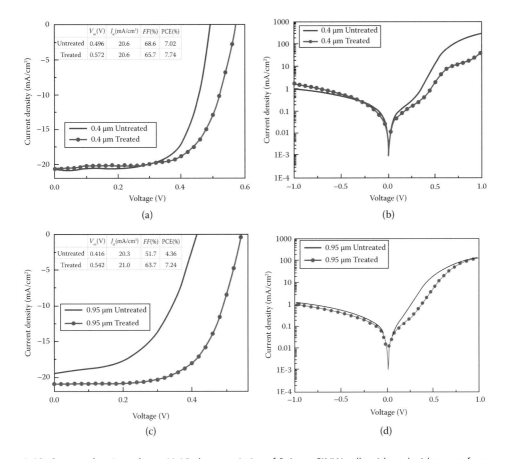

Figure 1.43 Current density-voltage (*J–V*) characteristics of 0.4-μm SiNW cells with and without surface treatment under (a) illumination and (b) dark condition. The corresponding results for 0.95-μm SiNW cells are shown in (c) and (d), respectively. (Reprinted with kind permission from Springer Science+Business Media: *Nanoscale Res. Lett.*, Thin film silicon nanowire/PEDOT:PSS hybrid solar cells with surface treatment, 11, 2016, 311, Wang, H., et al., Copyright 2016.)

A combination of Si semiconductor properties and such important factor distinguishing this material as strong light scattering in it allows novel properties to be reached. In particular, SiNW arrays are characterized by much higher efficiencies of many optical processes (reflection, absorption, photoluminescence, Raman scattering, nonlinear optical interactions) in comparison with c-Si. This opens up new possibilities for SiNW array application as sensors, thermoelectric, and photovoltaic elements.

Arrays, hybrids, and core-shell

REFERENCES

Adachi MM, Khorasaninejad M, Saini SS, and Karim KS (2013). Optical properties of silicon nanowires. In: *UV-VIS and Photoluminescence Spectroscopy for Nanomaterials Characterization* (Kumar CSSR, ed), pp. 357–385. Berlin, Heidelberg: Springer-Verlag.

Adu KW, Gutiérrez HR, Kim UJ, Sumanasekera GU, and Eklund PC (2005). Confined phonons in Si nanowires. *Nano Lett.* 5: 409–414.

Anaya J, et al. (2013). Study of the temperature distribution in Si nanowires under microscopic laser beam excitation. *Appl. Phys. A* 113: 167–176.

Anaya J, et al. (2014). Raman spectrum of Si nanowires: temperature and phonon confinement effects. A. *Appl. Phys. A* 114: 1321–1331.

Baek SH, Park JS, Jeong YM, and Kim JH (2016) Facile synthesis of Ag-coated silicon nanowires as anode materials for high-performance rechargeable lithium battery. *J. Alloys Compd.* 660: 387–391.

Becker M, et al. (2008). Nanowires enabling signal-enhanced nanoscale Raman spectroscopy. *Small* 4: 398–404.

Bestem'yanov KP, Gordienko VM, Ivanov AA, Konovalov AN, and Podshivalov AA (2004). Optical heterodyning study of the propagation dynamics of IR femtosecond laser pulses in a strongly scattering porous medium. *Quant. Electron* 34: 666–668.

Bett AJ, et al. (2014). Front side antireflection concepts for silicon solar cells with diffractive rear side structures. In: *The 29th European PV Solar Energy Conference and Exhibition*, 22–26 September 2014, pp. 987–991. Amsterdam, The Netherlands. München: WIP.

Bodena SA and Bagnall DM (2008). Tunable reflection minima of nanostructured antireflective surfaces. *Appl. Phys. Lett.* 93: 133108.

Boukai AI, Bunimovich Y, Tahir-Kheli J, Yu J-K, Goddard III WA, and Heath JR (2008). Silicon nanowires as efficient thermoelectric materials. *Nature* 451: 168–171.

Brahiti N, Bouanikb SA, and Hadjersi T (2012). Metal-assisted electroless etching of silicon in aqueous NH_4HF_2 solution. *Appl. Surf. Sci.* 258: 5628–5637.

Brönstrup G, Jahr N, Leiterer C, Csáki A, Fritzsche W, and Christiansen S (2010). Optical properties of individual silicon nanowires for photonic devices. *ACS Nano.* 4: 7113–7122.

Bunkov KV, et al. (2013). Dependence of the efficiency of Raman scattering in silicon nanowire arrays on the excitation wavelength. *Semiconductors* 47: 354–357.

Calaza C, et al. (2015). Bottom-up silicon nanowire arrays for thermoelectric harvesting, *Materials Today: Proceedings* 2: 675–679.

Campbell IH and Fauchet PM (1986). The effects of microcrystal size and shape on the one phonon Raman spectra of crystalline semiconductors. *Solid State Commun.* 58: 739–741.

Cao A, Sudhölter EJR, and de Smet LCPM (2014). Silicon nanowire-based devices for gas-phase sensing. *Sensors* 14: 245–271.

Chan CK, et al. (2008). High-performance lithium battery anodes using silicon nanowires. *Nat. Nano* 3: 31–35.

Chang HCh, Lai KYu, Dai YuA, Wang HH, Lin ChA, and He JH (2011). Nanowire arrays with controlled structure profiles for maximizing optical collection efficiency. *Energy Environ. Sci.* 4: 2863–2869.

Chiappini C (2014) MACE silicon nanostructures. In: *Handbook of porous silicon* (Canham L, ed.), pp. 171–186. New York: Springer Reference.

Choi JY, Alford TL, and Honsberg CB (2015). Fabrication of periodic silicon nanopillars in two-dimensional hexagonal array with enhanced control on structural dimension and period. *Langmuir* 31: 4018–4023.

Demichel O, et al. (2008). Photoluminescence of confined electron-hole plasma in core-shell silicon/silicon oxide nanowires. *Appl. Phys. Lett.* 93: 213104.

Demichel O, et al. (2011). Quantum confinement effects and strain-induced band-gap energy shifts in core–shell $Si-SiO_2$ nanowires. *Phys. Rev. B* 83: 245443

Dhindsa N and Saini SS (2016). Comparison of ordered and disordered silicon nanowire arrays: experimental evidence of photonic crystal modes. *Opt. Lett.* 41: 2045–2048.

Efimova A, et al. (2016). Enhanced photon lifetime in silicon nanowire arrays and increased efficiency of optical processes in them. *Opt. Quant. Electron.* 48: 232–240.

Georgobiani VA, Gonchar KA, Osminkina LA, and Timoshenko VYu (2015). Structural and photoluminescent properties of nanowires formed by the metal-assisted chemical etching of monocrystalline silicon with different doping level. *Semiconductors* 49: 1025–1029.

Ghosh R, Giri PK, Imakita K, and Fujii M (2014). Origin of visible and near-infrared photoluminescence from chemically etched Si nanowires decorated with arbitrary shaped Si nanocrystals. *Nanotechnology* 25: 045703.

Glybovsky SB, Tretyakov SA, Belov PA, Kivshar YS, and Simovski CR (2016). Metasurfaces: from microwaves to visible. *Phys. Rep.* 632: 1–72.

Golovan LA, et al. (2012a). Enhancement of the Raman scattering and the third-Harmonic generation in silicon nanowires. *ALT Proceedings* 1. doi: 10.12684/alt.1.81

Golovan LA, Gonchar KA, Osminkina LA, Timoshenko VYu, Petrov GI, and Yakovlev VV (2012b). Coherent anti-Stokes Raman scattering in silicon nanowire ensembles. *Laser Phys. Lett.* 9: 145–150.

Gonchar KA (2015). *Optical properties of scattering media based on silicon nanowires.* Ph.D. Thesis. Lomonosov Moscow State University.

Gonchar KA, Osminkina LA, Sivakov V, Lysenko V, and Timoshenko VYu (2014). Optical properties of nanowire structures produced by the metal-assisted chemical etching of lightly doped silicon crystal wafers. *Semiconductors* 48: 1613–1618.

Gonchar KA, Golovan LA, Gayvoronsky VYa, and Timoshenko VYu (2015). Investigation of optical properties of silicon nanowires prepared by metal-assisted chemical etching. In: *European materials research society (EMRS-2015 fall meeting),* 15–18 September 2015, p. 19. Warsaw, Poland.

Gonchar KA, Zubairova AA, Schleusener A, Osminkina LA, and Sivakov V (2016). Optical properties of silicon nanowires fabricated by environment-friendly chemistry. *Nanoscale Res. Lett.* 11: 357.

Hainey MF and Redwing JM (2016). Aluminum-catalyzed silicon nanowires: growth methods, properties, and applications. *Appl. Phys. Rev.* 3:040806.

Han H, Huang Z, and Lee W (2014). Metal-assisted chemical etching of siliconand nanotechnology applications. *NanoToday* 9: 271–304.

Hochbaum AI, et al. (2008). Enhanced thermoelectric performance of rough silicon nanowires. *Nature* 451: 163–167.

Hochbaum AI, Gargas D, Hwang YJ, Yang P. (2009). Single crystalline mesoporous silicon nanowires. *Nano Lett.* 9: 3550–3554.

Huang CT, et al. (2007). Er-doped silicon nanowires with 1.54 µm light-emitting and enhanced electrical and field emission properties. *Appl. Phys. Lett.* 91: 093133.

Huang JA, et al. (2013). Ordered Ag/Si nanowires array: wide-range surface-enhanced Raman spectroscopy for reproducible biomolecule detection. *Nano Lett.* 13: 5039–5045.

Huang Z, et al. (2009). Ordered arrays of vertically aligned [110] silicon nanowires by suppressing the crystallographically preferred <100> etching directions. *Nano Lett.* 9: 2519–2525.

Huang Z, Geyer N, Werner P, de Boor J, and Gösele U (2011). Metal-assisted chemical etching of silicon: a review. *Adv. Mat.* 23: 285–308.

Huang ZP, Wang RX, Jia D, Maoying L, Humphrey MG, and Zhang C (2012). Low-cost, large-scale, and facile production of Si nanowires exhibiting enhanced third-order optical nonlinearity. *ACS Applied Mat. Interfaces* 4: 1553–1558.

In HJ, Field CR, and Pehrsson PE (2011). Periodically porous top electrodes on vertical nanowire arrays for highly sensitive gas detection. *Nanotechnology* 22: 355501.

Johnson PM, Imhof A, Bret BPJ, Rivas JG, and Lagendijk A (2003). Time-resolved pulse propagation in a strongly scattering material. *Phys. Rev. E* 68 (2003): 016604.

Jung D, Cho SG, Moon T, and Sohn H (2016). Fabrication and characterization of porous silicon nanowires. *Electron. Mater. Lett.* 12: 17–23.

Jung JY, Guo Zh, Jee SW, Um HD, Park KT, and Lee JH (2010). A strong antireflective solar cell prepared by tapering silicon nanowires. *Opt. Expr.* 18: A286–A292.

Kazemi-Zanjani N, Kergrene E, Liu L, Sham T-K, and Lagugné-Labarthe F (2013). Tip-enhanced Raman imaging and nano spectroscopy of etched silicon nanowires. *Sensors* 13: 12744–12759.

Ke Y, et al. (2011). Single wire radial junction photovoltaic devices fabricated using aluminum catalyzed silicon nanowires. *Nanotechnology* 22: 445401.

Kelzenberg MD, et al. (2008). Photovoltaic measurements in single-nanowire silicon solar cells. *Nano Lett.* 8: 710–714.

Kempa TJ, et al. (2012). Coaxial multishell nanowires with high-quality electronic interfaces and tunable optical cavities for ultrathin photovoltaics. *Proc. Natl. Acad. Sci. U. S. A.* 109: 1407.

Khorasaninejad M and Saini SS (2010). Silicon nanowire optical waveguide (SNOW). *Opt. Exp.* 18: 23442.

Khorasaninejad M, Swillam MA, Pillai K, and Saini SS (2012b). Silicon nanowire arrays with enhanced optical properties. *Opt. Lett.* 37: 4194–4196.

Khorasaninejad M, Walia J, and Saini SS (2012a). Enhanced first-order Raman scattering from arrays of vertical silicon nanowires. *Nanotechnology* 23: 275706.

Ko M, Baek SH, Song B, Kang JW, Kim SA, and Cho CH (2016). Periodically diameter-modulated semiconductor nanowires for enhanced optical absorption. *Adv. Mater.* 28: 2504–2510.

Koynov S, Brandt MS, and Stutzmann M (2006). Black nonreflecting silicon surfaces for solar cells. *Appl. Phys. Lett.* 88: 203107–203107.

Krupenkin TN, Taylor JA, Schneider TM, and Yang S (2004). From rolling ball to complete wetting: the dynamic tuning of liquids on nanostructured surfaces. *Langmuir* 20: 3824–3827.

Kumar V, Saxena SK, Kaushik V, Saxena K, Shukla AK, and Kumar R (2014). Silicon nanowires prepared by metal induced etching (MIE): good field emitters. *RSC Adv.* 4: 57799–57803.

Kuyken B, et al. (2016). Nonlinear optical interactions in silicon waveguides. *Nanophotonics* 5: 1–16.

Lai RA, Hymel TM, Narasimhan VK, and Cui Y (2016). Schottky barrier catalysis mechanism in metal-assisted chemical etching of silicon. *Appl. Mater. Interfaces* 8: 8875–8879.

Lajvardi M, Eshghi H, Ghazi ME, Izadifard M, and Goodarzi A (2015). Structural and optical properties of silicon nanowires synthesized by Ag-assisted chemical etching. *Mat. Sci. Semicond. Process.* 40: 556–563.

Li Y, et al. (2015). A comparison of light-harvesting performance of silicon nanocones and nanowires for radial-junction solar cells. *Sci. Rep.* 5: 11532; doi: 10.1038/srep11532.

Liu X, Coxon PR, Peters M, Hoex B, Cole JM, and Fray DJ (2014). Black silicon: fabrication methods, properties and solar energy applications. *Energy Environ. Sci.* 7: 3223–3263.

Mehdipour H and Ostrikov K (2013). Size- and orientation-selective Si nanowire growth: thermokinetic effects of nanoscale plasma chemistry. *J. Am. Chem. Soc.* 135: 1912–1918.

Miyamoto Y and Hirata M (1976). Role of agents in filamentary growth of amorphous silicon. *Jpn. J. Appl. Phys.* 15: 1159–1160.

Morales AM and Lieber CM (1998). A laser ablation method for the synthesis of crystalline semiconductor nanowires. *Science* 279: 208–211.

Moutanabbir O, et al. (2011). Atomically smooth p-doped silicon nanowires catalized by aluminum at low temperature. *ACS Nano* 5: 1313.

Muskens OL, Rivas JG, Algra RE, Bakkers EPAM, and Lagendijk A (2008). Design of light scattering in nanowire materials for photovoltaic applications. *Nano Lett.* 8: 2638–2642.

Nebol'sin VA and Shchetinin AA (2003). Role of surface energy in the vapor–liquid–solid growth of silicon. *Inorg. Mater.* 39: 899–903.

Nebol'sin VA, Shchetinin AA, Dolgachev AA, and Korneeva VV (2005). Effect of the nature of the metal solvent on the vapor-liquid-solid growth rate of silicon whiskers. *Inorg. Mater.* 41: 1256–1259.

Noh J, Kim H, Kim B, Lee E, Cho H, and Lee W (2011). High-performance vertical hydrogen sensors using Pd-coated rough Si nanowires. *J. Mater. Chem.* 21: 15935–15939.

Oda K, Nanai Y, Sato T, Kimura S, and Okuno T (2014). Correlation between photoluminescence and structure in silicon nanowires gfabricated by metal-assisted etching. *Phys. Status Solidi A* 211: 848–855.

Osada Y, Nakayama H, Shindo M, Odaka T, and Ogata Y (1979). Growth and structure of silicon fibers. *J. Electrochem. Soc.* 126: 31–36.

Osminkina LA, et al. (2012). Optical properties of silicon nanowire arrays formed by metal-assisted chemical etching: evidences for light localization effect. *Nanoscale Res. Lett.* 7: 524.

Peng K, Y Xu, Wu Y, Yan Y, Lee ST, and Zhu J (2005). Aligned single-crystalline Si nanowire arrays for photovoltaic applications. *Small* 1: 1062–1067.

Peng KQ, et al. (2006). Fabrication of single-crystalline silicon nanowires by scratching a silicon surface with catalytic metal particles. *Adv. Funct. Mater.* 16: 387–394.

Peng KQ, Wang X, and Lee ST (2009). Gas sensing properties of single crystalline porous silicon nanowires. *Appl. Phys. Lett.* 95: 243112.

Qiu T, Wu XL, Shen JC, Ha PCT, and Chu PK (2006). Surface-enhanced Raman characteristics of Ag cap aggregates on silicon nanowire arrays. *Nanotechnology* 17: 5769.

Qu Y, Liao L, Li Y, Zhang H, Huang Y, and Duan X (2009). Electrically conductive and optically active porous silicon nanowires. *Nano Lett.* 9: 4539–4543.

Richter H, Wang ZP, and Ley Y (1981). The one phonon Raman spectrum in microcrystalline silicon. *Solid State Commun.* 39: 625–629.

Rodichkina SP, et al. (2015). Raman diagnostics of photoinduced heating of silicon nanowires prepared by metal-assisted chemical etching. *Appl. Phys. B.* 121: 337–344.

Rouquerol J, et al. (1994). Recommendations for the characterization of porous solids. *Pure Appl. Chem.* 66: 1739–1758.

Sadeghian RB and Islam MS (2011). Ultralow-voltage field-ionization discharge on whiskered silicon nanowires for gas-sensing applications. *Nat. Mat.* 10: 135–140.

Schmidt V, Wittemann JV, Senz S, and Gösele U (2009). Silicon nanowires: a review on aspects of their growth and their electrical properties. *Adv. Mater.* 21: 2681–2702.

Seo K, et al. (2011). Multicolored vertical silicon nanowires. *Nano Lett.* 11: 1851–1856.

Singh N, et al. (2015). Midinfrared supercontinuum generation from 2 to 6μm in a silicon nanowire. *Optica* 2: 797–802.

Sivakov V, et al. (2009). Silicon nanowire-based solar cells on glass: synthesis, optical properties, and cell parameters. *Nano Lett.* 9: 1549–1554.

Sivakov VA, et al. (2010a). Realization of vertical and zigzag single crystalline silicon nanowire architectures. *J. Phys. Chem. C.* 114: 3798–3803.

Sivakov VA, Voigt F, Berger A, Bauer G, and Christiansen SH (2010b). Roughness of silicon nanowire sidewalls and room temperature photoluminescence. *Phys. Rev. B* 82: 125446.

Sivakov V, Voigt F, Hoffmann B, Gerliz V, and Christiansen S (2011). Wet - chemically etched silicon nanowire architectures: formation and properties. In: *Nanowires - Fundamental Research* (Hashim A, ed.), pp. 45–80. Rijeka, Croatia: InTech. DOI: 10.5772/16736.

Sun X, Lin L, Li Z, Zhang Z, and Feng J (2009). Fabrication of silver-coated silicon nanowire arrays for surface-enhanced Raman scattering by galvanic displacement processes. *Appl. Surf. Sci.* 256: 916–920.

Theiß W (1997). Optical properties of porous silicon. *Surf. Sci. Rep.* 29: 91–192.

Timoshenko VYu, et al. (2011). Photoluminescence and Raman scattering in arrays of silicon nanowires. *Journal of Nanoelectronics and Optoelectronics* 6: 519–524.

Treuting RG and Arnold SM (1957). Orientation habits of metal whiskers. *Acta Met.* 5: 598.

Tsakalakos L, et al. (2007). Strong broadband optical absorption in silicon nanowire films. *J Nanophoton.* 1: 013552.

Venturi G, Castaldini A, Schleusener A, Sivakov V, and Cavallini A (2015). Electronic levels in silicon MAWCE nanowires: evidence of a limited diffusion of Ag. *Nanotechnology* 26: 425702.

Wagner RS and Ellis WC (1964). Vapor-liquid-solid mechanism of single crystal growth. *Appl. Phys. Lett.* 4: 89–90.

Wagner RS and Ellis WC (1965). The vapor-liquid-solid mechanism of crystal growth and its application to silicon. *Trans. Met. Soc. AIME* 233: 1053–1064.

Wanekaya AK, Chen W, Myung NV, and Mulchandani A (2006). Nanowire-based electrochemical biosensors. *Electroanalysis* 18: 533–550.

Wang H, Wang J, Hong L, Tan YH, Tan CS, and Rusli (2016). Thin film silicon nanowire/PEDOT:PSS hybrid solar cells with surface treatment. *Nanosc. Res. Lett.* 11: 311.

Wang N, et al. (1998). Transmission electron microscopy evidence of the defect structure in Si nanowires synthesized by laser ablation. *Chem. Phys. Lett.* 283: 368–372.

Wang Y, Schmidt V, Senz S, and Gösele U (2006). Epitaxial growth of silicon nanowires using an aluminium catalyst. *Nat. Nano.* 1: 186–189.

Wang XT, Shi WS, She GW, Mu LX, and Lee ST (2010). High-performance surface-enhanced Raman scattering sensors based on Ag nanoparticles-coated Si nanowire arrays for quantitative detection of pesticides. *Appl. Phys. Lett.* 96: 053104.

Weber WM, et al. (2006). Silicon nanowires: catalytic growth and electrical characterization. *Phys. Stat Sol. b*, 243: 3340–3345.

Whang SJ, Lee SJ, Yang WF, Cho BJ, Liew YF, and Kwong DL (2007). Complementary metal-oxide-semiconductor compatible Al-catalyzed silicon nanowires. *Electrochem. Solid-State Lett.* 10: E11–E13.

Williams EH, et al. (2012). Selective streptavidin bioconjugation on silicon and silicon carbide nanowires for biosensor applications. *J. Mater. Res.* 28: 68–77.

Wu C, et al. (2001). Near-unity below-band-gap absorption by microstructured silicon. *Appl. Phys. Lett.* 78: 1850–1852.

Wu Y, Cui Y, Huynh L, Barrelet CJ, Bell DC, and Lieber CM (2004). Controlled growth and structures of molecular-scale silicon nanowires. *Nano Lett.* 4: 433–436.

Yablonovitch E (1982). Statistical ray optics. *JOSA* 72: 899–907.

Yae S, Morii Y, Fukumuro N, and Matsuda H (2012). Catalytic activity of noble metals for metal-assisted chemical etching of silicon. *Nanoscale Res. Lett.* 7: 352.

Yi C, et al. (2010). Patterned growth of vertically aligned silicon nanowire arrays for label-free DNA detection using surface-enhanced Raman spectroscopy (July 2009 – June 2010) *Anal. Bioanal. Chem.* 397: 3143–3150.

Yin J, Qi X, Yang L, Hao G, Li J, and Zhong J (2011). A hydrogen peroxide electrochemical sensor based on silver nanoparticles decorated silicon nanowire arrays *Electrochim. Acta* 56: 3884–3889.

Yu N and Capasso F (2014). Flat optics with designer metasurfaces. *Nat. Mater.* 13: 139–150.

Yu P, et al. (2013). 13% efficiency hybrid organic/silicon nanowire heterojunction solar cell via interface engineering. *ACS Nano* 7: 10780–10787.

Zhang B, Wang H, Lu L, Ai K, Zhang G, and Cheng X (2008). Large-area silver-coated silicon nanowire arrays for molecular sensing using surface-enhanced Raman spectroscopy. *Adv. Funct. Mater.* 18: 2348–2355.

Zhang C, et al. (2013). Enhanced photoluminescence from porous silicon nanowire arrays. *Nanoscale Res. Lett.* 8: 277.

Zhang ML, et al. (2008). A surface-enhanced Raman spectroscopy substrate for highly sensitive label-free immunoassay. *Appl. Phys. Lett.* 92: 043116.

Zhu J, et al. (2009). Optical absorption enhancement in amorphous silicon nanowire and nanocone arrays. *Nano Lett.* 9: 279–282.

Arrays, hybrids, and core-shell

2 Inverted silicon nanopencil arrays

Fei Xiu and Johnny C. Ho

Contents

2.1 INTRODUCTION

Inspired by the rapid development of nanotechnology, fabrications of various subwavelength nanostructures have been successfully demonstrated to achieve many enhanced physical properties such as the antireflection performances and so on.[1–5] Specifically, as illustrated in Figure 2.1, many different types of photon management structures have been extensively explored, which include vertical nanowire, nanocone, nanowell, nanopyramid arrays, and so on, exhibiting impressive light-trapping properties with the minimal active material thickness required.[6–9] All of these structures with excellent light management properties can then be utilized in numerous optoelectronic devices, desiring the proper photon management. The typical applications involve solar cells, photodetectors, and phototransistors, among others.[1,10–12]

In general, there are three major principles of light-coupling mechanisms employed in these three-dimensional (3D) nanostructures in order to perform the effective photon trapping.[13–16] As shown in Figure 2.1, vertical nanowire arrays with a high aspect ratio display a remarkable antireflection characteristic, which can be attributed to the prolongation of the optical path length of light by simply increasing the frequency of reflected waves bouncing between the structures.[6] The light propagation nature within nanowire, nanocone, and nanorod structures has been further explored by the finite-difference time-domain (FDTD) analysis. Figure 2.1f and g depict the simulated solar-spectrum-weighted electrical field intensity $|E(r)|^2$ and corresponding generation rate $G(r)$ for Si nanowires and nanocones.[17] It is found that the antireflection properties are highly dependent on the geometry of nanostructures utilized, in which the better light absorption could be observed in nanoarrays with the higher aspect ratio. As compared with nanowires, the stronger electrical field intensity and the higher generation rate are observed in nanocones due to the suppressed light reflection, which is related to the unique tapered shape of nanocone structures.[2] As shown in Figure 2.2h, the time-averaged and normalized transverse mode electric (TE) field existing within ZnO nanorods indicates that nanorod structures can facilitate the light propagation and widen the field distribution for the enhanced light scattering, resulting in a significant improvement of the

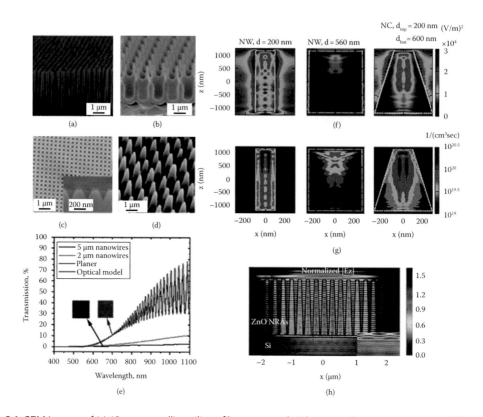

Figure 2.1 SEM images of (a) 10 μm crystalline silicon film patterned with inverted nanopyramids on a 700-nm-period. (b) 1-μm-height nanopillars on 1-μm-depth nanowells with a 710 nm well diameter. (c) Ordered silicon nanowire arrays made by deep reactive ion etching. (d) a-Si:H nanocones. (e) Transmission spectra of thin silicon window structures before (red) and after etching to form 2 μm (green) and 5 μm (black) nanowires. The spectra from an optical model for a 7.5 μm thin silicon window is in blue and matches very well with the planar control measurement. The insets are backlit color images of the membranes before and after etching. Clearly there is a large-intensity reduction and red shift in the transmitted light after the nanowires are formed, suggesting strong light trapping. (f) The solar-spectrum-weighted electrical field intensity |E(r)|2 and (g) solar-spectrum-weighted generation rate G(r) for three representative silicon nanowires and nanocones. From left to right, the systems shown are nanowire arrays with d = 200 nm, nanowire arrays with d = 560 nm and nanocone arrays with d$_{top}$ = 200 nm and d$_{bot}$ = 600 nm. (h) Time-averaged and normalized TE electric field, |Ez|, distribution at 543 nm simulated by FDTD analysis within ZnO nanorod structures. (Reproduced from Garnett, E. & Yang, P., Light trapping in silicon nanowire solar cells, *Nano Lett.*, 10, 1082–1087, 2010, Copyright 2010; Lin, Q., et al., Efficient light absorption with integrated nanopillar/nanowell arrays for three-dimensional thin-film photovoltaic applications, *ACS Nano*, 7, 2725–2732, 2013, Copyright 2013; Mavrokefalos, A., et al., Efficient light trapping in inverted nanopyramid thin crystalline silicon membranes for solar cell applications, *Nano Lett.*, 12, 2792–2796, 2012, Copyright 2012; Zhu, J., et al., Optical absorption enhancement in amorphous silicon nanowire and nanocone arrays, *Nano Lett.*, 9, 279–282, 2009, Copyright 2009; Wang, B. & Leu, P.W., Enhanced absorption in silicon nanocone arrays for photovoltaics, *Nanotechnology* 23, 194003, 2012, Copyright 2012; Tsai, D.-S., et al., Ultra-high-responsivity broadband detection of Si metal-semiconductor-metal Schottky photodetectors improved by ZnO nanorod arrays, *ACS Nano* 5, 7748–7753, 2011, Copyright 2011. With permission from The American Chemical Society and The Institute of Physics.)

corresponding responsivity.[18] Other tapered nanostructures, such as nanodomes as fabricated in Figure 2.2a–c, can as well demonstrate the enhanced light-coupling behavior for efficient photon trapping.[19] In any case, a nanocone-based structure with the tapered morphological shape is widely recognized as an ideal structure for efficient light harvesting because of its gradual reduction of effective refractive index between the structure and the air, as indicated in the superior omnidirectional photovoltaic performances in silicon nanocone solar cell devices presented in Figure 2.2d–l.[9]

Instead of high-aspect-ratio nanostructures with the efficient light-trapping properties, silica microspheres or nanospheres can also be utilized as another effective light management structure in order to enhance the light scattering, in which the scattering events are found to be closely related to the size of silica spheres

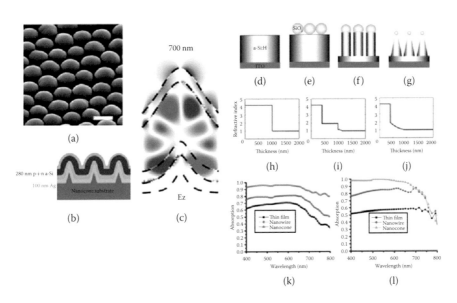

Figure 2.2 (a) Structure (scale bar = 500 nm), (b) schematic, and (c) electromagnetic simulation of an amorphous silicon nanodome cell showing the presence of an electric field component in the vertical direction, which indicates coupling into guided modes in the active layer. (d–g) Schematic illustration of 1 μm thick a-Si:H on ITO-coated glass substrate, a monolayer of silica nanoparticles on top of a-Si:H thin film, NW arrays, and nanocone (NC) arrays. The effective refractive index profiles of the interfaces between air and (h) a-Si:H thin film, (i) 600 nm a-Si:H NW arrays, and (j) 600 nm a-Si:H NC arrays. (k) Measured value of absorption on samples with a-Si:H thin film, NW arrays, and NC arrays as top layer over a large range of wavelengths at normal incidence. (l) Measured results of absorption on samples with a-Si:H thin film, NW arrays, and NC arrays as top layer over different angles of incidence (at wavelength λ= 488 nm). (Reproduced with permission from Zhu, J., et al., Nanodome solar cells with efficient light management and self-cleaning, *Nano Lett.*, 10, 1979–1984, 2010, Copyright 2010 American Chemical Society; Zhu, J., et al., Optical absorption enhancement in amorphous silicon nanowire and nanocone arrays, *Nano Lett.*, 9, 279–282, 2009, Copyright 2009 American Chemical Society.)

as the light diffusion centers.[20] Based on the FDTD simulations as displayed in Figure 2.3a and b,[20] it is noticed that when the size of silica microspheres is larger than the wavelength of incident light, strong interference patterns can then be observed at both near and far distances from the surface, achieving the effective diffuse light scattering. In contrast, the subwavelength silica spheres cannot be employed to diffuse the light propagation because all scattering events would be suppressed by the subwavelength structure except the zeroth-order diffraction. As a result, proper selection of the dimension of those spheres would determine the underlying photon scattering and propagation characteristics into the active device layer.

At the same time, there is a similar concept to realize the light guiding by forming whispering-gallery resonant modes (WGMs) within a spherical nanoshell structure, which refers to a type of wave traveling around the surface of a concave structure. WGMs would typically occur at particular resonant wavelengths of light for a given shell geometry with the specific cavity size and shape. The coupled light would then be trapped within the guide or void for timescales of the order of nanoseconds when it undergoes total internal reflection at the inner surface of a cavity, increasing the total light absorption due to the increased photon path length.[21] The prolonged optical paths in WGM would lead to the enhanced total absorption and electron-hole (e-h) pair generation.[22,23] For instance, as shown in Figure 2.3c–e, when the TiO$_2$ sphere with variable diameter is used as a model system, a broadband and omnidirectional light absorption enhancement is observed.[23] In detail, Figure 2.3f illustrates the simulated electric field (|E|) distribution established in the model, from which a very strong resonant enhancement can be seen. More important, when the geometry of the sphere (diameter of TiO$_2$) is carefully designed, these resonances and light trapping can be optimized for the maximum light absorption. This geometric effect has been further confirmed by the electric field distribution of a simulated sphere structure with the varying particle size, as given in Figure 2.3g and h.[23] The principle of this optical resonance can be reinforced by considering that the near field intensity and gradients lead to the locally increased e-h generation rates, being beneficial to many optical and optoelectronic utilizations.[24]

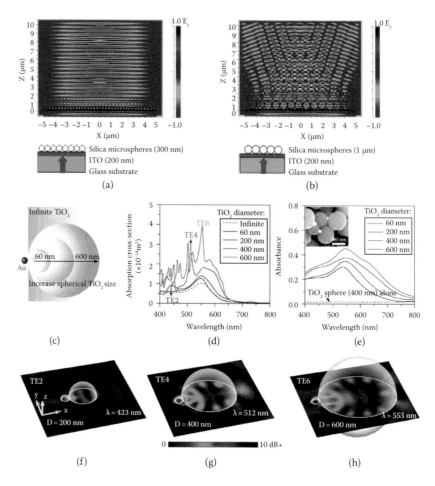

Figure 2.3 Calculated electric fields of light passing through the (a) silica microspheres (300 nm)/ITO/glass, (b) silica microspheres (1 μm)/ITO/glass. The thickness of ITO film is fixed at 200 nm. (c) Schematic of the WGM model system: a 60 nm Au NP is supported on an infinite TiO$_2$ plane or on TiO$_2$ spheres with various diameters. (d) Averaged absorption cross section calculated using FEM simulations. (e) Experimentally measured absorption spectra of isolated 60 nm Au NPs (Au monomers) supported on a TiO$_2$ particle with various diameters (Au mono-TiO$_2$). (f–h) Simulated electric field distribution of Au mono-TiO$_2$ for TiO$_2$ diameter of 200, 400, and 600 nm, respectively. The modes shown (TE2, TE4, and TE6) are identified in panel b. (Reproduced from Ko, Y.H. & Yu, J.S., Urchin-aggregation inspired closely-packed hierarchical ZnO nanostructures for efficient light scattering, *Opt. Express*, 19, 25935–25943, 2011, Copyright 2011; Zhang, J., et al., Engineering the absorption and field enhancement properties of Au–TiO$_2$ nanohybrids via whispering gallery mode resonances for photocatalytic water splitting, *ACS Nano*, 10, 4496–4503, 2016, Copyright 2016. With permission from The Optical Society of America and The American Chemical Society.)

Although the earlier discussion on different light management mechanisms has illustrated the technological potency of these various microstructures or nanostructures for the efficient broadband and omnidirectional light harvesting in practical applications of photovoltaic and so on, it is well known that the performance of optoelectronic devices (e.g., solar cells) is largely dependent on both the optical absorption behavior of the cells as well as the carrier collection dynamics of the devices, indicating the necessity of surface texturing with nanostructures as the efficient photon harvester and carrier collector. On the other hand, thinner devices with the consumption of less material are highly preferred for the cost-effective fabrication and subsequent operation; however, as depicted in Figure 2.4, when the device thickness is decreased, this may result in an incomplete absorption of photons,[25] highlighting the importance of employing ultrathin nanostructures as the effective light absorber for the thin devices. In particular, silicon is intrinsically weak for the optical absorption in infrared or near-infrared regions; therefore, a sufficient silicon substrate thickness is required for the efficient photon absorption in the long wavelength range for photovoltaics. As the silicon substrate thickness decreases, the light-trapping effect of nanostructures becomes more important to extend the light propagation for the effective photon harvesting.[26,27] For example, ultrathin nanocone-based monocrystalline silicon solar cells with a substrate thickness of only 10 μm were reported with the optical image shown in Figure 2.5a.[27] As indicated in the external quantum efficiency (EQE) data and *J-V* characteristics of both nanocone and planar control devices in Figure 2.5b and c, a 30.7% higher short-circuit current density is observed for the nanocone cells as compared with that of the planar control sample even coated with the optimized Si_3N_4 antireflection layer, suggesting the performance enhancement induced by the nanocone surface texturization.

Apart from the aforementioned nanostructures, our previous studies have investigated a novel nanostructure with the inverted pencil appearance, as given in Figure 2.6, the light-trapping capability of

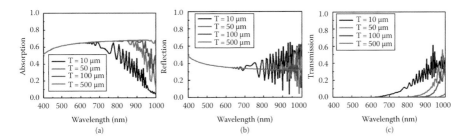

Figure 2.4 Simulation of optical properties for planar silicon with different substrate thickness. (a) Absorption, (b) reflection, and (c) transmission. (Reproduced with permission from from Lin, H., et al., Rational design of inverted nanopencil arrays for cost-effective, broadband, and omnidirectional light harvesting. *ACS Nano* 8, 3752–3760, 2014, Copyright 2014 American Chemical Society.)

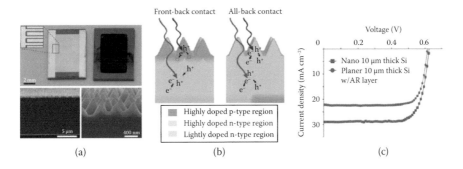

Figure 2.5 (a) Optical image of the back (top, left) and front (top, right) side of the 10 μm thick Si solar cell. The inset shows the optical microscope image of the interdigitated metal electrodes. SEM images of the cross-sectional view of the device (bottom, left) and cross-sectional view of the nanocones (bottom, right). The thin layer at the top of the nanocones is an 80 nm thick SiO_2 layer. (b) EQE data of the nanocone device and a planar control. (c) J–V characteristics of a nanocone device and a planar control sample. (Reproduced by permission from Macmillan Publishers Ltd. *Nat. Commun.*, Jeong, S., et al., All-back-contact ultra-thin silicon nanocone solar cells with 13.7% power conversion efficiency, 4, 345–50, 2013, copyright 2013.)

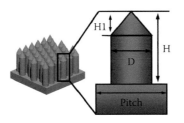

Figure 2.6 Schematic illustration of inverted nanopencil arrays studied in this chapter.

which could be comparable or even better than the nanocone structures.[25] In brief, we performed systematic studies on their low-cost and large-scale fabrication, optical behavior, light-trapping mechanism, design principle, and device applications, where all these findings would provide useful information for utilizing these inverted nanopencil arrays as promising active materials for photovoltaic and optoelectronic device applications. In this chapter, we will begin with a comprehensive review of our recently developed, simple, wet-chemistry-only fabrication scheme for the controllable hierarchy of nanopencil arrayed structures, followed by the detailed experimental and simulated results of their optical characteristics. The light-trapping mechanism and corresponding design principle of inverted nanopencils will be thoroughly discussed as well in order to obtain insights into the technological potency of these pencil arrays as efficient photon harvesters for photovoltaics. Other potential applications of these nanostructures in optoelectronics will also be explored. Finally, we will provide concluding summary and important remarks with perspectives on the future development of next-generation nanostructured silicon photovoltaics with optimized photovoltaic performances.

2.2 FABRICATION OF HIGHLY REGULAR, LARGE-SCALE INVERTED SILICON NANOPENCIL ARRAYS

Until now, numerous methods involving both "top-down" and "bottom-up" approaches have been developed for the controllable fabrication of 3D silicon nanostructures, which is crucial for the realization of enhanced light harvesting properties and PV applications. These approaches mainly include vapor–liquid–solid (VLS) growth, reactive ion etching (RIE), metal-assisted chemical etching, and so on.

Among many fabrication methods, the bottom-up schemes simply employ the selective material deposition to achieve nanostructured arrays. For instance, the metal-catalyzed VLS growth technique has been utilized to realize the fabrication of vertical silicon nanowire arrays with wire diameters ranging from ~5 nm all the way to several hundred nanometers and lengths from ~100 nm to tens of micrometers.[28–30] In this process, growth atoms are transported to a furnace from the gas precursor molecules or vapor containing the growth species, and metal nanoparticles react with the source atoms, catalyzing the growth of nanowires by forming a liquid metal droplet at an elevated temperature.[28] This procedure is known as the chemical vapor deposition (CVD), during which the catalytic seed floats on top of the nanowire as it grows and defines the diameter while the length is determined by the growth rate kinetics together with the growth time.[28] Silicon constituents can be provided by either silane (SiH_4) or tetrachlorosilane ($SiCl_4$), which is usually used as the gas precursor. As shown in Figure 2.7a, during the growth, the substrate temperature is held above the eutectic temperature of the Au–Si alloy once the vaporized Si atoms reach the substrate surface, which is covered with metal nanoparticles. This leads to the supersaturation of Au–Si liquid alloy droplets with silicon atoms, and then the growth of Si nanowires can be realized by the precipitation of Si from the catalytic alloys. To date, VLS is widely considered as an attractive method due to its process simplicity and potential suitability for the large-area device fabrication, and there have been many in-depth studies to accomplish various hierarchical silicon nanostructures based on the VLS technique.[31,32]

In addition, deep reactive ion etching (DIRE) is another effective and common method to fabricate vertical Si nanostructures via the top-down scheme through high-aspect-ratio anisotropic etching of the substrate.[33–36] In fact, the anisotropic DIRE of silicon has been already established as a mature technology in the standard semiconductor device fabrication to accomplish high-aspect-ratio structures for many different applications.[37] As depicted in Figure 2.7b, a typical DRIE process includes a unique Si etching using high-density plasma followed by the polymer deposition to prevent the already-carved features being further laterally etched;[38] however, during the etching, "scalloping" (or "ripple") morphology is easy to be formed on the sidewalls via a series of bites into the silicon, and it can be a potentially serious problem when the feature size and scallop size become comparable.[37] For example, this defect-rich surface can result in a serious limitation of the as-made nanostructures for PV applications due to the existence of these defects as scattering and carrier recombination centers to degrade the corresponding performance.[33] To ensure smooth sidewalls in the etching, the ripple effect must be minimized through controlling the process conditions, including optimizing the source power, gas flow rate, etching time, temperature, and so on.[39] Based on the optimized DRIE process, large-area silicon nanostructures, including nanowires,

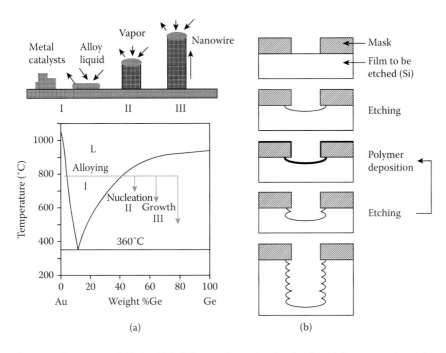

Figure 2.7 Schematic illustration of (a) the VLS NW growth mechanism (top) and the conventional Au-Ge binary phase diagram to demonstrate the composition and phase evolution during the NW growth process (bottom), (b) individual steps in the deep reactive ion etching (DRIE) process. (Reproduced from Wu, Y. & Yang, P., Direct observation of vapor-liquid-solid nanowire growth, *J. Am. Chem. Soc.*, 123, 3165–3166, 2001, Copyright 2001; Rhee, H., et al., Comparison of deep silicon etching using SF_6/C_4F_8 and SF_6/C_4F_6 plasmas in the Bosch process, *J. Vac. Sci. Technol. B*, 26, 576, 2008, Copyright 2008, with permission from The American Chemical Society and The American Institute of Physics.)

nanopillars, and nanowalls with smooth sidewalls, have been successfully achieved, indicating the great potential of DRIE for nanofabrication of 3D features.[39]

Despite DRIE being a well-established and an effective technique for the top-down nanostructure formation, wet etching by chemical solutions is still considered as the standard preferred process for the commercial solar cell surface texturization due to its low cost and simplicity. Although wet etching is isotropic in nature, there is a recent advanced development to introduce the etching directionality by coupling catalytic effects, which is known as the metal-assisted chemical etching (MacEtch). MacEtch is a simple and low-cost technique widely employed to obtain various highly regular silicon nanostructures that has attracted increasing attention in recent years for several advantages.[31,40] First, MacEtch is an effective approach to achieve various silicon nanostructures with the ability to control geometrical parameters precisely for PV devices, including the diameter, length, orientation, cross-sectional shape, doping type and level, and so on.[41] Second, as compared with DRIE, MacEtch is free of surface damage because there are no high-energy ions involved in the process. Third, as a more versatile technique, MacEtch can be used to realize features with the higher surface-to-volume ratio and, more important, it is readily scalable to the large-scale production. As shown in Figure 2.8, in a typical MacEtch process, silicon substrates are partly covered by a noble metal and then subjected to an etchant composed of HF and an oxidative agent, such as H_2O_2.[41] The noble metal plays a catalytic role in the etching reaction. Usually, silicon beneath the noble metal is etched much faster than the region without any noble metal coverage, resulting in the generation of pores, wires, or pillars in the processed substrate.

As a standard etching, MacEtch can produce sharp and well-controlled features; nevertheless, it is still not a suitable scheme for the controllable hierarchy of fine-designed nanostructures with tapered geometry, shape, and gradual change of material-filling ratio because the chemical etchants preferentially erode the underlying substrate equally in all directions. In our previous work, we have demonstrated a facial and anisotropic wet-chemistry-only fabrication scheme for the controllable hierarchy of tapered silicon

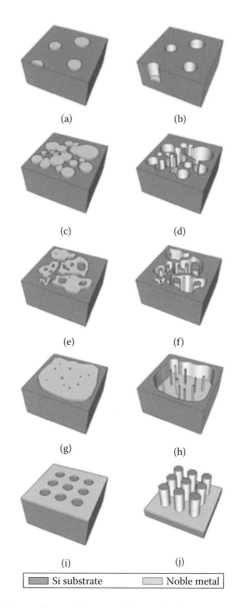

Figure 2.8 Process schematics of typical morphologies of the etched structures (right column) induced by metal catalysts with differently shaped noble metals (left column). (Reproduced from Huang, Z., et al: Metal-assisted chemical etching of silicon: A review. *Adv. Mater.* 2011. 23. 285–308, Copyright Wiley-VCH Verlag GmbH & Co. KGaA. With permission.)

nanoarrays with different geometrical morphologies, ranging from nanorods, nanocones to nanopencils over large areas, which is particularly attractive for applications requiring large-scale, low-cost, and efficient light management. Notably, for the first time, inverted nanopencil arrays were realized using this anisotropic wet-etching technique. The detailed fabrication method of various tapered silicon nanoarrays, especially nanopencil structures, via the aforementioned anisotropic wet-chemistry-only method as well as the corresponding etching mechanisms will be discussed in detail in the following sessions.

2.2.1 FABRICATION METHOD OF NANOPENCIL ARRAYS

As shown in Figure 2.9, silicon nanopillars with controllable diameters and periodicities were used as templates for the nanopencil fabrication, which were prepared via nanosphere lithography followed by the MacEtch process in HF/H$_2$O$_2$ solution.[42] First, a close pack monolayer of polystyrene (PS) was assembled

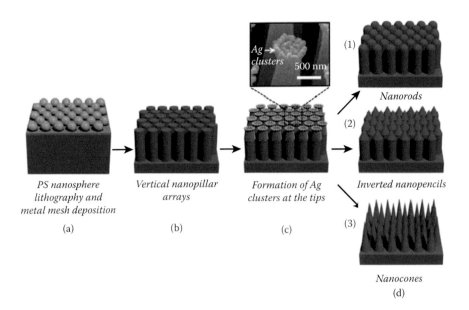

Figure 2.9 Process schematics for the formation of different morphological nanoarrays via the anisotropic wet-etching technique, including nanopillars, nanorods, inverted nanopencils, and nanocones. The inset shows the zoom-in SEM image of the nanopillar tips, demonstrating the selective deposition of Ag nanoclusters. (Reproduced from Lin, H., et al., Rational design of inverted nanopencil arrays for cost-effective, broadband, and omnidirectional light harvesting, *ACS Nano*, 8, 3752–3760, 2014, Copyright 2014, with permission from The American Chemical Society.)

on silicon (100) substrates employing the Langmuir–Blodgett (LB) method and treated by subsequent oxygen plasma etching to control the dimension and geometrical pitch of the spheres. These obtained spheres were then used as the mask, and Ti/Au metal mesh was thermally evaporated onto the silicon substrate, resulting in the formation of nanopillars with controllable geometries.

The as-made nanopillar templates were next treated with a mixture of ($AgNO_3$, HF, and HNO_3 or H_2O_2) etching solution, leading to the selective deposition of Ag clusters in the rim region of nanopillar tips, which performed the first function of this mixture.[42] At the same time, the ($AgNO_3$ + HF + HNO_3/H_2O_2) system would lead to the anisotropic etching of Si. In short, once the metal nanoclusters were formed, truncated pillars could be obtained by the subsequent HNO_3 washing to remove the cluster particles, and different morphological nanopillar arrays could be achieved by repeating this deposition/removal process multiple times. In this way, various nanoarrays, including nanopillars, nanorods, nanopencils, and nanocones, with different geometries could be successfully realized using this simple wet-chemistry technique, as indicated in Figure 2.10. It is worth mentioning that morphologies, dimensions, and aspect ratios of these nanoarrays could be finely controlled over large areas (e.g., > 1.5 cm × 1.5 cm), which would be particularly attractive for the efficient photon-trapping applications.[42]

2.2.2 ETCHING MECHANISM

The anisotropic wet-chemistry-only etching mechanism of silicon had been systematically studied in our work, and this etch anisotropy could be elucidated to originate from the introduction of $AgNO_3$ to the (HF + HNO_3/H_2O_2) system.[42] The mixture of ($AgNO_3$ + HF + HNO_3/H_2O_2) led to the selective deposition of Ag nanoclusters at the nanopillar tips, which acted as catalysts in the silicon etching. Chemical and structural characterizations given in Figure 2.11 and Figure 2.12 confirmed the presence of Ag nanoclusters site-selectively deposited at the rim region of the pillar tips. This observed site selectivity occurred because silicon atoms in the core region were less reactive as compared with those around the rim region (Figure 2.11a), due to the rim or edge effect, which was also observed in other material systems. Importantly, the nanoarrays exhibited the same crystallinity with the

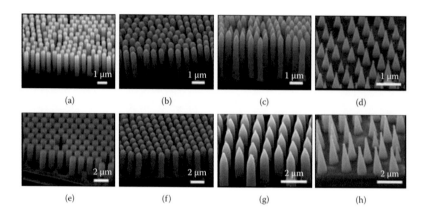

Figure 2.10 SEM images of the fabricated high aspect ratio (a and e) nanopillar, (b and f) nanorod, (c and g) nanopencil, and (d and h) nanocone arrays for the pitch of 0.6 and 1.3 μm, respectively. (Reproduced from Lin, H., et al., Developing controllable anisotropic wet etching to achieve silicon nanorods, nanopencils and nanocones for efficient photon trapping, *J. Mater. Chem. A*, 1, 9942, 2013. By permission of The Royal Society of Chemistry.)

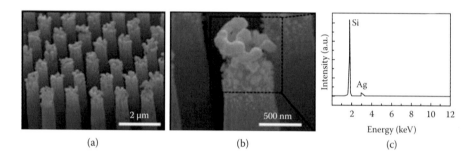

Figure 2.11 (a) SEM image of the Si nanopillar arrays after the selective deposition of Ag nanoclusters. (b) Zoom-in image of the Si nanopillar tips, demonstrating the selective coating of Ag nanoclusters (only on the tips but not on the base). (c) EDS spectrum of the tip region, confirming the presence and chemical content of pure Ag nanoclusters. (Reproduced from Lin, H., et al., Developing controllable anisotropic wet etching to achieve silicon nanorods, nanopencils and nanocones for efficient photon trapping, *J. Mater. Chem. A*, 1, 9942, 2013. By permission of The Royal Society of Chemistry.)

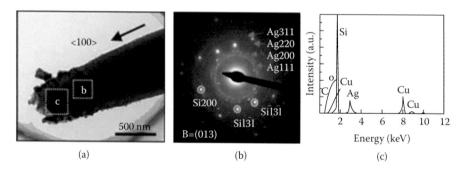

Figure 2.12 (a) TEM image of the representative Si nanopillar tip after selective deposition of Ag nanoclusters. (b) SAED pattern in the location "b" of the Si nanopillar tip, illustrating the existence of polycrystalline Ag nanoclusters selectively coated onto the Si nanopillar tips. Also, the diffraction pattern indicates the Si nanopillar axial direction being <100>, consistent with the orientation of the starting Si (100) substrates. (c) EDS spectrum of the location "c" of the Si nanopillar tip, again confirming the presence of pure Ag nanoclusters. (Reproduced from Lin, H., et al., Developing controllable anisotropic wet etching to achieve silicon nanorods, nanopencils and nanocones for efficient photon trapping, *J. Mater. Chem. A*, 1, 9942, 2013. By permission of The Royal Society of Chemistry.)

untreated silicon wafers, which is in distinct contrast to the other template-assisted bottom-up fabrication approaches.

To remove the Ag nanoclusters, nanopillar arrays were treated with HNO_3, leading to the conversion of nanopillars to truncated pillar structures. By repeating the Ag deposition and removal cycles, nanostructures with different morphologies in the tips could be realized. Schematic illustrations of the formation mechanism of Si nanostructures, including nanorods, nanopencils, and nanocones, via this anisotropic wet-etching method are described in Figure 2.13. Detailed process procedures for the fabrication of various nanoarrays with a pitch of 1.3 µm were studied, and SEM images depicting this process are demonstrated in Figure 2.14. For instance, the use of the $(AgNO_3 + HF + H_2O_2)$ solution system yielded nanorods with the dome head terminal after the third Ag deposition and acid washing cycle, as indicated in Figure 2.14a1–a3. This particular morphology was obtained because the presence of Ag nanoclusters modulated the oxidation kinetics of H_2O_2 in the etching process. On the other hand, pencil head terminals could be obtained after the fifth Ag deposition and acid washing cycles utilizing HNO_3 instead of H_2O_2 in the $(AgNO_3 + HF)$ solution mixture (Figure 2.14b1–b3). It was found that different etch anisotropies and aspect ratios of nanoarrays could be modulated by varying the ratio and the components of the $(AgNO_3, HF$ and $HNO_3)$ mixture, resulting in different morphologies. In the case of using extra $AgNO_3$ to HF, nanocone structures could be obtained because the additional $AgNO_3$ would speed up the formation of Ag nanoclusters enhancing the etch rate (i.e., Figure 2.14c1–c3).

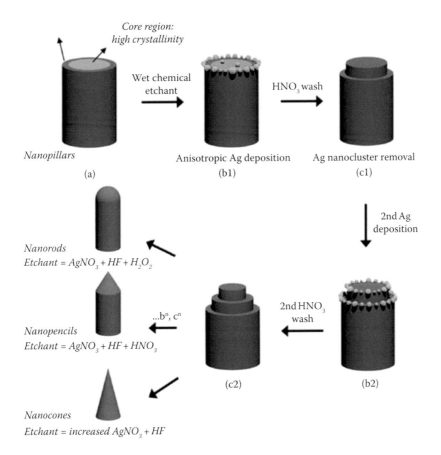

Figure 2.13 Schematic illustrations of the formation mechanism of silicon nanostructures with different morphologies. (Lin, H., et al., Developing controllable anisotropic wet etching to achieve silicon nanorods, nanopencils and nanocones for efficient photon trapping, *J. Mater. Chem. A*, 1, 9942, 2013. Reproduced by permission of The Royal Society of Chemistry.)

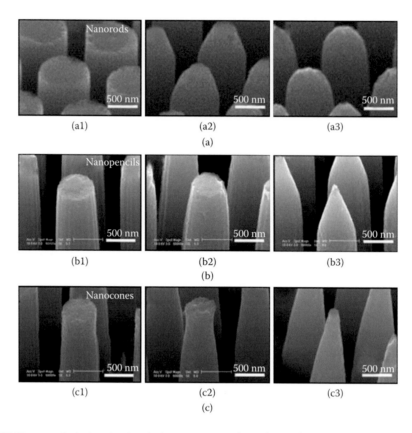

Figure 2.14 SEM images depicting the detailed process procedures for the fabrication of (a) nanorod arrays, (b) nanopencil arrays, and (c) nanocone arrays (all with the same pitch of 1.3 μm). (a1) Starting nanopillar array template; (a2) after the first Ag deposition and acid washing cycle. The etchant is a solution mixture of $AgNO_3$ (0.0002 M), HF (2 M), and H_2O_2 (0.005 M); (a3) after the third Ag deposition and acid washing cycle with the same etchant to obtain the round dome head terminals; (b1) after the first Ag deposition and acid washing cycle. The etchant is a solution mixture of $AgNO_3$ (0.0002 M), HF (2 M), and HNO_3 (0.005 M); (b2) after the second and (b3) fifth Ag deposition and acid washing cycles with the same etchant to obtain the pencil head terminals; (c1) after the first Ag deposition and acid washing cycle. The etchant is a solution mixture of $AgNO_3$ (0.0004 M) and HF (2 M); (c2) after the fourth and (c3) seventh Ag deposition and acid washing cycles with the same etchant with increasing $AgNO_3$ concentration up to 0.001 M (increment in 0.0001 M) in the final cycle in order to obtain the final shape of nanocones. (Lin, H., et al., Developing controllable anisotropic wet etching to achieve silicon nanorods, nanopencils and nanocones for efficient photon trapping, *J. Mater. Chem. A*, 1, 9942, 2013. Reproduced by permission of The Royal Society of Chemistry.)

2.3 LIGHT MANAGEMENT WITH INVERTED SILICON NANOPENCILS

After discussing their fabrications, we now focus on the enhanced physical properties of these inverted nanopencil arrays. Particularly, light-trapping mechanisms in the complex 3D nanophotonic structures have been comprehensively explored in various studies.[6,9,43] The light coupling, propagation, and absorption nature of these inverted nanopencil structures have been systematically investigated combining both experiments and simulations. All of these findings not only offer insight into the specific photon-capturing mechanism in complex 3D nanopencils but also provide efficient broadband and omnidirectional photon harvesters for the next generation of cost-effective ultrathin nanostructured photovoltaics.[25]

2.3.1 OPTICAL PROPERTIES

To quantitatively describe the optical characteristics of various nanoarrays fabricated via the wet-chemistry-only fabrication scheme as described in Figure 2.15, their antireflection performances were presented and

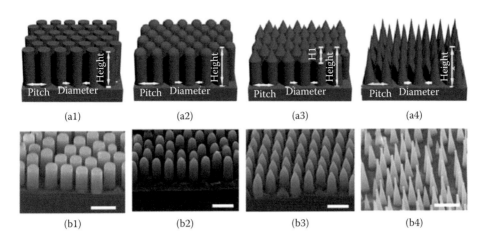

Figure 2.15 Schematic illustration of (a1) nanopillars, (a2) nanorods, (a3) inverted nanopencils, and (a4) nanocones studied in this chapter. The 45° tilted-angle-view SEM images of (b1) nanopillars, (b2) nanorods, (b3) inverted nanopencils, and (b4) nanocones. The structural pitches of all nanoarrays are 1.27 μm. (Reproduced with permission from Lin, H., et al., Rational design of inverted nanopencil arrays for cost-effective, broadband, and omnidirectional light harvesting, *ACS Nano*, 8, 3752–3760, 2014. Copyright 2014 American Chemical Society.)

evaluated in Figure 2.16.[25] The measurement was performed using ultraviolet-visible (UV-vis) spectroscopy equipped with an integrating sphere in the wavelength ranging from 400 to 1000 nm with a fixed angle of incidence at 0°. On the other hand, optical simulations were performed by the FDTD method in R-Soft software in order to verify and understand the experimental results. As it can be clearly observed, the experimental data indicate the comparable light-trapping performance between nanopencils and nanocones. The reflectance of the as-made inverted nanopencil arrays can be depressed below 5% over the entire wavelength range, demonstrating the best antireflection property among all the structures. A similar trend is observed from the simulated optical reflection spectra of the same nanostructures, where the inverted pencil structures with the tapered shapes exhibit excellent antireflection. By comparing Figure 2.16a and c, it is obvious that the pencil nanostructure with the smaller pitch of ~0.6 μm and the diameter of ~380 nm would exhibit a lower reflectance. It is suggested that this reduced reflectance comes from the depressed bottom surface reflection related to the reduced spacing among pencil arrays in the basal plane, indicating the effect of different structural pitch on the optical properties. The reflectance obtained from simulation of all the nanostructures is found to be slightly higher than those experimentally measured because of the perfectly smooth surface considered in the simulations, suggesting the significant effect of surface roughness induced by the wet-etching processes on the antireflection behavior. Besides, the effect of structural pitch on the optical properties of various nanostructures is further studied by simulation. As illustrated in Figure 2.17, the relatively low reflectance (<1%) could be observed when the structural pitch is between 0.3 and 0.8 μm. For the pitch below 0.3 μm, little light is coupled and propagated into the structure due to the small space between nanoarrays. When the pitch of nanostructures is larger than 0.8 μm, increased reflectance is observed because of the increased reflection from the enlarged bottom region.

In addition, antireflection properties of various nanostructures (nanopillar, nanorod, nanopencil, and nanocone) were also simulated as a function of height and diameter in order to explore their geometrical effect on the light-trapping properties. Reflection spectra integrated with the AM1.5G solar photon flux spectrum in the range of 400–1000 nm were plotted as two-dimensional (2D) contours, which is given in Figure 2.18 (i.e., features with 1.27 μm pitch) and Figure 2.19 (i.e., features with 0.6 μm pitch). It can be seen that the reflectance is dominantly dependent on the factors of aspect ratio and material filling ratio (MFR) based on a fixed pitch. For all the nanostructures with the small diameters, the higher reflectance is observed, which comes from the bottom reflection due to the relatively large open basal area. To achieve the efficient light absorption, this bottom reflection must be depressed and then larger diameters or MFR are needed. However, the larger diameter structures do not always bring the lower reflectance, especially for the nanopillar system. The reflectance of the pillars would get increased dramatically when

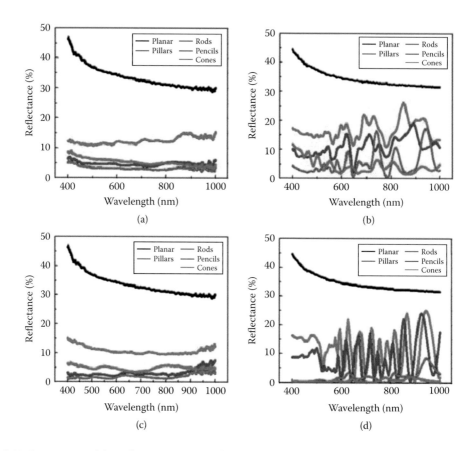

Figure 2.16 Comparison of the reflectance spectra of various nanostructured arrays between experimental measurement and optical simulation. (a) Experimental and (b) simulated spectra of different nanostructures with the pitch of 1.27 µm. (c) Experimental and (d) simulated spectra of different nanostructures with the pitch of 0.6 µm. (Reproduced with permissionfrom Lin, H., et al., Rational design of inverted nanopencil arrays for cost-effective, broadband, and omnidirectional light harvesting, *ACS Nano*, 8, 3752–3760, 2014. Copyright 2014 American Chemical Society.)

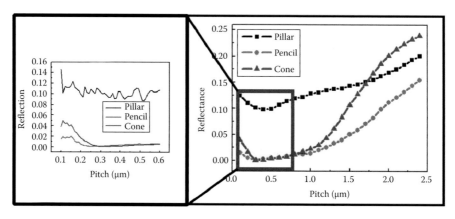

Figure 2.17 Simulated broadband-integrated reflectance of nanopillars, nanopencils, and nanocones with different pitch. The material filling (base-diameter-to-pitch) ratio is fixed at 0.6 µm while the pillar height is fixed at 2 µm. (Reproduced with permission from Lin, H., et al., Rational design of inverted nanopencil arrays for cost-effective, broadband, and omnidirectional light harvesting, *ACS Nano*, 8, 3752–3760, 2014. Copyright 2014 American Chemical Society.)

Arrays, hybrids, and core–shell

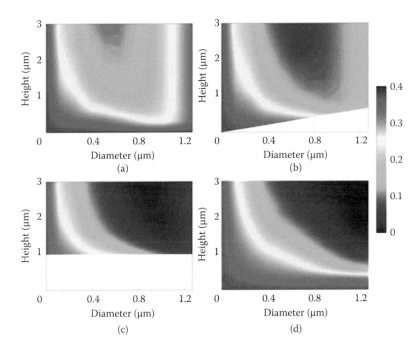

Figure 2.18 Two-dimensional simulated contours of broad-band-integrated reflectance of various nanostructures as a function of the base diameter and pillar height: (a) nanopillars, (b) nanorods, (c) inverted nanopencils, and (d) nanocones. Pitch is 1.27 μm in all panels. (Reproduced with permission from Lin, H., et al., Rational design of inverted nanopencil arrays for cost-effective, broadband, and omnidirectional light harvesting, *ACS Nano*, 8, 3752–3760, 2014. Copyright 2014 American Chemical Society.)

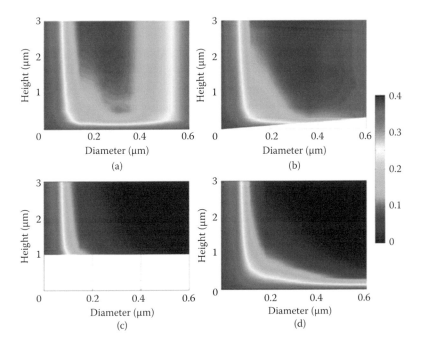

Figure 2.19 Two-dimensional simulated contours of broadband-integrated reflectance of various nanostructures with the pitch of 0.6 μm: (a) nanopillars, (b) nanorods, (c) inverted nanopencils, and (d) nanocones. (Reproduced with permission from Lin, H., et al., Rational design of inverted nanopencil arrays for cost-effective, broadband, and omnidirectional light harvesting, *ACS Nano*, 8, 3752–3760, 2014. Copyright 2014 American Chemical Society.)

Arrays, hybrids, and core–shell

the pillar diameter is increased larger than 1 μm, which is caused by the large reflective top surface area in pillar tips. As compared with the pillars, nanorods exhibit the lower reflectance and the extended antireflection band. The depressed light reflection of the nanorod structure results from its unique geometry, which could be explained by the refractive index matching between the hemispherical tips and air.[4] For nanocones with the tapered shape or the needle-like tips, their antireflection effect would become more profound and exciting; as a result, the nanocone-based structures have been considered as an ideal photon harvester with the reflective index matching with the air that suppresses light reflection significantly.[8,44,45] According to Figures 2.18d and 2.19d, the reflectance decreases as the cone height increases and the diameter approaches the pitch. Notably, the reflectance would be reduced to zero when the silicon substrate is fully covered by infinitely tall and continuous nanocone arrays. However, this ideal geometry is technically challenging and economically ineffective to be achieved. Nanopencil arrays also offer such a mechanism to couple light into the silicon substrate with the suppressed top surface reflection of tapered tips. More exciting, the pillar base of the nanopencil structure can be employed with scatter light along the in-plane dimension, which enhances the light traveling path for the absorption, providing the excellent light-trapping behavior. As indicated in Figure 2.18d, the excellent light-trapping behavior close to the perfect antireflection occurs for the pillar height larger than 2.2 μm and the diameter around 0.9 μm due to the minimized top surface reflection and maximized transmission of photons to basal pillars for the efficient absorption. In contrast to the nanocones, this optimized geometry of nanopencil arrays can be easily and cost-effectively achieved through our anisotropic wet-etching method. The reflectance 2D contour of nanopencils with the smaller pitch of 0.6 μm also demonstrates and confirms the similar trend, as demonstrated in Figure 2.19.

Apart from the geometrical effect of these nanopencils, it is also important to investigate the contribution of inverted nanopencil arrays to the light absorption for thinner silicon substrates for practical utilizations. Simulations of optical properties for nanopencils with different substrate thickness were performed in the diffraction mode in R-Soft software. The light absorption of nanopencils with the pitch of 1.27 μm and four different thicknesses is demonstrated in Figure 2.20, from which it can be observed that the 500 μm thick silicon surface texturized with nanopencils can even absorb the visible light almost completely. As the thickness of substrate decreases, the inverted nanopencil arrays still exhibit the excellent light-trapping and antireflection properties, although the slightly higher reflection and transmission are observed in the long-wavelength region. By comparing all these optical characteristics, it should be noted that the light absorption improvement of these thin substrates texturized with nanopencils is superior to that of thick planar silicon substrates, indicating the advantages of inverted nanopencil structure for solar cells with an ultrathin absorber layer.

The aforementioned optical properties were all performed with the normal incident light. Nonetheless, the angular-dependent optical absorption must be considered for practical photovoltaic usages for the maximization of energy conversion efficiency.[46] Specifically, the omnidirectional broadband photon flux absorption of various nanostructures over different angles of incidence was calculated and plotted by

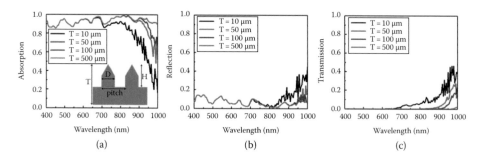

Figure 2.20 Simulated (a1) absorption, (a2) reflection, and (a3) transmission properties for inverted nanopencils with different substrate thickness. Nanopencils are with the pitch of 1.27 μm, base diameter of 800 nm, pillar height of 2 μm, and pencil tip height of 1 μm. (Reproduced with permission from Lin, H., et al., Rational design of inverted nanopencil arrays for cost-effective, broadband, and omnidirectional light harvesting, *ACS Nano*, 8, 3752–3760, 2014. Copyright 2014 American Chemical Society.)

the diffraction mode in R-Soft software, which is depicted in Figure 2.21. It is clear that inverted nanopencils exhibit the best absorption behavior over the entire angle range for 400–1000 nm wavelength, as indicated in the 3D absorption contours of Figure 2.21a1–a4. Notably, the higher absorption of pencil structures is demonstrated for the small incident angles and long-wavelength region, which is confirmed by the experimental broadband-integrated absorption spectra presented in Figure 2.21b,c. As presented in Figure 2.21b, the nanopencils and nanocones demonstrate the high absorption with over 95% absorption until the angle of incidence of 60° is reached. These measured excellent optical behavior and are comparable or even better than that of nanostructures, including nanocone, nanodome, and nanopyramids, fabricated by complicated techniques. For angles of incidence above 60°, a decrease of absorption for all the nanostructures is observed because fewer photons can enter the structures at such large angles, leading to the diminished probability of light trapping. These effects are also observed for nanostructures with the smaller pitch of 0.6 µm, as demonstrated in Figure 2.21c.

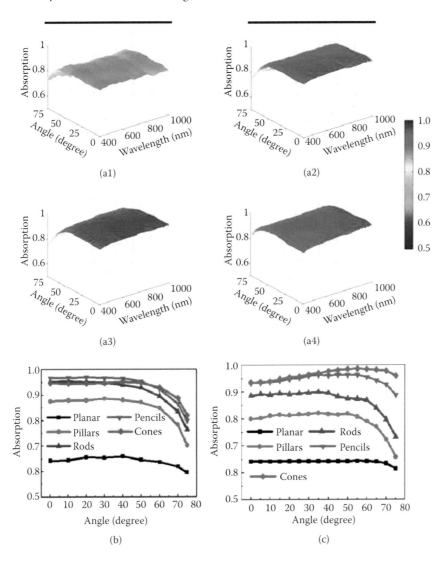

Figure 2.21 (a1–a4) Three-dimensional contours of absorption spectra of nanopillars, nanorods, inverted nanopencils, and nanocones with the pitch of 1.27 µm and the pillar height of 2 µm. (b) Experimental broadband-integrated absorption as a function of the incident angle of irradiation for different nanostructures with the pitch of 1.27 µm and (c) 0.6 µm, respectively. (Reproduced with permission from Lin, H., et al., Rational design of inverted nanopencil arrays for cost-effective, broadband, and omnidirectional light harvesting, ACS Nano, 8, 3752–3760, 2014. Copyright 2014 American Chemical Society.)

Arrays, hybrids, and core-shell

2.3.2 LIGHT-TRAPPING MECHANISM

To further assess the light-trapping mechanism of these complex 3D nanophotonic structures, the propagation nature of light in various nanoarrays was studied. The solar-spectrum-weighted (300–1100 nm) electrical field intensity ($|E|^2$) distribution of the electromagnetic (EM) wave on the nanostructures with the pitch of 1.27 and 0.6 μm was visualized and plotted in Figure 2.22.[25] The EM plane waves were guided to propagate downward from Z = 2.2 μm in simulations. Nanostructures with the height of 2 μm, MFR of ~0.6, and the pitch of 1.27 and 0.6 μm were evaluated and the simulated $|E|^2$ cross-sectional distribution is demonstrated in Figure 2.22a1–a4 and b1–b4, respectively. For nanopillars with the large flat top surface, strong electrical filed intensity can be observed on top of the structure, suggesting a significant reflection there (Figure 2.22a1). In comparison, the field intensity on top of the nanorod with a hemisphere tip (Figure 2.22a2) becomes weaker as compared with the pillar structure with the same aspect ratio. In the case of nanopencil (H1 = 1 μm) and nanocone structure with the tapered shape and small tips, the minimal top surface reflection is observed as shown in Figure 2.22a3 and a4. All these findings indicate that the top surface morphology plays an important role in the antireflection behavior of nanostructures, and suppressed reflectance can be realized by tuning the top surface morphology from the planar to the tapered shape. On the other hand, as contrasted with the other three nanostructures, stronger EM waves are noticed to be propagating at the center of the nanopencil structures, indicating the higher probability of light absorption there. However, the substantial field intensity is also observed near the basal region of large pitch structures (Figure 2.22a1–a4), suggesting the significant bottom surface reflection because of the large basal open area. Nonetheless, nanopencil structures have demonstrated the superior light absorption over nanocones with the same pillar height, base diameter, and MFR due to the larger effective MFR in the basal region of pencils for the enhanced light trapping and absorption. By comparing the field distribution of nanostructures with different structural pitches, nanostructures with a smaller pitch of 0.6 μm and MFR demonstrate the weaker field intensity near the bottom surface, as shown in Figure 2.22b1–b4. Interestingly, it is found that the field intensity almost vanishes near the basal plane of pencils and cones, signifying the effective propagation of incident light and the efficient absorption of the EM wave energy with the smaller pitch pencil and cone structures before the photons reaching the bottom region. Different with nanostructures with a pitch of 1.27 μm, the nanocone structure shows the strongest field intensity at the center of the feature among all the 0.6 μm pitch nanostructures studied

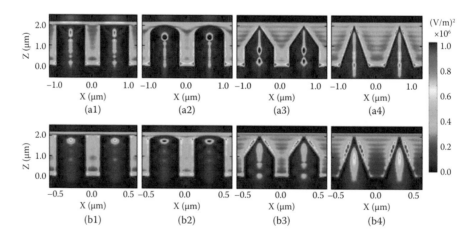

Figure 2.22 Two-dimensional solar-spectrum-weighted electrical field intensity contours of various nanostructures obtained by FDTD simulations. (a1–a4) Nanopillars, nanorods, inverted nanopencils, and nanocones with the pitch of 1.27 μm, base diameter (D) of 800 nm, and pillar height (H) of 2 μm, (b1–b4) nanopillars, nanorods, inverted nanopencils, and nanocones with the pitch of 0.6 μm, D of 380 nm, and H of 2 μm. The pencil tip height (H1) is 1 μm. (Reproduced with permission from Lin, H., et al., Rational design of inverted nanopencil arrays for cost-effective, broadband, and omnidirectional light harvesting, *ACS Nano*, 8, 3752–3760, 2014. Copyright 2014 American Chemical Society.)

here because of the minimized bottom reflection for the more efficient absorption. All these findings not only provide essential information about the light-trapping mechanism in complex 3D nanostructures but also illustrate the technological potency of these inverted nanopencil structures for next-generation cost-effective ultrathin nanostructured photovoltaics as the efficient broadband and omnidirectional photon harvesters.

2.4 APPLICATIONS IN OPTOELECTRONICS

For practical device applications, semiconductor devices consisting of periodic nanoarrays of radial p-n junctions as described in Figure 2.23a would consist of various advantages, including the minimized reflection, efficient light-trapping property, relaxed interfacial strain, single-crystalline synthesis on non-epitaxial substrates, and so on.[47] More important, as compared with the substrate junctions[48] presented in Figure 2.23b, respectively, this radial p–n junction can provide the shorter travel distance of excited carriers to the collection electrodes in a direction normal to the light absorption, resulting in the enhanced carrier collection efficiency. Above all, these high-aspect-ratio nanostructures with the radial p–n junction contain the physical shape, which is elongated in the direction of incident light but thin in the perpendicular dimension, enabling the enhanced light absorption as well as effective carrier collection.[47,49] In this case, when configuring these unique radial junctions into nanostructured silicon solar cells, the efficient dissociation and extraction of light-induced charge carriers would be achieved, providing that the minority carrier diffusion length is as short as the structural radius, which is of great importance to reduce the cost of solar cells by using the low-end and low-quality silicon material with short carrier diffusion lengths.

In practice, radial p–n junctions could be realized by BF_2 and P ion implantation. Figure 2.24a and b present the simulated boron and phosphorous profile in a nanowire structure after the BF_2 core and subsequent P shell implant.[49] Distribution of different dopant concentrations is described by the color gradient in the figure. It could be seen that a junction depth (at which both dopant concentrations are approximately equal) of 50 nm is obtained after P doping and a schematic description of the radial p–n junction in a nanowire could be found in Figure 2.23c. Current-voltage (I-V) characteristics of this core–shell silicon nanowire–based solar cell are tested under AM 1.5G illumination, as compared with that of the planar device, as shown in Table 2.1. The short-circuit current density, J_{sc}, in the nanowire-based device is observed to be 52% higher than that of the planar device, which can be attributed to the improved light absorption in nanowires and the enhanced carrier collection efficiency from the core–shell structure.[49]

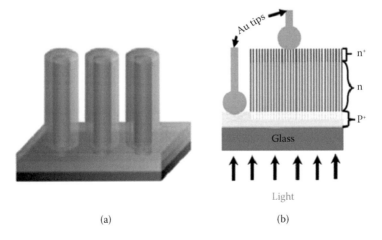

(a)   (b)

Figure 2.23 (a) Silicon nanowire solar cell structure with radial p–n junction. Single crystalline n-Si NW core is in brown; the polycrystalline p-Si shell is in blue; and the back contact is in black. (b) Schematic representation of the I-V curve measurements of SiNW-based substrate p-n junctions. (Reproduced with permission from Garnett, E.C. & Yang, P., Silicon nanowire radial p–n junction solar cells, *J. Am. Chem. Soc.*, 130, 9224–9225, 2008. Copyright 2008; Sivakov, V., et al., Silicon nanowire-based solar cells on glass: Synthesis, optical properties, and cell parameters, *Nano Lett.*, 9, 1549–1554, 2009. Copyright 2009 American Chemical Society.)

Arrays, hybrids, and core-shell

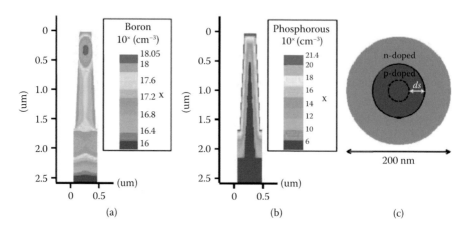

Figure 2.24 (a) Simulated boron profile in a nanowire after BF_2 core implant (rotation: 0°, 90°, 180°, 270°; dose: 2.5×10^{13} cm^{-2}, energy: 80 keV, tilt: 7° for each rotation) and 1 hour drive-in at 1000°C. (b) Simulated phosphorous profile in a nanowire after P shell implant (rotation: 0°, 90°, 180°, 270°; dose: 10^{15} cm^{-2}, energy: 7 keV, tilt: 7° for each rotation). The color gradient depicts distribution of different dopant concentrations in the vertical cross section of the wire. Junction depth (at which both dopant concentrations are approximately equal) is estimated to be 50 nm. (c) A schematic illustration of the radial p–n junction in a nanowire, indicating the estimated junction depth and depletion width d. (Reproduced from Li, Z., et al., Optical and electrical study of core–shell silicon nanowires for solar applications, *Opt. Express,* 19, A1057, 2011. With permission of Optical Society of America.)

Table 2.1 Characterization of both planar and core–shell silicon nanowire solar cells

	PLANAR Si	CORE–SHELL SiNW
J_{sc} (mA/cm²)	9.34	14.2
V_{oc} (V)	0.548	0.485
FF	72.8%	42.9%
R_s (Ω)	0.95	12.9
PCE	3.73%	2.95%
FF[a]	74.0%	63.9%
PCE[a]	3.79%	4.40%
J_{sc}[b] (mA/cm²)	13.3	20.3
PCE[b]	5.33%	4.21%
PCE[a,b]	5.39%	6.29%

Source: Reproduced from Li, Z., et al., Optical and electrical study of core–shell silicon nanowires for solar applications., *Opt. Express,* 19, A1057, 2011. With permission of Optical Society of America.

[a] In the absence of R_s.
[b] In the absence of 30% shading losses.

At the same time, a device physics model for radial p–n junction–based nanorod solar cells was also developed to explore their corresponding design principle.[50] In the model, the energy conversion efficiency of the cell versus the cell thickness and minority-electron diffusion length were evaluated for both conventional planar p–n junction cell and radial p–n junction nanorod cell, as given in Figure 2.25. In the planar geometry, J_{sc} is independent of the trap density in the depletion region, but it would decrease with the increase of quasineutral-region trap density, resulting in a low efficiency for this planer device. In contrast, the overall efficiency of a radial p-n junction can remain high even for very large trap densities in

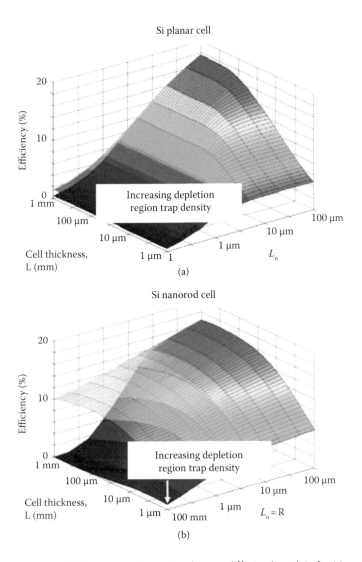

Si planar cell

(a)

Si nanorod cell

(b)

Figure 2.25 Efficiency versus cell thickness L and minority electron diffusion length L_n for (a) a conventional planar p–n junction silicon cell, (b) a radial p–n junction nanorod silicon cell. In all cases the top surface shown in the plot has a depletion region trap density fixed at 10^{14} cm^{-3} so that the carrier lifetimes $\tau_{n0},\ \tau_{p0}$ = 1 μs, whereas the bottom surface has a depletion region trap density equal to the trap density in the quasi-neutral region at each value of Ln. In the radial p–n junction nanorod case, the cell radius R is set equal to L_n, a condition that is near optimal. (Reproduced with permission from Kayes, B.M., et al., Comparison of the device physics principles of planar and radial p-n junction nanorod solar cells, *J. Appl. Phys.*, 97, 1–11, 2005. Copyright 2005, American Institute of Physics.)

quasi-neutral regions, providing that the trap density in the depletion region is relatively low.[50] Also, both experimental and theoretical findings have indicated that the design of radial p–n junction nanostructure-based devices could achieve significant improvement in their efficiencies only if the rate of carrier recombination in the depletion region is effectively suppressed. It means that for silicon solar cells, the carrier lifetimes in the depletion region must be shorter than ~10 ns; otherwise, only a modest improvement in efficiency could be realized. For instance, even though a higher J_{sc} is realized in the aforementioned nanowire-based solar cell than that of the planar device due to the improved light absorption, both fill factor and conversion efficiency are still found to be relatively low (i.e., fill factor of 42.9, efficiency of ~2.95%), attributable to the severe recombination effect caused by the ion implantation-induced defects.[49]

To alleviate this recombination effect, deposition of surface passivation layers is a commonly utilized approach with the aim to reduce interface traps and implement a field-effect passivation.[51–55]

Another effective way to reduce the surface recombination is by applying a mixture of nitric acid (HNO$_3$) and hydrofluoric acid (HF) to obtain a smooth and contamination-free surface, which is known as the chemical polishing etching (CPE) treatment.[56] In this process, large-area, 3D hierarchical structures combining micropyramids and nanowires, with the excellent antireflection behavior as displayed in Figure 2.26a, were obtained using the metal-assisted chemical etching for the fabrication of radial p–n junction solar cells. However, the hierarchical structure–based solar cell without CPE

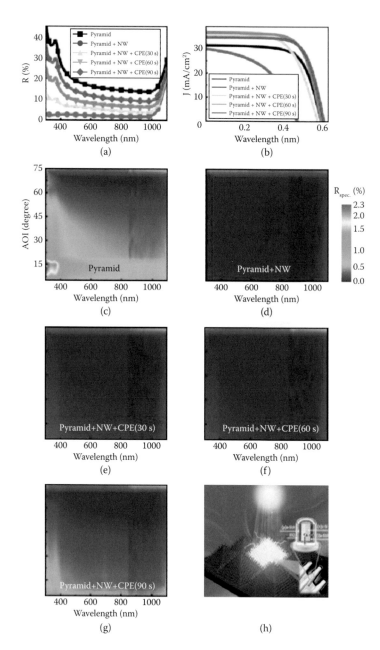

Figure 2.26 (a) Total reflectance spectra of micropyramid (MP) and hierarchical structures with different CPE durations over the wavelength regions of 300–1100 nm. (b) J-V characteristics of solar cells based on MP and hierarchical structures with different CPE durations. Specular reflectance spectra of (c) micropyramidal surfaces and (d–g) hierarchical surfaces with different CPE durations. (h) Schematic of hierarchical structures. (Reproduced with permission from Wang, H.-P., et al., Realizing high-efficiency omnidirectional n-type Si solar cells via the hierarchical architecture concept with radial junctions, *ACS Nano*, 7, 9325–9335, 2013. Copyright 2013 American Chemical Society.)

demonstrates the relatively low J_{sc} and conversion efficiency due to the severe recombination effect caused by rich surface defects induced by the etching process, as indicated in Figure 2.26b. After utilizing the CPE treatment, solar cells based on pyramids and NW hierarchical structures synergize several advantageous features, including the excellent broadband, omnidirectional light absorption, as well as improved carrier collection efficiency. As a result, their conversion efficiency is found to be 15.14%, which contributes to the highest reported efficiency among all n-type silicon nanostructured solar cells reported until now, indicating the great potency of this special hierarchical structure and CPE surface treatment for the next-generation of high-performance photovoltaics.[56]

2.5 SUMMARY AND REMARKS

In summary, due to the energy crisis, there is an urgent need to develop novel, cost-effective schemes to achieve efficient light absorption and carrier collection in order to further improve the energy conversion efficiency and to reduce the cost of solar panels. Recent studies have shown that inverted nanopencil arrays demonstrate the unique photon management ability, which may serve as the efficient light absorber material in the next generation of ultrathin nanostructured photovoltaics. In this chapter, a simple, wet-chemistry-only and large-scale fabrication scheme for the controllable hierarchy of nanopencil structures is introduced. In addition, we have systematically reviewed the light-coupling, propagation, and absorption nature of these nanopencil arrays by combining both experimental and simulation approaches. The light-trapping mechanism of nanopencil arrays was also discussed to provide the essential design principle of these nanopencils. It was revealed that the properly designed nanopencil arrays could demonstrate the excellent broadband and omnidirectional light absorption performance, which is comparable to the commonly accepted ideal but costly nanocone structures. What is more, behaviors and design guidelines of radial p–n junction nanostructured silicon solar cells with excellent photon-harvesting capability and efficient carrier collection were discussed.

Despite the rapid development of nanostructure-based silicon solar cells in the last decade, further research and studies are still needed to solve several important issues to reach an optimized photovoltaic performance. First, the recombination effect must be effectively suppressed without sacrificing the light-trapping capability. For the moment, proper doping techniques should also be developed to achieve atomically sharp and shallow doping profiles in radial p–n junctions without inducing severe defects. In addition, cost-effective fabrication approaches for controllable fabrication of large-scale, regular and 3D light management nanostructures are needed for cost-effective ultrathin nanostructured photovoltaics.

ACKNOWLEDGMENTS

This work was financially supported by the General Research Fund of the Research Grants Council of Hong Kong SAR, People's Republic of China (CityU 11204614); the Science Technology and Innovation Committee of Shenzhen Municipality (grant no. JCYJ20160229165240684); and the Natural Science Foundation of Jiangsu Province, People's Republic of China (grant no. BK20150955). It was also supported by a grant from the Shenzhen Research Institute, City University of Hong Kong.

REFERENCES

1. Liu, J. *et al.* Fabrication and photovoltaic properties of silicon solar cell with different diameters and heights nanopillars. *Energy Technol.* **26**, 2805–2811 (2012).
2. Tsui, K. H. *et al.* Low-cost, flexible, and self-cleaning 3D nanocone anti-reflection films for high-efficiency photovoltaics. *Adv. Mater.* **26**, 2805–2811 (2014).
3. Wei, W. R. *et al.* Above 11%-efficiency organic-inorganic hybrid solar cells with omnidirectional harvesting characteristics by employing hierarchical photon-trapping structures. *Nano Lett.* **13**, 3658–3663 (2013).
4. Narasimhan, V. K. & Cui, Y. Nanostructures for photon management in solar cells. *Nanophotonics* **2**, 187–210 (2013).
5. Li, X. Metal assisted chemical etching for high aspect ratio nanostructures: A review of characteristics and applications in photovoltaics. *Curr. Opin. Solid State Mater. Sci.* **16**, 71–81 (2012).

6. Garnett, E. & Yang, P. Light trapping in silicon nanowire solar cells. *Nano Lett.* **10**, 1082–1087 (2010).

7. Lin, Q., Hua, B., Leung, S.-F., Duan, X. & Fan, Z. Efficient light absorption with integrated nanopillar/nanowell arrays for three-dimensional thin-film photovoltaic applications. *ACS Nano* **7**, 2725–2732 (2013).

8. Mavrokefalos, A., Han, S. E., Yerci, S., Branham, M. S. & Chen, G. Efficient light trapping in inverted nano-pyramid thin crystalline silicon membranes for solar cell applications. *Nano Lett.* **12**, 2792–2796 (2012).

9. Zhu, J. *et al.* Optical absorption enhancement in amorphous silicon nanowire and nanocone arrays. *Nano Lett.* **9**, 279–282 (2009).

10. Cheng, Y. *et al.* Self-assembled wire arrays and ITO contacts for silicon nanowire solar cell applications. *Chin. Phys. Lett.* **28**, 035202 (2011).

11. Jesper, W. *et al.* InP nanowire array solar cells achieving 13.8% efficiency by exceeding the ray optics limit. *Science* **339**, 1057–1059 (2013).

12. Pudasaini, P. R. *et al.* High efficiency hybrid silicon nanopillar–Polymer solar cells. *ACS Appl. Mater. Interfaces* **5**, 9620–9627 (2013).

13. Lin, Q., Leung, S.-F., Tsui, K.-H., Hua, B. & Fan, Z. Programmable nanoengineering templates for fabrication of three-dimensional nanophotonic structures. *Nanoscale Res. Lett.* **8**, 268 (2013).

14. Jeong, S., Wang, S. & Cui, Y. Nanoscale photon management in silicon solar cells. *J. Vac. Sci. Technol.* **30**, 060801 (2012).

15. Wang, H.-P. *et al.* Periodic Si nanopillar arrays by anodic aluminum oxide template and catalytic etching for broadband and omnidirectional light harvesting. *Opt. Express* **20**, A94 (2012).

16. Fan, Z. *et al.* Ordered arrays of dual-diameter nanopillars for maximized optical absorption. *Nano Lett.* **10**, 3823–3827 (2010).

17. Wang, B. & Leu, P. W. Enhanced absorption in silicon nanocone arrays for photovoltaics. *Nanotechnology* **23**, 194003 (2012).

18. Tsai, D.-S. *et al.* Ultra-high-responsivity broadband detection of Si metal-semiconductor-metal Schottky photo-detectors improved by ZnO nanorod arrays. *ACS Nano* **5**, 7748–7753 (2011).

19. Zhu, J., Hsu, C.-M., Yu, Z., Fan, S. & Cui, Y. Nanodome solar cells with efficient light management and self-cleaning. *Nano Lett.* **10**, 1979–1984 (2010).

20. Ko, Y. H. & Yu, J. S. Urchin-aggregation inspired closely-packed hierarchical ZnO nanostructures for efficient light scattering. *Opt. Express* **19**, 25935–25943 (2011).

21. Wiersma, D. S. *et al.* Optics of nanostructured dielectrics. *J. Opt. A Pure Appl. Opt.* **7**, S190–S197 (2005).

22. Atwater, H. A. & Polman, A. Plasmonics for improved photovoltaic devices. *Nat. Mater.* **9**, 205–213 (2010).

23. Zhang, J. *et al.* Engineering the absorption and field enhancement properties of Au–TiO$_2$ nanohybrids via whispering gallery mode resonances for photocatalytic water splitting. *ACS Nano* **10**, 4496–4503 (2016).

24. Edman Jonsson, G., Fredriksson, H., Sellappan, R. & Chakarov, D. Nanostructures for enhanced light absorption in solar energy devices. *Int. J. Photoenergy* **11**, 477–493, (2011).

25. Lin, H. *et al.* Rational design of inverted nanopencil arrays for cost-effective, broadband, and omnidirectional light harvesting. *ACS Nano* **8**, 3752–3760 (2014).

26. Wang, P. & Menon, R. Optimization of periodic nanostructures for enhanced light-trapping in ultra-thin photovoltaics. *Opt. Express* **21**, 6274 (2013).

27. Jeong, S., Mcgehee, M. D. & Cui, Y. All-back-contact ultra-thin silicon nanocone solar cells with 13.7% power conversion efficiency. *Nat. Commun.* **4**, 345–50 (2013).

28. Wu, Y. & Yang, P. Direct observation of vapor-liquid-solid nanowire growth. *J. Am. Chem. Soc.* **123**, 3165–3166 (2001).

29. Gunawan, O. & Guha, S. Characteristics of vapor–liquid–solid grown silicon nanowire solar cells. *Sol. Energy Mater. Sol. Cells* **93**, 1388–1393 (2009).

30. Picraux, S. T., Dayeh, S. A., Manandhar, P., Perea, D. E. & Choi, S. G. Silicon and germanium nanowires—Growth, properties, and integration. *Low Dimens. Nanomater.* **62**, 35–43 (2008).

31. Kayes, B. M., Atwater, H. A. & Lewis, N. S. Comparison of the device physics principles of planar and radial p-n junction nanorod solar cells. *J. Appl. Phys.* **97**, (2005).

32. Park, W. I., Zheng, G., Jiang, X., Tian, B. & Lieber, C. M. Controlled synthesis of millimeter-long silicon nanowires with uniform electronic properties. *Nano Lett.* **8**, 3004–3009 (2008).

33. Morton, K. J., Nieberg, G., Bai, S. & Chou, S. Y. Wafer-scale patterning of sub-40 nm diameter and high aspect ratio (>50:1) silicon pillar arrays by nanoimprint and etching. *Nanotechnology* **19**, 345301 (2008).

34. Huang, Y., Duan, X. & Lieber, C. M. Nanowires for integrated multicolor nanophotonics. *Small* **1**, 142–147 (2005).

35. Sainiemi, L. *et al.* Rapid fabrication of high aspect ratio silicon nanopillars for chemical analysis. *Nanotechnology* **18**, 505303 (2007).

36. Rand, B. P., Genoe, J., Heremans, P. & Poortmans, J. Solar cells utilizing small molecular weight organic semiconductors. *Prog. Photovolt. Res. Appl.* **15**, 659–676 (2007).

Arrays, hybrids, and core–shell

37. Lang, W. Silicon microstructuring technology. *Mater. Sci. Eng. R Rep.* **17**, 1–55 (1996).
38. Rhee, H. *et al.* Comparison of deep silicon etching using SF_6/C_4F_8 and SF_6/C_4F_6 plasmas in the Bosch process. *J. Vac. Sci. Technol. B* **26**, 576 (2008).
39. Fu, Y. Q. *et al.* Deep reactive ion etching as a tool for nanostructure fabrication. *J. Vac. Sci. Technol. B Microelectron. Nanometer Struct.* **27**, 1520 (2009).
40. Li, X. & Bohn, P. W. Metal-assisted chemical etching in HF/H_2O_2 produces porous silicon. *Appl. Phys. Lett.* **77**, 2572–2574 (2000).
41. Huang, Z., Geyer, N., Werner, P., De Boor, J. & Gösele, U. Metal-assisted chemical etching of silicon: A review. *Adv. Mater.* **23**, 285–308 (2011).
42. Lin, H. *et al.* Developing controllable anisotropic wet etching to achieve silicon nanorods, nanopencils and nanocones for efficient photon trapping. *J. Mater. Chem. A* **1**, 9942 (2013).
43. Wang, B. & Leu, P. W. Enhanced absorption in silicon nanocone arrays for photovoltaics. *Nanotechnology* **23**, 194003 (2012).
44. Hua, B., Wang, B., Yu, M., Leu, P. W. & Fan, Z. Rational geometrical design of multi-diameter nanopillars for efficient light harvesting. *Nano Energy* **2**, 951–957 (2013).
45. Jeong, S. *et al.* Hybrid silicon nanocone-polymer solar cells. *Nano Lett.* **12**, 2971–2976 (2012).
46. Seshan, C. Cell efficiency dependence on solar incidence angle. *Conf. Rec. IEEE Photovolt. Spec. Conf.* 2102–2105 (2010). doi:10.1109/PVSC.2010.5616340.
47. Garnett, E. C., Yang, P. Silicon nanowire radial p–n junction solar cells. *J. Am. Chem. Soc.* **130**, 9224–9225 (2008).
48. Sivakov, V. *et al.* Silicon nanowire-based solar cells on glass: Synthesis, optical properties, and cell parameters. *Nano Lett.* **9**, 1549–1554 (2009).
49. Li, Z., Wang, J., Singh, N. & Lee, S. Optical and electrical study of core-shell silicon nanowires for solar applications. *Opt. Express* **19**, A1057 (2011).
50. Kayes, B. M., Atwater, H. A. & Lewis, N. S. Comparison of the device physics principles of planar and radial p-n junction nanorod solar cells. *J. Appl. Phys.* **97**, 1–11 (2005).
51. Schultz, O., Hofmann, M., Glunz, S. W. & Willeke, G. P. Silicon oxide silicon nitride stack system for 20% efficient silicon solar cells. *Conf. Rec. Thirty-First IEEE Photovolt. Spec. Conf.* 872–876 (2005). doi:10.1109/pvsc.2005.1488271.
52. Dingemans, G. & Kessels, W. M. M. Status and prospects of Al_2O_3 -based surface passivation schemes for silicon solar cells. *J. Vac. Sci. Technol. A* **30**, 1–27 (2012).
53. Kopfer, J. M., Keipert-Colberg, S. & Borchert, D. Capacitance-voltage characterization of silicon oxide and silicon nitride coatings as passivation layers for crystalline silicon solar cells and investigation of their stability against x-radiation. *Thin Solid Films* **519**, 6525–6529 (2011).
54. Wang, W.-C. *et al.* Surface passivation of efficient nanotextured black silicon solar cells using thermal atomic layer deposition. *ACS Appl. Mater. Interfaces* **5**, 9752–9759 (2013).
55. Kim, D. R., Lee, C. H., Rao, P. M., Cho, I. S. & Zheng, X. Hybrid Si microwire and planar solar cells: Passivation and characterization. *Nano Lett.* **11**, 2704–2708 (2011).
56. Wang, H.-P. *et al.* Realizing high-efficiency omnidirectional n-type Si solar cells via the hierarchical architecture concept with radial junctions. *ACS Nano* **7**, 9325–9335 (2013).

Arrays, hybrids, and core–shell

Single-crystal silicon nanopore and arrays

Zewen Liu and Yifan Wang

Contents

3.1 INTRODUCTION

3.1.1 SOLID-STATE NANOPORES

The research on nanopores, especially on solid-state nanopores, emerged and was promoted by the upsurge of genetic analysis around the world. Since the chain-termination method proposed by Sanger et al. in 1977 [1] first provided a feasible way to detect nucleic acid sequences, the DNA sequencing technologies have followed suit, and the development rate has equaled and now even outpaces Moore's law [2].

These DNA sequencing technologies provided an opportunity for finding genomic information to prevent, diagnose, and cure human diseases, and it led medical research and medical care to a new era. However, the sequencing cost was huge (~$10 million) by traditional Sanger sequencing technology due to the enormous amount (~3 billion base pairs) of DNA found in human genomes, which made it made it unfeasible for DNA sequencing to be a part of routine medical practice. In 2008, the cyclic-array sequencing technology (including wash-and-scan, PCR, and termination processes) took the first leap in the DNA sequencing field. The second-generation DNA sequencers (e.g., 454 Genome Sequencers, Illumina, HelioScope, and SOLiD) have gradually replaced the first-generation ones and reduced more than three-fold the cost in sequencing 1 megabase. More details about the second-generation sequencing technologies can be found in previous reviews.

To further reduce the cost, operation time, and equipment size of the DNA sequencing process, while continually increasing the contiguous read length, throughput, and accuracy, researchers have proposed different approaches for DNA sequencing [3]. This led to the emergence of the third-generation sequencing technologies, such as the real-time sequencing by synthesis technology and direct image technology. Nanopore-based DNA sequencing technology has become one of the most attractive and promising third-generation sequencing technologies because of its outstanding characteristics of label-free, amplification-free, great read length, and high throughput, which offer possibilities of high-quality gene sequencing applications, such as de novo sequencing, high-resolution analysis of chromosomal structure variation, and long-range haplotype mapping.

The basic principle of nanopore-based DNA sequencing is straightforward as in Coulter counters. As illustrated as Figure 3.1, the chip containing a nanopore separates the container, which is filled with electrolyte solution, into two chambers (trans and cis). When a bias voltage is applied across the chambers, there should (theoretically) be a constant ionic current appearing through the pore. The DNA molecule is negatively charged in electrolyte solution. Therefore, under the electric field force, the DNA molecule will be dragged though the nanopore.

As we all know, most DNA molecules consist of two biopolymer strands coiled with each other to form a double helix. The DNA strands are composed of four different kinds of units called nucleotides. Each nucleotide is composed of a nitrogen-containing nucleobase—cytosine (C), guanine (G), adenine (A), and thymine (T), respectively—as well as a sugar called deoxyribose and a phosphate group. As shown in Figure 3.2, four different kinds of bases on DNA have different size and spatial structure. When the DNA strands pass through a nanopore with size comparable to the DNA molecules, the ionic current will significantly decrease and vary with the different bases. Thus, they provide us with information about DNA sequence.

The nanopore-based DNA sequencing was first proposed by Church et al. in 1995 [4]. Subsequently, experiments of the electronic behavior of ssDNA passing through biological nanopore were carried out and marked the beginning of the nanopore-based sequencing field. Since then, various biological nanopores were used to sequence DNA and RNA molecules, for example, the octameric protein channel of MspA [5,6]. The biological nanopores have many advantages, such as the dimension reproducibility, the compatibility of genetic or chemical modification processes, and relatively slower DNA translocation velocity through the nanopores. However, there are also some undeniable drawbacks [7,8]: (a) the rigorous environmental demands

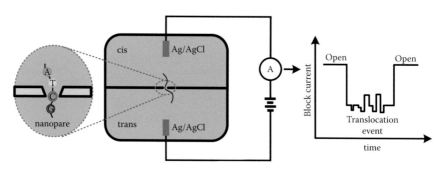

Figure 3.1 The principle of nanopore-based DNA sequencing.

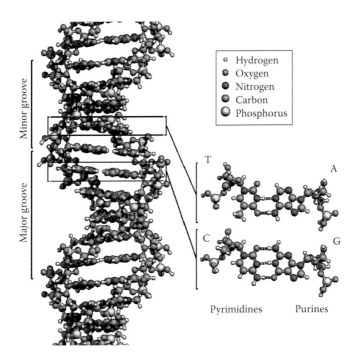

Figure 3.2 The structure of a DNA molecule and the structure of four different kinds of bases: adenine, cytosine, guanine, and thymine.

(e.g., temperature, electrolyte concentration, and pH) of biological nanopores to keep their biological activities; (b) the fragility of the lipid bilayer, which makes the biological nanopores break down easily; and (c) the incompatibility with the standard semiconductor fabrication process.

With the rapid development of solid-state nanopore fabrication technologies [9,10,11], solid-state nanopores have become an inexpensive and superior alternative to biological nanopores due to the following superiorities: (a) better robustness and durability; (b) superior mechanical, chemical, and thermal characteristics; (c) more easily handled shape and size fabrication process with nanometer precision; and (d) compatibility with semiconductor technology, allowing the integration with other nanodevices.

3.1.2 SILICON-BASED NANOPORES AND NANOPORE ARRAYS

A silicon-based nanopore is one of the most important categories of solid-state nanopore. Compared with traditional solid-state nanopores made of other material, such as Si_3N_4, polymer, and two-dimensional (2D) materials, the silicon-based nanopore has many of the following advantages:

1. The silicon-based nanopore is mainly fabricated by wet-etching technology and metal-assisted chemical/plasma etching, which determines the characteristic of low-cost, massive productivity.
2. The typical structure of the silicon-based nanopore is an inverted-pyramidal or conical structure, which is different from most cylindrical nanopores fabricated by focus ion/electron beam (FI/EB) drilling. This lateral asymmetrical structure shows a special rectification effect on the ionic current, which theoretically provides a new control method over the transport behavior of DNA molecules. Moreover, the inverted-pyramid nanopore can achieve better spatial resolution as compared with cylindrical nanopores and is more robust as compared with 2D material nanopores.
3. Silicon is the most widely used material in semiconductor technology and can be compatible with most of the semiconductor process. The subsequent processing and modification of the silicon-based nanopore is relatively easy to achieve. Thus, it meets the requirement of better biological compatibility.

Because of the aforementioned advantages, silicon-based nanopores and nanopore arrays have drawn increasing academic attention. In this section, we have briefly introduced the origin of nanopores and the classification of nanopores.

3.2 FABRICATION TECHNIQUES OF SINGLE-CRYSTAL SILICON NANOPORE

3.2.1 TECHNIQUE PRINCIPLE

3.2.1.1 Basic anisotropic wet etching of silicon

The formation of pyramidal silicon nanopore depends mainly on the anisotropic wet-etching process. Some wet etchants, such as potassium hydroxide (KOH) and tetramethylammonium hydroxide (TMAH), etch silicon at very different rates depending upon which crystal plane is exposed and the doping condition. This phenomenon is known as anisotropic wet etching or orientation-dependent etching. As illustrated in Figure 3.3, the wet etchants display etching rate selectivity much higher in <100> crystal directions than in <111> directions. Etching a (100) silicon surface through a rectangular hole in a masking material, for example a hole in a layer of silicon nitride, creates a pit with flat sloping (111)-oriented sidewalls and a flat (100)-oriented bottom. The (111)-oriented sidewalls have an angle to the surface of the wafer ((100)-oriented) of 54.74°, which is calculated by Equation 3.1:

$$\theta = \arctan\sqrt{2} = 54.74° \tag{3.1}$$

If the etching is continued "to completion," that is, until the flat bottom disappears, the pit becomes a trench with a V-shaped cross section. If the original rectangle is a perfect square, the pit when etched to completion displays a pyramidal shape. Thus, it provides a self-forming inverted-pyramid structure in (100)-oriented silicon wafer. It is notable that if the etchant displays a high etch rate selectivity between the <111> crystal directions and <100> directions, the structure formed by wet etching can be considered to be a self-limiting process.

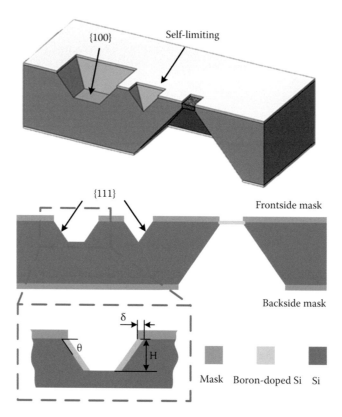

Figure 3.3 The anisotropic etching properties of a (100)-oriented silicon wafer. When the etching is not completed, the cross section shows a trapezoidal shape. When the etching process is completed, the cross section will change to a V shape. The pink material is etching mask and the blue material is silicon.

As the etchants make different oriented planes simultaneously, the undercut δ (illustrated in Figure 3.3) is described by Equation 3.2:

$$\delta = \frac{\kappa V_{100} \cdot t}{\sin \theta} = \frac{\kappa H}{\sin \theta} \tag{3.2}$$

where V_{100} is the etching rate in the <100> direction, t is the etching time, H is the etch depth, θ indicates the angle between the (111)-oriented and (100)-oriented planes, and κ displays the etching rare selectivity between the <111> crystal directions and <100> directions.

Different etchants have different anisotropic characteristics. Table 3.1 shows some major parameters of common anisotropic etchants for silicon.

3.2.1.2 Silicon-based nanopore formation principle by the anisotropic wet-etching method

In the last section, we briefly introduce the anisotropic wet etching characteristic of silicon from a macroscopic view, which is the basic theoretical formation principle of silicon-based nanopore. However, the theoretically optimum size of the nanopore is close to the atomic scale; thus, in this section we further analyze the silicon-based nanopore formation principle at nanoscale.

As illustrated in Figure 3.4, silicon crystallizes in a diamond cubic crystal structure with a lattice spacing (indicated by a) of 5.430710 Å. The space between each layer is only a quarter of the lattice spacing.

Without considering the influence of dangling bond and lattice defects, when the wet-etching process is completed (i.e., the (100)-oriented bottom disappears), the silicon atomic arrangement at the V-shaped cross section tip is illustrated in Figure 3.5.

From the atomic view, it can be indicated that if the silicon atoms are moved layer by layer from the bottom up, we can get small pores at nanoscale theoretically. The limit value of silicon-based nanopore can be calculated by following analysis. As illustrated in Figure 3.5, the nanopore will appear until the first three layers (Layer 1 through Layer 3) are removed. Due to the periodical arrangement and highly symmetrical characteristic of silicon diamond-crystal structure, the shape of the nanopore shows a periodical change with different layers. Figure 3.6 presents the schematic diagram of two kinds of different nanopore shape

Table 3.1 **Major parameters of common anisotropic etchants (KOH and TMAH) for silicon**

ETCHANT	OPERATING TEMPERATURE (°C)	R_{100} (μM/MIN)	$S = R_{100}/R_{111}$	MASK MATERIALS
KOH	80	1.5	400	Si_3N_4, SiO_2 (etches at 6.67 nm/min)
TMAH	80	0.6	37	Si_3N_4, SiO_2

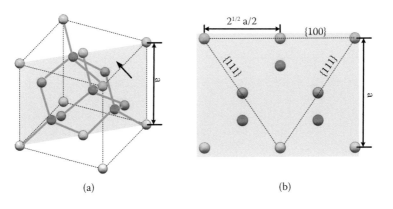

(a)   (b)

Figure 3.4 (a) Silicon crystal structure in a unit cell. (b) Silicon crystal structure in a unit cell viewed perpendicularly to the green plane.

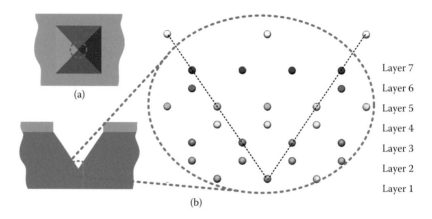

Figure 3.5 The inverted-pyramid formed by the wet-etching process, view from the top side (a) and cross section (b). The dashed red area shows the silicon atomic arrangement at the tip. Balls with different colors present the silicon atoms at different layers.

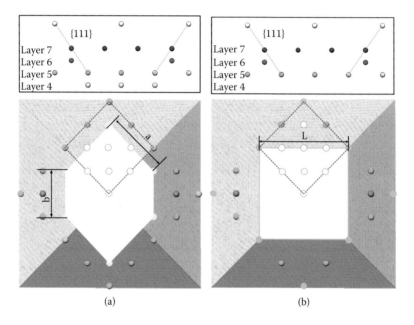

Figure 3.6 Schematic diagram of two kinds of nanopore shape generated by a different etched layer. Atom at the tip of inverted-pyramid structure belongs to Layer 1. The red dashed lines present the {111} planes, the squares with black dashed lines show unit cell of silicon crystal structure view perpendicular to the {100} planes.

generated by a different etched layer. As we know, diamond lattice consists of two interpenetrating face-centered cubic (FCC) Bravais lattices, displaced along the body diagonal of the cubic cell by one quarter the length of the diagonal (Figure 3.4); that is, the green, blue, and yellow atoms correspond to one of the two interpenetrating FCC lattices, and the red and pink atoms correspond to another of the two interpenetrating FCC lattices. The {100} planes with these two different kinds of atoms will also display different characteristics in the shape of a nanopore, which is shown in Figure 3.6. L_d indicates the feature size of the nanopore with the shape of a six-sided diamond.

$$L_d = 3 \cdot \frac{\sqrt{2}a}{2} \cdot n \qquad (3.3)$$

$$n = [\frac{l}{4}], l = 4,6,8......$$

(3.4)

$$L_{d\,min} = \frac{3\sqrt{2}}{2} a$$

(3.5)

where a is the silicon lattice spacing of 5.430710 Å; L_{dmin} is the limit feature size of the nanopore with a six-sided diamond shape; and l indicates the number of the layer at the bottom.

L_s indicates the feature size of the nanopore with the shape of a square:

$$L_s = \sqrt{2}a \cdot m$$

(3.6)

$$m = \frac{l-3}{2}, l = 5,7,9......$$

(3.7)

$$L_{d\,min} = \sqrt{2}a$$

(3.8)

where a is the silicon lattice spacing of 5.430710 Å; L_{smin} is the limit feature size of the nanopore with a six-sided diamond shape; and l indicates the number of the layer at the bottom.

3.2.2 FABRICATION PROCESS OF PYRAMIDAL SILICON NANOPORES AND NANOPORE ARRAYS

Silicon-based nanopores and nanopore arrays are mainly prepared by chemical etching methods, which can roughly be divided into two types: (a) the anisotropic wet-etching method and (b) the metal-assisted chemical/plasma (MaC/PE) etching method [10]. Both methods can achieve massive productivity; however, the nanopores prepared by MaC/PE always present a cylindrical or analogous conical structure. More details about the MaC/PE can be found in Ref. [1,10]. According to the topics in this chapter, we will introduce the anisotropic wet-etching method in detail in this section.

The fabrication process of silicon-based nanopores and nanopore arrays by anisotropic wet etching is carried out in (100) double-side-polished single crystalline silicon wafer. Figure 3.7 shows the basic flow of the fabrication process.

1. After being cleaned with standard RCA solvent, the wafer was coated with a layer of Cr (3000 Å) on the top side, to form the wet-etching mask, and a layer of Al/Si$_3$N$_4$ (7000 Å/3000 Å) was coated on the bottom side to provide the dry/wet-etching mask.
2. The etching windows were patterned using the normal photolithographic process. Remove the exposed mask materials, that is, Cr, Al, and Si$_3$N$_4$, by ceric ammonium nitrate solution, phosphoric acid, and dry etching (such as reactive ion etching [RIE] and inductive coupled plasma [ICP]), respectively;
3. Create an inverted-pyramid structure in the top side of the silicon wafer by the KOH wet etching process.
4. and (f) Silicon backside thinning at high etch rate by wet-etching and dry-etching process, respectively. Considering the etching efficiency and cost, the former demonstrates a fiercely competitive edge.
5. and (g) Nanopore open event at low etching rate by wet-etching and dry-etching process. Both methods can achieve low etching rate. Compared to the wet-etching process, the latter shows a lower etching rate, a higher control accuracy, but a coarser surface.

Cr was selected as the wet-etching mask layer because of its good adhesion with Si substrates, stable chemical properties (nearly no reaction in KOH solutions), and removability. The anisotropic wet-etching process obeys the Arrhenius equation; thus, the etching rate varies with the solution temperature.

The etchant's concentration will also influence the wet-etching performance. KOH solution shows a quite high etching selectivity between different silicon crystal planes. Moreover, as shown in Figure 3.8, 33 wt% KOH solution presents a higher etching rate in high-temperature environment (e.g., 70 °C) than other

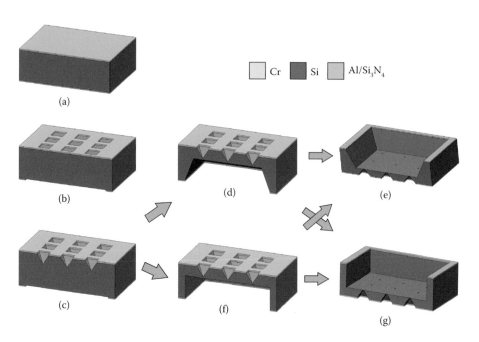

Figure 3.7 The fabrication process of silicon-based nanopores and nanopore arrays by anisotropic wet etching. (a) Create mask layers; (b) pattern the mask layers by lithography technology and wet/dry-etching process; (c) create inverted-pyramid structure in the top side of silicon wafer by KOH wet-etching process; (d) backside thinning by wet-etching process at high etch rate; (e) nanopore open event at low etching rate by wet etching; (f) backside thinning by ICP deep-etching process at high etch rate; (g) nanopore open event at low etching rate by ICP.

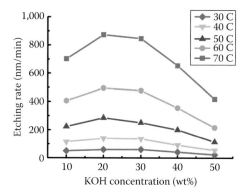

Figure 3.8 The silicon etching rate against KOH concentration and temperature.

concentration conditions, and quiet low etching rate under low temperature (e.g., 30°C), providing both efficiency and controllability over the fabrication process. Therefore, KOH solution with 33 wt% concentration turns out to be the most common choice for the anisotropic wet-etching process.

By the aforementioned wet-etching process (Figure 3.7), nanopores and nanopore arrays with square and rectangular shape can be obtained. The scanning electron microscope (SEM) images are shown in Figure 3.9.

3.2.3 PORE-SIZE REDUCTION TECHNIQUES

Inverted-pyramid silicon nanopores, which can theoretically provide higher spatial resolution, have been successfully obtained. However, controlling the nanopore array becomes increasingly difficult and the fabrication conditions become more demanding as a function of reducing the feature size of the nanopore. From experimental results, nanopores possessing a feature size of ~200 nm are much easier to fabricate than nanopores having feature sizes of ~40 nm.

Arrays, hybrids, and core–shell

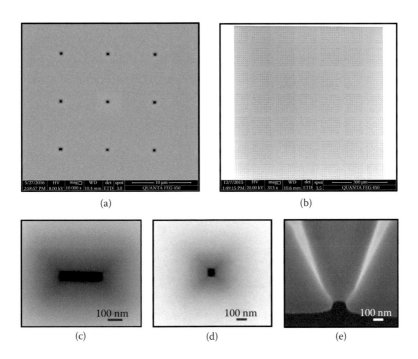

(a) (b)

(c) (d) (e)

Figure 3.9 The SEM image of nanopores and nanopore arrays. (a) and (c) present the nanopore and nanopore arrays opened by the wet-etching process (illustrated in Figure 3.7e); (b) and (d) are the nanopore and nanopore arrays opened by the dry-etching process (illustrated in Figure 3.7g); (e) shows the cross-section image of (d).

3.2.3.1 Material deposition methods

To balance quality and cost, a further treatment to reduce the size of the nanopore is necessary. Material deposition is considered to be the most direct and convenient way to reduce the size of prefabricated nanopores. This idea triggered the appearance of various deposition-induced nanopore shrinkage technologies. Using these techniques, nanopores with dimensions in the range of tens to hundreds of nanometers can be shrunk to sub-10 nm or even totally closed in a controllable way. In addition to modifying the nanopore geometry, some of these techniques can also improve the nanopore properties, like enhancing its mechanical strength and controlling the surface charge. Compared with the electron or ion beam sculpting, these shrinking techniques can process massive nanopores and nanopore arrays at a time, with an equivalent subnanometer resolution.

Material deposition methods can be generally subdivided into the following methods:

1. Conventional chemical vapor deposition (CVD) technology. The efficiency of the nanopore shrinkage process carried out by conventional CVD process can be very high. However, the nanopores' cross section will suffer a distortion, due to the low shape-preserving ability.
2. Atomic layer deposition (ALD) technology. ALD is a cyclic process carried out by dividing a conventional CVD process into an iterated sequence of self-saturating deposition cycles, which allows for very thin, excellent conformal films with control of thickness and composition of the films possible at the atomic level; thus, it achieves single Angstrom contraction precision and theoretically high shape-preserving ability. Furthermore, this technology can realize massive productivity. However, it is a relatively time-consuming process compared with CVD shrinking technology.
3. Conventional physical vapor deposition (PVD) technology. Evaporation and sputtering deposition methods both belong to the PVD technology. In most cases, the material deposited by PVD technology is metal. By these methods, we can get contracted and metalized nanopore at the same time, which can improve the noise performance of the nanopore.
4. Ion/electron-beam-assisted deposition method. A common problem associated with the nanopore shrinking techniques is that the entire membrane is always coated with the deposited materials, leading to the changing of its properties. Ion/electron-beam-assisted deposition method can restrict the deposition to only a specified area surrounding the nanopore.

It should be noticed that the shrinkage process introduced by material deposition may increase the length of the nanopore, especially cylindrical nanopores. For DNA analysis, this means that a longer section of the molecule is contained within the pore, and thus the analysis resolution is reduced. In addition, for a fixed pore diameter, the relative drop in conductance due to the presence of the molecule is lower for longer pore length. However, the length-extended phenomenon may not be obvious in the case of pyramidal nanopores. At beveled edges of the pyramidal tip, the deposited material might have suffered distortion and tune the effective length of the nanopore with minor change.

3.2.3.2 Dry oxidation–introduced shrinkage method

The dry oxidation process is a dedicated shrinkage method for silicon-based nanopores. Under high temperature (around 900°C) and sufficient oxygen supply, a layer of SiO_2 is developed on the surfaces of the Si pore. Simultaneously, the developed SiO_2 is softened and diffuses to the pore edge to find a structural morphology with lower surface free energy. Thus, the size of the Si nanopore will be reduced.

Compared with the other Si nanopore shrinking techniques, the oxidation-induced shrinking method possesses some remarkable advantages. First, it can process massive nanopore samples at a time and reduce their in-built residual stress during the high-temperature annealing. Second, the shrinking process is independent on the nanopore membrane thickness, which means the nanopore will be surely shrunk other than enlarge, no matter how thin the nanopore membrane is. Third, statistics of the pore dimensions indicate that the shrinking process does not broaden the pore size distribution; in most cases, the distribution even slightly decreased, as a result of a self-limiting shrinking effect. Most important, oxidation of the truncated pyramidal structure results in an hourglass-shaped nanopore with smooth arcs forming the bottle neck of the nanopore; thus, the active length of the nanopore will not increase during the shrinking process. As the "hourglass" is not symmetric along the y-axis, it permits unique ion transport properties, which is particularly interesting for various molecular sensing applications. In addition, with the relatively high aspect ratio of the pore (membrane thickness to bottleneck diameter), which would lower the electric field strength when an electric bias is applied across the membrane in a buffer solution, the velocity of a molecule passing through a nanopore would be slower. Hence, a better temporal resolution could be achieved during a biomolecular detection process.

3.3 SILICON-BASED NANOPORE FOR SINGLE-MOLECULE SEQUENCING

As mentioned in Section 1, the potential application in single-molecule sensing is the primary driver of solid-state nanopores' development. Therefore, in this section, we mainly focus on the detecting and controlling methods based on silicon-based nanopore sequencing devices.

3.3.1 DETECTING METHODS BASED ON NANOPORE SEQUENCING DEVICES

According to the different types of the signals, the detection methods can be roughly classified into two categories: the electrical detection methods and the optical readout methods. The detection principle of silicon-based nanopore sequencing devices is mainly based on electrical signal for its compatibility with IC technology, which makes it possible to integrate the signal processing circuit with a sensing unit to achieve a smaller, smarter, and more durable reuseable sequencing system. In this section, we will introduce several nanopore-based electrical methods to detect the sequence of a DNA molecule, as well as the major challenges of these methods.

3.3.1.1 Detection methods based on ionic blockade current

Measuring the ionic blockade current is the most common and original method to detect the sequence of a DNA molecule. Figure 3.10 schematically illustrates the basic principle of the detection method based on ionic blockade current. A membrane with a solid-state nanopore separates the container into two chambers, which are filled with electrolyte solution. A constant bias voltage applied astride the membrane will induce a steady-state ionic current through the pore. By adding DNA molecules into the negatively biased chamber, the electrophoresis force will drive the self-charged DNA molecule passing through the nanopore, and different bases theoretically cause base-specific ionic blockade current, providing evidence of the DNA structural information.

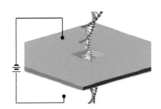

Figure 3.10 The electrical detection methods to sequence DNA molecules based on solid-state nanopores. The basic principle of DNA sequencing via ionic blockade current.

According to the detection principle, the size of an ideal nanopore should be small enough to generate remarkable signal changes in ionic current when a DNA molecule is passing through the pore. The effective length of the nanopores should be short enough to distinguish each single base of DNA molecule. It should be noted that the effective lengths of such conical and pyramidal nanopores are naturally shorter than that of cylindrical nanopores with the same membrane thicknesses. Thus, the ionic blockade current detection method sets high requirements for solid-state nanopore morphology and fabrication technologies. Another problem is that the current signal might not entirely indicate the actual DNA translocation events. It is because a DNA occluded event at the pore mouth may still generate significant reduction of ionic current.

According to the aforementioned challenges, the magnitude of changes in ionic blockade current seem insufficient. To further improve the detection accuracy, the DNA bases' duration time can be added as assisted information of the DNA sequence. In addition, nanopores modified by biomolecules (e.g., oligonucleotides and enzyme) can also help improve this situation. However, such a method puts forward a higher requirement of biocompatibility of solid-state nanopores.

3.3.1.2 Detection methods based on tunneling current

The tunneling-current-based detection method relies on the measurement of tunneling current signals, which is generated when DNA bases pass through a pair of voltage-biased tips in a very close distance, as displayed in Figure 3.11. The tunneling current signals are theoretically base-specific due to the different chemical and electronic structures of the four different DNA bases [12]. Instead of monitoring the ionic blockade current, which occurs due to the occupancy of bases in the entire nanopore channel, the tunneling current is slightly influenced by the adjacent base. Because it is controlled by base-electrode coupling and the energy of the molecular states, it provides an opportunity to achieve single-base sequencing resolution. This method might be the least expensive and fastest route for DNA sequencing.

With the development of the advanced nanofabrication technologies, the embedded-electrodes-sequencing method has drawn growing attention, because of its potential to realize portable and integratable sequencers. Several electrodes, such as Au, Pt, and C, have been adopted to detect the electronic properties of individual nucleotides and DNA strands. Furthermore, graphene and other 2D materials are considered to be the most promising alternative material of transelectrode membrane, due to their in-plant electronic conduction sensitivity to the immediate surface environment and transmembrane solution potential, as well as their extreme thin layer structure.

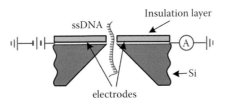

Figure 3.11 Schematic of tunnel-current detection method via a pair of nanoelectrodes fabricated at the edge of the nanopore.

Tunneling-current-based detection method is feasible, but significant challenges remain:

1. The DNA molecule should be specifically located and oriented when it is passing through the electrode pairs. Tunneling current is very sensitive to the electrode spacing and the nucleotides' orientations (perpendicular or parallel to the electrode surface), and it is exponentially affected by the distance between the bases and the electrodes, which will lead to orders of magnitude of fluctuations in the value of current.

2. The translocation rate of the DNA molecule should be slow enough (<0.1 ms/base) to meet the requirement of the high-bandwidth current readout instrument, as well as to minimize the inevitable noise.

3. The fabrication of tunneling-current-based sequencing device remains a formidable challenge. More economical and efficient methods to manufacture nanopores with precisely aligned tunneling electrodes are still needed.

4. The prediction of the tunneling current behavior caused by different bases of unknown DNA strand is difficult, because of the variation of voltage bias, ion condition, and DNA state fluctuation.

In general, more theoretical and experimental research, involving the fundamental factors that influence tunneling current and controlling methods to dominate the DNA molecules spatial manners are still needed.

3.3.1.3 Detection methods based on capacitance variation

Capacitive synthetic nanopores, as illustrated in Figure 3.12, are another potential alternative platform to detect and sequence the DNA structure. When a DNA molecule translocates through the capacitive nanopore, the capacitor will be polarized due to the unique electrostatic charge distribution of the DNA bases, thus triggering base-specific voltage signatures to identify the sequence of the DNA molecules [13].

A large number of simulation and experiment results indicate the feasibility of a capacitance-variation-based detection method; however, there are several challenges for the practical application of this method [14].

1. The bases of DNA molecules should be well oriented to the electrode, because the electrical signature is sensitive to the dipole moment, which associates with each unique base. Thus, the size of nanopore should be carefully controlled at the scale of the target DNA molecule to enhance the voltage signature by orienting the DNA bases along the pore axis.

2. The desired signature would be easily masked by the larger backbone signal because the DNA molecule is negatively charged and most of the charge is accounted for by the phosphate backbone.

3. A large fluctuation of ions at the entrance of the pore may contribute to the voltage fluctuation.

4. More economical and efficient methods to prepare capacitive synthetic nanopores remain a formidable challenge.

3.3.2 CONTROLLING METHODS FOR NANOPORE SEQUENCING DEVICES

The accurate solid-state nanopore-based single-base DNA sequencing cannot be fulfilled without a high degree of control over the molecule behavior through the nanopore. As mentioned earlier, a high velocity of DNA passing through a nanopore is one of the most serious obstacles to the further development and actual application of nanopore-based DNA sequencing technology. The reasonable value of the DNA translocation speed is considered to be 0.01–1 milliseconds per base. To take control of the DNA molecule translation process, various theories and approaches have been put forward in recent decades [15].

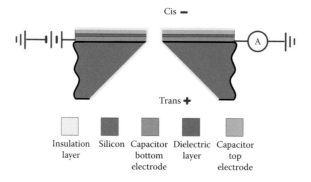

Figure 3.12 Schematic of MOS capacitive synthetic nanopore to detect DNA sequence.

3.3.2.1 Physical control of solution properties

The conventional physical control over the solution involves controlling the viscosity, voltage bias, and temperature. By increasing the viscosity of the solution, adding proper voltage bias between *cis* and *trans* chambers, and reducing temperature of the solution, the translocation speed dropped markedly.

However, the changes of these factors may lead to three kinds of problems:

1. The sequencing process might not be conducted under optimal conditions.
2. The readout signals at nanoampere level would be further weakened.
3. Variations in the translocation dynamics caused by nonspecific DNA–pore interactions may not be eliminated and mitigated.

Therefore, some more effective methods to control the DNA translocation behaviors are needed, such as the modification of nanopores with physical, biological, and chemical methods, the decoration of the DNA molecule, and the tweezers techniques.

3.3.2.2 Reduction of solid-state nanopore size

The translocation velocity of the particle transport through the nanopore shows a remarkable drop as the nanopore's feature size decreases. By decreasing the feature size of the nanopores, the interaction between the pore and the DNA molecule is enhanced. This interaction is caused by a series of thermally activated jumps over small energy barriers (~12 kBT) and influenced by factors such as the size of the nanopores, temperature, and DNA length. The small change in the diameter of a nanopore will also affect the translocation time due to the hydrodynamic coupling between the molecule and the nanopore. These experimental results indicate that it is key to fabricate nanopores with a well-controlled diameter. However, the DNA translocation behavior in a smaller pore is more complex and will introduce more collision, which might result in poor signal-to-noise ratio, compared with a larger nanopore.

3.3.2.3 Surface charge condition modulation of solid-state nanopores

As DNA molecules are self-charged in electrolyte solutions, the surface charge density conditions of a nanopore will affect the DNA translocation performance in the nanopore. Both positively charged and negatively charged nanopores can be used to modulate the DNA translocation process. Because the silicon surface can be modified by many kinds of materials, this method is suitable for pyramidal silicon nanopores [16].

By changing the surface charge density and distribution of the nanopore's sidewall, the interaction between the DNA molecule and the nanopore will be enhanced. Thus, it will result in a decrease of translocation velocities.

Instead of obtaining a nanopore with a fixed surface charge condition, the gate-bias voltage applied at the substrate of a nanopore can achieve real-time control over wall-surface charge density. This phenomenon led to the rectification property in asymmetrical nanochannels, especially in conical and pyramidal nanopores. Thus, it provides an effective way to modulate DNA molecule velocity through the nanopore. The modulation mechanism is also based on the manipulation of the surface charge condition by a varying gate voltage applied across the nanopore.

Compared with cylindrical nanopores under the same feature sizes, pyramidal structures have lower resistance and higher sensitivity to the variation of gate voltages, resulting in a stronger ionic current and better manipulation of DNA molecule behaviors. Therefore, the asymmetrical nanopores seem to be a better platform for DNA sequencing.

3.3.2.4 Hybrid biological–solid-state nanopores

As mentioned in Section 1, biological nanopores can achieve better control of DNA translocation time and performance. Therefore, hybrid biological–solid-state nanopores might theoretically provide a new platform with both enhanced robustness and unique biological qualities [17]. There are several ways to obtain hybrid biological–solid-sate nanopores.

1. Inserting a biological nanopore into a solid-state nanopore is the simplest and most direct method to obtain a hybrid nanopore, as shown in Figure 3.13a. Replace the lipid bilayer by solid material to achieve better robustness. However, the size of the solid-state nanopores should be precisely controlled to ensure the smooth and effective assembly process with the αHL protein nanopores.

Arrays, hybrids, and core-shell

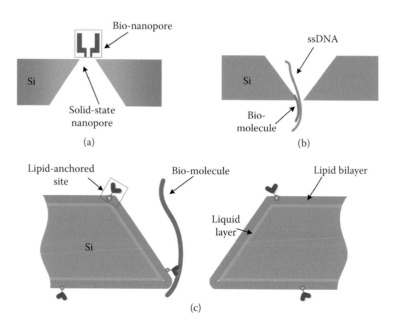

Figure 3.13 Technologies to modulate DNA molecule translocation behavior. (a) Inserting a biological nanopore into a solid-state nanopore. (b) Decorating the mouth edge of solid-state nanopore with biological molecule. (c) The cross-section of a lipid-bilayer-coated nanopore with lipid-anchored site (dashed box), which can anchor and translate the bio-molecule (red strand) through the nanopore.

2. Use biomolecule to decorate the solid-state nanopore, as presented in Figure 3.13b. Oligonucleotides, DNA origami, and enzymes are usually used to modify the mouth edge of solid-state nanopores. These enhanced pore–DNA interactions will lead to a decrease of DNA translocation velocity However, such a method put forward a higher requirement of biocompatibility of solid-state nanopores, where silicon nanopores show good potential.
3. Coat solid-state nanopores by lipid bilayer with fluid-anchored capture sites, which can tune the pore size in situ with the control of temperature condition, as illustrated in Figure 3.13c. By anchoring the analytes onto the lipid, the dominant factor that influences the translocation behavior changes from the low-viscosity aqueous electrolyte to the high-viscosity lipid bilayer coating, prolonging the translocation time.

3.3.2.5 Modification of the DNA

The modification of the DNA strands is another alternative approach to substantially reduce the DNA translocation velocity through the nanopore. By binding a biomolecule, such as enzymes or oligonucleotides, to DNA strands, the determinate factor of the translocate velocity changes from the electrophoretic force to the enzyme replicating or unzipping processes, which ratchets the DNA strands one nucleotide at a time and successively acts up to tens of thousands of nucleotides. Unlike decorating a nanopore, these methods reduce the demand of biological compatibility of solid-state nanopores. However, these methods show a higher requirement on the morphology of solid-state nanopores, which means the nanopore should be small enough to hinder/unzip the DNA-enzyme/oligomer complex, while remaining large enough for ssDNA to pass through.

3.3.2.6 Optical and magnetic tweezers

The methods mentioned earlier in this chapter may share a common problem that the actual position and the force exerted on the DNA molecule inside the nanopore are hard to predict, which can only be inferred by molecule dynamic (MD) simulation. The optical tweezers technology may provide a straightforward way to slow down the DNA translocation velocity. This method was initially used in the measurement of the forces exerted on the DNA molecule inside the nanopore during the voltage-driven process and was first published by Keyser's group in 2006. The DNA molecule is controlled by the tethered bead and laser tweezers, as shown in Figure 3.14a. By the control of the speed and direction of the optical tweezers, the DNA translocation velocity can be set at any desired value. This approach can theoretically achieve true

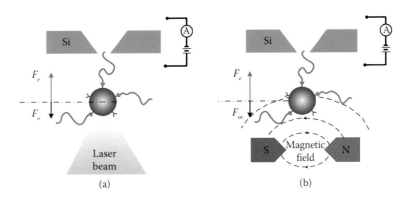

Figure 3.14 Optical and magnetic tweezers. (a) Left: a DNA-tethered bead is trapped near the solid-state nanopore by a tightly focused laser beam. Right: the electrical force F_{el} drives the DNA strand through the nanopore and the strand is straightened and controlled by the composite of optical force (F_{ot}) and F_{el}[18]. (b) Schematic of magnetic tweezers to control the translocation of a DNA-attached colloid by magnetic force [19].

three-dimensional spatial control of the DNA molecule and allow massively parallel detections. However, the ionic current is sensitive to the motion of the optical bead due to the absorption of the laser light by the solution, which results in a great influence on the ionic current measurement [20].

The magnetic tweezers technology utilizes an adjacent magnet to generate a magnetic-field gradient, thus inducing a constant force to control the DNA molecule attached to the magnetic colloid, as shown in Figure 3.14b. The sequencing process can be theoretically operated in a massively parallel fashion. However, magnetic tweezers cannot take total control of the DNA translocation behavior and a series of different stages might occur during the translocation process, because the resultant force, which consists of magnetic, electric, and DNA–pore interaction force, is roughly constant.

3.3.3 THE OPPORTUNITIES AND CHALLENGES

Over the past few decades, the solid-state nanopore-based sequencing technology has been promoting the development of a single-molecule real-time DNA sequencing field and providing a brand new future in the gene detection area. Although biological and solid-state nanopore-based DNA sequencing technologies have plenty of superiorities, substantial long fragments of DNA molecules have yet been sequenced with high accuracy. Several bottlenecks hold the further development of nanopore-based sequencing technologies, for instance, such as how to achieve massive and repeatable fabrication of high-quality solid-state nanopores at a low cost, how to improve the electrical detective methods for the high-resolution nanopore-based DNA sequencing, and how to take control of DNA translocation behavior and velocity through the nanopore. All these challenges demand high-quality nanopores with sufficiently small feature size, good pore shape, and proper materials. Meanwhile, all these challenges provide us with the impetus to seek a better, more efficient method to fabricate and decorate solid-state nanopore.

Therefore, in this chapter, we not only focused on the formation principle and preparation method of silicon-based nanopores and nanopore arrays; we also introduced the typical and advanced techniques applied in the field of solid-state nanopore-based DNA sequencing technology involving the detection methods for base-specific signals, the solid-state nanopore fabrication techniques, and the methods to modulate the DNA translocation behaviors. The nanopores with asymmetry or a 2D structure seem to be a better choice than the symmetrical ones due to the higher spatial resolution. Both biological and solid-state nanopores have their unique advantages in DNA sequencing application, and we believe that with the joint efforts of researchers in various fields, the hybrid biological–solid-state nanopores will finally play an important part in this sequencing platform. The field of solid-state nanopores-based DNA sequencing is still in its early stages. With the development of novel fabrication technology and advanced material, and with the combination of the nanofluidic and in situ electrical or optical readout devices, which are associated with reliable data processing and calibration methods, the parallel DNA sequencers based on solid-state nanopores will finally become an integrated and efficient nanosystem providing services for personalized medicine fields.

3.4 FUTURE APPLICATIONS AND SUMMARY

3.4.1 FUTURE APPLICATIONS

The future applications of silicon-based nanopores and nanopore arrays are not limited to the biosensing platform, which can be used to investigate a wide range of phenomena involving DNA, RNA, and proteins. The characteristics of silicon-based nanopores and nanopore arrays can also lead to other applications.

1. *Nanofabrication.* The pyramidal silicon nanopore arrays can be used as a temple for direct surface nanopatterning. Due to the continually decreasing feature size of integrated circuits, the surface nanopatterning has become one of the most attractive topics in the engineering fields. There are various techniques in fabricating nanoscale patterns on the surface of substrates, involving electron-beam lithography (EBL) technology, nanoimprinting and replica molding process, scanning probe microscope (SPM) writing technology, and templated-based methods. Among these technologies, the templated-based nanopatterning method with pyramidal silicon nanopore arrays has many advantages:
   a. High efficiency. Both preparation and patterning stages of pyramidal silicon nanopore arrays templates can produce multiple samples at a time.
   b. Low cost. The cost of equipment required in both preparation and patterning stages of pyramidal silicon nanopore arrays templates is low.
   c. High flexibility. The shape and size of the nanopore in the pyramidal silicon nanopore arrays templates can be controlled; thus, the corresponding size and shape of the surface structures transferred from the template are tunable.
2. *Liquid environment sensors.* Pyramidal silicon nanopore and nanopore arrays can act as a liquid environment sensor, because that nanopore with dozens to several nanometers of dimensions is sensitive to the ionic solution conductance. Different conditions in ionic solution, such as temperature, light intensity, and ionic concentration, will cause different changes in conductance of the ionic liquid. Therefore, pyramidal silicon nanopore and nanopore arrays can be used as a sensitive liquid environment sensor.
3. *Optical filter and other optical application.* For pyramidal silicon nanopore and nanopore arrays, the optical properties are still largely unknown. However, it has been discovered that nanopores in an opaque metal film, with sizes smaller than the wavelength of incident light, lead to a wide variety of unexpected optical properties such as strongly enhanced transmission of light through the pores and wavelength filtering [21]. These intriguing effects are probably caused by the interaction of the light with electronic resonances in the surface of the metal film, which can be controlled by changing the size and geometry of the pores.
4. *Simulator for theoretical physical study.* Pyramidal silicon nanopore and nanopore arrays might be used as simulators to explore new regimes for electrochemical, near-field optics, and nanoparticle translocation processes.

Besides the aforementioned potential application, pyramidal silicon nanopore and nanopore arrays might also find use in quantum and atom optics. Nanopore arrays with subwavelength are promising tools in the study of the quantum versus classical of physical nature. Larger solid-state pores can act as chip-based probes and are promising alternatives to conventional patch-clamping for electrophysiology on cells.

3.4.2 SUMMARY

Inspired by nanopore-based DNA sequencing, solid-state nanopores have been undergoing explosive growth over recent years, due to their remarkable properties, such as enhanced robustness and durability, the ability to tune the pore geometry and surface properties, and the compatibility with existing semiconductor and microfluidics fabrication techniques.

In this chapter, we focus on the pyramidal silicon-based nanopores and nanopore arrays with their origin, basic formation principle and fabrication process, electrical detecting methods, common controlling methods of the DNA molecule, and their potential future application. Table 3.2 demonstrates the silicon-based nanopore fabrication and further treatment methods. Compared with other kinds of solid-state nanopores, pyramidal silicon-based nanopores and nanopore arrays demonstrated the following advantages:
1. Massive productivity (high efficiency).
2. Low cost.

Table 3.2 The silicon-based nanopore fabrication and further treatment methods

METHOD	MATERIALS TESTED	FABRICATE PRECISION	INNER STRUCTURE OF THE NANOPORE	PARALLEL MASSIVE PRODUCTION	APPLICATIONS OF THE NANOPORES
Anisotropic wet etching	Si	Larger than several nanometers	Pyramidal	Yes	Nanopore-based sensors; masks for nanostencil lithography
Metal-assisted chemical etching	Si	Mainly determined by the metal particles	Cylindrical; cone; helical structure	Yes	Bio- and chemical sensor; porous silicon structures with intense photoluminescence; Si solar cells
Metal-assisted plasma etching	Common Si; flexible Si membranes, Si power crystals	Mainly determined by the metal particles	Cylindrical	Yes	Ionic switches; nanopore-based sensors; ionic logic circuitry; controllable molecular separation platforms
ALD-induced shrinking	SiN_x, SiO_2, Al_2O_3, TiO_2	~1 Å	Conformal deposition	Yes	Nucleic acid–based diagnostics; Electrofluidic application (ionic field effect transistor)
Evaporation-induced shrinking	SiN, SiO_2	Several nanometers	Associated with the structure of prenanopore	Yes	Nanopore-based sensors
Electro-deposition-induced shrinking	SiN_x	—	Mainly determined by the diffusion field in or close the pore	Yes	Nanopore-based sensors
Ion-beam-assisted deposition-induced shrinking	SiO_2	—	Cylinder, cone	No	Selective detection of biological organism and molecules
Electron-beam-assisted deposition-induced shrinking	SiO_2	Subnanometer (0.5nm)	Associated with the structure of prenanopore	No	Selective detection of biological organism and molecules
Dry oxidation–induced shrinking	Si	Nanometers	Associated with the structure of prenanopore	Yes	Nanopore-based sensors; masks for nanostencil lithography

Arrays, hybrids, and core–shell

3. High flexibility. The size and shape of pyramidal silicon nanopore can be tuned. The silicon surface is easily modified by material deposition process.
4. Asymmetric structure (high spatial resolution for molecule sensing), higher control over the molecule behavior due to the rectification characteristic.
5. Superior compatibility of the IC process.

The field of solid-state nanopores is still young and exciting, although much has already been achieved. Novel technologies and new materials will continuously push the limits and boundaries of the fabrication techniques. More precise control over pore geometry and higher yield will be realized with lower cost and less time. Multifunctional nanosystems with solid-state nanopores as the key components will become a hot research topic. For example, parallel DNA sequencers based on solid-state nanopores will combine with the nanofluidic and in situ electrical or optical readout devices to constitute an integrated and efficient nanosystem providing services for personalized medicine. The applications of solid-state nanopores will be significantly expanded. Besides acting as biosensing platforms, lithography nanostencils, light modulators, and ionic rectification devices, various solid-state nanopores will find their specific applications in many other fields.

REFERENCES

1. Sanger F, Nicklen S, and Coulson A R. DNA sequencing with chain-terminating inhibitors. *Proc. Natl. Acad. Sci. U. S. A.* **74**, 5463–5467 (1977).
2. Muers M. Technology: Getting Moore from DNA sequencing. *Nat. Rev. Genet.* **12**, 586–587 (2011).
3. Shendure J and Ji HL. Next-generation DNA sequencing. *Nat. Biotechnol.* **26**, 1135–1145 (2008).
4. Church G, Deamer DW, Branton D, Baldarelli R, and Kasianowicz J. *Characterization of individual polymer molecules based on monomer-interface interactions*, U.S. Patent 5795782 (1998).
5. Majd S, Yusko EC, Billeh YN et al. Applications of biological pores in nanomedicine, sensing, and nanoelectronics. *Curr. Opin. Biotechnol.* **21**, 439–476 (2010).
6. Branton D, Deamer DW, Marziali A et al. The potential and challenges of nanopore sequencing. *Nat. Biotechnol.* **26**, 1146–1153 (2008).
7. Majd S, Yusko E C, Billeh Y N et al. Applications of biological pores in nanomedicine, sensing, and nanoelectronics. *Curr. Opin. Biotechnol.* **21**, 439–476 (2010).
8. Venkatesan B M and Bashir R. Nanopore sensors for nucleic acid analysis. *Nat. Nanotechnol.* **6**, 615–624 (2011).
9. Rhee M and Burns MA. Nanopore sequencing technology: Nanopore preparations. *Trends Biotechnol.* **25**, 174–181 (2007).
10. Deng T, Li M, Wang Y et al. Development of solid-state nanopore fabrication technologies. *Sci. Bull.* **60**, 304–319 (2015).
11. Fologea D, Uplinger J, Thomas B et al. Slowing DNA translocation in a solid-state nanopore. *Nano Lett.* **5**, 1734–1737 (2005).
12. Di Ventra M, Pantelides ST, and Lang ND. First-principles calculation of transport properties of a molecular device. *Phys. Rev. Lett.* **84**, 979–982 (2000).
13. Heng JB, Aksimentiev A, Ho C et al. Beyond the gene chip. *Bell Labs Tech. J.* **10**, 5–22 (2005).
14. Gracheva ME, Aksimentiev A, and Leburton JP. Electrical signatures of single-stranded DNA with single base mutations in a nanopore capacitor. *Nanotechnology* **17**, 3160–3165 (2006).
15. Nakane J, Wiggin M, and Marziali A. A nanosensor for transmembrane capture and identification of single nucleic acid molecules. *Biophys. J.* **87**, 615–621 (2004).
16. Siwy ZS and Howorka S. Engineered voltage-responsive nanopores. *Chem. Soc. Rev.* **39**, 1115–1132 (2010).
17. Wanunu M and Meller A. Chemically modified solid-state nanopores. *Nano Lett.* **7**, 1580–1585 (2007).
18. Keyser UF, Koeleman BN, Van Dorp S et al. Direct force measurements on DNA in a solid-state nanopore. *Nat. Phys.* **2**, 473–477 (2006).
19. Peng HB and Ling XSS. Reverse DNA translocation through a solid-state nanopore by magnetic tweezers. *Nanotechnology* **20**, 185101 (2009).
20. Keyser UF. Controlling molecular transport through nanopores. *J. R. Soc. Interface* **8**, 1369–1378 (2011).
21. Genet C and Ebbesen TW. Light in tiny holes. *Nature* **445**(7123), 39–46 (2007).

Fabrication of three-dimensional Si quantum dot array by fusion of biotemplate and neutral beam etching

Seiji Samukawa

Contents

4.1 INTRODUCTION

As material size shrinks to a nanometer scale, the materials start to show specific electrical and optical characteristics based on the quantum confinement of carriers and/or excitons in nanoscale. Recently, superlattices with closely packed quantum dots (QDSLs) with high-uniformity and high-density have received great attention to develop high-performance optoelectronic devices, including lasers and solar cells [1–4]. In the superlattices with closely coupled QDs, discrete states of each QD merge to form broadened mini-bands to behave as brand-new materials [5–9]. For photovoltaic applications, such engineered QDs can be used as adjustable absorber layers with well-designed intermediate bandgap energy to build tandem solar cells [9–12]. QDs of III-V compounds formed by a bottom-up process such as Stranski–Krastanov growth based on self-organization [13] have been applied to optoelectronic devices, including high-performance QD lasers [1–4]. In the "bottom-up" process, there are some limitations in control of the density of QDs: The distance between QDs is too narrow to avoid the coupling of wave functions for high-gain lasers. On the other hand, in "top-down" process technologies, there are also limitations in fabrication of defect-free nanostructures through the process sequences such as photolithography, plasma etching, co-sputtering, and annealing [14–16].

To break through these problems with the bottom-up and top-down processes, an innovative process has been proposed, utilizing a damage-free neutral beam (NB) etching technology [17] combined with a biotemplate process [18]. The process was established by one of the authors (SS) and applied to fabricate uniform and closely packed arrays of nanodisks (NDs) and nanowires (NWs). So far, we applied the new method to Si, GaAs, and Ge and fabricated uniform and closely packed arrays of NDs/NWs with sub-10-nm in diameter. We employ "ferritins" to make a nano-etching mask as a biotemplate. The ferritins produced by DNA information have uniform 7-nm iron oxide (Fe_2O_3) cores in their cavities and a protein shell. A high-density, two-dimensional (2D) ordered array of ferritins on a substrate provides an array of well-aligned cores with controlled distances—roughly twice the protein shell thickness after selective protein shell elimination. This biotemplate technology leads us to generate a lattice of NDs/NWs that is an enabler of a new materials engineering technology.

In this chapter, we have reviewed our damage-free NB etching technology combined with a biotemplate process and reported some of our recent applications to produce NDs/NWs of various materials.

4.2 NEUTRAL BEAM SOURCES FOR DAMAGE-FREE ETCHING AND SURFACE OXIDATION

For the development of damage-free etching processes, NB sources have been investigated as substitutes for the conventional plasma sources. By using the NB sources, damage-free etching can be achieved by eliminating charged particles (ions and electrons) as well as ultraviolet (UV) light from the plasma source. During etching, energetic neutral beams, rather than ions, bombard the wafer surface where radicals absorb, resulting in the removal of the wafer materials, that is, etching. In particular, Samukawa has developed practical NB sources for high-performance etching. Figure 4.1 provides a schematic representation of an NB source that consists of an induction-coupled plasma (ICP) and parallel carbon plates [17, 19, 20]. The process chamber is separated from the plasma source chamber by a bottom electrode made of carbon, which has the low sputtering yield under high-energy bombardment and does not contaminate semiconductor devices. NB beams are extracted from the plasma through numerous high-aspect-ratio apertures in the bottom electrode. The NB sources employ pulse-time-modulated ICPs to generate large quantities of negative ions, which are more effectively neutralized than positive ions. This is because the detachment energy of electrons from negative ions is much smaller than the charge transfer energy of positive ions. For example, in previous-generation neutral beam sources, where neutral beams are obtained through the charge transfer of positive ions, the neutralization efficiency is at most 60% and the beam energy must be greater than 100 eV to obtain a large degree of neutralization of positive ions. On the other hand, an NB source using Cl2 plasma with pulsed operation can obtain a neutralization efficiency of almost 100% and high neutral beam flux of more than 1 mA/cm². At the same time, because of the presence of apertures, UV radiation from the plasma is shaded and/or absorbed in the bottom electrode; consequently,

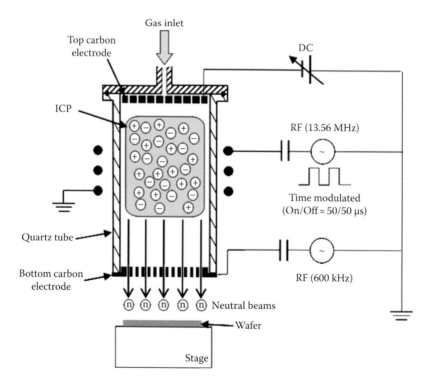

Figure 4.1 Schematics of the neutral beam system.

UV radiation damage can be suppressed. The NB process is a promising solution for next-generation nanoscale device fabrication, which requires damage-free etching conditions. Additionally, we also developed damage-free neutral beam oxidation (NBO) for the fabrication of ultrathin oxide films on Si and GaAs surfaces at a temperature of less than 300°C by using an energy-controlled oxygen-neutral beam [21–23]. In this system, the Si aperture was used for neutral beam generation to eliminate carbon contamination at the interface.

4.3 FABRICATION OF NANODISK STRUCTURES BY COMBINATION OF BIOTEMPLATE AND NEUTRAL BEAM ETCHING

It has been suggested that by 2020, Moore's law will break down and we will reach the physical limits of transistor operation. Work is therefore under way in several countries to develop nanodevices based on new principles using quantum effects. To fabricate quantum effect devices, it is essential that defect-free nanostructures (dots and wires) can be formed with precision down to the atomic layer level. Two approaches to the formation of quantum nanodots have hitherto been studied: a top-down approach that uses processes such as plasma etching, and a bottom-up approach that uses self-organization techniques based on processes such as molecular beam epitaxy. However, in top-down processing using a plasma process, the plasma emits ultraviolet rays and electrical charges accumulate at the substrate surface. This reduces the selectivity of the mask and underlying substrate material, and leaves a high density of defects deep within the processed surface, with the result that processing is limited to dimensions of several tens of nanometers. On the other hand, although the bottom-up process has fewer problems related to defects and the like, since the growth process involves lattice strains, it has problems such as nonuniformity and stress deformation of the arrangement and structure of the nanodots, which means that quantum effects can only be achieved with a limited range of materials and structures. To deploy these structures in nanodevices, we must be able to fabricate nanostructures without relying on more accurate materials. Therefore, the author proposed and is researching the formation of quantum nanodots of less than 10 nm in size by means of a top-down process using a low-energy neutral beam capable of defect-free processing. An advantage of the top-down process is that it can form nanostructures with an arrangement that can be uniformly controlled no matter what combination of materials is used. Instead of photolithography, we used a biotemplate as proposed by Yamashita [18] as an etching mask with dots of a few nanometers in size. As shown in Figure 4.2, the biological supermolecule (protein) ferritin has a diameter of 12 nm and a 7 nm internal cavity. There is a negative charge inside this cavity, and when ferritin is put into a solution containing dissolved Fe ions, these Fe-positive ions are introduced into the cavity of ferritin molecules to form iron oxide cores. These iron cores are 7 nm in diameter. Ferritin molecules containing these iron cores are selectively placed in a 2D arrangement on a silicon oxide film, and the protein is then removed by UV/ozone or heat processing, leaving behind the 7-nm iron cores on the substrate for use as an etching

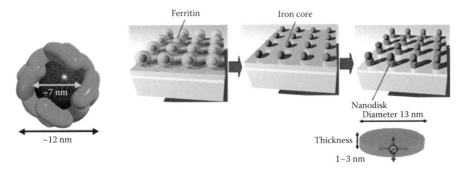

Figure 4.2 The neutral beam etching process using the biological supermolecule (protein) ferritin, whose self-organizing properties result in 2D crystallization. The etching mask is made from iron cores encapsulated within the ferritin molecules. By using iron cores with a diameter of 7 nm as etching masks, we can form defect-free ultrafine structures with a size of less than 10 nm.

Arrays, hybrids, and core–shell

mask [22]. Finally, a Cl$_2$-based neutral beam can etch any kind of surface materials using the etching mask of 7 nm iron cores. The authors are using this process to develop quantum-effect devices with a quantum nanodisk structure. A nanodisk is a nanoscale cylindrical structure whose height (thickness) is smaller than its diameter. Figure 4.3 shows nanodisk structures of silicon, germanium, gallium arsenide, and graphene with a diameter of about 10 nm that were produced in this way. As this figure shows, the sub-10-nm quantum nanodisks are formed in an array configuration with uniform spacing. Figure 4.4 shows the

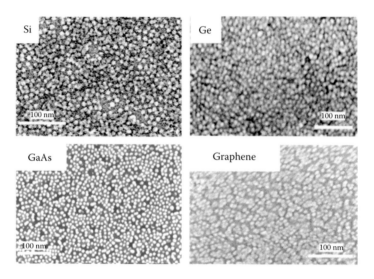

Figure 4.3 SEM image of defect-free nanodisk structures (diameter 7 nm) formed with a uniform density and regular arrangement by using ferritin iron cores as an etching mask for silicon, germanium, gallium arsenide, and graphene.

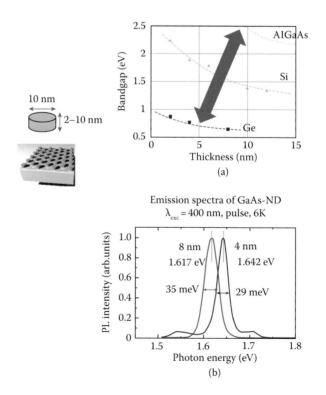

Figure 4.4 (a) Variation of band gap energy with disk thickness in nanodisk structures of silicon, germanium, graphene, and aluminum gallium arsenide, and (b) photoluminescence of gallium arsenide nanodisk structures.

photoluminescence from GaAs/AlGaAs quantum dots and the precise control of bandgap energy in these nanodisk structures with different materials when varying the thickness of these dots while keeping the diameter fixed at 10 nm. It can be seen that the bandgap can be controlled over a wide range with high precision by varying the nanodisk size and material. No other quantum dot fabrication techniques can offer this kind of flexible and precise bandgap control, Also, this is the first time that photoluminescence has been observed from gallium arsenide quantum dots fabricated by a top-down process, and from time-resolved measurements it has been confirmed that this light originates from the quantum dots themselves and not from defects [24,25]. These results show that the benefits of the top-down process are achieved for any material, with defects fully suppressed at the surface interface of sub-10-nm quantum dots formed by a neutral beam process. The authors are currently developing a high-efficiency quantum dot solar cell and quantum dot laser with a flexible band structure [24,25].

4.4 FABRICATION OF SILICON NANODISKS

4.4.1 PROCESS FLOW

To measure optical transmittances, we prepared 2D arrays of Si-NDs with SiO_2 or SiC matrix on quartz substrate. The fabrication of a 2D Si-ND array structure using the two types of biotemplate of ferritin and Listeria-Dps (Lis-Dps) and damage-free NB etching is schematically shown in Figure 4.5a

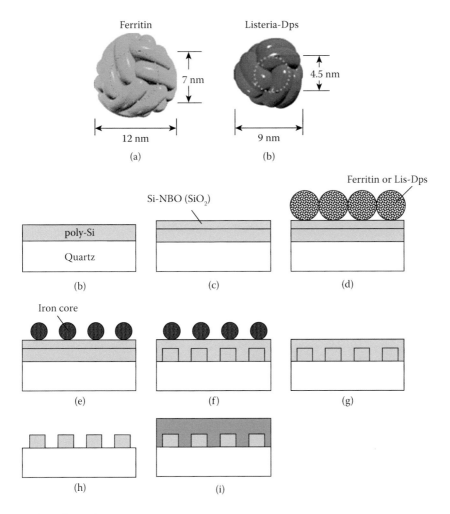

Figure 4.5 (a–i) Flow of sample preparation.

through i. The fabrication steps are as follows. First, 2- to 12-nm-thick poly-Si and 3-nm-thick Si-NBO (SiO_2) layers were fabricated on a 10×10 mm^2 quartz substrate, as shown in Figure 4.5b and c. The poly-Si layer was prepared using molecular beam epitaxy with a controlled deposition rate of 0.05 nm/min followed by annealing in argon atmosphere at 600°C for 16 h. By in situ monitoring the poly-Si deposition thickness, we could precisely control the thickness of the Si-ND. Then a 3-nm SiO_2 layer was fabricated using our developed neutral-beam oxidation process at a low temperature of 300°C as a surface oxide (Si-NBO) [21–23]. Second, a new biotemplate of a 2D array of ferritin (a 7.0 nm diameter iron oxide core in the cavity) or Lis-Dps (a 4.5 nm diameter iron oxide core) molecules was formed, as shown in Figure 4.5d. Lis-Dps is a Dps protein that is synthesized from Listeria bacteria. It has a spherical protein shell with an external diameter of 9.0 nm and a cavity diameter of 4.5 nm, and it biomineralizes iron as a hydrate iron oxide core (diameter: 4.5 nm) in the cavity and stores it. By using iron oxide core as the 4.5 nm diameter etching mask, we would like to realize a much smaller diameter size of the Si-ND structure. Next, protein shells were removed by heat treatment in oxygen atmosphere at 500°C for 1 h to obtain a 2D array of iron oxide cores as the etching masks, as shown in Figure 4.5e. Etching was carried out by a combination of NF_3/H-radicals treatment at 100°C and anisotropic etching of poly-Si using Cl_2 NB, respectively, as shown in Figure 4.5f. During the etching process by changing the NF_3/H-radicals treatment time to 15 min and 40 min, we could change the diameter of ND due to the side etching of Si-NBO. Then, the 2D iron oxide core array was removed using hydrochloric solution to obtain a 2D Si-ND array structure, as shown in Figure 4.5g. Because the native oxide grows between Si-NDs, we call this structure simply a "Si-ND array structure with a SiO_2 matrix." For fabrication of the Si-NDs with SiC matrix, finally, the SiO_2 matrix was removed by the NF_3/H-radicals treatment and then a 3-nm-thick SiC was deposited on Si-NDs by sputtering system as shown in Figure 4.5h and i.

Figures 4.6a and 4.7a show a top-view SEM image of the 2D Si-ND array structure fabricated using Ferritin and Lis-Dps as the etching mask, which consisted of a closely packed array of 7 nm and 4.5 nm diameter iron oxide cores on the Si-NBO surface. As the figure shows, this array structure had a high density (7×10^{11} cm^{-2} and 1.4×10^{12} cm^{-2} respectively), uniform size (Si-ND diameter: 6.4 nm and 10 nm) and well-ordered arrangement (quasi-hexagonal ordered arrays). The density of the 2D Si-ND array structure fabricated using Lis-Dps increased to two times that of the array fabricated using ferritin, while the diameter of the Si-ND shrunk to 61% of that obtained by using ferritin with 40 min NF_3/H-radicals treatment. To confirm regularity in this array, we measured the center-to-center distance between adjacent Si-NDs, as shown in the SEM image in Figure 4.6b and Figure 4.7b. These figures show that the standard deviation of the center-to-center distance was less than 10%. The results show that the 2D Si-ND array structure fabricated using a Ferritin and Lis-Dps etching mask with a 4.5 nm and 7 nm diameter iron

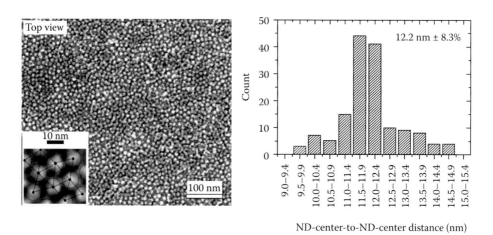

Figure 4.6 (a) SEM picture of a 2D array of Si-NDs. (b) Distribution of ND-center-to-ND-center distances.

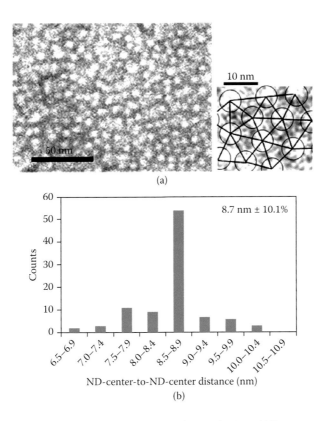

Figure 4.7 (a) SEM image of a 2D Si-NDs array using Lis-Dps. (b) Distribution of ND-center-to-ND-center distances.

oxide core formed a 2D superlattice structure with a high density and well-ordered arrangement, making it a suitable quantum dot structure.

4.4.2 PHOTOLUMINESCENCE IN SILICON NANODISK

To analyze defects at the interface between a Si-ND and the SiO_2 matrix, high-sensitivity electron spin resonance (ESR) analysis was performed to quantify the presence of any paramagnetic defects. We used Si-NBO grown onto a p-type Si wafer with a (111) surface orientation and a high resistivity of more than 1000 $\Omega \cdot$cm. ESR measurement was performed on the samples at 4 K using a Bruker-ESP300E spectrometer. The microwave frequency and power were approximately 9.62 GHz and 0.1 mW, respectively. The paramagnetic (P_b) center density at the Si-ND/Si-NBO (SiO_2) interface was 5.0×10^{10} cm^{-2}. Incidentally, the surface areas of the (7 nm in diameter) Si-NDs with thicknesses of 4 nm and 8 nm were 1.6×10^{-12} cm^2 and 2.5×10^{-11} cm^2, respectively. Therefore, we were able to estimate the number of defects in 4-nm- and 8-nm-thick Si-NDs at 0.08 and 0.16, respectively. In achieving these results, we found that the interface between the Si-NDs and the SiO_2 matrix was almost completely defect-free.

Silicon is basically an indirect band gap semiconductor and thus its optical absorption and radiative recombination efficiencies are markedly low. On the other hand, with a QD structure the surface effect and size effect, which modify the carrier wave function, can improve radiative recombination efficiency [26–29]. However, in the conventional fabrication processes, a large amount of defect-induced and interfacial local energy levels can be formed easily. As a result, the majority of experiments on Si nanostructures have shown extremely slow decaying PL with decay times ranging from several tens of nanoseconds to microseconds. This can be attributed to carriers recombining at the defect states rather than the nanostructure itself [28]. With this in mind, we researched photo-exited emissions from the Si-ND array structure. Figure 4.8 shows time-integrated PL spectra of a 2D array of 8-nm-thick Si-NDs with a SiO_2 interlayer for an excitation power of 50 mW. We performed the measurement from 10K to 300K. The result shows two PL emissions centered at wavelengths of 665 nm (E_1: 1.86 eV) and 555 nm (E_2: 2.23 eV), and the highest peak of PL intensity was

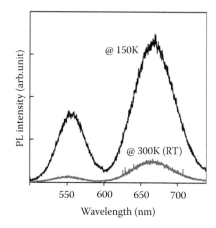

Figure 4.8 Time-integrated PL spectra of a 2D array of 8-nm-thick Si-NDs with a SiO$_2$ interlayer for an excitation power of 50 mW.

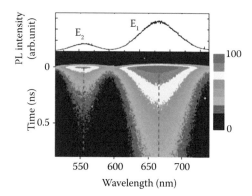

Figure 4.9 Time-integrated and time-resolved PL spectra of a 2D array of 8-nm-thick Si-NDs measured at a temperature of 150K.

observed at 150K. Moreover, PL emission of 2D array of Si-NDs structure also could be observed even at room temperature. Figure 4.9 shows time-integrated and time-resolved PL spectra of a 2D array of 8-nm-thick Si-NDs measured at a temperature of 150K. While the decay characteristic of two PL bands is slow in the case of conventional self-assembled QD (i.e., the microsecond region originating in local states in Si and SiO$_2$ induced by a defect or interface state [28]), both peaks in our structure show short lifetimes of 3 ns or less. Therefore, we can rule out assigning the emission bands to defect-related emissions.

4.4.3 CONTROL OF BANDGAP ENERGY BY GEOMETRIC PARAMETERS OF NANODISK

Figure 4.10a shows the results of E_g as a function of Si-ND thickness. We found that, for the 2D array of Si-NDs, E_g can be controlled from 2.3 to 1.3 eV when the ND thickness is changed from 2 to 12 nm. We also saw that, when the poly-Si thin film thickness changes from 2 to 8 nm, E_g varies from 1.6 to 1.1 eV, which is the E_g of bulk Si. From these results, as shown in Figure 4.10a, we know that the controllable Eg range of a 2D array of Si-NDs is much larger than that of poly-Si thin film. Figure 4.10b shows the results of E_g as a function of Si-ND diameter controlled by the biotemplates and NF$_3$/H-radicals treatment. By shrinking the Si-ND diameter from 12.5 to 6.4 nm, the E_g increased from 1.9 to 2.1 eV even at a ND thickness of 4 nm. We found that the E_g could be controlled by both Si-ND thickness and diameter. The diameter from 6.4 to 12.5 nm allowed the E_g to be changed in the range of 0.1 eV, while controlling the thickness from 2 to 12 nm allowed the gap to change in the range of 0.5 eV. This result also suggests that stronger quantum confinement occurs in the thickness direction. These results made it clear that independently changing the geometric parameters of thickness and diameter in our proposed Si-ND array structure

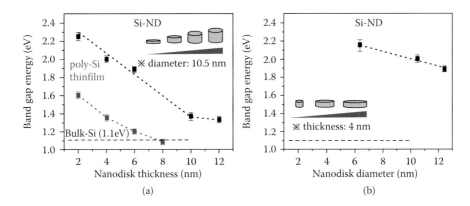

Figure 4.10 Bandgap control by changing (a) thickness and (b) diameter.

enables the optical bandgap energy to be precisely designed within a wide range. This wide controllable range of E_g is very suitable for developing all-Si tandem solar cells. For the three cells of all-Si tandem solar cells, the E_g of the top and middle cells requires 2.0 and 1.5 eV, respectively [30].

4.5 APPLICATION TO SOLAR CELLS

We have developed a sub-10 nm Si nanodisk (Si-ND) structure using a biotemplate and damage-free NB etching [18,19]. In this study, we first investigated the controllable range of E_g and optical absorption characteristic of Si-NDs by changing the geometric parameters of Si-ND and matrix material, and we discussed the mechanism of band gap energy by comparing the experimental result with the simulation result [31]. Second, we observed an enhancement of conductivity in Si-NDs by formation of mini-bands. Within the envelope function theory and Anderson Hamiltonian method, we also calculated electronic structures and the current transport, which theoretically proved that mini-bands enhanced the conductivity. Finally, we verified the high optical absorption and conductivity properties by fabricating p-i-n solar cells with Si-NDs and clarified carrier generation and carrier collection in our Si-ND structure [31].

4.5.1 CONTROL OF ABSORPTION COEFFICIENT BY MINI-BAND FORMATION

We prepared 2D arrays of Si-NDs with a SiC or SiO_2 interlayer whose ND thickness, diameter, and average ND-center-to-ND-center distance corresponded to 4, 10, and 12 nm. Figure 4.11a presents the results for the absorption coefficients of the Si-NDs with SiC, Si-NDs with SiO_2, and SiC interlayer on the substrate, and Figure 4.11b has the $(\alpha h \nu)^{1/2}$ versus photon energy ($h\nu$). The absorption coefficient of the 2D Si-ND array with the SiC interlayer was higher than that of the SiO_2 matrix because of the higher photon absorption in the SiC layer. The absorption edge of the Si-ND array with a SiC interlayer was also blue shifted more than the SiO_2 interlayer. Moreover, even though the E_g of the SiC layer was 3.0 eV, the absorption coefficient from 2 eV to 3 eV also increased when using SiC as the interlayer compared with using SiO_2 and only SiC film. These results would have originated in variations in the band offset that changed the localization of the wave function and the electronic states in Si-ND; the typical conduction band offset of a Si/SiO_2 interface is 3.2 eV and that of a Si/SiC interface is 0.5 eV (32). Therefore, this indicates that the coupling of wave functions strengthens and forms a wide mini-band due to the lower bandgap energy of the SiC interlayer, which enhanced photon absorption in the structure. We estimated the one-sun (photon intensity: AM1.5, 1000W/m^2) illuminated photon absorbance from E_g (2.0 eV) to 5.0 eV in one layer of the Si-ND array (Figure 4.2) to better understand the efficiency of photon utilization. We found that when the interlayer was changed from SiO_2 to SiC, the photon absorbance in the 2D array of Si-NDs increased from 14% to 16% of sunlight from 2.0 eV to 5.0 eV. We also prepared a 2D array of small-diameter (6.4 nm) Si-NDs with a SiC interlayer to investigate the optical absorption properties. In this case, the one-sun illuminated photon absorbance from E_g (2.0 eV) to 5.0 eV in one layer of the Si-ND array is also shown in Figure 4.12. The photon absorbance in the 2D array of Si-NDs was clearly increased to 21% due to a

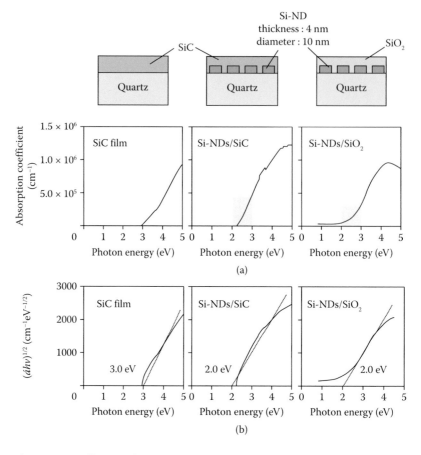

Figure 4.11 (a) Absorption coefficients of Si-NDs with SiC, Si-NDs with SiO$_2$, and SiC thin film, and (b) Tauc plot ($\acute{a}hv$)$^{1/2}$ versus photon energy (hv) for each sample.

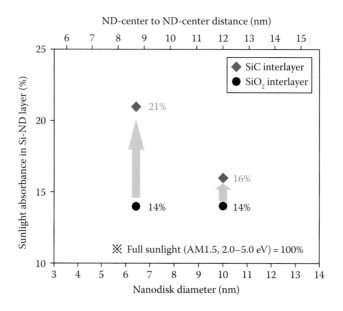

Figure 4.12 One-sun illuminated photon absorbance from E_g (2.0 eV) to 5.0 eV in a 2D array of Si-NDs with SiC and SiO$_2$ matrix.

combination of the small diameter of the Si-ND array and the SiC interlayer. It means that shrinking the Si-ND diameter simultaneously increased the optical absorption coefficient.

These interesting phenomena can perhaps be attributed to the lateral coupling of the NDs. There are not only strongly coupled wave functions at the discrete initial quantum level, but they also relax selection rules to induce additional transitions, which commonly means an increased state density that leads to enhanced total absorption. We solved one-band Schrödinger equations with classic envelope function theory to obtain the electronic structure of our Si NDs and the electron spatial probability (square wave function)

$$-\nabla\left(\frac{\hbar}{2m^*}\nabla\phi\right)+V\phi=E\phi, \tag{4.1}$$

where $\hbar$, m^*, V, E, and ϕ correspond to Planck's constant divided by 2π, the effective mass, the position-dependent potential energy, the quantum levels, and the electron envelope function. The finite element method was used to finely describe our complex structures. Electron spatial probability in three neighboring NDs is indicated by the logarithmic axis in Figure 4.13. Using the SiC interlayer effectively enhanced wave function coupling more than using the SiO_2 interlayer because the energy potential well at the SiC/Si interface is lower than that in the SiO_2 interlayer (Figure 4.13a). Moreover, wave function coupling is also effectively enhanced by decreasing the diameter (Figure 4.13b). A possible reason is that, by decreasing diameters, the center-to-center distance between NDs is decreased so that wave functions more easily spread to the neighboring NDs and are coupled with each other. This enhanced wave function coupling will enhance total photon absorption and conductivity in the 2D array of Si-ND structures.

Then, we measured optical absorption coefficient in three-dimensional (3D) array of Si-NDs to investigate the effect of 3D mini-band formation on the optical absorption. Figure 4.14 shows absorption coefficients in 2D and 3D array of Si-NDs samples prepared on transparent quartz substrates. The absorption coefficient in 3D array was almost the same as that in 2D array, the calculated E_g of both samples was 2.2 eV, and the total photon absorbance in 3D array was increased to 30%. Following the results, we calculated a width of mini-band in 2D and 3D array of Si-NDs by Equation 4.1, and a change in mini-band width between the samples was estimated as 3.85 meV (single layer: 0.95 meV; four layers: 4.80 meV). Therefore, it seems that the change of 3.85 meV in mini-band width is not large enough to affect the photon absorption.

4.5.2 ENHANCEMENT OF ELECTRON CONDUCTIVITY DUE TO MINI-BAND FORMATION IN SILICON NANODISK SUPERLATTICES

Theoretical discussion was carried out to investigate enhancement of the conductivity due to mini-band formation. Our developed top-down nanotechnology achieves great flexibility in designing parts of the quantum structure, such as the independently controllable diameter and thickness, high aspect ratio, different matrix materials, and so on. The finite element method very suitably describes complex quantum structures. Within the envelope function theory, the electronic structure and wave function are presented as Equation 4.1. Here, we mainly consider the matrix material, realistic geometry structure, and the number of stacking layers. Results are shown in Figure 4.15. A distinct feature is that, due to the higher band-offset of Si/SiO_2 interface, electron wave functions are more strongly confined in the Si-NDs in SiO_2 matrix. Thus, they result in the higher quantum levels. In addition, in the same geometry alignment, the stronger confinement means the weaker coupling of wave function and narrower mini-band.

In Figure 4.16, by stacking our NDs from one layer to ten layers, the mini-band gradually broadens; and at around four to six layers, the broadening width seems to saturate. The probability of wave function diffusing into barrier exponentially reduces with distance, which indicates that wave function coupling saturates as the number of layers increases. Perhaps four- or six-layer NDs are enough to maximize the advantage of mini-bands.

With the Anderson Hamiltonian model, Chang et al. considered interdot coupling to deduce the tunneling current density, as shown in detail in [33].

As shown in Figure 4.17, calculated results also reveal that the wider mini-band in SiC matrix brings a better transport property than that in SiO_2 matrix. A simplified, but not too distorted, explanation is that

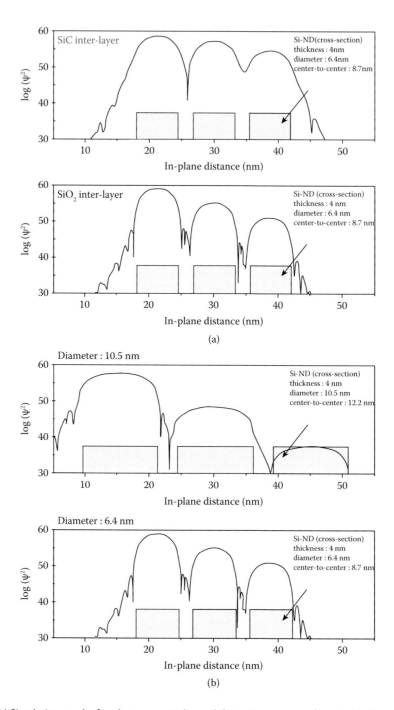

Figure 4.13 (a, b) Simulation results for electron spatial possibilities (square wave function) in three lateral coupled NDs in accordance with our samples structures.

mini-band formation broadens the resonance levels to increase the joint-state density. The carrier transport in this two-barrier structure mainly depends on the resonant tunneling. Mini-band formation broadens resonant peak to allow more states to approach the maximum, which results in the enhanced current. Thus, the wider mini-band means a higher current density and lower threshold voltage, as shown in the Si-NDs in the SiC matrix. In addition, the 2D array of Si-NDs in the SiC matrix has a lower mini-band level, E_0, which also shifts the I–V curves to a lower threshold voltage. This tendency closely matches

Arrays, hybrids, and core–shell

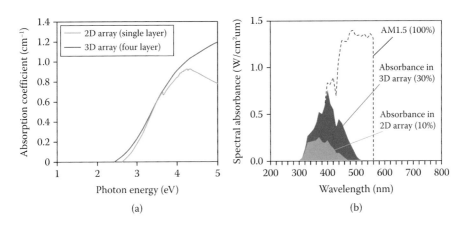

Figure 4.14 (a) Absorption coefficients of 2D and 3D arrays of Si-NDs with SiC matrix. (b) One-sun illuminated photon absorbance from E_g to 5.0 eV in 2D and 3D arrays of Si-NDs.

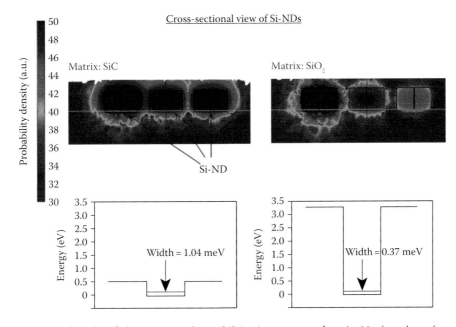

Figure 4.15 Calculated results of electron spatial possibilities (square wave function) in three lateral coupled NDs and mini-band width in a 2D array of Si-NDs.

our experimental results, and due to the larger tunneling resistance in the SiO_2 interlayer, the threshold voltage (V) is further increased in realistic I-V curves. Moreover, the conductivity in 2D and 3D arrays of Si-NDs were enhanced due to the same mechanism of broadening of the wave functions and formation of wider mini-bands. These were also very consistent with the trend in our experimental results, so it is clarified that mini-band formation in both in-plane and out-of-plane could enhance the carrier transport in QDSL. The enhanced conductivity is very significant for electronic/optoelectronic devices, which indicates high charge-injection efficiency in lasers and carriers collection efficiency in solar cells.

4.5.3 SILICON NANODISK SUPERLATTICES FOR HIGH-EFFICIENCY MULTISTACKED SOLAR CELLS

Quantum dot solar cells are very promising devices to break the frustrating Schockley–Queisser limit. Many exciting results have been reported to verify the key operating mechanism, the preparation of various

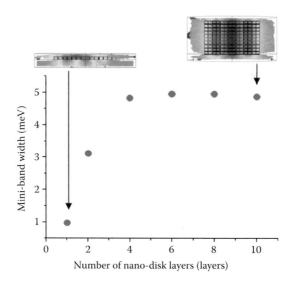

Figure 4.16 Calculated results of mini-band width in a 3D array of Si-NDs.

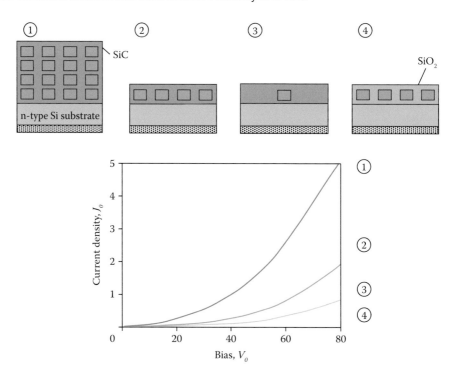

Figure 4.17 Simulation results for I-V properties of our sample structures.

materials, and the fabrication of prototype devices. However, these researches, unsurprisingly, is still insufficient to optimize these novel devices after exciting developments with conventional cells. One critical problem has been the lack of ideal QDSLs and basic understanding of their electrical/optical properties, especially key optical absorption and electrical conductivity.

The 2D array of Si-NDs fabricated by using a combination of an etching mask of Listeria-Dps (Lis-Dps) iron cores and a SiC interlayer was the most effective structure to enhance conductivity and optical absorption in the Si-ND layer. The SiC interlayer could enhance the conductivity in the 2D array of Si-NDs due to the strong coupling of wave functions. By using a 2D array of an etching mask with a Lis-Dps iron core, a 2D array of Si-NDs with an ND-center-to-ND-center distance of 8.7 nm could be achieved with enhanced optical absorption due to the strong overlap of wave functions. Hence, we designed solar cell structures with a 2D array of Si-ND layers fabricated by using a Lis-Dps biotemplate and a SiC interlayer (Figure 4.18). The diameter, space between NDs, and average ND-center-to-ND-center distance corresponded to 6.4, 2.3, and 8.7 nm. We fabricated Si-ND solar cells with different Si-ND thicknesses of 2 and 4 nm to clarify how the Si-ND layer contributed to the performance of solar cells. Four types of structures were prepared (Figure 4.18) to clarify the effects of the Si-ND array structure: a solar cell with a 2-nm-thick-SiC/2-nm-thick-Si-ND/2-nm-thick-SiC layer, a 2-nm-thick-SiC/4-nm-thick-Si-ND/2-nm-thick-SiC layer, a 2-nm-thick SiC layer, and a p–n junction without a Si-ND structure. The p–n junction solar cell without any structures inserted could be used as a sample, which is established as a criterion to evaluate current generation in Si-NDs and carrier collection through them. All samples were fabricated on a 400-nm-thick, 1–1.5-Ω p-type Si substrate. After that, 30-nm-thick undoped epitaxial (epi)-Si was grown by electron beam evaporation at 600°C. An n-type phosphorus-doped Si emitter was formed with a method of diffusion by rapid thermal annealing [34, 35]. We deposited 70-nm-thick indium tin oxide by sputtering to use it as an antireflective film. The back-surface field contact was formed by printing it with aluminum paste and subsequent annealing, and the front finger contact was formed by firing silver paste through an ITO film. There is a cross-sectional SEM image of Si-ND array solar cells in Figure 4.18. Figure 4.19 shows the photon absorbance for each layer in the actual structure of a solar cell (Figure 4.18a and b) as a function of the light wavelength calculated from the absorption coefficient in each layer. The results indicate that the SiC/4-nm-thick Si-ND/SiC structure (Figure 4.18a) has a more efficient absorption of photons that entered from the top of the solar cell than the SiC thin film.

We measured the cell properties by using a characterization system (Jasco, YQ-250BX) with an AM1.5 solar simulator (100 mW/cm²) as a light source at room temperature (RT). Figure 4.20a plots the one-sun

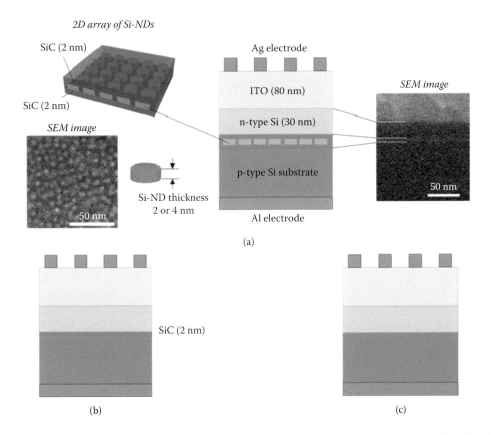

Figure 4.18 Schematics and SEM image of structures of solar cells: (a) our fabricated Si-ND/SiC solar cells, (b) SiC thin film solar cells, and (c) single-layer p–n junction solar cells.

Arrays, hybrids, and core–shell

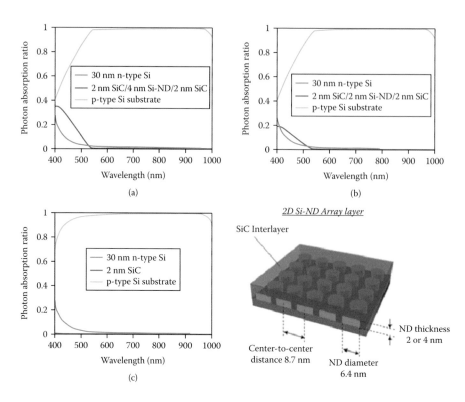

Figure 4.19 Photon absorbance for each layer in actual solar cell structure (Figure 4.8a, b, and c) as a function of light wavelength calculated from absorption coefficient in each layer.

Table 4.1 One-sun illuminated cell parameters of three different solar cells measured at RT

TYPE OF SOLAR CELL	SATURATION CURRENT, J_{SC} (MA/CM²)	OPEN CIRCUIT VOLTAGE, V_{OC} (V)	FILL FACTOR, *FF*	CONVERSION EFFICIENCY, η (%)
p–n junction	33.4	0.601	0.634	12.7
2 nm SiC	29.0	0.544	0.340	5.4
2 nm SiC/2 nm Si-ND/2 nm SiC	29.9	0.539	0.578	9.3
2 nm SiC/4 nm Si-ND/2 nm SiC	31.3	0.556	0.724	12.6

illuminated I-V curves of Si-ND solar cells with 2-nm- and 4-nm-thick Si-NDs, a solar cell with 2-nm-thick SiC film, and a p–n junction solar cell. Here, the contributions and reduction in cell performance in inserted layers (Si-ND or SiC thin film) could be evaluated by comparing them with the cell performance of a standard sample of a p–n junction solar cell. The solar cell with the 2-nm-thick SiC film had a conversion efficiency (η) of 5.4% with an open-circuit voltage (V_{oc}) of 0.544 V, short-circuit current density (J_{sc}) of 29.0 mAcm⁻², and a fill factor (*FF*) of 34%, due to large resistivity in SiC thin film that resulted in reductions in current and *FF* (Table 4.1). Interestingly, despite the two types of quantum dot solar cells having the same SiC film thickness (2 nm), the cell with 2-nm-thick-SiC/4-nm-thick-Si-ND/2-nm-thick SiC had superior performance with V_{oc} of 0.556 V, J_{sc} of 31.3 mAcm⁻², *FF* of 72.4%, and η of 12.6% (Table 4.1). The cell parameters including conversion efficiency had good values close to those of the p–n junction solar cell by comparing them with those of the solar cell with SiC film. This demonstrated that the photo-generated carriers in the substrate were efficiently

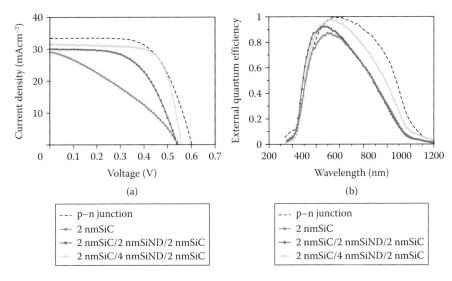

Figure 4.20 Properties of solar cells. (a) One-sun illuminated I-V curves of Si-ND solar cells with 2-nm- and 4-nm-thick Si-NDs and solar cell with 2-nm-thick SiC film measured at RT. (b) External quantum efficiency of three different solar cell structures.

transported through the Si-ND structure without generating any reductions in cell performance. Figure 4.20b plots the external quantum efficiency (EQE) of the three different solar cell structures. The spectrum EQE of the Si-ND solar cell had a good response at ~620 nm, while the EQE peak of photon absorption when 2-nm-thick Si-NDs (E_g: 2.2 eV) were used was observed at a shorter wavelength (570 nm) than that (620 nm) when 4-nm-thick Si-NDs (E_g: 2.0 eV) were used. That is, the EQE peak was blue shifted by increasing E_g due to the decreasing thickness of Si-NDs. From the results of the absorption coefficient in the Si-ND structure, the largest possible contribution of the Si-ND structure to the total current density can be calculated as

$$J_{Si-ND} = \frac{q}{hc} \cdot \int I_{AM1.5}(\lambda) \cdot A_{Si-ND}(\lambda) \cdot \lambda \cdot d\lambda \tag{4.2}$$

where $I_{AM1.5}$, A_{Si-ND}, q, and c correspond to the solar spectral irradiance of AM1.5, the absorption ratio in the Si-ND layer (see Figure 4.20), the elementary charge, and the speed of light. The contribution of the Si-ND layer can be estimated at 1.2 mA/cm² when using 4-nm-thick Si-NDs. Although the absorption of photons in the Si substrate contributed to a large fraction of the current, this suggests that a Si-ND array also can play a key role in the efficient absorption of photons as well as in power generation in solar cells.

However, the EQE spectra of the Si-ND solar cells decreased at wavelengths from 600 to 1100 nm, which might result in lower J_{sc} and V_{oc} than that for the p–n junction solar cell. We concluded from the bandgap energy of each layer (Si-ND: 2.0eV and bulk Si: 1.1eV) that the EQE from 600 to 1100 nm belonged to carriers generated in the Si substrate. Therefore, this could be attributed to still low conductivity vertically even in the 2D array of Si-NDs. This indicates that the electrons generated in the p-type Si substrate could not be completely extracted to the n-type Si layer through the Si-ND layer compared with the p–n junction solar cell. Namely, the conductivity in our single-layer Si-ND array structure was still lower than that in the bulk Si p–n junction. We speculated that this problem could be solved by generating mini-bands both vertically and laterally using a 3D Si-ND array structure.

4.6 SUMMARY

Recently, quantum dot superlattices (QDSLs) with uniform dot size, uniform interdot spacing, and high-density QDs have been widely used to develop new-generation devices, such as QD lasers, QD solar cells, and QD electronic devices. Quantum size effect in QDs provides an opportunity to engineer the bandgap energy (E_g) by adjusting the size of the QD. Carrier wave functions with close-packed and well-aligned

Arrays, hybrids, and core-shell

QDs overlap one another and discrete confined energy levels merge into broadened mini-bands. Such nanoengineered QDs in photovoltaic applications have great potential as novel adjustable absorber layers to achieve intermediate band solar cells and tandem solar cells. Moreover, for Si-QDSLs, the SiC for the interlayer is one of the most promising materials to be used to form mini-bands that anticipate enhanced photon absorption and electrical conductivity in Si-QD solar cells. Carrier wave functions more easily spread into the SiC interlayer to form mini-bands to achieve Si-QD solar cells because of its lower E_g compared with other insulating materials such as SiO_2 and SiN. However, it is difficult to fabricate defect-free and sub-10 nm nanostructures by using conventional methods such as plasma processes with photolithography techniques and co-sputtering techniques with annealing and molecular beam epitaxy processes. Furthermore, a fabrication process with a high degree of control is also required for different QD materials in different applications. Our aim in this paper was to develop a uniform and closely packed array of sub-10-nm Si structure as realistic QDSL using a new process that combines a biotemplate and damage-free NB etching to address these problems, and investigated and controlled optical and electronic properties due to quantum effects for future devices. We first investigated the controllable range of E_g and optical absorption characteristic of Si-NDs by changing the geometric parameters of Si-ND and matrix material, and we discussed the mechanism by comparing the experimental result with the simulation result. It was clarified that independently changing the geometric parameters of thickness and diameter in our proposed Si-ND array structure enables the optical E_g to be precisely designed within a wide range from 1.3 eV to 2.3 eV. It was demonstrated that the coupling of wave functions strengthens and forms a wide mini-band due to the lower E_g of the SiC or decreasing of the center-to-center distance between NDs, which enhanced photon absorption in the structure. Then, an enhancement of conductivity in Si-NDs by formation of mini-bands was investigated. Within the envelope function theory and Anderson Hamiltonian method, we also calculated electronic structures and the current transport, which theoretically proved that mini-bands enhanced the conductivity. Finally, the highly optical absorption and electrical conductivity were verified by fabricating p-i-n solar cells with Si-NDs, and effective carrier generation and transport in our Si-ND structure were clarified.

REFERENCES

1. S. Baskoutas and A. F. Terzis. Size-dependent band gap of colloidal quantum dots. *J. Appl. Phys.* **99**, 013708 (2006).
2. H. Yu, J. Li, R. A. Loomis, P. C. Gibbons, L. W. Wang and E. W. Buhro. Cadmium selenide quantum wires and the transition from 3D to 2D confinement. *J. Am. Chem. Soc.* **125**, 16168 (2003).
3. A. Kongkanand, K. Tvrdy, K. Takechi, M. Kuno and P. V. Kamat. Quantum dot solar cells. Tuning photoresponse through size and shape control of CdSe-TiO$_2$ architecture. *J. Am. Chem. Soc.* **130**, 4007 (2008).
4. B. Pejova and I. Grozdanov. Three-dimensional confinement effects in semiconducting zinc selenide quantum dots deposited in thin-film form. *Mater. Chem. Phys.* **90**, 35 (2005).
5. L. Goldstein, F. Glas, J. Y. Marzin, M. N. Charasse and G. Le Roux. Growth by molecular beam epitaxy and characterization of InAs/GaAs strained-layer superlattice. *Appl. Phys. Lett.* **47**(10), 1099–1101 (1985).
6. A. Luque and A. Marti. Increasing the efficiency of ideal solar cells by photon induced transitions at intermediate levels. *Phys. Rev. Lett.* **78**, 5014 (1997).
7. A. A. Konakov and V. A. Burdov. Optical gap of silicon crystallites embedded in various wide-band amorphous matrices: role of environment. *J. Phys. Condens. Matter.* **22**, 215301 (2010).
8. L. H. Thamdrup, F. Persson, H. Bruu, A. Kristensen and H. Flyvbjerg. Experimental investigation of bubble formation during capillary filling of SiO$_2$ nanoslits. *Appl. Phys. Lett.* **91**, 163505 (2007).
9. G. Conibeer, M. A. Green, R. Corkish, Y. Cho, E. C. Cho, C. W. Jiang, T. Fangsuwannarak, E. Pink, Y. Huang, et al. Silicon nanostructures for third generation photovoltaic solar cells. *Thin Solid Films* **511/512**, 654 (2006).
10. E. C. Cho, S. Park, X. Hao, D. Song, G. Conibeer, S. C. Park and M. A. Green. Silicon quantum dot/crystalline silicon solar cells. *Nanotechnology* **19**, 245201 (2008).
11. Y. Okada, R. Oshima and A. Takata. Characteristics of InAs/GaNAs strain-compensated quantum dot solar cell. *J. Appl. Phys.* **106**, 024306 (2009).
12. R. B. Laghumavarapu, M. El-Emawy, N. Nuntawong, A. Moscho, L. F. Lester and D. L. Huffakerb. Improved device performance of InAs/GaAs quantum dot solar cells with GaP strain compensation layers. *Appl. Phys. Lett.* **91**, 243115 (2007).

Arrays, hybrids, and core–shell

13. R. P. Raffaelle, S. L. Castro, A. F. Hepp and S. G. Bailey. Quantum dot solar cells. *Prog. Photovolt. Res. Appl.* 10, 433 (2002).

14. G. Conibeer, M. A. Green, E. C. Cho, D. König, Y. H. Cho, T. Fangsuwannarak, G. Scardera, E. Pink, Y. Huang, et al. Silicon quantum dot nanostructures for tandem photovoltaic cells. *Thin Solid Films* **516**, 6748 (2008).

15. X. J. Hao, A. P. Podhorodecki, Y. S. Shen, G. Zatryb, J. Misiewicz and M. A. Green. Effects of Si-rich oxide layer stoichiometry on the structural and optical properties of Si QD/SiO$_2$ multilayer films. *Nanotechnol.* **20**, 485703 (2009).

16. Y. Kurokawa, S. Tomita, S. Miyajima, A. Yamada and M. Konagai. Photoluminescence from silicon quantum dots in Si quantum dots/Amorphous SiC superlattice. *Jpn. J. Appl. Phys.* **46**, L833 (2007).

17. S. Samukawa, K. Sakamoto and K. Ichiki. Generating high-efficiency neutral beams by using negative ions in an inductively coupled plasma source. *J. Vac. Sci. Technol. A* **20**, 1566 (2002).

18. I. Yamashita. Fabrication of a two-dimensional array of nano-particles using ferritin molecule. *Thin Solid Films* 393, **12** (2001).

19. S. Samukawa, K. Sakamoto and K. Ichiki. Generating high-efficiency neutral beams by using negative ions in an inductively coupled plasma source. *J. Vac. Sci. Technol. A* **20**, 1566 (2002).

20. S. Samukawa, K. Sakamoto and K. Ichiki. High-efficiency low energy neutral beam generation using negative ions in pulsed plasma. *Jpn. J. Appl. Phys.* **40**, L997 (2001).

21. M. Yonemoto, T. Ikoma, K. Sano, K. Endo, T. Matsukawa, M. Masahara and S. Samukawa. Low temperature, beam-orientation-dependent, lattice-plane-independent, and damage-free oxidation for three-dimensional structure by neutral beam oxidation. *Jpn. J Appl. Phys.* **48**, 04C007 (2009).

22. C. H. Huang, M. Igarashi, M. Wone, Y. Uraoka, T. Fuyuki, M. Takeguchi, I. Yamashita and S. Samukawa. Two-dimensional Si-nanodisk array fabricated using bio-nano-process and neutral beam etching for realistic quantum effect devices. *Jpn. J. Appl. Phys.* **48**, 04C187 (2009).

23. A. Wada, K. Sano, M. Yonemoto, K. Endo, T. Matsukawa, M. Masahara, S. Yamasaki and S. Samukawa. High-performance three-terminal fin field-effect transistors fabricated by a combination of damage-free neutral-beam etching and neutral-beam oxidation. *Jpn. J. Appl. Phys.* **49**, 04DC17 (2010).

24. M. Igarashi, M. F. Budiman, W. Pan, W. Hu, Y. Tamura, M. E. Syazwan, N. Usami and S. Samukawa. Effects of formation of mini-bands in two-dimensional array of silicon nanodisks with SiC interlayer for quantum dot solar cells. *Nanotechnology* **24**, 015301 (2013).

25. Y. Tamura, T. Kaizu, T. Kiba, M. Igarashi, R. Tsukamoto, A. Higo, W. Hu, C. Thomas, M. E. Fauzi, et al. Quantum size effects in GaAs nanodisks fabricated using a combination of the bio-template technique and neutral beam etching. *Nanotechnology* **24**, 285301 (2013).

26. L. Pavesi, L. D. Negro, C. Mazzoleni, G. Franzo and F. Priolo. Optical gain in silicon nanocrystals. *Nature* **408**, 440 (2000).

27. K. Kusová, O. Cibulka, K. Dohnalova, I. Pelant, J. Valenta, A. Fucíkova, K. Zídek, J. Lang, J. Englich, et al. Brightly luminescent organically capped silicon nanocrystals fabricated at room temperature and atmospheric pressure. *ACS Nano* **4**, 8, 4495 (2010).

28. W. Boera, H. Zhangb and T. Gregorkiewicz. Optical spectroscopy of carrier relaxation processes in Si nanocrystals. *Mater. Sci. Eng. B* **159–160**, 190 (2003).

29. F. Trojánek, K. Neudert, M. Bittner and P. Malý. Picosecond photoluminescence and transient absorption in silicon nanocrystals. *Phys. Rev. B* **72**, 075365 (2005).

30. I. A. Walmsley. Looking to the future of quantum optics. *Science* **29**, 1211 (2008).

31. M. Igarashi, M. F. Budiman, W. Pan, W. Hu, Y. Tamura, M. E. Syazwan, N. Usami and S. Samukawa. Effects of formation of mini-bands in two-dimensional array of silicon nanodisks with SiC interlayer for quantum dot solar cells. *Nanotechnology* **24**, z015301 (2013).

32. G. Conibeer, M. A. Green, E. C. Cho, D. König, Y. H. Cho, T. Fangsuwannarak, G. Scardera, E. Pink, Y. Huang, et al. Mansfield, silicon quantum dot nanostructures for tandem photovoltaic cells. *Thin Solid Films* **516**, 6748 (2008).

33. D. M. T. Kuo, G. Y. Guo and Y. C. Chang. Tunneling current through a quantum dot array. *Appl. Phys. Lett.* **79**, 3851 (2001).

34. W. Pan, N. Usami, K. Hara, N. Arifuku, M. Matsui and S. Matsushima. *PVSEC21, 2D-1P-04* (Fukuoka, Japan, 2011).

35. S. Yamada, Y. Kurokawa, S. Miyajima, A. Yamada and M. Konagai. High open-circuit voltage oxygen-containing silicon quantum dots superlattice solar cells, in Proceeding of the 35th IEEE Photovoltaic Specialists Conference, 317, Honolulu, HI (2010).

Arrays, hybrids, and core–shell

Systems of silicon nanocrystals and their peculiarities

Vladimir A. Terekhov, Sergey Y. Turishchev, and Evelina P. Domashevskaya

Contents

5.1 INTRODUCTION

Systems with silicon nanocrystals revealing peculiarities in atomic and electronic structure are studied by means of techniques sensitive to physical and chemical states, local atomic surroundings, and electronic structures. These techniques include scanning and high-resolution transmission electron microscopy, X-ray diffraction, ultrasoft X-ray emission spectroscopy (USXES), X-ray photoelectron spectroscopy (XPS), and X-ray absorption near-edge structure spectroscopy (XANES). The last two techniques were applied with the use of highly brilliant synchrotron radiation.

In the energies corresponding to the absorption near the XANES threshold interval that are 50–100 eV from the main absorption edge, photoelectrons have a small kinetic energy. This allows a large number of surrounding atoms at great distances from the absorbing atom to be involved in the scattering process.

Therefore, the fine structure of XANES spectra is the interference result of the primary photoelectron wave exiting the absorbing atom and the secondary one reflected from neighboring atoms. As a result, the interference effects of the interaction of the primary and reflected waves depend on the relative position of the absorbing atom and the surrounding atoms. And despite the difficulties of interpretation, XANES spectra fine structure and its qualitative change in the transition from one compound to another are used to study the local surroundings of the absorbing atom. In the dipole approximation, the X-ray absorption near-edge structure reflects the distribution of the local (the atoms of a given element) partial density of free electronic states (e.g., s, p, d, f) in the conduction band of any material under study with complex composition according to the equation:

$$\mu(E) \sim \nu^3 \sum_f |M_{fi}|^2 \delta (E_f - E_i - h\nu),$$

where $M_{fi} = \int \varphi_f^* H' \varphi_i dr$ is a matrix element of electron transition probability from the core level with the wave function φ_i to the conduction band with the wave function φ_f and eigenvalue of E_f, the H' is a perturbation operator, and $h\nu$ is the energy of the perturbation quantum.

USXES allows us to determine local partial density of the occupied states in the valence band of the investigated material:

$$I(E) \sim \nu^3 \sum_j |M_{ij}|^2 \delta (E_i - E_j - h\nu),$$

where $M_{ij} = \int \varphi_i^* H' \varphi_j dr$ is a matrix element of electron transition probability from the valence band with the wave function φ_j and eigenvalue of E_j to the vacancy in the core level with the wave function φ_i, H' is a perturbation operator, and $h\nu$ is the energy of the emitted X-ray quantum.

Investigated structures with silicon nanocrystals are nanoporous silicon, massives of silicon nanocrystals in silicon oxide matrix obtained in thin films by Si^+ ion implantation, SiOx thermal decomposition, or that formed in multilayered nanoperiodical structures, silicon nanowires, Al-Si composite films, and free silicon nanopowders.

All the investigated systems revealed certain atomic and electronic structure peculiarities, which have led to observation of physical properties noticeably different from the bulk materials such as visible photoluminescence at room temperature. It is shown that the influence of silicon nanocrystals formation conditions on their size and shape stipulated peculiarities of atomic and electronic structure and optical properties.

The presented results show that interaction peculiarities between nanometer size wavelength synchrotron radiation with nanostructures are caused by photon elastic scattering processes on silicon nanocrystals.

5.2 ELECTRONIC STRUCTURE AND OPTICAL PROPERTIES OF SILICON NANOPOWDERS

Attention to silicon nanopowders is due to scientific interest in silicon nanocrystals for their potential practical application. For example, in the last decade, researchers all over the world made efforts to design new light-emitting diodes on the basis of silicon nanocrystals. It is known that with a decrease of the particle size, there appears an uncertainty in momentum of the charge carriers localized inside the particles. Hence, direct electron transitions can take place as a result of recombination of the electron-hole pair, which cannot be realized in the crystalline silicon (indirect band gap material), where the participation of photons is necessary for the recombination. Moreover, silicon nanocrystals are prospective materials if used in the permanent memory cells that have several advantages compared with the usual ones, such as the higher recording density and lower voltage for the recording process. A practical interest in nanopowders is also due to the possibility of making nanocomposite materials on their basis.

Silicon nanopowders (Si-npwd) were obtained by evaporation of silicon ingot with a strong electron beam with the energy of electrons in the beam 1.4 MeV. Evaporation was performed under different conditions in the atmosphere of argon and nitrogen [1] (Si-npwd(Ar) and Si-npwd(N$_2$) correspondingly).

An advantage of this method as compared with the other ones is the possibility of obtaining the great amount of nanocrystals with different properties as a result of variation of the experimental conditions. The obtained samples were investigated with the use of transmission electron microscopy technique, Raman spectroscopy, photoluminescence (PL), USXES, and XANES.

The TEM data show that particles have spherical form (Figure 5.1). The range of these particle sizes is quite large. In Figure 5.1, particles with nanometer size could be revealed. Combined, they form spherical particles in the range of 10–20 nm and bigger spherical formations in the submicron range. At the same time, ordered arrangement of atoms with the 3.1 Å interplanar spacing that corresponds to (111) silicon planes is observed in the smaller particles (10–20 nm) by the high-resolution TEM data (Figure 5.2).

Raman spectra of the particles obtained in the atmosphere of argon (Figure 5.3) have a peak shifted toward less wave numbers by approximately 3.5 cm^{-1} relative to the single-crystalline silicon that corresponds to the diameter of the particles about 3–4 nm. To determine the size of the particles, a technique of convolution for the density of the effective vibration states described in [2] and [3] was applied. The width of the peak at the half-height is rather large and is about 8 cm^{-1}, which corresponds to a large scattering of particle size. For the powders obtained in the atmosphere of nitrogen-Si-npwd(Ar), the peak of Raman spectra is rather broad and asymmetric with a shoulder slowly dropping down to the value of 480 cm^{-1}, which corresponds to the presence of the amorphous phase.

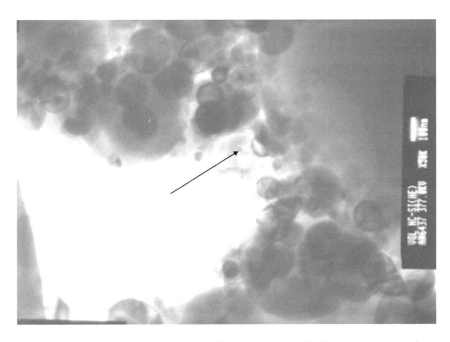

Figure 5.1 Transmission electron microscopy pattern of silicon nanopowder (the arrow is pointed to one of the particle in nanometer range).

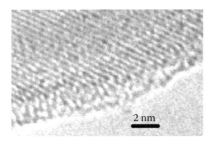

Figure 5.2 High-resolution transmission electron microscopy image of a silicon nanopowder particle with the ordered arrangement of atoms.

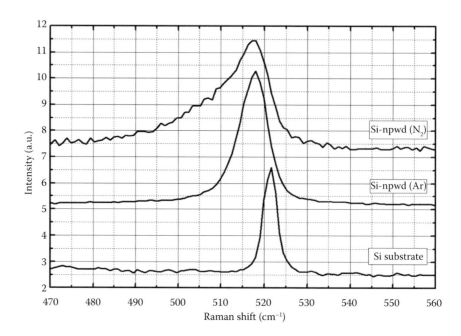

Figure 5.3 Raman spectra of crystalline silicon and silicon nanopowders obtained in argon atmosphere (Si-npwd[Ar]) and nitrogen atmosphere (Si-npwd[N_2]).

PL spectrum of silicon nanoparticles (Figure 5.4) represents a broad peak centered at the wavelength of 580 nm that corresponds to the recombination energy of 2.1 eV, whereas the value of the band gap for silicon is 1.1 eV. Such a strong difference is due to quantum-size effect. PL spectrum in a dependence of the particle radius was calculated in the effective mass approximation, and, as a result, the mean radius of the particle was determined as 1.8 nm. Comparison was made as with the results of theoretical models obtained in [4] in the atomic orbitals (LCAO) approximation for the clusters with free bonds, occupied with hydrogen atoms as with the experimental data on the photoluminescence of silicon nanoclusters with different sizes [5]. Results of these calculations and experiments give the diameter of the particles about 3 nm for position of PL peak equal to 2.1 eV.

Silicon nanopowders were also investigated with the use of USXES technique as well as XANES technique. Because these methods are sensitive to the local atomic structure, they allow for distinguishing silicon atoms located in the structures with a different degree of ordering and with different kinds of atoms in their nearest environment. Figure 5.5 represent Si $L_{2,3}$ emission spectra of nanopowders obtained in argon and nitrogen (Si-npwd(Ar) and Si-npwd(N_2) correspondingly) as well as the reference spectra of c-Si (crystalline silicon), a-Si (amorphous silicon), and SiO_2 (silicon dioxide). Comparison of these spectra indicates a predominance of the crystalline silicon phase in these powders. This is supported by the presence of two maxima in the density of states at 92 eV and 89.4–90 eV with a dip between them in the spectra of powders just as for the reference c–Si. At the same time, in the spectrum of the powder obtained in nitrogen Si-npwd(N_2), and especially in the powder obtained in Ar–Si-npwd(Ar) as compared with the reference c–Si, a higher intensity can be observed in the range of 69–90 eV as well as in the range of 93–96 eV, and this can indicate the presence of an amorphous phase in the powder. This phase is characterized by smoothing of the density of states and an increase of the density of states in the upper part of the valence band (Figure 5.5). Besides, the appearance of the well-expressed maximum with the energy of ~95 eV (clearly noticeable for Si-npwd(Ar) powder) can imply a partial oxidation of the powder.

For the quantitative estimation of the phase composition of nanopowders by the X-ray emission spectra, a special computer program of the component analysis was applied [6]. The essence of the method implies simulation of the experimental spectra on the basis of the reference spectra (Figure 5.5) for the assumed amorphous and crystalline silicon phases that could be presented in the powders. Results of the simulation are presented in the same figures as thin solid lines in comparison with the experimental data (dotted line).

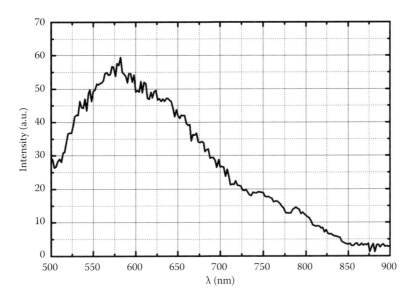

Figure 5.4 Photoluminescence spectra of silicon nanopowders.

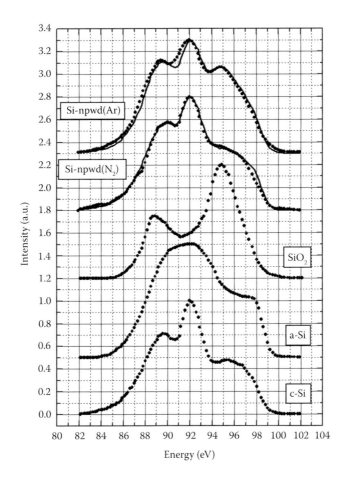

Figure 5.5 Si L$_{2,3}$ ultrasoft X-ray emission spectra of silicon nanopowders obtained in argon and nitrogen atmospheres Si-npwd(Ar) and Si-npwd(N$_2$) correspondingly and reference samples: c-Si (crystalline silicon), a-Si (amorphous silicon), and SiO$_2$ (silicon dioxide). Solid line: simulated spectra.

Results of simulation of the experimental spectra for nanopowders using this method (Figure 5.5) showed their rather good agreement, thus allowing us to make a conclusion about the presence of crystalline silicon, amorphous silicon, and silicon oxide in the powders (Table 5.1). Si-npwd(N_2) nanopowder involves quite a lot of amorphous silicon besides the crystalline silicon phase and very little silicon oxide, whereras the powder obtained in argon involves less amorphous silicon but quite a lot of SiO_2. Thus, results obtained with the use of the USXES technique indicate the predominance of the crystalline phase of silicon in investigated nanopowders.

Because emission spectra were excited by electrons with the energy of 3 keV that corresponds to the depth of analysis of about 60 nm, the results of investigations by this method represent the phase content in the bulk of nanopowder particles. To analyze the features in the composition and structure of near-surface layers (less than 5 nm) for the particles of the powder, the XANES technique was applied. Figure 5.6 represent Si $L_{2,3}$ XANES spectra of these powders obtained at synchrotron center SRC of the University of Wisconsin-Madison (USA). XANES spectra of the reference samples of c-Si, a-Si, and SiO_2 (10 nm thermally grown oxide film) are presented in the same figure. As it follows from Figure 5.6 (XANES spectra for c–Si and a–Si) for single-crystalline and amorphous silicon, one can observe rather abrupt edge at energies about 100 eV. Furthermore, at E > 105 eV in the structure of Si $L_{2,3}$ edge of single-crystalline silicon, just as in the case of amorphous Si, one can observe the peaks characteristic of native SiO_2 (Figure 5.6)

Table 5.1 **Analysis of the phase composition of silicon nanopowders**

	c-Si, %	a-Si, %	SiO_2, %	Δ, %
Si-npwd(N_2)	70	30	0	4
Si-npwd(Ar)	57	16	27	7

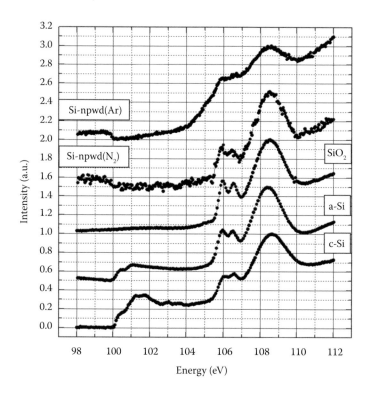

Figure 5.6 Si $L_{2,3}$ X-ray absorption spectra of silicon nanopowders obtained in argon and nitrogen atmospheres Si-npwd(Ar) and Si-npwd(N_2) correspondingly and reference samples: c-Si (crystalline silicon), a-Si (amorphous silicon), and SiO_2 (10 nm thermally grown oxide film).

Arrays, hybrids, and core–shell

as a result of the formation of tetrahedron bonds of silicon atoms with oxygen. The presence of the thin native oxide layer on the surface of the reference samples of c–Si and a–Si does not prevent observation of XANES structure characteristic of the elementary silicon. XANES for thermally grown SiO_2 film (10 nm thickness) at the energy range above 105 eV reveals fine structure corresponding to SiO_2; at the same time, the elementary silicon part (100–104 eV) does not exist for this sample (Figure 5.6).

For investigated nanopowders, fine structures peculiar to silicon oxide exist at XANES spectra, but the absorption edge peculiar to elementary silicon is absent (Figure 5.6). Hence, the absence of the elementary silicon edge in XANES spectra of Si-npwd(N_2) and Si-npwd (Ar) means the presence of more thick surface oxide covered the particles composing nanopowders than natural silicon oxide on c–Si and a-Si.

Thus, investigation of the silicon nanopowders, obtained by evaporation of silicon ingot with a powerful electron beam, allows us to state that obtained nanopowders contain nanocrystals with the average size of about 3–4 nm (by Raman and photoluminescence data), amorphous silicon, and silicon oxide (by USXES and XANES data). The thickness of the oxide-covered nanocrystals forming nanopowders is greater than that of the native oxide layer on the surface of crystalline or amorphous silicon. The obtained powders demonstrate photoluminescence in the visible range, thus opening up possibilities for the design of light-emitting structures on the basis of these powders.

5.3 INVESTIGATIONS OF THE ELECTRON ENERGY STRUCTURE AND PHASE COMPOSITION OF POROUS SILICON WITH DIFFERENT POROSITY

At the present time, there exist a number of models describing visible PL in porous silicon (por-Si). These are the models of quantum confinement [7,8], of surface passivation [9], photoluminescence due to the presence of Si–SiO$_2$ boundaries [10,11], and some others. However, none of them can explain all of the observed experimental facts. One can assume that most likely the data on PL of por-Si could be explained by some "superposition," a set of the most complete models.

Porosity of por-Si—the ratio of the empty volume, that is, the volume of pores, to the total volume of the porous layer—directly depends on the technology of formation of the porous silicon. Obviously, the properties of this material and, first of all, its photoluminescence properties quite strongly depend on the method of formation. It is clear that the electron energy structure of por-Si depends on the applied technology as well. Hence, the investigations of the electron energy structure and optical properties as well as the regularities of the changes of porous layer composition on the conditions of porous silicon formation seem to be actual. To analyze the influence of the silicon atoms with different local surroundings, it is necessary to use methods sensitive to local atomic neighboring such as ultrasoft X-ray spectroscopy.

Porous silicon was obtained on the substrate of a single-crystalline silicon (100) and (111) of n-type conductivity using the laboratory procedure of electrochemical etching in alcoholic solution of hydrofluoric acid. The samples of por-Si demonstrated rather bright and reproducible PL. Just before etching of single-crystalline Si, a standard washing of the original silicon wafers was performed, providing a low concentration of residual metal impurities and organic compounds in the surface layers of silicon. Etching was made in galvanostatic mode. A plate of chemically stable stainless steel was used as a cathode. Etching was performed for different intervals of time in order to elucidate the influence of porosity and phase composition of the surface layers of por-Si on the structure of its energy bands.

The following samples were obtained: single-crystalline silicon substrates <111> (with a resistivity of 0.35 Ohm·cm) doped with phosphorus were etched for 1, 2, 3, 5, and 10 min. Porosity of the samples etched for 5 and 10 min was of about 45% and 75%, respectively. Single-crystalline silicon substrates <111> doped with antimony (resistivity of 0.01 Ohm·cm) was etched for 1, 3, 5, and 10 min, which allowed us to obtain the layers of 80% porosity for the sample etched at 10 min. All of the samples were kept in the atmosphere for about 2 weeks. This exposure was necessary in order to provide investigations of various samples at approximately the same ageing time.

Ultrasoft X-ray emission spectroscopy was applied for the study of electron structure. This method provides information on the occupied electron states in valence band and unoccupied states in the conduction band with rather high energy resolution that can be simply interpreted. X-ray emission spectra allow us to determine local partial density of the occupied states in the valence band of the investigated material [12]:

$$I(E) \sim v^3 \sum_j |M_{ij}|^2 \delta (E_i - E_j - h\nu) \tag{5.1}$$

where $M_{ij} = \int \varphi_i^* H' \varphi_j dr$ is a matrix element of electron transition probability from the valence band with the wave function φ_j and eigenvalue of E_j to the vacancy in the core level with the wave function φ_I, H' is a perturbation operator, and $h\nu$ is the energy of emitted X-ray quantum.

X-ray photoeffect quantum yield spectra are proportional to the absorption coefficient, and their XANES spectra represent the distribution of the local partial density of states in the conduction band [12]:

$$\mu(E) \sim v^3 \sum_f |M_{fi}|^2 \delta (E_f - E_i - h\nu) \tag{5.2}$$

Ultrasoft X-ray emission Si $L_{2,3}$ spectra of silicon were obtained with X-ray spectrometer-monochromator RSM-500. The depth of the analyzed layer was 10–60 nm, and its variation was performed by the change of the kinetic energy of electrons exciting the spectrum from 1 to 3 keV. The operating vacuum in the X-ray tube and the volume of spectrometer during spectra survey was of 2×10^{-6} Torr.

XANES Si $L_{2,3}$ spectra were obtained with Russian-German beamline of BESSY II synchrotron (Berlin). The X-ray-optic scheme of XANES measurements includes four mirrors with gold coating and four gold-coated gratings with 600 lines per millimeter. Energy resolution was of 0.03 eV. Thickness of the analyzed layer for the investigated samples determined by the electron yield depth did not exceed 10 nm. Vacuum in the analytical chamber was continuously kept at $5 \times 10^{-9} - 10^{-10}$ Torr during the survey of spectra.

Figure 5.7 represents the data on photoluminescence where the PL excitation source was a pulse laser operating at 337 nm wavelength of emission. An atomic force microscope (AFM) image of the surface for the sample of porous silicon etched for 10 min is also given in Figure 5.7.

To analyze the obtained spectra XANES data for single-crystalline silicon, amorphous silicon and silicon oxide were applied; they are presented in Figure 5.8.

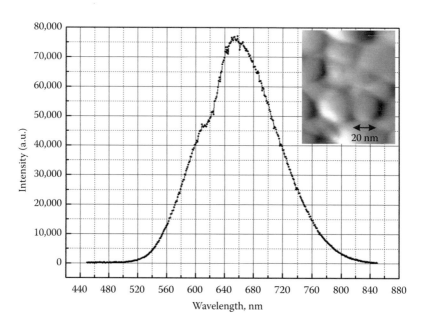

Figure 5.7 Photoluminescence spectrum of the porous silicon sample obtained by 10 min etching of single-crystalline silicon <111> wafer doped with phosphorus. Inset: AFM image of the sample surface.

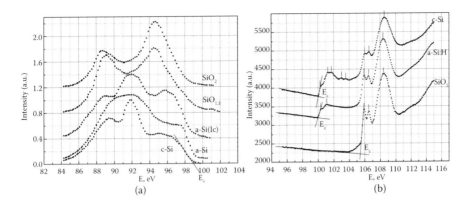

Figure 5.8 Si $L_{2,3}$ USXES (a) used for simulation and XANES spectra (b) of the reference samples.

Analysis of the phase composition was performed with the use of a special technique of computer simulation envisaging simulation of a complicated X-ray emission band of a sample by combining X-ray emission bands of the reference materials [6]. To determine phase composition of the investigated porous silicon samples, their spectra were simulated with the use of the spectra for reference "phases", which can likely be present in the porous layer, namely single-crystalline silicon (c-Si), amorphous silicon (a-Si), low-coordinated silicon (a-Si(lc)) (this phase with a coordination number of ~ 2,5-3 was observed in amorphous films of Si [13]), and two kinds of silicon oxide: sub-oxide $SiO_{1.3}$ and silicon dioxide SiO_2. Ultrasoft X-ray emission spectra of the reference phases are presented in Figure 5.8.

5.3.1 RESULTS OF INVESTIGATIONS BY THE USXES TECHNIQUE

Figure 5.9 represents ultrasoft X-ray emission Si $L_{2,3}$ spectra of the porous silicon samples as a function of time of formation varying from 1 to 10 min (dashed line) with an increasing porosity. Note that for the samples with etching time of 1 and 2 min the spectra of valence states were obtained for the depth of analysis up to 35 nm. This is because at the initial stages of etching the substrate of c-Si participated in the formation of the spectrum for a greater depth of analysis, and thus it was indistinguishable from the substrate spectrum. For the larger etching time, the depth of analysis was up to 60 nm. The absence of the data on the density of states for the samples etched for 10 min at the depth of analysis of 10–35 nm is due to a low reproducibility of the experiment. It is due to the presence of various products of reactions on the surface of the sample and instability for some of them under exposure of the electron beam used for excitation of X-ray spectra. The depth of the analyzed layer in Figure 5.9 is given near the corresponding spectra.

Comparison of these spectra with one of the original silicon c-Si allows to determine that just at the first several minutes of por-Si formation considerable changes are observed in the density of valence states of silicon. At the depth of analysis equal to 10 nm with an increase of etching time, a relative reduction of the density of states is observed in the range of peak with the energy of 89.6 eV, in such a way that for a 2-min etching it almost disappears while the density of states near E_v ($E_v - E \sim 3$ eV) increases as compared with the results for single-crystalline silicon. Since the peak of the density of states at 89.6 eV is due to the splitting of electron states as a result of ns–ns interactions [14], the decrease of its intensity in the surface layers of por-Si can be a result of loosening up this interaction for a part of silicon atoms due to a strong dissolution of silicon crystal and the appearance of rather thin Si wires at the surface of por-Si. At the same time, an increase of the density of states in the range of 96 eV is due to the change in the character of hybridization of s- and p-states of silicon as a result of the change of the coordination number for the atoms on the surface of wires.

Besides, a total increase in the density of states for the range of 94–98 eV, which is especially noticeable in the surface layers of por-Si, can be related with the formation of amorphous phase in the surface layers, as can be seen from comparison with the spectrum of amorphous silicon (Figure 5.8). Then with an increase of the etching time up to 3–5 min the intensity of the first peak at 89.6 eV in the surface layers is almost recovered, though with a reduced intensity of the low-energy slope region. At the same time the

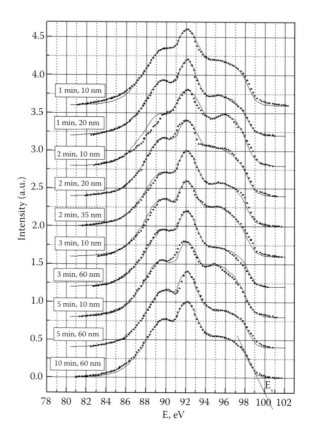

Figure 5.9 USXES Si L$_{2,3}$ spectra of porous silicon for the time of porous layer formation from 1 to 10 min at different depth of analysis (Ev is the valence band maximum). Solid line: simulated spectra.

increased density of states in the range of 94–98 eV is kept. An increase of the depth of analysis up to 35 nm and then up to 60 nm decreases manifestation of these effects due to enhanced contribution of c-Si. However, for the large etching time, these changes in the density of states are quite noticeable at the depth of analysis of 60 nm.

Let us discuss the features in the density of states in the surface layers of por-Si formed under 2-min etching of silicon that is mostly different from c-Si. Such shoulder shape of the spectrum in the range of 86–91 eV was observed in [15] in the compounds or complexes of SiAsx kind. These compounds are characterized by the changes in the nearest neighboring silicon atom by arsenic, an antimony resulting in an increase of length of Si–Si bonds. Therefore, the use of the reference sample a-Si(lc) for the analysis of the "phase" composition resulted in a considerable deviation of the simulated spectrum from the experimental one. For low-coordinated silicon, along with the decrease of coordination number a decrease of the length of the Si–Si bond takes place [13], resulting not in a decrease of the maximum but in its moving away by ~0.5 eV (Figure 5.8), as a result of enhancement of ns–ns interaction. At the same time, the presence of a rather clearly expressed maximum in por-Si at E ≈ 96 eV just as in low-coordinated silicon a-Si(lc) means a similar character of hybridization for s- and p-states. It means that in this case we can note the "extension" of silicon–silicon bonds in the wires inside the surface layer of por-Si under the etching of silicon.

Besides a qualitative analysis of the effect of por-Si structural features on the distribution of the density of states, we performed the quantitative analysis of the phase composition in the surface layers of por-Si with the use of mathematical spectra simulation. Results of the investigations of phase composition for all of the samples are presented in Table 5.2, together with the depth of the analyzed layer.

Comparative analysis of the experimental and simulated spectra means that for all of the samples except for the surface layer of the por-Si sample obtained at a 2-min etching, good agreement can be observed.

Table 5.2 Analysis of the phase composition of porous silicon samples

ETCHING TIME, MIN	DEPTH OF ANALYSIS, NM	PHASE COMPOSITION (%)				
		c-Si	a-Si	a-Si(lc)	$SiO_{1.3}$	SiO_2
1	10	45	25	26	0	4
1	20	76	8	14	2	0
2	10	19	8	73	0	0
2	20	49	21	19	0	11
2	35	59	8	30	0	3
3	10	57	11	16	0	16
3	60	70	24	5	1	0
5	10	54	6	22	0	17
5	60	76	24	0	0	0
10	60	60	27	9	3	0

Under formation of the porous silicon, an amorphous layer appears on the surface of the remaining wires of single-crystalline silicon. Silicon oxide (in the form of dioxide and, possibly, suboxide of $SiO_{1.3}$) often observed in the spectra is probably due to a partial oxidation of the por-Si surface during its storage in the air after its removal from the electrochemical cell until it is mounted in the vacuum chamber of the X-ray spectrometer.

A large amount of the crystalline silicon in the porous layer indicates the presence of big wires preserving their crystalline structure. Obviously, with an increase of the etching time, the number of such wires decreases. It is supported by some reduction of content for this phase. The presence of the phase of low-coordinated silicon indicates the presence of the structure forms in a porous layer, where a considerable part of silicon atoms are characterized by reduced coordination number. It is also known that silicon wires of large diameter are coated with small ones of nanometer size [16–18]. These objects are formed during the etch of more large wires, and they have the broken crystal structure not only on the surface but also in the bulk because they are of the nanometer size as compared with the wires of micrometer size. According to Table 5.2, the contribution of the a-Si(lc) phase in the composition of por-Si surface layers can attain several dozens of percents and, in our opinion, the process of etching of the walls in the big "macro"-pores deep inward, the wires should take place in the surface layers of por-Si with a formation of nc-Si.

The presence of the phase of amorphous silicon according to the results of simulation of the experimental spectra means formation of the phases with a complete failure of ordering in the silicon structure in the process of etching. According to L. M. Peter et al. [19] and D. N. Goryachev et al. [20], due to the instability of Si^{2+} ions formed in the process of etching and their following disproportionation (mutual exchange of electrons), formation of a "secondary" silicon presumably in amorphous state and Si^{2+} ions takes place. Thus, the question is concerned not only with the etching of silicon in the bulk of substrate but also with the formation of a new amorphous or low-crystalline layer on its surface [19,20]. In our opinion, the presence of phases of amorphous and low-coordinated silicon in the surface layer of por-Si at the depth of 10 nm as at 60 nm means that during etching with pore formation, there is redeposition of secondary atomic-like silicon with a formation of nanosize clusters, which cover the surface of the porous layer (nonetched parts and wires), as well as an amorphous phase that correlates with the data of [19,20].

An increase of the etching time results in a certain decrease of the contribution of amorphous component into the energy spectrum of the surface layer of 10 nm thickness with a simultaneous increase of the contribution for oxide components. It was assumed that a decrease of a-Si content is related to the interaction between oxygen and amorphous silicon. Moreover, because the amount of low-coordinated silicon is also reduced, we assume oxidation of low-coordinated silicon along with the oxidation of a-Si.

Arrays, hybrids, and core–shell

5.3.2 RESULTS OF XANES INVESTIGATIONS

Figure 5.10 represents XANES spectra of the investigated samples for Si substrate orientation of <100> and <111>. Energy positions of the main spectral features of the investigated samples are given in Table 5.3 together with the position of the conduction-band bottom relative to the Si 2p-level obtained from the experimental data.

Let us consider in detail the position and the structure of absorption edges in the reference spectra of absorption presented in Figure 5.8. For single-crystalline silicon, one can observe an abrupt edge with a characteristic "step" as well as two clearly expressed double maxima in the range of 101.2–101.7 eV and 102.2–102.7 eV. The distance between the split peaks approximately corresponds to the spin-orbital splitting of the core Si $L_{2,3}$ level. A more simple structure with a "step" and a single maximum is peculiar to the Si $L_{2,3}$ edge of amorphous silicon. The latter can be due to the smoothing of the density of states

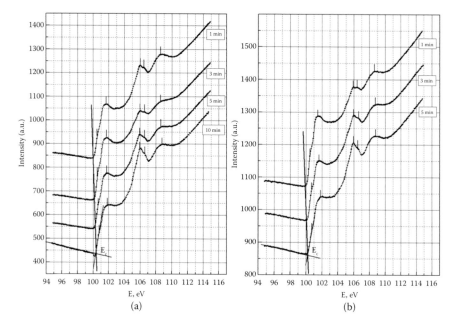

Figure 5.10 Si $L_{2,3}$ XANES spectra of porous silicon obtained on c-Si substrates with <100> (a) and <111> (b) orientation for different etching time (Ec is the conduction band bottom).

Table 5.3 The main spectral features of Si $L_{2,3}$ XANES for porous silicon samples obtained on c-Si substrate under different etching time

ETCHING TIME, MIN	E_C EDGE, eV	ENERGY OF SPECTRAL FEATURE, eV		
		<100>		
1	100.1	101.5	105.9/106.4	108.5
3	100.2	101.6	106.0/106.4	108.6
5	100.3	101.6	105.9/106.5	108.5
10	100.4	101.6	106.0/106.5	108.6
		<111>		
1	100.1	101.4	106.1/106.5	108.6
3	100.2	101.5	106.0/106.6	108.7
5	100.3	101.6	106.0/106.6	108.6

as a result of disordering in a-Si [15], as a result of the manifestation of Si–H bonds on the surface due to a-Si:H formation. Furthermore, at E > 104eV in the structure of the Si $L_{2,3}$ edge of amorphous silicon, just as in the case of single-crystalline Si, one can observe the peaks characteristic of SiO_2 (Figure 5.8).

Comparing these edges with XANES Si $L_{2,3}$ spectra of porous silicon, one can note a more simple structure of the edge in por-Si as compared with c-Si and a weakly expressed structure connected with the oxidation (E > 104 eV). Though position of this structure coincides well with that of SiO_2 (Figures 5.8 and 5.10), its relative intensity is considerably different.

Comparing XANES spectra of porous silicon and amorphous silicon within the range of 100–104 eV, one can observe their rather good similarity. At the same time a fine structure of XANES spectrum characteristic of c-Si is absent in the spectra of por-Si (energy range of 101–102 eV and 103–103.5 eV). This result shows that there is a layer of amorphous silicon on the surface of por-Si. This confirms the presence of the amorphous layer according to X-ray emission spectroscopy. However, due to a greater depth of analysis when surveying of emission spectra, we obtained superposition of the spectra from the silicon single-crystal and amorphous layer. In the case of XANES spectra, less depth of analysis revealed an almost pure amorphous layer with the traces of oxidation.

The peaks in absorption spectrum in the range of 106–107 eV and 108 eV for the sample of single-crystalline silicon (Figure 5.8b) correspond to the natural oxide growing on the surface of c-Si. Doublet feature (106–107 eV) for the first of the peaks is due to spin-orbit splitting of the Si 2p core level. Similar peaks with a close position and structure appear in a-Si:H as well, as a result of natural oxidation in the air. As for porous silicon the structure of XANES spectra in the range of 105–110 eV (Figure 5.10) quite noticeably differs from that one in the spectra of c-Si and a-Si:H. This structure is of quite low intensity despite the fact that the samples were kept about 2 weeks in the air after their preparation before making the experiments. It means that the oxidation at the surface of porous silicon for this period of time is quite low. The structure of XANES spectra in this energy range for porous silicon is characterized by a weaker maximum in the range of ~108–109 eV than in the range of ~106 eV. It should be noted that this ratio is observed for all of por-Si samples independent of the time of etching. Moreover, with an increase of the etching time, one can see some relative increase in the contrast of the XANES maximum at ~106 eV. Because the reconstruction of XANES spectra in this energy range can be due only to the change of the nearest neighboring for silicon atoms by oxygen ions, the obtained results suggest that the result of oxidation of por-Si surface does not respond to SiO_2 oxide. According to the data of X-ray emission and XPS measurements, the surface oxide layer represents a mixture of the oxide SiO_2 and suboxide $SiO_{1,3}$ [6,21,22]. Therefore, it is possible to consider that XANES spectra of porous silicon also indicates the formation of suboxides SiO_x, where $x < 2$, on the surface of por-Si.

It is easily seen from Figure 5.10 and Table 5.3 that the conduction band minimum is shifted to the higher energies, and this shift is independent of the crystal orientation of the original substrate, but it depends on the etching time. This shift attains up to 0.3 eV, and it can be a justification of the quantum confinement effects in porous silicon resulting in the increase of the band gap in this material. In our opinion, quantum confinement occurs in cluster-like structures formed on the surface of silicon wires in por-Si layers. A possibility of formation of such structures was considered in a view of the features of valence electrons energy spectra. This idea is in agreement with the data of XANES [23]: the energy position of conduction band minimum varies with the change of the mean size of silicon clusters.

Thus, the results indicate that porous silicon is a complex multiphase system consisting of crystalline silicon with different types of breakdowns in crystalline structure. In the process of electrochemical etching, redeposition of silicon ions occurs on the surface of porous silicon, resulting in the formation of amorphous phase covering Si wires, formed in the porous layer. These wires can involve low-coordinated silicon. The surface of wires is coated with amorphous silicon and silicon oxide with a reduced oxidation degree.

Finally, we propose the model of photoluminescence in porous silicon based on the following ideas. According to our experimental results, the band gap of porous silicon increases with the increase of etching time and, hence, porosity. Thus, we can directly observe quantum confinement effect proposed in various models of PL in porous silicon. However, the estimations of the band gap E_g mean that PL should be observed in the nearest IR range, but not in the visible part of the spectrum. This contradiction can be explained by the fact that X-ray emission and XANES spectra represent the distribution of electron states in a certain volume of porous layer involving Si wires of different diameter as well as clusters of different sizes. The values of the limiting energies E_v and E_c are related with electron energy spectrum in nc-Si wires characterized by a certain size distribution. Meanwhile, nc-Si wires only with a certain size contribute to a visible PL because due to the quantum confinement effect, the restrictions imposed on the change of electron quasi-momentum are eliminated.

On the other hand, the results indicate the influence of the oxide phases on the distribution of the density of states both in the valence and conduction bands (Figures 5.9 and 5.10). This is in agreement with PL models assuming the presence of the radiation centers participating in PL of por-Si at the defect boundary of Si–SiO$_x$ (Si–O or Si–OH bonds) [10,11]. Previously it was found that under exposure of por-Si, prepared in a similar way, in the air for a few months, the intensity of PL increases almost by 10 times [21], while the position of the main PL peak does not change. At the same time, according to X-ray spectral data, an increase of the oxide content in the surface layers of por-Si, mainly in the form of suboxide SiO$_{1.3}$ [6,21], can be observed. Thus, the shift of E_c by 0.4 eV with an increase of porosity due to quantum confinement effect observed in our experiments stipulates the value of the band gap Eg $\approx$ 1.5 eV. Nevertheless, this value can determine only the lower limit of the energy of PL quantum in such complex heterophase structures as porous silicon. The maximum of PL intensity usually arranged at $\approx$1.8 eV (according to the optical data) is related with electron transitions in heterojunctions at the boundaries between nanocrystalline wire and the amorphous layer covering the wires, nc-Si/a-Si (Figure 5.11). Note that for hydrogenated amorphous silicon, the optical band gap attains 1.8–2 eV. As a result, there appears a possibility of efficient PL in the visible range. Moreover, radiation centers participating in the formation of PL in por-Si can appear at the defect boundary of (Si–O or Si–SiOH bonds). This idea is confirmed by the presence of SiO$_x$ phase on the surface of nanosize silicon wires.

Therefore, comparing results of analysis on XANES and USXES spectra, it is possible to conclude that photoluminescence in porous silicon can be related with several mechanisms connected with the boundary phenomena in such complex systems as porous silicon (Figure 5.11).

It was found that in porous silicon demonstrating visible photoluminescence, the phases of amorphous silicon and silicon suboxide with low oxidation degree were formed on the surface of nanocrystalline Si wires. Increase of porosity results in the shift of conduction-band bottom and increase of the band gap due to the quantum confinement effect. On the basis of experimentally observed regularities in the distribution of the local partial density of states as well as the changes in phase composition, it was determined that photoluminescence of porous silicon is a result of competition between several mechanisms of radiation transitions of electrons.

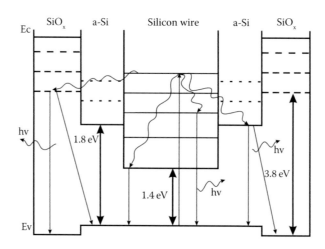

Figure 5.11 Proposed model of photoluminescence in porous silicon. Figures indicate the values of the band gaps for the corresponding phases. Arrows: possible optic transitions in por-Si.

5.4 ATOMIC AND ELECTRONIC STRUCTURE PECULIARITIES OF SILICON WIRES FORMED ON SUBSTRATES WITH VARIED RESISTIVITY ACCORDING TO ULTRASOFT X-RAY EMISSION SPECTROSCOPY

Beginning from the time when visible PL from porous silicon was first obtained [7], problems associated with development of silicon nanostructures with light-emitting properties stable over the course of time have been a subject of intense scientific and practical interest [8,24]. Recently developed technologies can form arrays of nanostructured crystalline silicon nanowires, including ones with nanometer transverse dimensions, and this is of indubitable interest for various applications in optoelectronics, photonics, and photovoltaics [25]. Here, the problems associated with the stability of the exhibited properties, primarily light-emitting properties, are among the most important and require detailed study of the specific features of the atomic and electronic structure, interatomic interactions, and phase composition. Therefore, non-destructive methods of X-ray spectroscopy with varied thickness of the layer being analyzed, which possess increased sensitivity to the electronic structure and local environment of a given sort of atoms, are exceedingly in demand and highly informative [26–30]. The present study is concerned with specific features of the atomic and electronic structure and phase composition of Si nanowire arrays by the method of USXES. Samples of nanowire Si massives were produced by the method of metal-assisted wet chemical etching (MAWCE) [25]. Crystalline silicon c-Si (100) wafers with a diameter of 100 mm and high (10^{20} cm^{-3}, resistivity <0.005 Ω/cm]) and low (10^{15} cm^{-3}, resistivity ~1–5 Ω/cm) degrees of doping with boron (B). Here we name the sample formed on a substrate with high-resistivity (20-min etching) low-density silicon nanowires (LD SiNW), while we name the sample formed on a substrate with low resistivity (60-min etching) high-density silicon nanowires (HD SiNW). At the first stage, silver nanoparticles were deposited onto the surface of a Si wafer submerged for 30 seconds in an aqueous solution of AgNO$_3$ and HF (0.2 and 5 mol, volume ratio 1 : 1). At the second stage, the wafers covered with silver nanoparticles were submerged in a solution containing 5 mol of hydrofluoric acid and hydrogen peroxide (30%). The sample formed on the high-resistivity substrate was etched in the solution for 20 min, whereas that on the low-resistivity substrate was etched for 60 min. The residual Ag nanoparticles were removed by washing in nitric acid (65%) for 15 min. The atomic structure and morphology of silicon nanowires were analyzed by scanning electron microscopy (Carl Zeiss Ultra 55, FESEM) and transmission electron microscopy (FEI CM200FEG, CM 20). USXES spectra provide information on the local partial density of occupied electron states in the valence band of a material under study. Si L$_{2,3}$ spectra were examined with an RSM-500 ultrasoft X-ray spectrometer–monochromator [12]. The depths of analysis were 10, 35, 60, and 120 nm at accelerating voltages applied to the X-ray tube of 1, 2, 3, and 6 kV, respectively. The working residual pressure in the X-ray tube and in the working volume of the spectrometer was ~10^{-4} Pa. The energy resolution was ~0.3 eV. Figure 5.12a and b show typical cross-sectional SEM images of HD-SiNW and LD-SiNW samples on c-Si with the inherent porous structure of silicon nanowires. An ordered growth of the nanowires preserving the crystallographic orientation of the substrate was observed. The length of the nanowires was about 5 μm after 60 min of etching of the HD-SiNW sample and 20 μm for LD-SiNW after 20 min of etching. The TEM data (insets in Figure 5.12) clearly demonstrate that the nanowire crystals of the HD-SiNW sample have a highly developed porous structure as compared with LD-SiNW, with silicon nanocrystals situated on the side walls of a nanowire. This porous structure of nanowires is typical of wire-like silicon crystals produced by the MAWCE method on a heavily doped silicon substrate [31]. Figure 5.13a shows Si L$_{2,3}$ spectra of the HD-SiNW sample recorded for various depths of the informative layer (the experimental spectra are represented by points). The Si L$_{2,3}$ spectra of the LD-SiNW sample are shown in Figure 5.13b. Comparison of these spectra with each other and with spectra of reference samples of c-Si and SiO$_2$, also shown in Figure 5.13, easily shows that there are important differences between the Si L$_{2,3}$ spectra of two arrays, which are primarily due to the different thicknesses of the oxide layer. It also seems possible to note the gradual changes in the distribution of spectral features with increasing analysis depth. The spectrum of the nanowires array of sample HD-SiNW at an analysis depth of 10 nm shows two peaks at energies of 89 and 94.7 eV and a satellite at an energy of 77 eV (Figure 5.13a), all characteristic of the SiO$_2$ spectrum,

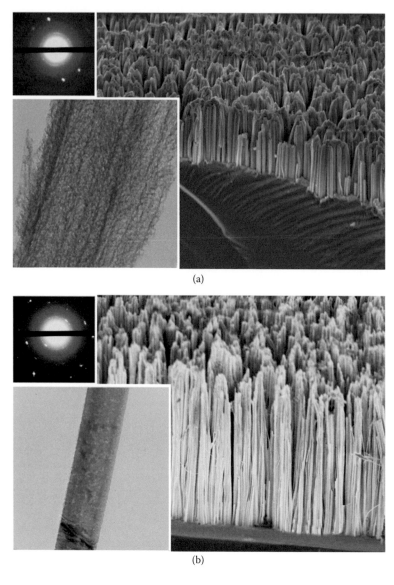

(a)

(b)

Figure 5.12 SEM images of silicon nanowires in samples (a) HD-SiNW and (b) LD-SiNW. Insets: microdiffraction and TEM data.

which indicate that silicon oxide is predominant in the composition of a 10 nm surface nanolayer. Later, the ratio between the peak intensities remains nearly unchanged at 35 nm, but the dip between these peaks becomes less pronounced and there appears a weakly pronounced shoulder at E ~ 92 eV. With the analysis depth increasing to 120 nm, the contribution of the feature at E ~ 92 eV in the USXES Si $L_{2,3}$ spectrum increases, indicating the influence exerted by elementary silicon on the spectrum. Figure 5.13b shows Si $L_{2,3}$ spectra of another array of silicon wires from the LD-SiNW sample. First, it should be noted that there are significant differences in the distribution of the main spectral features from the aforementioned array of the HDSiNW sample. At an analysis depth of 10 nm, a major peak of crystalline silicon appears, with a comparable intensity, in the Si $L_{2,3}$ spectrum between the peaks of silicon oxide (89 and 94.7 eV). This means, in the first place, that the oxide layer on the surface of nanowires in this array is substantially thinner. As the analysis depth increases to 35 nm and, further, to 120 nm, the contribution of the intensity of the major peak of crystalline silicon increases and becomes predominant. Thus, for the HD-SiNW nanowire array, the 92 eV feature corresponding to crystalline silicon is hardly seen at an analysis depth of ≥ 35 nm,

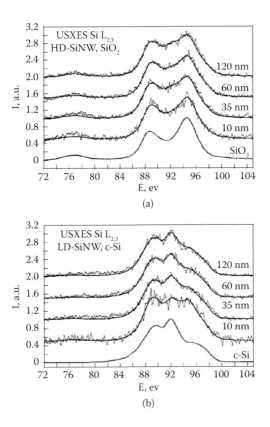

Figure 5.13 Si $L_{2,3}$ series in the USXES spectra of samples (a) HD-SiNW and (b) LD-SiNW, recorded at different depths of analysis, and spectra of reference SiO_2 and c-Si. Solid line: simulated spectra.

whereas, for the other array, LD-SiNW, this feature clearly dominates beginning at an analysis depth of 35 nm or greater. To quantitatively describe the variation of the phase composition across the sample thickness, we simulated experimental spectra with reference sample spectra by our method that was described in detail in [6]. The results obtained in a simulation by this procedure are presented in Table 5.4. In Figure 5.13, the simulated spectra are drawn by a fine solid line. As reference spectra of intermediate silicon oxides SiOx, we used the Si $L_{2,3}$ spectra from [32]. On the whole, the contribution of these reference spectra shows the possible presence of silicon suboxides in the surface layers of silicon nanowire arrays. The good coincidence of the experimental and simulated spectra indicates that the simulated results are sufficiently adequate and

Table 5.4 Phase composition of LD-SiNW and HD-SiNW samples at various depths of the informative layer

SAMPLE	DEPTH OF ANALYSIS, NM	c-Si, MOL %	SiO_2, MOL %	SiO_X, MOL %	ERROR, %
HD-SiNW	10	0	35	65	8
	35	13	37	50	8
	60	19	34	47	6
	120	24	42	34	6
LD-SiNW	10	45	22	33	6
	35	67	13	20	10
	60	72	8	20	7
	120	74	11	15	10

the data on the phase composition of silicon nanowire arrays are reliable. It follows from Table 5.4 that the HD-SiNW sample has a thin (~10 nm) surface oxide layer containing no elementary silicon. In addition, at all analysis depths, the main contribution to the phase composition of the morphologically developed surface of the HD-SiNW sample is made by silicon oxides SiOx and SiO_2. In the other array of silicon nanowires, in the LD-SiNW sample, crystalline silicon is found in an amount that is comparable with the amount of the oxide even in the 10 nm layer, increases at a depth of 35 nm, and becomes predominant at larger analysis depths up to 120 nm.

On the whole, the results of the phase analysis based on simulation of the experimental spectra with the use of reference spectra are in good agreement with the electron microscopic data. The nanowires in an LD-SiNW sample produced on a high-resistivity silicon wafer are mostly composed of crystalline silicon covered with a thin (<10 nm) oxide layer. The surface of the other nanowire array in the HD-SiNW sample formed on a low-resistivity silicon wafer is covered by a substantially (one to two orders of magnitude) thicker layer of silicon oxides compared with the LD-SiNW sample. This can be attributed to the strongly developed surface of the nanowire arrays in this sample. Thus, the method of ultrasoft X-ray emission spectroscopy was used to demonstrate that there are significant differences in the electronic structure and phase composition of silicon nanowire array samples formed at different times of etching of substrates with different doping levels. Analysis of the phase composition at informative layer depths from 10 to 120 nm demonstrated that the morphologically more developed HD-SiHW sample is substantially more strongly subject to oxidation and has a thicker layer of surface oxides, with a thickness exceeding by more than an order of magnitude the thickness of the oxide layer in the nanowire array of the LD-SiNW sample.

5.5 SILICON NANOCRYSTALS IN SiO_2 MATRIX OBTAINED BY ION IMPLANTATION UNDER CYCLIC DOSE ACCUMULATION

Materials containing the nanocrystals draw serious attention from researchers due to the unusual physical properties. These objects are characterized by the quasi-atomic energy structure of their electron states. However, the main regularities of the changes in their electron spectra and other physical properties under transition from the bulk crystalline materials to the nanosize objects have, to date, been studied insufficiently. Specific features of the interaction between such crystals and a substance of the ambient matrix, which is used for quantum confinement of the current carriers, passivation of the cluster surface, and stabilization of their properties with time are not investigated in detail.

On the other hand, investigations of manufacturing methods and impacts on semiconductor structures involving silicon nanocrystals (nc-Si) prove to be very prospective for a number of reasons. First, silicon is the main material for microelectronics at the present time as well as for the near future. Second, a decrease in the dimensions of the elements is the main tendency in present-day microelectronics. And this fact inevitably results in a move into the nanolectronic world. Finally, an ability of silicon nanocrystals to emit rather intensive visible light at room temperature, unlike the bulk single-crystalline silicon, allows for design of the elements of microchips that are able to process information using the principles of optoelectronics [33–35].

Nowadays, nc-Si is usually formed due to the effect of self-organization under decomposition of the oversaturated solid solution of Si in SiO_2. Ion implantation also proves to be a widespread method of obtaining nc-Si. Conditions for obtaining silicon nanocrystals seem to be quite critical—the required doses of ions are of the order of ~10^{17} cm^{-2} at the temperature of postimplantation anneal of ~1100°C. Recently some new methods of exposure have appeared that allow for stimulating the processes of nanocrystals formation or directly modifying their properties [36–39].

Different methods of the investigations for studying nc-Si/SiO_2 nanostructures obtained by ion implantation can be applied such as high-resolution electron microscopy (HREM), AFM, optical spectroscopy, and Raman spectroscopy. All of these methods provide different information on the substructure and optical properties of the investigated samples. Depth profiling methods are of special interest for analyzing these structures. Because the distribution of the interstitial silicon is nonuniform during ion implantation, the composition of the implanted layer is not usually uniform in depth. In a number of cases, it is important to get information on the composition of the implanted layer not only averaged over the layer but on the distribution of layers with different compositions near the surface as well as at a certain distance from the surface.

We determined for the first time the contribution of different phases (nc-Si and SiO_2) of the surface layer of semiconductor-dielectric nanocomposite into electron energy spectrum near the conduction-band bottom using the method of XANES spectroscopy.

For the first series of samples, silicon ions with the energy of E = 140 keV and total dose of $\Phi = 10^{17}$ cm^{-2} were implanted into the films of silicon oxide (with a thickness of 510 nm) obtained by thermal oxidation of silicon wafers in humid oxygen. The density of ion current was less than 5 µA/cm². Accumulation of the implantation dose was performed either in one stage or cyclically but in such a way that each time the same doses were accumulated. Each time after implantation the samples were subjected to the anneal in dry nitrogen (1100°C) in such a way that the total time of annealing was equal to 2 hours. Thus, the following samples were made:

Sample 1. Dose of 10^{17} cm^{-2}, single-stage annealing in dry nitrogen for 2 hours.

Sample 2. Dose of 5×10^{16} cm^{-2}, annealing in dry nitrogen for 1 hour, two cycles.

Sample 3. Dose of 3.3×10^{16} cm^{-2}, annealing in dry nitrogen for 40 min, three cycles.

According to the data of photoluminescence, Raman scattering, and high-resolution transmission electron microscopy [33], Si nanocrystals of 2–5 nm size were observed in the implanted layer for all of the investigated samples. Moreover, their average concentration was reduced with the increase of the total number of annealing cycles; that is, the splitting of the Si implantation dose resulted in a decrease of the total efficiency of nanocrystal formation.

The second series of samples (samples 11, 12, and 13) was made in a similar way, but original oxide was additionally annealed in the air at 1100°C for 3 hours before ion implantation in order to heal the defects introduced under the "humid" oxidation and to form more dense oxide matrix.

XANES investigations were performed at SRC synchrotron of Madison University (Madison, WI). The vacuum in the operation chamber was of 10^{-9} Torr. The instrumental broadening was about 0.05 eV. The depth of analysis was less than 5 nm. To obtain XANES spectra, we used current sample technique measurements.

Analysis of the shape and position of the main energy features for XANES Si $L_{2,3}$ spectra were performed by comparison of the experimental spectral data with the known spectra of the standard samples such as crystalline silicon c-Si covered with native oxide, amorphous a-Si:H, and thermally grown silicon dioxide SiO_2. Si $L_{2,3}$ XANES of standard samples are represented in Figure 5.14.

The spectra of c-Si and a-Si:H are rather different in the energy range near the main absorption edge of 100–104 eV. For the crystalline silicon one can see along with the main absorption edge two doublet maxima at 101.2–101.7 eV and 103–103.6 eV in this energy range. The difference between a-Si:H and c-Si is connected with smoothing and smearing of the density of states function near the bottom of the conduction band in the amorphous material. The features related to silicon oxide are observed at the energies above 104 eV for all of the spectra, and the main sharp absorption edge of this phase is located at ~105.5 eV (Figure 5.14). It should be noted that formation of the excited states in SiO_2 spectra takes place mainly inside SiO_4 tetrahedron—the basic structure element of the oxide. Within the confines of the quasi-molecular approach, the basic features of the fine structure for SiO_2 are connected with transitions to the excited states of the cluster $(SiO_4)^{4-}$:a_1,t_2,e and t_2+e [40,41]. That's why it is possible to assert that the fine structure of XANES spectra provides information on the nearest local environment for silicon atoms.

Thus, a comparison of XANES spectra for the analyzed samples with those for reference samples allows us not only to identify the presence of two phases in the near-surface layer of the Si-SiO₂ system but also to estimate qualitatively the degree of perfection of their atomic structure by the presence or absence of the fine structure in the spectra.

Si $L_{2,3}$ XANES spectra for the first series of the implanted samples are presented in Figure 5.15. Similar data for the samples with more dense oxide are given in Figure 5.16.

For the first series of the samples obtained by split (cyclical) implantation, two edges can be observed: the first one at ~100 eV is related with elementary silicon, whereas the second one at E > 104 eV is related to SiO_2 matrix (both edges are separated by the energy gap of about 4 eV).

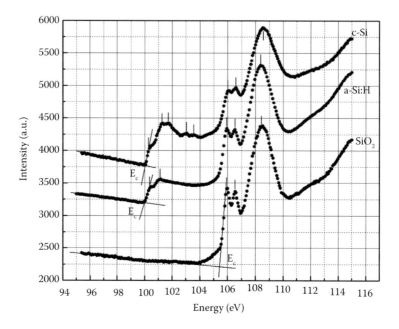

Figure 5.14 Si $L_{2,3}$ XANES spectra of the standard samples: single-crystalline silicon c-Si with native oxide, amorphous a-Si:H, and thermally grown silicon dioxide SiO_2.

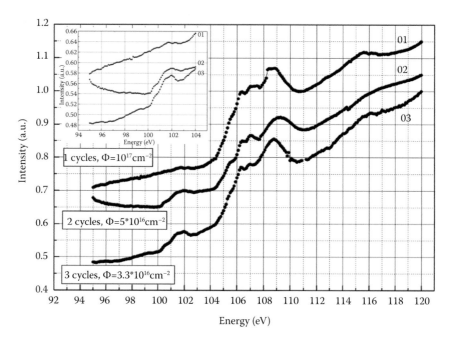

Figure 5.15 Si $L_{2,3}$ XANES spectra of thin SiO_2 films containing Si nanocrystals obtained under different number of accumulation cycles of the total implantation dose (F = 10^{17} cm^{-2}). Insert: the elementary silicon (nc-Si) absorption edges.

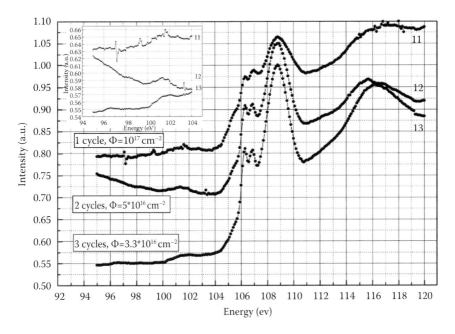

Figure 5.16 Si L$_{2,3}$ XANES spectra for thin films of SiO$_2$ subjected to the additional annealing before ion implantation, containing Si nanocrystals obtained at the different variants of cyclical accumulation of the total implantation dose (F = 10^{17} cm^{-2}). Insert: the elementary silicon (nc-Si) absorption edges.

However, in the Si L$_{2,3}$ XANES spectrum for Sample 1 with a single accumulation of the dose a 10^{17} cm^{-2} silicon edge at ~100 eV is not observed though the states above 104 eV related to SiO$_2$ can be clearly distinguished. It means that the nanocrystals in the surface layer of the sample are practically absent. One should also remember that the depth of analysis for the XANES technique in this case is of about 5 nm.

The appearance of a distinct silicon edge at the energies above 100 eV in Samples 2 and 3 is due to the formation of inclusions of the elementary silicon in the surface layers of oxide matrix. The observed well-distinguished fine structure in the range of 100–103 eV means an ordered arrangement of Si atoms in these inclusions, that is, formation of silicon nanocrystals.

Analyzing the relative intensities of the spectral features characteristic of silicon and its oxide in XANES spectra represented in Figure 5.15, it should be noted that the double accumulation of the total dose and time of annealing results in the increase of the intensities ratio for the edges of nc-Si and SiO$_x$, while for threefold accumulation this ratio is higher than for a single one but lower as compared with the double accumulation. It means that the splitting of the process of synthesis into two or three cycles increases concentration of silicon nanocrystals formed just near the surface. Moreover, the increase depends nonmonotonously on the number of cycles. Besides, under splitted ion irradiation, less damage of the oxide matrix takes place, which is confirmed by a less distorted structure of the spectrum in the energy range of 105–110 eV, corresponding to oxide as compared with the spectrum of standard SiO$_2$ (compare Figures 5.14 and 5.15). Here we designate the phase of oxide by the formula of SiO$_x$ because the shape of the spectrum in the energy range above 104 eV is considerably distorted in comparison with the spectrum of standard SiO$_2$, given in Figure 5.14. This indicates rather strong distortions of silicon-oxygen octahedrons in SiO$_2$ matrix under the implantation of silicon ions inside it that cannot be restored even during the following 2-hour annealing at 1100°C in dry nitrogen.

How can these results, in some respects unusual ones, be explained? We have already noted that with an increase of the number of cycles, the mean (or depth-integrated) concentration of Si nanocrystals is monotonously reduced [33]. However, these data are not contradictory. The matter is that unlike of [33], in our case a thin subsurface layer is probed and its thickness is small as compared with the run path R$_p$ of the ions (R$_p$ ≈ 200 nm). Due to the bell-shaped original (just before annealing) distribution

Arrays, hybrids, and core-shell

of the implanted silicon ions, their concentration near the surface is low if no diffusion is taken into account. Because the probability of nanocrystal nucleation depends on the original concentration quadratically [42], then the concentration of the latter ones in this area must be quite low. Generally, in the process of anneals a diffusion spreading of silicon profiles takes place that increases concentration of the excess atoms for this element near the surface. However, for Sample 1 such a spreading is minimal due to a high concentration of the nanocrystals formed near R_p that prove to be capture centers for atoms. For Sample 2 (two-stage process of the implantation-annealing) concentration of the nanocrystals near R_p is less, diffusion spreading is expressed more clearly, and, therefore, their concentration near the surface is higher than in the case of Sample 1. As for Sample 3, this tendency should be expressed even more distinctly, but it is not true since under rather high splitting of the implantation dose even near the surface a factor that causes a decrease of the integral concentration of nc-Si comes into force. It was already mentioned earlier that the probability of nc-Si formation is proportional to the squared concentration of the free silicon. In the case of splitting of the procedure into three cycles, the implantation dose in each cycle is so small that the cycle amount of the free silicon is insufficient, even taking into account a diffusion influx of the atoms; perhaps the concentration gradient providing the diffusion in this case is still less. This argument can qualitatively explain the obtained result. It is also clear that the splitting of ion irradiation with the inclusion of intermediate anneals should result in less "damage" of the matrix.

For the second series of samples, two edges are also observed in XANES spectra (Figure 5.16). However, in the presence of a very weak jump of absorption at the first edge (elementary silicon), the structure of the absorption edge in the range of >104 eV, characteristic of the oxide, becomes similar to that one at the absorption edge of stoichiometric SiO_2 (compare Figures 5.14 and 5.16). It means that an increase of the structure perfection in original oxide matrix resulted in a weaker radiation damage or a higher degree of reconstruction of the layer structure in the process of annealing. In this case even under single-stage dose accumulation (Sample 11), the shape of the spectra in this region becomes just the same as in the samples with "loose" oxide matrix under threefold dose accumulation (Sample 3). It should be noted that for the sample of 11 prominent absorption edge can be observed in the range corresponding to the absorption edge of elementary silicon just as in the case of Sample 1 from the first series. At the same time, the absorption edge of the elementary silicon for the samples of 12 and 13 can be observed, although it is considerably less expressed as compared with Samples 2 and 3. Nevertheless, a weak resemblance of the fine structure in the range of 100–103 eV means a slight ordering in silicon inclusions.

Thus, results of the investigations for the samples of the second series obtained with the use of additional annealing in the air of the oxide matrix show that cyclical accumulation of the dose can promote formation of nanocrystalline inclusions near the surface of a sample similar to the results of investigations for the first series of samples. However, under implantation of silicon into more dense oxide (Samples 12 and 13) the probability of formation of nc-Si inclusions in the surface layers of the samples is reduced. It follows from comparison of the relative intensities for absorption edges of the elementary silicon and SiO_2 in XANES spectra for the samples of the second series with similar spectra for the samples of the first series with more "loose" oxide matrix. At the same time photoluminescence data [42] show that the total amount of nc-Si in the case of applying the additional thermal treatment of the oxide matrix, on the contrary, increases. This fact is in complete agreement with the explanation presented earlier: the higher the concentration near R_p, the lower the diffusion influx of Si atoms to the surface and, hence, the higher the total concentration of nc-Si.

The results of synchrotron investigations of XANES spectra in $Si–SiO_2$ nanostructures demonstrate the following:

- Cyclical accumulation of the implantation dose proved to be more efficient for the formation of nanocrystalline silicon in a thin (as compared with the ion run path) surface layer of SiO_2 matrix.
- Relative content of the nanocrystalline phase nc-Si in the surface layer is reduced if more dense oxide (subjected to the additional anneal in the air before ion implantation) is used as a matrix.
- Under cyclical dose accumulation the oxide matrix is less damaged than under single-step ion irradiation with the same total dose.

5.6 XANES, USXES, AND XPS INVESTIGATIONS OF ELECTRON ENERGY AND ATOMIC STRUCTURE PECULIARITIES OF THE SILICON SUBOXIDE THIN FILMS SURFACE LAYERS CONTAINING Si NANOCRYSTALS

The possibility of thin layers formation with well-expressed PL in the framework of silicon technologies is of a great interest for technologists and researchers working in the field of SiO$_2$ films containing silicon nanocrystals (SiO$_2$:nc-Si). Redundant silicon in dioxide matrix could be obtained by the annealing of the SiO$_x$ films formed from SiO powder [43–45]. Under high-temperature annealing of these films, SiO$_x$ decomposition to Si + SiO$_2$ should take place with a simultaneous self-organization of silicon atoms in clusters and/or nanocrystals. The point of the study was identification of silicon nanocrystals obtained under annealing of SiO$_x$ layers and estimation of their size and embedding depth.

The films with a thickness of about 350 nm were formed with the use of SiO$_x$ molecular-beam deposition in a vacuum onto (111) silicon substrates and following high-temperature anneal. Films deposition was performed at 250°C, 300°C, and 350°C substrate temperatures (T$_s$). The following anneal was performed for 2 hours under 900°C–1100°C (T$_a$).

PL spectra were taken at the room temperature in the 350–900 nm wavelength range with nitrogen pulse laser excitation (25 Hz pulse recurrence frequency) at λ = 337 nm.

USXES provides information about local partial density of electron states in the valence band. USXES data were obtained with the use of laboratory X-ray spectrometer-monochromator RSM-500 by electron beam spectra excitation with 3 keV energy and 2 mA X-ray tube current that corresponds to a 60 nm depth of analysis. The operating vacuum in spectrometer volume was 10^{-6} Torr. The energy resolution was 0.3 eV.

A XANES investigation near silicon L$_{2,3}$ level was performed at the Mark V beamline of SRC synchrotron radiation facility (University of Wisconsin-Madison, Stoughton, WI). The operating vacuum in the experimental chamber was 10^{-11} Torr, and instrumental broadening was 0.05 eV. The sample current measurement technique was used to detect XANES spectra.

XPS investigation for SiO$_2$:nc-Si samples was performed with an ultra-high vacuum Omicron Multiprobe setup (10^{-11} Torr pressure) with Mg Kα (1254 eV) X-ray source and constant absolute resolution 0.3 eV. For surface cleaning and investigations at different depths of analysis, the samples were etched by Ar$^+$ ion beam with up to 5 keV energies. Similar investigations but without ion etching were performed at the same synchrotron with the use of HERMON beamline under 700 eV photon energy (customized chamber with cylindrical mirror analyzer).

X-ray diffraction investigations were performed at laboratory DRON-3 diffractometer with the use of Cu Kα radiation.

5.6.1 PHOTOLUMINESCENCE SPECTRA

Figure 5.17 represents PL spectra of the investigated samples: initial unannealed SiOx films with different substrate temperatures (a); 900°C–1100°C annealed films with different substrate temperatures: 250°C (b), 300°C (c), and 350°C (d). Initial film deposited at minimal substrate temperature of 250°C is characterized by broad PL band in the 350–750 nm range and maximum at ~575 nm. Increasing the substrate temperature is accompanied by a steady PL intensity decreasing in the 600–700 nm range and appearance of the shoulder in the 500 nm range. High-temperature annealing of SiO$_x$ films leads to practically complete quenching of the PL with 575 nm maxima and to the appearance of 700–800 nm and 400–500 nm PL bands, which are more clearly expressed in the films deposited at T$_s$ = 300 and 350°C and annealed at T$_a$ = 1100°C.

5.6.2 ULTRASOFT X-RAY EMISSION SPECTRA

Si L$_{2,3}$ USXES interpretation is made according to dipole approximation. Thus, these spectra represent density of occupied s, d states distribution in the valence band of the sample surface layer [12]. Investigations of the electron energy spectra of the valence electrons by USXES data for SiO$_x$ films allows us to determine

Arrays, hybrids, and core-shell

that just after deposition there is a large amount of elementary silicon in the film. The evidence of this fact is the appearance of the maximum with the energy corresponding to elementary silicon (~ 92 eV) in Si $L_{2,3}$ emission spectra. At the same time on both sides of this peak two maxima typical for silicon oxides (~ 89.4 and 94.7 eV) appeared [6].

More detailed analysis with the use of modeling [6] of the spectrum of unannealed film shows that it contains about 43% of crystalline silicon (c-Si), ~ 15% of amorphous silicon (a-Si), and 42% of SiO_2 oxide (Table 5.5). Agreement of model spectra with the experimental ones determines accuracy of the reference phase content in the analyzed layer.

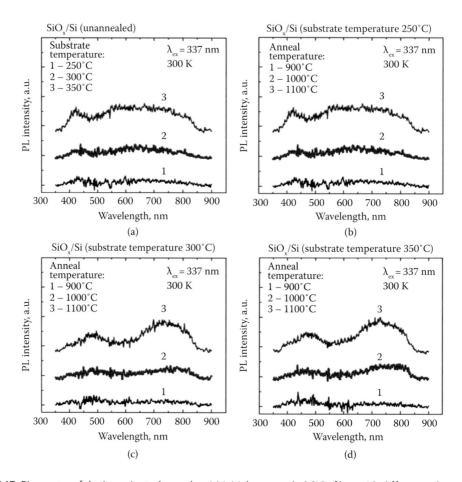

Figure 5.17 PL spectra of the investigated samples: (a) initial unannealed SiO^x films with different substrate temperatures; (b) 900–1100°C annealed films with different substrate temperatures 250°C, (c) 300°C, and (d) 350°C.

Table 5.5 SiO_2:nc-Si/Si films with T_s = 250°C and initial powder SiO phase composition by USXES data

	c-Si, %	a-Si, %	SiO_2, %	$SiO_{1.3}$, %	Δ, %
Initial SiO powder	22	18	43	17	5
Unannealed film	43	15	42	0	7
1000°C anneal	34	13	53	0	13
1100°C anneal	23	0	77	0	7

Source: Terekhov, A., et al., *J. Electr. Spectr. Rel. Phen.*, 114, 895, 2001.
Note: Δ is the analysis accuracy.

Investigation of the Si $L_{2,3}$ emission spectra of the "SiO" powder used for films deposition shows considerable decomposition of this powder to Si and SiO_2 at the storage stage already by the presence of the weak expressed maximum at 92 eV (Table 5.5).

With increase of the film's anneal temperature, the relative intensity of the elementary crystalline silicon maximum (~92 eV) is gradually reduced with the increase of the relative contribution of SiO_2 peaks. However, after anneal at 1100°C a weak crystalline silicon feature in spectrum (23%) is observed.

5.6.3 X-RAY PHOTOELECTRON SPECTRA

XPS investigations of the film's composition were performed by measuring the binding energy of silicon core levels Si 2p and oxygen O 1s. Results show that on the surface of the unannealed film (T_s = 250°C) only silicon oxide is observed, but with the oxidation degree lower than for silicon dioxide. This means that we have the SiO_x phase. The evidence of this fact is that the BE value of the Si 2p core level (102.6 eV) is intermediate between elementary silicon (99.5 eV) and silicon dioxide (103.3 eV) [46].

For the sample annealed at 1100°C XPS, data analysis reveals a considerable shift of Si 2p core level to high energies—up to 105 eV. At the surface of the annealed films, Si 2p BE appear to be 1–1.5 eV higher than for SiO_2 [46]. The fact that the binding energy for the O 1s core level at the same time slightly increased the shift of the Si 2p level could be explained only as an increase of the oxidation degree. Shifts of Si 2p and O 1s levels to higher energies were observed previously [47] in thin SiO_2 films obtained by sol-gel method and annealed in nitrogen and were connected with multisegmented structures formation from silicon-oxygen tetrahedrons.

In addition, XPS measurements of the same Si 2p and O 1s core levels were performed for T_s = 300°C sample after a 1100°C anneal with the use of layer-by-layer ion etching. According to these measurements, elementary silicon was detected after ion gun etching at about 60 nm depth (Figure 5.18).

Thus, XPS data indicate that in SiO_2:nc-Si/Si film, elementary silicon is found in the SiO_2 layer at the depth ≥60 nm (that is in a good agreement with USXES data, i.e., film upper layer < 60 nm is a dioxide layer).

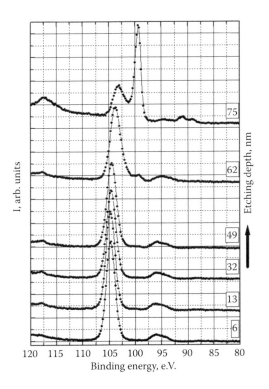

Figure 5.18 Si 2p XPS of SiO_2:nc-Si/Si (T_s = 300°C) after 1100°C subjected to layer by layer ion etching.

Arrays, hybrids, and core–shell

5.6.4 X-RAY DIFFRACTION INVESTIGATIONS

X-ray diffraction investigations confirm the presence of nanocrystalline silicon in SiO_x film with the average lateral size of nanocrystals calculated by Scherrer equation about 20 nm in the initial film and up to 60 nm in film annealed at 1100°C. At the same time nanocrystals have prevailed with an orientation parallel to the (111) plane of the substrate.

5.6.5 X-RAY ABSORPTION NEAR-EDGE STRUCTURE SPECTRA

XANES spectra in the range of the Si $L_{2,3}$ edge provide information about the distribution of local partial density of electron energy states above conduction-band bottom in ~5 nm depth layer [48]. For the investigated SiO_2:nc-Si/Si samples, XANES spectra are presented in Figure 5.19. XANES spectra taken with different grazing angles for reference samples of c-Si with the natural oxide on the surface and thermal SiO_2 60 nm film on single crystalline silicon substrate are presented in Figure 5.20. We use these reference spectra for comparison with XANES spectra of investigated SiO_2:nc-Si/Si structures. A total of 100–104 eV spectral features are related to elementary silicon, while features at E >104 eV are related to SiO_2.

   For the initial SiO_x film obtained with substrate temperature T_s = 250°C in the range of elementary silicon absorption edge 100–104 eV, a weak spectral feature is observed (Figure 5.19a — lower spectrum a). After a 900°C anneal, the XANES feature in this range appears to have more contrast, but it is inverted relative to the normal absorption structure and became lower than the background. With the increase of the annealing temperature, this "inversed" structure appears to be more expressed and still remains negative relative to the background; that is, the dip in the electron yield becomes more profound. Thus, in these films instead of the normal electron emission at hv ≥ 100 eV caused by the presence of elementary silicon formations, we observed anomalous yield in the energy range corresponding to elementary silicon structure ("inversed intensity"). In our previous work [48], it was shown that the presence of elementary silicon clusters formed in thermally grown SiO_2 films by silicon ion implantation led to normal Si $L_{2,3}$ absorption edge formation at ~100 eV.

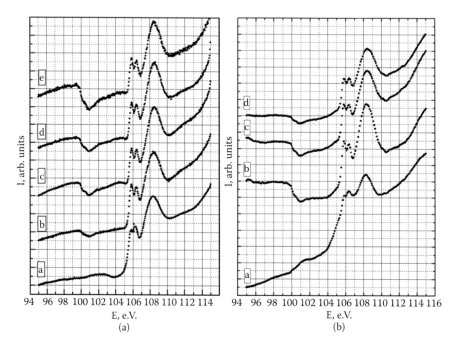

Figure 5.19 (a) Si $L_{2,3}$ XANES spectra of SiO_2:nc-Si/Si films. a–d: obtained with T_s = 250°C; a, T_a = 0°C; b, T_a = 900°C; c, T_a = 1000°C; d, T_a = 1100°C. e: film with T_s = 350°C and T_a = 1100°C. Synchrotron radiation grazing angle is 90°. (b) Si $L_{2,3}$ XANES spectra of SiO_2:nc-Si/Si film obtained with T_s = 250°C and T_a = 1100°C for different grazing angles of the primary beam: a, 10°; b, 30°; c, 60°; d, 90°.

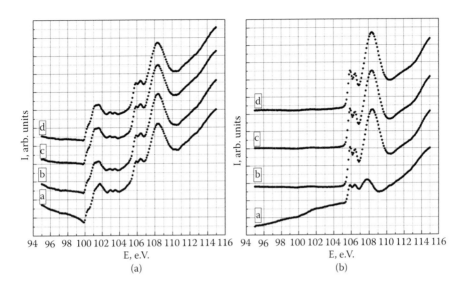

Figure 5.20 (a) c-Si XANES spectra obtained for different grazing angles of the primary beam: a, 10°; b, 30°; c, 60°; d, 90°. (b) SiO$_2$ (60 nm film) XANES spectra obtained for different grazing angles of the primary beam: a, 10°; b, 30°; c, 60°; d, 90°.

For the interpretation of this "inversed intensity" phenomenon, it should be taken into account the large value of the absorption coefficient in the ultrasoft X-ray range [49]; that is why for XANES registration the X-ray photoeffect electron quantum yield χ near the absorption edge is most frequently detected [50].

According to M. A. Rumsh et al. [50], electron quantum yield χ at the given grazing angle θ is proportional to the absorption coefficient μ, but at the same time it depends on reflection coefficient R:

$$\chi = \frac{[1 - R(\theta)]hc}{4E\lambda}\frac{\mu}{\sin\theta}, \tag{5.3}$$

where E is the average energy used for electron generation, λ is the wavelength, h is the Planck constant, and c is the speed of light. With grazing angles considerably greater than critical one for the total external reflection values R(θ) approaches zero and quantum yield dependence on the quantum energy repeats spectral dependence of μ. This is usually used for XANES measurements of various objects in the ultrasoft X-ray range.

Anomalous quantum yield spectra for the investigated films could be due to unusual behavior of the effective reflection coefficient R(θ) in these films. Because R(θ) should depend on grazing angle, we observed XANES spectral behavior for different θ. Figure 5.19b represents XANES spectra behavior with different grazing angles $\theta = 90°, 60°, 30°$, and $10°$ for the film obtained at T$_s$ = 250°C and annealed at 1100°C. As one can see, with the increase of θ, the absolute intensity of the inversed part (100–104 eV) of the XANES spectra first increases, appearing to be more expressed at the 30° grazing angle. And only at about $\theta = 10°$ the spectrum takes "normal" appearance but with a flat absorption edge and weakly expressed features typical for elementary silicon.

Anomalous quantum yield spectra and its behavior modification with the grazing angle decrease in the investigated SiO$_x$:nc-Si/Si films can be explained by diffraction or interference effects occurring in the considered wavelength range. This fact is not accounted for by Equation 5.3. For Bragg reflection appearance of photons with 100–105 eV energies in a considered system, the particles should be ordered—periodically or quasi-periodically. But the weak dependence of the inversed peak intensity with grazing angles argues against the proposition of Bragg diffraction at silicon nanocrystals. Moreover, the effect of XANES "inversed intensity" near C Kα edge was observed by our group previously [51] in the disordered system–layered amorphous SiC$_x$ films.

According to XPS data SiO$_x$ film up to a 60 nm depth corresponds to silicon dioxide. Under this layer, another layer containing a large amount of silicon nanocrystals is located. Let us assume that anomalous behavior of the X-ray external photoeffect quantum yield is caused by X-rays interference from the layer contained in nc-Si. In this case, estimation of the silicon nanocrystals thickness (d) providing interference path difference that is necessary for intensity, minimum formation can be made by the following equation:

$$\tfrac{1}{2}\lambda = 2d \sin\theta \qquad (5.4)$$

For the wavelength values from 11.9 to 12.4 nm in χ anomalous behavior range and grazing angles of 30°–90°, this equation makes $d = 3 – 6.2$ nm.

To explain electron yield anomalous behavior in the Si L$_{2,3}$ edge spectral range of elementary silicon, it should be taken into account that refraction coefficient n in the X-ray range is a little lower than 1 and determined by the following equation [52]:

$$n = 1 - i\alpha - \beta, \qquad (5.5)$$

where α and β are the atom's ability to absorb and scatter, respectively. In the E < 100 eV range, the photon energy is not enough for L$_{2,3}$ silicon-level ionization (as for elementary Si as for SiO$_2$). At the same time, the elementary silicon absorption coefficient is 0.3×10^5 cm^{-1} (hν = 99 eV [49]), and for SiO$_2$ this coefficient is one order of magnitude lower [53], so radiation with this energy passes all the structure up to the substrate.

Electrons that escaped into the vacuum and were registered in our experiments are generated by initial quantum beam as they are by a backscattered one. With hν < 100 eV, the absorption is not great; backscattered photons are forming in the whole film volume and electron yield is proportional to general photon intensity near the surface (≤ 5 nm).

At the $100 \leq$ hν ≤ 104 eV, a part of the photons begins to be absorbed by silicon nanocrystals in the SiO$_2$ layer volume and film appears to be optically inhomogeneous in the considered energy range. In this case the backflow will be determined by reflection process from nanocrystals.

Earlier [54] noticeable XANES inversion effects as a result of interference of the X-rays reflected from the layer boundaries were observed near the Si L$_{2,3}$ edge for the LiF/Si/LiF multilayer structure with a thin silicon layer. Therefore, one can assume that silicon nanocrystals in SiO$_2$ layer represent nanosized structures with two boundaries SiO$_2$/Si/SiO$_2$ and absolute reflectance coefficient R is determined by the well-known equation:

$$R = \frac{R_{12} + R_{21}e^{2i\Delta}}{1 + R_{12}e^{2i\Delta}}, \qquad (5.6)$$

where R_{12} is the SiO$_2$/Si boundary reflection coefficient and R_{21} is for the Si/SiO$_2$ boundary. In this case, n_1 is the refractive coefficient of SiO$_2$, and n_2 is the refractive coefficient of silicon nanocrystals and phase shift $\Delta = \dfrac{2\pi d n_2}{\lambda} \cos\theta$ depends on nanocrystals size d and their refractive coefficient n_2. In a dependence of nanocrystal size, there could appear effective reflection of electromagnetic wave and in this case the radiation takes part in surface photoemission. On the other hand, considerable decay of the reflected electromagnetic wave as a result of the interference could lead to a general reduction of the overall electromagnetic field intensity in the surface layers. In the latter case the decrease of the electron yield relative to hν < 100 eV and hν > 104 eV should be observed as the result of photoemission reduction. According to V. Paillard et al. [3], a refractive coefficient value of nanocrystals material will fluctuate in the 100–104 eV range because of the absorption coefficient variation and corresponding small fluctuations will appear in phase shift Δ and the final decay of the electromagnetic field. We have already noted that for radiation intensity reduction as the result of interference in thin nc-Si layer of SiO$_2$ matrix, the thickness of this layer should be 3.6–6.2 nm. This estimation of the particle sizes is in good agreement with those made by PL maximum position (1.7 eV), observed in our samples (Figure 5.17). According to PL peak dependence on nanoparticles sizes [55], this position corresponds to the average particle size of ~4 nm. This PL interpretation (with maximum at 1.7 eV) is in agreement with PL data for silicon nanocrystals in the SiO_2 matrix [56]. At the same time, the obtained nanocrystal size estimation contradicts that of X-ray diffraction data. The latter gives sizes of ~20–60 nm in a dependence of anneal temperature. This contradiction could be eliminated if one can assume that in our case silicon nanocrystals are formed as flat disks with greater lateral dimensions (20–60 nm) and small thickness ~5 nm. In this case, due to nanosize thickness, one-dimensional constraint of the charge carriers takes place in nc-Si particle, providing PL of the considered structure and the conditions for wave interference, with λ = 12.4 nm corresponding to the main Si $L_{2,3}$ absorption edge energy.

In conclusion, it should be noted that formation and evolution of the light-emitting layers obtained from SiO powder films has a complicated nature. First, immediately after deposition, there are amorphous and nanocrystalline phases of elementary silicon in the investigated film. Second, under anneal, the redundant silicon is spent in the process of phase decomposition for the formation of 3–5 nm thickness flat Si nanocrystals preferably oriented parallel to (111) substrate and luminescent in the 700–800 nm range. At the same time, lateral sizes (20–60 nm) of the particles are by one order of magnitude greater than their width. "Inversed intensity" phenomenon observed in silicon XANES spectra is caused by synchrotron radiation interference on flat silicon particles in the SiO_2 matrix.

5.7 SYNCHROTRON INVESTIGATION OF THE MULTILAYER NANOPERIODICAL Al_2O_3/SiO/Al_2O_3/SiO…Si STRUCTURES FORMATION

The possibility of the formation of thin layers with the well-expressed PL in the framework of silicon technologies is of a great interest for technologists and researchers working on the field of SiO_2 films containing silicon nanoparticles. The redundant silicon in the silicon dioxide matrix can be obtained by annealing the SiO_x films formed from SiO powder [43–45,57]. Under the high temperature of annealing these films, SiO_x decomposition to Si + SiO_2 should take place with the simultaneous self-organization of silicon atoms in nanoclusters and/or nanocrystals. That is why the formation of structures containing periodical massives of the silicon nanoclusters/nanocrystals in the dielectric layers attracts serious attention. The problem of silicon nanoparticles massives formation is a rather complicated technological task mostly because of the production of silicon nanoparticles with fixed sizes. One of the possible solutions here is the formation of the multilayered nanoperiodical structures (MNS) with fixed thicknesses of nanolayers containing silicon nanoparticles located between nanolayers of different materials (Al_2O_3, for example). As the layers containing Si nanoparticles, one could take silicon oxide nanolayers decomposed under the high-temperature annealing. In the latter case, required sizes of silicon nanoparticles are specified by two factors: the thickness of the silicon oxide nanolayers and the presence of other material limiting nanolayers.

The MNS formation study and their following annealing necessitate the use of nondestructive methods sensitive to the phase composition and the surface structure. As is known, the XANES technique provides information about the local partial density of free electron states near the conduction-band bottom. This technique is sensitive to the local atomic environment (for the specific kind of atom—Si or Al in our case) in surface nanolayers. In this work, the results of XANES spectroscopy study with the use of synchrotron radiation are presented for the MNS Al_2O_3/SiO/Al_2O_3/SiO…Si.

The investigated MNS Al_2O_3/SiO/Al_2O_3/SiO…Si samples were formed with the use of the layer-by-layer deposition of SiO and Al_2O_3 nanolayers on to silicon (100) substrates by the resistive evaporation and the electron beam evaporation, respectively. The accelerating voltage of the electron beam was 6 kV. The residual pressure was less than 10^{-5} Torr, substrates temperature was 150°C, and substrate degassing was performed at ~200°C. The thickness of Al_2O_3 layers in all structures was 5 nm while thicknesses of SiO layers were 4, 7, and 10 nm with nine of the total layers pair. The annealing was performed at 500°C, 700°C, 900°C, and 1100°C in the atmosphere of dry nitrogen for 120 min.

XANES investigations near Si and Al $L_{2,3}$ core levels were performed at Mark V beamline [58] of SRC synchrotron radiation facility (University of Wisconsin-Madison, Stoughton, WI). The operating vacuum in the XAB experimental chamber was 10^{-10} Torr, and instrumental broadening was about 0.05 eV. The sample current measurement technique was used to detect XANES spectra in the total electron yield (TEY) mode. According to work [59] performed at the same synchrotron radiation (SR) facility, the depth of analysis was ~5 nm for silicon $L_{2,3}$ XANES. All of the MNS samples were measured at SR grazing angle $\theta = 30°$. References were measured at $\theta = 90°$.

As a reference for Si $L_{2,3}$ XANES measurements, we used the single-crystalline silicon plate covered by the native oxide, the thermally grown silicon dioxide thin film of 10 nm thickness, and the amorphous silicon sample obtained by PCVD from atmosphere containing silane. Let us consider in detail the position and the structure of XANES features in reference spectra of absorption presented in Figure 5.21a. For the single-crystalline silicon, one can observe the rather abrupt edge with the characteristic "step" as well as two clearly expressed double maxima in the range of 101.2–101.7 eV and 102.2–102.7 eV. The distance between the split peaks corresponds to the spin-orbital splitting of the core Si $L_{2,3}$ level. More simple structure with the "step" and the single maximum is peculiar to the Si $L_{2,3}$ edge of the amorphous silicon. The latter can be due to the smoothing of the density of states as a result of the disordering in an amorphous material [60]. Furthermore, at $E > 104\mathrm{eV}$ in the structure of the Si $L_{2,3}$ edge of the amorphous silicon, just as in the case of the single-crystalline Si, one can observe spectral features of SiO_2. The inlay in Figure 5.21a demonstrates XANES data in the energy range of 100–105 eV. This energy range is peculiar to the elementary silicon absorption edge according to F. C. Brown et al. [49].

Comparison of XANES Si $L_{2,3}$ data for references and Si $L_{2,3}$ XANES for the initial SiO thin 18 nm film is given in Figure 5.21. One can indicate at the elementary silicon formation after the thermal anneal. Elementary silicon absorption edges in the inlay of Figure 5.21 represent the presence of the elementary silicon in the 5 nm surface layer of the annealed initial film with temperatures 900°C and 1100°C. So one can assume the SiO decomposition as Si + SiO_2. Also it should be noted that Si $L_{2,3}$ XANES for the initial

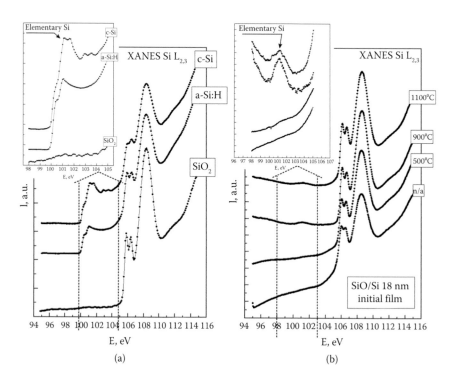

Figure 5.21 Si $L_{2,3}$ XANES spectra of the reference samples c-Si, a-Si, and thermally grown SiO_2 (a), initial 18 nm SiO film annealed at different temperatures (b). Insets: the elementary silicon absorption edges.

18 nm SiO film is close to the reference of SiO$_2$, most likely because of the total oxidation of the ~5 nm surface layer of the sample.

XANES spectra for the Al$_2$O$_3$/SiO = 5 nm/4 nm MNS samples are shown in Figure 5.22a. First, the low intensity for the XANES signal can be noted for all the structures annealed up to the temperatures lower than 1100°C. This is because the upper layer is the 5 nm Al$_2$O$_3$. Because this upper layer thickness is comparable with the electrons escaped (from the SiO layer) depth for XANES data, a considerable decrease of the spectra intensity takes place. Nevertheless, the main spectral features peculiar to the silicon oxide (hν ~ 106 and hν ~ 108.5 eV) are observed. Moreover, the wide maximum in the elementary silicon X-ray absorption energy range (hν = 100–102 eV) is detected. The minor increasing of the Si L$_{2,3}$ spectrum intensity mostly in the "oxide" part is taking place for the sample annealed under 1100°C. It can be connected with the partial cracking of the upper layer under the high-temperature annealing. The detailed elementary silicon absorption part is presented in the inlay for the Figure 5.22a. The fine structure of the absorption edges peculiar to the ordered (crystalline) silicon is absent if compared with Figure 5.21a. The latter fact means that most of the silicon atoms formed under SiO → Si + SiO$_2$ decomposition and situated between the Al$_2$O$_3$ layers are in the disordered state for all the samples annealed with temperatures up to 1100°C.

According to the registered Si L$_{2,3}$ data, the increase of the SiO layer thickness up to 7 nm had no influence on the elementary silicon formation (Figure 5.22b). At the same time this increasing of the SiO layer thickness considerably changed the "oxide" part of the spectra (Figure 5.22b). For the unannealed structure as well as for the annealed ones at temperatures up to 900°C under the synchrotron radiation grazing angle 30° instead of the maximum at the energy of 108 eV (Figures 5.21 and 5.22a), the intensity dip is observed. We named this phenomenon the "inversed intensity." As this inversion observation cannot be determined by the absorption, one can assume that according to Equation 5.3 for the quantum yield, this phenomenon can be connected with the specific behavior of the reflection coefficient R

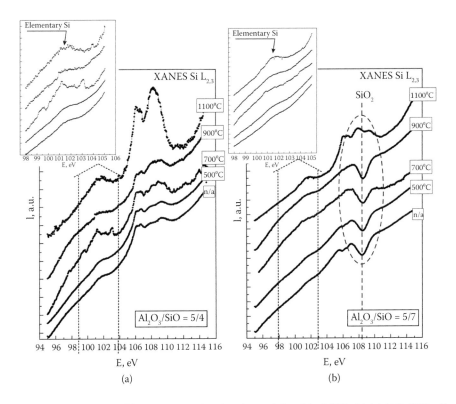

Figure 5.22 Si L$_{2,3}$ XANES spectra of the investigated surface layers (~5 nm) for MNS with (a) Al$_2$O$_3$/SiO = 5 nm/4 nm ratio and (b) Al$_2$O$_3$/SiO = 5 nm/7 nm ratio with the inversed intensity phenomena outlined at energies ~108,5 eV. Insets: the elementary silicon absorption edges.

Arrays, hybrids, and core–shell

at the multilayered structure. Therefore, we registered XANES spectra under the different synchrotron radiation grazing angles: 10°, 60°, and 90° (Figure 5.23). These data did not reveal any unexpected intensity inversions. So this very specific feature (Figures 5.22b and 5.23) was observed only under a 30° grazing angle. Obviously this kind of XANES spectra behavior in the 5/7 structures cannot be connected with the λ = 11.4 nm (hν = 108.5 eV) radiation interference at the double-layered Al$_2$O$_3$/SiO (see inlay for the Figure 5.23) since the interference must be observed under wide-angle range [54,61]. On the other hand, the well-expressed angle dependence can be observed under the radiation diffraction on the multilayered structure with the 12 nm parameter (Figure 5.22b and inlay of the Figure 5.23). According to the Bragg law, $n\lambda$ = 2dsinθ the estimation of the parameter d under λ = 11.4 nm and θ = 30° gives values d = 11.4 nm that is close to the technology MNS parameter of d = 12 nm for the 5/7 samples.

The absence of the inversion phenomenon for the 5/7 MNS sample annealed at the 1100°C (Figure 5.22b) is most likely the evidence of the deformation or the destruction of the layered structure because of the highest annealing temperature. This structural change may prevent the Bragg diffraction observation phenomenon. Under this annealing temperature the elementary silicon concentration is noticeably increasing as well as the SiO decomposition contribution.

For the Al$_2$O$_3$/SiO = 5 nm/10 nm MNS samples, the same result of nanoclusters formation was obtained for observation of the elementary silicon absorption edges. At the same time the oxide part of Si L$_{2,3}$ XANES for these MNS appears closer to reference SiO$_2$ due to the highest value of silicon oxide nano-layer thickness (Figure 5.23).

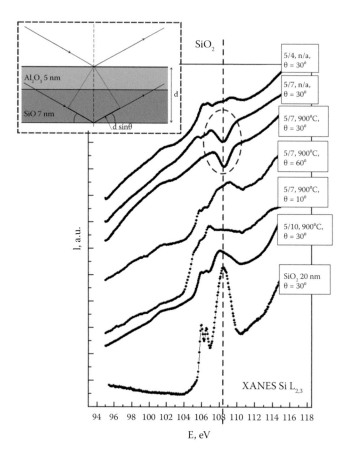

Figure 5.23 Si L$_{2,3}$ XANES spectra of the investigated surface layers (~5 nm) for the MNS with different Al$_2$O$_3$/SiO ratios and registered at different synchrotron radiation grazing angles θ as well as XANES Si L$_{2,3}$ for 20 nm reference SiO$_2$ film. Inset: schematic pattern of the outlined inversed intensity phenomena explanation for the MNS with Al$_2$O$_3$/SiO=5 nm/7 nm ratio at θ = 30°.

Arrays, hybrids, and core–shell

Al $L_{2,3}$ XANES investigations for Al_2O_3 layers of all the MNS were additionally performed. Figure 5.24 represents the Al $L_{2,3}$ data registered for the 5/4 MNS as well as reference data for the natural oxide of the metallic Al [62] and for the aluminum silicate sillimanite s-Al_2SiO_5 [63]. The analysis of this spectra showed that for the unannealed structure and for ones annealed at 500°C, the observed features are peculiar to Al_2O_3. Under annealing at 700°C, the unknown additional spectral feature A‴ is observed showing changes in the aluminum oxide interlayer electronic structure. Under the following increase of the annealing temperatures, another additional feature A″ appeared at the energies ~78.3 eV. According to C. Weigel et al. [63], this feature is peculiar to the aluminum silicate formation (Figure 5.24) by its energy position. It means that under the high annealing temperatures, the interaction between the aluminum and the silicon oxides leads to the additional phase formation at the layers boundary. Also it can be clearly seen that peaks marked as B are not essentially changed, indicating the absence of the destruction for the Al_2O_3 under high annealing temperatures. This fact confirms the formation of aluminum silicates at the layers' boundary.

Thus, the possible silicon nanocluster formation is shown in surface layers of multilayered nanoperiodical $Al_2O_3/SiO/Al_2O_3/SiO…Si$ structures under their annealing at the temperatures from 500°C to 1100°C. The possibility of the aluminum silicate formation in the MNS layers boundary is revealed as the results of the interaction between the aluminum and the silicon oxides under high annealing temperatures. The phenomenon of the Bragg diffraction is shown as the result of the synchrotron radiation interaction with the layered MNS structure.

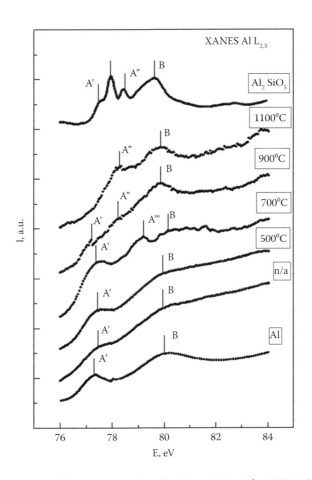

Figure 5.24 Al $L_{2,3}$ XANES spectra of the investigated surface layers (~5 nm) for MNS with Al_2O_3/SiO = 5 nm/4 nm ratio and reference data for the natural oxide of the metallic Al [61] and for the aluminum silicate sillimanite s-Al_2SiO_5 [54].

Arrays, hybrids, and core–shell

5.8 X-RAY ABSORPTION NEAR-EDGE STRUCTURE ANOMALOUS BEHAVIOR IN STRUCTURES WITH BURIED LAYERS CONTAINING SILICON NANOCRYSTALS

Silicon nanocrystals (nc-Si) formation in dielectric matrix that is luminescent in visible red and near infrared ranges (1.4eV–1.8 eV) is one of the modern directions for the opto- and nanoelectronics [64]. The luminescent band shift to the short wavelengths with overlapping of the whole visible range can extend application possibilities of silicon structures for different optoelectronics devices. In several papers (see, for example, Refs. [65–69]), it was shown that simultaneous implantation of silicon and carbon ions into SiO_2 films leads to photoluminescence in the range from near infrared to ultraviolet wavelengths due to the formation of silicon nanocrystals as well as carbon and silicon carbide nanocrystals. In the study of Belov et al. [70], the luminescence extension to the visible and ultraviolet spectral range was realized by the carbon ion implantation into SiO_x films on silicon substrates where the nanocrystalline silicon phase was formed as the nonstoichiometric oxide decomposition $SiO_x \rightarrow Si + SiO_2$ under high-temperature annealing. By means of the X-ray photoelectron spectroscopy technique (XPS), it was found [69, 70] that silicon atoms with the Si2p core level binding energy values close to crystalline SiC (100.8 eV) [71] were located over the 70–170 nm surface layers of investigated samples. But the absence of the SiC sharp peak in the XPS data mentioned earlier as well as the presence of low-oxidation-degree oxides at the same depth with silicon binding energies of about 100.5–101.2 eV [72] argues with the proposition of silicon carbide formation. In the present work we attempted to investigate the same samples by means of the XANES technique that is highly sensitive to the local surroundings of given atoms. Along with the confirmation of the SiC phase formation, we detected spectral features that are demonstrating the influence on XANES spectra of chemical bonds of those Si atoms that are located much deeper than the probing depth of the technique used. These spectral features are of common interest due to the application of the XANES technique in the nanostructured systems studies.

Nonstoichiometric silicon oxide films (SiO_x) with the thickness of about 300 nm were formed on KDB-0.005 (111) and KDB-12 (100) silicon substrates with the use of the technique described in [70]. Part of the samples was subjected to the ion implantation of carbon. After that, irradiated samples as well as initial ones were annealed at 1100°C in the atmosphere of nitrogen for 2 hours. Implantation doses (6×10^{16}, 9×10^{16} и 1.2×10^{17} cm^{-2}) and carbon ions energy (40 keV) were the same as in [70].

Samples were investigated by the XANES technique with the use of the Synchrotron Radiation Center's Aladdin storage ring synchrotron radiation (University of Wisconsin-Madison, Stoughton, WI). Spectra registration near the Si $L_{2,3}$ absorption edge with 0.05 eV instrumental broadening was performed at the Mark V beamline. Spectra registration near the K absorption edge of silicon was performed at the DCM beamline with 0.9 eV instrumental broadening. The sample drain current detection was used for XANES spectra registration under synchrotron radiation photon energies variations. Herewith the spectral dependence of the Auger and photoelectron yield is detected from the sample's surface. In case of the common XANES technique, this yield is proportional to the X-ray absorption coefficient in a thin surface layer with the thickness determined by the atoms' energy structure and their surroundings [50]. As will be shown later under nanostructures investigation, the spectrum shape can depend on the structure and composition of layers that are deeper than the XANES technique's regular probing depth, from which secondary electrons are not escaping directly.

To analyze the chemical state of silicon in the investigated structures formed before and after carbon ion implantation as well as annealed ones, let us consider XANES data presented in Figure 5.25 in the energy range of the silicon K edge. For the comparison, spectra of "reference" samples are presented in Figure 5.26: silicon plate, a single crystal of cubic silicon carbide β-SiC, and amorphous films of SiO_2 with 10 nm and 100 nm thickness obtained by silicon thermal oxidation. The layer thickness from which photoelectrons and Auger electrons are emitting in the considered energy range is about 65 nm according to Kasrai et al. [59]. Besides the main elemental silicon maximum under 1841 eV in spectra of reference silicon samples, we observed the 1847 eV feature that is connected with the silicon oxide presence.

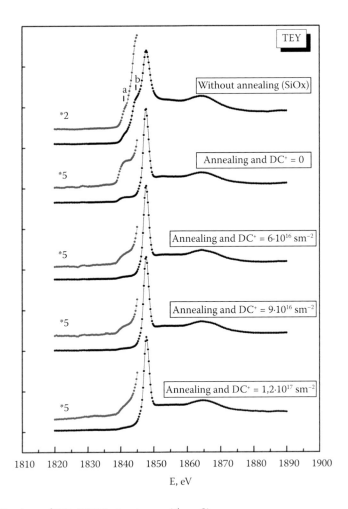

Figure 5.25 XANES K edges of SiO$_x$/Si(111) structures with nc-Si.

For the thin 10 nm SiO$_2$ film, elemental silicon peak observation (feature at the Figure 5.26) is caused by c-Si substrate signal that is not detectable for the 100 nm SiO$_2$ film.

The main peak for the investigated unannealed SiO$_x$ films is caused by the SiO$_2$ phase that follows from the comparison of spectra given at the Figures 5.25 and 5.26. The shoulder (feature) "a" peculiar to the K absorption edge of elemental silicon (1841 eV) is observed at the low energy range of the given spectra, indicating at the c-Si phase inclusions in considered films. Relatively intensive shoulder "b" at 1844.2 eV is observed as well. We associate feature "b" observation with silicon atoms' presence in the initial film with intermediate oxidation degree since the energy position of this shoulder is close to the mean value for elemental Si and SiO$_2$ phases' maxima.

Relative intensities for the low-energy shoulders "a" and "b" mentioned earlier are noticeably decreased after high-temperature annealing of films. This is the evidence of elemental silicon and nonstoichiometric oxide content decreasing within the analyzed layer. It was established earlier [61] that the thermal treatment equal to one considered in the present work led to the stoichiometric SiO$_2$ formation in the 60 nm surface layer caused by the residual oxygen in the annealing atmosphere.

After C$^+$ ion implantation followed by annealing, the more noticeable decrease of the elemental silicon feature (Figure 5.25) is observed. This is in a good agreement with XPS results that demonstrated elemental silicon depth of occurrence $\geq$60 nm for annealed SiO$_x$ films without carbon implantation [73] and >70 nm

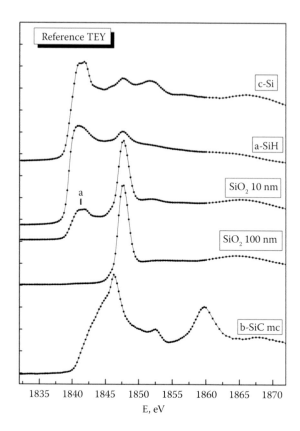

Figure 5.26 XANES K edges of reference samples.

after carbon ion implantation according to [70]. According to L. Pavesi et al. [64], the increase of the carbon implantation dose leads to elemental silicon content decrease in the analyzed layer that is caused by reaction with implanted atoms. The shape changes of the low-energy spectra tail (under energies lower than 1843.5 eV) should be noted as well with the increase of the implantation dose. Instead of the plateau at 1840–1843.5 eV that is well expressed in the nonimplanted sample, the spectral shape is transformed into the nearly linear slope. Moreover, the spectral part intensity at ~1843.5 is increased under ultimate implantation dose (1.2×10^{17} cm^{-2}). This transformation can be connected with the silicon carbide formation because SiC XANES at the Si K edge is maximum at 1846 eV and continuous shoulder falling down with energies up to 1840 eV (Figure 5.26).

Let us proceed to the results of silicon $L_{2,3}$ absorption spectra investigations. Figure 5.27 represents XANES spectra for the investigated structures formed on (111) substrates before and after ion implantation of carbon, and Figure 5.28 represents XANES Si $L_{2,3}$ data for measured "references": c-Si, SiO_2 and β-SiC. At the Si $L_{2,3}$ absorption edge, XANES (TEY) sampling depth is known as about 5 nm according to [59]. Because "reference" samples of c-Si and β-SiC had a natural SiO_2 layer on their surface, the observed spectral features at 106 and 108 eV (Figure 5.28) are caused by this oxide layer. As it can be easily seen from Figure 5.27 in the case of the initial film, we observed spectrum that is typical for the stoichiometric SiO_2. As it was shown earlier for the Si K edges, this is the evidence of the near-surface thin SiO_x layer oxidation to SiO_2 that took place while the samples had been stored in the ambient air.

After annealing the nonimplanted film in the energy range starting at 100 eV (that corresponds to the absorption edge of the elemental silicon), the abnormal "inversed" spectrum shape is observed: instead of the usual sample drain current rising with the increase of photon energies, the decrease of spectrum relative intensity is observed, forming the "dip" in the 100–104 eV range. The same behavior of the spectrum shape was observed earlier [61] for sputtered and annealed SiO_x films. In [61], we explained

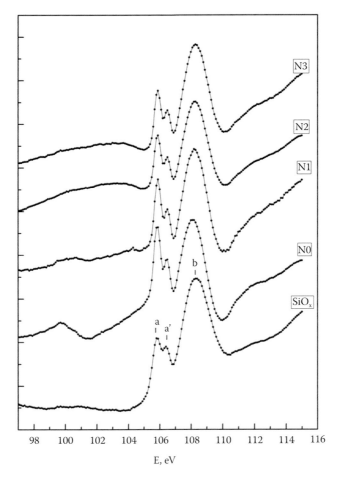

Figure 5.27 Si $L_{2,3}$ XANES for SiO_x/Si(111) structures with nc-Si. SiO_x, the initial film; N0, the film without the implantation and after the anneal; N1, the film after the implantation with the dose of 6×10^{16} cm^{-2} and the following anneal; N2, the film after the implantation with the dose of 9×10^{16} cm^{-2} and the following anneal; N3, the film after the implantation with the dose of 1.2×10^{17} cm^{-2} and the following anneal.

this phenomenon by actual proportionality of the electron yield from the sample that is registering in the XANES technique to the X-rays beam electromagnetic filed in the near-surface layer from which electrons are emitted. The electromagnetic filed intensity in this layer is determined not only by the X-ray quanta flux falling but by the backscattered one from deeper layers. In the case of nanosized inclusions, presence in the considered layer that is different from the matrix by its structure and composition, the backscattered beam amplitude and phase, in their turn, are determined by the following processes: the photons' elastic scattering, the absorption of the falling and backscattered beams, the reflection (generally multiple) on internal heteroboundaries, and the possible interference. In the case of the nanocomposite structure, any of these factors can be found in a strong dependence on the elemental and phase composition, the nano-sized inclusions concentration, size and morphology, and even more their depth distribution. Under those quanta energies where ionization of the certain atomic shell appears to be possible (where sample drain current rising is observed in the case of homogenous single-phase samples), in nanocomposite systems, the TEY rise is possible as well as the TEY decrease depending on the relative contribution of different factors (mentioned earlier). Apparently in the case of the investigated system, the main role of the Si $L_{2,3}$ range dip formation plays weakening of quanta back flow near the surface as the result of the initial and back-scattered beam absorption intensification by silicon nanocrystals that are located under the surface SiO_2 layer (with energies enough for Si $L_{2,3}$ shell ionization). Moreover, in the XANES dip formation for the

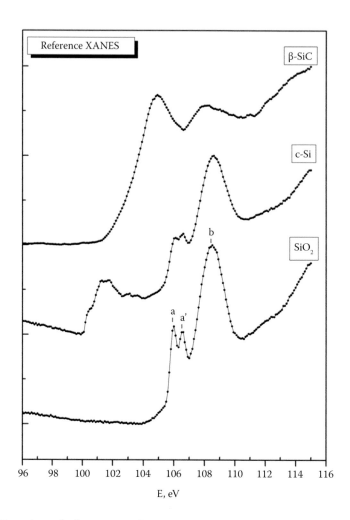

Figure 5.28 XANES $L_{2,3}$ edges of reference samples.

absorption edge range of atoms that are forming nanoparticles, the phenomenon of the anomalous elastic scattering [74] can make a significant contribution. In the case of the photon energy coincidence with core level ionization energy, this phenomenon's cross section has a sharp minimum.

The XANES spectra inversion disappeared in the case of the samples obtained on (111) substrates and measured after the carbon implantation with and without annealing (Figure 5.27). Small traces of the inversion remained under the lowest carbon dose as the small feature observed at the 99–101 eV range. The inversion disappearance indicates composition changes (and optical properties as well) at the films phase. Herewith the elemental silicon that is contained in nanocrystals is bonding with carbon.

Let us turn to the XANES spectra analysis for films formed on silicon substrates with the (100) orientation given in Figure 5.29. For C⁺ implanted samples, these spectra are different from the ones taken for the films obtained on the (111) silicon substrates in the 100–105 eV energy range. As it was demonstrated for (111) substrates under the lowest C⁺ implantation dose (6×10^{16} cm⁻²), the inversion phenomenon connected with the elemental silicon formation disappeared in the 100–105 eV energy range with only some traces left at ~100 eV as a kink. But in contrast to the (111) substrate orientation case, the "dip" (inversion) is observed in the 105–112 eV energy range. The peaks observed in this range for reference samples and the initial SiOx film formed on the silicon substrate are caused by Si bonds with oxygen in SiO_2. With the C⁺ implantation dose increasing, this inversion becomes less expressed and under the 1.2×10^{17} cm⁻² dose the spectra get their normal shape in the considered energy range. Under the 6×10^{16} cm⁻² carbon implantation dose, the inversion appears to be less expressed; under a decrease of the radiation grazing angle,

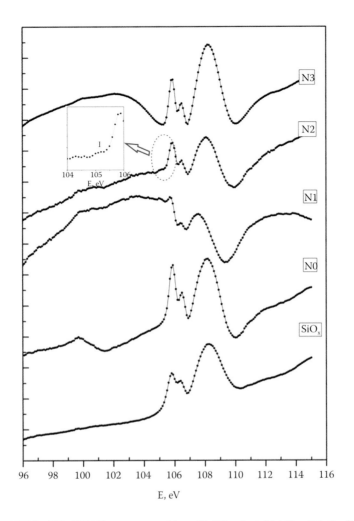

Figure 5.29 Si $L_{2,3}$ XANES for SiO_x/Si(100) structures with nc-Si. SiO_x, the initial film; N0, the film without the implantation and after the anneal; N1, the film after the implantation with the dose of 6×10^{16} cm^{-2} and the following anneal; N2, the film after the implantation with the dose of 9×10^{16} cm^{-2} and the following anneal; N3, the film after the implantation with the dose of 1.2×10^{17} cm^{-2} and the following anneal.

the spectrum shape is getting close to the "regular" one (Figure 5.30). Because the variation of the radiation grazing angle leads to a variation of the effective depth for radiation interaction with the films material, the latter observation confirms that the observed spectrum inversion phenomenon is connected with the structural and phase condition of films layers that are placed outside the oxidized one formed as the result of the anneal. It should be noted that for the sample with the lowest implantation dose under all grazing angles used for Si $L_{2,3}$ spectra registration (Figures 5.29 and 5.30), the noticeable feature is observed under quanta energy ~105 eV that corresponds to the silicon carbide spectrum main maximum position (Figure 5.28). This is more evidence of the latter compound presence in studied SiO_2 film. The spectral feature appears in a smaller degree for the sample with the greater carbon implantation dose of 9×10^{16} cm^{-2} (Figure 5.29). This can be caused by the predominant formation of the carbon inclusions instead of the SiC.

By the same method, we used earlier to explain the 100–105 eV spectral feature inversion phenomenon observation, the inversion in the 105–112 eV range is caused by nanostructure peculiarities that affect the backscattered beam behavior [75]. X-ray quanta absorption with energies relevant to the Si $L_{2,3}$ edge of silicon atoms bonded with oxygen and carbon leads to the attenuation of backscattered X-ray photons, which weakening electrons yield intensity in its turn and registering as the "dip" in the spectrum. It should be noted that reference spectra extremes that are caused by β-SiC phase are placed in the same range as

Arrays, hybrids, and core-shell

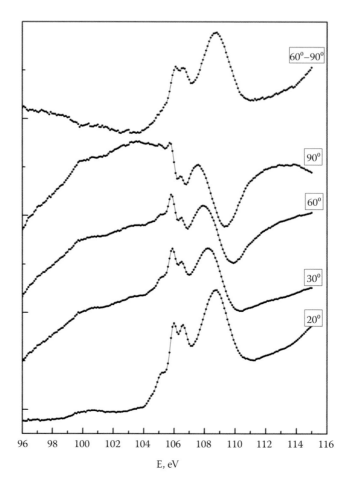

Figure 5.30 Si $L_{2,3}$ XANES spectra for the sample N1 (the implantation dose 6×10^{16} cm^{-2}) formed on the (100) substrate registered at different grazing angles and the difference spectrum for 60° and 90° registration.

silicon dioxide ones (see Figure 5.29). The important role of deep layers in the observed phenomenon is visually demonstrated by the difference curve for the spectra registered under 90° and 60° grazing angles (Figure 5.30). This curve almost coincides with the "regular" SiO_2 absorption spectrum (Figure 5.28). This fact is connected to the low contribution of the depths with formed nanoparticles into the difference spectra.

The fact that we did not observe the spectrum inversion anytime we registered one from nanostructures shows that the inversion can take place only under certain criteria that are connected with structural-phase conditions of the backscattering X-ray quanta layer. One of these criteria can be the nanosized inclusions comparability to the distance between them. For example, we did not observe any inversions near the Si $L_{2,3}$ absorption edge in [48], where studied SiO_2:nc-Si structures were obtained under the Si^+ ion implantation into SiO_2. In latter structures nc-Si volume ratio was relatively low (~10%), so distances between nanocrystals were quite long. Besides, silicon nanocrystals in SiO_2:nc-Si structures were formed not only in films depths but in the near-surface layers so the absence of the intermediate "pure" SiO_2 layer took place. Thus, specific criteria that lead to the inversion phenomenon are the subject for more detailed studies.

What is the reason for such spectra behavior in dependence on substrate orientation? In [76], we demonstrated that in similar samples, silicon nanocrystals are not oriented chaotically and their predominant orientation coincides with the substrates used. Apparently, crystal orientation affects the optical properties of the considered nanostructures due to the anisotropy of optical constants that lead to intensity differences that are observed for backscattered X-rays.

Arrays, hybrids, and core–shell

Our results reveal that under XANES data interpretations for nanostructured systems, it is necessary to consider the contribution of the (falling) direct X-ray beam as well as the backscattered one. On the one hand, this fact makes spectra interpretation more complicated. On the other hand, this fact creates additional diagnostic possibilities for the structure of such systems and morphology analysis by the nondestructive XANES technique. The practical realization of these opportunities requires additional research.

Obtained results confirm the formations of silicon carbide nanosized inclusions under the carbon ion implantation of SiO_x films.

5.9 SPECIFIC FEATURES OF THE ELECTRONIC AND ATOMIC STRUCTURES OF SILICON SINGLE CRYSTALS IN THE ALUMINUM MATRIX

It is widely known that the properties of nanostructured silicon substantially differ from those of the bulk material. In particular, nanostructured silicon exhibits luminescent properties. For example, at room temperature, visible photoluminescence is observed in porous silicon [7] and in dielectric films of silicon oxide or nitride containing silicon nanocrystals [77,78]. In recently published works, nanostructured silicon was produced by magnetron evaporation of a target consisting of at least 55% of aluminum and at most 45% of silicon [79]. The selective removal of aluminum makes it possible to obtain nanostructured silicon in which the sizes of silicon particles depend on the silicon content in the initial aluminum matrix. The use of such nanostructured silicon as a material for lithium ion accumulators makes it possible to increase the capacity and the number of recharging cycles as compared to those of accumulators with a graphite anode and to avoid the destruction of the sample after a large number of recharging cycles [80]. We studied two series of samples obtained by the deposition of an aluminum plus silicon film onto a substrate of single-crystal silicon (111) by magnetron evaporation of a complex target. In the first series of samples, we used a target consisting of 45 at %Si and 55 at %Al, and in the second series, 30 at %Si and 70 at %Al. The film thickness was on the order of 0.5 μm. The selective removal of aluminum was performed in orthophosphoric acid at a temperature of 50°C [79,80].

The morphology of the surface layer of the nanocomposite before and after the etching was examined with a JEOL JSM_6380LV scanning electron microscope. The phase composition of films and the mean size of silicon nanocrystals were determined from X-ray diffraction data obtained on a PANanalytical Empyrean diffractometer (Cu$K\alpha$ radiation). The specific features of the electron energy distribution in the valence band of the composites were studied using the emission spectra obtained on a RSM-500 ultrasoft X-ray spectrometer monochromator [81,82]. The electron energy distribution in the conduction band was studied using XANES spectra obtained at the SRC synchrotron (University of Wisconsin–Madison, Stoughton, WI).

According to the electron microscopy data (Figure 5.31a), the surface of the initial structure of the first series has inhomogeneities with the sizes of 30–40 nm. The removal of aluminum leads to a noticeable change in the morphology of the composite, which manifests itself in the transition to a coral-like structure with the characteristic diameter of elements of 25–30 nm (Figure 5.31b). The diffractometric analysis of the Al–Si composite revealed the presence of broadened reflections from both the aluminum and silicon phases in the initial film. After the etching, reflections of pure aluminum disappear (Figure 5.32). From the broadening of the reflection of silicon (220), the mean sizes of nanocrystals were calculated. Estimation of the mean sizes of nanocrystals give the values of 25 and 20 nm for the first and second series (i.e., containing more and less silicon), respectively. The obtained experimental X-ray emission Si $L_{2,3}$ spectra of samples are presented in Figure 5.33. For comparison, the spectrum of bulk crystalline silicon and the spectrum of silicon theoretically calculated by the method of orthogonalized plane waves (OPW) in [83] are shown. The experimental spectra of the Al–Si composite resemble in shape the spectrum of crystalline silicon: they are observed in the energy range from 80 to 100 eV and have two clear maxima at the same emission energy (89.6 and 92 eV). However, there are certain distinctions. The spectra of the samples after the removal of aluminum slightly differ from the spectrum of the single crystal (Figure 5.33): they have a higher intensity in the region 94–96 eV. The spectra of silicon in the aluminum matrix noticeably differ in the whole photon energy range of 82–92 eV: in the photon energy range of 82–86 eV

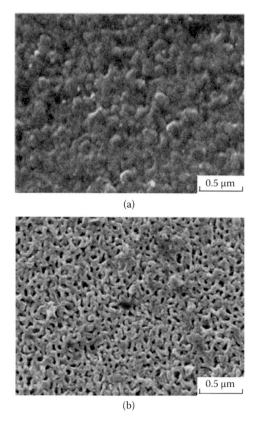

(a)

(b)

Figure 5.31 Structure of the Al–Si nanocomposite (the first series) according to scanning microscopy data: (a) initial structure and (b) after etching of aluminum.

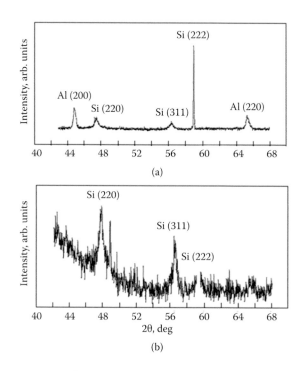

Figure 5.32 X-ray diffraction patterns of (a) initial sample and (b) after etching of aluminum.

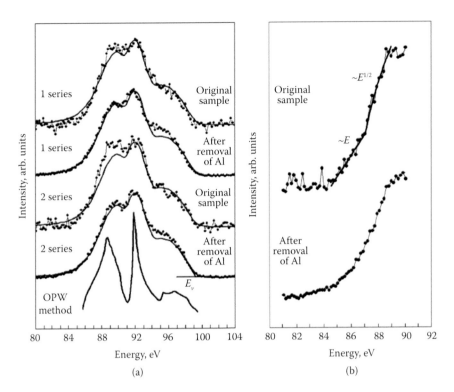

Figure 5.33 (a) Si L$_{2,3}$ spectra of Al–Si composite films (shown by points) and the spectrum of crystalline silicon (solid line). Shown at the bottom is the spectrum theoretically calculated by the OPW method in the energy range of 80–100 eV. (From Klima, J., *J. Phys. C Solid State Phys.*, 3, 70, 1970. With permission.) (b) Si L$_{2,3}$ spectra of Al–Si composite films (30% Si, the second series) near the bottom of the valence band (shown by points) and the approximation (solid lines).

(near the bottom of the valence band), the intensity of the spectra of nanostructured samples is noticeably lower than in the single crystal while, near the maximum of the spectrum (at hv = 89.6 eV), it is noticeably higher. Of special interest is the sharp decrease in the intensity of the emission spectra of the initial samples near the bottom of the valence band: at the energies of 84–87 eV (Figure 5.33). The Si L$_{2,3}$ spectrum near the bottom of the valence band of nanostructured silicon after the removal of aluminum practically coincides with the spectrum of single crystal c-Si. At the same time, the spectrum of silicon found in the aluminum matrix has no such long tail and, near the bottom of the valence band (in the region of 84–87 eV), the intensity linearly depends on the energy: I(E) ~ E. Above this energy and almost up to the first maximum of the density of states (E = 87–89 eV), the intensity increases with the energy as I(E) ~ v2(E − E0)1/2, where E0 = 86.7 eV. The dependence I(E) ~ E1/2 near the bottom of the valence band was theoretically predicted for the L$_{2,3}$ spectra [84], which is illustrated by the Si L$_{2,3}$ spectrum theoretically calculated from the band structure [83]. However, in the experimental Si L$_{2,3}$ spectra of c-Si near the bottom of the band, E0, instead of the dependence I(E) ~ E1/2, we observe a wide tail, as we clearly see from the Si L$_{2,3}$ spectra of c-Si and nanostructured silicon after the removal of aluminum (Figure 5.33).

According to the Tombulian data [84], the tail in the dependence ~E1/2 is caused by Auger broadening of levels near the bottom of the valence band, which is not taken into account in the one electron approximation. The sharp decrease in the intensity in the Si L$_{2,3}$ spectra of the initial nano-composites in which silicon nanocrystals are found indicates that Auger broadening of levels near the bottom of the valence band disappears. Because the probability of Auger process is determined by the electron–electron interaction whose matrix element involves the wave functions of interacting electrons [85], the disappearance of the Auger tail is possible if the wave functions of these electrons are localized. The reason for the localization may be the fact that silicon nanocrystals are found in the

Arrays, hybrids, and core-shell

aluminum matrix and do not interact with one another. As a result, the energy spectrum of valence electrons of silicon is the sum of all states belonging to all nanocrystals slightly differing in the sizes and not interacting with one another. Because the valence band width for each nanocrystal depends on its sizes, different states near the bottom of the valence band can be localized on different nanocrystals. This localization mechanism is confirmed by the fact that, after the removal of aluminum, the Si $L_{2,3}$ spectrum of particles becomes identical to the spectrum of bulk silicon (Figure 5.33), because nanocrystals come in contact and begin to interact with one another (Figure 5.31). It is well known that, for the local electronic state [86], an exponential decrease in the density of these states is observed near the boundary of the band. At the same time, according to [87,88], in the region of localized states, the intensity of the X-ray spectrum is proportional to the logarithm of the density of states. In this case, the exponential decrease in the density of electronic states must produce a linear decrease in the intensity in the X-ray spectrum, which is observed in the experiment. For the analysis of the specific features of the atomic and electronic structure of the surface layers (~5 nm) of Al–Si nanocomposites, synchrotron studies of the free electron density of states were performed from the XANES spectra near the Si $L_{2,3}$ edge of absorption on the near environment of absorbing atoms and the character of their ordering [89,90]. Unfortunately, we failed to obtain the Si $L_{2,3}$ spectrum for the initial sample because of the absence of noticeable signal in the given region of synchrotron radiation photon energy. This fact evidences at a very low silicon concentration over the composite surface. However, after the removal of aluminum, we obtained a sufficiently intensive Si $L_{2,3}$ spectrum (Figure 5.34) near the bottom of the conduction band E_c. Figure 5.34 also presents reference spectra for the c-Si single crystal and a-Si:H film. The comparison of these spectra shows that, on the surface of nanograined silicon after removal of aluminum, we observe the distribution of the density of states of the same character as in amorphous silicon. Moreover, the analysis of the XANES spectra reveals noticeable tails in the density of states below E_c (99.2–100 eV), and these tails are expressed much more strongly than in the spectrum of amorphous silicon. This may suggest a reconstruction of the electronic structure of silicon inclusions in the composites under study.

Thus, complex studies of aluminum–silicon composites have shown that silicon nanoparticles in the aluminum matrix are nanocrystals whose mean size depends on the amount of silicon contained in the aluminum matrix, and it does not change after etching of aluminum. The surface of each nanocrystal is covered with a layer of amorphous silicon. The absence of the interaction between silicon nanocrystals in the aluminum matrix results in the appearance of localized states near the bottom of the valence band. After the removal of aluminum, remaining nanocrystals interact with one another. In this case, below the conduction band, noticeable tails in the density of states are revealed.

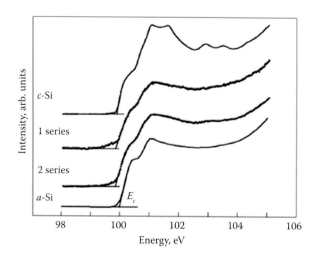

Figure 5.34 XANES spectra of amorphous and crystalline silicon and Al–Si composite films after removal of aluminum near the Si $L_{2,3}$ edge.

REFERENCES

1. M. D. Efremov, V. A. Volodin, D. V. Marin, et al., *JETP Lett.* 80, 544–547 (2004).
2. I. H. Campbell, P. M. Fauchet, *Solid State Commun.* 58, 739–741 (1986).
3. V. Paillard, P. Puech, M. A. Laguna, R. Carles, *J. Appl. Phys.* 86, 1921–1924, (1999).
4. C. Delerue, G. Allan, M. Lannoo, *Phys. Rev. B.* 48, 11024–11036 (1993).
5. G. Leodux, O. Guillois, D. Porterat, C. Reynaud, F. Huisken, B. Kohn, V. Paillard, *Phys. Rev. B.* 62, 15942–15951 (2000).
6. V. A. Terekhov, V.M. Kashkarov, E.Y. Manukovskii, A.V. Schukarev, E.P. Domashevskaya, *J. Electron Spectros. Relat. Phenom.* 114–116, 895–900 (2001).
7. L. T. Canham, *Appl. Phys. Lett.* 57, 1046 (1990).
8. G. C. John, V. A. Singh, *Phys. Rev. B* 50, 5329 (1994).
9. K. M. Yung, S. Shin, D. L. Kwong, *J. Electrochem. Soc.* 140, 3046 (1993).
10. A. N. Obraztsov, V. Y. Timoshenko, H. Okushi, H. Watanabe, *Semiconductors* 33, 322 (1999).
11. X.-M. Bao, X. He, T. Gao, et. al., *Solid State Commun.* 109, 169 (1999).
12. T. M. Zimkina, V. A. Fomichev, *Ultrasoft X-ray spectroscopy*, LGU, Leningrad, 1971.
13. A. I. Mashin, A. F. Khokhlov, E. P. Domashevskaya, V. A. Terekhov, N. I. Mashin, *Semiconductors* 35, 956 (2001).
14. V. I. Nefedov, Y. V. Salyn, E. P. Domashevskaya, Y. A. Ugai and V. A. Terekhov, *J. Electron Spectros. Relat. Phenom.* 6, 231 (1975).
15. V. A. Terekhov, E. P. Domashevskaya, *Izv. Akad. Nauk SSSR, Seriya Fizicheskaya* 49, 1531 (1985).
16. J. L. Gole, F. P. Dudel, D. Grantier, D. A. Dixon, *Phys. Rev. B* 56, 2137 (1997).
17. P. Li, G. Wang, Y. Ma, R. Fang, *Phys. Rev. B* 58, 4057 (1998).
18. T. Y. Gorbach, G. Y. Rudko, P. S. Smertenko et. al., *Semicond. Sci. Technol.* 11, 601 (1996).
19. L. M. Peter, D. J. Blackwood, S. Pons, *Phys. Rev. Lett.* 62, 308 (1989).
20. D. N. Goryachev, L. V. Belyakov, O. M. Sreseli, *Semiconductors* 34, 1090 (2000).
21. T. V. Torchynska, M. M. Rodrigues, G. P. Polupan, L. I. Khomenkova, N. E. Korsunskaya, V. P. Papusha, L. V. Scherbina, E. P. Domashevskaya, V. A. Terekhov, S. Y. Turishchev, *Surf. Rev. Lett.* 9, 1047 (2002).
22. E. P. Domashevskaya, V. M. Kashkarov, E. Y. Manukovskii, A. V. Schukarev, V. A. Terekhov, *J. Electron Spectros. Relat. Phenom.* 88–91, 969 (1998).
23. T. Van Buuren, L. N. Dinh, L. L. Chase, W. J. Siekhaus, I. Jimenez, L. J. Terminello, M. Grush, T. A. Callcott, J. A. Carlisle, Advances in microcrystalline and nanocrystalline semiconductors—1996, Collins RW, Ed, *Pittsburgh Mater. Res. Soc. Symposium Proc.* 452, 171–175 (1997).
24. K. H. Jung, S. Shin, D. L. Kwong, *J. Electrochem. Soc.* 140, 3046 (1993).
25. V. A. Sivakov, F. Voigt, A. Berger, et al., *Phys. Rev. B* 82, 125446 (2010).
26. E. P. Domashevskaya, O. A. Golikova, V. A. Terekhov, S. N. Trostyanskii, *J. Non-Cryst. Solids* 90, 135 (1987).
27. A. S. Shulakov, *Cryst. Res. Technol.* 23, 835 (1988).
28. V. A. Terekhov, S. N. Trostyanskii, A. E. Seleznev, E. P. Domashevskaya, *Poverkhnost Fiz. Khim. Mekh.* 5, 74 (1988).
29. S. Y. Turishchev, V. A. Terekhov, V. M. Kashkarov, et al., *J. Electron Spectrosc. Relat. Phenom.* 156–158, 445 (2007).
30. V. A. Terekhov, V. M. Kashkarov, S. Y. Turishchev, et al., *J. Mater. Sci. Eng. B* 147, 222 (2008).
31. V. A. Sivakov, G. Bronstrup, B. Pecz, et al., *J. Phys. Chem. C* 114, 3798 (2010).
32. G. Wiech, H. O. Feldhutter, A. Simunek, *Phys. Rev. B* 47, 6981 (1993).
33. G. A. Kachurin, V. A. Volodin, D. I. Tetel'baum, D. V. Marin, A. F. Leier, A. K. Gutakovskii, A. G. Cherkov, A. N. Mikhailov, *Semiconductors* 39, 552 (2005).
34. T. Shimizu-Iwayama, K. Fujita, S. Nakao, K. Saitoh, T. Fujita, N. Itoh, *J. Appl. Phys.* 75, 7779 (1994).
35. P. Mutti, G. Ghislotti, S. Bertoni, L. Bonoldi, G. F. Cerofolini, L. Meda, E. Grilli, M. Guzzi, *Appl. Phys. Lett.* 66, 851 (1995).
36. A. Mimura, M. Fujii, S. Hayashi, D. Kovalev, F. Koch, *Phys. Rev. B* 62, 12625 (2000).
37. J. Zhao, D. S. Mao, Z. X. Lin, X. Z. Ding, B. Y. Jiang, Y. H. Yu, X. H. Liu, *Appl. Phys. Lett.* 74, 1403 (1999).
38. V. G. Kesler, S. G. Yanovskaya, G. A. Kachurin, A. F. Leier, L. M. Logvinsky, *Surf. Interface Anal.* 33, 914 (2002).
39. I. E. Tyschenko, L. Rebohle, R. A. Yankov, W. Skorupa, A. Misiuk, G. A. Kachurin, *J. Luminesc.* 80, 229 (1999).
40. A. S. Vinogradov, E. O. Filatova, T. M. Zimkina, *Pis'ma v JETF* 15, 84 (1989).
41. L. Windt, *Appl. Opt.* 30, 15 (1991).
42. D. I. Tetel'baum, O. N. Gorshkov, A. P. Kasatkin, A. N. Mikhailov, A. I. Belov, D. M. Gaponova, S. V. Morozov, *Phys. Solid State* 47, 13 (2005).

43. L. X. Yi, J. Heitmann, R. Scholz, M. Zacharias, *Appl. Phys. Lett.* 81, 4248 (2002).
44. T. Inokuma, Y. Wakayama, T. Muramoto, R. Aoki, Y. Kurata, S. Hasegawa, *J. Appl. Phys.* 83, 2228 (1998).
45. H. Rinnert, M. Vergnat, A. Burneau, *J. Appl. Phys.* 89, 237 (2001).
46. V. I. Nefedov, *XPS spectroscopy of chemical compounds: Handbook,* Chemistry, Moscow, 1984.
47. O. M. Kanunnikova, *Perspektivnye Mater.* 6, 88 (2006).
48. V. A. Terekhov, S. Y. Turishchev, V. M. Kashkarov, E. P. Domashevskaya, A. N. Mikhailov, D. I. Tetel'baum, *Physica E Low Dimens. Syst. Nanostruct.* 38, 16 (2007).
49. F. C. Brown, O. P. Rustgi, *Phys. Rev. Lett.* 28, 497 (1972).
50. M. A. Rumsh, A. P. Lukirskii, V. N. Shchemelev, *Izv. Akad. Nauk SSSR, Seriya Fizicheskaya* 25,1060 (1961).
51. V. A. Terekhov, E. I. Terukov, I. N. Trapeznikova, V. M. Kashkarov, O. V. Kurilo, S. Y. Turishchev, A. B. Golodenko, E. P. Domashevskaya, *Semiconductors* 39, 830 (2005).
52. A. V. Vinogradov, I. A. Brytov, I. Y. Grudskii, et al., *X-ray mirror optics, Engineering,* Leningrad, 1989.
53. D. H. Tomboulian, D. E. Bedo, *Phys. Rev.* 104, 590 (1956).
54. M. Watanabe, T. Ejima, N. Miyata, T. Imazono, M. Yanagihara, *Nucl. Sci. Tech.* 17, 257 (2006).
55. G. Ledoux, J. Gong, F. Huisken, O. Guillois, C. Reynaud, *Appl. Phys. Lett.* 80, 4834 (2002).
56. I. A. Kamenskikh, D. N. Krasikov, O. A. Shalygina, et al., *HASYLAB at DESY Annual report.* HASYLAB and DESY, Hamburg, Germany, 2007, 721.
57. A. V. Ershov, D. I. Tetelbaum, I. A. Chugrov, A. I. Mashin, A. N. Mikhailov, A. V. Nezhdanov, A. A. Ershov, I. A. Karabanova, *Semiconductors* 45, 731 (2011).
58. http://www.src.wisc.edu/facility/list/Port_043.pdf.
59. M. Kasrai, W. N. Lennard, R. W. Brunner, G. M. Bancroft, J. A. Bardwell, K. H. Tan, *Appl. Surf. Sci.* 99, 303 (1996).
60. V. A. Terekhov and E. P. Domashevskaya, *Izv. Akad. Nauk SSSR, Seriya Fizicheskaya* 49, 1531 (1985).
61. V. A. Terekhov, S. Y. Turishchev, K. N. Pankov, I. E. Zanin, E. P. Domashevskaya, D. I. Tetelbaum, A. N. Mikhailov, A. I. Belov, D. E. Nikolichev, S. Y. Zubkov, *Surf. Interface Anal.* 42, 891 (2010).
62. H. Piao, N. S. McIntyre, *Surf. Interface Anal.* 31, 874 (2001).
63. C. Weigel, G. Calas, L. Cormier, L. Galoisy, G. S. Henderson, *J. Phys. Condens. Matter* 20, 135219 (2008).
64. L. Pavesi, R. Turan (Eds.), *Silicon nanocrystals: fundamentals, synthesis and applications,* WILEY-VCH Verlag GmbH & Co. KGaA, Weinheim, 2010, 613.
65. J. Zhao, D. S. Mao, Z. X. Lin, *Appl. Phys. Lett.* 73, 1838 (1998).
66. O. Gonzalez-Varona, A. Perez-Rodriguez, B. Garrido, *Nucl. Instrum. Methods Res. B* 904, 161–163 (2000).
67. A. Perez-Rodriguez, O. Gonzalez-Varona, B. Garrido, P. Pellegrino, J. R. Morante, C. Bonafos, M. Carrada, A. J. Claverie, *Appl. Phys.* 94, 254 (2003).
68. D. I. Tetelbaum, A. N. Mikhaylov, V. K. Vasiliev, et al., *Surf. Coat. Technol.* 203, 2658 (2009).
69. A. V. Boryakov, D. E. Nikolitchev, D. I. Tetelbaum, A. I. Belov, A. V. Ershov, A. N. Mikhaylov, *Phys. Solid State* 54, 394–403 (2012).
70. A. I. Belov, A. N. Mikhaylov, D. E. Nikolitchev, A. V. Boryakov, A. P. Sidorin, A. P. Gratchev, A. V. Ershov, D. I. Tetelbaum, *Semiconductors* 44, 1450–1456 (2010).
71. G. Dufour, F. Rochet, *Phys. Rev. B* 56, 4266–4282 (1997).
72. F. J. Himpsel, F. R. McFeely, A. Taleb-Ibrahimi, J. A. Yarmoff, G. Hollinger, *Phys. Rev. B* 38, 6084–6096 (1988).
73. V. A. Terekhov, S. Y. Turishchev, K. N. Pankov, I. E. Zanin, E. P. Domashevskaya, D. I. Tetelbaum, A. N. Mikhailov, A. I. Belov, D.E. Nikolichev, *J. Surf. Invest.* 5, 958–968 (2011).
74. A. N. Hoperskii, V. A. Yavna, *X-ray photon anomalous elastic scattering by atom,* SKNC VSh, Rostov-On-Don. [In Russian].
75. A. V. Vinogradov, *X-ray mirror optics,* Mashinostroenie, Leningrad. Otdelenie, 1989. [In Russian].
76. V. A. Terekhov, D. I. Tetelbaum, I. E. Zanin, K. N. Pankov, D. E. Spirin, A. N. Mikhailov, A. I. Belov, A. V. Ershov, *Izv. Vys. Uch. Zav. Mater. El. Tech.* 4, 54–59 (2012). (In Russian).
77. V. Y. Bratus', V. A. Yukhimchuk, L. I. Berezhinsky, M. Y. Valakh, I. P. Vorona, I. Z. Indutnyi, T. T. Petrenko, P. E. Shepelyavyi, I. B. Yanchuk. *Semiconductors* 35 (7), 821 (2001).
78. E. S. Demidov, N. A. Dobychin, V. V. Karzanov, M. O. Marychev, V. V. Sdobnyakov, S. V. Khazanova, *Vestn. Nizhegorodskogo Univ.* 5, 298 (2010).
79. S. K. Lazarouk, D. A. Sasinovich, P. S. Katsuba, V. A. Labunov, A. A. Leshok, V. E. Borisenko. *Semiconductors* 41 (9), 1109 (2007).
80. S. K. Lazaruk, A. A. Leshok, P. S. Katsuba, *Low temperature technique for amorphous porous silicon formation,* Proceedings of the VIII International Conference "Amorphous and Microcrystalline Semiconductors," St. Petersburg, 2012 (St. Petersburg State Polytechnical University, St. Petersburg, 2012), p. 124.
81. A. S. Shulakov, A. P. Stepanov. *Poverkhnost,* 10, 146 (1988).
82. T. M. Zimkina, V. A. Fomichev, *Ultrasoft X-ray spectroscopy,* Leningrad State University, Leningrad, 1971. [in Russian]

83. J. Klima, *J. Phys. C: Solid State Phys.* 3, 70 (1970).

84. D. G. Tombulian, *Xrays*, Ed. by M. A. Blokhin, *Inostrannaya Literatura*, Moscow, 1960. [in Russian]

85. T. A. Carlson, *Photoelectron and Auger Spectroscopy*, Plenum, New York, 1975.

86. N. F. Mott and E. A. Davis, *Electronic Processes in NonCrystalline Materials,* Vol. 1, Oxford University Press, Oxford, UK, 1971.

87. S. K. Balagurov, N. Yu. Karpova, V. A. Terekhov, S. N. Trostyanskii, and E. P. Domashevskaya, *Sov. Phys. Solid State* 33 (10), 1712 (1991).

88. V. A. Terekhov, Poverkhnost, N 4–5, 167 (1997).

89. V. A. Terekhov, S. Yu. Turishchev, V. M. Kashkarov, E. P. Domashevskaya, A. N. Mikhailov, and D. I. Tetel'baum, *Physica E* (Amsterdam) 38, 16 (2007).

90. V. A. Terekhov, S. Yu. Turishchev, K. N. Pankov, I. E. Zanin, E. P. Domashevskaya, A. N. Mikhailov, D. I. Tetelbaum, A. N. Mikhailov, A. I. Belov, D. E. Nikolichev, and S. Yu. Zubkov, *Surf. Interface Anal.* 42, 891 (2010).

Arrays, hybrids, and core–shell

6 Silicon/polymer composite nanopost arrays

Xueyao Liu, Wendong Liu, and Bai Yang

Contents

6.1 INTRODUCTION

Silicon nanostructures have played an important role in the development of the semiconductor industry. They have proved to be promising building blocks for devices in the fields of optoelectronics, energy conversion, nanoelectronics, energy storage, and bio (chemical) sensors [1]. Silicon nanopost arrays have aroused great attention, owing to their reproducible and facile fabrication process, and adjustable structure parameters. Attributed to the silicon nanopost arrays' efficient light trapping assisted by the interactions of light within the space between the posts, they exhibit increased light-harvesting properties in contrast to flat surfaces, which makes them ideal for solar energy applications. Moreover, periodic silicon nanopost arrays are two-dimensional photonic crystal (2DPC), which presents vivid structural colors due to their specific photonic band gap. They are therefore considered to be potential candidates for building biological or chemical sensors. Furthermore, owing to the high surface area, they have been studied for use as substrates for cell adhesion and gene delivery.

Composite materials possess the combination of the properties of each component or even better properties when the individual components cooperate synergistically and exert superior performance to all the components alone. Silicon/polymer nanopost arrays are therefore gifted with the advantages of both silicon nanopost arrays and polymers. Responsive polymer nanostructures are expected to be of great importance for the smart surfaces or channels, biointerfaces, diagnostics, microfluidic devices, and sensors realization [2–6]. The modification of intelligent responsive polymers, such as light-driven configuration tunable, thermoresponsive polymers, and polymers with variable wettability and functions, suggests great potential for the composite silicon/polymer nanopost arrays for multifunctional devices.

Herein we will review the fabrication methods of silicon/polymer composite nanopost arrays, the consequent properties brought by the composite structures, and the related application areas of the silicon/polymer composite nanopost arrays, including sensing, separation, biointerfaces, and photoelectric devices. Finally, we'll provide a summary and perspective of the developing direction of silicon/polymer composite nanopost arrays, including the development trend of fabrication approaches and the broadening of the application fields in the near future.

6.2 FABRICATION METHODS OF SILICON/POLYMER COMPOSITE NANOPOST ARRAYS

Despite the paucity of published works on silicon/polymer composite nanopost arrays, we can still be inspired by the fabrication methods of silicon nanopost arrays and polymer growth on nanostructure separately. Benefitting from the diversity of the fabrication methods of the two sections alone, countless silicon/polymer composite nanopost arrays fabrication methods can be derived. In this section, we will mainly discuss the fabrication of silicon nanopost arrays and methods to grow polymer on the silicon nanopost arrays herein. Later, the other existing form of silicon/polymer composite nanopost arrays will be discussed briefly.

6.2.1 FABRICATION OF SILICON NANOPOST ARRAYS

Up to now, myriad approaches have been explored to develop silicon nanopost arrays. Classified by the structure order, the silicon nanopost arrays can be divided into hierarchical Si nanopost arrays and Si nanopost arrays. As for the fabrication methods, based on the etching medium component category, we can divide them into dry etching and wet etching. The silicon nanopost arrays' fabrication methods will be reviewed concisely herein.

6.2.1.1 Primary silicon nanopost arrays' fabrication

6.2.1.1.1 Wet etching

Wet etching is performed in liquid, sorted to metal-assisted chemical etching and electrochemical wet etching, as sketched in Figure 6.1 [7].

As for electrochemical wet etching, the electrolyte together with anode and cathode will generate a charged double layer near the silicon surface, resulting in the nanostructure formation. The pattern shape is determined by the mask layer or the substrate's covering layer [7]. The process of obtaining patterned macropores with electrochemical etching (ECE) with photolithography has emerged since 1990 [8]. Since then, it has developed based on the gradual progress of the device and materials. Schuster et al. [9] localized electrochemical reactions

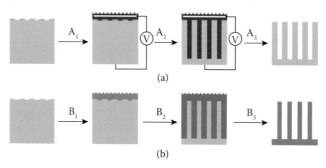

Figure 6.1 Scheme for wet etching: (a) electrochemical etching and (b) MACE. (A1) Load the pattern mask, immerse the silicon surface into HF, and connect the system with cathode and anode. (A2) Apply the anodic bias and etch. (A3) Sample and mask removal. (B1) Load the metal mask and immerse the sample in HF. (B2) Etching catalyzed by metal. (B3) Remove the sample and mask.

on conducting materials to a submicrometer scale, which is attributed to the ultrashort voltage pulses between the tool electrode and the workpiece in the electrochemical environment. Bassu et al. enabled high-complexity Si nanostructure creation with high accuracy even with the aspect ratio up to 100 [10].

Metal-assisted chemical etching (MACE) is simple, inexpensive, and the morphology parameter can be finely controlled. The first MACE applied on Si was reported in 1997, and the porous silicon was obtained by stain etching in the mixture of HNO_3, HF, and H_2O with the help of the predeposited Al layer [1]. However, the thickness of the as-prepared porous Si was limited to a rather thin scope (less than 1.5 nm). The most widely used MACE nowadays is proposed by Li and Bohn [11]: the Si will be etched in the mixture of HF, H_2O_2, and EtOH with the etching process catalyzed by patterned noble metal layer deposited onto the Si surface. As a result, the Si surface will be etched into nanopost or nanopore arrays. Huang et al. [1] reviewed MACE in detail, including the basic mechanism and various influence factors, such as noble metal, etchant, temperature, illumination, and intrinsic properties of Si on MACE. To ensure the high reproducibility and controllable fabrication of Si nanopost, ordered structural fabrication is essential. Template-based MACE endows the structure high ordering and excellent controllability. The most common templates include nanosphere, AAO, block-copolymer, and mask obtained by interference lithography [1]. Based upon the template category, the obtained nanoposts are endowed with morphologically distinguished characteristics.

6.2.1.1.2 Dry etching

Unlike wet etching, dry etching is processed in a reacting chamber under vacuum. The Si substrate will be covered by a layer of etching mask, usually photoresist or nanospheres. Assisted by UV photolithography or electron-beam lithography, the pattern can therefore be transferred to the substrate. Finally, under the mask's shield, the Si will be etched to Si nanopost with the help of usually (deep) reactive ion etching ([D] RIE). (D) RIE is conducted by chemically reactive plasma, including SF_6, O_2, CHF_3, CF_4, and C_4F_8, which can react with the uncovered Si. The anisotropic etching thereby leaves Si nanopost morphology.

As for the aspect ratio consideration, the mask category can influence the structure aspect ratio severely. Polystyrene nanospheres taken as etching mask will result in a relatively low aspect ratio as a result of the fast erosion speed [12]. Therefore, during the pattern transfer process, silica [13] or metallic nanoparticles [14,15], which can withstand long-period etching, are also employed as an etching mask in the nanosphere lithography (NSL) in order to achieve a higher aspect ratio. Inorganic hardmask can increase the aspect ratio; however, it also suffers the problems of sidewall roughening, pattern distortion, and difficulty in mask removal. Aiming at high aspect ratio, the Bosch process, alternating etching and passivation gases, is a good choice. Its disadvantage is the inevitable scalloping on the sidewall caused by the isotropic etching. To fix this problem, the cryogenic etching (–110°C) can smooth the sidewall, resulting from the thin passivation layer's protection [16]. However, the requirement on the equipment limits the application prospect of this fabrication method.

Black silicon etching, a kind of maskless etching, is also a feasible way to fabricate silicon nanopost arrays [17]. Native oxides, dust, and so on will act as micromasks in that case. The direct etching results in the spike appearance. Wafer-scale patterning can be achieved and is attributed to the maskless etching. Nevertheless, randomly distributed structures cannot avoid the disadvantage of poor reproducibility and difficulty in designing the structure subjectively.

6.2.1.2 Hierarchical silicon nanopost arrays' fabrication

To mimic the natural surface structures, much effort has been put into creating hierarchical nanostructure. Stemming from the hierarchical structure provided by the synergistic contribution of interfacial properties in wetting, electrical conduction, adhesion, and light trapping, these structures exhibit superior device performance [18]. Hierarchical silicon nanopost arrays are also reported by many groups in recent years.

As most hierarchical structures are not on one order of magnitude, majority reported methods to fabricate the hierarchical Si nanopost arrays are the combination of two or three primary structure fabrication methods. Combining DRIE and galvanic etching, Yuan et al. obtained micro-nano-hierarchical silicon surfaces [19]. The first step of DRIE combining photolithography results in the formation of micropillars. Afterward, assisted by the silver nanoparticles deposited on the micropillars, MACE was performed and the hierarchical silicon surfaces were achieved. Ho et al. reported hierarchical silicon nanopost arrays by the combination of MACE, NSL, and conventional lithography method [18]. They used NSL and conventional lithography to get metal mesh on the silicon surface, which will act as the etching mask in the MACE step. In summary, most hierarchical structures reported are prepared through the combination of different etching methods, with various scales of size obtained from each step.

6.2.2 METHODS TO GROW POLYMER ON THE SILICON NANOPOST ARRAYS

In this section, we will discuss the methods to grow polymer brush on the silicon nanopost arrays. Polymer brush is attached to the substrate surface on one end, and the other end is suspended in the solvents. Owing to its composition, density, and length-control advantages, polymer brush is widely applied to build functional surfaces. Either by covalently attaching or physical adsorption, polymer brush can be grafted onto the surface. The physical adsorption includes electrostatic interactions, hydrophobic or hydrophilic interactions, and hydrogen bonding, which suffer instability problems. Covalent attachment embraces "grafting to" and "grafting from" two pathways. The former one is conducted by grafting the prepared polymer (end-functionalized) to the modified surface. The latter one is performed by growing polymer directly from the initialized surface.

To develop controlled/living radical polymerization (CRP) methods is the goal of polymer chemistry. A variety of CRP methods, including atom transfer radical polymerization (ATRP), degenerative transfer (DT), reversible addition fragmentation chain transfer (RAFT), nitroxide-mediated polymerization (NMP), and stable free radical polymerization (SFRP), have been developed to generate polymer brushes in a controllable manner. All of them, as shown in Figure 6.2 [20], include activation (rate constant k_{act}) and deactivation (rate constant k_{deact}), except DT and RAFT, for which the steps will be simplified to just the exchange process (rate constant k_{exch}). The free radicals propagate (rate constant k_p) and terminate (rate constant k_t). Though termination occurs, it still behaves as a nearly living and controlled system. The most common surface modification to polymerize is surface initialized ATRP (SI-ATRP), which maintains all characteristics of conventional ATRP.

SI-ATRP is catalyzed by a redox-active transition metal complex, at a lower oxidation state that reacts with initiator generating radicals and propagates. The radicals can terminate and be deactivated with a higher oxidation state deactivator. The polymerization rate depends on the propagation and deactivation rate constants and the concentration ratio of two oxidation states [21]. Distinguished by the composition, the SI-ATRP-enabled polymer brush can be classified as a homogeneous brush, a block copolymer brush, and a mixed type. Depending on the surface topography, the polymer brush can be gradient polymer brushes, patterned brushes, spherical brushes, and asymmetric brushes. Moreover, SI-ATRP can be regulated by many external stimuli, such as applied potential or current during electrochemistry, chemical reducing reagents, and irradiation in photoinitialized reactions, which makes it easy to control the whole process and adaptable to any substrate.

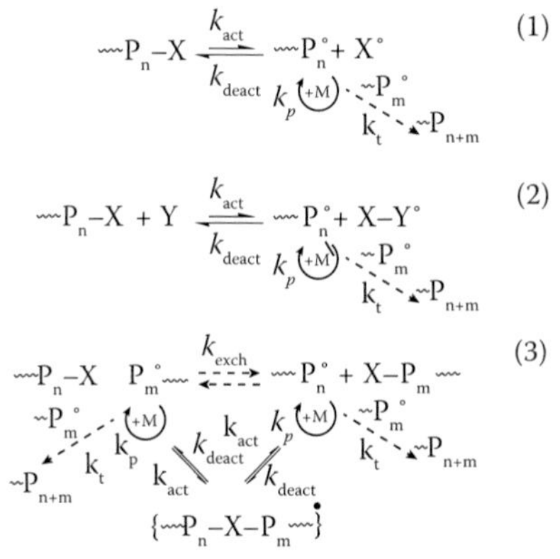

Figure 6.2 Scheme for CRP methods.

With the development of the polymer brush exploration, it has been applied widely and revolutionized the field of surface polymerization. As for the sensing performance, integrated with other structures, the polymer brush provides more surface area for interaction with the analytes, reducing the sensor's detection limits at the same time [22]. Patterned polymer brushes can conjugate protein and form complex protein patterns, which will guide the cell's adhesion behavior and regulate the cell's adhesion manner [23]. Assisted by the stimuli-responsive characteristic of the modified polymer brush, the hierarchical structure can synergistically exploit the advantages of both structured surface and polymer brush. Yang et al. grafted PNIPAM onto PET nanocones and achieved thermotunable wetting behavior of the surface [24].

6.2.3 OTHER FORMS OF SILICON/ POLYMER COMPOSITE NANOPOST ARRAYS

The composite form of silicon nanopost arrays with polymer can be diverse. With the exception of growing polymer brush from the silicon nanopost arrays, there are also many other existing forms for the combination of the silicon nanostructure and the polymer. The polymer can be spin-coated into the space among the silicon nanopost, which is usually seen in solar cell preparation. Moreover, the polymer can merely appear on the top of each nanopost. Taking advantage of PVA's physisorb onto hydrophobic surfaces, Hatton and Aizenberg painted the tip of silicon nanoposts (fluorosilane-treated previously) with hydrophilic patches [25]. This is attributed to the Cassie state wetting behavior. Additionally, using multilayer etching method (Nanosphere-Si-Polymer-Si surface), Frommhold et al. obtained high-aspect-ratio silicon nanopost arrays with polymer remaining on each independent nanopost top [26].

6.3 VARIOUS PROPERTIES OF SILICON/POLYMER COMPOSITE NANOPOST ARRAYS

The same as pure silicon or polymer-based nanopost arrays, silicon/polymer composite nanopost arrays also present their own advantages and excellent properties since much work has been done based on this kind of hybrid structure. In this part, we will provide a brief introduction about the wetting, optical, and stimuli-responsive properties of these hybrid nanopost arrays.

6.3.1 WETTABILITY

Wetting is a feature phenomenon that exists in nature, and it is greatly dependent on the surface's composition and structures in microscale or nanoscale. In recent years, wettability has become one of the most important properties of structured surfaces and has drawn great attention since series of suprawetting surfaces were successfully obtained [26]. Nanopost arrays fabricated on silicon substrate can perform suprawetting behavior because the ordered structures endowed the surfaces with specific roughness, which can meet requirements of suprawetting [26]. Meanwhile, liquid droplets on pure polymer-based nanopost arrays can also display suprawetting behavior because the polymer nanocone not only provides the roughness but also the feature surface energy, which can be easily regulated via changing different polymers with specific properties [27]. However, these monocomponent nanopost arrays possess their own drawbacks: the surface property of silicon is simplex, whereas the polymer nanostructure is difficult to control, which is caused by the flexibility of polymer. Thus, silicon/polymer composite nanopost arrays have a great advantage to form surfaces with suprawettability, even though there are few works focused on using this hybrid structure for controllable wetting. This hybrid structure can inherit the advantages of silicon and polymer, and it can successfully overcome the defects of these two kinds of materials with the help of the complementary nature of organic and inorganic materials.

Jiang et al. [28] prepared an oil superwetting surface based on the silicon/polymer hybrid structure, as shown in Figure 6.3. In this work, mixed silane agents of HFMS and ATMS were anchored onto the surface of the silicon structure via silane coupling chemistry, and then the PNIPAMs were grafted onto the surface by SI-ATRP (Figure 6.3a). Because of underwater oleophilic HFMS and thermoresponsive PNIPAM, the wettability of the hybrid surface can be switched easily between superoleophobicity and superoleophilicity. Meanwhile, temperature-induced underwater low-adhesive superoleophobicity (20°C) and high-adhesive superoleophobicity (60°C) surfaces were obtained by modifying the silicon structure with pure PNIPAM brushes (Figure 6.3b). This work proves that the silicon/polymer composite nanopost

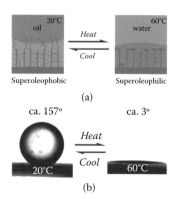

Figure 6.3 Underwater temperature-responsive switch between superoleophobicity and superoleophilicity. (a) By tuning surrounding temperature, PNIPAM chains could conceal or expose oleophilic heptadecafl uoro-decyltrimethoxysilane (HFMS) by virtue of a thermoresponsive conformational change of the PNIPAM molecules. Thus, underwater wettability of 1, 2-dichloroethane (DCE) on silicon nanowire arrays (SiNWAs) modified with a mixed brush of PNIPAM and HFMS can switch between superoleophobicity and superoleophilicity. (b) At 20°C, the underwater contact angle of 2 μL DCE is about 157°. When increasing temperature to 60°C, the underwater contact angle of DCE decreases to only about 3°.

arrays provide a way to fabricate smart surfaces with specific wettability and adhesivity, and these hybrid structures will be further applied in controllable adhesion, mixture separation, and miniature reactors with more research into this feature structure.

6.3.2 OPTICAL PROPERTY

Nanopost arrays are typical two-dimensional (2D) structures that possess the same optical properties as the ordered 2D structures in nature and artificial elliptical hemisphere arrays, silver nanohole arrays, nanovolcano arrays, hollow nanocone arrays, and asymmetric half-cone/nanohole arrays. Since these 2D structures possess periodically arranged media with differing dielectric constants in two directions, which is similar to 2D photonic crystals, the periodic potential is replaced by a periodic dielectric, giving rise to selective transmission or reflection of particular light wavelengths, resulting in the generation of structure color, antireflection, and improved permeability [29–31].

Silicon/polymer hybrid nanopost arrays can also perform specific optical properties because the dielectric constant can be easily regulated via modifying the polymer film grafted onto the silicon nanopost. Yang et al. [32] reported a silicon/polymer composite nanopost array fabricated by the combination of colloidal lithography and SI-ATRP. The composite nanopost arrays possess a core/shell structure; the core is silicon nanopost (Figure 6.4a and b), and the shell is constructed by PHEMA brushes. Because the nanopost arrays finely inherited the orderliness of the colloidal crystal mask, the obtained silicon/polymer composite nanopost arrays can perform an obvious structure color. The color can be easily modulated via regulating the thickness of the polymer brush film (Figure 6.4c–h), because the increase in thickness can induce a great change of the reflective index of the space between the adjacent silicon nanopost, leading to the change of the reflection peak of the light causing the color conversion. This work provides guidance for using silicon/polymer hybrid nanopost arrays for regulation of particular light wavelength, and this kind of hybrid nanopost array can be further applied to other optical relative fields with further changing of the polymer or introducing other materials with feature-reflective index, which may greatly modulate the index variation of silicon and polymer layering in the future.

6.3.3 STIMULI RESPONSIVENESS

Another significant property of silicon/polymer composite nanopost arrays is the responsiveness under external stimuli when the stimuli-responsive polymer grafted. Thus, the hybrid nanopost arrays can present

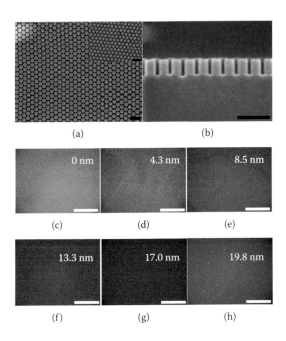

Figure 6.4 The top-view (a) and cross-sectional (b) scanning electron microscopy (SEM) images of the silicon nanopost arrays prepared by colloidal lithography. Inset of (a) shows SEM images of the 2D PS colloidal crystals used as masks in the colloidal lithography process. All scale bars are 500 nm. The photographs (c–h) of the composite nanopost arrays with different shell thickness of PHEMA. The concentration of water vapor in the air is 0.43%. All scale bars are 500 μm.

various responsive behaviors by regulating the composition of polymer to meet the specific requirement of researchers.

Yu et al. [33] demonstrated a light-driven switch of superhydrophobic adhesion based on the silicon/polymer hybrid structure. As shown in Figure 6.5a, a layer of cyclized rubber negative photoresist was covered onto micro-nanopost-arrayed silicon wafer as a transitional layer and then a hydrophobic azo-polymer (PA8AB6, poly[4-(8-acryloyloxy) octyloxy-4′-hexyloxy] azobenzene) was spin-coated onto micro-nanopost arrays as a light-responsive layer. This composite structure–based surface can perform a fully light-driven switch of superhydrophobic adhesion of a water droplet when exposing the hybrid structure to UV-Vis light, which greatly attributed to the light-responsive property of the azo-polymer (Figure 6.5b–e). This kind of noncontact real-time controllable switching of a droplet's adhesion might inspire and facilitate the designs and applications in novel microfluidic devices.

Yang et al. [34] also did much valuable work based on the stimuli-responsive property of silicon/polymer hybrid nanopost arrays. As shown in Figure 6.6a, a Janus hybrid nanopost array with controllable wettability difference was successfully obtained. The Janus silicon post structure was modified with hydrophilic self-assembled monolayer on one side and switchable PNIPAm on the other side. Owing to the temperature-driven switchable wettability of the PNIPAm brush, the surface could switch between anisotropic and isotropic wetting at different temperature, which can be introduced into the microfluidic channel to control the motion of microfluid (Figure 6.6b and c). And these thermal-responsive hybrid Janus nanopost arrays would provide a new strategy to control the flow and motion of fluids in microfluidic channels and chips for the flowing work.

As mentioned earlier, the wetting, optical, and stimuli-responsive properties of the silicon/polymer nanopost arrays are greatly benefited by the feature property of the silicon and polymer constructed hybrid structure. With the help of the variety of polymer and the fast development of microfabrication techniques, silicon/polymer composite nanopost arrays will be attracting the attention of researchers, and series of hybrid nanopost arrays which possess specific properties will be further fabricated and broadly used in the future.

Arrays, hybrids, and core–shell

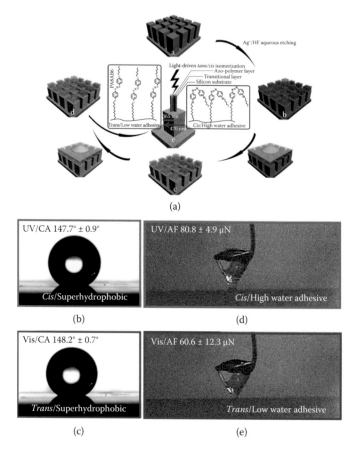

Figure 6.5 (a) Schematic illustration of the preparation process of the micro-nanocomplex azo-polymer coating and the light-driven *trans/cis* isomerization. (b and c) Profile of the water droplet on the array after UV light and visible light irradiation. (d) Optical image of the maximum-deformed water droplet on the array after UV light irradiation. The surface is in *cis* state, which shows high water adhesion (the average AF, 80.8 ± 4.9 µN). (e) Optical image of the maximum-deformed water droplet on the array after visible light irradiation. The surface is in *trans* state, which shows low water adhesion (the average AF, 60.6 ± 12.3 µN).

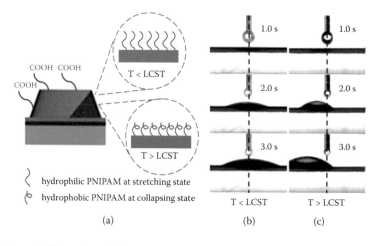

Figure 6.6 (a) A schematic illustration of the PNIPAM/MHA Janus pillar surface wettability at different temperatures. (b and c) The photographs of the water drop taken at different times after the water drop contacted the surface of PNIPAM/MHA Janus array at a temperature below and above the LCST of PNIPAM brush. The red dotted line is the center axis of the syringe needle.

6.4 APPLICATIONS OF SILICON/POLYMER COMPOSITE NANOPOST ARRAYS

In the recent decades, we've seen silicon nanopost arrays applied in various areas, such as antireflective layer, DNA or other biomolecule separation, enhanced-fluorescence-based detection of analytes, biointerfaces, and wetting behavior–related applications. Combination with polymer will offer the entire structure more opportunity and prospects in application for the advantages of both silicon nanopost arrays and polymer provide. However, silicon/polymer composite nanopost arrays are still reported by relatively few groups. Therefore, it is essential and desirable to explore the application potential for the composite structures. Herein, we'll sum up not only silicon nanopost array–based composite structures but also some successful silicon nanowire/polymer composite structure application based on the few works reported in recent decades. Classified by the specific applied areas, we'll interpret the application of composite structures from sensing, separation, biointerfaces, and hybrid solar cell aspects.

6.4.1 SENSING TO SOME SPECIAL MOLECULES AND MATERIALS

The characteristic properties of polymers confer the composite structures' unique superiority in sensing performance owing to the silicon nanopost arrays' own structure parameters' facile manipulation and 2D photonic crystal's (2DPC) reflectance band gap.

Yang et al. fabricated silicon/polymer composite nanopost arrays and further applied them in water-vapor sensing. The fabrication procedure and general sensing process are as shown in Figure 6.7a [32]. Combining NSL and RIE, they prepared silicon nanopost arrays. Afterward, the structure was turned into a composite structure assisted by the SI-ATRP method. Interestingly, with the grafted polymer thickness increase, the composite structure's reflectance peak gradually red shifts, which is caused by the efficient refractive index change of the entire structure. The whole structure performs a vivid color change when the grafted polymer thickness changes. Furthermore, due to the hydrophilic property of the grafted poly(2-hydroxyethyl methacrylate) (PHEMA), the composite structure can therefore sense water vapor. PHEMA can swell upon the absorbed water, resulting in the efficient refractive index's increase and the reflectance peak's red shift. Increased concentration of water vapor results in the reflectance peak's red shift gradually, demonstrating the composite structure's humidity-sensing performance as shown in Figure 6.7b. The as-prepared structure exhibits high stability and reproducibility, which ensures its application as a humidity sensor.

6.4.2 SEPARATION OF BUILDING BLOCKS WITH DIFFERENT WETTING PROPERTIES

Based on the chemically modified molecule's wetting property difference on the opposite sides, the composite structure can separate building blocks with different wetting properties. With the help of incline metal deposition-assisted selective modification, Yang et al. functionalized the silicon nanopost arrays with molecules of distinct surface energies, 16-Mercaptohexadecanoic acid (MHA) and trichloro(1H,1H,2H,2H-perfluorooctyl)silane (PFS) [35]. As for the microfluidic system, the flow behavior

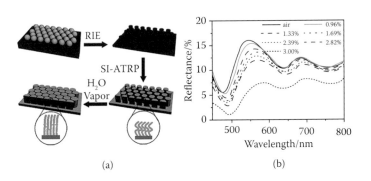

Figure 6.7 (a) Scheme for the fabrication procedure of composite nanopost arrays employed as a water-vapor sensor. (b) The composite structure's reflectance spectra when exposed to water vapor of a series of concentrations. The concentration of water vapor in air is 0.43%.

in a microchannel is intensely influenced by the surrounding walls' surface properties for the large surface-to-volume ratio. Therefore, the flow direction of fluid inside the microchannel can be directed by regulating the wettability of the walls [36]. Based on this, the as-prepared Janus composite nanopost arrays acted as a valve for microfluidic system, realizing gas–liquid separation owing to their anisotropic wetting property. Now that functional molecules modification has already proved to fulfill the goal of separation, how does it make sense to functionalize the silicon nanopost arrays with polymers?

Polymers, owing to their multiple chain units and possible various configurations under different external conditions, can be stimuli-responsive; therefore, this confers the device with additional intelligent response superiority. As mentioned in the last section, which is also the expanded work of Janus Si nanopost arrays modified by molecules reported by Yang's group [34], they selectively modified the Si nanopost arrays with poly(N-isopropylacrylamide) [PNIPAM] on one side and MHA on the other side. Owing to the thermoresponsive characteristic of the composite structure to exhibit anisotropic and isotropic (two distinct wetting behaviors) under different temperatures (above and below the (lower critical solution temperature) LCST of PNIPAM), the composite structure can be integrated into microfluidic system to build a thermoresponsive separator. Figure 6.8 is the scheme for the microfluidic system fabrication and the fluorescence microscope pictures during the process of injecting fluorescent dye solution into the microfluidic channel under temperatures above and below the LCST of PNIPAM. The channel can separate the liquid to one side when the temperature is above the LCST of PNIPAM attributed to the hydrophobicity of PNIPAM in that case, which is the opposite of MHA's hydrophilicity. The direction the solution flowed is the MHA-modified direction.

To our knowledge, there are a variety of intelligent stimuli-responsive polymers whose stimuli categories can be multifarious. Therefore, even though reported works on the related composite structures employed as separators are still very few in number, the future exploration of this area is very attractive.

6.4.3 BIOINTERFACES

Nature itself is a masterpiece that all the material scientists adore. Educated by the natural topography features and topographical interactions, countless biomimetic or derived three-dimensional (3D) biointerfaces gradually emerge in the public's vision. Herein, we mainly shed light on the biointerfaces derived from the

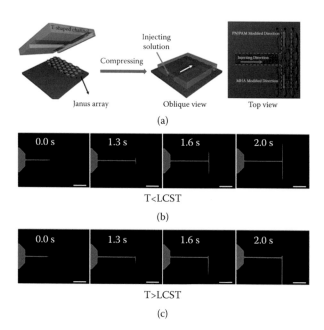

Figure 6.8 (a) Scheme for the T-shaped microfluidic channel's fabrication. (b and c) The fluorescence microscope pictures of the Rhodamine aqueous solution injected into a T-shaped microfluidic channel taken at different time when the temperature is below and above the LCST of PNIPAM. The scale bars are 1000 μm, and the downward direction is the MHA-modified direction.

Si nanopost or nanowire (NW). The biointerfaces discussed here will be classified by their polymers modified on the Si nanopost arrays: biopolymer derived biointerfaces and chemical-responsive polymer-derived biointerfaces.

6.4.3.1 Biopolymer-derived biointerfaces

Depending on the specific applied purpose of biointerfaces, the biopolymer can be classified to functional proteins, DNAs or RNAs, and specific-recognition molecules such as antibodies and aptamers.

As for the cell-adhesion requirement, the substrate needs coating with adhesive protein, including arginine-glycine-aspartic acid (RGD) peptide groups, collagens, and fibronectins, in order to facilitate the cell adhesion [37]. When it comes to the demand of antiadhesion, polyethylene glycol (PEG) will play a role.

Resting on the ability of silicon nanowires to penetrate the cell's membrane and release surface-bound molecules, the composite structure can accomplish gene delivery without chemical modification, releasing the genes directly into the cell. Park et al. reported introducing a series of biological effectors, including DNAs, RNAs, peptides and proteins, into almost any cell category [38]. The molecules were bound to the nanostructure by simply dispensing solution atop the substrate until the solvent is totally evaporated without washing. The modification is nonspecific and noncovalent, saving much effort in typical bioassay. Based upon this work, they explored the adaptability for immune cells for the possibility that immune cells may sense NWs as foreign substances that could be activating or lead apoptosis [39]. Therefore, they further studied the NW-based delivery modality's applicability for immune cells. According to the results, NWs can effectively deliver biomolecules into primary immune cells with minimum invasion, realizing systematical analysis of cell circuits and functional responses in normal and malignant hematopoietic cells from human and mouse. Especially, the NW-assisted gene silencing provides insight into the molecular circuitry contributing to heterogeneity.

Modification of specific-recognition biopolymers like antibody and aptamers endows the composite structure's ability to isolate target-recognized cells from the mixture of cells within the blood. Lee et al. functionalized the Si NWs with streptavidin, which possess high-affinity binding with biotin-labeled CD4+ T lymphocytes [40]. The CD4+ T lymphocytes were therefore picked out and the composite nanostructure is therefore considered to be promising for use as a cell separator. However, the problem herein is that the target has to be labeled, which is likely to be complex and costly. Wang et al. reported epithelial-cell adhesion molecule antibody (anti-EpCAM), whose cell-capture efficiency is outstanding, and modified silicon nanopost arrays employed as a 3D nanostructured surface for cancer-cell capturing [41]. They modified streptavidin onto the silicon nanopost arrays with the help of N-hydroxysuccinimide (NHS)/maleimide chemistry. The biotinylated anti-EpCAM is therefore introduced to the nanostructure for the subsequent cell capture. As shown in Figure 6.9, the silicon nanopost arrays provide more anchored sites for the interactions with cellular surface and results in significantly improved cell-capture efficiency compared with unstructured surface.

6.4.3.2 Chemical responsive polymer-derived biointerfaces

Intelligent responsive polymers facilitate the formation of smart surfaces. Their stimuli-responsive characteristic brings the composite structure turn-on/off function.

PNIPAM-modified Si NWs can control the adhesion and detachment of cancer cells by tuning the temperature [42]. The antibodies were connected to PNIPAM-Si NWs through hydrophobic interaction (between antibody anchored BSA and PNIPAM) when the temperature is above PNPAM's LCST, realizing the target cell's adhesion. When the temperature is below LCST, the PNIPAM becomes hydrophilic and the target cells detach simultaneously. The high surface area offered by the 3D nanostructure enables much higher adhesion as well as detachment efficiency as compared with unstructured PNIPAM-Si.

6.4.4 APPLICATION AS A FUNCTIONAL INTERFACE FOR PHOTOELECTRIC DEVICES

Owing to the advantages of low cost, relatively low toxicity, being easily solution processable, and having high absorption coefficients of greater than 10^5 cm^{-1}, polymer-based solar cells have attracted

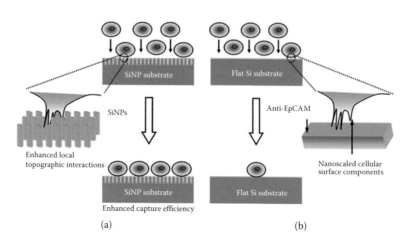

Figure 6.9 Schematic illustration for how anti-EpCAM-coated Si nanopost arrays can be employed to achieve significantly enhanced capture of EpCAM-positive cells (i.e., CTCs) from a cell suspension in contrast to an anti-EpCAM-coated unstructured (i.e., flat Si) substrate. (a) Interdigitation of nanoscale cellular surface components and SiNPs enhances local topographic interactions, resulting in vastly improved cell-capture efficiency. (b) Lack of local topographic interactions between cells and flat Si substrate compromises the respective cell-capture efficiency.

researchers' attention. However, they suffer from poor exciton diffusion lengths that are far less than that needed by polymers. Combination with silicon nanopost arrays enables a pore radius smaller than the exciton diffusion length to obtain maximum exciton harvesting, higher adsorption by polymer due to the large surface area, enhanced hole mobility, and increased charge carrier's direct path for the straight pores and promoted charge transfer [43]. Based on these advantages, the composite structure of Si nanopost arrays combined with polymers is considered to be a promising candidate for photoelectric devices fabrication.

Gowrishankar et al. reported amorphous-silicon (a-Si:H) nanoposts/polymer hybrid solar cells [43]. The polymer (P3HT or MEH-PPV) was spin-coated onto the nanopost arrays and heated afterward to infiltrate into the nanostructure completely. According to the results, there is strong energy transfer in the composite structure while inefficient hole transfer also occurs resulting in small currents. Nevertheless, the as-prepared device achieved improved efficiency compared with a bilayer device and an almost perfect charge-carrier extraction under short-circuit conditions. Attempts have also been made to form a hybrid between Si NWs and polymer to build solar cells by similar methods, achieving enhanced efficiency as well [44,45]. However, the defective transport properties of conjugated polymers result in a polymer layer that is too thick to make holes collected by external electrode, owing to the long transfer distance. The polymer layer should be as thin as possible. Sun et al. fabricated hybrid solar cells by combining Si nanopost arrays with thin-layer P3HT [46]. Through well-controlled regulation of the Si nanopost arrays' density, dipped in PCl$_5$ solution for different times, the polymer can reach the bottom of nanopost under appropriate conditions. They found that the P3HT layer thickness plays a critical role on Jsc, Voc as well as FF, and optimum thickness, which will define the final device efficiency. The as-prepared solar cell with P3HT layer of 10 nm achieved PCE of 9.2%, which improved considerably, compared with the previously reported silicon nanostructure/polymer hybrid solar cells (mostly less than 6%).

6.5 SUMMARY AND PERSPECTIVES

In this chapter, we have provided an overview of the silicon/polymer composite nanopost arrays: their fabrication, the corresponding unique properties, and various applications. The composite nanostructure can be presented in many formations: grafting polymers from nanoposts, spin-coating polymers onto the nanoposts, and depositing polymers onto the top of the individual nanopost. The silicon nanopost arrays can be fabricated through wet etching and dry etching with facile and adjustable approaches. Assisted by SI-ATRP or spin-coating or covalently modification of functional molecules, the polymer can hybrid with silicon nanopost arrays. On one hand, depending on the polymer's function, the composite structure exhibits various stimuli-responsive properties. On the other hand, the silicon nanopost's 2D PC property endows the composite structure a characteristic reflectance band gap, and the rough surface of silicon nanopost confers the composite structure's superwetting ability. All of these unique properties are supposed to be used in the application of sensing, separation, biointerfaces, and photoelectric devices.

The reported silicon/polymer composite nanopost arrays, however, are still quite few in number. The fabrication methods of both silicon nanopost arrays and polymer modification are well documented. Nevertheless, researchers are far less concerned with fabrication approaches to the composite nanostructures. Undoubtedly, there is some other existent form of composite nanostructures and corresponding fabrication methods we didn't embrace here and that remains to be discovered, which encourages more research efforts in this area. It is noteworthy that few composite nanopost arrays were reported to be employed as sensors and separators. Even so, the stimuli-responsive polymers are varied, providing a fundamental guarantee for the sensing requirements. Most biointerfaces typically modify the nanostructures with functional biomolecules. The chemically modified composite nanostructures are paid much less attention, and the polymer modified is mostly stereotyped. Much more passion and concern need to be put into this field. To conclude, the silicon/polymer composite nanopost arrays in sight are still just a tip of the iceberg. More efforts are needed to have a better view of this area and develop a perfect blueprint of the composite nanostructure material's future.

REFERENCES

1. Huang Z, Geyer N, Werner P, de Boor J, Gösele U. (2011). Metal-assisted chemical etching of silicon: A review. *Adv. Mater.* 23:285–308.
2. del Campo A, Arzt E. (2008). Fabrication approaches for generating complex micro-and nanopatterns on polymeric surfaces. *Chem. Rev.* 108:911–945.
3. Hou X, Jiang L. (2009). Learning from nature: Building bio-inspired smart nanochannels. *ACS Nano* 3:3339–3342.
4. Lahann J, et al. (2003). A reversibly switching surface. *Science* 299:371–374.
5. Barbey R, et al. (2009). Polymer brushes via surface-initiated controlled radical polymerization: Synthesis, characterization, properties, and applications. *Chem. Rev.* 109:5437–5527.
6. Stuart MAC, et al. (2010). Emerging applications of stimuli-responsive polymer materials. *Nat. Mater.* 9:101–113.
7. Elbersen R, Vijselaar W, Tiggelaar RM, Gardeniers H, Huskens J. (2015). Fabrication and doping methods for silicon nano- and micropillar arrays for solar-cell applications: A review. *Adv. Mater.* 27:6781–6796.
8. Lehmann V, Föll H. (1990). Formation mechanism and properties of electrochemically etched trenches in n-type silicon. *J. Electrochem. Soc.* 137:653–659.
9. Schuster R, Kirchner V, Allongue P, Ertl G. (2000). Electrochemical micromachining. *Science* 289:98–101.
10. Bassu M, Surdo S, Strambini LM, Barillaro G. (2012). Electrochemical micromachining as an enabling technology for advanced silicon microstructuring. *Adv. Funct. Mater.* 22:1222–1228.
11. Li X, Bohn PW. (2000). Metal-assisted chemical etching in HF/H_2O_2 produces porous silicon. *Appl. Phys. Lett.* 77:2572–2574.
12. Li W, Hu M, Ge P, Wang J, Guo Y. (2014). Humidity sensing properties of morphology-controlled ordered silicon nanopillar. *Appl. Surf. Sci.* 317:970–973.
13. Hsu CM, Connor ST, Tang MX, Cui Y. (2008). Wafer-scale silicon nanopillars and nanocones by Langmuir–Blodgett assembly and etching. *Appl. Phys. Lett.* 93:133109.
14. Chen JK, et al. (2012). Using colloid lithography to fabricate silicon nanopillar arrays on silicon substrates. *J. Colloid Interface Sci.* 367:40–48.
15. Ghoshal T, Senthamaraikannan R, Shaw MT, Holmes JD, Morris MA. (2012). "In situ" hard mask materials: A new methodology for creation of vertical silicon nanopillar and nanowire arrays. *Nanoscale* 4:7743–7750.
16. Tachi S, Tsujimoto K, Okudaira S. (1988). Low-temperature reactive ion etching and microwave plasma etching of silicon. *Appl. Phys. Lett.* 52:616–618.
17. Jansen H, de Boer M, Legtenberg R, Elwenspoek M. (1995). The black silicon method: A universal method for determining the parameter setting of a fluorine-based reactive ion etcher in deep silicon trench etching with profile control. *J. Micromech. Microeng.* 5:115–20.
18. Lin H, et al. (2014). Hierarchical silicon nanostructured arrays via metal-assisted chemical etching. *RSC Adv.* 4:50081–50085.

19. He Y, Jiang C, Yin H, Chen J, Yuan W. (2011). Superhydrophobic silicon surfaces with micro–nano hierarchical structures via deep reactive ion etching and galvanic etching. *J. Colloid Interface Sci.* 364:219–229.

20. Matyjaszewski K, Xia J. (2001). Atom transfer radical polymerization. *Chem. Rev.* 101:2921–2990.

21. Li B, Yu B, Ye Q, Zhou F. (2015). Tapping the potential of polymer brushes through synthesis. *Acc. Chem. Res.* 48:229–237.

22. Hu W, Liu Y, Chen T, Liu Y, Li C. (2015). Hybrid ZnO nanorod-polymer brush hierarchically nanostructured substrate for sensitive antibody microarrays. *Adv. Mater.* 27:181–185.

23. Li Y, et al. (2012). Polymer brush nanopatterns with controllable features for protein pattern applications. *J. Mater. Chem.* 22:25116-25122.

24. Liu W, et al. (2014). Bioinspired polyethylene terephthalate nanocone arrays with underwater superoleophobicity and anti-bioadhesion properties. *Nanoscale* 6:13845–13853.

25. Hatton BD, Aizenberg J. (2012). Writing on superhydrophobic nanopost arrays: Topographic design for bottom-up assembly. *Nano Lett.* 12:4551–4557.

26. Frommhold A, Robinson APG, Tarte E. (2012). High aspect ratio silicon and polyimide nanopillars by combination of nanosphere lithography and intermediate mask pattern transfer. *Microelectron. Eng.* 99:43–49.

27. Huang CF, Lin Y, Shen YK, Fan YM. (2014). Optimal processing for hydrophobic nanopillar polymer surfaces using nanoporous alumina template. *Appl. Surf. Sci.* 305:419–426.

28. Liu H, Zhang X, Wang S, and Jiang L. (2015). Underwater thermoresponsive surface with switchable oil-wettability between superoleophobicity and superoleophilicity. *Small* 11:3338–3342.

29. Xu H, Lu N, Qi D, Hao J, Gao L, Zhang B, Chi L. (2008). Biomimetic antireflective Si nanopillar arrays. *Small* 4:1972–1975.

30. Chen JK, Zhou GY, Huang CF, Ko FH. (2013). Using nanopillars of silicon oxide as a versatile platform for visualizing a selective immunosorbent. *Appl. Phys. Lett.* 102:251903.

31. Chen JK, Zhou GY, Chang CJ, Cheng CC. (2014). Label-free detection of DNA hybridization using nanopillar arrays based optical biosensor. *Sensors Actuators B* 194:10–18.

32. Li Y, et al. (2011). Fabrication of silicon/polymer composite nanopost arrays and their sensing applications. *Small* 7:2769–2774.

33. Li C, et al. (2012). In situ fully light-driven switching of superhydrophobic adhesion. *Adv. Funct. Mater.* 22:760–763.

34. Wang T, et al. (2015). Janus Si micropillar arrays with thermal-responsive anisotropic wettability for manipulation of microfluid motions. *ACS Appl. Mater. Interfaces* 7:376–382.

35. Wang T, et al. (2014). Anisotropic Janus Si nanopillar arrays as a microfluidic one-way valve for gas–liquid separation. *Nanoscale* 6:3846–3853.

36. Takei G, Nonogi M, Hibara A, Kitamori T, Kim H. (2007). Tuning microchannel wettability and fabrication of multiple-step Laplace valves. *Lab Chip* 7:596–602.

37. Liu X, Wang S. (2014). Three-dimensional nano-biointerface as a new platform for guiding cell fate. *Chem. Soc. Rev.* 43:2385–2401.

38. Shaleka AK, et al. (2010). Vertical silicon nanowires as a universal platform for delivering biomolecules into living cells. *Proc. Natl. Acad. Sci. U. S. A.* 107:1870–1875.

39. Shalek AK, et al. (2012). Nanowire-mediated delivery enables functional interrogation of primary immune cells: Application to the analysis of chronic lymphocytic leukemia. *Nano Lett.* 12:6498–6504.

40. Kim ST, et al. (2010). Novel streptavidin-functionalized silicon nanowire arrays for CD4+ T lymphocyte separation. *Nano Lett.* 10:2877–2883.

41. Wang S, et al. (2009). Three-dimensional nanostructured substrates toward efficient capture of circulating tumor cells. *Angew. Chem.* 121:9132–9135.

42. Liu H, et al. (2013). Hydrophobic interaction-mediated capture and release of cancer cells on thermoresponsive nanostructured surfaces. *Adv. Mater.* 25:922–927.

43. Gowrishankar V. (2008). Exciton harvesting, charge transfer, and charge-carrier transport in amorphous-silicon nanopillar/polymer hybrid solar cells. *J. Appl. Phys.* 103:064511.

44. Zhang F, Sun B, Song T, Zhu X, Lee S. (2011). Air stable, efficient hybrid photovoltaic devices based on poly(3-hexylthiophene) and silicon nanostructures. *Chem. Mater.* 23:2084–2090.

45. Shiu SC, Chao JJ, Hung S, Yeh CL, Lin CF. (2010). Morphology dependence of silicon nanowire/poly(3,4-ethylenedioxythiophene):Poly(styrenesulfonate) heterojunction solar cells. *Chem. Mater.* 22:3108–3113.

46. Zhang F, Han X, Lee S, Sun B. (2012). Heterojunction with organic thin layer for three dimensional high performance hybrid solar cells. *J. Mater. Chem.* 22:5362–5368.

7 Vertical silicon nanostructures via metal-assisted chemical etching

Seyeong Lee, Dong-Hee Kang, Seong-Min Kim, and Myung-Han Yoon

Contents

7.1 INTRODUCTION

Metal-assisted chemical etching (MACE) is a top-down fabrication method based on wet chemical etching and has been successfully used for preparing a variety of vertical silicon nanostructures. MACE allows for not only facile control over the geometric factors of vertical silicon nanostructures, but also large-batch fabrication in a cost-effective manner. Therefore, vertical silicon nanostructures fabricated via MACE have been proposed for various applications, such as photovoltaics, sensors, and bio-interfaces. Unlike typical wet etching processes, MACE uses a mixture of an acid (e.g., HF) and an oxidizing agent (e.g., H_2O_2) in the presence of a thin layer of metal catalyst (e.g., Au, Ag, and Pt) pre-patterned on the target silicon substrate. During the vertical etching process, the oxidizing agent is reduced by electrons transferred from silicon through the metal catalyst layer. Thus, the silicon surface just beneath the patterned metal layer is oxidized to silicon dioxide, which is subsequently dissolved by hydrofluoric acid (HF) contained in the etchant mixture. Accordingly, the geometric parameters of the resulting vertical silicon nanostructures can be delicately controlled by varying metal catalyst type, metal patterns, etchant composition, and etching reaction conditions used in MACE.

As many researchers across various fields could benefit from versatile fabrication and application of vertical silicon nanostructures, a number of remarkable studies have recently been published regarding the fundamental MACE mechanism, the fine control over silicon nanostructure properties (e.g., geometric parameters), and various applications demonstrated with these nanostructures. Therefore, we suppose that a comprehensive review of vertical silicon nanostructures fabricated via MACE should be very beneficial not only for scientists researching on the nanoscale silicon wet-etching, but also for engineers interested in practical applications of geometry-controlled vertical silicon nanostructures for a variety of purposes.

In this chapter, the following topics are covered: (1) the proposed mechanism of MACE, (2) the effects of various experimental parameters on the resultant vertical silicon nanostructure properties (e.g., catalyst metal, dopant type and doping level in silicon, and etchant composition), (3) the fine control of silicon nanostructure geometry relying on a variety of mask patterning methods, (4) a concise overview of important applications of vertical silicon nanostructures, and (5) brief prospective comments on vertical silicon nanostructure–related research in the future.

7.2 BASIC MECHANISM OF MACE

MACE differs from other conventional silicon etching techniques in that the silicon etching rate beneath a thin noble metal film–based catalyst layer is much faster than that at the silicon surface without metal coverage. Therefore, silicon nanostructures can be spontaneously formed on the silicon surface submerged in an appropriate etching solution because of the different silicon etching rate depending on the existence of metal coverage. In brief, the commonly accepted mechanism of MACE involves (1) the reduction of oxidant at the noble metal surface, (2) the oxidation of silicon interfaced with the catalyst metal, (3) the removal of oxidized silicon in the presence of HF in the etchant mixture, and (4) the mass transport of reactants and products near the nanostructured surface. Note also that the electroless metal deposition (EMD) with metal salts such as $AgNO_3$ and $CuCl_2$ has been successfully employed to fabricate vertical silicon nanostructures instead of directly-deposited metallic films or nanoparticles. In the following, we will first describe the MACE process in terms of EMD, silicon etching, and mass transport.

7.2.1 ELECTROLESS METAL DEPOSITION

EMD occurs via spontaneous metal growth in a solution without the supply of an external electrical power source. Morinaga et al. (1994) originally described the EMD on silicon substrates using an aqueous solution containing HF and $CuCl_2$. Currently, the most commonly used method for fabricating vertical silicon nanostructures involves the use of solutions containing HF and $AgNO_3$. This is due to the relatively easy solution preparation and simple fabrication process as Peng et al. (2006b) successfully demonstrated that single crystalline silicon nanowires can be formed simply by using the mixture solution of HF and $AgNO_3$.

The silver deposition on a silicon substrate is accomplished by a single redox reaction at the silicon–solution interface, involving cathodic (most likely hydrogen evolution and Ag ion reduction) and anodic (electrons release from or hole consumption in silicon) processes. Each possible reaction can be expressed including the corresponding standard redox potential with respect to the standard hydrogen electrode (SHE). The possible cathodic reactions are:

$$2H^+ + 2e^- \rightarrow H_2 \uparrow \qquad E^0 = 0.00 \text{ V/SHE} \tag{7.1}$$

$$Ag^+ + e^- \rightarrow Ag \qquad E^0 = +0.79 \text{ V/SHE} \tag{7.2}$$

and the possible anodic reactions are:

$$Si + 2F^- \rightarrow SiF_2 + 2e^- \qquad E^0 = -1.2 \text{V/SHE} \tag{7.3}$$

$$Si + 2F^- + 2H^+ \rightarrow SiF_2 \qquad E^0 = -1.2 \text{V/SHE} \tag{7.4}$$

where the redox reaction or charge exchange is driven by the potential difference in a paired reaction in the solution and the Fermi level of the silicon substrate. When Ag ions approach the silicon surface, they capture electrons from silicon and metallic Ag film is deposited on the silicon surface. While the Ag layer with strong catalytic activity could function as active sites for the previously described cathodic reactions, the reduced Ag strongly attracts electrons from silicon and becomes negatively charged because of its higher electronegativity than silicon. Therefore, Ag ions in the solution tend to pull electrons from the nearest Ag nuclei rather than from the bare silicon surface, leading to the Ag nucleus growth while the underlying silicon keeps losing electrons and forming oxide (SiO_2) (Chattopadhyay et al. 2002; Peng et al. 2003, 2006b; Kolasinski 2005; Hadjersi 2007; Tsujino and Matsumura 2007).

7.2.2 SILICON ETCHING

Generally, MACE is conducted in an aqueous solution of HF and hydrogen peroxide (H_2O_2). As mentioned previously, the silicon etching rate in the HF (40%)–H_2O_2 (35%)–H_2O solution (1/5/10, v/v) is very low and increases from 8 to 15 nm/min when the etching solution temperature increases from 20°C to 40°C in the absence of a metal catalyst. In the presence of a noble metal in contact with the silicon surface, both the reduction of H_2O_2 at the interface between the metal and solution (cathodic reaction) and the dissolution of silicon at the interface between the silicon and metal (anodic reaction) occur simultaneously. For such systems, the cathodic reaction is generally accepted as the reduction of H_2O_2 at the metal as described below, where NHE denotes the normal hydrogen electrode:

$$H_2O_2 + 2H^+ \rightarrow 2H_2O + 2H^+ \qquad E^0 = +1.76 \text{ V / NHE} \tag{7.5}$$

As a strong oxidant, H_2O_2 allows for hole injection into silicon at the metal–silicon interface, and the metal dramatically enhances the silicon etching rate by catalyzing the reduction of H_2O_2. The reduction of H^+ into H_2 was proposed to accompany the cathodic reaction with the reduction of H_2O_2 in the presence of Pt deposits, since gas evolution occurred during silicon etching. However, Chartier et al. (2008) considered that gas evolution may be related to the dissolution of silicon at the anodic site. It is well known that H_2 is evolved during the electrochemical dissolution of silicon in aqueous HF solutions with a high HF/HNO_3 ratio, leading to the porous silicon formation. This could be related to the fact that porous silicon should be initially formed by the dissolution of silicon, as proposed by Geyer et al. (2012). It is assumed that if a porous silicon layer is underneath the metal layer, then the reduction of H_2O_2 could occur not only at the metal–solution interface but also at the metal–silicon interface. This hypothesis was investigated by using a chromium (Cr) layer (which does not catalyze the reduction of H_2O_2) deposited on top of the catalytic noble metal surface to block the reduction of H_2O_2 at the metal–solution interface. In this case, the etching rate is similar to the case without Cr. Therefore, this prior study concluded that while the reduction of H_2O_2 at the metal–silicon interface is possible, this cannot occur if there exists no porous silicon at the interface.

Although anodic reactions, during which silicon is oxidized and dissolved, have been proposed in many previous models, these can be categorized into three reactions as shown below:
(1) Direct dissolution of silicon into the tetravalent state (Reaction 7.1)

$$Si + 4H^+ + 4HF \rightarrow SiF_4 + 4H^+ \tag{7.6}$$

$$SiF_4 + 2HF \rightarrow H_2SiF_6 \tag{7.7}$$

(2) Direct dissolution of silicon into the divalent state (Reaction 7.2)

$$Si + 4HF_2^- \rightarrow SiF_6^{2-} + 2HF + H_2 \uparrow + 2e^- \tag{7.8}$$

Arrays, hybrids, and core–shell

(3) Silicon oxide formation followed by dissolution of oxide (Reaction 7.3)

$$Si + 2H_2O \rightarrow SiO_2 + 4H^+ + 4e^- \tag{7.9}$$

$$SiO_2 + 6HF \rightarrow H_2SiF_6 + 2H_2O \tag{7.10}$$

There are two primary differences between Reactions 7.1, 7.2, and 7.3, due to silicon oxide formation at the silicon surface before the dissolution of silicon and H_2 generation during the dissolution of silicon. It still remains an open question whether the silicon oxide formation occurs during etching, due to the difficulty of in situ or ex situ investigations of this phenomenon. Chartier et al. (2008) suggested the following mixed reaction for this behavior, composed of divalent and tetravalent dissolution for the anodic reaction in MACE:

$$Si + 6HF + nH^+ \rightarrow H_2SiF_6 + nH^+ + \left[\frac{4-n}{2}\right]H_2 \tag{7.11}$$

Here, n is the number of holes per dissolved silicon atom. Taking into account the cathodic reaction of H_2O_2, the balanced overall reaction was proposed as follows:

$$Si + 6HF + \frac{n}{2}H_2O_2 \rightarrow H_2SiF_6 + nH_2O + \left[\frac{4-n}{2}\right]H_2 \tag{7.12}$$

In this case, ρ is defined as $(HF)/([HF]+[H_2O_2])$, for the maximum etching rate of ~80%. Assuming that the maximum rate is related to the stoichiometry of the reaction, they suggested n = 3 in the above overall redox reaction (Lee et al. 2008; Zhang et al. 2008b; Li 2012).

7.2.3 MASS TRANSPORT

In addition to understanding the reactions among the constituent chemicals in MACE, the mass transport of reactants to and the continuous removal of byproducts from the catalyst should be considered for fabricating vertical silicon nanostructures with well-defined geometric parameters. When the metal particles act as catalyst, the electrolyte and reactants in the etching solution can reach the interface between the metal and silicon. Therefore, the influence of mass transport of the reactants and byproducts on the overall MACE reaction is negligible. However, the diffusion of reactants and byproducts becomes a rate-determining factor if the noble metal catalyst used in the form of an extended film for MACE. Three possible diffusion models are presented in Figure 7.1. Model 1 is where the diffusion of reactants and byproducts occurs through small pores in the metal, which form in the region without the metal coverage if the thickness of the metal film is below 15 nm. Model 2 shows that the diffusion of reactants and byproducts takes place in a thin permeable channel formed at the metal–silicon interface. Lastly, Model 3 describes the case that silicon atoms diffuse through the metal film, oxidize at the surface of the metal, and then finally dissolve as $(SiF_6)^{2-}$.

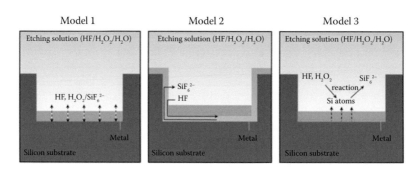

Figure 7.1 Suggested diffusion models in metal-assisted chemical etching.

Arrays, hybrids, and core–shell

Geyer et al. (2012) explained the details of the diffusion process in each model based on MACE experimental results. When the thin metal film contains additional small pores, the resultant vertical silicon nanostructures are composed of ordered silicon wires and whiskers with the diameter of 5–15 nm. In the case of thick metal films (>50 nm), however, only the ordered wires formed, which excludes Model 1 in the case of vertical silicon nanostructure fabrication by MACE.

The validity of Models 2 and 3 can be verified by investigating the etching rate as a function of the size and thickness of metal film, since the diffusion length is determined by the lateral film size in Model 2 and by the film thickness in Model 3. At the constant film thickness of 40 nm, the samples with the later size of 390 nm and 710 nm exhibited the etching rates of 35 and 20 nm/min, respectively. Although the film thickness was changed, the etching rate remained constant. Therefore, these results clearly show the validity of Model 2. In conclusion, the silicon oxidation and dissolution occur at the metal–silicon interface, even if the lateral size of metal films is on the micrometer scale. For metal films with the width of 3–12 µm, the etching rate is faster at the edges than in the middle of the film because of the long diffusion path beneath the metal film. If the diffusion length is too long, the metal film bends as the reactants are consumed before the silicon underneath the entire metal film is laterally etched (Peng et al. 2003; Chartier et al. 2008).

7.3 EXPERIMENTAL PARAMETERS

Vertical silicon nanostructures fabricated by MACE can be tuned to obtain a variety of mechanical, physical, and optical properties by varying their fabrication conditions. The major parameters that can strongly influence such structural properties are catalyst metal, silicon substrate type, and etching solution conditions. In this section, we will describe the effects of these parameters on the etching reactions and resultant silicon nanostructures including other factors.

7.3.1 METAL CATALYST

The deposition of a noble metal film onto the silicon surface can be performed by a variety of methods. Initially, EMD and metal particles were used as catalysts in MACE (Figure 7.2a and b). Since it is difficult to control the defined geometry of vertical silicon nanostructures in this manner, vacuum deposition methods such as thermal evaporation, sputtering, and e-beam evaporation are more advantageous. By using vacuum deposition, metal films can be formed in a particular pattern that can determine the spatial arrangement of the resultant vertical silicon nanostructure arrays (Figure 7.2c) (Peng et al. 2005; Chartier et al. 2008; Hildreth et al. 2009).

The type of used metal also influences the properties of fabricated vertical silicon nanostructures. Typically, MACE employs noble metals such as Ag, Au, Pt, and Pd. The straight pores are formed in the case of using isolated Ag or Au particles, whereas the pores with somewhat different shapes are formed using Pt particles. For example, straight or helical pores are induced by EMD-derived Pt particles, whereas Pt particles prepared from electroless plating or sputtering tend to move randomly. Additionally, in the case of using Pt, the reaction rate during the etching process and the porosity in the consequent surface are enhanced compared with Au or Ag, as shown in Figure 7.3. Although the variations in MACE reaction among these noble metals are not fully investigated yet in the literature, they may be attributed to the different catalytic activity of each noble metal with regard to H_2O_2 reduction. Indeed, as more holes are injected, the etching rate becomes faster and the diffusion of holes to the side walls of the etched silicon structure (which induced porous layer formation) is more frequent (Peng et al. 2003, 2006a; Chattopadhyay and Bohn 2004; Lee et al. 2008; Chiappini et al. 2010).

7.3.2 SILICON SUBSTRATE TYPE

Commercial silicon wafers can be classified in terms of dopant type, doping level, and crystal orientation. Doping level and dopant type seemingly affect both the rate of silicon etching and the morphology of the resultant vertical silicon nanostructures, although the exact mechanisms are not clear yet. While there has been a report that the pore size and etching depth are slightly different between highly boron doped p-type

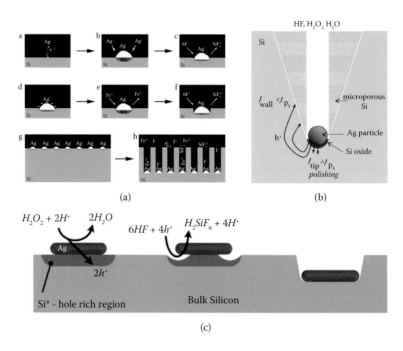

Figure 7.2 The dependence of metal shape on catalytic reduction of H_2O_2. (a) Metal nanoparticles from electroless metal deposition. (Reproduced from Peng, K., et al.: Uniform, axial-orientation alignment of one-dimensional single-crystal silicon nanostructure arrays. *Angewandte Chemie International Edition.* 2005. 44. 2737. Copyright Wiley-VCH Verlag GmbH & Co. KGaA. With permission.) (b) Spherical metal particles. (Reprinted from *Electrochimica Acta*, 53, Chartier, C., et al., Metal-assisted chemical etching of silicon in HF–H_2O_2, 5509–5516, Copyright 2008, with permission from Elsevier.) (c) Metal strip. (Reprinted with permission from Hildreth, O.J., et al., Effect of catalyst shape and etchant composition on etching direction in metal-assisted chemical etching of silicon to fabricate 3D nanostructures, *ACS Nano*, 3, 4033–4042, 2009. Copyright 2009 American Chemical Society.)

silicon (0.01–0.03 Ω·cm) and low boron doped p-type silicon (1–10 Ω·cm), there is another report showing the conflict result that the etching depth in low boron doped p-type silicon (10 Ω·cm) is 1.5 times higher than that of highly boron doped p-type silicon (0.01 Ω·cm) under the identical condition. Regarding the effects of dopant type on silicon etching rate, p-type silicon substrates are etched more slowly than n-type silicon substrates in the case of (100) and (111) orientations. Mikhael et al. (2011) have shown that the etching depth on the low phosphorous doped n-type silicon is the highest, then low boron doped p-type silicon and highly phosphorous doped n-type silicon (which are almost the same) follows, and that on the highly boron doped p-type silicon is the lowest (Figure 7.4a). Furthermore, the etched vertical silicon nanostructures tend to show the morphology transition from microporous to mesoporous as the doping level increases. As shown in Figure 7.3, whether the resulting vertical silicon nanostructures are micro- or mesoporous is also affected by the composition of the etching solution and the metal type. The apparent trend of porosity formation seems to be related to the resistivity of silicon substrate (thus, doping level), and the similar trend was observed in the fabrication of vertical silicon nanostructures (Figure 7.4b and c). Although this tendency is apparent in numerous studies, other conditions should be considered carefully for fabricating porous vertical silicon nanostructures. Zhang et al. (2008) reported the similar result with highly boron doped p-type silicon (100) substrates (0.003–0.005 Ω·cm) and highly boron doped p-type silicon (111) substrates (0.004–0.008 Ω·cm) using aqueous solutions of HF (4.8 M)/H_2O_2 (0.15 M) with Ag particles. The origin that highly doped silicon substrates exhibit micro- or mesoporous structures after MACE processing may be attributed to the diffusion of holes at the metal–silicon interface to the region without the metal coverage during the exposure to etching solution. Similarly, the top surface area of vertical silicon nanostructures fabricated with MACE is typically more porous than other regions due to the prolonged exposure time.

Chen et al. (2008) demonstrated that there is a preferential crystallographic orientation of vertical silicon nanostructures formed by MACE using HF/H_2O_2/AgNO$_3$. As shown in Figure 7.5,

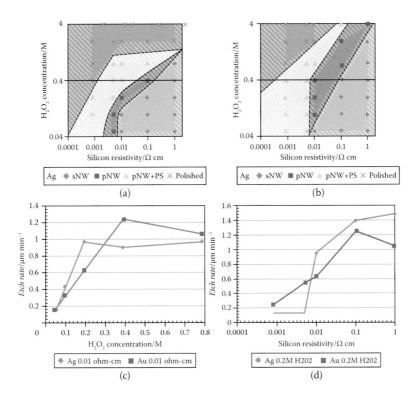

(a)

(b)

(c)

(d)

Figure 7.3 The influences of various H_2O_2 concentration and silicon substrate resistivity. Phase diagrams showing the different silicon nanostructure morphologies when (a) Ag and (b) Au are used as a catalyst. The etching rates varying as a function of (c) H_2O_2 concentration and (d) silicon substrate resistivity. (Reproduced from Chiappini, C., et al.: Biodegradable porous silicon barcode nanowires with defined geometry. *Advanced Functional Materials.* 2010. 20. 2231–2239. Copyright Wiley-VCH Verlag GmbH & Co. KGaA. With permission.)

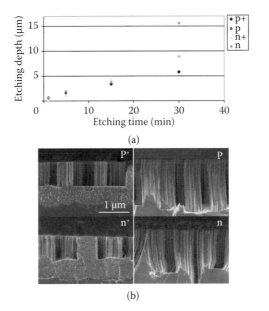

(a)

(b)

Figure 7.4 (a) The etching depth at different etching time in four types of silicon substrates p$^+$, p, n$^+$, and n etched using the etching solution of HF/H_2O_2 (4 M/0.88 M). (b) Cross-sectional SEM images of four types of silicon substrates p$^+$, p, n$^+$, and n processed using the etching solution of HF/H_2O_2 (4 M/0.1 M). (Reprinted with permission from Mikhael, B., et al., New silicon architectures by gold-assisted chemical etching, *ACS Applied Materials & Interfaces*, 3, 3866–3873, 2011. Copyright 2011 American Chemical Society.)

Arrays, hybrids, and core-shell

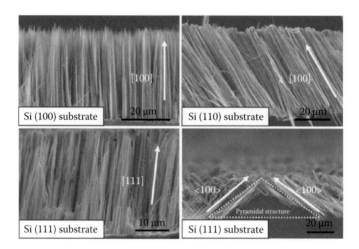

Figure 7.5 Cross-sectional SEM images of vertical silicon nanowires formed on (a) (100), (b) (110), and (c, d) (111) silicon substrates. (d) Certain regions on a (111) silicon substrate are formed under [100] anisotropic etching. (Reproduced from Chen, C.-Y., et al.: Morphological control of single-crystalline silicon nanowire arrays near room temperature. *Advanced Materials*. 2008. 20. 3811–3815. Copyright Wiley-VCH Verlag GmbH & Co. KGaA. With permission.).

vertical silicon nanostructures etched in this condition follow the [100] orientation on silicon (100), (110), and (111) substrates. Such an orientation dependence could be related to the silicon lattice configuration at the reaction site and the passivation with hydrogen (H)-terminated silicon atoms (Zhang et al. 2008b; Zhong et al. 2011; Balasundaram et al. 2012; Smith et al. 2013).

7.3.3 ETCHING SOLUTION

MACE etching solutions are typically composed of HF, H_2O_2, and H_2O (with ethanol added occasionally). The concentrations of H_2O_2 and HF affect the etching rate and the morphologies of the resulting vertical silicon nanostructures. As shown in Figure 7.3, the etching rate generally increases as the H_2O_2 concentration increases, since H_2O_2 acts as a hole donor and oxidant during the etching process up to a certain saturation point. When Pt particles are employed as catalyst, straight pores are formed with etching solutions containing low HF concentration (HF [50%]: H_2O_2 (30%): H_2O = 2 : 1: 8, v: v: v). The cone-shaped porous structures with the diameter of >1 μm are also formed near the straight pore. In the case of the fourfold increase in HF concentration (HF [50%]: H_2O_2 = 10:1, v: v), the size of porous structure near the straight pore dramatically decreases below 100 nm. According to Liu et al. (2012), the oxidation speed of silicon around Ag particles increases with the H_2O_2 concentration, resulting in the accelerated horizontal etching rate of silicon. As the H_2O_2 concentration in the etching solution reaches 20%, more silicon is oxidized to SiO_2 around the Ag nanoparticles and then dissolved by HF, leading to the increased horizontal etching rate (Figure 7.6a). This escalated horizontal etching causes the 20% of the resultant vertical silicon nanowires to possess a diffusion configuration with low nanowire density and enlarged spacing. Increasing the H_2O_2 concentration up to 30% enables the horizontal etching rate to overcome the gravity of Ag nanoparticles and shift their positions. In this case, their vertical positions can be deviated from the initial location during the etching process, and the consequent vertical silicon nanowires exhibit randomly oriented porous structures. Ethanol, when used as a surfactant, can improve the wetting of water-based solutions on the silicon surface. As shown in Figure 7.6b, the rise in ethanol concentration induces the reduction in etching rate. This result is associated with the sequential transition from solid (no porous layers) to *reemerging* polished surface as indicated by the presence of a porous surface morphological phase (in between the porous nanowire and porous bottom surface) and a polishing phase (Hadjersi 2007; Tsujino and Matsumura 2007; Chartier et al. 2008; Chiappini et al. 2010; Mikhael et al. 2011).

Arrays, hybrids, and core–shell

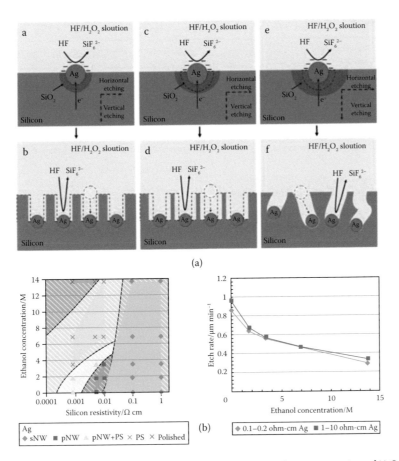

(a)

(b)

Figure 7.6 (a) A scheme of MACE with Ag nanoparticles while increasing the concentration of H_2O_2 (10%, 20%, and 30%). (With kind permission from Springer Science+Business Media: *Nanoscale Research Letters,* Fabrication and photocatalytic properties of silicon nanowires by metal-assisted chemical etching: Effect of H_2O_2 concentration, 7, 2012, 663, Liu, Y., et al.) (b) Phase diagram (left) and etching rate (right) of different silicon nanostructure morphologies as a function of silicon resistivity and ethanol concentration. (Reproduced from Chiappini, C., et al.: Biodegradable porous silicon barcode nanowires with defined geometry. *Advanced Functional Materials.* 2010. 20. 2231–2239. Copyright Wiley-VCH Verlag GmbH & Co. KGaA. With permission.).

7.3.4 OTHER EXPERIMENTAL PARAMETERS

MACE processes can also be controlled by temperature or light exposure. Previous studies have shown that the etching rate is accelerated as the solution temperature is increased. While the difference in etching depth between the dark and room-light conditions is less than 5% for both p- and n-type silicon substrates, the difference becomes 150% (i.e., 1.5 times) in the case of using a 20 W bulb. These results can be explained in terms of the number of photo-excited holes; under the low-intensity illumination, the number of photo-excited holes is negligible compared with that originated from the reduction of H_2O_2, but both numbers become comparable to each other under the high-intensity illumination.

In certain cases, the catalyst layer can be composed of two types of metal. In 2011, Kim et al. studied the MACE process using a Au/Ag bilayer system. Since Au is inert to the oxidative dissolution by etching solutions, the upper Au layer can effectively block the undesired structural disintegration in the underlying Ag layer as well as the tapering in fabricated vertical silicon nanowires. Yeom et al. (2014) reported the MACE reaction employing a titanium (Ti) layer which was deposited between Au and silicon as an effective adhesion layer. In the Au/Ti bilayer system, Ti is instantly dissolved in the etching solution, affording the Au layer an intimate contact with the underlying silicon substrate, so that silicon could be etched without any delay (Chen et al. 2010; Hildreth et al. 2012).

Arrays, hybrids, and core–shell

7.4 METHODS FOR NANOSTRUCTURE GEOMETRY CONTROL

The geometric control of silicon nanostructures is an important challenge, as the nanoscale morphology plays a vital role in the properties and further applications of the resulting silicon nanowires. Among the known fabrication methods for silicon nanowires, MACE is relatively straightforward in terms of nanoscale geometric control. As described in Section 7.2 on the basic mechanism of MACE, the etching reaction occurs at the metal–silicon interface. Therefore, the selective coverage of catalyst layer via patterning a metal film on silicon determines the overall nanostructure geometry and distribution. In this section, the control over silicon nanostructure geometry based on various metal layer lithography methods is mainly discussed.

7.4.1 METAL FILM DEWETTING

The control over the geometric parameters of vertical silicon nanostructures can be mainly achieved by depositing the metal catalyst film with size-defined pores at designated locations. First, the simple fabrication of vertical silicon nanowire arrays can be done using metal film dewetting processes without complicated lithography-based pattering (Fang et al. 2006; Chartier et al. 2008; Yang et al. 2008). Liu et al. (2013) employed the Ag deposition at different film thickness and annealing temperature conditions in order to tune the patterned Ag film properties. The dewetting process of thin metal films involves hole nucleation and growth phenomenon on solid silicon substrates. The most common origin for this heterogeneous nucleation is grain boundary grooving, which may arise from the surface free from metal coverage and the metal–substrate interface. The morphology of dewetted Ag film as a function of film thickness (9–14 nm) at 150°C shows the clear distinction depending on Ag film thicknesses (Figure 7.7). The 9-nm-thick flat Ag film was completely converted to one with the nanoparticle-like morphology, while the bicontinuous morphology was obtained from the 11-nm-thick Ag film. Note also that at a given film thickness, the metal film morphology can be modified further by controlling the annealing temperature. The nanostructure geometry control based on metal dewetting is relatively simple and inexpensive in MACE. However, it is difficult to precisely control the nanowire diameter and the nanowire-to-nanowire distance with this method.

7.4.2 NANOSPHERE LITHOGRAPHY

Nanosphere lithography is an inexpensive and simple nanofabrication technique, which relies on a monolayer of hexagonally close packed nanosphere arrays, and has been successfully combined with MACE for fabricating vertical silicon nanowire arrays with defined nanostructure geometry (Peng et al. 2007; Wang et al. 2010; Mikhael et al. 2011). To accurately determine silicon nanowire dimension and spacing, a self-assembled monolayer of nanosphere arrays was deposited onto silicon substrates, followed by the typical MACE process (Huang et al. 2007). A schematic describing the fabrication procedure is presented in Figure 7.8a. First, a self-assembled monolayer of closely-packed polystyrene nanospheres was formed on a clean silicon substrate. A reactive ion etching (RIE) process is applied to reduce the diameter of nanospheres, thereby generating the interparticle spacing. Next, a metal catalyst layer is deposited by thermal evaporation onto the silicon surface covered with a nonclosely packed nanosphere array. Owing to this

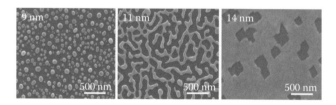

Figure 7.7 SEM images showing surface morphologies of silicon substrates prepared via MACE with different Ag film thicknesses annealed at 150°C for 10 min. (With kind permission from Springer Science+Business Media: *Nanoscale Research Letters,* Lithography-free fabrication of silicon nanowire and nanohole arrays by metal-assisted chemical etching, 8, 2013, 155, Liu, R., et al.)

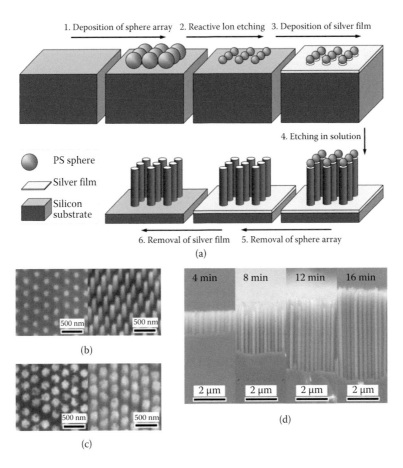

Figure 7.8 (a) A schematic diagram of MACE process combined with nanosphere lithography. (b) and (c) are top- and tilted-view of vertical silicon nanowire arrays with the diameter of 100 and 180 nm, respectively. (d) Cross-sectional SEM images of height-varied vertical silicon nanowires after etching for 4–16 min. (Reproduced from Huang, Z., et al.: Fabrication of silicon nanowire arrays with controlled diameter, length, and density. *Advanced Materials*. 2007. 19. 744–748. Copyright Wiley-VCH Verlag GmbH & Co. KGaA. With permission.)

polystyrene monolayer mask, the resulting metal film features a hexagonal array of holes. Subsequently, MACE is conducted using a mixture of water, HF, and H_2O_2. Finally, the fabrication of geometry-controlled vertical silicon nanostructure arrays is finalized by removing PS nanospheres and metal film with a $CHCl_3$ solution and boiling aqua regia, respectively. Figure 7.8b and c shows that the diameter of fabricated silicon nanowires can be tuned by using polystyrene nanospheres with different sizes. Another geometric parameter, vertical nanostructure height, is related to the etching time. The height of vertical silicon nanowire array varies linearly with the duration of etching reaction, which enables the good control over the length of the etched nanowire (Figure 7.8d). Nanosphere lithography is the most popular method for patterning the holes in metal films and it is successfully combined with MACE to fabricate a wide range of nanowire diameters and densities; nanosphere diameter and spacing after the RIE process reflect nanowire diameter and nanowire-to-nanowire distance, respectively. Note that the center-to-center distance among nanowires is determined by the initial nanosphere diameter. Thus, the diameter and areal density of vertical silicon nanowire arrays are coupled to each other, and it is difficult to control these parameters in an independent manner (Yeom et al. 2014).

Recently, an unconventional but simple approach toward independently controlling these geometric parameters was reported, in which a nonclosely packed nanosphere array was uniformly deposited on silicon substrates (Lee et al. 2015). They introduced the polyelectrolyte multilayer as an adhesive layer, which enables nanospheres to be attached in a uniform monolayer with random spacing (Figure 7.9a).

Arrays, hybrids, and core-shell

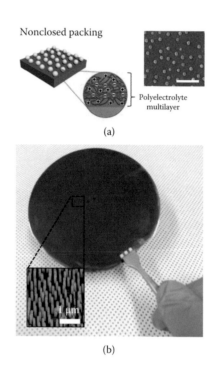

Nonclosed packing

Polyelectrolyte
multilayer

(a)

(b)

Figure 7.9 (a) A schematic of randomized nonclosely packed nanosphere lithography using polyelectrolyte multilayer-based adhesives. (b) Optical photograph of a 4-inch silicon wafer fully modified with vertical silicon nanowire arrays. (Reproduced from Lee, S., et al., Polyelectrolyte multilayer-assisted fabrication of nonperiodic silicon nanocolumn substrates for cellular interface applications, *Nanoscale*, 7(35), 14627–14635, 2015. By permission of The Royal Society of Chemistry.)

In this case, the diameter and areal density of PS nanospheres on the silicon surface can be directly translated into the diameter and density of the vertical silicon nanowires, respectively, without the aid of hexagonal cross-packing and RIE etching of nanospheres. Furthermore, the polyelectrolyte adhesion layer facilitates the simple and cost-effective fabrication of vertical silicon nanowire arrays on very large substrates, for example, over 3-inch to 4-inch silicon wafers (Figure 7.9b).

7.4.3 AAO LITHOGRAPHY

Practically, it is not straightforward to uniformly cover large silicon substrates with a self-assembled monolayer of nanospheres. In contrast, an anodic aluminum oxide (AAO) membrane can be conveniently formed by the anodization of aluminum, which enables a diverse range of nanopore diameters (10–350 nm) and densities (5×10^8 to 3×10^{10} pores/cm^2). Huang et al. (2008) employed AAO membranes as a mask for patterning metal catalyst layers, thereby forming vertical silicon nanostructures with controlled diameters. Both the average pore diameter in the AAO membrane and the thickness of metal catalyst film can affect the resulting silicon nanowire geometry. Huang et al. (2010) demonstrated that the silicon nanowire diameter can be precisely controlled with a precision of 10 nm by fine tuning the nanopore diameter in the AAO membrane (Figure 7.10a and b). The fabrication procedure for AAO-assisted vertical silicon nanowires are illustrated in Figure 7.10c. The AAO membrane was first transferred to the surface of the silicon substrate and then Cr (10 nm) and Au (30 nm) metal films were deposited through the AAO nanopores by thermal evaporation. Subsequently, the AAO membrane was removed and a thin layer of Au was evaporated onto the surface. After forming metal catalyst nanopatterns, MACE was conducted using a solution containing 4.6 M HF and 0.44 M H$_2$O$_2$. In this method, the diameter and density of the resulting vertical silicon nanowires can be modulated by carefully manipulating the pore size in the AAO membrane. Furthermore, the use of Pt and Au in conjunction with AAO will also be of great interest for fabricating vertical silicon nanostructures with finely controlled diameters (Kim et al. 2011b; Márquez et al. 2011).

Arrays, hybrids, and core–shell

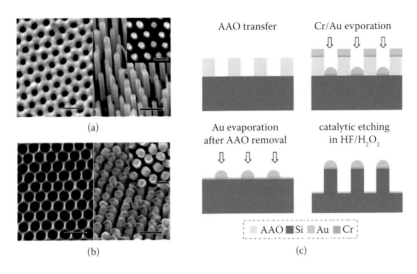

Figure 7.10 SEM images of the anodic aluminum oxides (AAOs) with average pore diameters (left) of (a) 40 nm and (b) 80nm, respectively, and the corresponding vertical nanowire arrays (right). (c) A schematic of vertical silicon nanowire fabrication using AAO. (Reprinted with permission from Huang, J., et al., Fabrication of silicon nanowires with precise diameter control using metal nanodot arrays as a hard mask blocking material in chemical etching, *Chemistry of Materials*, 22(13), 4111–4116, 2010. Copyright 2010 American Chemical Society.)

7.4.4 LASER INTERFERENCE LITHOGRAPHY

Laser interference lithography is a maskless patterning technique for producing a regular array of fine features using two or more coherent light sources. By adjusting the exposure process, the photoresist openings with various cross-sectional shapes, including circles, ovals, and rectangles, can be generated (Hildreth et al. 2009; Chang et al. 2010a). Choi et al. (2008) reported the fabrication of vertical silicon nanostructures based on the combination of laser interference lithography and MACE as the main experimental processes (Figure 7.11a). First, a silicon wafer is coated with a photoresist and then is exposed to the laser-induced interference pattern. The circular-shaped photoresist dots are remained on the silicon surface after the photoresist at the exposed areas are removed by dipping in a developer solution. Subsequently, the photoresist-patterned substrate is treated by oxygen plasma etching in order to remove any residual photoresist and reduce the lateral size of nanoscale photoresist patterns. Next, a Au metal catalyst film was deposited by thermal evaporation onto the patterned silicon substrate, which is then dipped in an etchant solution at room temperature. This technique can produce perfectly periodic nanostructures over a very large area, and independently control the diameter and nanostructure-to-nanostructure spacing down to 200 nm or less. In particular, the density of vertical silicon nanowire arrays can be controlled by varying the exposure angle during interference lithography (Figure 7.11b–d).

7.4.5 BLOCK-COPOLYMER LITHOGRAPHY

Another way to fabricate metal catalyst patterns is through block-copolymer lithography. The direct self-assembly of block copolymers has shown great potential for sub-20-nm lithography due to its high resolution scalability, cost effectiveness, and compatibility with the conventional fabrication processes for integrated circuits. Initially, Chang et al. (2009) used one type of block copolymer to form a metal catalyst pattern for MACE. An ordered structure of polystyrene-*block*-polyferrocenyldimethylsilane (PS-*b*-PFS) could be obtained by spin coating and vacuum annealing on top of silicon dioxide/silicon substrates. Vertical silicon nanowire arrays with high areal density and high aspect ratio (up to 200) were successfully fabricated with this method. Bang et al. (2011) demonstrated the simple and large-scale production of vertical silicon nanowire arrays by combining block-copolymer templates and MACE techniques. Figure 7.12a shows the overall fabrication procedure for porous vertical silicon nanowire arrays using this method. Block copolymer–containing iron salts were spin coated onto the silicon surface and treated with oxygen plasma to form dotted patterns of iron oxide. Subsequently, a Ag catalyst film was deposited on the silicon surface,

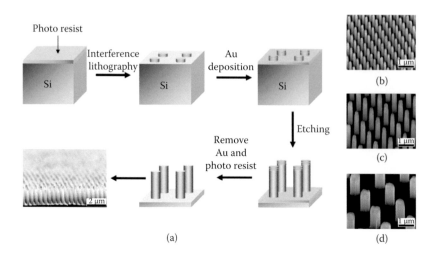

Figure 7.11 (a) A schematic illustration of the fabrication procedure for vertical silicon nanostructures by combining interference lithography and MACE. SEM images of silicon nanowire arrays with different areal densities: (b) 4×10^6 mm^{-2}, (c) 1×10^6 mm^{-2}, and (d) 3.5×10^5 mm^{-2}. (Reprinted with permission from Choi, W.K., et al., Synthesis of silicon nanowires and nanofin arrays using interference lithography and catalytic etching, *Nano Letters*, 8(11), 3799–3802, 2008. Copyright 2008 American Chemical Society.)

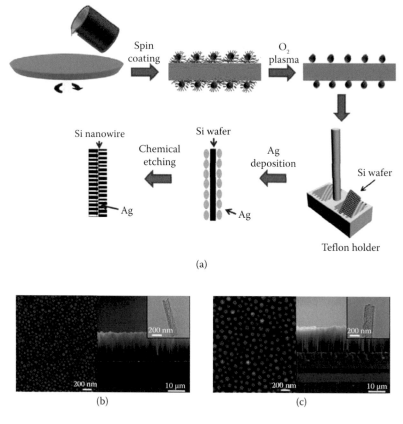

Figure 7.12 (a) A schematic diagram of fabricating vertical silicon nanowire arrays by combining block copolymer template and MACE. (b, c) AFM images of iron oxide–dotted arrays (left) and cross-sectional SEM images of vertical silicon nanowire arrays (right) prepared from different molecular weight of block-copolymer templates (insets show a magnified TEM images of nanowire). (Reproduced from Bang, B.M., et al., Mass production of uniform-sized nanoporous silicon nanowire anodes viablock copolymer lithography, *Energy & Environmental Science*, 4(9), 3395–3399, 2011. By permission of The Royal Society of Chemistry.)

followed by the immersion of silicon wafers in an etchant solution. Such a block-copolymer template enabled the simple and large-scale production of vertical silicon nanowire arrays. In this case, the average diameter of vertical silicon nanowires was adjusted by the molecular weight of block copolymer used in the template pattern. Figure 7.12b and c shows that iron oxide–dotted arrays generated using a block-polymer template with the molecular weight of 59 and 136 kg/mol resulted in vertical silicon nanowires with the diameter of 190 and 240 nm, respectively.

7.5 APPLICATIONS OF VERTICAL SILICON NANOSTRUCTURES

Vertical silicon nanostructures cover a variety of interesting applications in the field of nanomaterial technology. For instance, vertical silicon nanostructures have been successfully combined with various energy conversion and storage materials, leading to many practical applications such as solar cells, photoelectrochemical cells (PECs), and batteries (Li 2012; Han et al. 2014). In particular, the development of solution-based cost-effective high-throughput fabrication of vertical silicon nanostructures via MACE would have several advantages compared with the vacuum-based counterparts. In this section, we describe the diverse applications of vertical silicon nanostructures fabricated by the MACE process in comparison with those prepared by vapor–liquid–solid (VLS) growth or gas-based dry etching.

7.5.1 ENERGY CONVERSION AND STORAGE APPLICATIONS

It has been proposed that vertical silicon nanostructures could be very promising for solar energy conversion (Fang et al. 2008; Liu et al. 2012). The conventional silicon solar cells rely on the planar junction between semiconducting layers, and require a large amount of high-purity silicon to fully adsorb the incident sun light because of the indirect band gap of silicon material. In order to minimize charge recombinations, the silicon semiconductor can collect a broad region of sun light absorption and exhibit the short diffusion lengths of charged carriers. Unlike the conventional planar-type solar cells, however, vertically aligned silicon nanostructures show the reduced sensitivity to impurities and the decreased optical reflectance. Vertical silicon nanostructures prepared by MACE were successfully employed for a photovoltaic cell (Peng et al. 2005b). They fabricated the large-area freestanding vertical silicon nanowire arrays on mono- and polycrystalline silicon substrates, which exhibit excellent anti-reflection in the wavelength range of 400–1000 nm (Figure 7.13a). This remarkably low optical reflectance comes from the subwavelength scale of the corresponding vertical silicon nanowire arrays, as well as their ultrahigh density and large surface area. The combination of p-type silicon nanowires with n-type silicon to form p–n junctions (performed by $POCl_3$ diffusion in a quartz tube furnace) showed the power conversion efficiency of 9.31% and 4.73% on monocrystalline and polycrystalline silicon substrates, respectively. Garnett and Yang (2008) demonstrated the vertical silicon n–p core–shell nanowire solar cells. In this case, they fabricated n-type silicon nanowires with the diameter of 50–100 nm and deposited the 150-nm-thick p-type amorphous silicon on top using low-pressure chemical vapor deposition, followed by the crystallization by rapid thermal annealing. The overall efficiency in this vertical silicon n–p core shell nanowire array was approximately 0.46%. It is interesting to note that vertical silicon nanostructures fabricated by Peng et al. (2005b) can exhibit different efficiency values. This low efficiency could be attributed to high series resistance in the thick polycrystalline shell, and the enhanced recombination at the surface–interlayer junction.

Peng et al. (2008b) also developed PECs based on vertical silicon nanostructures fabricated by MACE. Compared with those prepared with VLS growth, vertical silicon nanowire arrays produced using MACE offer the equivalent electrical properties and the robust nanowire structures as the integral parts of silicon substrates. Moreover, the nanowire surface is very rough due to the lattice faceting of silicon during the metal-induced solution etching to form highly antireflective surfaces. They fabricated n-type vertical silicon nanostructures as the photoelectrode and used them a Pt mesh while the counter electrode was immersed in the corresponding redox electrolyte solution (Figure 7.13b). The representative photocurrent density in silicon nanostructure–based PECs was tested with different silicon photoelectrodes, including a polished silicon wafer. The vertical silicon nanostructures featured the increased photovoltage and photocurrent compared with the polished planar silicon substrate. The open-circuit voltage of vertical silicon nanostructure–based solar cells was significantly increased from 0.43% to 0.73%, implying a low surface recombination velocity between the silicon nanostructure and the electrolyte.

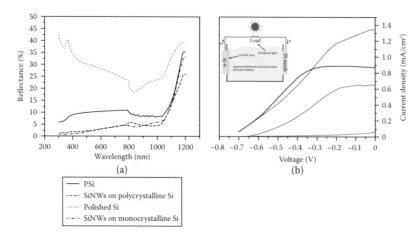

Figure 7.13 (a) Hemispherical reflectance measurement of vertical silicon nanowire arrays, porous silicon (PSi), and polished crystalline silicon. (Reproduced from Peng, K., et al.: Aligned single-crystalline Si nanowire arrays for photovoltaic applications. *Small*. 2005. 1. 1062–1067. Copyright Wiley-VCH Verlag GmbH & Co. KGaA. With permission.) (b) Photocurrent densities of PEC solar cells equipped with different silicon photoelectrodes fabricated by various etching solution: black, HF–AgNO$_3$ for 45 min; blue, Ag–HF–H$_2$O$_2$ for 45 min; red, Ag–HF–H$_2$O$_2$ for 45 min without silver removal; and green, polished silicon. Inset is a schematic illustration of a photovoltaic PEC setup. (Reprinted with permission from Peng, K., et al., Silicon nanowire array photoelectrochemical solar cells, *Applied Physics Letters*, 92(16), 163103, 2008. Copyright 2008, American Institute of Physics.)

Li et al. (2013) discovered the superior purification of dirty silicon and fabricated PECs using this type of silicon. In the current silicon-based photovoltaics industry, the development of solar-cell grade silicon is very challenging due to the tradeoff between the material cost and the conversion efficiency. In this case, they examined various silicon wafers with different grades to fabricate vertical silicon nanostructures using MACE. The metallurgical silicon with upgrading from 99.74% to 99.99% purity induced the silicon purity increase from 99.999772% to 99.999899% after MACE, which is almost identical to other high-purity silicon materials. The high amount of metal impurities could be removed during etching while the quality of the resultant silicon nanostructures was improved. To verify the advantages of upgraded metallurgical silicon nanostructures, they also fabricated PECs using this material. The PEC characterization showed that the photocurrent was increased by 35% in the upgraded vertical silicon structures.

One of the most promising energy storage devices based on vertical silicon nanostructures is lithium ion battery (Chang et al. 2010; Huang and Zhu 2010; Wu and Cui 2012). It is well known that for lithium ion batteries, silicon-based anode materials exhibit very high charge storage capacity (theoretical values up to ~4200 mAh/g) compared with the conventional graphite anode materials (limited to 370 mAh/g). However, the bulk silicon cannot accommodate the large volume change during lithium insertion and extraction. It was demonstrated that the lithium ion batteries having vertical silicon nanostructures fabricated by MACE display the excellent charge storage performance (Peng et al. 2008). Figures 7.14a and b show the voltage profiles during the galvanostatic charging/discharging cycles and SEM image of vertical silicon nanostructures after several these cycles, respectively. Although the first discharging is longer than the following discharging process, both the charging and discharging capacities remained constant with minimal fading during subsequent cycles. Therefore, the anode based on vertical silicon nanowires has higher capacity and longer cycling stability than the conventional planar silicon anode. In addition, the porosity of nanowire surface affected excellent cycling retention compared to nonporous surface (Bang et al. 2011).

Thermoelectric devices could potentially convert the temperature gradient to electrical energy; the conversion efficiency is dependent on the thermoelectric figure of merit (ZT) of the material itself, which is a function of the Seebeck coefficient, the electrical resistivity, the thermal conductivity, and the absolute temperature of system. Hochbaum et al. (2008) employed MACE to fabricate vertical silicon nanostructures for use as scalable thermoelectric materials. These nanostructures have the same Seebeck coefficient and electrical resistivity values as bulk silicon; however, their thermal conductivity can be controlled by the average diameter of

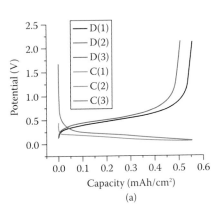

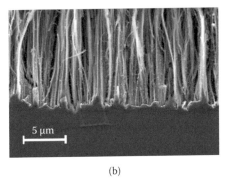

(a)

(b)

Figure 7.14 (a) Voltage-capacity profiles during galvanostatic charging/discharging with the anode of vertical silicon nanowire arrays. (b) A cross-sectional SEM image of vertical silicon nanowires after several charging/discharging cycles. (Reprinted with permission from Peng, K., et al., Silicon nanowires for rechargeable lithium-ion battery anodes, *Applied Physics Letters*, 93(3), 33105, 2008. Copyright 2008, American Institute of Physics.)

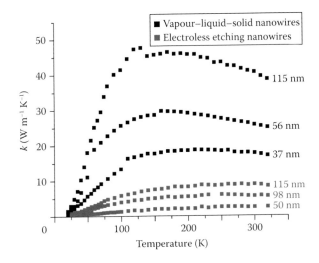

Figure 7.15 The temperature-dependent k (thermal conductivity) using vertical silicon nanowire arrays with various densities at the different fabrication methods. (Reprinted by permission from Macmillan Publishers Ltd. *Nature*, Hochbaum, A.I., et al., Enhanced thermoelectric performance of rough silicon nanowires, 451(7175), 163–167, 2008, copyright 2008.)

vertical silicon nanowires (Figure 7.15). The conventional bulk silicon has high thermal conductivity and low ZT (~0.01 at 300 K) values. One advantage that silicon nanostructures offer for thermoelectric applications is the large difference in mean free path lengths between electrons and phonons at room temperature. Moreover, the silicon nanowires with the rough surface reduced the thermal conductivity compared with those with the smooth surface of nanowires fabricated by VLS growth at equivalent diameters. The roughness of a silicon nanowire surface acts as a secondary scattering phase, which may contribute to the enhanced rate of diffusive reflection or phonon backscattering. Therefore, these processes have been predicted to affect the increased thermal conductivity of silicon nanowires (Yuan et al. 2012; Ghossoub et al. 2013; Krali and Durrani 2013).

7.5.2 BIOLOGICAL APPLICATIONS

Over the past few decades, a variety of silicon nanomaterials have been developed. Since silicon is well known for its biocompatibility and relative abundance, there have been many recent studies focusing on the application of silicon nanomaterials to biomedical engineering in addition to optics, electronics,

Arrays, hybrids, and core-shell

and energy devices. Among various silicon nanomaterials, vertical silicon nanostructures supported on a two-dimensional bottom substrate have shown strong potential as a new cellular engineering platform and opened up a novel prospect in the field of biomedical engineering. For example, vertical silicon nanowire arrays were successfully used for a noninvasive cellular interface for investigating the responses of mammalian cells to various physical stimuli, the direct intracellular delivery, the circulating tumor cell capture, and so on. In combination with excellent biocompatibility, the fine geometry control make vertical silicon nanowire arrays popular in cellular engineering applications, and MACE, which enables the facile laboratory-scale fabrication of such nanostructures, has drawn much attention in the related research fields.

In general, the cellular physical interaction with extracellular environments is mediated by mechanotransduction, which consists of various mechanisms converting mechanical stimuli to intracellular activities. Since cellular traction force is crucial to mechanotransduction, the measurement of cellular traction force would be very useful for understanding the cellular- or subcellular-level mechanisms of wound healing, embryogenesis, angiogenesis, histogenesis, invasion, cancer metastasis, and so on. In 2009, Li et al. reported a biophysical method for quantifying the maximum cell traction forces in three different cell types including normal mammalian cells, benign cells, and malignant cells (Figure 7.16). This platform is expected to be potentially useful for oncology, disease diagnosis, drug development, and tissue engineering because cancer cells exhibit significantly larger traction force than normal cells. In addition to these cell traction force measurements, vertical silicon nanostructures can serve as a useful tool for exerting a specific traction force by tuning their geometric parameters.

Vertical nanostructures have shown the capability of effectively capturing suspended cells compared with conventional flat substrates. Many previous studies have shown that vertical silicon nanostructures remarkably improve the efficiency of capturing circulating tumor cells and T-lymphocytes (Figure 7.17). The effective capture of circulating tumor cells, which are rarely detected from blood, is very encouraging for the cancer diagnose research. This phenomenon may be attributed to the fact that cells are able to recognize the nano topography of external environments.

Gene delivery, in which foreign DNA is introduced into host cells, is a critical step in genetic engineering. Therefore, the effective gene delivery requires high delivery efficiency in various cell types, the minimal toxicity, and the prolonged stability in a medium or circulation system. Furthermore, the capability of performing multiple continuous deliveries is needed while exogenous genes are steadily supplied over the

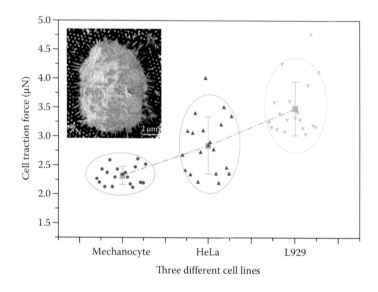

Figure 7.16 The measure of maximum and average cell traction force of mechanocyte, HeLa, and L929 cells. The vertical bar represents the standard deviation. Inset is the top-view SEM image of a mechanocyte on vertical silicon nanowire substrate. (Reprinted with permission from Li, Z., et al., Quantifying the traction force of a single cell by aligned silicon nanowire array, *Nano Letters*, 9(10), 3575–3580, 2009. Copyright 2009 American Chemical Society.)

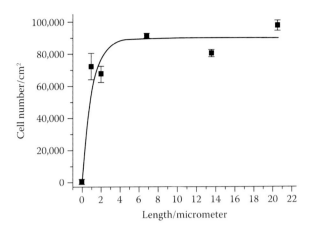

Figure 7.17 Quantitative evaluation of the capture yield for CCRF-CEM cells using silicon nanowire arrays with different lengths. (Reproduced from Chen, L., et al.: Aptamer-mediated efficient capture and release of T lymphocytes on nanostructured surfaces. *Advanced Materials*. 2011. 23. 4376–4380. Copyright Wiley-VCH Verlag GmbH & Co. KGaA. With permission.)

duration of the desired biological process. A conventional method for gene delivery involves using viral vectors due to their practical advantages. However, viral vectors have several potential risks such as immunogenicity, inflammatory responses, and potential insertion mutagenesis. Recently, nonviral vectors based on biocompatible nanomaterials, for instance, inorganic nanoparticles, carbon nanomaterials, artificial liposomes, and polypeptides, have been proposed to combat these issues. Despite the serious efforts to search for nonviral gene delivery methods, low transfection efficiency and cell-type dependence remained as issues. These limitations come from the fact that the intracellular delivery is mainly governed by endocytosis, an active cellular transport process. Vertical silicon nanostructures, which have shown minimal invasion to cellular membranes, may serve as an alternative delivery tool for overcoming the abovementioned limitations (Figure 7.18). Indeed, several reports have been published claiming the high gene delivery efficiency

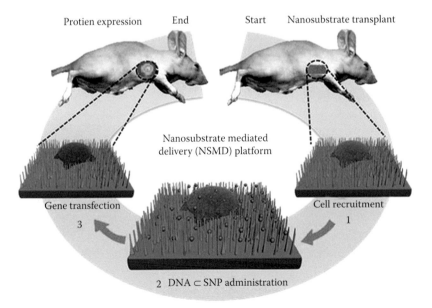

Figure 7.18 A schematic illustration of the unique mechanism governing the nanosubstrate-mediated delivery approach for both in vivo and in vitro experiments. (Reprinted with permission from Peng, J., et al., Molecular recognition enables nanosubstrate-mediated delivery of gene-encapsulated nanoparticles with high efficiency, *ACS Nano*, 8(5), 4621–4629, 2014. Copyright 2014 American Chemical Society.)

Arrays, hybrids, and core–shell

using vertical silicon nanostructures. Furthermore, the efficiency of direct intracellular gene delivery using vertical silicon nanostructures has been improved in combination with nanowire geometric control and/or surface chemical modification (Zhang et al. 2008a; Chen et al. 2011; Piret et al. 2011; Liu et al. 2013, 2014a; Lee et al. 2014; Pan et al. 2014; Peng et al. 2014; Chiappini et al. 2015a, 2015b). Nonetheless, there are still some controversial points: whether this enhanced delivery efficiency is enabled by the penetration of vertical silicon nanostructures into cellular membranes, or whether the curvature of cellular membranes (due to the topography of vertical silicon nanostructures) enhances the endocytosis activity.

7.5.3 SENSOR APPLICATIONS

Surface-enhanced Raman spectroscopy (SERS) is a powerful analytical tool where the plasmon resonance on a nanostructured metallic surface enhances the Raman signals. In order to employ the surface plasmon effect for fingerprinting in biosensing and imaging applications, SERS substrates must be stable, reproducible, and controllable. Vertical silicon nanostructures fabricated by MACE are potentially useful as SERS substrates due to the scalable fabrication of and fine geometric control over such nanostructured substrates. Large-area Ag-coated vertical silicon nanowire arrays were successfully used for this purpose (Zhang et al. 2008a). In this research, the morphologies of silicon nanowire arrays and the type of Ag plating solution used are two key factors for determining the magnitude of SERS signal enhancement and the detection sensitivity. Controlling the etching time and etchant temperature affected the morphologies of final vertical silicon nanostructures. The optimized forest-like vertical nanowires could be the most efficient SERS substrates; as shown in Figure 7.19a, Ag-coated vertical

(a)

(b)

Figure 7.19 (a) SEM image of the silver-coated vertical silicon nanowire arrays. (Reproduced from Zhang, B., et al.: Large-area silver-coated silicon nanowire arrays for molecular sensing using surface-enhanced raman spectroscopy. *Advanced Functional Materials*. 2008. 18. 2348–2355. Copyright Wiley-VCH Verlag GmbH & Co. KGaA. With permission.) (b) SEM images of the vertical silicon nanowire arrays decorated with silver nanoparticles via redox reaction. (The source of the material Yang, J., et al., High aspect ratio SiNW arrays with Ag nanoparticles decoration for strong SERS detection, *Nanotechnology*, 25(46), 465707, 2014 is acknowledged.)

silicon nanowire arrays exhibit rough fractal-like nanostructures. This unique feature induced the large enhancement in electromagnetic fields and showed good stability and reproducibility for SERS applications. In addition, Yang et al. (2014) reported Ag nanoparticle-decorated vertical silicon nanowire arrays for the extremely sensitive SERS detection. By combining interference lithography with MACE, they could obtain the well-ordered vertical silicon nanowire arrays. Subsequently, the as-fabricated nanowire substrates were dipped into an aqueous "Ag-deposition" solution composed of HF and $AgNO_3$, finally resulting in vertical silicon nanowire arrays decorated with Ag nanoparticles (Figure 7.19b). These well-ordered silicon nanowires resulted in the enhanced SERS signal intensity and improved uniformity due to the optimized height and periodic nanostructure configuration.

7.6 SUMMARY AND PERSPECTIVE

In summary, we reviewed the fabrication of vertical silicon nanostructures based on MACE. Compared with other conventional fabrication methods for vertical silicon nanostructures, MACE is a facile and cost-effective solution-based process which also enables the wafer-scale production of such nanostructures. The mechanism and procedure of MACE can be described in the following steps: (1) the reduction of the oxidant on the surface of noble metal catalyst, (2) the injection of holes, which were generated by the reduction of the oxidant, into the silicon–metal interface, (3) the oxidation of silicon subjected to hole injection and the removal of oxidized silicon using etchant (including HF), and (4) the mass transport of reactants and byproducts. Additionally, we discussed the influence of specific noble metal catalyst, silicon crystal structure, and etching solution on the geometry and morphology of the resulting vertical silicon nanostructures. Even though the detailed effects of etching conditions is difficult to clarify due to a number of different environmental conditions involved, we reviewed a good number of studies about the influence of specific experimental conditions on MACE. By choosing a certain method of metal catalyst patterning in conjunction with MACE, it is possible to precisely control the nanowire diameter, height, density, position, and shape. Well-ordered or randomly-distributed vertical silicon nanowire arrays can be selected depending on the specific type of their applications. We suppose that vertical silicon nanowire arrays produced by MACE may serve as a potentially useful platform in a variety of fields covering energy, biology, and environmental applications, not only from an academic but also industrial perspective.

REFERENCES

Balasundaram, Karthik, Jyothi S. Sadhu, Jae Cheol Shin, Bruno Azeredo, Debashis Chanda, Mohammad Malik, Keng Hsu, et al. 2012. Porosity Control in Metal-Assisted Chemical Etching of Degenerately Doped Silicon Nanowires. *Nanotechnology* 23 (30): 305304. doi:10.1088/0957-4484/23/30/305304.

Bang, Byoung Man, Hyunjung Kim, Jung-Pil Lee, Jaephil Cho, and Soojin Park. 2011. Mass Production of Uniform-Sized Nanoporous Silicon Nanowire Anodes Viablock Copolymer Lithography. *Energy & Environmental Science* 4 (9): 3395–3399. doi:10.1039/C1EE01898A.

Céline Chartier, Stéphane Bastide, and Claude Lévy-Clément. 2008. Metal-Assisted Chemical Etching of Silicon in HF–H_2O_2. *Electrochimica Acta* 53 (17): 5509–5516. doi:10.1016/j.electacta.2008.03.009.

Chang, Shih-Wei, Vivian P. Chuang, Steven T. Boles, Caroline A. Ross, and Carl V. Thompson. 2009. Densely Packed Arrays of Ultra-High-Aspect-Ratio Silicon Nanowires Fabricated Using Block-Copolymer Lithography and Metal-Assisted Etching. *Advanced Functional Materials* 19 (15): 2495–2500. doi:10.1002/adfm.200900181.

Chang, Shih-Wei, Vivian P. Chuang, Steven T. Boles, and Carl V. Thompson. 2010a. Metal-Catalyzed Etching of Vertically Aligned Polysilicon and Amorphous Silicon Nanowire Arrays by Etching Direction Confinement. *Advanced Functional Materials* 20 (24): 4364–4370. doi:10.1002/adfm.201000437.

Chang, Shih-Wei, Jihun Oh, Steven T. Boles, and Carl V. Thompson. 2010b. Fabrication of Silicon Nanopillar-Based Nanocapacitor Arrays. *Applied Physics Letters* 96 (15): 153108. doi:10.1063/1.3374889.

Chattopadhyay, Soma, and Paul W. Bohn. 2004. Direct-Write Patterning of Microstructured Porous Silicon Arrays by Focused-Ion-Beam Pt Deposition and Metal-Assisted Electroless Etching. *Journal of Applied Physics* 96 (11): 6888–6894. doi:10.1063/1.1806992.

Chattopadhyay, Soma, Xiuling Li, and Paul W. Bohn. 2002. In-Plane Control of Morphology and Tunable Photoluminescence in Porous Silicon Produced by Metal-Assisted Electroless Chemical Etching. *Journal of Applied Physics* 91 (9): 6134–6140. doi:10.1063/1.1465123.

Chen, Chia-Yun, Chi-Sheng Wu, Chia-Jen Chou, and Ta-Jen Yen. 2008. Morphological Control of Single-Crystalline Silicon Nanowire Arrays near Room Temperature. *Advanced Materials* 20 (20): 3811–3815. doi:10.1002/adma.200702788.

Chen, Huan, Hui Wang, Xiao-Hong Zhang, Chun-Sing Lee, and Shuit-Tong Lee. 2010. Wafer-Scale Synthesis of Single-Crystal Zigzag Silicon Nanowire Arrays with Controlled Turning Angles. *Nano Letters* 10 (3): 864–868. doi:10.1021/nl903391x.

Chen, Li, Xueli Liu, Bin Su, Jing Li, Lei Jiang, Dong Han, and Shutao Wang. 2011. Aptamer-Mediated Efficient Capture and Release of T Lymphocytes on Nanostructured Surfaces. *Advanced Materials* 23 (38): 4376–4380. doi:10.1002/adma.201102435.

Chiappini, Ciro, Xuewu Liu, Jean Raymond Fakhoury, and Mauro Ferrari. 2010. Biodegradable Porous Silicon Barcode Nanowires with Defined Geometry. *Advanced Functional Materials* 20 (14): 2231–2239. doi:10.1002/adfm.201000360.

Chiappini, Ciro, Jonathan O. Martinez, Enrica De Rosa, Carina S. Almeida, Ennio Tasciotti, and Molly M. Stevens. 2015b. Biodegradable Nanoneedles for Localized Delivery of Nanoparticles in Vivo: Exploring the Biointerface. *ACS Nano* 9 (5): 5500–5509. doi:10.1021/acsnano.5b01490.

Ciro Chiappini, Enrica De Rosa, Jon Martinez, Xuewu Liu, Joe Steele, Molly Stevens, and Ennio Tasciotti. 2015a. Biodegradable Silicon Nanoneedles Delivering Nucleic Acids Intracellularly Induce Localized in Vivo Neovascularization. *Nature Materials* 14 (5): 532–539. doi:10.1038/nmat4249.

Fang, Hui, Xudong Li, Shuang Song, Ying Xu, and Jing Zhu. 2008. Fabrication of Slantingly-Aligned Silicon Nanowire Arrays for Solar Cell Applications. *Nanotechnology* 19 (25): 255703. doi:10.1088/0957-4484/19/25/255703.

Fang, Hui, Yin Wu, Jiahao Zhao, and Jing Zhu. 2006. Silver Catalysis in the Fabrication of Silicon Nanowire Arrays. *Nanotechnology* 17 (15): 3768. doi:10.1088/0957-4484/17/15/026.

Garnett, Erik C., and Peidong Yang. 2008. Silicon Nanowire Radial P–n Junction Solar Cells. *Journal of the American Chemical Society* 130 (29): 9224–9225. doi:10.1021/ja8032907.

Geyer, Nadine, Bodo Fuhrmann, Zhipeng Huang, Johannes de Boor, Hartmut S. Leipner, and Peter Werner. 2012. Model for the Mass Transport during Metal-Assisted Chemical Etching with Contiguous Metal Films As Catalysts. *The Journal of Physical Chemistry C* 116 (24): 13446–13451. doi:10.1021/jp3034227.

Hadjersi, Toufik. 2007. Oxidizing Agent Concentration Effect on Metal-Assisted Electroless Etching Mechanism in HF-Oxidizing Agent-H2O Solutions. *Applied Surface Science* 253 (9): 4156–4160. doi:10.1016/j.apsusc.2006.09.016.

Han, Hee, Zhipeng Huang, and Woo Lee. 2014. Metal-Assisted Chemical Etching of Silicon and Nanotechnology Applications. *Nano Today* 9 (3): 271–304. doi:10.1016/j.nantod.2014.04.013.

Hildreth, Owen J., Andrei G. Fedorov, and Ching Ping Wong. 2012. 3D Spirals with Controlled Chirality Fabricated Using Metal-Assisted Chemical Etching of Silicon. *ACS Nano* 6 (11): 10004–10012. doi:10.1021/nn303680k.

Hildreth, Owen James, Wei Lin, and Ching Ping Wong. 2009. Effect of Catalyst Shape and Etchant Composition on Etching Direction in Metal-Assisted Chemical Etching of Silicon to Fabricate 3D Nanostructures. *ACS Nano* 3 (12): 4033–4042. doi:10.1021/nn901174e.

Hochbaum, Allon I., Renkun Chen, Raul Diaz Delgado, Wenjie Liang, Erik C. Garnett, Mark Najarian, Arun Majumdar, and Peidong Yang. 2008. Enhanced Thermoelectric Performance of Rough Silicon Nanowires. *Nature* 451 (7175): 163–167. doi:10.1038/nature06381.

Huang, Jinquan, Sing Yang Chiam, Hui Huang Tan, Shijie Wang, and Wai Kin Chim. 2010. Fabrication of Silicon Nanowires with Precise Diameter Control Using Metal Nanodot Arrays as a Hard Mask Blocking Material in Chemical Etching. *Chemistry of Materials* 22 (13): 4111–4116. doi:10.1021/cm101121c.

Huang, Rui, and Jing Zhu. 2010. Silicon Nanowire Array Films as Advanced Anode Materials for Lithium-Ion Batteries. *Materials Chemistry and Physics* 121 (3): 519–522. doi:10.1016/j.matchemphys.2010.02.017.

Huang, Zhipeng, Xuanxiong Zhang, Manfred Reiche, Lifeng Liu, Woo Lee, Tomohiro Shimizu, Stephan Senz, and Ulrich Gösele. 2008. Extended Arrays of Vertically Aligned Sub-10 Nm Diameter [100] Si Nanowires by Metal-Assisted Chemical Etching. *Nano Letters* 8 (9): 3046–3051. doi:10.1021/nl802324y.

Kim, Jungkil, Hee Han, Young Heon Kim, Suk-Ho Choi, Jae-Cheon Kim, and Woo Lee. 2011a. Au/Ag Bilayered Metal Mesh as a Si Etching Catalyst for Controlled Fabrication of Si Nanowires. *ACS Nano* 5 (4): 3222–3229. doi:10.1021/nn2003458.

Kim, Jungkil, Young Heon Kim, Suk-Ho Choi, and Woo Lee. 2011b. Curved Silicon Nanowires with Ribbon-like Cross Sections by Metal-Assisted Chemical Etching. *ACS Nano* 5 (6): 5242–5248. doi:10.1021/nn2014358.

Kolasinski, Kurt W. 2005. Silicon Nanostructures from Electroless Electrochemical Etching. *Current Opinion in Solid State and Materials Science* 9 (1–2): 73–83. doi:10.1016/j.cossms.2006.03.004.

Krali, Emiljana, and Zahid A. K. Durrani. 2013. Seebeck Coefficient in Silicon Nanowire Arrays. *Applied Physics Letters* 102 (14): 143102. doi:10.1063/1.4800778.

Kuiqing Peng, Juejun Hu, Yunjie Yan, Yin Wu, Hui Fang, Ying Xu, ShuitTong Lee, and Jing Zhu. 2006a. Fabrication of Single-Crystalline Silicon Nanowires by Scratching a Silicon Surface with Catalytic Metal Particles. *Advanced Functional Materials* 16 (3): 387–394. doi:10.1002/adfm.200500392.

Kuiqing Peng, Yunjie Yan, Shangpeng Gao, and Jing Zhu. 2003. Dendrite-Assisted Growth of Silicon Nanowires in Electroless Metal Deposition. *Advanced Functional Materials* 13 (2): 127–132. doi:10.1002/adfm.200390018.

Lee, Chia-Lung, Kazuya Tsujino, Yuji Kanda, Shigeru Ikeda, and Michio Matsumura. 2008. Pore Formation in Silicon by Wet Etching Using Micrometre-Sized Metal Particles as Catalysts. *Journal of Materials Chemistry* 18 (9): 1015. doi:10.1039/b715639a.

Lee, Sang-Kwon, Dong-Joo Kim, GeeHee Lee, Gil-Sung Kim, Minsuk Kwak, and Rong Fan. 2014. Specific Rare Cell Capture Using Micro-Patterned Silicon Nanowire Platform. *Biosensors & Bioelectronics* 54 (April): 181–188. doi:10.1016/j.bios.2013.10.048.

Lee, Seyeong, Dongyoon Kim, Seong-Min Kim, Jeong-Ah Kim, Taesoo Kim, Dong-Yu Kim, and Myung-Han Yoon. 2015. Polyelectrolyte Multilayer-Assisted Fabrication of Non-Periodic Silicon Nanocolumn Substrates for Cellular Interface Applications. *Nanoscale* 7 (35): 14627–14635. doi:10.1039/c5nr02384j.

Li, Xiaopeng, Yanjun Xiao, Jin Ho Bang, Dominik Lausch, Sylke Meyer, Paul-Tiberiu Miclea, Jin-Young Jung, Stefan L. Schweizer, Jung-Ho Lee, and Ralf B. Wehrspohn. 2013. Upgraded Silicon Nanowires by Metal-Assisted Etching of Metallurgical Silicon: A New Route to Nanostructured Solar-Grade Silicon. *Advanced Materials* 25 (23): 3187–3191. doi:10.1002/adma.201300973.

Li, Xiuling. 2012. Metal Assisted Chemical Etching for High Aspect Ratio Nanostructures: A Review of Characteristics and Applications in Photovoltaics. *Current Opinion in Solid State and Materials Science,* Photonic Nanostructure Materials, Processing and Characterization, 16 (2): 71–81. doi:10.1016/j.cossms.2011.11.002.

Li, Zhou, Jinhui Song, Giulia Mantini, Ming-Yen Lu, Hao Fang, Christian Falconi, Lih-Juann Chen, and Zhong Lin Wang. 2009. Quantifying the Traction Force of a Single Cell by Aligned Silicon Nanowire Array. *Nano Letters* 9 (10): 3575–3580. doi:10.1021/nl901774m.

Liu, Dandan, Changqing Yi, Chi-Chun Fong, Qinghui Jin, Zuankai Wang, Wai-Kin Yu, Dong Sun, Jianlong Zhao, and Mengsu Yang. 2014. Activation of Multiple Signaling Pathways during the Differentiation of Mesenchymal Stem Cells Cultured in a Silicon Nanowire Microenvironment. *Nanomedicine: Nanotechnology, Biology and Medicine* 10 (6): 1153–1163. doi:10.1016/j.nano.2014.02.003.

Liu, Dandan, Changqing Yi, Kaiqun Wang, Chi-Chun Fong, Zuankai Wang, Pik Kwan Lo, Dong Sun, and Mengsu Yang. 2013a. Reorganization of Cytoskeleton and Transient Activation of Ca2+ Channels in Mesenchymal Stem Cells Cultured on Silicon Nanowire Arrays. *ACS Applied Materials & Interfaces* 5 (24): 13295–13304. doi:10.1021/am404276r.

Liu, Ruiyuan, Fute Zhang, Celal Con, Bo Cui, and Baoquan Sun. 2013b. Lithography-Free Fabrication of Silicon Nanowire and Nanohole Arrays by Metal-Assisted Chemical Etching. *Nanoscale Research Letters* 8 (1): 155. doi:10.1186/1556-276X-8-155.

Liu, Yousong, Guangbin Ji, Junyi Wang, Xuanqi Liang, Zewen Zuo, and Yi Shi. 2012. Fabrication and Photocatalytic Properties of Silicon Nanowires by Metal-Assisted Chemical Etching: Effect of H2O2 Concentration. *Nanoscale Research Letters* 7 (1): 663. doi:10.1186/1556-276X-7-663

Marc Ghossoub, Krishna Valvala, Myunghoon Seong, Bruno Azeredo, Keng Hsu, Joythi S. Sadhu, Piyush Singh, and Sanjiv Sinha. 2013. Spectral Phonon Scattering from Sub-10 Nm Surface Roughness Wavelengths in Metal-Assisted Chemically Etched Si Nanowires. *Nano Letters* 13 (4): 1564–1571. doi:10.1021/nl3047392.

Márquez, Francisco, Carmen Morant, Vicente López, Félix Zamora, Teresa Campo, and Eduardo Elizalde. 2011. An Alternative Route for the Synthesis of Silicon Nanowires via Porous Anodic Alumina Masks. *Nanoscale Research Letters* 6 (1): 495. doi:10.1186/1556-276X-6-495.

Mikhael, Bechelany, Berodier Elise, Maeder Xavier, Schmitt Sebastian, Michler Johann, and Philippe Laetitia. 2011. New Silicon Architectures by Gold-Assisted Chemical Etching. *ACS Applied Materials & Interfaces* 3 (10): 3866–3873. doi:10.1021/am200948p.

Morinaga, Hitoshi, Makoto Suyama, and Tadahiro Ohmi. 1994. Mechanism of Metallic Particle Growth and Metal-Induced Pitting on Si Wafer Surface in Wet Chemical Processing. *Journal of The Electrochemical Society* 141 (10): 2834–2841. doi:10.1149/1.2059240.

Pan, Jingjing, Zhonglin Lyu, Wenwen Jiang, Hongwei Wang, Qi Liu, Min Tan, Lin Yuan, and Hong Chen. 2014. Stimulation of Gene Transfection by Silicon Nanowire Arrays Modified with Polyethylenimine. *ACS Applied Materials & Interfaces* 6 (16): 14391–14398. doi:10.1021/am5036626.

Peng, Jinliang, Mitch André Garcia, Jin-sil Choi, Libo Zhao, Kuan-Ju Chen, James R. Bernstein, Parham Peyda, et al. 2014. Molecular Recognition Enables Nanosubstrate-Mediated Delivery of Gene-Encapsulated Nanoparticles with High Efficiency. *ACS Nano* 8 (5): 4621–4629. doi:10.1021/nn5003024.

Arrays, hybrids, and core-shell

Peng, Kuiqing, Hui Fang, Juejun Hu, Yin Wu, Jing Zhu, Yunjie Yan, and ShuitTong Lee. 2006b. Metal-Particle-Induced, Highly Localized Site-Specific Etching of Si and Formation of Single-Crystalline Si Nanowires in Aqueous Fluoride Solution. *Chemistry – A European Journal* 12 (30): 7942–7947. doi:10.1002/chem.200600032.

Peng, Kuiqing, Jiansheng Jie, Wenjun Zhang, and Shuit-Tong Lee. 2008a. Silicon Nanowires for Rechargeable Lithium-Ion Battery Anodes. *Applied Physics Letters* 93 (3): 33105. doi:10.1063/1.2929373.

Peng, Kuiqing, Xin Wang, and Shuit-Tong Lee. 2008b. Silicon Nanowire Array Photoelectrochemical Solar Cells. *Applied Physics Letters* 92 (16): 163103. doi:10.1063/1.2909555.

Peng, Kuiqing, Yin Wu, Hui Fang, Xiaoyan Zhong, Ying Xu, and Jing Zhu. 2005a. Uniform, Axial-Orientation Alignment of One-Dimensional Single-Crystal Silicon Nanostructure Arrays. *Angewandte Chemie (International Ed. in English)* 44 (18): 2737–2742. doi:10.1002/anie.200462995.

Peng, Kuiqing, Ying Xu, Yin Wu, Yunjie Yan, Shuit-Tong Lee, and Jing Zhu. 2005b. Aligned Single-Crystalline Si Nanowire Arrays for Photovoltaic Applications. *Small* 1 (11): 1062–1067. doi:10.1002/smll.200500137.

Peng, Kuiqing, Mingliang Zhang, Aijiang Lu, Ning-Bew Wong, Ruiqin Zhang, and Shuit-Tong Lee. 2007. Ordered Silicon Nanowire Arrays via Nanosphere Lithography and Metal-Induced Etching. *Applied Physics Letters* 90 (16): 163123. doi:10.1063/1.2724897.

Piret, Gaëlle, Elisabeth Galopin, Yannick Coffinier, Rabah Boukherroub, Dominique Legrand, and Christian Slomianny. 2011. Culture of Mammalian Cells on Patterned Superhydrophilic/superhydrophobic Silicon Nanowire Arrays. *Soft Matter* 7 (18): 8642–8649. doi:10.1039/C1SM05838J.

Smith, Zachary R., Rosemary L. Smith, and Scott D. Collins. 2013. Mechanism of Nanowire Formation in Metal Assisted Chemical Etching. *Electrochimica Acta* 92 (Spring): 139–147. doi:10.1016/j.electacta.2012.12.075.

Tsujino, Kazuya, and Michio Matsumura. 2007. Morphology of Nanoholes Formed in Silicon by Wet Etching in Solutions Containing HF and H2O2 at Different Concentrations Using Silver Nanoparticles as Catalysts. *Electrochimica Acta,* ELECTROCHEMICAL PROCESSING OF TAILORED MATERIALS Selection of papers from the 4th International Symposium (EPTM 2005) 3–5 October 2005, Kyoto, Japan, 53 (1): 28–34. doi:10.1016/j.electacta.2007.01.035.

Wang, Hsin-Ping, Kun-Yu Lai, Yi-Ruei Lin, Chin-An Lin, and Jr-Hau He. 2010. Periodic Si Nanopillar Arrays Fabricated by Colloidal Lithography and Catalytic Etching for Broadband and Omnidirectional Elimination of Fresnel Reflection. *Langmuir* 26 (15): 12855–12858. doi:10.1021/la1012507.

Wee Kiong Choi, Tze Haw Liew, Mohamed Khalide Dawood, Henry I. Smith, Carl V. Thompson, and Minghui Hong. 2008. Synthesis of Silicon Nanowires and Nanofin Arrays Using Interference Lithography and Catalytic Etching. *Nano Letters* 8 (11): 3799–3802. doi:10.1021/nl802129f.

Wu, Hui, and Yi Cui. 2012. Designing Nanostructured Si Anodes for High Energy Lithium Ion Batteries. *Nano Today* 7 (5): 414–429. doi:10.1016/j.nantod.2012.08.004.

Yang Jing, Jiabao Li, Qihuang Gong, Jinghua Teng, and Minghui Hong. 2014. High Aspect Ratio SiNW Arrays with Ag Nanoparticles Decoration for Strong SERS Detection. *Nanotechnology* 25 (46): 465707. doi:10.1088/0957-4484/25/46/465707.

Yeom, Junghoon, Daniel Ratchford, Christopher R. Field, Todd H. Brintlinger, and Pehr E. Pehrsson. 2014. Decoupling Diameter and Pitch in Silicon Nanowire Arrays Made by Metal-Assisted Chemical Etching. *Advanced Functional Materials* 24 (1): 106–116. doi:10.1002/adfm.201301094.

Yi Min Yang, Paul K. Chu, Zhengwei Wu, Shihao Pu, Takfu Hung, Kaifu Huo, Guixiang Qian, Wenjun Zhang, and Xinglong Wu. 2008. Catalysis of Dispersed Silver Particles on Directional Etching of Silicon. *Applied Surface Science* 254 (10): 3061–3066. doi:10.1016/j.apsusc.2007.10.055.

Yuan, Guodong, Rüdiger Mitdank, Anna Mogilatenko, and Saskia F. Fischer. 2012. Porous Nanostructures and Thermoelectric Power Measurement of Electro-Less Etched Black Silicon. *The Journal of Physical Chemistry C* 116 (25): 13767–13773. doi:10.1021/jp212427g.

Zhang, Baohua, Haishui Wang, Lehui Lu, Kelong Ai, Guo Zhang, and Xiaoli Cheng. 2008a. Large-Area Silver-Coated Silicon Nanowire Arrays for Molecular Sensing Using Surface-Enhanced Raman Spectroscopy. *Advanced Functional Materials* 18 (16): 2348–2355. doi:10.1002/adfm.200800153.

Zhang, Ming-Liang, Kui-Qing Peng, Xia Fan, Jian-Sheng Jie, Rui-Qin Zhang, Shuit-Tong Lee, and Ning-Bew Wong. 2008b. Preparation of Large-Area Uniform Silicon Nanowires Arrays through Metal-Assisted Chemical Etching. *The Journal of Physical Chemistry C* 112 (12): 4444–4450. doi:10.1021/jp077053o.

Zhipeng Huang, Hui Fang, and Jing Zhu. 2007. Fabrication of Silicon Nanowire Arrays with Controlled Diameter, Length, and Density. *Advanced Materials* 19 (5): 744–748. doi:10.1002/adma.200600892.

Zhong, Xing, Yongquan Qu, Yung-Chen Lin, Lei Liao, and Xiangfeng Duan. 2011. Unveiling the Formation Pathway of Single Crystalline Porous Silicon Nanowires. *ACS Applied Materials & Interfaces* 3 (2): 261–270. doi:10.1021/am1009056.

Arrays, hybrids, and core–shell

8 Silicon nanowire and nanohole arrays

Changying Zhao and Xing Fang

Contents

8.1 INTRODUCTION

Nanowires and nanoholes are two kinds of typical nanostructures. Commonly, these nanostructures have cross-sectional dimensions (wires and holes) that can be tuned from a few nanometers to hundreds of nanometers, with lengths spanning from hundreds of nanometers to several micrometers. A large number of nanowires can be distributed on substrates to form nanowire arrays, and a mass of nanoholes can also be arranged in slabs to form nanohole arrays.

Nanowires are known as one-dimensional (1D) nanostructures because they are confined in two dimensions, that is, cross-sectional planes, thus allowing electrons, holes, or photons to propagate freely along the third dimension. There are two types of nanowire arrays: horizontal and vertically aligned nanowire arrays. Horizontal nanowire elements lie on the substrate plane, whereas vertically aligned nanowires are oriented perpendicular to the substrate. Their high aspect ratio allows for the bridging of the nanoscopic and macroscopic world. As an important class of nanostructures, nanowires have emerged since the 1990s. These were initially called "nanowhiskers" (Yazawa et al. 1991), later "nanowires" (Xia et al. 2003). Since their introduction in the 1990s, nanowires have been extensively studied and much insight has been gained into tuning their optical and electrical properties by controlling their size and dimensions. Furthermore, the 1D nature of nanowires permits materials synthesis in traditionally inaccessible compositional regions. Such nanowires with tunable optical and electronic structures hold great promise in photovoltaics, solid-state lighting, and solar-to-fuel energy conversion. Moreover, the ability to form diverse heterostructures sets nanowires apart from other nanomaterials (such as quantum dots and carbon nanotubes) and represents a substantial advantage for the development of increasingly powerful and unique nanoscale electronic and optoelectronic devices (Li et al. 2006). Considering that the nano–macro interface of nanowires is fundamental to the integration of nanoscale building blocks in electrical or optoelectronic device applications, they were immediately recognized as one of the essential building blocks for nanophotonics (Yan et al. 2009).

On the other hand, nanoholes are not regarded as 1D nanostructures strictly, because of that the substantial medium is free for propagating electrons, holes, or photons. However, walls of holes largely enhance scattering of electrons, holes, or photons, which also make their unique optical and electrical properties. Moreover, nanoholes can be etched in the ultrathin slabs that have two-dimensional (2D) nanostructure nature. One can control or optimize structural parameters and dimensions to tune optical and electrical properties of nanoholes as well. Therefore, nanoholes are versatile nanostructures that are applicable for photodetectors, chemical and gas sensors, waveguides, light-emitting diodes, microcavity lasers, solar cells, and nonlinear optical converters. Similar to nanowires, an integrated photonic platform using nanohole building blocks will achieve advanced functionalities at dimensions compatible with on-chip technologies.

Both nanowire and nanohole arrays have been widely studied for different applications (Li et al. 2006; Wei and Jie 2015). Nanowires and nanoholes have promising thermoelectric material's figure of merit and inspire different thermoelectric converter device concepts including nanowire device concepts, p–n junction thermoelectric generators, and integrated spot coolers (Schierning 2014). In life sciences, nanowires and nanoholes are able to influence cellular adhesion, morphology, migration, proliferation, and differentiation. They can be taken as tools for the delivery of biomolecular cargoes into mammalian cells and used in drug delivery and for biosensing (Elnathan et al. 2014).

In photovoltaic applications (which are mainly concerned in this chapter), silicon nanowire and nanohole arrays exhibit unique optical and electrical characteristics for building photovoltaic devices with high performance to cost ratios over traditional planar junction bulk silicon structures. By optimizing the structural parameters, such as wire and hole diameter and array periodicity, antireflection even superior to the optimized antireflective coatings can be realized. In the meantime, excellent light confinement is easily achievable for silicon nanowire and nanohole arrays using much less materials compared to their bulk silicon counterparts (Fang et al. 2014). For a high-performance solar cell, maximized absorption of the solar radiation, effective carrier generation, and collection must be simultaneously satisfied. From the electrical aspect, the radial p–n junction configuration formed around the wires significantly reduces the minority carrier collection length along the radial direction, providing outstanding tolerance to material qualities. Owing to these charming properties, silicon nanowire array–based solar cells have been attracting extensive attention. The power conversion efficiency has also made huge progress from <1% to the present >12% in less than 10 years (Li et al. 2014).

Before understanding the optical and electrical properties of silicon nanowire and nanohole arrays, it is necessary to know their fabrication techniques and processes. Fabrication with the good quality of nanowires and nanoholes is a prerequisite for guaranteeing competitive optical and electrical properties that are predicted by theories and numerical simulations. Moreover, fast, low-cost, and matured fabrication techniques will largely promote large-scale applications of nanowires and nanoholes.

8.2 FABRICATION OF SILICON NANOWIRE AND NANOHOLE ARRAYS

Controllable nanostructure fabrication routes can usually be categorized into two approaches: "top-down" and "bottom-up" (Bandaru and Pichanusakorn 2010; Hobbs et al. 2012). In the top-down approach, both nanowire and nanohole arrays are derived from a bulk substrate and are obtained by progressive sculpting or etching to carve nanowires and nanoholes out of the bulk material (Lieber and Wang 2011). Nanowire arrays can also be obtained by the bottom-up approach. The bottom-up fabrication approach relies on packing the atoms and molecules along energetically preferential directions, assembling nanowire arrays from smaller units (Wang et al. 2006; Borgstrom et al. 2007).

There are numerous fabrication methods for nanowires or nanoholes, including top-down approaches using reactive ion etching (Fu et al. 2009; Li et al. 2013) and metal-assisted chemical etching (MACE)

(Huang et al. 2011), and bottom-up approaches using epitaxial-based vapor–liquid–solid (VLS) growth (Wagner and Ellis 1964; Schwarz and Tersoff 2009; Lerose et al. 2010) and metal–organic vapor-phase epitaxy (Dick et al. 2005; Joyce et al. 2011), all of which are to some extent capable of controlling the properties of the resulting nanowire and nanohole structures. As such, nanowire and nanohole arrays can be generated by electron-beam lithography (Pevzner et al. 2010), nanosphere lithography (NSL) (Haynes and Van Duyne 2001; Li et al. 2011a, 2011b), dip-pen nanolithography (Banholzer et al. 2009), and nanoimprint lithography (Mårtensson et al. 2004; Luo and Epps 2013). Here, we review some common fabrication methods in top-down and bottom-up approaches for nanowire and nanohole research.

8.2.1 TOP-DOWN FABRICATION OF NANOWIRE AND NANOHOLE ARRAYS

In top-down fabrication approaches, a bulk material is typically patterned by a series of masking steps. Controllable geometries and patterns can be created by using various lithography methods, including photo, colloidal, nanoimprint, and e-beam lithography, which are frequently employed to produce nanowire and nanohole arrays (Whang et al. 2003). Then, etching is an essential step to selectively remove materials from the patterned substrate, leading to the formation of nanowires and nanoholes. Etching can be carried out either in liquid or gas phase, that is, wet or dry etching, respectively. A dry etching method such as reactive-ion etching is one of the common techniques for fabricating nanowires and nanoholes, yet the achievable depth is limited, and an undesirable side effect is ion-induced damage to walls of nanowires and nanoholes (Amir et al. 2007; Fu et al. 2009; Pevzner et al. 2010). Alternatively, one of the most popular top-down wet-etching techniques for fabricating silicon nanowires and nanoholes is the MACE approach, involving the fabrication of anisotropic, high aspect ratio, and porous as well as nonporous structures (Peng et al. 2005a; Huang et al. 2011; Kim et al. 2011; Geyer et al. 2013).

Nowadays, in top-down fabrication schemes, both nanowire and nanohole arrays of well-defined architectures with consistent diameter, length, and period are typically fabricated through a two-step MACE method as shown in Figure 8.1.

First, a lithographically structured noble metal thin film is deposited on the substrate. For nanohole arrays, it is convenient to use NSL (Ellinas et al. 2011; Mikhael et al. 2011; Yeom et al. 2014) to generate discrete noble metal nanosphere array as the mask film in Figure 8.1a. The lithographical noble metal thin film of nanowire arrays as shown in Figure 8.1c can be obtained by typical lithographic approaches including laser interference lithography (Johannes de et al. 2010; Mai et al. 2012), e-beam lithography (Hildreth et al. 2009), block copolymer lithography (Chang et al. 2009; Huang et al. 2013; Luo and Epps 2013), and NSL. Among these patterning techniques, NSL is also one of the routinely employed methods for the fabrication of nanowire arrays.

Second, etching in hydrogen fluoride (HF) solution plus oxidant is performed as shown in Figure 8.1b and d. As an etching technique, MACE is carried out progressively relying on thin layer of noble metal

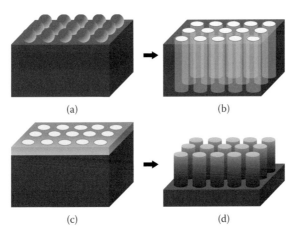

(a) (b)

(c) (d)

Figure 8.1 Schematic of nanowire and nanowire arrays fabricated by MACE. (a, b) Nanohole arrays etched by a discrete noble metal nanosphere array, and (c, d) nanowire arrays etched by a patterned noble metal film.

(e.g., Ag or Au) on silicon in a solution of HF and an oxidant (e.g., H_2O_2) (Li and Bohn 2000; Peng et al. 2006). The noble metal acts as a nanoscale cathode, on which reduction of the oxidant generates holes. These holes diffuse through the metal catalyst and are injected into the valence band of the silicon that is in contact with the metal, causing the silicon to be oxidized and finally dissolved by HF (Chern et al. 2010; Li 2012).

The MACE technique has many advantages over other fabrication techniques: (1) its versatility offers controllability in the synthesis path, including the noble metal catalyst type (Ag, Au, Pt, Pd, and Cu), metal pattern, etchant composition, etching rate, and intrinsic properties of the Si substrate (doping type, doping level, and crystal orientation); (2) it offers a wide selection of shapes for the resulting silicon nanostructures, since the etched structure replicates the shape of the deposited metal film; (3) it enables the fabrication of porous and nonporous nanowire and nanohole arrays, where the ratio of the oxidant and acid in solution, wafer resistivity, and catalyst type determine the overall porosity and the surface roughness of the resulting nanowire and nanohole arrays (Chiappini et al. 2010; Huang et al. 2011).

8.2.2 BOTTOM-UP FABRICATION OF NANOWIRE ARRAYS

A typical "bottom-up" nanowire synthesis process involves VLS or vapor–solid growth mechanisms in which nanoparticles are used as catalysts to continuously feed the 1D material growth (Qian et al. 2005). This approach can not only fabricate silicon nanowires but also have the ability to grow a wide range of nanowire materials, including group IV, III–V, and II–VI core/shell, superlattice, and branched structures, while simultaneously allowing precise control of the nanowire composition, morphology, and electrical properties through fine tuning of growth parameters that enhance or suppress the axial and radial growth processes (Wei and Charles 2006; Lieber 2011; Lieber and Wang 2011; Yang 2011).

The VLS mechanism was first suggested by Wagner and Ellis (1964) at Bell Laboratories in the early 1960s to explain gas-phase silicon whisker growth in the presence of a liquid gold droplet. But it was not until the late 1990s that the concept was promoted (Hu et al. 1999; Shi et al. 2000). The VLS growth mechanism has been reexamined and developed over the past two decades by Lieber (Yang and Lieber 1996; Morales and Lieber 1998), Yang (Wu and Yang 2001), Samuelson (Björk et al. 2002), and other research groups.

Understanding the VLS mechanism has led to nanowire growth with precise control over length, diameter, growth direction (Kuykendall et al. 2004), morphology, and composition. Nanowire diameter, typically from several nanometers to hundreds of nanometers, is determined by the size of the metal alloy droplets. Consequently, nanowire arrays with uniform size can be readily obtained by using monodispersed metal nanoparticles (Hochbaum et al. 2005). The length of these nanowires can be easily controlled from nanometers to millimeters. Precise orientation control during nanowire growth can be achieved by applying conventional epitaxial crystal growth techniques to this VLS process, known as VLS epitaxy. This process is particularly powerful in the controlled synthesis of high-quality nanowire arrays and single-wire devices (He et al. 2005). Moreover, within the past 10 years, great progress has been made in the precise positional control of nanowire growth using advanced lithography processes.

Importantly, owing to the nature of free-standing nanowire growth, these nanowires and their heterojunctions are usually dislocation-free and can readily accommodate large lattice mismatches, unlike thin-film growth (Ertekin et al. 2005). Therefore, semiconductor nanowire growth also allows direct integration of optically active semiconductors (e.g., III–V compounds) onto silicon substrates (Bakkers et al. 2004; Svensson et al. 2008). This capability represents a significant advantage over conventional thin-film technology. In high-efficiency photovoltaics and solid-state lighting, the precise control and full-range tunability over the composition of doped and alloyed nanowires are of particular interest.

In the VLS approach, the process relies on the deposition of metallic nanoparticle catalysts of defined diameter on a substrate as shown in Figure 8.2a. The substrate is heated in a tube furnace above the eutectic temperature of the relevant metal–semiconductor system in the presence of the semiconductor reactant vapor-phase source. The anisotropic growth is promoted by the presence of a liquid alloy–solid interface as shown in Figure 8.2b. Based on the Si–Au binary phase diagram, for instance, silicon (e.g., from the decomposition of SiH_4) and Au will generate a liquid alloy when the temperature is higher than the eutectic point (Duan et al. 2013). The liquid surface has a large accommodation coefficient and is therefore a preferred site of deposition for incoming silicon vapor. After the liquid alloy becomes supersaturated

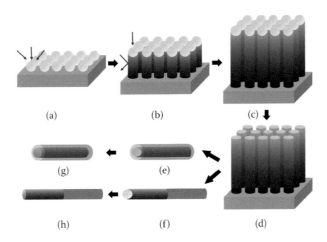

Figure 8.2 Schematic of nanowire arrays and heterostructures grown by the VLS approach. (a–d) Nanowire array synthesis through catalyst-mediated axial growth, (e, g) switching of the source material results in nanowire radial heterostructures; and (f, h) conformal deposition of different materials leads to the formation of axial nanowire heterostructures.

with silicon, silicon nanowire growth occurs by precipitation at the solid–liquid interface as shown in Figure 8.2c, because the growth temperature is higher than the eutectic point but lower than the melting point of the nanowire material. Moreover, it is convenient to use this mechanism to create heterostructures, including co-axial (Lauhon et al. 2002; Qian et al. 2008) and longitudinal (Gudiksen et al. 2002; Algra et al. 2008) variations. Typically, direct overgrowth on the side wall of the nanowire leads to co-axial heterostructures as shown in Figure 8.2g, whereas sequential nanowire VLS growth produces longitudinal heterostructures as shown in Figure 8.2h.

In summary, the VLS approach has two significant advantages over other fabrication techniques: (1) it is suitable for nanowires with the small diameters, large surface areas, and relatively smooth surfaces of nanowire materials that may be difficult to obtain from their top-down aggressively scaled counterparts and (2) it truly sets the bottom-up nanowire system apart, which is the ability to obtain heterostructures during growth, including radial core/shell heterostructures and axial superlattice heterostructures that are very challenging to match, or are even unobtainable, via top-down lithographic means.

8.3 OPTICAL PROPERTIES OF SILICON NANOWIRE AND NANOHOLE ARRAYS

As an abundant element on the earth, silicon is nontoxic and compatible with the semiconductor-manufacturing technology (Tsakalakos 2008; Lin and Povinelli 2009). Currently, 80%–90% of the photovoltaic market is based on crystalline silicon solar cells (Catchpole and Polman 2008). The use of expensive silicon wafers accounts for more than 40% of the cost of a solar module (Shah et al. 1999; Green 2007; Catchpole and Polman 2008), which impels the development of the thin-film solar cell for reducing the material cost. New designs and technologies have been proposed to make up the relatively low absorption due to smaller thicknesses, and nanowire and nanohole arrays are one of the most promising ways to address this issue.

More and more researchers are convinced that better nano and microphotonics such as nanowires and nanoholes for photon management inside solar cell is the key challenge and has great potential for cost-effective photovoltaic energy conversion (Lewis 2007; Polman and Atwater 2012; Vynck et al. 2012). Compared to traditional silicon solar cells, for example, silicon wafer and film-based ones, the solar cells constructed on silicon nanowire and nanohole arrays have some major potential advantages, which are totally tracing the key requirements of developing high efficiency and low-cost photovoltaic devices. One is the excellent light management with reduced materials consumption, which is first indicated by efficient antireflection (Peng et al. 2005b; Hu and Chen 2007; Li et al. 2009b). Thus, reduced light reflection without employing high cost and complex antireflection schemes can be achieved. Besides antireflection, effective scattering and the formation of guided resonance for the incident light in the silicon nanowire and nanohole arrays can remarkably improve the light absorption efficiency with reduced materials consumption, reducing the material cost for related solar cells (Lin and Povinelli 2009; Li et al. 2009a). On the other hand, for nanowire array–based devices, remarkable electrical properties allow the low-quality material silicon, further saving the material cost while keeping efficient carrier collection, which is discussed in detail in the next section.

Some nanowire arrays are fabricated and show their high conversion efficiencies for photovoltaic applications (Garnett and Yang 2008, 2010; Boettcher et al. 2010; Kelzenberg et al. 2010), as well as nanohole arrays (Chen et al. 2012; Meng et al. 2012; Yahaya et al. 2013). Furthermore, from the view of light trapping, nanohole arrays have better spectral absorption than the nanowire arrays (Han and Chen 2010), and this is further supported by more researchers (Xiong et al. 2010; Wang et al. 2011; Chen et al. 2012; Meng et al. 2012; Yahaya et al. 2013; Fu-Qiang et al. 2015).

8.3.1 STRUCTURAL PARAMETERS OF NANOWIRE AND NANOHOLE ARRAYS

It is worth noting that only relying on experimental studies makes it difficult to accurately depict the optical characteristics of silicon nanowire and nanohole arrays due to the parasitic absorption by substrates and surface defects created during the fabrication (Tsakalakos et al. 2007; Sivakov et al. 2009). Accordingly, simulation methods are preferred for clearly manifesting the optical processes, which can contribute to carrier generation. However, in simulations, the optical processes in nanowire and nanohole arrays are complicated because absorption characteristics vary with structural parameters and incident conditions.

In the view of light scattering, if objects have comparable characteristic dimensions to the wavelength of the incident light, light scattering would be significantly enhanced. Then, the optical path length is elongated, resulting in enhanced confinement to the respective light waves (Li et al. 2009a, 2009b) as visually sketched in Figure 8.3b. If the wavelength does not match the characteristic dimensions of the objects, for example, the wavelength is much longer as shown in Figure 8.3a or shorter than the characteristic dimensions as shown in Figure 8.3c, light propagation will rarely be affected or be strongly reflected by the objects (Li et al. 2012). Besides the improved light trapping based on light scattering, several other ideas are proposed to understand the enhanced absorption in silicon subwavelength structures, including guided resonance and leaky mode–based models (Lin and Povinelli 2009; Cao et al. 2010; Sturmberg et al. 2011; Wang and Leu 2012). Some conclusions such as how structural parameters influence the optical properties of nanowire and nanohole arrays are reviewed in the following sections.

8.3.1.1 Length

Low light transmission can be realized if the wires and holes are long enough. However, other issues such as low fabricability, poor wire quality, high fabrication cost, and prolonged fabrication period are associated

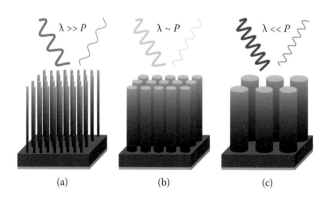

Figure 8.3 Schematized representative optical processes between incident light and silicon nanowire arrays with (a) the wavelength (λ) >> the array periodicity (P), (b) $\lambda \sim$ P, and (c) $\lambda \ll$ P.

with long silicon nanowires and nanoholes. An optimized length of wires and holes exists considering the tradeoff between maximizing the light trapping and minimizing the fabrication cost. Furthermore, the main absorption mechanisms will change with the length. For the hexagonally aligned wire arrays with periodicity and diameters of 600 and 500 nm, light scattering dominates the absorption enhancement when the wire length is shorter than 1000 nm or longer than 5000 nm, and the guided resonance-induced absorption enhancement is evidenced in an intermediate length range (Li et al. 2012). Considering that absorption enhancement weakens dramatically as the length is larger than 2000 nm, it usually sets the length of nanowires and nanoholes to about 2330 nm. This value is compatible with the thickness of thin silicon solar cells (Yamamoto et al. 1999, 2004) and is adopted by many researchers in simulations (Hu and Chen 2007; Lin and Povinelli 2009; Xiong et al. 2010; Wangyang et al. 2013).

8.3.1.2 Periodicity and diameter

Periodicity and diameter are structural parameters in the radial plane, and they affect light management in the plane vertical to the length. Various periodicities and diameters dominate different mechanisms of light absorption enhancement in silicon nanowire and nanohole arrays as well as the length, and subwavelength-scale periodicity and diameter even have a potential to break the conventional light-trapping limit (Yu et al. 2010c; Basu Mallick et al. 2011; Kupec and Witzigmann 2012). Hu and Chen first investigated nanowire arrays with diameters ranging from 50 to 80 nm and the periodicity fixed at 100 nm, and much lower reflectivity than thin films only occurs in the high-frequency regime (>2 eV) (Hu and Chen 2007). The role of the periodicity on absorption for nanowire arrays is further examined by Lin and Povinelli. The ultimate efficiency (absorption over the whole incident spectrum) increases with the period from 100 to 600 nm, which benefits from the guided resonance modes in nanowire arrays (Lin and Povinelli 2009). In photovoltaic applications, the ultimate efficiency is vague when the periodicity gets larger than 600 nm (Li et al. 2009a; Wangyang et al. 2013). Therefore, high-performance silicon nanowire and nanohole array–based photovoltaic devices should have the optimal array periodicity matching the wavelength region in which the photon number is relatively large. To find optimal periodicity and diameters, the optical absorption of nanowire and nanohole arrays with different cross-sectional shapes are numerically investigated in a wide periodicity ranging from 100 to 1500 nm. Optimal periodicity appears around 600 nm for all arrays, whereas optimal diameters are quite different for each array to achieve the maximum ultimate efficiency (Fang et al. 2014). When the substrate participates in the multireflection and absorption in nanowire arrays, the numerical results predicted that the ultimate efficiency reaches the peak when the periodicity is around 500 nm, which is less than free-standing arrays in the vacuum (Li et al. 2009a).

8.3.1.3 Patterns

By comparing the absorption spectra of the squarely and hexagonally arranged nanowire arrays, it is found that the array symmetry has little impact on the optical behaviors (Li et al. 2012). Apart from periodic arrays, aperiodic nanowire and nanohole arrays are found to have good broadband absorption enhancement. Because symmetry can sometimes prevent the guided resonance in the film structure from coupling outside, aperiodicity could be helpful to enhance broadband absorption (Yu et al. 2010b). Based on numerical electromagnetic simulations, Bao et al. demonstrated that in vertically aligned nanowire arrays with random position, diameter, and length, there is further enhanced light absorption compared to their ordered counterparts (Bao and Ruan 2010). For nanowire arrays with various optimized diameter distributions, Sturmberg et al. numerically demonstrated that the ultimate absorption efficiency can be 28% larger than an array with a uniform diameter. The mechanism of absorption enhancement in nanowire arrays is attributed to enhanced multiple scattering (Bao and Ruan 2010) or better matching of solar spectra using leaky-mode resonance (LMR) (Cao et al. 2009; Sturmberg et al. 2012). For aperiodic nanohole arrays, Vynck et al. studied the 2D slab with short-range positioned hole pattern and pointed out that amorphous patterns with short-range correlation could achieve better broadband absorption enhancement in a relatively broad frequency range (Vynck et al. 2012). Later, a new scheme to form nanohole arrays with short-range correlation was proposed by Oskooi et al., in which appropriate tiny displacements of holes are made from the ordered nanohole arrays (Oskooi et al. 2012). Semiconductor solar cells with correlated disordered nanoholes have larger absorption efficiency than ordered counterparts, which is subsequently validated by experimental results (Burresi et al. 2013; Pratesi et al. 2013). Besides, Fang et al. systematically investigated the absorption in three types of disordered nanohole arrays, that is, random positions, nonuniform radii, and amorphous patterns, and found that nanohole arrays with an amorphous pattern have the best light-trapping functionality for photovoltaics based on the statistic results of numerous cases (Fang et al. 2015). Recently, they also found that additional thin blocks on substrates can further enhance absorption in nanowire arrays (Fang and Zhao 2017). Moreover, the optical behavior of the hybrid nanowire–nanohole arrays indicated that nanowires are distributed in each square nanohole and higher absorption enhancement can be achieved (Wangyang et al. 2013). But, from the complicated fabrication's view, these hybrid nanostructures have no attraction for actual applications.

Then, absorption theories and common simulation methods in nanowire and nanohole arrays are introduced. The research progress in nanowire and nanohole arrays largely benefits from the advanced simulation methods and the improvement of computers ability (Mokkapati and Catchpole 2012).

8.3.2 APPROXIMATE THEORIES AND SIMULATION METHODS

Based on the characteristics of optical problems in nanostructures, one can solve them as either the source problem or the eigenvalue problem. For the source problem, an imposed illumination is given and the electromagnetic fields within the computational domain are determined. Then, optical properties for a particular geometry, wavelength, and incidence are obtained. In the eigenvalue problem, oscillatory solutions of the Maxwell equations in the absence of an excitation are computed. The solutions of the eigenvalue problem known as the modes are used to explain the behavior of nanostructures (Kupec and Witzigmann 2012).

There are different numerical methods for calculating optical parameters such as fields, reflectivity, and absorptivity in nanowire and nanohole arrays. The simulation methods include transfer matrix method (Hu and Chen 2007; Lin and Povinelli 2009; Han and Chen 2010; Xiong et al. 2010), finite-difference time-domain (FDTD) method (Meng et al. 2012; Yahaya et al. 2012, 2013), rigorous coupled-wave analysis (RCWA) (Lagos, Sigalas, and Niarchos 2011; Wangyang et al. 2013), finite-element method (Li et al. 2009a; Tok et al. 2011; Umut Tok and Şendur 2013), discrete dipole approximation (Draine and Flatau 2008; Loke et al. 2011), and so on. In general, it is necessary to perform a full numerical algorithm to calculate fields and optical properties of arrays. However, when the characteristic dimensions such as periodicity and diameter have particular restricted conditions, the theories such as effective medium theory (EMT) and the frame of LMRs are accurate enough to obtain and analyze optical behaviors in nanowire and nanohole arrays.

8.3.2.1 Effective medium theory

When the periodicity and diameter are much shorter than incident wavelengths, the silicon nanowire arrays can be viewed as effective media with the effective dielectric constant given by (Bergman 1978; Yang et al. 2009; Xiong et al. 2010)

$$[\varepsilon_{NW}] = \begin{pmatrix} \varepsilon_x & 0 & 0 \\ 0 & \varepsilon_x & 0 \\ 0 & 0 & \varepsilon_z \end{pmatrix}, \varepsilon_z = f\varepsilon_{Si} + (1-f)\varepsilon_{air}, \frac{\varepsilon_x - \varepsilon_{air}}{\varepsilon_x + \varepsilon_{air}} = \frac{\varepsilon_{Si} - \varepsilon_{air}}{\varepsilon_{Si} + \varepsilon_{air}} f \tag{8.1}$$

where ε_{Si} and ε_{air} are dielectric constants of silicon and the vacuum, respectively, and f is the filling ratio for the arrays. Similarly, the effective dielectric constant of nanohole arrays is given by

$$[\varepsilon_{NH}] = \begin{pmatrix} \varepsilon_x & 0 & 0 \\ 0 & \varepsilon_x & 0 \\ 0 & 0 & \varepsilon_z \end{pmatrix}, \varepsilon_z = f\varepsilon_{Si} + (1-f)\varepsilon_{air}, \frac{\varepsilon_x - \varepsilon_{Si}}{\varepsilon_x + \varepsilon_{Si}} = \frac{\varepsilon_{air} - \varepsilon_{Si}}{\varepsilon_{Si} + \varepsilon_{air}}(1-f) \tag{8.2}$$

After the effective dielectric constant of arrays is obtained, the absorption A, reflection R, and transmission T of an effective-medium slab in the vacuum at normal incidence can be directly calculated by the following equation:

$$T = \left| \cos\left(n_e k_0 L\right) - \frac{i\left(n_e^2 + 1\right)}{2n_e} \sin\left(n_e k_0 L\right) \right|^{-2}, \quad R = \left| \frac{n_e^2 - 1}{2n_e} \sin\left(n_e k_0 L\right) \right|^2 T, \quad A = 1 - R - T \qquad (8.3)$$

where $n_e = \sqrt{\varepsilon_x}$ is the effective refractive index and k_0 is the wavenumber of light in vacuum.

The EMT has been successfully employed in heavily doped silicon nanowires for applications in infrared range (Liu et al. 2013; Liu and Zhang 2013; Wang et al. 2013) considering that infrared wavelength is much longer than the periodicity of nanowire arrays. It is worth noting that the analysis based on the EMT may be not accurate when the period is comparable to the wavelength in silicon or when the air or silicon rods are close to each other.

8.3.2.2 Leaky-mode resonances

When the length of nanowires is longer than incident wavelength, and the array is sparse so that fields between a nanowire and neighboring nanowires have less interaction, one can describe modes in nanowires as modes in optical fibers and microscale dielectric resonators based on classical waveguide theory. By solving Maxwell's equations with the appropriate boundary conditions (Snyder and Love 1983), the excitation of leaky modes occurs in an infinitely long dielectric cylinder of radius a when the following condition is satisfied:

$$\left(\frac{1}{\kappa^2} - \frac{1}{\gamma^2}\right)^2 \left(\frac{\beta m}{a}\right)^2 = k_0^2 \left[n^2 \frac{J_m'(\kappa a)}{\kappa J_m(\kappa a)} - n_0^2 \frac{H_m'(\gamma a)}{\gamma H_m(\gamma a)} \right] \left[\frac{J_m'(\kappa a)}{\kappa J_m(\kappa a)} - \frac{H_m'(\gamma a)}{\gamma H_m(\gamma a)} \right] \qquad (8.4)$$

where $\gamma(\kappa)$ and $n_0(n)$ are the transverse wave vector and refractive index outside (inside) the cylinder, respectively, β and k_0 are the wave vectors along the cylindrical axis and in free space, respectively, J_m and H_m are the mth-order Bessel function of the first kind and Hankel function of the first kind, respectively, and the prime denotes differentiation with respect to related arguments.

This waveguide theory has well explained the spectral tunability and selectivity of absorption resonances in single nanowire (Cao et al. 2009), horizontal nanowire arrays (Cao et al. 2010), and vertical nanowire arrays (Wang and Leu 2012). However, in dense nanowire arrays or more common nanowire and nanohole arrays, a complete mode analysis is necessary to give the whole modes and their coupling effects (Sturmberg et al. 2011; Wen et al. 2012; Michallon et al. 2014).

To analyze optical behaviors in general nanowire and nanohole arrays, two of the most widely used numerical methods are described in the following sections. One is RCWA that is a high-efficient algorithm for periodic arrays, and the other is FDTD that is a power tool for simulating any complex electromagnetic problem.

8.3.2.3 Rigorous coupled-wave analysis

RCWA is an efficient and accurate numerical tool especially suitable for solving the electromagnetic field in periodic nanostructures (Moharam et al. 1995a, 1995b; Lalanne 1997). With better algorithms using the correct Li's factorization (Li 1996), the convergence of RCWA has been further accelerated and the memory requirement is significantly reduced. It is a semi-analytical method where the wave equation is solved analytically in the longitudinal direction. To implement the method, structures are divided into layers that are uniform in the longitudinal direction. The transverse problem is solved in reciprocal space by expressing the field as a sum of spatial harmonics. Thus, the wave equation transforms into a set of ordinary differential equations known as coupled-wave equations. The coupled-wave equation in the ith layer is expressed as

$$\frac{d}{dz^2} \begin{bmatrix} \mathbf{s}_x^{(i)} \\ \mathbf{s}_y^{(i)} \end{bmatrix} - \Omega_i^2 \begin{bmatrix} \mathbf{s}_x^{(i)} \\ \mathbf{s}_y^{(i)} \end{bmatrix} = 0 \qquad (8.5)$$

Arrays, hybrids, and core–shell

where **s** is the vector of spatial harmonic amplitudes of electric fields, and the coefficient matrix in Equation 8.5 is

$$\Omega_i^2 = \mathbf{P}_i\mathbf{Q}_i, \ \mathbf{P}_i = \begin{bmatrix} \mathbf{K}_x\varepsilon_{r,i}^{-1}\mathbf{K}_y & \varepsilon_{r,i} - \mathbf{K}_x\varepsilon_{r,i}^{-1}\mathbf{K}_x \\ \mathbf{K}_y\varepsilon_{r,i}^{-1}\mathbf{K}_y - \varepsilon_{r,i} & -\mathbf{K}_y\varepsilon_{r,i}^{-1}\mathbf{K}_x \end{bmatrix}, \mathbf{Q}_i = \begin{bmatrix} \mathbf{K}_x\varepsilon_{r,i}^{-1}\mathbf{K}_y & \varepsilon_{r,i} - \mathbf{K}_x\varepsilon_{r,i}^{-1}\mathbf{K}_x \\ \mathbf{K}_y\varepsilon_{r,i}^{-1}\mathbf{K}_y - \varepsilon_{r,i} & -\mathbf{K}_y\varepsilon_{r,i}^{-1}\mathbf{K}_x \end{bmatrix} \quad (8.6)$$

where **K** is the diagonal matrix of wave vector components defined in terms of reciprocal lattice vectors of an unit cell, and **ε** and **μ** are coefficient matrixes of permittivity and permeability expanded by Fourier series, respectively.

Coupled-wave equations are classical eigenvalue equations that have eigenvectors characterizing the configurations of spatial harmonics and eigenvalues describing longitudinal behavior in terms of a complex propagation constant that incorporates loss, gain, and coupling between modes. The solution of the coupled-wave equation is in the form as

$$\psi(\tilde{z}) = \begin{bmatrix} \mathbf{s}_x(\tilde{z}) \\ \mathbf{s}_y(\tilde{z}) \\ \mathbf{u}_x(\tilde{z}) \\ \mathbf{u}_y(\tilde{z}) \end{bmatrix} \begin{bmatrix} \mathbf{W} & \mathbf{W} \\ -\mathbf{V} & \mathbf{V} \end{bmatrix} \begin{bmatrix} e^{-\lambda\tilde{z}} & \mathbf{0} \\ \mathbf{0} & e^{\lambda\tilde{z}} \end{bmatrix} \begin{bmatrix} \mathbf{c}^+ \\ \mathbf{c}^- \end{bmatrix} \quad (8.7)$$

where **u** is the vector of spatial harmonic amplitudes of magnetic fields, **W** and λ are eigenvectors and eigenvalues of the matrix Ω^2, respectively, and **v** = **QW**λ^{-1}. Besides, **c**$^+$ and **c**$^-$ are proportionality constants corresponding to forward and backward propagating waves, respectively, which can be obtained through the boundary conditions by matching tangential fields at the segment interfaces. Therefore, after the boundary conditions are given, the overall solutions of spatial harmonics are obtained by Equation 8.7, which can transform into fields in the real space finally.

RCWA seems more tedious and complicated in the formulation than other methods, but its implementation is surprisingly simple and compact. Numerous researchers have used RCWA to simulate optical properties of periodic nanostructure arrays (Mallick et al. 2010; Peters et al. 2010; Tham and Heath 2010; Fang et al. 2014).

8.3.2.4 Finite-difference time-domain method

FDTD has an advantage for simulating large and complicated nanostructures, because it does not require to solve a large set of linear equations and can be parallelized to execute on high-performance computing systems (Adams et al. 2007; Yu et al. 2010). A simulation model with the FDTD method for simulating the optical process in nanowire arrays is shown in Figure 8.4a. In FDTD algorithm, a perfectly matched layer that absorbs light with any incident angle and any frequency is crucial to mimic infinitely extended space (Berenger 1994) and is described as gray layers in the top and bottom of the model. A source plane shown in yellow is added to stimulate the optical process in the simulation domain. Then, optical responses are recorded by monitors indicated by green and red planes. Besides, periodic boundaries are set on the walls of the model. FDTD discretizes the spatial derivatives of differential operators of Maxwell equations by expanding the field on a grid consisting of the Yee cells (Kane 1966), and the temporal derivatives by time stepping are simultaneously integrated.

To clarify the fundamental thought in FDTD, the electromagnetic problem in a linear, homogeneous, nondispersion medium is taken as an example, and the curl equations of Maxwell equations in the time domain are expressed as

$$\nabla \times \vec{E}(t) = -\mu\frac{\partial \vec{H}(t)}{\partial t}, \nabla \times \vec{H}(t) = \varepsilon\frac{\partial \vec{E}(t)}{\partial t} \quad (8.8)$$

To approximate the time derivatives using central finite differences, the electric and magnetic fields are staggered in time, and the discretization form is given as

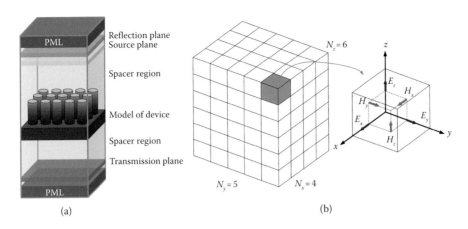

Figure 8.4 (a) Schematic of simulation model of FDTD; and (b) grid structures and the Yee cell.

$$\nabla \times \vec{E}\,|_{t} = -\mu \frac{\vec{H}\,|_{t+\Delta t/2} - \vec{H}\,|_{t-\Delta t/2}}{\Delta t}, \nabla \times \vec{H}\,|_{t+\Delta t/2} = \varepsilon \frac{\vec{E}\,|_{t+\Delta t} - \vec{E}\,|_{t}}{\Delta t} \tag{8.9}$$

Thus, the update equations are derived by solving the above equations for the fields at the future time value as below, and fields in the whole simulation time can be obtained theoretically

$$\vec{H}\,|_{t+\Delta t/2} = \vec{H}\,|_{t-\Delta t/2} - \frac{\Delta t}{\mu}\left(\nabla \times \vec{E}\,|_{t}\right), \vec{E}\,|_{t+\Delta t} = \vec{E}\,|_{t} + \frac{\Delta t}{\varepsilon}\left(\nabla \times \vec{H}\,|_{t+\Delta t/2}\right) \tag{8.10}$$

Actual FDTD programs have much more steps, and detailed FDTD method can be learned from an enormous library of literature of FDTD. There are many references on this method including books devoted entirely to FDTD (Taflove and Hagness 2000; Kantartzis and Tsiboukis 2006; Polycarpou 2006). There are many commercial software and free-available source codes based on FDTD, which also promote this method to widely apply to nanostructure arrays (Bao and Ruan 2010; Fang et al. 2015).

8.4 ELECTRICAL PROPERTIES OF SILICON NANOWIRE AND NANOHOLE ARRAYS

Although sometimes more efficient light management can be achieved using nanowire and nanohole arrays, one must see the accompanying challenges in fabricating the corresponding high-quality heterojunctions which can effectively separate photogenerated carriers and achieve acceptable open-circuit voltages. Without excellent electrical structures, the absorbed photon energy will be finally converted to heat with no acceptable electricity output. Fortunately, there is the possibility of employing the radial p–n junction configuration for nanowires, which can be easily engaged into the wires to guide the carrier collection along the radial direction rather than the light-trapping direction in the traditional solar cells with a planar p–n junction (Kayes et al. 2005; Tian et al. 2007). Therefore, the carrier collection capability of the related solar cells is not strongly dependent on the materials quality, further saving the material cost while keeping efficient carrier collection. Moreover, this advantage in electrical properties exists theoretically in compact nanohole arrays as well.

The general structure of p–n junction in nanowires is described in Figure 8.5. Energy levels and carriers in the half of vertical plane are also sketched. Similar to the plane solar cells, the p–n junction is divided into two regions: a p-type base, which is shown in green and comprises the core of the nanowire, and a shallow n-type emitter (blue) at the outside surface. An electron–hole pair (ehp) is generated through the absorption of an incident photon. The current collected at the base and emitter contacts depends on the rate at which the minority carriers arrive at those contacts and consequently depends strongly on the minority carrier diffusion. Minority carriers generated in the emitter or base that can reach the space charge layer by diffusion will be swept across the junction by the built-in electric field and contribute to the photocurrent.

Arrays, hybrids, and core-shell

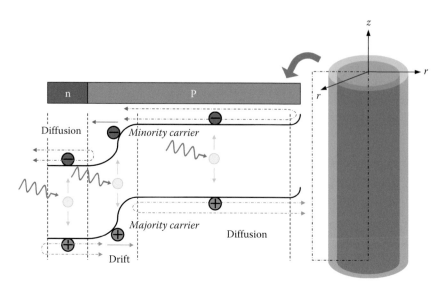

Figure 8.5 Schematic of p–n junction in a nanowire.

To numerically investigate the electrical properties of nanowire and nanohole arrays, the optical simulation models should be built to find the generation of electron–hole pairs in arrays by calculating the absorbed optical energy. The electrical simulation then determines how many of these photogenerated electron–hole pairs get collected at the electrical contacts and contribute to the output electrical power. The key performance metrics for a solar cell are short-circuit current (J_{sc}), open-circuit voltage (V_{oc}), fill factor (FF), and photovoltaic efficiency (η).

To calculate the absorption as a function of space and frequency, one needs to know the electric field E intensity and the imaginary part of the permittivity ε. The number of absorbed photons per unit volume can then be calculated by dividing this value by the energy per photon, and the generation rate is the integration of g over the simulation spectrum

$$g = \frac{-0.5|E|^2 \, \mathrm{Im}(\varepsilon)}{\hbar} \tag{8.11}$$

The quantum efficiency of a solar cell, QE(λ), is defined by

$$QE(\lambda) = \frac{P_{abs}}{P_{in}} \tag{8.12}$$

where λ is the wavelength, and P_{in} and P_{abs} are the powers of the incident light and absorbed light within the silicon nanowire and nanohole arrays, respectively.

Assuming that all absorbed photons generate electron–hole pairs and contribute to photocurrent, the short-circuit current density and the open-circuit voltage are given by (Shockley and Queisser 1961)

$$J_{SC} = e \int \frac{\lambda}{hc} QE(\lambda) I_{AM1.5}(\lambda) d\lambda, \quad V_{OC} = \frac{k_B T_a}{e} \ln\left(\frac{J_{SC}}{J_0} + 1 \right) \tag{8.13}$$

where e is the charge on an electron, h is Planck's constant, c is the speed of the light in the vacuum, $I_{AM1.5}(\lambda)$ is the spectral irradiance of AM 1.5 direct and circumsolar spectrum, k_B is Boltzmann's constant, T_a is the ambient temperature, and J_0 is the reverse saturation current density.

To evaluate the performance of the solar cell, we typically measure the photovoltaic energy conversion efficiency, that is,

$$\eta = \frac{FF \times V_{OC} \times J_{SC}}{S_{AM1.5}} \tag{8.14}$$

where FF is the fill factor, V_{oc} is the open-circuit voltage, J_{sc} is the short-circuit current, and $S_{AM1.5}$ is the incident power from the AM1.5 solar model. The fill factor is related to the maximum power point, where the power product JV is at its maximum value

$$FF = \frac{P_{max}}{V_{OC} J_{SC}} \tag{8.15}$$

Then, a brief summary is given regarding the radial p–n junction built around silicon nanowires and concluded with several key points for guiding to improve the electrical properties (Li et al. 2014): (1) The radial p–n junction is more tolerant to low material qualities, but the open-circuit voltage of nanowires is still sensitive to defects. It is predicted that the minority lifetime should be kept longer than 1 μs for achieving an acceptable V_{oc} > 0.6 eV. (2) The radial p–n junction is relatively insensitive to surface recombination velocity (SRV), so passivation of the wire surface to reduce SRV to <10^3 cm/s is necessary for realizing high device performance. (3) The radial distribution of the doping profile leads to the asymmetric distributions of the space charge density and electric field strength, and accordingly the doping concentrations and thicknesses of the p and n regions necessitate careful adjustment. (4) To keep the electric field strength strong enough (i.e., on the order of 10^7 V/m) for effective carrier separation, the doping levels >10^{18} cm^{-3} are suggested for the cases with comparable core and shell thicknesses. Although higher doping levels make stronger electric fields, the enhanced Auger recombination is detrimental to the open-circuit voltage. Accordingly, the doping concentration of 10^{18} cm^{-3} is recommended.

8.5 SUMMARY

The past decade has seen tremendous progress in the research field of nanowire and nanohole arrays. It has been demonstrated that these nanostructures can indeed have many different functionalities such as light trapping, light emission, lasing, waveguiding, nonlinear optical mixing, and so on. However, it is still necessary to make quantitative comparisons with conventional thin-film technology in terms of efficiency, fabrication cost, and stability, which is regarded as the benchmark assessing whether this new class of nanostructures is a viable candidate for future generations of photonic technologies. Moreover, a systematic evaluation of the environmental and health implications of the large-scale production of these materials is urgently required as well.

Nowadays, both nanowire and nanohole arrays with high quality can be prepared by numerous fabrication schemes. Some fabrication methods have been explored to control the specific properties of the resulting nanowire and nanohole structures. In these fabrication techniques, MACE method and VLS method are two kinds of most widely used to produce high-quality nanowires and nanoholes.

The recent progress in investigating optical properties and related mechanisms of nanowire and nanohole arrays is reviewed. The influence of structural parameters of arrays on light trapping is discussed in detail, then approximate theories and common simulation method for figuring out the optical process in nanowire and nanohole arrays are introduced. In nanowire and nanohole arrays, the absorption enhancement is achieved by light scattering and excitation of guide-mode resonances, which is considered to be applicable to guide the optical design for high-efficiency nanostructured photoelectric devices including solar cells and photodetectors.

The general p–n junction in nanowires is demonstrated to discuss the electric properties of the nanostructures. Electrical performance is accessed based on light energy absorbed in arrays. Nanowires have the superior carrier collection, but they have inferior open-circuit voltage to the traditional planar junction. Following that, some guidelines for fabricating a working p–n junction are concluded, presenting the key electrical parameters such as the tolerable surface and bulk defect densities and appropriate doping concentrations.

Arrays, hybrids, and core-shell

ACKNOWLEDGMENTS

This work is supported by the Natural Science Foundation of China (Nos: 51636004, 51476097, and 51306111) and Shanghai Key Fundamental Research Grant (No.: 16JC1403200).

REFERENCES

Adams, S., J. Payne, and R. Boppana. 2007. Finite difference time domain (FDTD) simulations using graphics processors. Paper read at DoD High Performance Computing Modernization Program Users Group Conference, 18–21 June 2007.

Algra, Rienk E., Marcel A. Verheijen, Magnus T. Borgstrom, Lou-Fe Feiner, George Immink, Willem J. P. van Enckevort, Elias Vlieg, and Erik P. A. M. Bakkers. 2008. Twinning superlattices in indium phosphide nanowires. *Nature* 456 (7220):369–372. http://www.nature.com/nature/journal/v456/n7220/suppinfo/nature07570_S1.html.

Amir, Sammak, Azimi Soheil, Mohajerzadeh Shams, Khadem-Hosseini Bahar, and Fallah-Azad Babak. 2007. *Silicon nanowire fabrication using novel hydrogenation-assisted deep reactive ion etching.* Paper read at Semiconductor Device Research Symposium, 2007 International, 12–14 Dec 2007.

Bakkers, Erik P. A. M., Jorden A. van Dam, Silvano De Franceschi, Leo P. Kouwenhoven, Monja Kaiser, Marcel Verheijen, Harry Wondergem, and Paul van der Sluis. 2004. Epitaxial growth of InP nanowires on germanium. *Nature Materials* 3 (11):769–773.

Bandaru, P. R., and P. Pichanusakorn. 2010. An outline of the synthesis and properties of silicon nanowires. *Semiconductor Science and Technology* 25 (2):024003.

Banholzer, Matthew J., Lidong Qin, Jill E. Millstone, Kyle D. Osberg, and Chad A. Mirkin. 2009. On-wire lithography: Synthesis, encoding and biological applications. *Nature Protocols* 4 (6):838–848.

Bao, Hua, and Xiulin Ruan. 2010. Optical absorption enhancement in disordered vertical silicon nanowire arrays for photovoltaic applications. *Optics Letters* 35 (20):3378–3380.

Basu Mallick, Shrestha, Nicholas P. Sergeant, Mukul Agrawal, Jung-Yong Lee, and Peter Peumans. 2011. Coherent light trapping in thin-film photovoltaics. *MRS Bulletin* 36 (06):453–460.

Berenger, Jean-Pierre. 1994. A perfectly matched layer for the absorption of electromagnetic waves. *Journal of Computational Physics* 114 (2):185–200. doi: http://dx.doi.org/10.1006/jcph.1994.1159.

Bergman, David J. 1978. The dielectric constant of a composite material—A problem in classical physics. *Physics Reports* 43 (9):377–407. doi: http://dx.doi.org/10.1016/0370-1573(78)90009-1.

Björk, M. T., B. J. Ohlsson, T. Sass, A. I. Persson, C. Thelander, M. H. Magnusson, K. Deppert, L. R. Wallenberg, and L. Samuelson. 2002. One-dimensional heterostructures in semiconductor nanowhiskers. *Applied Physics Letters* 80 (6):1058–1060. doi: http://dx.doi.org/10.1063/1.1447312.

Boettcher, Shannon W., Joshua M. Spurgeon, Morgan C. Putnam, Emily L. Warren, Daniel B. Turner-Evans, Michael D. Kelzenberg, James R. Maiolo, Harry A. Atwater, and Nathan S. Lewis. 2010. Energy-conversion properties of vapor-liquid-solid–erown silicon wire-Array photocathodes. *Science* 327 (5962):185–187.

Borgstrom, Magnus T., George Immink, Bas Ketelaars, Rienk Algra, and P. A. M. BakkersErik. 2007. Synergetic nanowire growth. *Nature Nanotechnology* 2 (9):541–544.

Burresi, Matteo, Filippo Pratesi, Kevin Vynck, Mauro Prasciolu, Massimo Tormen, and Diederik S. Wiersma. 2013. Two-dimensional disorder for broadband, omnidirectional and polarization-insensitive absorption. *Optics Express* 21 (102):A268–A275.

Cao, Linyou, Pengyu Fan, Alok P. Vasudev, Justin S. White, Zongfu Yu, Wenshan Cai, Jon A. Schuller, Shanhui Fan, and Mark L. Brongersma. 2010. Semiconductor nanowire optical antenna solar absorbers. *Nano Letters* 10 (2):439–445.

Cao, Linyou, Justin S. White, Joon-Shik Park, Jon A. Schuller, Bruce M. Clemens, and Mark L. Brongersma. 2009. Engineering light absorption in semiconductor nanowire devices. *Nature Materials* 8 (8):643–647.

Catchpole, K. R., and Albert Polman. 2008. Plasmonic solar cells. *Optics Express* 16 (26):21793–21800.

Chang, Shih-Wei, Vivian P. Chuang, Steven T. Boles, Caroline A. Ross, and Carl V. Thompson. 2009. Densely packed arrays of ultra-high-aspect-ratio silicon nanowires fabricated using block-copolymer lithography and metal-assisted etching. *Advanced Functional Materials* 19 (15):2495–2500. doi: http://dx.doi.org/10.1002/adfm.200900181.

Chen, Ting-Gang, Peichen Yu, Shih-Wei Chen, Feng-Yu Chang, Bo-Yu Huang, Yu-Chih Cheng, Jui-Chung Hsiao, Chi-Kang Li, and Yuh-Renn Wu. 2012. Characteristics of large-scale nanohole arrays for thin-silicon photovoltaics. *Progress in Photovoltaics: Research and Applications.*

Chern, Winston, Keng Hsu, Ik Su Chun, Bruno P. de Azeredo, Numair Ahmed, Kyou-Hyun Kim, Jian-min Zuo, Nick Fang, Placid Ferreira, and Xiuling Li. 2010. Nonlithographic patterning and metal-assisted chemical etching for manufacturing of tunable light-emitting silicon nanowire arrays. *Nano Letters* 10 (5):1582–1588. doi: http://dx.doi.org/10.1021/nl903841a.

Chiappini, Ciro, Xuewu Liu, Jean Raymond Fakhoury, and Mauro Ferrari. 2010. Biodegradable porous silicon barcode nanowires with defined geometry. *Advanced Functional Materials* 20 (14):2231–2239. doi: http://dx.doi.org/10.1002/adfm.201000360.

Dick, Kimberly A., Knut Deppert, Thomas Mårtensson, Bernhard Mandl, Lars Samuelson, and Werner Seifert. 2005. Failure of the vapor–liquid–solid mechanism in Au-assisted MOVPE growth of InAs nanowires. *Nano Letters* 5 (4):761–764. doi: http://dx.doi.org/10.1021/nl050301c.

Draine, Bruce T, and Piotr J Flatau. 2008. Discrete-dipole approximation for periodic targets: theory and tests. *JOSA A* 25 (11):2693–2703.

Duan, Xiaojie, Tian-Ming Fu, Jia Liu, and Charles M. Lieber. 2013. Nanoelectronics-biology frontier: From nanoscopic probes for action potential recording in live cells to three-dimensional cyborg tissues. *Nano Today* 8 (4):351–373. doi: http://dx.doi.org/10.1016/j.nantod.2013.05.001.

Ellinas, Kosmas, Athanasios Smyrnakis, Antonia Malainou, Angeliki Tserepi, and Evangelos Gogolides. 2011. "Mesh-assisted" colloidal lithography and plasma etching: A route to large-area, uniform, ordered nano-pillar and nanopost fabrication on versatile substrates. *Microelectronic Engineering* 88 (8):2547–2551. doi: http://dx.doi.org/10.1016/j.mee.2010.12.073.

Elnathan, Roey, Moria Kwiat, Fernando Patolsky, and Nicolas H. Voelcker. 2014. Engineering vertically aligned semiconductor nanowire arrays for applications in the life sciences. *Nano Today* 9 (2):172–196.

Ertekin, Elif, P. A. Greaney, D. C. Chrzan, and Timothy D. Sands. 2005. Equilibrium limits of coherency in strained nanowire heterostructures. *Journal of Applied Physics* 97 (11):114325. doi: http://dx.doi.org/10.1063/1.1903106.

Fang, Xing, Minhan Lou, Hua Bao, and C. Y. Zhao. 2015. Thin films with disordered nanohole patterns for solar radiation absorbers. *Journal of Quantitative Spectroscopy and Radiative Transfer* 158:145–153. doi: http://dx.doi.org/10.1016/j.jqsrt.2015.01.002.

Fang, Xing, C.Y. Zhao, and Hua Bao. 2014. Radiative behaviors of crystalline silicon nanowire and nanohole arrays for photovoltaic applications. *Journal of Quantitative Spectroscopy and Radiative Transfer* 133:579–588.

Fang, Xing, C.Y. Zhao. 2017. Grading absorption and enhancement in silicon nanowire arrays with thin blocks. *Journal of Quantitative Spectroscopy and Radiative Transfer* 194:7–16.

Fu, Y. Q., A. Colli, A. Fasoli, J. K. Luo, A. J. Flewitt, A. C. Ferrari, and W. I. Milne. 2009. Deep reactive ion etching as a tool for nanostructure fabrication. *Journal of Vacuum Science & Technology B* 27 (3):1520–1526. doi: http://dx.doi.org/10.1116/1.3065991.

Fu-Qiang, Zhang, Peng Kui-Qing, Sun Rui-Nan, Hu Ya, and Lee Shuit-Tong. 2015. Light trapping in randomly arranged silicon nanorocket arrays for photovoltaic applications. *Nanotechnology* 26 (37):375401.

Garnett, Erik, and Peidong Yang. 2010. Light trapping in silicon nanowire solar cells. *Nano Letters* 10 (3):1082–1087.

Garnett, Erik C., and Peidong Yang. 2008. Silicon nanowire radial p–n junction solar cells. *Journal of the American Chemical Society* 130 (29):9224–9225.

Geyer, Nadine, Bodo Fuhrmann, Hartmut S. Leipner, and Peter Werner. 2013. Ag-mediated charge transport during metal-assisted chemical etching of silicon nanowires. *ACS Applied Materials & Interfaces* 5 (10):4302–4308. doi: http://dx.doi.org/10.1021/am400510f.

Green, Martin A. 2007. Thin-film solar cells: review of materials, technologies and commercial status. *Journal of Materials Science: Materials in Electronics* 18 (1):15–19.

Gudiksen, Mark S., Lincoln J. Lauhon, Jianfang Wang, David C. Smith, and Charles M. Lieber. 2002. Growth of nanowire superlattice structures for nanoscale photonics and electronics. *Nature* 415 (6872):617–620.

Han, Sang Eon, and Gang Chen. 2010. Optical absorption enhancement in silicon nanohole arrays for solar photovoltaics. *Nano Letters* 10 (3):1012–1015.

Haynes, Christy L., and Richard P. Van Duyne. 2001. Nanosphere lithography: A versatile nanofabrication tool for studies of size-dependent nanoparticle optics. *The Journal of Physical Chemistry B* 105 (24):5599–5611. doi: http://dx.doi.org/10.1021/jp010657m.

He, R., D. Gao, R. Fan, A. I Hochbaum, C. Carraro, R. Maboudian, and P. Yang. 2005. Si nanowire bridges in microtrenches: Integration of growth into device fabrication. *Advanced Materials* 17 (17):2098–2102. doi: http://dx.doi.org/10.1002/adma.200401959.

Hildreth, Owen James, Wei Lin, and Ching Ping Wong. 2009. Effect of catalyst shape and etchant composition on etching direction in metal-assisted chemical etching of silicon to fabricate 3D nanostructures. *ACS Nano* 3 (12):4033–4042. doi: http://dx.doi.org/10.1021/nn901174e.

Hobbs, Richard G., Nikolay Petkov, and Justin D. Holmes. 2012. Semiconductor nanowire fabrication by bottom-up and top-down paradigms. *Chemistry of Materials* 24 (11):1975–1991. doi: http://dx.doi.org/10.1021/cm300570n.

Hochbaum, Allon I., Rong Fan, Rongrui He, and Peidong Yang. 2005. Controlled growth of Si nanowire arrays for device integration. *Nano Letters* 5 (3):457–460. doi: http://dx.doi.org/10.1021/nl047990x.

Hu, Jiangtao, Min Ouyang, Peidong Yang, and Charles M. Lieber. 1999. Controlled growth and electrical properties of heterojunctions of carbon nanotubes and silicon nanowires. *Nature* 399 (6731):48–51.

Hu, Lu, and Gang Chen. 2007. Analysis of optical absorption in silicon nanowire arrays for photovoltaic applications. *Nano Letters* 7 (11):3249–3252.

Huang, Yinggang, Tae Wan Kim, Shisheng Xiong, Luke J. Mawst, Thomas F. Kuech, Paul F. Nealey, Yushuai Dai, et al. 2013. InAs nanowires grown by metal–organic vapor-phase epitaxy (MOVPE) employing PS/PMMA diblock copolymer nanopatterning. *Nano Letters* 13 (12):5979–5984. doi: http://dx.doi.org/10.1021/nl403163x.

Huang, Zhipeng, Nadine Geyer, Peter Werner, Johannes de Boor, and Ulrich Gösele. 2011. Metal-assisted chemical etching of silicon: A review. *Advanced Materials* 23(2):285–308. doi: http://dx.doi.org/10.1002/adma.201001784.

Johannes de, Boor, Geyer Nadine, V. Wittemann Jörg, Gösele Ulrich, and Schmidt Volker. 2010. Sub-100 nm silicon nanowires by laser interference lithography and metal-assisted etching. *Nanotechnology* 21 (9):095302.

Joyce, Hannah J., Qiang Gao, H. Hoe Tan, C. Jagadish, Yong Kim, Jin Zou, Leigh M. Smith, Howard E. Jackson, Jan M. Yarrison-Rice, Patrick Parkinson, and Michael B. Johnston. 2011. III–V semiconductor nanowires for optoelectronic device applications. *Progress in Quantum Electronics* 35 (2–3):23–75. doi: http://dx.doi.org/10.1016/j.pquantelec.2011.03.002.

Kane, Yee. 1966. Numerical solution of initial boundary value problems involving Maxwell's equations in isotropic media. *IEEE Transactions on Antennas and Propagation* 14 (3):302–307. doi: http://dx.doi.org/10.1109/TAP.1966.1138693.

Kantartzis, Nikolaos V., and Theodoros D. Tsiboukis. 2006. Higher order FDTD schemes for waveguide and antenna structures. *Synthesis Lectures on Computational Electromagnetics* 1 (1):1–226. doi: http://dx.doi.org/10.2200/S00018ED1V01Y200604CEM003.

Kayes, Brendan M., Harry A. Atwater, and Nathan S. Lewis. 2005. Comparison of the device physics principles of planar and radial p-n junction nanorod solar cells. *Journal of Applied Physics* 97 (11):114302-114302-11.

Kelzenberg, Michael D., Shannon W. Boettcher, Jan A. Petykiewicz, Daniel B. Turner-Evans, Morgan C. Putnam, Emily L. Warren, Joshua M. Spurgeon, Ryan M. Briggs, Nathan S. Lewis, and Harry A. Atwater. 2010. Enhanced absorption and carrier collection in Si wire arrays for photovoltaic applications. *Nature Materials* 9 (3):239–244.

Kim, Jungkil, Hee Han, Young Heon Kim, Suk-Ho Choi, Jae-Cheon Kim, and Woo Lee. 2011. Au/Ag bilayered metal mesh as a Si etching catalyst for controlled fabrication of Si nanowires. *ACS Nano* 5 (4):3222–3229. doi: http://dx.doi.org/10.1021/nn2003458.

Kupec, Jan, and Bernd Witzigmann. 2012. Computational electromagnetics for nanowire solar cells. *Journal of Computational Electronics* 11 (2):153–165.

Kuykendall, Tevye, Peter J. Pauzauskie, Yanfeng Zhang, Joshua Goldberger, Donald Sirbuly, Jonathan Denlinger, and Peidong Yang. 2004. Crystallographic alignment of high-density gallium nitride nanowire arrays. *Nature Materials* 3 (8):524–528. http://www.nature.com/nmat/journal/v3/n8/suppinfo/nmat1177_S1.html.

Lagos, N, MM Sigalas, and D Niarchos. 2011. The optical absorption of nanowire arrays. *Photonics and Nanostructures-Fundamentals and Applications* 9 (2):163–167.

Lalanne, Philippe. 1997. Improved formulation of the coupled-wave method for two-dimensional gratings. *Journal of the Optical Society of America A* 14 (7):1592–1598. doi: http://dx.doi.org/10.1364/JOSAA.14.001592.

Lauhon, Lincoln J., Mark S. Gudiksen, Deli Wang, and Charles M. Lieber. 2002. Epitaxial core-shell and core-multishell nanowire heterostructures. *Nature* 420 (6911):57–61. http://www.nature.com/nature/journal/v420/n6911/suppinfo/nature01141_S1.html.

Lerose, Damiana, Mikhael Bechelany, Laetitia Philippe, Johann Michler, and Silke Christiansen. 2010. Ordered arrays of epitaxial silicon nanowires produced by nanosphere lithography and chemical vapor deposition. *Journal of Crystal Growth* 312 (20):2887–2891. doi: http://dx.doi.org/10.1016/j.jcrysgro.2010.07.023.

Lewis, Nathan S. 2007. Toward cost-effective solar energy use. *Science* 315 (5813):798–801.

Li, Junshuai, HongYu Yu, and Yali Li. 2012. Solar energy harnessing in hexagonally arranged Si nanowire arrays and effects of array symmetry on optical characteristics. *Nanotechnology* 23 (19):194010.

Li, Junshuai, HongYu Yu, She Mein Wong, Gang Zhang, Xiaowei Sun, Patrick Guo-Qiang Lo, and Dim-Lee Kwong. 2009a. Si nanopillar array optimization on Si thin films for solar energy harvesting. *Applied Physics Letters* 95 (3):033102.

Li, Junshuai, HongYu Yu, She Mein Wong, Xiaocheng Li, Gang Zhang, Patrick Guo-Qiang Lo, and Dim-Lee Kwong. 2009b. Design guidelines of periodic Si nanowire arrays for solar cell application. *Applied Physics Letters* 95 (24):243113.

Li, Liang, Tianyou Zhai, Haibo Zeng, Xiaosheng Fang, Yoshio Bando, and Dmitri Golberg. 2011a. Polystyrene sphere-assisted one-dimensional nanostructure arrays: Synthesis and applications. *Journal of Materials Chemistry* 21 (1):40–56. doi: http://dx.doi.org/10.1039/C0JM02230F.

Li, Lifeng. 1996. Use of fourier series in the analysis of discontinuous periodic structures. *JOSA A* 13 (9):1870–1876.

Li, X., and P. W. Bohn. 2000. Metal-assisted chemical etching in HF/H2O2 produces porous silicon. *Applied Physics Letters* 77 (16):2572–2574. doi: http://dx.doi.org/10.1063/1.1319191.

Li, Xiuling. 2012. Metal assisted chemical etching for high aspect ratio nanostructures: A review of characteristics and applications in photovoltaics. *Current Opinion in Solid State and Materials Science* 16 (2):71–81. doi: http://dx.doi.org/10.1016/j.cossms.2011.11.002.

Arrays, hybrids, and core–shell

Li, Yali, Qiang Chen, Deyan He, and Junshuai Li. 2014. Radial junction Si micro/nano-wire array photovoltaics: Recent progress from theoretical investigation to experimental realization. *Nano Energy* 7:10–24.

Li, Yat, Fang Qian, Jie Xiang, and Charles M Lieber. 2006. Nanowire electronic and optoelectronic devices. *Materials Today* 9 (10):18–27.

Li, Yue, Guotao Duan, Guangqiang Liu, and Weiping Cai. 2013. Physical processes-aided periodic micro/nano-structured arrays by colloidal template technique: Fabrication and applications. *Chemical Society Reviews* 42 (8):3614–3627. doi: http://dx.doi.org/10.1039/C3CS35482B.

Li, Yue, Naoto Koshizaki, and Weiping Cai. 2011b. Periodic one-dimensional nanostructured arrays based on colloidal templates, applications, and devices. *Coordination Chemistry Reviews* 255 (3–4):357–373. doi: http://dx.doi.org/10.1016/j.ccr.2010.09.015.

Lieber, Charles M. 2011. Nanoscale science and technology: Building a big future from small things. *MRS Bulletin* 28 (7):486–491. doi: http://dx.doi.org/10.1557/mrs2003.144.

Lieber, Charles M., and Zhong Lin Wang. 2011. Functional nanowires. *MRS Bulletin* 32(2):99–108. doi: http://dx.doi.org/10.1557/mrs2007.41.

Lin, Chenxi, and Michelle L. Povinelli. 2009. Optical absorption enhancement in silicon nanowire arrays with a large lattice constant for photovoltaic applications. *Optics Express* 17 (22):19371–19381.

Liu, X. L., L. P. Wang, and Z. M. Zhang. 2013. Wideband tunable omnidirectional infrared absorbers based on doped-silicon nanowire arrays. *Journal of Heat Transfer* 135 (6):061602.

Liu, X. L., and Z. M. Zhang. 2013. Metal-free low-loss negative refraction in the mid-infrared region. *Applied Physics Letters* 103 (10):103101.

Loke, Vincent LY, M Pinar Mengüç, and Timo A Nieminen. 2011. Discrete-dipole approximation with surface interaction: Computational toolbox for MATLAB. *Journal of Quantitative Spectroscopy and Radiative Transfer* 112 (11):1711–1725.

Luo, Ming, and Thomas H. Epps. 2013. Directed block copolymer thin film self-assembly: Emerging trends in nanopattern fabrication. *Macromolecules* 46 (19):7567–7579. doi: http://dx.doi.org/10.1021/ma401112y.

Mai, Trong Thi, Chang Quan Lai, H. Zheng, Karthik Balasubramanian, K. C. Leong, P. S. Lee, Chengkuo Lee, and W. K. Choi. 2012. Dynamics of wicking in silicon nanopillars fabricated with interference lithography and metal-assisted chemical etching. *Langmuir* 28 (31):11465–11471. doi: http://dx.doi.org/10.1021/la302262g.

Mallick, Shrestha Basu, Mukul Agrawal, and Peter Peumans. 2010. Optimal light trapping in ultra-thin photonic crystal crystalline silicon solar cells. *Optics Express* 18 (6):5691–5706. doi: http://dx.doi.org/10.1364/OE.18.005691.

Mårtensson, Thomas, Patrick Carlberg, Magnus Borgström, Lars Montelius, Werner Seifert, and Lars Samuelson. 2004. Nanowire arrays defined by nanoimprint lithography. *Nano Letters* 4 (4):699–702. doi: http://dx.doi.org/10.1021/nl035100s.

Meng, Xianqin, Valérie Depauw, Guillaume Gomard, Ounsi El Daif, Christos Trompoukis, Emmanuel Drouard, Cécile Jamois, Alain Fave, Frédéric Dross, and Ivan Gordon. 2012. Design, fabrication and optical characterization of photonic crystal assisted thin film monocrystalline-silicon solar cells. *Optics Express* 20 (104):A465–A475.

Michallon, Jérôme, Davide Bucci, Alain Morand, Mauro Zanuccoli, Vincent Consonni, and Anne Kaminski-Cachopo. 2014. Light trapping in ZnO nanowire arrays covered with an absorbing shell for solar cells. *Optics Express* 22 (104):A1174–A1189.

Mikhael, Bechelany, Berodier Elise, Maeder Xavier, Schmitt Sebastian, Michler Johann, and Philippe Laetitia. 2011. New silicon architectures by gold-Assisted chemical etching. *ACS Applied Materials & Interfaces* 3 (10):3866–3873. doi: http://dx.doi.org/10.1021/am200948p.

Moharam, M. G., T. K. Gaylord, Eric B. Grann, and Drew A. Pommet. 1995a. Formulation for stable and efficient implementation of the rigorous coupled-wave analysis of binary gratings. *Journal of the Optical Society of America A* 12 (5):1068–1076. doi: http://dx.doi.org/10.1364/JOSAA.12.001068.

Moharam, M. G., T. K. Gaylord, Drew A. Pommet, and Eric B. Grann. 1995b. Stable implementation of the rigorous coupled-wave analysis for surface-relief gratings: enhanced transmittance matrix approach. *Journal of the Optical Society of America A* 12 (5):1077–1086. doi: http://dx.doi.org/10.1364/JOSAA.12.001077.

Mokkapati, S., and K. R. Catchpole. 2012. Nanophotonic light trapping in solar cells. *Journal of Applied Physics* 112:101101.

Morales, Alfredo M., and Charles M. Lieber. 1998. A laser ablation method for the synthesis of crystalline semiconductor nanowires. *Science* 279 (5348):208–211. doi: http://dx.doi.org/10.1126/science.279.5348.208.

Oskooi, Ardavan, Pedro A. Favuzzi, Yoshinori Tanaka, Hiroaki Shigeta, Yoichi Kawakami, and Susumu Noda. 2012. Partially disordered photonic-crystal thin films for enhanced and robust photovoltaics. *Applied Physics Letters* 100 (18):181110. doi: http://dx.doi.org/10.1063/1.4711144.

Peng, Kuiqing, Hui Fang, Juejun Hu, Yin Wu, Jing Zhu, Yunjie Yan, and ShuitTong Lee. 2006. Metal-particle-induced, highly localized site-specific etching of Si and formation of single-crystalline Si nanowires in aqueous fluoride solution. *Chemistry—A European Journal* 12 (30):7942–7947. doi: http://dx.doi.org/10.1002/chem.200600032.

Peng, Kuiqing, Yin Wu, Hui Fang, Xiaoyan Zhong, Ying Xu, and Jing Zhu. 2005a. Uniform, axial-orientation alignment of one-dimensional single-crystal silicon nanostructure arrays. *Angewandte Chemie International Edition* 44 (18):2737–2742. doi: http://dx.doi.org/10.1002/anie.200462995.

Peng, Kuiqing, Ying Xu, Yin Wu, Yunjie Yan, Shuit-Tong Lee, and Jing Zhu. 2005b. Aligned single-crystalline Si nanowire arrays for photovoltaic applications. *Small* 1 (11):1062–1067. doi: http://dx.doi.org/10.1002/smll.200500137.

Peters, Marius, Marc Rüdiger, Benedikt Bläsi, and Werner Platzer. 2010. Electro–optical simulation of diffraction in solar cells. *Optics Express* 18 (104):A584–A593.

Pevzner, Alexander, Yoni Engel, Roey Elnathan, Tamir Ducobni, Moshit Ben-Ishai, Koteeswara Reddy, Nava Shpaisman, Alexander Tsukernik, Mark Oksman, and Fernando Patolsky. 2010. Knocking down highly-ordered large-scale nanowire arrays. *Nano Letters* 10 (4):1202–1208. doi: http://dx.doi.org/10.1021/nl903560u.

Polman, Albert, and Harry A. Atwater. 2012. Photonic design principles for ultrahigh-efficiency photovoltaics. *Nature Materials* 11 (3):174–177.

Polycarpou, Anastasis C. 2006. Introduction to the finite element method in electromagnetics. *Synthesis Lectures on Computational Electromagnetics* 1 (1):1–126. doi: http://dx.doi.org/10.2200/S00019ED1V01Y200604CEM004.

Pratesi, Filippo, Matteo Burresi, Francesco Riboli, Kevin Vynck, and Diederik S. Wiersma. 2013. Disordered photonic structures for light harvesting in solar cells. *Optics Express* 21 (S3):A460–A468. doi: http://dx.doi.org/10.1364/OE.21.00A460.

Qian, Fang, Silvija Gradečak, Yat Li, Cheng-Yen Wen, and Charles M. Lieber. 2005. Core/multishell nanowire heterostructures as multicolor, high-efficiency light-emitting diodes. *Nano Letters* 5 (11):2287–2291. doi: http://dx.doi.org/10.1021/nl051689e.

Qian, Fang, Yat Li, Silvija Gradecak, Hong-Gyu Park, Yajie Dong, Yong Ding, Zhong Lin Wang, and Charles M. Lieber. 2008. Multi-quantum-well nanowire heterostructures for wavelength-controlled lasers. *Nature Materials* 7 (9):701–706. http://www.nature.com/nmat/journal/v7/n9/suppinfo/nmat2253_S1.html.

Schierning, Gabi. 2014. Silicon nanostructures for thermoelectric devices: A review of the current state of the art. *Physica Status Solidi A* 211 (6):1235–1249.

Schwarz, K. W., and J. Tersoff. 2009. From droplets to nanowires: Dynamics of vapor-liquid-sSolid growth. *Physical Review Letters* 102 (20):206101.

Shah, Arvind, P Torres, R Tscharner, N Wyrsch, and H Keppner. 1999. Photovoltaic technology: The case for thin-film solar cells. *Science* 285 (5428):692–698.

Shi, W. S., H. Y. Peng, Y. F. Zheng, N. Wang, N. G. Shang, Z. W. Pan, C. S. Lee, and S. T. Lee. 2000. Synthesis of large areas of highly oriented, very long silicon nanowires. *Advanced Materials* 12 (18):1343–1345. doi: http://dx.doi.org/10.1002/1521-4095(200009)12:18<1343::AID-ADMA1343>3.0.CO;2-Q.

Shockley, William, and Hans J Queisser. 1961. Detailed balance limit of efficiency of p-n junction solar cells. *Journal of Applied Physics* 32 (3):510–519.

Sivakov, V., G. Andrä, A. Gawlik, A. Berger, J. Plentz, F. Falk, and S. H. Christiansen. 2009. Silicon nanowire-based solar cells on glass: Synthesis, optical properties, and cell parameters. *Nano Letters* 9 (4):1549–1554. doi: http://dx.doi.org/10.1021/nl803641f.

Snyder, AW, and J Love. 1983. *Optical Waveguide Theory*. Springer Science & Business Media, London, UK.

Sturmberg, Björn C. P., Kokou B. Dossou, Lindsay C. Botten, Ara A. Asatryan, Christopher G. Poulton, C. Martijn de Sterke, and Ross C. McPhedran. 2011. Modal analysis of enhanced absorption in silicon nanowire arrays. *Optics Express* 19 (105):A1067–A1081.

Sturmberg, Björn C. P., Kokou B. Dossou, Lindsay C. Botten, Ara A. Asatryan, Christopher G. Poulton, Ross C. McPhedran, and C. Martijn de Sterke. 2012. Nanowire array photovoltaics: Radial disorder versus design for optimal efficiency. *Applied Physics Letters* 101 (17):173902.

Svensson, C. Patrik T, Mårtensson Thomas, Trägårdh Johanna, Larsson Christina, Rask Michael, Hessman Dan, Samuelson Lars, and Ohlsson Jonas. 2008. Monolithic GaAs/InGaP nanowire light emitting diodes on silicon. *Nanotechnology* 19 (30):305201.

Taflove, Allen, and Susan C. Hagness. 2000. *Computational Electrodynamics*. Artech house, Norwood, MA.

Tham, Douglas, and James R. Heath. 2010. Ultradense, deep subwavelength nanowire array photovoltaics as engineered optical thin films. *Nano Letters* 10 (11):4429–4434. doi: http://dx.doi.org/10.1021/nl102199b.

Tian, Bozhi, Xiaolin Zheng, Thomas J. Kempa, Ying Fang, Nanfang Yu, Guihua Yu, Jinlin Huang, and Charles M. Lieber. 2007. Coaxial silicon nanowires as solar cells and nanoelectronic power sources. *Nature* 449 (7164):885–889.

Tok, Rüstü Umut, Cleva Ow-Yang, and Kürsat Sendur. 2011. Unidirectional broadband radiation of honeycomb plasmonic antenna array with broken symmetry. *Optics Express* 19 (23):22731–22742.

Tsakalakos, Loucas. 2008. Nanostructures for photovoltaics. *Materials Science and Engineering R* 62 (6):175–189.

Tsakalakos, Loucas, Joleyn Balch, Jody Fronheiser, Min-Yi Shih, Steven F. LeBoeuf, Matthew Pietrzykowski, Peter J. Codella, Bas A. Korevaar, Oleg V. Sulima, and Jim Rand. 2007. Strong broadband optical absorption in silicon nanowire films. *Journal of Nanophotonics* 1 (1):013552-013552-10.

Umut Tok, Rüştü, and Kürşat Şendur. 2013. Engineering the broadband spectrum of close-packed plasmonic honeycomb array surfaces. *Journal of Quantitative Spectroscopy and Radiative Transfer* 120:70–80.

Vynck, Kevin, Matteo Burresi, Francesco Riboli, and Diederik S Wiersma. 2012b. Photon management in two-dimensional disordered media. *Nature Materials* 11 (12):1017–1022.

Wagner, R. S., and W. C. Ellis. 1964. Vapor-liquid-solid mechanism of single crystal growth. *Applied Physics Letters* 4 (5):89–90. doi: http://dx.doi.org/10.1063/1.1753975.

Wang, Baomin, and Paul W Leu. 2012. Tunable and selective resonant absorption in vertical nanowires. *Optics Letters* 37 (18):3756–3758.

Wang, Fei, Hongyu Yu, Junshuai Li, Shemein Wong, Xiao Wei Sun, Xincai Wang, and Hongyu Zheng. 2011. Design guideline of high efficiency crystalline Si thin film solar cell with nanohole array textured surface. *Journal of Applied Physics* 109 (8):084306-084306-5.

Wang, Han, Xianglei Liu, Liping Wang, and Zhuomin Zhang. 2013. Anisotropic optical properties of silicon nanowire arrays based on the effective medium approximation. *International Journal of Thermal Sciences* 65:62–69.

Wang, Yewu, Volker Schmidt, Stephan Senz, and Ulrich Gosele. 2006. Epitaxial growth of silicon nanowires using an aluminium catalyst. *Nature Nanotechnology* 1 (3):186–189.

Wangyang, Peihua, Qingkang Wang, Xia Wan, Kexiang Hu, and Kun Huang. 2013. Optical absorption enhancement in silicon square nanohole and hybrid square nanowire-hole arrays for photovoltaic applications. *Optics Communications* 294:377–383.

Wei, Lu, and M. Lieber Charles. 2006. Semiconductor nanowires. *Journal of Physics D: Applied Physics* 39 (21):R387.

Wei, Lu, and Xiang Jie. 2015. *Semiconductor Nanowires From Next-Generation Electronics to Sustainable Energy.* The Royal Society of Chemistry, Cambridge, UK.

Wen, Long, Xinhua Li, Zhifei Zhao, Shaojiang Bu, XueSong Zeng, Jin-hua Huang, and Yuqi Wang. 2012. Theoretical consideration of III–V nanowire/Si triple-junction solar cells. *Nanotechnology* 23 (50):5055202.

Whang, Dongmok, Song Jin, and Charles M. Lieber. 2003. Nanolithography using hierarchically assembled nanowire masks. *Nano Letters* 3 (7):951–954. doi: http://dx.doi.org/10.1021/nl034268a.

Wu, Yiying, and Peidong Yang. 2001. Direct observation of vapor–liquid–solid nanowire growth. *Journal of the American Chemical Society* 123 (13):3165–3166. doi: http://dx.doi.org/10.1021/ja0059084.

Xia, Y., P. Yang, Y. Sun, Y. Wu, B. Mayers, B. Gates, Y. Yin, F. Kim, and H. Yan. 2003. One-dimensional nanostructures: Synthesis, characterization, and applications. *Advanced Materials* 15 (5):353–389. doi: http://dx.doi.org/10.1002/adma.200390087.

Xiong, Zhiqiang, Fangyuan Zhao, Jiong Yang, and Xinhua Hu. 2010. Comparison of optical absorption in Si nanowire and nanoporous Si structures for photovoltaic applications. *Applied Physics Letters* 96 (18):181903. doi: http://dx.doi.org/10.1063/1.3427407.

Yahaya, Nor Afifah, Noboru Yamada, Yukio Kotaki, and Tadachika Nakayama. 2013. Characterization of light absorption in thin-film silicon with periodic nanohole arrays. *Optics Express* 21 (5):5924–5930.

Yahaya, Nor Afifah, Noboru Yamada, and Tadachika Nakayama. 2012. Light trapping potential of hexagonal array silicon nanohole structure for solar cell application. *Advanced Materials Research* 512:90–96.

Yamamoto, K, M Yoshimi, Y Tawada, Y Okamoto, A Nakajima, and S Igari. 1999. Thin-film poly-Si solar cells on glass substrate fabricated at low temperature. *Applied Physics A* 69 (2):179–185.

Yamamoto, Kenji, Akihiko Nakajima, Masashi Yoshimi, Toru Sawada, Susumu Fukuda, Takashi Suezaki, Mitsuru Ichikawa, Yohei Koi, Masahiro Goto, and Tomomi Meguro. 2004. A high efficiency thin film silicon solar cell and module. *Solar Energy* 77 (6):939–949.

Yan, Ruoxue, Daniel Gargas, and Peidong Yang. 2009. Nanowire photonics. *Nature Photonics* 3 (10):569–576.

Yang, Jiong, Xinhua Hu, Xin Li, Zheng Liu, Zixian Liang, Xunya Jiang, and Jian Zi. 2009. Broadband absorption enhancement in anisotropic metamaterials by mirror reflections. *Physical Review B* 80 (12):125103.

Yang, Peidong. 2011. The chemistry and physics of semiconductor nanowires. *MRS Bulletin* 30 (2):85–91. doi: http://dx.doi.org/10.1557/mrs2005.26.

Yang, Peidong, and Charles M. Lieber. 1996. Nanorod-superconductor composites: A pathway to materials with high critical current densities. *Science* 273 (5283):1836–1840. doi: http://dx.doi.org/10.1126/science.273.5283.1836.

Yazawa, M., M. Koguchi, and K. Hiruma. 1991. Heteroepitaxial ultrafine wire-like growth of InAs on GaAs substrates. *Applied Physics Letters* 58 (10):1080–1082. doi: http://dx.doi.org/10.1063/1.104377.

Yeom, Junghoon, Daniel Ratchford, Christopher R. Field, Todd H. Brintlinger, and Pehr E. Pehrsson. 2014. Decoupling diameter and pitch in silicon nanowire arrays made by metal-assisted chemical etching. *Advanced Functional Materials* 24 (1):106–116. doi: http://dx.doi.org/10.1002/adfm.201301094.

Yu, W., R. Mittra, X. Yang, Y. Liu, Q. Rao, and A. Muto. 2010a. High-performance conformal FDTD techniques. *IEEE Microwave Magazine* 11 (4):43–55. doi: http://dx.doi.org/10.1109/MMM.2010.936496.

Yu, Zongfu, Aaswath Raman, and Shanhui Fan. 2010b. Fundamental limit of light trapping in grating structures. *Optics Express* 18 (103):A366–A380.

Yu, Zongfu, Aaswath Raman, and Shanhui Fan. 2010c. Fundamental limit of nanophotonic light trapping in solar cells. *Proceedings of the National Academy of Sciences* 107 (41):17491–17496.

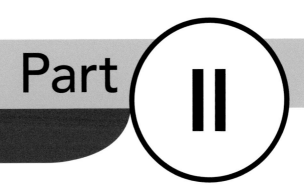

Functional materials

Silicon-based core–shell nanostructures

Mallar Ray, Sayak Dutta Gupta, and Atrayee Hazra

Contents

9.1 INTRODUCTION

During the last two decades, a variety of core–shell structures involving nano-Si as one of the components have been synthesized and investigated. Such structures are important both from the standpoint of understanding basic physics and chemistry as well as for diverse applications. Attempts to understand the controversial luminescence photophysics of ultra-small Si nanocrystals (NCs) have been primarily based on investigation of these NCs capped with a variety of materials, thereby forming core–shell nanostructures of some sort (English et al. 2002; Dasog et al. 2013; Wang et al. 2015; Wen et al. 2015). On the other hand, applications of Si nanostructures either as Li-ion battery anodes (Hwang et al. 2012a; Wu et al. 2013) or as active material of solar cells (Tang et al. 2011; Adachi et al. 2013) involve coating nano-Si with appropriate materials. The importance of such core–shell materials can therefore be hardly overemphasized. In this article, we will try to deal with the preparation, properties and applications of the major variants of Si-based core–shell nanostructures with an emphasis on understanding the interface of the core and the shell.

Conventionally, composite nanomaterials constructed with some inner material (core), which is surrounded fully or partially with some outer layer material (shell), both at nanoscale, are broadly defined as core–shell nanoparticles (NPs) (Wang et al. 2013). The imposition that both the core and the shell must be of nanodimensions leaves out systems where Si NPs are embedded in a bulk host. All other forms of nano-Si, where the surface of the NCs is either terminated by some ligand or surrounded by some other NP system, should comprise Si-based core–shell NPs. Such core–shell geometry is the simplest motif in two-component systems. In comparison with the conventional single-component systems, complex systems offer the advantage of coupling and tuning the size-dependent properties of two systems to meet specific requirements.

Of the different varieties of core–shell Si nanostructures, Si–Si oxide core–shell nanostructures are most common, because the unsaturated surface bonds of a Si nanostructure have uncanny affinity toward oxygen under ambient conditions, unless they are terminated by stabilizing species. Nearly ubiquitous presence of oxide in Si nanosystems has been a major hindrance in understanding the luminescence mechanism from Si NCs. Several attempts have been made to overcome this limitation by capping Si with different organic molecules as well as with inorganic NPs. We therefore have different types of Si-based core–shell nanostructures.

9.1.1 CLASSIFICATION OF CORE–SHELL NANOSTRUCTURES

Although it is possible to have core–shell nanosystems where Si comprises the shell, in the present discussion, we will primarily concentrate on nanostructures where the core is Si NC. The wide variety of such structures having a Si NC core is usually classified based on the nature of the shell material. Most common among them is a Si-organic core–shell nanostructure, where the surface of Si NC is functionalized by an organic group. Whether such organic molecule–terminated Si NCs classify as conventional core–shell material may be debated because tailoring Si surfaces with organic moieties involves the formation of stable covalent bonds (Si–C). These structures are usually referred to as surface functionalized Si NCs and not as core–shell nanostructures. Nevertheless, we will discuss some important properties and methods of preparation of surface functionalized Si NCs, because they closely resemble typical core–shell nanostructures in modulating the overall properties and in the arrest of spontaneous surface oxidation.

Within the category of Si-inorganic core–shell nanostructures, a limited number of studies have been made on Si-metal and Si-semiconductor core–shell nanostructures. Metal–semiconductor and semiconductor–semiconductor core–shell nanostructures are extremely interesting systems to study charge and energy transfer between the materials of the core and the shell. Such systems are also important for understanding resonant interactions between semiconductor excitons and metallic plasmons. We will try to discuss the fundamentals of such interactions with few representative examples of Si-metal and Si-semiconductor core–shell nanostructures. From the materials perspective, therefore, Si-based core–shell nanostructures may be classified as Si-organic and Si-inorganic, where the inorganic material may be a metal or a semiconductor or a typical oxide-like insulator.

Another way of classifying core–shell nanostructures is based on the dimensionality of the system. Most common are the spherical core–shell nanostructures – so-called 0D quantum dots. For Si-based nanosystems, a substantial amount of work has also been done on core–shell nanowires and nanotubes, which

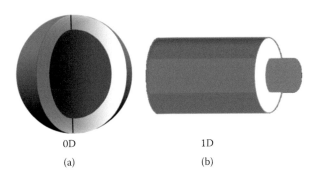

Figure 9.1 Schematic of core–shell nanostructure: (a) spherical (0D) and (b) nanowire (1D).

constitute the 1D nanostructures. A significant part of this discussion will focus on such 0D and 1D core–shell Si nanostructures. We will try to understand the differences in the overall properties and subsequently the different applications of these 0D and 1D core–shell structures. Schematic of 0D and 1D core–shell structures is shown in Figure 9.1.

Besides size, shape of nanostructures also plays significant role in determining the properties at the nanoscale. However, unlike other direct bandgap semiconductors (CdSe, CdS, ZnO, etc.) or metal nanostructures, synthesis of Si NCs having well-defined shapes like spherical, cuboids, and pyramidal is yet to be achieved. Both solid state and chemical routes of preparation usually lead to the formation of structures with minimum surface energy. Consequently, the classification of Si NCs according to shape cannot be done at this stage.

9.1.2 SIGNIFICANCE OF CORE–SHELL NANOSTRUCTURES

The significance of core–shell nanostructures is in the increased and tunable functionalities. The individual properties of the core (Si NC) and the shell may be coupled to derive multi-functionality. It is also possible to change, enhance or reduce a particular feature, like emission or absorption of the NCs by suitably choosing the shell material and thickness. It is possible to tune the luminescence spectrum of Si NCs through the entire visible electromagnetic spectrum, simply by coating the NCs with different surface groups (Dasog et al. 2014). Thermal conductivity of Si NCs can be significantly reduced by encapsulating Si nanowire with a Ge shell (Hu et al. 2011). Due to the low in vitro toxicity of Si NCs (Choi et al. 2009), core–shell structures having Si NCs at the core have also been widely suggested for numerous biological applications like fluorescent bio-imaging,(He et al. 2009) and contrast agents in magnetic resonance imaging (Erogbogbo et al. 2010).

Si in its nanocrystalline form remains one of the most attractive materials for lithium-ion batteries because of its highest known capacity (4200 mAh/g). But pristine Si NCs are unsuitable due to short cycle life and the limitation for scalable electrode fabrication. Here too, core–shell variant of Si NCs is preferred and has been widely studied (Hwang et al. 2012a). Of course, Si is still the mainstay of microelectronic industry, and the major application of Si NC should be ideally in photonic and photovoltaic. However, indirect bandgap of bulk Si has been a serious impediment for the application of Si in photonic communication. Recent developments on engineering NCs, a vast majority of which are in fact core–shell nanostructures, are starting to change the picture, and some nanostructures now approach or even exceed the performance of equivalent direct bandgap materials (Priolo et al. 2014). Here again, we will see how the shell of a Si NC plays a determining role in rendering the NCs amenable for opto-electronic devices.

From a more fundamental view point, core–shell nanostructures present a class of material where a host of basic physics- and chemistry-related issues can be best understood. For example—what is the nature of a crystal and amorphous interface? We have a substantial knowledge about crystal–crystal boundaries and their role in determining material properties but very little till date is known about crystal–amorphous interface and their role. Core–shell nanostructures of Si, particularly the Si NC-oxide structures present an immense possibility in this direction. Also, we know from basic thermodynamics that lattice matching is a basic requirement for epitaxial growth of one material on the surface of another to form a stable structure.

However, at nanoscale, it is possible to construct highly lattice mismatched yet stable core–shell structures (Zhang et al. 2010). How is this possible? What is/are the force/s that holds two mismatched structures together as a single entity?

In addition to the above questions, there are other phenomena like charge and energy transfer at nanoscale which need to be first understood so that they can be utilized for various purposes.

9.1.3 SCOPE OF THIS ARTICLE

In the following sections, we will focus on each of the above-mentioned aspects in relation to Si-based core–shell nanostructures. We will first describe briefly the different preparation strategies for Si–organic, Si–metal, Si–semiconductor, and Si–insulator (oxide) nanomaterials. The idea here is to provide a brief overview of the schemes involved in developing different core–shell nanostructures. Details of synthetic procedure and techniques will not be discussed. Following this, we will try to illustrate the broad structural features of the variety of nanostructures involving Si at the core. Here, we will deal both with 1D and 0D systems. Subsequently, the spectroscopic features, with emphasis on absorption and emission characteristics of the Si-based core–shell nanostructures will be discussed. We will deal with the coupled interaction of the core and shell materials and delve into some of the fundamental questions of interface and interactions. Finally, the reader will be provided with an exposure to the variety of applications that are either realized or may be potentially achieved with these core–shell nanostructures.

9.2 STRATEGIES FOR SYNTHESIS

There are many strategies for the development of Si-based core–shell nanostructures. All the conventional methods for synthesizing Si NCs, like vapor condensation, chemical methods and solid state routes, may be employed with some modification for the formation of a shell on the NC surface. We will discuss about the general principles that may be adopted for the development of different types of shells highlighting the difficulties in each method.

9.2.1 ORGANIC CAPPING OF SILICON NANOSTRUCTURES

Organically capped Si NCs are not classified as core–shell structures in literature. The idea is that in a conventional core–shell material, there should be a well-defined shell material having an interface with the core with which it forms a composite. In organic-capped NCs, usually an organic molecule attaches to an unsaturated surface bond of a NC, thereby stabilizing or passivating it. In such a case, it is not possible to isolate the shell material and the interface. Organic molecule, shell, and interface—all become synonymous. Nevertheless, we will present a short summary of the surface-functionalized Si NCs. This is because the surface-functionalized Si NCs very closely resemble typical core–shell nanostructures in more ways than one. Most importantly, the overall properties of a Si nanostructure can be tuned by changing the surface functional group.

Surface termination methodologies of Si NCs with organics are very different from that of other semiconductor NCs. In contrast to ligand exchange, which is by far the most common technique used for surface modification of other semiconductor NCs, hydrosilylation is the most important method for Si NCs. Hydrosilylation is a process by which unsaturated surface bonds across a Si NC are passivated by H, that is, Si–H bonds are formed. The Si–H bonds on NC surfaces can subsequently be replaced with suitable groups to prepare alkyl-terminated, amine-terminated, and carboxyl-terminated Si NCs. All that needs to be done to replace the Si–H bonds with a suitable organic ligand is to supply appropriate thermal or mechanical or photochemical energy in the presence of the organic moiety. Si–H bonds are then replaced with stable Si–C bonds on Si NC surface. Table 9.1 summarizes the different methods used for the preparation of H-terminated Si NCs, followed by the functionalization techniques. In all the cases listed in Table 9.1, the initial core that is developed consists of H-terminated Si NCs.

Hydrosilylation, as discussed earlier, is the most common technique for the preparation of hydride-terminated Si NCs, which are eventually substituted by different organic moieties to prepare organic-capped Si nanostructures. Inverse micelle method is generally the adopted hydrosilylation technique, which is explained in Figure 9.2a–c. Hydride-terminated Si NCs can be synthesized in TOAB micelles using

Table 9.1 Techniques of capping Si NCs with different organic functional groups

CORE/SHELL	SYNTHESIS TECHNIQUE AND BASIC RAW MATERIAL USED				REFERENCE
	CORE		SHELL		
	METHOD	REAGENT/S	METHOD	REAGENT/S	
Si/COOMe	Ultra-sonic fractionation of catalyzed electrochemically etched Si wafer (Si NC-H)	HF and H_2O_2	Thermal hydrosilylation	Methyl 4-pentenoate	Rogozhina et al. 2006
Si/COOH			Base hydrolysis	Water–methanol with 7.5 wt% NaOH	
Si/Oc	Annealing of SiOx powders followed by HF etching (Si NC-H)	SiO	Thermal alkylation reaction	1-octene, toluene	Liu et al. 2006
Si/Bu	Electrochemical etching of Si wafer (Si NC-H)	HF	Alkylation reaction	Bromine, Butyllithium	Carter et al. 2005
Si/R	Inverse micelle synthesis (Si NC-H)	Tetraoctylammonium bromide (TOAB), $SiCl_4$, $LiAlH_4$, MeOH quenching	Alkylation	1-decene, 1-tetradecene 1-hexadecene	Vasic et al., 2008
Si/RNH_2	Inverse micelle synthesis (Si NC-H)	TOAB, $Si(OCH_3)_4$, $LiAlH_4$	Amination	alkyl-amine, hex-5-enylamine, dodec-11-enylamine	Vasic et al. 2009b
Si/RNH-Ru(bpy)$_2$(spb)$^{2+}$			Amine ester coupling reaction	[Ru(bpy)$_2$(spb)]Cl$_2$; bpy = 2, 2'-bipyridine; spb = 4-(p-N-succinimidylcarboxyphenyl)-2,2'-bipyridine	Vasic et al. 2009a
Si/N_3R	Inverse micelle synthesis (Si NC-H)	TOAB, $SiCl_4$, $LiAlH_4$	Azidation	11-azido-undec-1-ene, 6-azido-hex-1-ene, 3-azidoprop-1-ene	PhD Thesis Rosso-Vasic 2008
Si/$O_3SiC_8H_{17}$	Reduction of $SiCl_4$ (Si NC-Cl)	Sodium napthalide	Alkoxylation, Hydroxylation, Silanization	Methanol, Glyme, Octyltrichlorosilane	Zou et al., 2006

Note: Bu = Butyl; Oc = octyl; R = alkyl group.

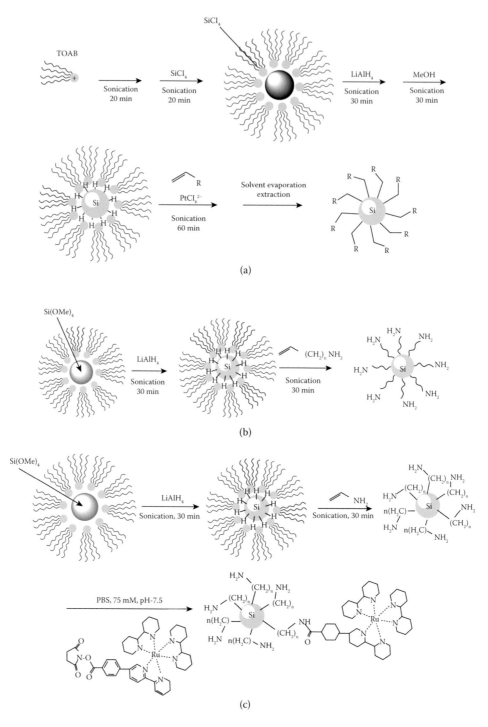

Figure 9.2 The chemical schemes for the synthesis of (a) alkyl (R)-capped Si NCs (reproduced from Vasic, M.R., et al., Alkyl-functionalized oxide-free silicon nanoparticles: Synthesis and optical properties, *Small*, 2008, 4, 1835–1841, Copyright Wiley-VCH Verlag GmbH & Co. KGaA, with permission); (b) (CH₂)ₙNH₂ (n = 1,4,9)-capped Si NCs (reproduced from Vasic, M.R., et al., 2009b, Amine-terminated silicon nanoparticles: Synthesis, optical properties and their use in bioimaging, *J. Mater. Chem.*,19, 5926–5933, by permission of The Royal Society of Chemistry); and (c) Ru(bpy)₂(spb)]-capped Si NCs from amine-terminated NCs (Reprinted with permission from Vasic, M.R., et al., 2009a, Efficient energy transfer between silicon nanoparticles and a Ru-polypyridine complex, *J. Phys. Chem. C*, 113, 2235–2240, Copyright 2009 American Chemical Society).

SiCL$_4$ and LiAlH$_4$ in toluene medium by inverse micelle method as shown in Figure 9.2a. The as-prepared Si NC–H is alkyl terminated by the addition of alkene in the presence of the catalyst H$_2$PtCl$_6$. Si(OCH$_3$)$_4$ may also be used in place of SiCL$_4$ for the inverse micelle synthesis as shown in Figure 9.2b. Alkyl-amine termination can be achieved in the same manner using the corresponding alkyl amines in the presence of some catalyst. The amine-terminated Si NCs can be further functionalized with [Ru(bpy)$_2$(spb)]Cl$_2$ using standard amine ester coupling reactions as shown in the Figure 9.2c.

9.2.2 SILICON–METAL CORE–SHELL NANOSTRUCTURES

If Si NCs are capped by an inorganic nanomaterial, then it classifies as a classical core–shell nanostructure. An assortment of inorganic materials has been used to coat Si NCs. However, most of these inorganic shells are either insulating or at the most semiconducting oxides. Getting a metal shell on Si NC is very difficult. This is because the lattice mismatch of metals with Si is large. Therefore, epitaxial growth of a metallic shell on Si NC surface is thermodynamically not favorable. An epitaxial shell growth can tolerate only a moderate lattice mismatch (<5%), and it is nearly impossible to achieve such a situation with metals. Moreover, thermodynamic stability also demands that in a typical core–shell structure, the atomically larger material should sit at the core, whereas the atomically smaller material constitutes the shell. This condition too is difficult to satisfy for any Si–metal combination where the Si NC forms the core. A reverse situation, that is, a metallic core with a Si shell, is more favorable, and there are some available reports on such core–shell nanostructures (Mohapatra et al. 2008; Wang et al. 2014b; Sugimoto et al. 2016).

An interesting method to synthesize a Si NC shell on noble metal NPs like Au, Ag, and Pt involves reduction of Au/Ag/Pt salts by Si NCs. H-terminated Si NCs are well-known reducing agents. Simple addition of metal salts in colloidal H-terminated Si NCs can reduce metal salts and initiate nucleation and growth of noble metal NPs. At the same time, Si NCs participate in a self-limiting shell growth on the noble metal NPs (Sugimoto et al. 2016). Scanning transmission electron microscopy-energy dispersive spectroscopy (STEM-EDS) mapped images of a Ag–Si and Au–Si core–shell nanostructure formed in this route are shown in Figure 9.3. Au/Ag NPs with Si NC shells are extremely important candidates for understanding and exploiting the interaction of metallic plasmons and semiconductor excitons that is discussed later.

Despite being thermodynamically unfavorable, there are couple of recent reports where nanostructures with Si NC core and metal shell have been synthesized. Two completely different techniques have been employed to achieve this structure. Solid Si target immersed in appropriate noble metal salt, when subjected to two pulsed lasers, can give rise to core–shell nanostructures with Si NC as the core. An energetic laser ablates the solid target, the other laser with a longer pulse width supposedly promotes a continuous dynamic crystallization of the Si noble metal core–shell NPs. Liu et al. (2015) used two synchronously controlled lasers to reduce single-crystal Si and simultaneously grow noble metal shell. The schematic of the set up used to prepare is shown in Figure 9.4.

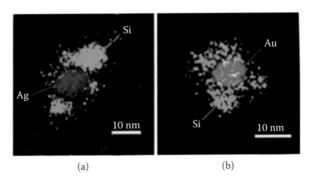

Figure 9.3 STEM-EDS mapping of Ag-Si (a) and Au-Si (b) core–shell NPs. (Reproduced from Sugimoto, H., et al., 2016, Silicon nanocrystal-noble metal hybrid nanoparticles, *Nanoscale*, 8, 10956–10962. By permission of The Royal Society of Chemistry.)

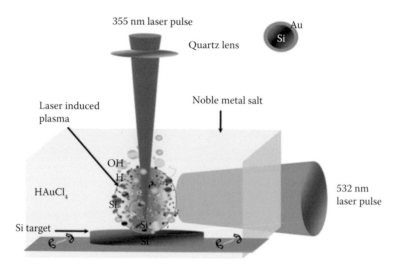

Figure 9.4 Schematic of process for producing Si NC core noble metal shell nanostructure. (Reproduced with permission from Liu, P., et al., 2015, Fabrication of Si/Au core/shell nanoplasmonic structures with ultrasensitive surface-enhanced Raman scattering for monolayer molecule detection, *J. Phys. Chem. C*, 119(2), 1234–1246. Copyright 2015 American Chemical Society.)

In another method, spherical Si NCs are first synthesized by successive etching-oxidation-etching of mechanically milled crystalline Si. The Si NCs are then oxide etched with buffered HF and immediately allowed to react with evolving Au NPs, under UV exposure at elevated temperature (~80°C). Surface of freshly HF-etched Si NCs is abundant with H and F terminations along with dangling bonds. Evolving Au NPs, prepared by standard citrate reduction, are continuously brought in proximity of the Si NCs in aqueous solution by thermal agitation. UV excitation facilitates charge transfer between the Si NCs and the Au NPs, which consequently charges both the nanostructures. UV excitation also exposes more dangling bonds on the Si NC surface, which are preferred cites for Au NP nucleation and growth. The proximal Au NPs therefore adhere to the Si NC surface and are held there due to simple coulombic force (Basu and Ray 2014; Ray et al. 2014). A typical Au NP-coated Si NC prepared in this route is shown in Figure 9.5.

Both the above methods—synchronous laser ablation of solid Si target in the presence of metal salt, and nucleation and growth of Au NPs at the dangling bonds of etched Si NCs—are very new techniques for

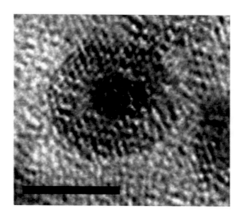

Figure 9.5 TEM image of a Si–Au core–shell nanostructure. The scale bar is 5 nm. (Reproduced from Ray, M., et al., 2014, Highly lattice-mismatched semiconductor–metal hybrid nanostructures: Gold nanoparticle encapsulated luminescent silicon quantum dots, *Nanoscale*, 6, 2201–2210. By permission of The Royal Society of Chemistry.)

getting metal NP coverage on Si NCs. Both these techniques need further development for controlled and reproducible development of these exciting hybrid core–shell nanostructures.

9.2.3 SILICON–SEMICONDUCTOR CORE–SHELL NANOSTRUCTURES

Another promising inorganic shell material is a semiconductor. When we think about a semiconductor in the nano-regime, the most interesting aspect that comes to our mind is perhaps its band structure variation with size, shape, and other factors that tune the optical and electrical properties. Due to the size-dependent variation in band structure and the tunability endowed by the core and the shell, such hetero-nanostructures with Si core have huge potential applications in optoelectronics, tunnel FET, renewal energy domain, and so on. For example, if a wider band gap material is used as the shell of a Si NC, the light absorption efficiency increases. Development of such nanostructures is also relatively easier because it is possible to select semiconductors with minimal lattice mismatch. Normal epitaxial growth in this case is the preferred technique for developing a semiconductor shell on Si NC.

In case of spherical core–shell nanostructures with Si NC core, the strategy is to etch the native oxide of the NC surface followed by appropriate surface modification. The shell is then grown on the modified surface by a suitable process. For example, Wang et al. (2014a) synthesized Si NC–CdS core–shell structures by a successive ion layer adhesion and reaction (SILAR) technique. Si NC cores were first synthesized by hydrogen reduction of $(HSiO_{1.5})_n$, and the surface oxide was etched by HF solution. To enhance dispersibility, the surfaces of Si NCs were modified with 1-decene and were extracted in an organic phase. In the SILAR shell growth process, sulfur precursor was adopted as the first addition in the growth of the CdS shell on Si NCs. The schematic of development of CdS shell is illustrated in Figure 9.6.

However, in the category of Si–semiconductor nanostructures, the dominant structures are 1D, where Si nanowire cores are coated by some semiconducting shell. The most common method employed here is fabrication of Si nanowires by vapor–liquid–solid (VLS) techniques followed by coating the nanowire with a coaxial shell by chemical vapor deposition (CVD).

In a typical 1D growth of Si nanowires by VLS mechanism, silane (SiH_4) is used as the Si reactant and assisted by a metal catalyst. Au is by far the most important catalyst and is ubiquitously used for Si nanowire growth. Gas precursor molecules (SiH_4) are introduced into a furnace or a CVD reactor at high temperatures. At high temperatures, the metal catalyst forms liquid alloy droplets by adsorbing SiH_4 vapor components. By varying the temperature and pressure, it is possible to make the liquid

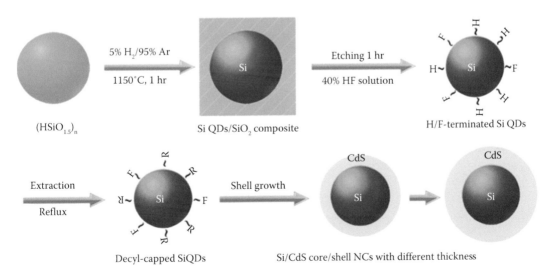

$(HSiO_{1.5})_n$ — 5% H_2/95% Ar, 1150°C, 1 hr → Si QDs/SiO$_2$ composite — Etching 1 hr, 40% HF solution → H/F-terminated Si QDs

Extraction Reflux → Decyl-capped SiQDs — Shell growth → Si/CdS core/shell NCs with different thickness

Figure 9.6 Synthetic strategy of core–shell Si–CdS nanocrystals. (Reproduced from Wang, G., et al., 2014, Type-II core–shell Si–CdS nanocrystals: Synthesis and spectroscopic and electrical properties, *Chem. Commun.*, 50, 11922–11925. By permission of The Royal Society of Chemistry.)

Functional materials

droplet supersaturated, that is, the actual concentration of the components becomes higher than the equilibrium concentration. Consequently, the component (in our case Si) is precipitated at the liquid–solid interface to achieve minimum free energy of the alloy system. This initiates 1D crystal growth and it continues as long as the vapor components are supplied. This catalytic seed floats on top of the nanowire as it grows. The diameter of the nanowire is, therefore, determined by the catalyst size, whereas the growth rate kinetics together with the growth time defines the nanowire length. The position of the catalyst on the substrate determines the position of the 1D structures, as the liquid phase is confined to the area of the precipitated solid phase. Figure 9.7 illustrates the mechanism of nanowire growth by VLS technique.

Si–Ge core–shell nanowires (Figure 9.8) can be developed by a similar two-step process involving CVD. The axial growth of intrinsic (i) Si nanowires is usually carried out at elevated temperatures (~550°C) by CVD in the presence of Au catalyst. After reducing the temperature to ambient conditions, radial growth of doped Ge shell layer may be carried out at high temperature (~500°C) in the presence of PH_3. The reduction of temperature and the presence of PH_3 assist in retarding the axial growth during the shell formation. The shell thickness is usually controlled by controlling the growth time. For p-type or n-type doping, diborane (B_2H_6) and phosphine (PH_3) are usually used as dopants (Fukata et al. 2012). Further modification of these Si–Ge structures is possible with Si_XGe_{1-X} shell on Si nanowires. The only modification during synthesis would be the presence of both SiH_4 and GeH_4 precursor (Dillen et al. 2016).

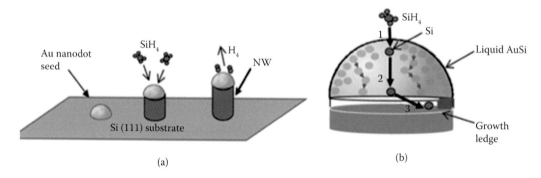

(a) (b)

Figure 9.7 Schematic illustration of VLS growth mechanism for Si nanowire using Au catalyst. (a) Schematic of Si nanowire growth from a liquid Au catalyst seed which floats on the nanowire. (b) Enlarged view illustrating the three kinetic steps for nanowire growth: (1) silane decomposition at the vapor–liquid interface, (2) Si atom diffusion through the AuSi liquid, and (3) nanowire crystallization by Si incorporation into a step at the growing liquid–solid interface of the nanowire. (Reproduced with permission from Picraux, S.T., et al., 2010, Silicon and germanium nanowires: Growth, properties, and integration, *JOM*, 62(4), 35–43. Copyright 2010 TMS.)

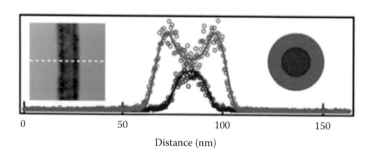

Distance (nm)

Figure 9.8 Elemental mapping cross section of Si–Ge core–shell nanowires indicating a 21-nm-diameter Si core (blue circles) surrounded by a 10-nm Ge shell (red) grown by VLS method. (Reproduced by permission from Macmillan Publishers Ltd. *Nature.* Lauhon, L.J., et al., 2002, Epitaxial core–shell and core–multishell nanowire heterostructures, 420, 57–61, copyright 2002.)

Following exactly similar method, p–i–n coaxial Si nanowires consisting of a p-type Si nanowire core capped with i- and n-type Si shells can also be fabricated (Tian et al. 2007). Other than epitaxially grown almost uniform shell, hierarchical branched nanostructures have received much attention. One of the main advantages of such structure is their large surface areas that enhance functionality. Si–ZnO core–shell hierarchical structure development includes three steps shown in Figure 9.9. First, p-Si nanowires should be grown by VLS method. On these Si nanowires, ZnO seed can be coated by atomic layer deposition at a temperature lower than that for core synthesis. Further growth of ZnO nanorods on the seeds may be carried out at a lower temperature than seeding temperature (Devika et al. 2010). Another hierarchical structure Si–InGaN can also be synthesized by placing Si nanowires inside halide CVD furnace in the presence of $GaCl_3$ and $InCl_3$ along with N_2 as carrier gas (Hwang et al. 2012b).

CdS and CdSe are arguably the two most common II–VI group semiconductor NCs and are efficient light emitters in the visible range. Core–shell nanowires of Si–CdSSe with modulated composition can be synthesized by a similar technique with some modification. Oxide-etched Si wafer, used as the source for Si, and CdS and CdSe powders, used as sources of alloying constituents, are placed together inside a horizontal tube furnace along with Au NPs deposited on quartz (Pan et al. 2008). The schematic of the setup is shown in Figure 9.10. The tube chamber is evacuated and back-flushed with the carrier gas—Ar. The growth of nanowire is then initiated at high temperature (~1080°C), and the growth is controlled by heating rate, gas flow rate, peak temperature, pressure, and other formation parameters.

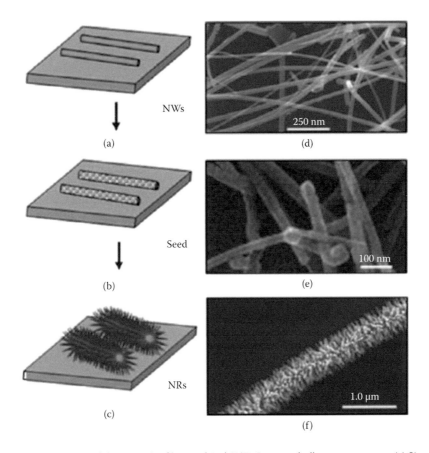

Figure 9.9 Schematic diagram of the growth of hierarchical Si/ZnO core–shell nanostructures: (a) Si nanowire core structures, (b) ZnO-seeded Si nanowire core, and (c) ZnO nanorod-coated Si wires. (d–f) Their corresponding FESEM images. (Reproduced from Devika, M., et al.: Heteroepitaxial Si/ZnO hierarchical nanostructures for future optoelectronic devices. *ChemPhysChem*. 2010. 11. 809–814. Copyright Wiley-VCH Verlag GmbH & Co. KGaA. With permission.)

Functional materials

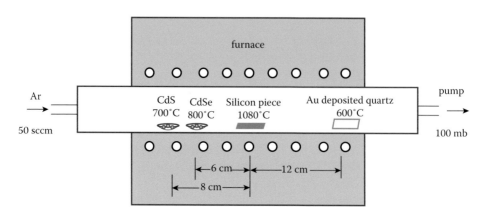

Figure 9.10 Experiment setup for growing the core–shell nanowires of Si-CdSSe with modulated composition. (Reproduced with permission from Pan, A.L., et al., 2008, Si-CdSSe core/shell nanowires with continuously tunable light emission, *Nano Lett.*, 8(10), 3413–3417. Copyright 2008 American Chemical Society.)

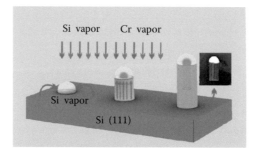

Figure 9.11 Schematic for preparation of chromium silicide-Si core–shell nanopillars. (Reproduced with permission from Chang, M.T., et al., 2009, Core–shell chromium silicide–silicon nanopillars: A contact material for future nanosystems, *ACS Nano*, 3, 3776–3780. Copyright 2009 American Chemical Society.)

Till now we were discussing about synthesis strategies for Si–semiconductor core–shell nanostructures as primary building blocks of various devices. But to implement a device, we need connections. This connecting scheme must not include any further modification of any property of these building blocks. To serve this purpose, chromium silicide–Si core–shell nanopillars would be ideal. These nanopillars can be fabricated by VLS technique as well. For this purpose, a Si substrate with Au catalyst on surface is placed inside a molecular beam epitaxy (MBE) chamber and the growth is mediated by Au catalyst. The chamber contains electron gun as the source for Si and chromium effusion cell. The synthesis strategy is shown in Figure 9.11 (Chang et al. 2009).

The growth of different types of Si–semiconductor core–shell nanostructures is therefore dominated by VLS technique using a CVD, which we have examined briefly. The same technique also forms the basis for development of other Si-based core–shell nanostructures which are discussed in the next sections.

9.2.4 SILICON–SILICON OXIDE CORE–SHELL NANOSTRUCTURES

Silicon–Si oxide, crystalline–amorphous core–shell nanostructures provide substantial potential for exploring fundamental physics and chemistry at the nanoscale and also for realizing novel device applications. Si NCs, unless deliberately terminated by some other material, are in most cases surrounded by a disorganized amorphous shell—Si oxide (Figure 9.12). This shell surrounding the NC is permeated with innumerable surface defects, which are introduced by the respective synthesis technique. The surface and the edge structures in turn play a key role in controlling the overall behavior of the nanosystem. Understanding the interface of such crystal–amorphous nanostructure is therefore of paramount importance.

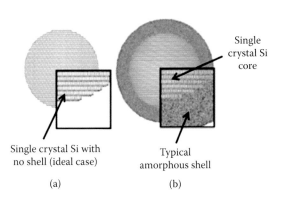

Single crystal Si core

Single crystal Si with no shell (ideal case)

Typical amorphous shell

(a) (b)

Figure 9.12 Simulated models of (a) an ideal single crystalline Si core and (b) Si NC surrounded by an amorphous oxide shell. (Reproduced with permission from Guénolé, J., et al., 2013, Plasticity in crystalline-amorphous core–shell Si nanowires controlled by native interface defects, *Phys. Rev. B*, 87, 045201. Copyright 2013 by the American Physical Society.)

Although most of the Si–Si oxide and Si–organic core–shell nanostructures are essentially crystalline–amorphous systems, they are seldom studied to understand the role of the crystal–amorphous interface. This is because there is hardly any control over the amorphous content which in many cases appears spontaneously. For example, top-down mechanical milling of bulk crystalline Si or laser-induced decomposition of SiH_4 can produce Si NCs as discussed earlier. Depending upon the medium in which reactions are carried out, the surface of the Si NCs gets terminated with either hydrogen or fluorine (in case of fluorine-based preparation) or most commonly oxygen. Even trace amount of oxygen present in reactor chambers can attach to the unsaturated surface bonds of Si NCs forming a thin, passivating oxide shell. Also, oxygen preferentially replaces hydrogen or fluorine from the surface of Si NCs resulting in the formation of amorphous surface oxide on the NC surface as shown in Figure 9.13, which is an electron microscopy image of a spherical Si NC surrounded by native amorphous oxide. Therefore, the easiest strategy to synthesize a crystalline–amorphous, core–shell Si NC is to simply expose Si NCs to oxygen. Unless, the NC surface is functionalized by some stabilizing agents that passivate all the dangling bonds, oxidation will proceed on the surface, thereby creating crystalline–amorphous core–shell nanostructure. Figure 9.14 shows a simulated model that clearly demonstrates how the amorphous shell over a Si NC can grow with increasing oxygen introduction.

Spherical Si NCs with about a 1-nm-thick amorphous SiO_2 layer can be synthesized by a high-temperature aerosol reaction (Littau et al. 1993). The process essentially involves gas-phase pyrolysis (i.e., decomposition by thermal energy) of SiH_4. Formation of NCs subsequent to the decomposition of SiH_4 is explained by standard nucleation and growth model. Here, larger NCs grow from a "monomer" created by fast initial decomposition.

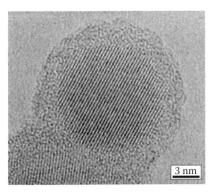

3 nm

Figure 9.13 HR-TEM image of single crystalline Si surrounded by an amorphous oxide shells prepared by laser pyrolysis of silane. (Reproduced with permission from Hofmeister, H., et al., 1999, Lattice contraction in nanosized silicon particles produced by laser pyrolysis of silane, *Eur. Phys. J. D.*, 9, 137–140. Copyrights 1999 The European Physical Journal.)

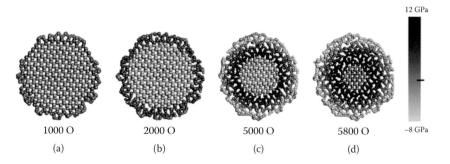

Figure 9.14 Molecular dynamic simulation demonstrating the development of thick amorphous oxide shell at the expense of oxidation and subsequent amorphization of the Si core. Shown are the morphology after (a) 1000, (b) 2000, (c) 5000, and (d) 5800 oxygen atoms are inserted. (Reproduced with permission from Dalla Torre, J., et al., 2002, Study of self-limiting oxidation of silicon nanoclusters by atomistic simulations, *J. Appl. Phys.*, 92, 1084–1094. Copyright 2002 American Institute of Physics.)

NCs with thicker amorphous oxide can be prepared by deliberate external oxidation of mechanically milled Si (Ray et al. 2009b). Oxidation and amorphization of Si NCs can be done chemically by treating the Si NCs b with standard Piranha solution (a mixture of 1 part of hydrogen peroxide and 2 parts of sulfuric acid) or simply by introducing oxygen into Si NC colloid. One way to control the thickness of the amorphous oxide layer is to etch the oxide with buffered HF and subsequently expose the Si NCs to controlled flow of oxygen allowing some desired thickness of amorphous shell to develop. But this process inevitably leads to the formation of a highly distributed core and shell sizes.

Crystalline–amorphous boundary in Si NC system has been somewhat controlled in axial core–shell nanowires developed by CVD using SiH$_4$ as the precursor molecule. Development of core–shell nanowires is a two-step process, which involves (i) metal nanocluster catalyzed Si nanowire core growth followed by (ii) homogeneous deposition of the amorphous Si shell. In this case, the control of crystalline and amorphous phases is based upon control of radial versus axial growth. Controlled promotion of axial growth is achieved by the addition and activation of reactant at the catalyst site along a pre-defined axis and not on the nanowire surface. Correspondingly, it is possible to drive conformal shell growth by altering conditions to favor homogeneous vapor-phase deposition on the nanowire surface. SiH$_4$ molecules catalytically decompose on the surface of Au nanocluster leading to nucleation and directed 1D growth of Si crystal (Figure 9.15a). 1D growth is maintained as reactant decomposition on the gold catalyst is strongly preferred (Figure 9.15b).

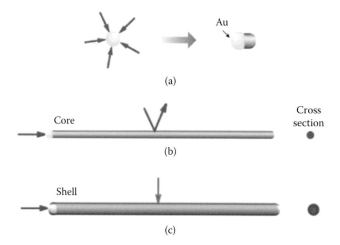

Figure 9.15 Schematic illustration of the controlled growth of crystalline–amorphous core–shell nanowire utilizing VLS method and CVD. (Reproduced by permission from Macmillan Publishers Ltd. *Nature*, Lauhon, L.J., et al., 2002, Epitaxial core–shell and core–multishell nanowire heterostructures, 420, 57–61, copyright 2002.)

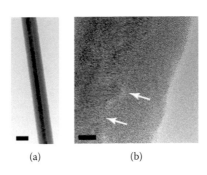

(a)   (b)

Figure 9.16 Clear HRTEM images showing crystalline Si nanowire surrounded by amorphous oxide shell. Scale bars 50 nm and 5 nm in (a) and (b), respectively. (Reproduced by permission from Macmillan Publishers Ltd. *Nature*, Lauhon, L.J., et al., 2002, Epitaxial core–shell and core–multishell nanowire heterostructures, 420, 57–61, copyright 2002.)

Finally, synthetic conditions are altered to induce homogeneous reactant decomposition on the nanowire surface (Figure 9.15c) (Lauhon et al. 2002). This mechanism of controlled growth of core–shell crystalline–amorphous nanowire is shown in Figure 9.15 and the resulting structure is shown in Figure 9.16.

Core–shell nanowires of Si–Si oxide have also been reported to be prepared by simple evaporation of Si monoxide in a thermal CVD (Lim et al. 2014). The nanowires with diameters ranging from 10 to 30 nm have been synthesized by controlling the substrate temperature in the absence of any catalyst. However, this method is yet to gain sufficient control over size and shape of the core and the shell.

Electrospinning, a well-known technique for making ultra-long nanofibers, has also been employed to prepare Si–SiO_2 core–shell nanowires. Polyacrilonitrile (PAN) fibers, prepared by electrospinning, were coated with Si using SiH_4 by CVD. Heating the Si-coated PAN nanowires in air at a high temperature oxidized the carbon core to CO_2 and also prepared the oxide shell on Si nanowire core (Wu et al. 2012).

One thing to be pointed out here is that development of well-defined and controllable crystalline–amorphous Si–oxide nanostructure is the pre-requisite for understanding the hitherto ambiguous nature of crystal–amorphous interface at the nanoscale. With the development of these controllable structures and with the advancement of modern characterization and analytical techniques, it should not be far till we develop a clear picture of the modification of electronic structure and the consequences thereof at the crystal–amorphous interface.

9.3 STRUCTURAL FEATURES

Before one can embark on any discussion about properties or applications of core–shell nanostructures, it is imperative to understand the structural features—the microstructural characteristics of the nanostructures. We need to know the shape, the dimensions of the core and the shell and, most importantly, the interface between the core and the shell. But how do we know? How do we "see" the core–shell structures which are so small? Of course, optical microscopes are hopeless in this regard as they can at the best reveal features that are 200 nm or more. The obvious answer would then be—electron microscopy. Using fast moving electrons of a scanning or a transmission electron microscope (SEM or TEM), it is possible to resolve features that are in the nanometric regime. Here again SEM falls short of TEM as feature sizes of few nanometers are difficult to see even with the most advanced SEM having a field emission gun source. The nearly ubiquitous method to "see" core–shell nanostructures of Si is therefore TEM—that too only when operated under high resolution (HR) mode. Consequently, we shall concentrate most of our discussions on HR-TEM investigations of Si-based core–shell nanostructures in this section.

9.3.1 ELECTRON MICROSCOPY OF CORE–SHELL NANOSTRUCTURES

Transmission electron microscopy is the most appropriate and hence the most commonly used technique to visualize core–shell nanostructures. However, there are several limitations in imaging and interpreting

HR-TEM images, particularly in understanding core–shell nanostructures, which are seldom discussed in literature. Some of these problems concern with the physics of image formation in a TEM and cannot be discussed in details here. However, it is very important to have an idea of the sources of errors which may otherwise lead to wrong conclusions. First, the low-scattering factor of Si makes it difficult to image Si with an electron beam. Because the first stage of TEM image formation involves diffraction of electron waves by a specimen, the low scattering factor of Si sets a natural limitation to obtain high diffraction contrast. Second, it is difficult to simultaneously image the different planes of Si NCs having different lattice spacing. This again is associated with the mechanism of image formation in a TEM by two-stage Fourier transform. The problem becomes even more complex if there are randomly oriented Si NCs as it would be very difficult to simultaneously image all NCs at a given focus setting. Third, in a particular image acquiring process, image settings are optimized for resolving the fringes of a given Si NC. This requires aligning the electron beam with the zone axis of the crystal structure under observation. For bulk crystalline material, this alignment is relatively easy because the imaging is sensitive to any deviation from the zone axis. However, for nanostructures, it is very difficult to verify whether the imaging is along a zone axis (Kohno et al. 2003). Studies of Si and metal NPs have shown that resolvable lattice fringes can be achieved for a wide range of tilt angles away from the zone axis resulting in erroneous lattice spacing values (Malm et al. 1997; Du et al. 2003). The fourth problem associated with HR-TEM imaging of individual core–shell NCs is image delocalization, where fringes can be extended beyond the true sizes of the core and the shell. This effect can result in overestimating particle sizes and can be significant when the defocus setting is far from optimal (Ziegler et al. 2002). Fifth problem is that of drift due to interaction of electron beam with nanostructures. Finally, the effect of beam heating can produce Si NCs in amorphous specimens (Du et al. 2003). Sufficient care needs to be taken to ensure that the specimen is not overheated particularly when 200 or 300 kV electron beam is used. Imaging Si based core–shell nanostructures by HR-TEM is therefore absolutely nontrivial. One needs to be aware of the sources of errors besides being careful. Nevertheless, HR-TEM imaging along with selected area electron diffraction pattern (SAEDP) remains the sole method for understanding and studying detailed structural features of core–shell nanosystems.

9.3.1.1 Surface functionalized Si NCs

Although TEM is the most versatile technique to image core–shell nanostructures, it is almost useless in examining any organic shell. This is because the organics like alkenes, amines, and esters, which are used routinely to cap Si, are made of light elements like H, C, O, and N and hence have very little contrast in TEM. Researchers generally use the well-known defocus technique to enhance the contrast of TEM images. However, this technique reduces the resolution of the images significantly and also creates some artifacts, thus making it difficult to interpret the images (Sur 2014). The TEM images shown in Figure 9.17a–d for octane-, acetal-, amino-propyl-, and heptane-capped Si NCs, respectively, clearly reveal that the capping molecules have no manifestation in the TEM images. Only the Si NC core in all the cases appears as dark spots that can be resolved to exhibit the fringes representing the atomic planes of Si in HR images as shown in the insets of Figure 9.17a and c. The organic capping is simply lost in the nearly homogeneous background contrast. Understanding the structural features of the organic shell and the interface with the Si NC surface is an open problem. Till now, the presence and effect of organic capping is inferred indirectly from other characterization techniques like emission and absorption. But, we still do not have any direct means to see a Si NC surrounded by an organic shell.

9.3.1.2 Si–semiconductor core–shell nanostructures

Many studies involving direct band gap semiconductor and metal core–shell nanostructures reveal excellent images where the core and the shell fringes can be resolved under HR-TEM. However, this is not the case when Si is involved. The Si–CdS core–shell nanostructures cannot be clearly visualized under TEM because CdS and Si have similar image contrasts. As we see in Figure 9.18, which is one of the best reported works on such core–shell structures, the core and the shell are not distinctly decipherable.

While dealing with Si, we cannot ignore the possibility of having an oxide layer on the surface of Si NC even in a core–shell structure. An additional step of HF etching is often performed to ensure removal

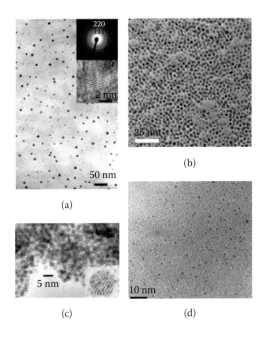

Figure 9.17 TEM images of (a) Octene-capped Si-NCs (reproduced with permission from Liu, S.M., et al., 2006, Enhanced photoluminescence from Si nano-organosols by functionalization with alkenes and their size evolution, *Chem. Mater.,* 18, 637–642, Copyright 2006 American Chemical Society); (b) acetal-capped Si-NCs. (reproduced with permission from Dasog, M., et al., 2014, Size vs surface: Tuning the photoluminescence of freestanding silicon nanocrystals across the visible spectrum via surface groups, *ACS Nano,* 8(9), 9636–9648, Copyright 2014 American Chemical Society); (c) 1-heptene-capped Si NCs (reproduced from Tilley, R.D., et al., 2005, Micro-emulsion synthesis of monodisperse surface stabilized silicon nanocrystals, *Chem. Commun.,* 14, 1833–1835, by permission of The Royal Society of Chemistry); and (d) 3-amino-propyl-terminated (reproduced from Vasic, M.R., et al., 2009, Amine-terminated silicon nanoparticles: Synthesis, optical properties and their use in bioimaging, *J. Mater. Chem.,* 19, 5926–5933, by permission of The Royal Society of Chemistry).

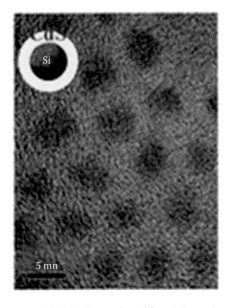

Figure 9.18 Core–shell nanostructures of Si–CdS. (Reproduced from Wang, G., et al., 2014, Type-II core–shell Si–CdS nanocrystals: Synthesis and spectroscopic and electrical properties, *Chem. Commun.,* 50, 11922–11925. By permission of The Royal Society of Chemistry.)

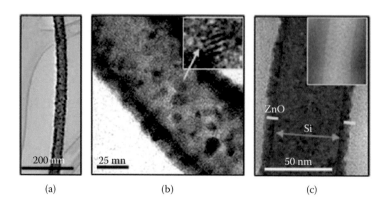

Figure 9.19 (a) Low- and (b) high-magnification TEM images of ZnO-seeded Si nanowires (inset shows the HRTEM image of a ZnO nanoparticle). (c) EF-TEM image of ZnO-seeded Si nanowire (the inset shows its thickness mapping image). (Reproduced from Devika, M., et al.: Heteroepitaxial Si/ZnO hierarchical nanostructures for future optoelectronic devices. *ChemPhysChem.* 2010. 11. 809–814.Copyright Wiley-VCH Verlag GmbH & Co. KGaA. With permission.)

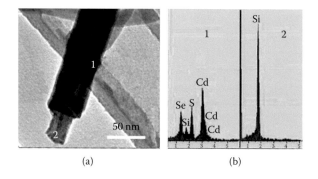

Figure 9.20 (a) TEM image of a broken core–shell wire from one of the Si-CdSSe samples after strong sonication; (b) dot EDX from the shell region (marked with "1") and the exposed core end (marked with "2") of the wire shown in image (a), respectively. (Reproduced with permission from Pan, A.L., et al., 2008, Si-CdSSe core/shell nanowires with continuously tunable light emission, *Nano Lett.*, 8(10), 3413–3417. Copyright 2008 American Chemical Society.)

of surface oxide. Compound semiconductor shells like InGaN can be developed on the surface of Si NC by HF etching (Hwang et al. 2012b). There are few cases when surface oxide layer is absent. For Si–ZnO core–shell nanostructures, absence of oxide layer is evident from bright field TEM and energy filtered TEM (EF-TEM) as shown in Figure 9.19a–c. No oxide removal step is needed in this case. When Zn seeds are deposited on Si nanowires, reaction with oxygen atoms present with oxide layer introduces stable ZnO structures (Devika et al. 2010).

In Figure 9.20a, we can see TEM image of broken Si–CdSSe hybrid nanostructure showing a coaxial core–shell structure. The EDX (Figure 9.20b) indicates the presence of Si only in exposed Si core and Cd, S, and Se in the shell region but there is no trace of oxide. In this case, absence of oxide layer is due to subsequent deposition of Cd, S, and Se inside the same chamber in a CVD process. The interface of this core–shell structure also appears to be very sharp. The lattice spacing of $CdS_{1-X}Se_X$ is in between CdS and CdSe, which follows Vegard's law (Pan et al. 2008).

From a structural point of view, possibly the best TEM images for Si NC semiconductor core–shell structures have so far been obtained for Si–Ge core–shell nanowires as shown in Figure 9.21, where the individual fringes of Si and Ge are resolvable. Although there are physical limitations in imaging Si NC-based core–shell nanostructures with a TEM, the Ge–Si nanowires show that much of these limitations could be reasonably addressed by careful sample preparation and appropriate imaging.

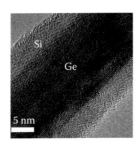

Figure 9.21 HR-TEM image of Ge–Si core–shell nanowires.(Reproduced with permission from Lu, W., et al., 2005, *Proc. Nat. Acad. Sci. U.S.A.*, 102, 10046 10051. Copyright 2005 National Academy of Sciences, U.S.A.)

9.3.1.3 Si–metal core–shell nanostructures

We have seen that development of a stable metal shell on Si NC core generally demands materials with moderate (<5%) lattice mismatch, and it is difficult to find metals that satisfy this condition. Moreover, most metals are heavier and atomically larger than Si. Therefore, capping Si with a metal means forming a shell that is made of heavier and atomically larger atoms, which is not a thermodynamically preferred scenario. Nevertheless, such thermodynamically less probable situations have been achieved at the nanoscale where Au and Ag NPs have been used to cap Si NCs. Structures of such core–shell nanostructures are reasonably well understood from HR-TEM images as shown in Figure 9.22a–d.

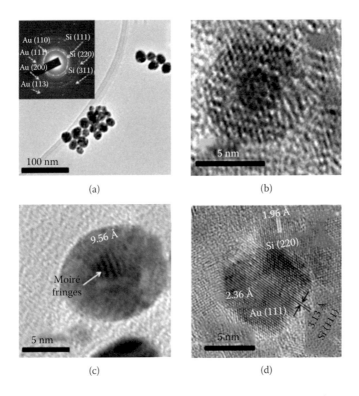

Figure 9.22 HR-TEM images of Si NC–Au NP core–shell nanostructures. (a) Low magnification image with SAEDP shown as inset, (b) high-magnification image of a single core–shell structure, (c) formation of Moiré patterns due to accordant overlap of two crystalline lattices, and (d) unfolding of the shell following centrifugation. (Reproduced from Ray, M., et al., 2014, Highly lattice-mismatched semiconductor–metal hybrid nanostructures: Gold nanoparticle encapsulated luminescent silicon quantum dots, *Nanoscale*, 6, 2201–2210. By permission of The Royal Society of Chemistry.)

Bright field low-magnification image as shown in Figure 9.22a provides very little information. But the corresponding SAEDP is always very important for core–shell structures of two crystalline materials. The diffraction pattern should have clear signatures of the core and the shell. Along with their presence, the pattern also affirms the nanocrystalline character of the core and shell—the diffused halo and scattered spots in a ring-like pattern is typical of any NC. However, it is difficult to conclude from the SAEDP that the two NCs form core and shell. A high-magnification image as shown in Figure 9.22b is critical. Seeing is believing—such images with distinct core and shell nearly unambiguously establish the formation of Si–Au core–shell nanostructure.

A very important point here is to note that representative fringes of either Au or Si are not always seen for core–shell structures, particularly in the core region. Instead as can be seen in Figure 9.22c, the core of the particle exhibits distinct alternate bright and dark bands. These regularly spaced bands are quite different from the dark fringes/stripes that are commonly seen in Au NPs due to twinning and point defects. The observed bands are the Moiré patterns formed due to the superposition of two misfit crystalline lattices of Si and Au. In fact, one can calculate the spacing of the Moiré fringes for the Moiré pattern using the expression:

$$D = \frac{d_2 d_1}{d_2 - d_1}$$

where D is the spacing between two successive bands and d_1 and d_2 are the misfit crystalline lattices of the overlapping planes. In case of structure shown in Figure 9.22c, the spacing observed is ~9.56 Å, which agrees with the calculated value of 9.43 Å corresponding to the overlap of Si (111) and Au (111) planes thereby indicating that the nanostructure shown is the image of a Si NC fully entrapped by Au NP forming a typical core–shell nanostructure.

An easy way to find out the stability of the shell over the core is to subject the nanostructures to mechanical agitation. For example, centrifugation of Si–Au core–shell nanostructures leads to partial unfolding of the Au shell shown in Figure 9.22d, where two-faceted Au NPs with (111) exposed planes are found to reside on a Si NC. Besides the Si NC with (111) exposed plane lying beneath the two Au NPs, other fringes corresponding to Si (200) plane are also identifiable. Such TEM micrographs strengthen the claim of formation of Au NP shell on Si NC.

Appearance of Moiré patterns in TEM images for overlap of two crystalline lattices of the core and the shell can also be observed for Ag NP shell on Si NC (Figure 9.23). The observed parallel bands are the translational and mixed (translation + rotation) Moiré patterns, formed due to accordant and non-accordant superposition of two misfit crystalline lattices of Si and Ag, respectively.

Unlike the distinct core–shell structures of Si and Au discussed above, another type of core–shell Si–Au nanosystem has been visualized by HR-TEM images. Laser ablation in liquid produces Si–Au core–shell structures, which appears very different from what we have seen before. Figure 9.24a–d shows that these structures comprise of a much larger Si NC (~100–500 nm) having a wrinkled surface coated with tiny Au particles. The wrinkles or the net like patterns form a thin slice that can be somewhat controlled by changing the formation parameters.

Structural investigation by HR-TEM of Si–metal nanosystems reveals the morphology of the core and the shell. However, it is difficult to infer about the mechanism of shell growth from such static micrographs. We are familiar with epitaxial growth of shell on a core. In very simple terms, epitaxial growth can be visualized as attachment of atoms. During shell growth, attachment of atoms of different materials having considerable lattice mismatch should initiate lattice strain. The strain extends from the interface to certain depths inside the core as well as the shell. This in turn should destabilize the shell. Then, how does the shell grow? Does it follow the customary epitaxial pathway like atom–atom stitching or adhesion by means of some interaction? Or is it an attachment of a pre-formed cluster of tiny Au NPs that promotes shell growth? These questions are still not well understood. We believe that in addition to static HR-TEM, an in situ dynamic imaging by electron microscopy can provide important insight to this interesting phenomenon.

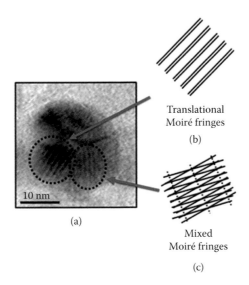

(a)

Translational
Moiré fringes

(b)

Mixed
Moiré fringes

(c)

Figure 9.23 (a) High-magnification image of one Si NC–Ag NP core–shell structure having translational and rotational Moiré fringes due to encapsulation of Si NC by lattice mismatched Ag NPs. Schematic illustrations of the formation mechanism of translational and mixed Moiré fringes are shown in (b) and (c). (Courtesy of T. S. Basu and M. Ray, University of Konstanz, Germany and IIEST, Shibpur, India.)

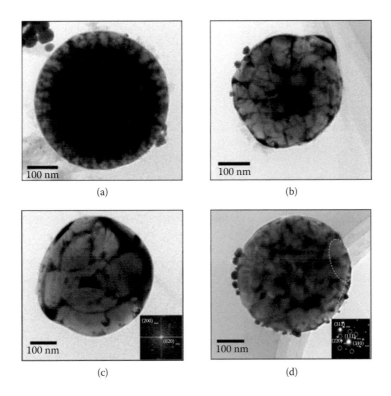

(a)

(b)

(c)

(d)

Figure 9.24 Bright-field TEM image of Si–Au core–shell nanostructures, having (a) a complete core–shell structure, (b) the net-shape wrinkled surface around the particle, (c) the wrinkled surface, with an SAED pattern showing the dark piece circled by white square can be indexed to Au crystal, and (d) the wrinkled slices and tiny particles around the surface, with the corresponding SAED pattern. (Reproduced with permission from Liu, P., et al., 2015, Fabrication of Si/Au core/shell nanoplasmonic structures with ultrasensitive surface-enhanced Raman scattering for monolayer molecule detection, *J. Phys. Chem. C*, 119(2), 1234–1246. Copyright 2015 American Chemical Society.)

Functional materials

9.3.1.4 Si–Si oxide core–shell nanostructures

Of the different forms of Si-based core–shell systems, most common is the Si–SiO$_2$ core–shell nanostructures due to the extraordinary high affinity between Si and O$_2$. Highly crystalline Si NCs may be produced by various routes discussed above, namely vapor condensation, chemical methods and solid-state methods. However, as we already know, all the NCs are invariably passivated with an amorphous SiO$_2$ shell unless utmost care is taken to terminate the same with other less probable species to yield the other variants of Si core–shell systems.

Spherical Si NC, which is the most common form of Si nanostructures, has an amorphous SiO$_2$ shell, which results in a crystalline–amorphous Si–SiO$_2$ core–shell structure. The morphology of the core and the shell and the interface between them are best understood from HR-TEM images. Figure 9.13 shows a typical SiO$_2$ shell covering a spherical core. However, such distinct core and shell of Si NC and SiO$_2$, respectively, are not always visible under an electron microscope. The reason lies in the sample preparation method. Oxide-coated Si NCs are either formed in a matrix of solid amorphous oxide or in colloidal suspensions. To form TEM samples, the colloidal suspensions are usually deposited on carbon-coated grids. Upon evaporation of the solvent on a TEM grid, the amorphous oxide coverage on the adjacent NCs merges together to form extended agglomerates having an appearance of crystalline islands in amorphous background as seen in Figure 9.25a–c. The images show randomly oriented, highly crystalline spherical NCs with visible lattice fringes having particle sizes as low as 2 nm. Both the spherical and rod-like Si NC appear to be embedded in a thick SiO$_2$ matrix that is formed by merging of the individual oxide shells.

Highly crystalline Si nanowires are also known to develop an amorphous SiO$_2$ layer. A Si core of 9.2 nm diameter on development of the amorphous shell resulted in a 1D Si–SiO$_2$ core–shell structure with diameter of 42 nm as seen in Figure 9.26.

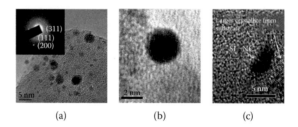

(a)　　　　　　　(b)　　　　　　　(c)

Figure 9.25 HR-TEM images of Si NCs. (a) A large number of particles with sizes less than 5 nm are identifiable. Inset shows the corresponding SAEDP, (b) an isolated spherical Si NC with diameter ~2 nm, and (c) Si NC derived from porous Si. The aspect ratio of the particle captured is ~2.4. (Reproduced with permission from Ray, M., et al., 2010, Luminescent core-shell nanostructures of silicon and silicon oxide: Nanodots and nanorods, *J. Appl. Phys.*, 107, 064311. Copyright 2010 American Institute of Physics.)

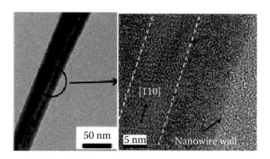

Figure 9.26 Crystal–amorphous, Si–Si oxide core–shell nanowire under HR-TEM. (Reproduced with permission from Adachi, M.M., et al., 2010, Optical properties of crystalline-amorphous core–shell silicon nanowires, *Nano Lett.*, 10, 4093–4098. Copyright 2010 American Chemical Society.)

As discussed before, a control over the growth of the amorphous SiO_2 shell over the crystalline Si core is of paramount importance for understanding the crystalline–amorphous interface and hence for application of the core–shell system. Controlled growth of oxide layer has been attempted in a vapor–liquid–solid deposition system, and the resulting structure has been observed by a TEM. The growth of SiO_2 over crystalline Si core for 10, 20, and 40 min as monitored by TEM shows that thickness of the SiO_2 increases linearly with increase in the growth time, whereas the core thickness remains unchanged (Figure 9.27).

Along with materials properties of Si, the unique role of this material in the electronic industry may be attributed to the remarkable properties of the Si–SiO_2 interface. Although Si is crystalline and the

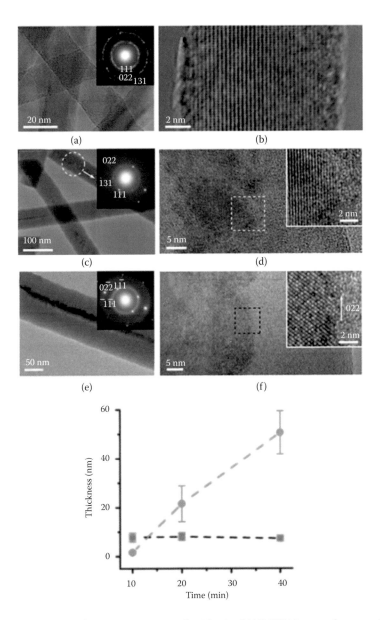

Figure 9.27 (a) TEM and SAEDP of Si nanowires grown for 10 min, (b) HR-TEM image of a nanowire grown for 10 min, (c) TEM and SAED of nanowires grown for 20 min, (d) HR-TEM images of a nanowire grown for 20 min, and (e) TEM and SAED of nanowires grown for 40 min, (f) HR-TEM images of a nanowire grown for 40 min, and (below panel) statistics of the core radius (red) and shell thickness (green) versus growth time. (Reproduced with permission from Cui, L.F., et al., 2009, Crystalline-amorphous core–shell silicon nanowires for high capacity and high current battery electrodes, *Nano Lett.*, 9(1), 491–495. Copyright 2009 American Chemical Society.)

oxide is amorphous, in the bulk form, the interface is essentially perfect, with an extremely low density of dangling bonds or other electrically active defects. However, the actual atomic structure of this interface is still not clearly understood. Of course at the nanoscale, the structure of the Si–oxide interface becomes more and more important as surface to volume ratio increases. Yet, we have very little idea till date about what exactly is happening at the interface. Experiments have been carried out to estimate the number of Si atoms at the interface having intermediate oxidation states (Himpsel et al. 1988; Banaszak Holl et al. 1994; Djurabekova and Nordlund 2008). Theoretical models have been made to explain the experimental results (Pasquarello et al. 1996), but the actual picture had remained largely elusive. It is generally accepted that primary connection between Si core and SiO_2 shell occurs via Si^{+2} (Tu and Tersoff 2000), but a lot remains to be understood regarding the true atomic structure and energetics of this extremely important crystal–amorphous interface.

9.3.2 SCANNING PROBE MICROSCOPY

Scanning probe techniques involving atomic force microscopy (AFM) and scanning tunneling microscopy (STM) have been extensively used for studying the characteristics of nanostructures. There are few but significant reports where AFM in the phase contrast mode has been used to understand core–shell structures as well. Core–shell nanostructures under phase contrast microscopy usually manifest as two phase material representative of core–shell structures. Phase imaging provides a map of stiffness variation on the sample surface such that a stiffer region appears brighter thereby providing a means for differentiating phases with different elastic moduli (Magonov et al. 1997). Later works have shown that phase contrast, in many cases, is independent of variations in the elastic moduli, particularly when the energy used in surface deformation is elastically recovered (García and Pérez 2002). Recent works attribute contrast variations with the changes in viscoelasticity (Lei et al. 2007). Even if we disregard the physics of contrast variation, it may be asserted that phase imaging reveals the existence of two distinct phases forming core and shell and hence could be a good tool to study the structural features of these nanostructures. In Figure 9.28, the darker core represents crystalline Si, whereas the brighter shell represents amorphous Si oxide.

A key factor in AFM of core–shell structures is the choice of substrates. An atomically flat substrate is naturally the ideal choice. However, the most common atomically flat substrate that is easily available or made in the laboratory is electropolished Si. Si on Si will definitely not be a good choice as there will be no phase contrast. Cleaved mica in this regard provides a good alternative but is useless for any conducting measurements as it is highly insulating. There are special techniques to deposit atomically flat Au or Ag on flat Si substrates, which should be used for imaging in conducting mode.

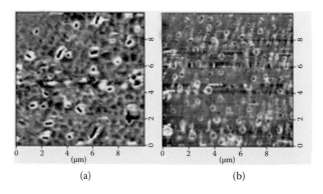

(a)   (b)

Figure 9.28 Phase contrast AFM images of Si–Si oxide core–shell nanostructures. (a) rod-like core–shell nanostructures produced by exfoliation of porous Si and (b) spherical structures produced by deliberate etching-oxidation etching of ball-milled Si. (Reproduced with permission from Ray, M., et al., 2010, Luminescent core–shell nanostructures of silicon and silicon oxide: Nanodots and nanorods,. *J. Appl. Phys.*, 107, 064311. Copyright 2010 American Institute of Physics.)

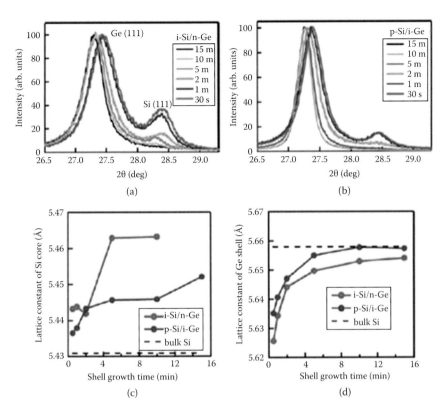

Figure 9.29 Evolution of shell captured in peak shift in the XRD profile for (a) i-Si/n-Ge and (b) p-Si/i-Ge core–shell nanowires. Variation of lattice parameters with shell growth for (c) the Si in the i-Si/n-Ge and p-Si/i-Ge core-shell nanowiress and (d) the Ge shell in the i-Si/n-Ge (Reproduced with permission from Fukata, N., et al., 2012, Characterization of impurity doping and stress in Si/Ge and Ge/Si core–shell nanowires, *ACS Nano*, 6(10), 8887–8895. Copyright 2012 American Chemical Society.)

9.3.3 X-RAY DIFFRACTION OF CORE–SHELL NANOSTRUCTURES

Besides electron microscopy and scanning probe techniques, X-ray diffraction (XRD) can also be used to infer about structural features of core–shell nanostructures. Along with shell growth and doping, the lattice parameters of core and shell suffer minor changes due to interfacial stress generation. Increase in shell thickness should in such a case lead to peak shift in XRD profile. The progressive development of shell in i-Si/n-Ge and p-Si/i-Ge core–shell nanowires has been studied by the shift of Ge (111) and Si (111) peaks as shown on Figure 9.29a and b. The tensile stress applied by the Ge shell on the Si core results in increased lattice parameter of the core, whereas the shell lattice parameter also increases during shell growth due to the compressive stress initiated at the core which weakens with shell growth. The variations of lattice parameters with shell growth are shown in Figure 9.29c and d.

Before ending the discussion of structural characterization of Si-based core–shell nanostructures, it is worth mentioning that grazing incidence X-ray diffraction using a synchrotron source is a very powerful tool for understanding the core–shell structure. Synchrotron-based studies can reveal interesting information about microscopic chemical composition, sizes along different dimensions, strain distribution, and many other structural features. However, this discussion requires separate attention and is not dealt with in this chapter.

9.4 SPECTROSCOPY OF SILICON-BASED CORE–SHELL NANOSTRUCTURES

It is well known that optical properties of Si NCs are of paramount importance. In fact, the worldwide interest in Si NCs was triggered after the discovery of light emission from nanostructured porous

Functional materials

Si because it opened up the possibility of developing Si-based photonic devices where communication between components can be done by photons instead of electrons. Nearly three decades have passed since then and a huge amount of research has been carried out in this filed. Yet, the development of a stable and device-integrable Si NC-based light source remains largely elusive. There is still no conclusive understanding about the exact mechanism of light emission. As discussed in other chapters of this book, broadly there are two competing theories: (i) quantum confinement of electron-hole pair and relaxation of k-selections rules in Si NCs and (ii) interfacial or surface-related defect states or surface oxide-related transitions.

In the absence of a stable capping of Si NCs, it is impossible to prevent oxidation of the NC under ambient conditions. Therefore, most of the significant works on understanding the optical properties of Si NCs are based on core–shell structures. A major problem in such case is the modification of the intrinsic property of Si NC in the presence of the shell. Presence of a shell not only modifies the overall behavior of the core–shell nanostructure but also affects the properties of the Si NC core. Hence, it is imperative to understand the coupled interaction of the core and the shell which manifest in a myriad of ways depending on the shell material as well as the structure and geometry of the nanostructure. Investigation of core–shell interaction reveals exciting physics like plasmon–exciton coupling in case of metal–Si nanostructures and at the same time renders these hetero-nanostructures amenable for other applications where the individual and coupled properties of the core and the shell can be modulated and used. In this section, we will discuss on these aspects in light of the emission and absorption features with focus on the luminescence mechanism of Si-based core–shell nanostructures.

9.4.1 ABSORPTION CHARACTERISTICS

Although our primary focus in this section will be on emission characteristics, we start our discussion with absorption features of Si NC core–shell structures. UV-visible and IR absorption are ubiquitously used to characterize colloidal nanostructures, often without realizing the full potential of these techniques. In order to absorb a particular frequency of incident radiation, the energy of the radiation needs to correspond to an allowed energy transition of the system. Energies of the UV and visible spectra usually correspond to electronic transition energies, whereas molecular transitions fall in the IR regime of the electromagnetic spectrum. Energies lying within the band gap of a system cannot be absorbed. According to the quantum confinement effect in NPs, particles with smaller sizes have a larger band gap. This relates to a blue shift in the absorption spectra with decreasing particle sizes. The changes in molecular bonding are captured by shifts in the IR spectrum. Therefore, UV–visible and IR spectra bear signatures of many exciting features of the size-dependent properties of core–shell nanostructures.

9.4.1.1 UV-visible spectroscopy

9.4.1.1.1 Surface functionalized Si NCs

We have already noted that Si NCs functionalized by organic molecule do not classify as conventional core–shell materials. Nevertheless, we have discussed the general strategy for their preparation and dealt about their structural characterization. Here, we will present a very brief discussion on their optical properties so that we can relate better to the typical core–shell nanostructures.

A simple way to demonstrate quantum confinement effect in Si NCs is to cap the oxide-free NCs of varying sizes with some organic ligand and observe the absorption and emission characteristics. For the same capping, we can then expect the absorption and emission peaks to progressively blue shift with decreasing size. A consistent blue shift of the absorption features with decreasing size was demonstrated for silanized Si NCs as shown in Figure 9.30, providing strong support toward the quantum confinement model. Then, why should we doubt this model? It is easy to find out counter evidences, but that does not explain why absorption bands of similarly capped Si NCs blue shift with decreasing size.

According to the quantum confinement model, all similar sized Si NCs, terminated with any species, should have similar spectra. However, Si NCs of same size and shape terminated with different species are known to exhibit difference in absorption peaks as shown in Figure 9.31.

The spectra in Figure 9.31 show that increasing the alkyl chain length results in a blue shift of the absorption bands. This has been attributed to two factors: (i) presence of NH_2 in the vicinity of Si NCs for small NC systems results in narrowing of energy gap (HUMO-LUMO gap) and (ii) less oxidation of Si

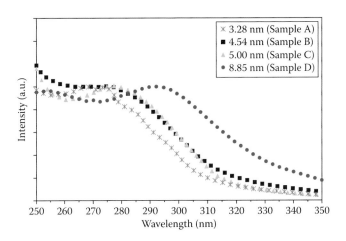

Figure 9.30 UV–vis absorption spectra of silanized Si NCs [Si–$O_3SiC_8H_{17}$] with varying Si core sizes. (Reproduced with kind permission from Springer Science+Business Media: *J. Cluster Sci.*, Size and spectroscopy of silicon nanoparticles prepared via reduction of $SiCl_4$, 17, 2006, 565–578, Zou, J., et al., copyright 2006.)

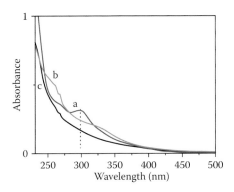

Figure 9.31 Absorption spectra of (a) Si-$C_3H_6NH_2$, (b) Si-$C_6H_{12}NH_2$, and (c) Si-$C_{11}H_{22}NH_2$NPs. (Reproduced from Vasic, M.R., et al., 2009, Amine-terminated silicon nanoparticles: Synthesis, optical properties and their use in bioimaging, *J. Mater. Chem.*, 19, 5926–5933. By permission of The Royal Society of Chemistry.)

in systems terminated with longer chains (Vasic et al. 2009b). Therefore, according to this result, surface states and interfacial states play the pivotal role in determining the optical property of Si NCs.

The above two absorption features and the arguments therein put the debate regarding the origin of visible luminescence in Si NCs in a perspective. There are two competing mechanisms: (i) quantum confinement and (ii) surface/interface states. There are probably as many experiments and "evidences" in favor of the former as for the later. We now know that both quantum confinement and surface/interface states (in most cases oxide related, because oxides are almost always present in Si NCs) play significant role in the process of light emission from these species. The question therefore is which mechanism dominates and under which circumstance? If both mechanisms are present, then how do we distinguish and control them individually? We will soon see how core–shell nanostructures help us in answering these questions.

9.4.1.1.2 Semiconductor shell on Si NCs

Development of a shell on a Si NC surface has many effects on the Si core and the overall nanostructure. Because there is almost always some lattice mismatch between the materials of the core and the shell, growth of a shell invariably results in a strained structure which in turn changes the band alignment and band gap. The outcomes of these changes are manifested in their optical and electrical properties. For Si–CdS spherical core–shell structures, the absorption peak continuously red shifts with increasing shell thickness as shown in Figure 9.32.

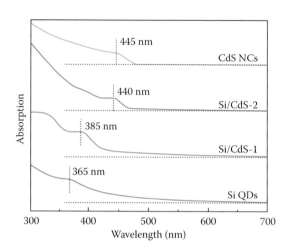

Figure 9.32 Absorption spectra of Si NCs, CdS, Si/CdS-1, and Si/CdS-2 NCs. (Reproduced with permission from Wang, X., et al,. (2014), Silicon/hematite core/shell nanowire array decorated with gold nanoparticles for unbiased solar water oxidation, *Nano Lett.*, 14(1), 18–23. Copyright 2014 American Chemical Society.)

Here, it appears that the overall absorption of the composite nanostructure shifts from that of Si to that of CdS as the shell thickness is increased and the band edge varies between the band edge of Si NC and CdS. The long tail-like feature of the absorption profile, corresponding to proposed subband gap absorption in NCs, remains after shell growth. This is somewhat expected—the overall property of the core–shell is somewhere in between the core and the shell and shifts toward that of the shell with increasing shell thickness.

9.4.1.1.3 Si-metal core–shell nanostructures

Absorption for semiconductor and metal is quite different. For semiconductor NCs, the absorption arises from its band gap, discrete energy levels and unavoidable defect states; whereas, for metallic NPs, the absorption features are dominated by localized surface plasmon resonance (LSPR). What would happen when a photon interacts with a structure having both semiconductor as well as metal nanostructures? For Si NCs, upon photon incidence, an electron from valence band will jump to any energy level in conduction band. This transition is referred as band-to-band transition that requires photons with energy either equal to or greater than the band gap energy. Defect states that may lie inside the band gap also take part in absorption. Transition from valance band to these states will require energy less than band gap energy and as a whole we may expect an absorption feature extending over a range of energy. In case of metallic NPs, absorption is associated with plasmon oscillation. When the frequency of incident electromagnetic wave matches with the natural frequency of oscillation of the plasmons of the NPs, that is, at resonance, maximum absorption will occur. But then what form will the absorption feature acquire upon metallic shell growth on Si NC? It is quite different from either constituent as can be seen from Figure 9.33a due to coupling of semiconductor exciton and metal plasmon. When metallic NPs and high refractive index semiconductor NCs are in close proximity, change in surrounding refractive index will change the natural frequency of oscillation of the surface plasmon, and hence, the LSPR peak of Au NPs changes its position. This is evident from the result shown in Figure 9.33a. This shift may also include the contribution of increase in overall particle size. Now, like the life time observed for semiconductor electrons or excitons, there is a dephasing time for plasmons after which plasmonic oscillations lose their coherence. Introduction of electromagnetic coupling with adjacent particles, electron–electron scattering, and elastic scattering by defects at surface may introduce this dephasing. When electrons participating in plasmon oscillation are perturbed as a result of core–shell formation, decrease in dephasing time takes place and it results in increase in full-width at half maxima (FWHM) of LSPR peak. This widening of LSPR peak as shown in Figure 9.33a suggests the formation of hybrid structure. Upon UV radiation, electron transfer from Si NCs to Au shell results in change in plasmon oscillation frequency leading to the blue shift in peak position. This shifted LSPR peak reverts back to its original position after withdrawal of UV, as

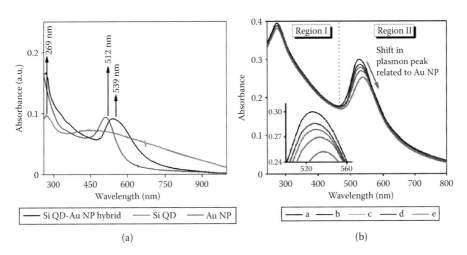

Figure 9.33 (a) UV-visible absorption spectra of the Si NC–Au NP core–shell along with that of bare Si NCs and Au NPs in aqueous solutions. (Reproduced with permission from Ray, M., et al., 2014, Highly lattice-mismatched semiconductor–metal hybrid nanostructures: Gold nanoparticle encapsulated luminescent silicon quantum dots, *Nanoscale*, 6, 2201–2210. Copyright 2014 *Nanoscale*.) (b) Absorption spectra of deaerated Si NC–Au NP colloids observed after UV illumination for 18 min ("a" in the graph). The spectra (b–e) were taken after stopping the UV illumination at intervals of 2, 6, 10, and 15 min, respectively. Inset shows the variation of the plasmon peak only. (Reproduced with permission from Basu, T.S. and Ray, M., 2014, Charge transfer induced encapsulation of Si quantum dots by atomically larger and highly lattice-mismatched Au nanoparticles, *J. Phys. Chem. C*, 118, 5041–5050. Copyright 2014 American Chemical Society.)

shown in Figure 9.33b. Thus, from simple UV-visible absorption experiments, it is possible to draw inference about exotic phenomena like plasmon–exciton coupling and charge transfer in Si–metal core–shell nanostructures.

9.4.1.1.4 Si–Si oxide core–shell nanostructures

Oxide shell on Si NC core is the most common of the core–shell structures. However, the absorption features of these structures are apparently uninteresting. Usually, such samples exhibit a curve without any well-defined band edge or any sharp absorption peak as shown in Figure 9.34a and b. Absorbance gradually increases with decreasing wavelength, from the not so well-defined absorption "on-set" wavelength. As shown in Figure 9.32, standard semiconductor quantum dots like CdS show a clear absorption onset below (in terms of energy) which the absorbance is zero. However, for Si NCs, particularly the one with an oxide shell, there is a long tail extending in longer wavelengths, which is usually attributed to sub-band gap transitions. We have observed that absorption spectrum of oxide-capped Si NCs usually shows a rather weak feature around 260–290 nm (Lin and Chen 2009; Ray et al. 2009a). This feature has been often attributed to the Γ–Γ direct band gap transition or to the presence of ultra-small (1–2 nm) Si NCs. In fact for Si NCs with oxide shell, it has been remarked that more Si NCs might have core diameters in this range after subtracting their oxide layer. So the absorption at about 260–290 nm might be referred to the Si NCs with core diameters less than 2 nm. The Si–Si oxide nanostructures represented in the absorption profiles shown in Figure 9.34a and b were synthesized in two completely different routes and by two different groups. Yet, the nature (weak feature around 260–290 nm) and the argument to explain the feature are remarkably the same in these two cases as well as for all the absorption curves shown before.

The effect of the presence of amorphous oxide shell surrounding crystalline Si nanowires in optical absorption has been studied in detail by Adachi et al. (2010). Figure 9.35 shows the effective optical absorption of Si nanowires with varying diameters. The absorption here is defined as 1 – T – R, where T is the total transmission and R is the total reflectance. For comparison, the absorption and total reflectance of a thin film of 400 nm thick a-Si are also shown. Interestingly, here we see that following the basic understanding of quantum confinement model, the absorption feature and the edge shift toward higher energy with decreasing size of the nanowire core. However, for the sizes of nanowires (15–180 nm) under

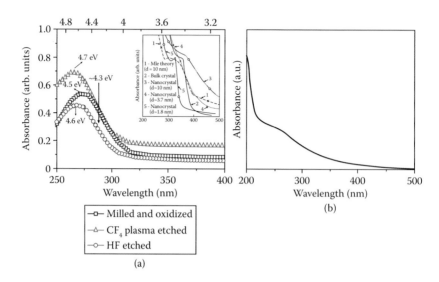

Figure 9.34 UV-vis absorption spectra of (a) colloidal suspensions of un-etched (black), CF$_4$ plasma etched (red) and HF etched (blue) Si–Si oxide core–shell NCs. The absorption edges calculated from the point of inflection for all the three samples are ~4.3 eV and are indicated by an arrow. (Reproduced with permission from Ray, M., et al., 2010, Free standing luminescent silicon quantum dots: Evidence of quantum confinement and defect related transitions, *Nanotechnology*, 21, 505602. Copyright 2010 American Institute of Physics.) The inset is the absorption spectra of several Si NC samples along with the absorption spectrum of bulk Si and a Mie theory calculation for d = 10 nm. (Reproduced with permission from Wilcoxon, J.P., et al., 1999, Optical and electronic properties of Si nanoclusters synthesized in inverse micelles, *Phys. Rev. B*, 60, 2704–2714. Copyright 1999 by the American Physical Society.) (b) Si–Si oxide core–shell NCs in water. (Reproduced from Lin, S.W. and Chen, D.H.: Synthesis of water-soluble blue photoluminescent silicon nanocrystals with oxide surface passivation. *Small*. 2009. 5. 72–76. Copyright Wiley-VCH Verlag GmbH & Co. KGaA. With permission.)

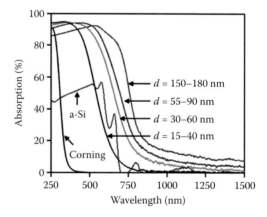

Figure 9.35 Absorption with varying total diameters, d of Si core of Si–Si oxide nanowire. The absorption of a thin film of a-Si with a thickness of 400 nm is shown for comparison. The absorption edge of Si nanowires shifts to longer wavelengths for increasing total diameter well beyond the absorption limit of thin film amorphous Si. (Reproduced with permission from Adachi, M.M., et al., 2010, Optical properties of crystalline-amorphous core–shell silicon nanowires, *Nano Lett.*, 10, 4093–4098. Copyright 2010 American Chemical Society.)

consideration, confinement effect should be very weak. The shift in absorption edge is not due to any confinement effect but well explained by increase in the filling ratio (i.e., the area ratio between nanowires and substrate). Therefore, we see that despite cases where absorption profile blue shifts with decreasing size, it is difficult to attribute such shifts to standard quantum confinement at least in case of Si–Si oxide core–shell nanowires.

9.4.1.2 Fourier transform infra-red spectroscopy

Infra-red radiations correspond to molecular orbital transitions. Hence, the peaks in the IR absorption spectrum correspond to different molecular bonds present in the system, particularly those on the surface of the NC. Typical FT-IR characteristics of Si NCs capped by different species are shown below in Figure 9.36, with each peak corresponding to a particular bond in the system. Understanding of such bonding becomes very important for core–shell nanostructures. For example, in xylene-capped Si NCs (Figure 9.37), FT-IR study revealed the presence of supramolecular bonds that have very little influence on the luminescence properties of uncapped Si NCs.

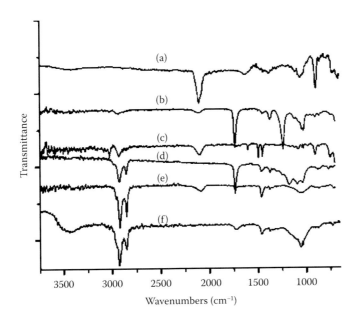

Figure 9.36 FT-IR spectra of Si NCs with a range of surface groups (a) hydrogen, (b) vinyl acetate, (c) styrene, (d) ethyl undecylenate, (e) 1-dodecene, and (f) undecanol. (Reproduced with permission from Hua et al., 2005., *Langmuir*. 21: 6054-6062 Copyright 2005 American Chemical Society.)

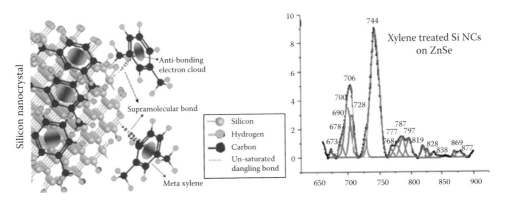

Figure 9.37 Supramolecularly bonded xylene-capped Si NCs and the corresponding FT-IR spectrum. (Reproduced with permission from Mandal et al., 2012, *J Phys. Chem. C*. 116 (27), 14644-14649. Copyright 2012 American Chemical Society.)

Functional materials

9.4.1.3 Raman spectroscopy

In addition to UV-visible and IR, Raman spectroscopy also provides important insights into the interfacial characteristics of core–shell nanostructures. Utilizing Raman spectrum, the stress created in the shell region of Si–Ge core–shell structure was examined as shown in Figure 9.38 (Fukata et al. 2012). With increase in growth time, that is, with increase in shell thickness, the optical peak position of Ge shifts (Figure 9.38a). The variation of the peak position with shell thickness, shown in Figure 9.38b, can be divided into three distinct regions. In the initial stage, we can see decrement in shift which is due to the compressive stress applied by Si NC core. Then, the increase in shift indicates that the effect of compressive stress is lowered. Finally, it becomes like a plateau after a certain shell growth time. For intrinsic Ge shell, the peak reaches the peak position of bulk Ge indicating no effect of core. But for n-type Ge shell, doping of P introduces a contraction of lattice parameter and so the position will not be same as intrinsic bulk Ge.

Along with modification of shell lattice, vibrational modes of dopant and the core can also be realized. These vibrational modes have been observed more prominently for Si–SiGe structure than Si–Ge. The highest intensity peak in Figure 9.39 corresponds to core Si and the position shifts upon shell growth. Other peaks present indicate the Ge–Ge, Ge–Si, and Si–Si interaction (from left to right, respectively) in shell region (Dillen et al. 2016).

9.4.2 STEADY-STATE LUMINESCENCE IN CORE–SHELL NANOSTRUCTURES OF SILICON

Photoluminescence from Si nanostructures is arguably the most widely explored characteristic of all forms of nano-Si. As mentioned earlier and discussed in details elsewhere in this book, the issue of light emission mechanism from Si nanostructures still awaits a consensus (Godefroo et al. 2008; Dasog et al. 2013; Hartel et al. 2013; Kolasinski 2013; Lu et al. 2013; Wolf et al. 2013). A detailed discussion on the state-of-art understanding in this regard is presented elsewhere. Here we simply point out the wide variety of reports regarding PL peaks of Si NCs passivated by different materials and the corresponding sizes in order to understand the role of core–shell nanostructures in this regard. Figure 9.40 provides a summary of variation of PL maxima with particle size of Si NCs, passivated by different ligands, extracted from few significant reports over the past two decades. The wide variety of reports in itself is a testimony for the complexity associated with understanding the mechanism and the ensuing debate.

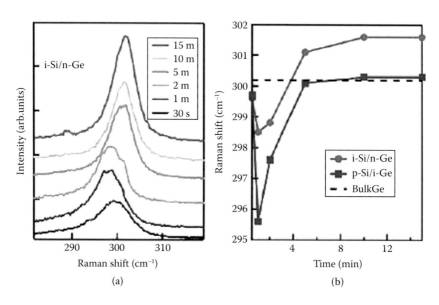

Figure 9.38 (a) Ge optical phonon peaks observed for i-Si/n-Ge core–shell nanowires and (b) Raman shift of Ge optical phonon peak as a function of the shell growth time. (Reproduced with permission from Fukata, N., et al., 2012, Characterization of impurity doping and stress in Si/Ge and Ge/Si core–shell nanowires, *ACS Nano*, 6(10), 8887–8895. Copyright 2012 American Chemical Society.)

Functional materials

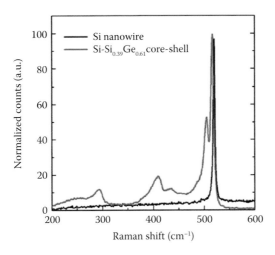

Figure 9.39 Comparison of Raman spectra between Si nanowires with no shell (black line) and Si–Si$_x$Ge$_{1-x}$core–shell. (Reproduced with permission from Dillen, D.C., et al., 2016, Coherently strained Si–SixGe1–x core–shell nanowire heterostructures, *Nano Lett.*, 16(1), 392–398. Copyright 2016 American Chemical Society.)

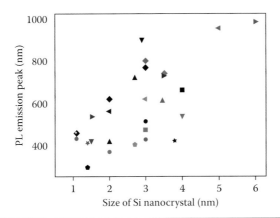

■ JPC 97, 1224 (1993); O$_2$ passivated	◓ Nano Lett. 80, 163 (2003); OH passivated
◀ PRB 54, 5029 (1996); O$_2$ passivated	▲ Science 296, 1203 (2003); O$_2$ passivated
▶ PRB 54, 5029 (1996); O$_2$ passivated	▲ Chem. Mater. 24, 393 (2003), alkyl passivated
● PRB 60, 2704 (1999); O$_2$ passivated	▼ Chem. Mater. 24, 393 (2003), alkyl passivated
▼ PRB 60, 2704 (1999); O$_2$ passivated	◀ Chem. Mater. 24, 393 (2003), alkyl passivated
▶ PRL 82, 197 (1999); O$_2$ passivated	▷ Chem. Mater. 24, 393 (2003), alkyl passivated
◀ PRL 82, 197 (1999); O$_2$ passivated	■ APL 83, 3473 (2003); N$_2$ passivated
● Nature 408, 440 (2000); O$_2$ passivated	● APL 83, 3473 (2003); O$_2$ passivated
▲ PRB 62, 15942 (2000); O$_2$ passivated	◆ Nat. Nanotechnol. 3, 174 (2008); O$_2$ passivated
▼ JACS 123, 3743 (2001); alkoxide linkage	◆ ACS Nano 7, 2676 (2013); dodecyl passivated
◆ APL 80, 4834 (2002); O$_2$ passivated	◆ ACS Nano 7, 2676 (2013); dodecyl passivated
● Nano Lett. 80, 163 (2003); H passivated	● ACS Nano 7, 2676 (2013); TAOB passivated
● Nano Lett. 80, 163 (2003); O$_2$ passivated	★ ACS Nano 7, 2676 (2013); NH$_4$Br passivated
✱ Nano Lett. 80, 163 (2003); O$_2$ passivated	◓ Nano Lett. 13, 2516 (2013); alkylamine passivated
	■ Nano Lett. 13, 2516 (2013); NH$_4$Br passivated

Figure 9.40 Reported variation of PL maxima with particle size of Si NC, passivated by different ligands, extracted from few significant reports over the past two decades. (Courtesy of T. S. Basu and M. Ray, University of Konstanz, Germany and IIEST, Shibpur, India.)

9.4.2.1 Surface functionalized Si NCs

We have seen while dealing with absorption features of Si NCs that the strongest evidence in favor of quantum confinement is provided by functionalized Si NCs, where varying sizes of NCs, terminated similarly, show shifts in optical characteristics in accordance with this model.

Highly monodispersed Si NCs with sizes of ~1, ~2, ~3, and ~4 nm and with a very narrow size distribution have been reported to emit a blue peak at 450 nm, a green peak at 520 nm, a red band at 640 nm, and an infrared band at 740 nm, respectively, as can be seen from the Figure 9.41a. Luminescence from alkyl-capped Si NCs also blue shift on decrease in particle size (Figure 9.41b). Particles of size ~6 nm show emission wavelength longer than 800 nm, whereas decrement of size to less than 3 nm shifts the peak to 560 nm (Liu et al. 2006). The same blue shift in PL peak is also observed in silanized Si NCs on decrease in the particle size from 8.85 to 3.28 nm, shown in Figure 9.41c (Zou et al. 2006). All these examples tend to provide reasonable support for quantum confinement model.

The surface functionalized Si NCs also exhibit visible tunable colors (Figure 9.42). Tuning the luminescence of Si NCs over the entire visible spectra with high relative PL quantum yields using different surface functionalization has been reported. But this can be achieved for the same system by functionalizing the NCs with different surface groups and not by size selection of the same sample of Si NCs—a feature that has worked against the quantum confinement model.

To add to the complexity of the issue, Wolf et al. (2013) have used scanning tunneling spectroscopy to show the quantum confinement effects in Si NCs capped by different capping ligands. They also showed that the effect is present in the PL spectra for Si NCs functionalized with dodecene or trioctylphosphine oxide (TOPO) but not for nitrogen-containing species. Veinot and coworkers (Dasog et al. 2013) have proposed that the nitrogen defect or impurity site provides the mechanism of the blue emission in Si NCs. Another contemporary study on Si nanorods shows that as-prepared nanorods are not luminescent, but etching with hydrofluoric acid to remove residual surface oxide followed by thermal hydrosilylation induces bright red PL (Lu et al. 2013). Based on all these observations, arguments, and counter

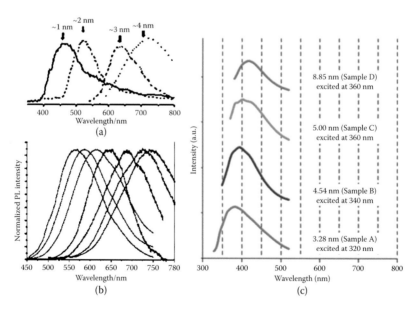

Figure 9.41 Blue shift in PL spectra of Si NCs with decreasing size of NC core, (a) with sizes from ~1 to ~4 nm (reproduced with permission from Kang, Z., et al., 2007, A polyoxometalate-assisted electrochemical method for silicon nanostructures preparation: From quantum dots to nanowires, *J. Am. Chem. Soc.*, 129(17), 5326–5327. Copyright 2007 American Chemical Society); (b) alkyl-capped Si NCs. (reproduced with permission from Liu, S.M., et al., 2006, Enhanced photoluminescence from Si nano-organosols by functionalization with alkenes and their size evolution, *Chem. Mater.*, 18, 637–642. Copyright 2006 American Chemical Society); and (c) silanized Si NCs (reproduced with kind permission from Springer Science+Business Media: *J. Cluster Sci.*, Size and spectroscopy of silicon nanoparticles prepared via reduction of SiCl₄, 17, 2006, 565–578, Zou, J., et al., copyright 2006).

Figure 9.42 PL from 3 to 4 nm Si–NCs functionalized with various surface groups and dispersed in toluene, under UV illumination. From left to right the Si–NC functionalization: blue, dodecylamine; blue–green, acetal; green, diphenylamine; yellow, TOPO; orange, dodecyl (air); red, dodecyl (inert). (Reproduced with permission from Dasog, M., et al., 2014, Size vs surface: Tuning the photoluminescence of freestanding silicon nanocrystals across the visible spectrum via surface groups, *ACS Nano*, 8(9), 9636–9648. Copyright 2014 American Chemical Society.)

arguments, it is now believed that blue luminescence from Si NCs is due to oxide and other defects or surface species, whereas red PL is mainly due to quantum confinement—a fact that many researchers are probably converging.

9.4.2.2 Si–semiconductor core–shell nanostructures

The PL emission characteristics of the semiconductor-based Si NC core–shell nanostructures can potentially be very informative. However, in the absence of reasonable study in this area, it is difficult to generalize. A change in band structure of the composite with respect to the individual components is the most important feature that has evolved so far. For example, with change in shell composition "x" for $CdS_{1-X}Se_X$, the peak of PL spectrum as well as the shape changes as shown in Figure 9.43. This is because the band gaps along with other band features are modulated by change in the composition (Pan et al. 2008).

The Si–CdS core–shell nanostructures are characterized by double peaks in the PL spectrum. Features of PL spectra evolve with growth of the shell as shown in Figure 9.44. Here we clearly see that the emission of CdS and Si NCs are both modified when they are combined. This is expected because electronic wave functions at the interface interact to produce new states. So, we now have a different band structure of the entire system—this is not the band structure of either Si NC or CdS or their linear combination. To find out the band structure of this composite system, we need to know all the interactions, which is extremely difficult from a purely theoretical stand point. Experiment shows that even slight perturbations at the interface, or shell structure, or defect distribution may significantly affect the PL characteristics and hence the band structure.

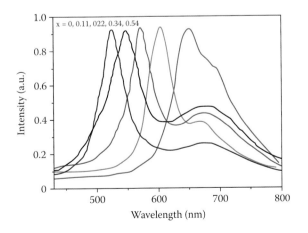

Figure 9.43 Micro-PL spectra from the as-grown single Si–$CdS_{1-x}Se_x$ core–shell nanowires with x = 0, 0.11, 0.22, 0.34, and 0.54, respectively. (Reproduced with permission from Pan, A.L., et al., 2008, Si-CdSSe core/shell nanowires with continuously tunable light emission, *Nano Lett.*, 8(10), 3413–3417. Copyright 2008 American Chemical Society.)

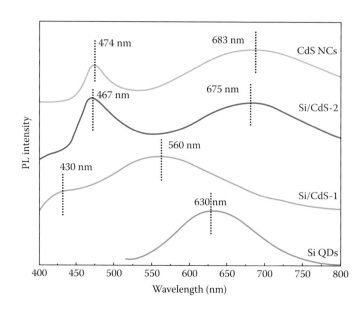

Figure 9.44 PL spectra of Si NCs, CdS, Si/CdS-1 and Si/CdS-2 NCs. (Reproduced from Wang, G., et al., 2014, Type-II core–shell Si–CdS nanocrystals: Synthesis and spectroscopic and electrical properties. *Chem. Commun.*, 50, 11922–11925. By permission of The Royal Society of Chemistry.)

9.4.2.3 Si–metal core–shell nanostructures

On absorbing energy equal to or higher than the band gap, valence electrons are excited to the conduction band. After a certain time (~life time), electrons again come back to their lower energy state via radiative recombination—a process that gives PL. Of course, the energy difference between the levels must be in the visible range to produce visible light. The very fact that nano-Si emits visible light shows there must be some gap which is more than the bulk band gap 1.1 eV. Also, the transitions across this energy gap must be dominant enough to produce visible PL. What we do not clearly understand is whether this energy gap is due to the widened indirect band gap which has now become pseudo direct? Or is this gap formed by the surface states? Or are they inside the surface oxide? Or is it the direct band gap of Si playing a dominant role when reduced to nanoscale?

Now, if there is a metal shell on Si NC, then what do we expect? The emission characteristic of bare Si NC is expected to be affected as seen in Figure 9.45a for Si–Au core–shell nanostructures. We will see in the next section that left to ambient atmosphere Si NCs get progressively oxidized which results in continuous blue shift of the PL spectrum (Ray et al. 2014). The first major observation related to PL is that this time-dependent shift in the spectrum can be arrested when Si NCs are capped by Au NPs, shown in Figure 9.45b. Also, the low energy peak centered around 750 nm is absent in case of Au-capped Si NCs, indicating almost no trace of oxide layer as a result of Au encapsulation (Basu and Ray 2014).

Metal encapsulation of Si NC, therefore, helps in stabilizing the emission characteristics of Si NCs. Additionally, disappearance of one peak and arresting time evolution of the other tentatively suggest that the low energy peak in this case is probably related to oxide and so is the time-dependent shift of the high energy peak. However, to reach to any assertive conclusion in this regard, more studies on metal encapsulated Si NCs are required.

9.4.2.4 Si–Si oxide core–shell nanostructures

As has been repeatedly mentioned, oxide-covered Si NCs are the most common variety of Si NCs. In fact, we have already discussed about oxide and their assumed role in blue PL. Localized defects at the interface of Si NC core and the oxide shell are suspected to play the major role. A study by Godefroo et al. (2008) based on measurements of PL in pulsed magnetic fields up to 50 T shows that defects at the interface of Si embedded in SiO₂ are the dominant source. However, there seems to be one factor which is still not clearly understood. Usually, the Si NCs embedded in an oxide matrix tend to emit in the red region,

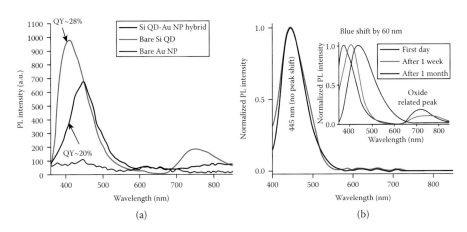

Figure 9.45 (a) PL spectra of bare Si NCs (red), Au NP-encapsulated Si NCs (black) and that of citrate protected Au NPs (blue) under 325nm UV excitation. The quantum yields (QYs) of the luminescent species are indicated. (Reproduced with permission from Ray, M., et al., 2014, Highly lattice-mismatched semiconductor–metal hybrid nanostructures: Gold nanoparticle encapsulated luminescent silicon quantum dots, *Nanoscale*, 6, 2201–2210. Copyright 2014 *Nanoscale*.) (b) PL emission of Si NC–Au NP system shows no peak shift. Inset panel, showing PL spectra of Si NCs dispersed in DI water exhibit continuous blue-shift of PL peaks with time of exposure to the atmosphere. (Reproduced with permission from Basu, T.S. and Ray, M., 2014, Charge transfer induced encapsulation of Si quantum dots by atomically larger and highly lattice-mismatched Au nanoparticles, *J. Phys. Chem. C*, 118, 5041–5050. Copyright 2014 American Chemical Society.)

whereas free-standing Si–Si oxide core–shell nanostructures generally emit in the blue region as shown in Figure 9.46a and b. Additionally, it is a common feature to observe that emission energy of the NCs is dependent on excitation energy for the colloidal suspensions, but the same is usually not observed for Si NCs embedded on a substrate.

Ray et al. (2010a) have tried to explain this in terms of discretization of phonon density of states. In colloidal suspensions of the NCs, the lattice waves within the nanostructures cannot propagate from one particle to another, that is, the phonons become confined inside the particle. The solid–liquid interface offers a boundary for the propagation of phonons, whereas for Si NCs on some matrix or on a substrate, phonons can easily propagate throughout the solid and consequently the restriction on phonon density of states in this case is much less compared to the former. It is therefore proposed that in case of the Si NCs in a matrix or on a substrate, the radiative recombination paths are preceded by relaxation of photo-excited carriers at the band edges or at the interface states. This relaxation energy is normally emitted in form of phonons in bulk materials, that is, $h\nu_{emission} = h\nu_{excitation} - h\nu_{phonon}$ (where ν_{phonon} is the phonon frequency). In bulk materials, all possible ν_{phonon} are present, and hence, there is no restriction in the phonon energy that can be emitted during thermal relaxation of the photo-generated carriers at the band edge or at the interface states. With increasing excitation frequency, the frequency of emitted phonon will increase, resulting in a fixed value of emitted photon frequency ($\nu_{emission}$). However, when these nanostructures are dispersed in colloidal suspensions, the lattice waves cannot propagate from one particle to another, that is, the phonons become confined inside the particle, both in the core and in the shell. Formation of discrete phonon density of states in colloidal suspensions implies that phonons with arbitrary frequencies cannot be emitted. Hence, ν_{phonon} can now assume some discrete values only and that accounts for the blue shift in PL peak with increase in $\nu_{excitation}$. The photoexcited carriers relax by emitting phonons corresponding to some discrete states, resulting in radiative recombination of relatively hot carriers. Appearance of features indicated in PL spectrum and the linear trend of the excitation versus emission (not shown here) also support this model. This is because, $\nu_{emission} = \nu_{excitation} - \nu_{phonon}$, and thus, for a fixed ν_{phonon}, $\nu_{excitation}$ versus $\nu_{emission}$ will be linear with an intercept equal to the emitted phonon energy. In addition, for higher values of excitation energy, we expect relaxation of the photo-generated carriers at different electronic states, determined by the discrete phonon states followed by radiative recombination that accounts for multiple peaks in the PL spectra as shown in Figure 9.46a. Colloidal Si NCs–SiO$_2$ core–shell systems are therefore characterized by multiple peaks and excitation-dependent emission peak shifts.

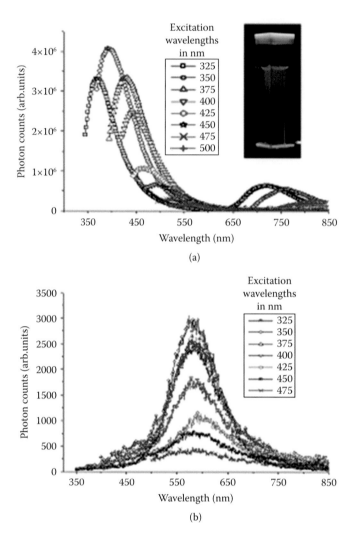

Figure 9.46 PL from (a) Si-Si oxide core–shell nanostructures in colloidal solution and (b) Si-Si oxide nanostructures grown on substrate. (Reproduced with permission from Ray, M., et al., 2010, Luminescent core–shell nanostructures of silicon and silicon oxide: Nanodots and nanorods, *J. Appl. Phys.*, 107, 064311. Copyright 2010 American Institute of Physics.)

Another typical feature of Si–Si oxide core–shell nanostructure worth mentioning is the continuous blue shift of the PL peak. Several groups have observed a gradual but consistent blue shift of the PL spectrum of Si NC–Si oxide core–shell nanostructures—a process that is apparently ceaseless when these nanostructures are exposed to air or water.

Thus, the chemical coverage of the surface of the NC, that is, oxide (Si–O–Si) and interface between the oxide and the crystallite surface, plays a dominant role in providing the necessary radiative recombination centers in generating PL from oxidized Si NC samples. This is also the most probable reason for the blue shift of the PL peak with time. Blue shift with aging has been explained in terms of reduction in particle size that is attributed to oxidation at the NC surface (Cooke et al. 2004). Exposure to water/air causes a reduction of Si–H bonds with a simultaneous growth of Si–O–Si bonds (Maruska et al. 1993), reducing the crystallite size in consequence. However, it is expected that reduction in crystallite size due to modification of surface and interface would slow down and finally cease when all the Si–H bonds are replaced by Si–O–Si bonds. This would imply that blue shifting should stop after certain time. But the shift of the *S*-band PL peaks is found to continue even after 3 years as shown in the inset of Figure 9.47 (Canham et al. 1991), which keeps the issue an

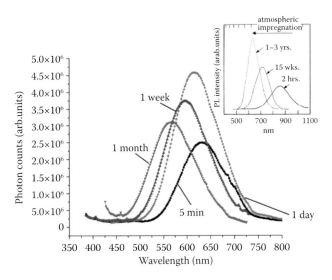

Figure 9.47 Continuous blue shift of PL peak of Si NC–Si oxide core–shell nanostructures. (Courtesy of M. Ray, IIEST, Shibpur, India.)

open-ended research question. It is often argued that oxidation of the Si NC is an invasive process, where NCs exposed to oxygen atmosphere continue to consume oxygen and convert Si–Si bonds to Si–O–Si bonds. The proposition on invasive oxidation is supported theoretically by molecular dynamic simulation demonstrating the development of thick amorphous oxide shell and subsequent amorphization of the Si core as demonstrated in Figure 9.14 (Dalla Torre et al. 2002). Needless to mention, such propositions warrant further investigations and therefore mark some active areas of research that needs to be explored.

9.5 APPLICATIONS OF CORE–SHELL NANOSTRUCTURES

The interest in core–shell nanostructures of Si is primarily due to the potential applications of these exciting structures. The promise of Si-based nanostructures is even more as they can be easily integrated with the existing Si-based technology. Optoelectronics and photovoltaic are arguably the two most important areas where applications of Si NC-based core–shell structures have been investigated and demonstrated. Besides, applications of Si nanostructures have been widely explored in lithium-ion batteries, and finally, there are some recent reports on bio-medical applications. Here, we present a very brief overview of some of these applications of core–shell nanosystems of Si that have been realized in the lab scale.

With regard to these applications, Si nanowire-based core–shell structures out perform all other forms of Si nanostructures. Large-area Si nanowire-based p–n junction diode arrays were developed by Zhu and coworkers in 2004 in Tshinghua University, China (Q. Peng et al. 2004). These nanowires were grown on Si substrates and had an oxide passivation, thereby representing Si–Si oxide core–shell system. Four years later, Lieber and coworkers (Hu et al. 2008) of Harvard developed core–shell Ge–Si nanowire field effect transistors in the sub-100 nm channel length regime which yielded scaled transconductances of 5.3 and 6.2 mS/μm and scaled on-currents of 1.8 and 2.1 mA/μm, respectively (Figure 9.48).

Subsequently, several device structures based on Si–Si oxide core–shell nanostructures have been made, and reasonably good device performance has been demonstrated. Ferrari and coworkers (Colli et al. 2009) of Cambridge University developed top-gated Si nanowire transistors by preparing all terminals (source, drain, and gate) on top of the nanowire as shown in Figure 9.49. The natural oxide shell on the Si nanowire transistor was reported to have negligible leakage.

Crossed crystalline–amorphous core–shell nanowire devices and arrays with metal contacts have been developed (Figure 9.50), and switching behavior in them has been demonstrated (Dong et al. 2008). Room-temperature electrical measurements on single Si/a-Si × Ag nanowire devices exhibit bistable switching between high (off) and low (on) resistance states with well-defined switching threshold voltages, on/off

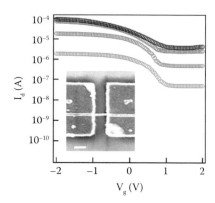

Figure 9.48 Typical transport data from a 100 nm channel device Inset: SEM image of the device. Scale bar is 100 nm. (Reproduced with permission from Hu, Y., et al., 2008, Sub-100 nanometer channel length Ge/Si nanowire transistors with potential for 2 THz switching speed, *Nano Lett.*, 8(3), 925–930. Copyright 2008 American Chemical Society.)

Figure 9.49 SEM image of a short-channel (400 nm) top-gated Si nanowire FET. Scale bar: 200 nm. (Reproduced with permission from Colli, A., et al., 2009, Top-gated silicon nanowire transistors in a single fabrication step, *ACS Nano*, 3(6), 1587–1593. Copyright 2009 American Chemical Society.)

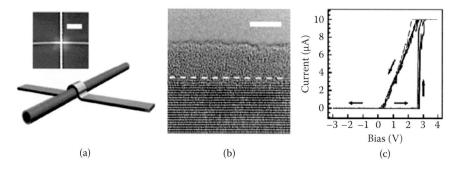

(a)   (b)   (c)

Figure 9.50 (a) A single switch is formed at the cross point of a Si(blue)/a-Si (cyan) core–shell nanowire and a metal nanowire (gray). (Inset) SEM image of a Si/a-Si nanowire (horizontal) crossed Ag–metal nanowire (vertical) device; scale bar is 1 μm. (b) HR-TEM image of aSi/a-Si core–shell nanowire. Scale bar is 5 nm. (c) Current versus voltage sweeps where arrows indicate the voltage-scanning direction. The initial cycle is in red and subsequent three cycles are in black. (Reproduced with permission from Dong, Y., et al., 2008, Si/a-Si core/shell nanowires as nonvolatile crossbar switches, *Nano Lett.*, 8(2), 386–391. Copyright 2008 American Chemical Society.)

ratios greater than 10^4, and current rectification in the "on" state. This structure represents a highly scalable and promising nanodevice element for assembly and fabrication of dense nonvolatile memory and programmable nanoprocessors.

The application of single core–shell Si p–i–n nanowires as nanoscale avalanche photodetectors (APDs) has also been reported by Lieber and his group at Harvard University (Figure 9.51). The p–i–n nanowires

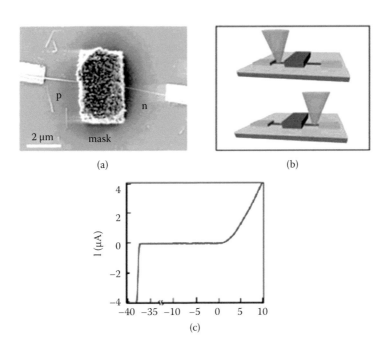

Figure 9.51 (a) SEM image of a p-i-n Si nanowire device with the i-region masked by a thick high optical density polymer. (b) Schematics of localized laser excitation of the p- and n-type regions. (c) Current versus applied voltage curve at room temperature. (Reproduced with permission from Yang, C., et al., 2006, Single p-type/intrinsic/n-type silicon nanowires as nanoscale avalanche photodetectors, *Nano Lett.*, 6, 2929–2934. Copyright 2006 American Chemical Society.)

were grown in three sequential steps, and spatially resolved photocurrent measurements were carried out. Photomultiplication as well as pure multiplication factors for electrons and holes were determined and found to be comparable to planar Si APDs (Yang et al. 2006).

The application of core–shell Si nanostructures in photovoltaic has remarkable promise. But comparatively little has been realized till date. Nanostructure-based solar cells are extremely promising for the third-generation low-cost highly efficient light harvesting devices. Two types of nanowire-based solar cells, that is, with axial or radial p–n junction, are under intensive investigation. A very promising approach of nanowire-based core–shell structure is a heterojunction with intrinsic thin layer configuration, in which crystalline silicon (c-Si) nanowires are used as a core and hydrogenated amorphous silicon (a-Si:H) as a shell (Jia et al. 2013). Many such attempts have been made but it has remained difficult to achieve the efficiency comparable to bulk crystalline Si. Third-generation approaches in photovoltaics attempt to increase device efficiencies while maintaining low cost. This can be achieved by circumventing the Shockley–Queisser limit for single band gap photovoltaic devices by using multiple energy threshold approaches. Such an approach can be realized either by incorporating multiple energy levels in tandem or intermediate band devices; or by modifying the incident spectrum on a cell by converting either high energy or low energy photons to photons more suited to the cell band gap; or by using an absorber which is heated by the solar photons with power extracted by a secondary structure (Conibeer 2009). Silicon-based core–shell nanostructures in this regard have been widely researched.

An array of disordered Si nanowires surrounded by a thin transparent conductive oxide is reported to have low diffuse and specular reflection over a broad wavelength range. These anti-reflective properties together with enhanced infrared absorption in the core–shell nanowire facilitate enhancement in external quantum efficiency using two different active shell materials: amorphous silicon and nanocrystalline silicon. The core–shell nanowire device exhibits a short circuit current enhancement of 15% with an amorphous Si shell and 26% with a Si NC shell compared to their corresponding planar devices (Adachi et al. 2013). Individual core–shell Si nanowire photovoltaic elements have demonstrated their potential for self-powered functional nanoelectronic systems (Tian et al. 2007).

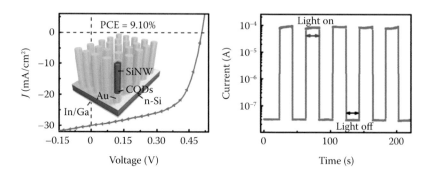

Figure 9.52 Performance of core–shell heterojunction of Si nanowire arrays and carbon quantum dots. (Reproduced with permission from Xie, C., et al., 2014, Core–shell heterojunction of silicon nanowire arrays and carbon quantum dots for photovoltaic devices and self-driven photodetectors, *ACS Nano*, 8(4), 4015–4022. Copyright 2014 American Chemical Society.)

A Si nanowire array–carbon quantum dot core–shell heterojunction photovoltaic device has been recently reported (Xie et al. 2014) (Figure 9.52). The device exhibited excellent rectifying behavior with very high power conversion efficiency. The heterojunction could function as a high-performance self-driven visible light photodetector operating in a wide switching wavelength with good stability, high sensitivity, and fast response speed.

In addition to the above, several other significant attempts have been made to marginally improve the conversion efficiency of a wide variety of photovoltaic devices like the typical dye synthesized solar cell, plasmonic solar cells, tandem solar cells, and hot-carrier solar cells. However, actual large-scale application of any of these technologies is yet to be achieved.

Silicon by virtue of its unprecedented theoretical capacity near 4000 mAh/g is one of the most promising candidates to serve as anode in lithium-ion batteries especially for large-scale energy storage applications. The capacity of Si is almost 10-fold compared to the in use graphite anodes in the cells. However, Si suffers from high volume changes during lithium insertion and extraction. This causes pulverization and capacity fading and limits the cycle life of Si anode to just hundreds of cycles. In order to circumvent this intrinsic bottleneck, research was directed toward the development of Si-based core–shell structures.

Si–SiO$_2$ core–shell nanostructures are one of the primary contenders to be used as anodes in the batteries. The oxide shell prevents the expansion of the outer surface of the Si NC core and the expanding inner surface is not exposed to the electrolyte creating a stable solid–electrolyte interface. These core–shell structures are known to cycle over 6000 times in half cells besides retaining 85% of their initial capacities (Wu et al., 2012). High electrochemical performance can also be achieved when these crystalline–amorphous core–shell structures are directly prepared on stainless steel or copper current collectors, where the crystalline Si cores function as a stable mechanical support and an efficient electrical conducting pathway while amorphous shells store Li+ ions (Cui et al. 2009; Lim et al. 2014). Stable solid–electrolyte interface layer on Si electrode surface can also be prepared by incorporation of conductive coatings like polypyrrole-Fe which significantly improves its capacity and cycling performance (Zhou et al. 2016). Due to the high energy density and long cycle life, Si–Sn-based core–shell nanostructures are also good candidates for this application in Li-ion battery (Choi et al. 2011). A substantial amount of literature is already available on the application of Si NC-based core–shell materials for Li-ion battery anodes (see review on this: Wu and Cui 2012), and we believe this application will soon see the light of the day in a commercial scale, once a cost-effective method for producing such nanostructures is developed.

Hydrogen fuel production is another good alternative to oil and other nonrenewable fuel. Photo-catalytic water splitting is an efficient way out to generate hydrogen ions from water and then utilize them to produce electrical power. To do this, we need a catalyst which may be any inorganic material that will split water. Semiconductors like Si with low band gap energy compared to the visible range (Peng et al. 2005; Jiang et al. 2013), while InGaAs or ZnO nanostructures with tunable bandgap energy covering from UV to infrared region of solar spectrum (Kuykendall et al. 2007), are most useful as they accommodate wider portion of visible band of solar spectrum in splitting. Core–shell nanostructures of this kind are therefore excellent candidates for this application. Not only the range and absorption efficiency but these core–shell nanostructures also increase the separation between photo-generated electrons and holes, resulting in decrease of recombination probability (Paracchino et al. 2011; Reece et al. 2011). Another advantage of such 1D core–shell structures is that in such structures, the directions of light absorption and photo-generated carrier collection directions are decoupled (Peng et al. 2005; Yu et al. 2009). One such hybrid structure is Si–ZnO core–shell nanowire array (Ji et al. 2013). Instead of 1D core–shell structure, for hierarchical Si–InGaAs core–shell heterostructures the catalyst–electrolyte interfacial area increases, resulting in higher conversion (Hwang et al. 2012b).

Another evolving area of application of core–shell Si NCs is in the bio-medical field. Functionalization of ultra-small semiconductor NPs to develop new luminescent probes that are optically bright, stable in aqueous environments, and sized comparably to small organic fluorophores would be of considerable utility for myriad applications in biology. However, most luminescent semiconductor quantum dots are toxic and harmful for human health. Therefore, light-emitting Si NC-based structures are very promising candidates for their low inherent toxicity and bio-inertness. Ruckenstein and coworkers reported the use of red-emitting Si NCs grafted with polyacrylic acid for fixed cell labeling (Li and Ruckenstein 2004). Luminescent micelle-encapsulated Si NCs have been used as luminescent labels for pancreatic cancer cells (Erogbogbo et al. 2008).

Boron-doped Si nanowires have been used to create highly sensitive, real-time sensors for biological and chemical species. Amine- and oxide-functionalized Si nanowires were shown to exhibit pH-dependent conductance that is linear over a large dynamic range. Biotin-modified Si nanowires were used to detect streptavidin, antigen-functionalized Si nanowires showed reversible antibody binding and concentration-dependent detection in real time, and detection of the reversible binding of the metabolic indicator Ca^{2+} was also demonstrated. The small size and capability of these semiconductor nanowires for sensitive, label-free, real-time detection of a wide range of chemical and biological species could be exploited in array-based screening and in vivo diagnostics.

Recently, a Si NC-based multimodal probe that combines the optical properties of Si quantum dots with the superparamagnetic properties of iron oxide nanoparticles to create biocompatible magnetofluorescent nanoprobes has been reported. In this probe, multiple nanoparticles of each type are co-encapsulated within the hydrophobic core of biocompatible phospholipid–polyethyleneglycol (DSPE-PEG) micelles. Enhanced cellular uptake of these probes in the presence of a magnetic field was demonstrated in vitro. Their luminescence stability in a prostate cancer tumor model microenvironment was also demonstrated in vivo as shown in Figure 9.53 (Erogbogbo et al. 2010).

In general, the instability of luminescence in Si NCs and its continuous oxidation in oxidizing environment pose two major challenges for realizing the applications in biological systems unless the surface of the NC is passivated with an appropriate shell with desired properties so as to make the NCs stable and surface passivated.

Figure 9.53 In vivo luminescence stability of Si NC-based multimodal probe in a prostate cancer tumor model microenvironment. (Reproduced with permission from Erogbogbo, F., et al., 2008, Biocompatible luminescent silicon quantum dots for imaging of cancer cells, *ACS Nano*, 2(5), 873–878. Copyright 2010 American Chemical Society.)

9.6 CONCLUSIONS

Silicon NC-based core–shell structures are extremely important, both from the standpoint of understanding fundamental physics associated with an interface at the nanoscale and its impact on electronic structure, as well as for realizing novel applications. Based on some preliminary discussions on Si–organic, Si–metal, Si–semiconductor, and Si–insulator (oxide) core–shell nanostructures, we see that both fundamental understanding and applications have indeed advanced in the last three decades but there are still many unanswered questions and ill-defined propositions. Synthesis or fabrication of core–shell nanostructures with Si NC sitting at the core has made significant progress. Particularly, the vapor–liquid–solid deposition methods are now routinely used to develop core–shell nanostructures of different geometry and composition. However, control of shape of 0D quantum dots or that of the cross section of 1D nanowires is still to be achieved. Nearly, all processes lead either to spherical structures or nanowires with circular cross sections. Consequently, the effect of shape on the properties is still not clear. With the development of the characterization techniques, structure and property of different variants of Si NC-based core–shell structures are now thoroughly investigated. Yet, the exact understanding of the interface between the core and the shell is still very challenging. The specific interaction between the properties of the shell and that of the NC core that affects the electronic structure of the composite is not properly understood. We believe, proper characterization and modelling of the nebulous boundary between the core and the shell will solve many unanswered questions and pave the path for large-scale commercial applications of these extremely promising engineered band gap materials.

REFERENCES

Adachi MM, Anantram MP, Karim KS. (2010). Optical properties of crystalline-amorphous core-shell silicon nanowires. *Nano Lett.* 10:4093–4098.

Adachi MM, Anantram MP, Karim KS. (2013). Core-shell silicon nanowire solar cells. *Sci. Rep.* 3:1546.

Banaszak Holl MM, Lee S, McFeely FR. (1994). Core-level photoemission and the structure of the Si/SiO$_2$ interface: A reappraisal. *Appl. Phys. Lett.* 65:1097–1099.

Basu TS, Ray M. (2014). Charge transfer induced encapsulation of Si quantum dots by atomically larger and highly lattice-mismatched Au nanoparticles. *J. Phys. Chem. C.* 118:5041–5050.

Canham LT, Houlton MR, Leong WY, Pickering C, Keen JM. (1991). Atmospheric impregnation of porous silicon at room temperature. *J. Appl. Phys.* 70:422–431.

Carter RS, Harley SJ, Power PP, Augustine MP. (2005). Use of NMR spectroscopy in the synthesis and characterization of air- and water-stable silicon nanoparticles from porous silicon. *Chem. Mater.* 17:2932–2939.

Chang MT, Chen CY, Chou LJ, Chen LJ. (2009). Core–shell chromium silicide–silicon nanopillars: A contact material for future nanosystems. *ACS Nano.* 3:3776–3780.

Choi J, Zhang Q, Reipa V, Wang NS, Stratmeyer ME, Hitchins VM, Goering PL. (2009). Comparison of cytotoxic and inflammatory responses of photoluminescent silicon nanoparticles with silicon micron-sized particles in RAW 264.7 macrophages. *J Appl. Toxicol.* 29(1):52–60.

Choi NS, Yao Y, Cui Y, Cho J. (2011). One dimensional Si/Sn—based nanowires and nanotubes for lithium-ion energy storage materials. *J. Mater. Chem.* 21(27):9825–9840.

Colli A, Tahraou A, Fasoli A, Kivioja JM, Milne WI, Ferrari AC. (2009). Top-gated silicon nanowire transistors in a single fabrication step. *ACS Nano.* 3(6):1587–1593.

Conibeer G. (2009). Third Generation Photovoltaics in Nanoscale Photonic and Cell Technologies for Photovoltaics II, *Proc. of SPIE* Vol. 7411.

Cooke DW, Muenchausen RE, Bennett BL, Jacobsohn LG, Nastasi M. (2004). Quantum confinement contribution to porous silicon photoluminescence spectra. *J. Appl. Phys.* 96:197–203.

Cui LF, Ruffo R, Chan CK, Peng H, Cui Y. (2009). Crystalline-amorphous core–shell silicon nanowires for high capacity and high current battery electrodes. *Nano Lett.* 9(1):491–495.

Dalla Torre J, et al. (2002). Study of self-limiting oxidation of silicon nanoclusters by atomistic simulations. *J. Appl. Phys.* 92:1084–1094.

Dasog M, Reyes GB, Titova LV, Hegmann FA, Veinot JGC. (2014). Size vs surface: Tuning the photoluminescence of freestanding silicon nanocrystals across the visible spectrum via surface groups. *ACS Nano.* 8(9):9636–9648.

Dasog M, et al. (2013). Chemical insight into the origin of red and blue photoluminescence arising from freestanding silicon nanocrystals. *ACS Nano.* 7(3):2676–2685.

Devika M, Reddy NK, Pevzner A, Patolsky F. (2010). Heteroepitaxial Si/ZnO hierarchical nanostructures for future optoelectronic devices. *ChemPhysChem.* 11:809–814.

Dillen DC, Wen F, Kim K, Tutuc E. (2016). Coherently strained Si–SixGe1–x core–shell nanowire heterostructures. *Nano Lett.* 16(1):392–398.

Djurabekova F, Nordlund K. (2008). Atomistic simulation of the interface structure of Si nanocrystals embedded in amorphous silica. *Phys. Rev. B.* 77:115325.

Dong Y, Yu G, McAlpine MC, Lu W, Lieber CM. (2008). Si/a-Si core/shell nanowires as nonvolatile crossbar switches. *Nano Lett.* 8(2):386–391.

Du Xw, Takeguchi M, Tanaka M, Furuya K. (2003). Formation of crystalline Si nanodots in SiO_2 films by electron irradiation. *Appl. Phys. Lett.* 82:1108–1110.

English DS, Pell LE, Yu Z, Barbara PF, Korgel BA. (2002). Size tunable visible luminescence from individual organic monolayer stabilized silicon nanocrystal quantum dots. *Nano Lett.* 2(7):681–685.

Erogbogbo F, Yong K.T, Hu R, Law W.C, Ding H, Chang C.W, et al. (2010). Biocompatible magnetofluorescent probes: Luminescent silicon quantum dots coupled with superparamagnetic iron (III) oxide. *ACS Nano.* 4(9):5131–5138.

Erogbogbo F, Yong KT, Roy I, Xu G, Prasad PN, Swihart MT. (2008). Biocompatible luminescent silicon quantum dots for imaging of cancer cells. *ACS Nano.* 2(5):873–878.

Fukata N, et al. (2012). Characterization of impurity doping and stress in Si/Ge and Ge/Si core–shell nanowires. *ACS Nano.* 6(10):8887–8895.

García R, Pérez R. (2002). Dynamic atomic force microscopy methods. *Surf. Sci. Rep.* 47:197–301.

Godefroo S, et al. (2008). Classification and control of the origin of photoluminescence from Si nanocrystals. *Nat. Nanotechnol.* 3:174–178.

Guénolé J, Godet J, Brochard S. (2013). Plasticity in crystalline-amorphous core-shell Si nanowires controlled by native interface defects. *Phys. Rev. B.* 87:045201.

Hartel AM, Gutsch S, Hiller D, Zacharias M. (2013). Intrinsic nonradiative recombination in ensembles of silicon nanocrystals. *Phys. Rev. B.* 87(3):0354281–0354287.

He Y, et al. (2009). Photo and pH stable, highly-luminescent silicon nanospheres and their bioconjugates for immunofluorescent cell imaging. *J. Am. Chem. Soc.* 131(12):4434–4438.

Himpsel FJ, McFeely FR, Taleb-Ibrahimi A, Yarmoff JA, Hollinger G. (1988). Microscopic structure of the SiO_2/Si interface. *Phys. Rev. B.* 38:6084–6096.

Hofmeister H, Huisken F, Kohn B. (1999). Lattice contraction in nanosized silicon particles produced by laser pyrolysis of silane. *Eur. Phys. J. D.* 9:137–140.

Hu M, Giapis KP, Goicochea JV, Zhang X, Poulikakos D. (2011). Significant reduction of thermal conductivity in Si/Ge core–shell nanowires. *Nano Lett.* 11(2):618–623.

Hu Y, Xiang J, Liang G, Yan H, Lieber CM. (2008). Sub-100 nanometer channel length Ge/Si nanowire transistors with potential for 2 THz switching speed. *Nano Lett.* 8(3):925–930.

Hua F, Swihart MT, Ruckenstein E. (2005) Efficient Surface Grafting of Luminescent Silicon Quantum Dots by Photoinitiated Hydrosilylation. *Langmuir.* 21: 6054-6062.

Hwang TH, Lee YM, Kong BS, Seo JS, Choi JW. (2012a). Electrospun core–shell fibers for robust silicon nanoparticle-based lithium ion battery anodes. *Nano Lett.* 12(2):802–807.

Hwang YJ, Wu CH, Hahn C, Jeong HE, Yang P. (2012b). Si/InGaN core/shell hierarchical nanowire arrays and their photoelectrochemical properties. *Nano Lett.* 12:1678–1682.

Ji J, et al. (2013). High density Si/ZnO core/shell nanowire arrays for photoelectrochemical water splitting. *J. Mater. Sci.: Mater.* 24:3474–3480.

Jia G, Eisenhawer B, Dellith J, Falk F, Thøgersen A, Ulyashin A. (2013). Multiple core–shell silicon nanowire-based heterojunction solar cells. *J. Phys. Chem. C.* 117:1091–1096.

Jiang J, Li S, Jiang Y, Wu Z, Xiao Z, Su Y. (2013). Mechanism of optical absorption enhancement of surface textured black silicon. *J. Mater. Sci.: Mater. Electron.* 24(2):463–466.

Kang Z, et al. (2007). A polyoxometalate-assisted electrochemical method for silicon nanostructures preparation: From quantum dots to nanowires. *J. Am. Chem. Soc.* 129(17):5326–5327.

Kohno H, Ozaki N, Yoshida H, Tanaka K, Takeda S. (2003). Misleading fringes in TEM images and diffraction patterns of Si nanocrystallites. *Cryst. Res. Technol.* 38:1082–1086.

Kolasinski KW. (2013). The mechanism of photohydrosilylation on silicon and porous silicon surfaces. *J. Am. Chem. Soc.* 135:11408–11412.

Kuykendall T, Ulrich P, Aloni S, Yang P. (2007). Complete composition tunability of InGaN nanowires using a combinatorial approach. *Nat. Mater.* 6:951–956.

Lauhon LJ, Gudiksen MS, Wang D, Lieber CM. (2002). Epitaxial core–shell and core–multishell nanowire heterostructures. *Nature.* 420:57–61.

Lei CH, Ouzineb K, Dupont O, Keddie JL. (2007). Probing particle structure in waterborne pressure-sensitive adhesives with atomic force microscopy. *J. Colloid Interface Sci.* 307:56–63.

Li ZF and Ruckenstein E. (2004). Water-soluble poly(acrylic acid) grafted luminescent silicon nanoparticles and their use as fluorescent biological staining labels. *Nano Lett.* 4(8):1463–1467.

Lim KW, et al. (2014). Catalyst-free synthesis of Si-SiOx core-shell nanowire anodes for high-rate and high-capacity lithium-ion batteries. *ACS Appl. Mater. Interfaces.* 6:6340–634.

Lin SW and Chen DH. (2009). Synthesis of water-soluble blue photoluminescent silicon nanocrystals with oxide surface passivation. *Small.* 5(1):72–76.

Littau KA, Szajowski PJ, Muller AJ, Kortan AR, Brus LE. (1993). A luminescent silicon nanocrystal colloid via a high-temperature aerosol reaction. *J. Phys. Chem.* 97:1224–1230.

Liu P, Chen H, Wang H, Yan J, Yang G. (2015). Fabrication of Si/Au core/shell nanoplasmonic structures with ultrasensitive surface-enhanced Raman scattering for monolayer molecule detection. *J. Phys. Chem. C.* 119(2):1234–1246.

Liu SM, Yang Y, Sato S, Kimura K. (2006). Enhanced photoluminescence from Si nano-organosols by functionalization with alkenes and their size evolution. *Chem. Mater.* 18:637–642.

Lu W, Xiang J, Timko BP, Wu Y, Lieber CM. (2005). One-dimensional hole gas in germanium/silicon nanowire heterostructures. *Proc. Nat. Acad. Sci. U.S.A.* 102:10046–10051.

Lu X, Hessel CM, Yu Y, Bogart TD, Korgel BA. (2013). Colloidal luminescent silicon nanorods. *Nano Lett.* 13:3101–3105.

Magonov SN, Elings V, Whangbo MH. (1997). Phase imaging and stiffness in tapping-mode atomic force microscopy. *Surf. Sci.* 375:L385–L391.

Malm J-O and O'Keefe MA. (1997). Deceptive "lattice spacings" in high-resolution micrographs of metal nanoparticles. *Ultramicroscopy.* 68:13–23.

Mandal AK, Ray M, Rajapaksa I, Mukherjee S, Datta A. (2012) Xylene-Capped Luminescent Silicon Nanocrystals: Evidence of Supramolecular Bonding. *J Phys. Chem. C.* 116 (27), 14644–14649.

Maruska HP, Namavar F, Kalkhoran NM. (1993). Energy bands in quantum confined silicon light-emitting diodes. *Appl. Phys. Lett.* 63:45–47.

Mohapatra S, Mishra YK, Avasthi DK, Kabiraj D, Ghatak J, Varma S. (2008). Synthesis of gold-silicon core-shell nanoparticles with tunable localized surface plasmon resonance. *Appl. Phys. Lett.* 92:103105.

Pan AL, et al. (2008). Si-CdSSe core/shell nanowires with continuously tunable light emission. *Nano Lett.* 8(10):3413–3417.

Paracchino A, Laporte V, Sivula K, Grätzel M, Thimsen E. (2011). Highly active oxide photocathode for photoelectrochemical water reduction. *Nat. Mater.* 10:456–461.

Pasquarello A, Hybertsen MS, Car R. (1996). Structurally relaxed models of the Si(001)–SiO$_2$ interface. *Appl. Phys. Lett.* 68:625–627.

Peng K, Xu Y, Wu Y, Yan Y, Lee S, Zhu J. (2005). Aligned single-crystalline Si nanowire arrays for photovoltaic applications. *Small.* 1(11):1062–1067.

Peng KQ, Huang ZP, Zhu J. (2004). Fabrication of large-area silicon nanowire p–n junction diode arrays. *Adv. Mater.* 16(1):73–76.

Picraux ST, Dayeh SA, Manandhar P, Perea DE, Choi SG. (2010). Silicon and germanium nanowires: Growth, properties, and integration. *JOM.* 62(4):35–43.

Priolo F, Gregorkiewicz T, Galli M, Krauss TF. (2014). Silicon nanostructures for photonics and photovoltaics. *Nat. Nanotechnol.* 9:19–32.

Ray M, et al. (2014). Highly lattice-mismatched semiconductor–metal hybrid nanostructures: Gold nanoparticle encapsulated luminescent silicon quantum dots. *Nanoscale.* 6:2201–2210.

Ray M, et al. (2010a). Luminescent core-shell nanostructures of silicon and silicon oxide: Nanodots and nanorods. *J. Appl. Phys.* 107:064311.

Ray M, Hossain SM, Klie RF, Banerjee K, Ghosh S. (2010b). Free standing luminescent silicon quantum dots: Evidence of quantum confinement and defect related transitions. *Nanotechnology.* 21:505602.

Ray M, et al. (2009a). Blue-violet photoluminescence from colloidal suspension of nanocrystalline silicon in silicon oxide matrix. *Solid State Commun.* 149:352–356.

Ray M, Sarkar S, Bandyopadhyay NR, Hossain SM, Pramanick AK. (2009b). Silicon and silicon oxide core-shell nanoparticles: Structural and photoluminescence characteristics. *J. Appl. Phys.* 105:074301.

Reece SY, et al. (2011). Wireless solar water splitting using silicon-based semiconductors and earth-abundant catalysts. *Science.* 334:645–648.

Rogozhina EV, Eckhoff DA, Gratton E, Braun PV. (2006). Carboxyl functionalization of ultrasmall luminescent silicon nanoparticles through thermal hydrosilylation. *J. Mater. Chem.* 16:1421–1430.

Sugimoto H, Fujii M, Imakita K. (2016). Silicon nanocrystal-noble metal hybrid nanoparticles. *Nanoscale.* 8:10956–10962.

Sur UK. (2014). Imaging of organic and biological materials by in-focus transmission electron microscopy. *Curr. Sci.* 106:17–19.

Tang J, Huo Z, Brittman S, Gao H, Yang P. (2011). Solution-processed core–shell nanowires for efficient photovoltaic cells. *Nat. Nanotechnol.* 6:568–572.

Tian B, et al. (2007). Coaxial silicon nanowires as solar cells and nanoelectronic power sources. *Nature.* 449:885–889.

Tilley RD, Warner JH, Yamamoto K, Matsuib I, Fujimoric H. (2005). Micro-emulsion synthesis of monodisperse surface stabilized silicon nanocrystals. *Chem. Commun.* 14:1833–1835.

Tu Y and Tersoff J. (2000). Structure and energetics of the Si-SiO$_2$ interface. *Phys. Rev. Lett.* 84:4393–4396.

Vasic MR. (2008). Azide-terminated silicon nanoparticles: Functionalization using Click Chemistry and Uptake by Growing Yeast Cells. In: *Synthesis and Photophysics of Functionalized Silicon Nanoparticles.* pp. 75–88, Wageningen Universiteit, Wageningen, The Netherlands.

Vasic MR, Spruijt E, Lagen BV, Cola LD, Zuilhof H. (2008). Alkyl-functionalized oxide-free silicon nanoparticles: Synthesis and optical properties. *Small.* 4(10):1835–1841.

Vasic MR, Cola LD, Zuilhof H. (2009a). Efficient energy transfer between silicon nanoparticles and a Ru-polypyridine complex. *J. Phys. Chem. C.* 113:2235–2240.

Vasic MR, et al. (2009b). Amine-terminated silicon nanoparticles: Synthesis, optical properties and their use in bioimaging. *J. Mater. Chem.* 19:5926–5933.

Wang H, Chen LY, Feng YH, Chen HY. (2013). Exploiting core–shell synergy for nanosynthesis and mechanistic investigation. *Acc. Chem. Res.* 46:1636–1646.

Wang G, et al. (2014a). Type-II core–shell Si–CdS nanocrystals: Synthesis and spectroscopic and electrical properties. *Chem. Commun.* 50:11922–11925.

Wang L, et al. (2015). Ultrafast optical spectroscopy of surface-modified silicon quantum dots: Unravelling the underlying mechanism of the ultrabright and color-tunable photoluminescence. *Light: Sci. Appl.* 4:e245.

Wang X, et al. (2014b). Silicon/hematite core/shell nanowire array decorated with gold nanoparticles for unbiased solar water oxidation. *Nano Lett.* 14(1):18–23.

Wen X, et al. (2015). Tunability limit of photoluminescence in colloidal silicon nanocrystals. *Sci. Rep.* 5:12469.

Wilcoxon JP, Samara GA, Provencio PN. (1999). Optical and electronic properties of Si nanoclusters synthesized in inverse micelles. *Phys. Rev. B.* 60:2704–2714.

Wolf O, Dasog M, Yang Z, Balberg I, Veinot JGC, Millo O. (2013). Doping and quantum confinement effects in single Si nanocrystals observed by scanning tunneling. *Nano Lett.* 13:2516–2521.

Wu H, et al. (2012). Stable cycling of double-walled silicon nanotube battery anodes through solid–electrolyte interphase control. *Nat. Nanotechnol.* 7:310–315.

Wu H, Cui Y. (2012). Designing nanostructured Si anodes for high energy lithium ion batteries. *Nano Today.* 7:414–429.

Wu H, et al. (2013). Stable Li-ion battery anodes by in-situ polymerization of conducting hydrogel to conformally coat silicon nanoparticles. *Nat. Commun.* 4:1943.

Xie C, et al. (2014). Core–shell heterojunction of silicon nanowire arrays and carbon quantum dots for photovoltaic devices and self-driven photodetectors. *ACS Nano.* 8(4):4015–4022.

Yang C, Barrelet CJ, Capasso F, Lieber CM. (2006). Single p-type/intrinsic/n-type silicon nanowires as nanoscale avalanche photodetectors. *Nano Lett.* 6:2929–2934.

Yu H, Chen S, Quan X, Zhao H, Zhang Y. (2009). Silicon nanowire/TiO$_2$ heterojunction arrays for effective photoelectrocatalysis under simulated solar light irradiation. *Appl. Catalysis B: Environ.* 90:242–248.

Zhang J, Tang Y, Lee K, Ouyang M. (2010). Nonepitaxial growth of hybrid core-shell nanostructures with large lattice mismatches. *Science.* 327(5973):1634–1638.

Zhou J, et al. (2016). Core-shell coating silicon anode interfaces with coordination complex for stable lithium-ion batteries. *ACS Appl. Mater. Interfaces.* 8(8):5358–5365.

Ziegler A, Kisielowski C, Ritchie RO. (2002). Imaging of the crystal structure of silicon nitride at 0.8 Ångström resolution. *Acta Mater.* 50:565–574.

Zou J, Sanelle P, Pettigrew KA, Kauzlarich SM. (2006). Size and spectroscopy of silicon nanoparticles prepared via reduction of SiCl$_4$. *J. Cluster Sci.* 17(4):565–578.

Porous silicon as template for magnetic nanostructures

Petra Granitzer and Klemens Rumpf

Contents

10.1 INTRODUCTION

10.1.1 VERSATILITY OF POROUS SILICON

Porous silicon was first discovered by Uhlir in 1956 [1] and 2 years later by Turner [2] during electropolishing experiments but without any further note. Nearly four decades later, porous silicon became very important after the discovery of its possibility to emit light in the visible in 1990 [3, 4]. Besides photoluminescence, also electroluminescence has been investigated intensely [5], especially with respect to optoelectronic applications.

Luminescent porous silicon offers structures in the nanometer scale (2–4 nm) and is called microporous after the IUPAC (international union of pure and applied chemistry) notation. The cause of the luminescence has been under discussion for a long time and has been attributed to siloxene [6], surface states [7], OH groups adsorbed on structural defects in SiO_2 [8], and quantum confinement effects [3] which finally have been figured out to be the origin of the visible light emission.

In the middle of the 1990s, the formation process was under intensive discussion and various morphologies have been produced in the mesoporous (5–50 nm) as well as in the macroporous (>50 nm) regime, for example, by Lehmann [9] and Föll [10, 11]. Especially macroporous silicon with pore diameters in the micrometer regime attained attention and such structures have been used to produce photonic crystals [12], whereas in this case as a preliminary step a photolithographic mask was necessary and after anisotropic KOH (potassium hydroxide) etching which results in pyramidal pits the samples have been anodized in an

HF (hydrofluoric acid) electrolyte [13]. Photonic crystals allowing a refractive index change in the vertical direction regarding the orientation of the pores with structures in user-specified wavelength ranges can be designed by varying the lattice period [14].

Besides optical applications, porous silicon has also been under intense investigation concerning gas sensors or biosensors [15–17]. Also such sensing is often based on optical detection due to a variation of the refractive index [16] or, for example, by using porous silicon interferometer for label-free, ultrasensitive detection of biomolecules [18].

Since the finding of the biocompatibility of porous silicon [19], various studies have been and still are carried out with respect to its applicability in biomedicine, for example, as biosensors for DNA or bacteria detection, for drug delivery, cancer therapy, or tissue engineering [20]. Reports have shown that porous silicon offers good biocompatibility [21], for example, supporting the growth in vitro of several cell types such as hepatocytes [22], neurons [23], and osteoblasts [24]. Besides the intrusion of anticancer therapeutics, proteins, and peptides, the use of porous silicon as dietary supplement has also been potentially discussed [25].

Since the nanostructured silicon offers a broadly tunable morphology and also a big surface area [26], it is a suitable system for the infiltration of various materials into its pores. Besides many organics, also the deposition of metals is under intense investigation. First, metal deposition has been performed to improve the conductivity of the material with regard to electroluminescence and also to influence the emitting light [27, 28]. Metal deposition has been carried out by many different groups, either to investigate the deposition mechanism of various metals [29–31] or to produce an applicable system such as copper deposited within porous silicon for heat sink technology [32].

The deposition of ferromagnetic materials into the pores of porous silicon has been investigated by various authors to figure out, on the one hand, the metal deposition mechanism and, on the other hand, the arising magnetic properties of the nanocomposite system [33–36]. The deposition of magnetic metals has been carried out in mesoporous as well as macroporous silicon [37,38]. A further approach to create a semiconducting/magnetic nanocomposite is the infiltration of synthesized iron oxide nanoparticles into porous silicon [39–41].

10.1.2 FORMATION OF POROUS SILICON

The fabrication of porous silicon can be carried out by wet etching [42], dry etching [43], or laser ablation [44]. However, in this chapter, only wet etching processes will be discussed. The dissolution of silicon takes place by supply of electronic holes from the bulk silicon, whereas holes are depleted between the pores. A very common process to produce porous silicon in every possible morphology is anodization of a silicon wafer. In the case of n-type silicon, generally the required holes are generated by illumination of the sample during the etching process. The resulting morphology depends on the applied current density, the electrolyte composition, concentration and temperature, as well as the doping of the wafer substrate. The relationship between doping density and pore density is demonstrated in Figure 10.1 [45].

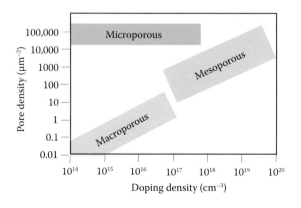

Figure 10.1 Porous silicon showing different morphologies for different silicon doping densities. The pore diameters range between 2 nm (micropores) and a few micrometers (macropores). The appearance of a certain morphology at a certain doping density depends additionally on the anodization conditions, for example, current density, bath concentration. (After Lehmann, V., et al., *Mater. Sci. Eng.*, 69/70, 11–22, 2000.).

An advantage of the anodization process is the tunability of the pore diameters in a broad range and with high accuracy without any prestructuring of the substrate.

Microporous silicon is generally achieved by anodization of low and moderate doped silicon, whereas in the case of an n-type silicon backside illumination is necessary to generate holes.

The formation of mesoporous silicon is often achieved by using highly doped silicon and applying high currents which leads to electrical breakdown conditions [46] and allows anodization without illumination also in the case of n-type silicon.

The thickness of the porous layer depends on the etching time as shown in Figure 10.2 for a porous silicon morphology with an average pore diameter of 60 nm. In this case, the growth rate is about 4 µm/min. Without agitation, the electrolyte becomes exhausted at the pore tips due to insufficient exchange of fluor ions, which results in a self-limitation of the silicon dissolution process [47].

Macropores can be achieved by anodization of low and moderate doped silicon, whereas in the case of prestructuring of the wafer by lithography arrangements of high regularity can be achieved as shown by Lehmann [48]. Many groups used different formation techniques in using various HF-based electrolytes, such as aqueous, nonaqueous, or organic ones, to fabricate macropores [11,49,50]. Ordered out-of-plane macropores with interconnections of smaller in-plane pores have been fabricated by controlled electrochemical etching under backside illumination of an n-type silicon wafer in exploiting breakdown voltage and H_2O_2 as oxidizing agent [51]. A further example of macropores forming a three-dimensional (3D) architecture with interconnections between the pores is shown in Figure 10.3.

A further often employed wet etching method is metal-assisted chemical etching (MACE), which is performed either by depositing the metal on the silicon in using an additional lithographic mask or by self-organization in using an electrolyte with the addition of a metal salt. Most common metals are Au, Ag, Pt, and Pd [52,53]; however, Cu has also been used to catalyze the silicon surface [54]. MACE is generally performed in an HF electrolyte containing an oxidizing agent such as H_2O_2 [55]. The morphology of the resulting porous structure depends on the type of noble metal if isolated particles are used, whereas the shape and size of the pores as well as the distance between them are determined by the shape and size of the metal particles since the silicon beneath the metal offers a much faster etching rate than the uncovered silicon [55]. Furthermore, the resulting structures are influenced by the used oxidizing agent as well as the electrolyte composition and concentration [55].

Stain etching, which is also an electroless method to achieve nanostructured silicon, was first carried out in the late 1950s and 1960 [56,57]. Besides many groups, Kolasinski [58] performed stain etching in various electrolytes and recently demonstrated that the etching process is initiated by hole injection into the valence band which is the rate determining step. The stain etching mechanism can be explained by hole injection into the valence band of the silicon substrate in the presence of the oxidizing agent at the surface. Each hole which is injected attacks one silicon atom [59]. Generally, the etching of the pores occurs randomly and thin porous layers are formed with a uniform lateral porosity [60].

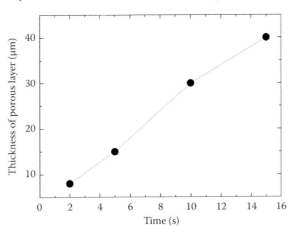

Figure 10.2 Dependence of thickness of the porous silicon layer on the etching time.

Functional materials

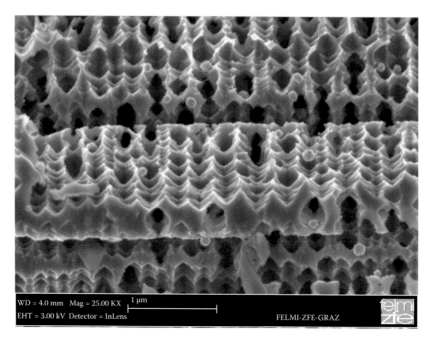

WD = 4.0 mm   Mag = 25.00 KX   1 μm
EHT = 3.00 kV   Detector = InLens   FELMI-ZFE-GRAZ

Figure 10.3 Cross-sectional SEM image of macropores grown perpendicular to the surface (diameter ~800 nm) as well as parallel to the surface (diameter ~200 nm) forming a 3D architecture.

A further electroless technique for etching of silicon is galvanic etching, which is carried out by deposition of a metal layer on the backside of the wafer. The nanostructuring of the wafer is caused by exposing it to an oxidizing agent containing HF electrolyte, which causes that the metal catalyzes hole injection from the oxidant [60].

10.1.3 ELECTRODEPOSITION WITHIN POROUS MEDIA

Another strongly used material, which is also achieved by self-organization, is porous alumina [61]. It is produced by etching of an alumina foil in oxalic or sulfuric acid solution resulting in closely packed honeycomb-like structures. Using a two-step fabrication process, a high regularity of the pore arrangement can be achieved [62]. Therefore, porous alumina templates are often used as templates for metal deposition, especially magnetic ones [63]. Sophisticated multilayered structures are investigated with regard to the domain wall motion and the coupling between them [64]. Porosified InP (indium phosphide) membranes have been used as template for metal deposition investigating the applicability as magnetoelectric sensors with a high sensitivity [65]. Magnetic nanostructures are of special interest due to the properties ascribed to their size, shape, and arrangement. Such nanostructures deposited within porous media offer specific properties, especially due to their size such as single domain or superparamagnetic behavior or particular domain wall motions and magnetization reversal. Also the magnetic interactions between them are of interest and worth investigation, especially with respect to their nanotechnological applicability (e.g., high-density data storage or biomedical diagnostics and therapeutics).

10.2 FABRICATION OF THE SEMICONDUCTING/MAGNETIC NANOCOMPOSITE

Nanostructured systems play a key role in today's nanoscience research, especially due to the new arising physical properties when a structure becomes smaller than a characteristic length scale. The effects of confinement and proximity affect the interdependence between physical characteristics and the structure size of the patterned materials. The fabrication of nanopatterned materials is widespread in many realms such as physics, chemistry, material science, and biology. The fabrication of nanostructured systems can be carried out by prestructuring in using lithography or focused ion beam methods or by self-organization which is a suitable technique due to low costs and saving of time. On the one hand, assembling of particles by self-organization [66] and, on the other hand, the formation of porous materials is intensely used [67].

Since the morphology of porous silicon can be tuned, it is an eligible system to be used as template for the incorporation of another material to produce a nanocomposite. In the following section, the fabrication of a semiconducting/magnetic composite system is discussed.

10.2.1 FABRICATION OF POROUS SILICON WITH QUASI-REGULAR PORE ARRANGEMENT

To use porous silicon as template for magnetic materials and for the exploitation of the arising magnetic properties, it is favorable to achieve ordered pore arrangements. The used templates in the meso-/macro-porous regime are produced by anodization of a highly n-doped silicon wafer in a 10 wt% hydrofluoric acid solution. The anodization has been performed under electrical breakdown conditions [68]. Average pore diameters between 30 and 100 nm have been produced by applying current densities between 50 and 125 mAcm^{-2}. The distances between the pores are formed concomitant to the pore diameter. In Table 10.1, the average pore diameters, the associated mean pore distances, and the applied current densities are listed. Figure 10.4 [69] shows a composed sketch of top view and cross-sectional scanning electron micrographs (SEMs) of a sample with an average pore diameter of 60 nm.

When the applied current density decreases, the pore diameters also decrease, whereas the distance between the pores is increased. Concerning the regularity of the arising morphology, one can say that larger pore diameters offer a better regularity than smaller ones, which is related to the formation process of the porous structures in this regime. The precondition for smooth pore growth is that the distance between the pores should not exceed twice the thickness of the space charge region [70]. By decreasing the pore diameter, it is found that the pore distance increases, and thus this condition is not fulfilled anymore which enables side pore growth due to the presence of free charge carriers between the pores. This increased dendritic pore growth results in a higher irregularity of the pore arrangements. Figure 10.5 (a–d) shows top view SEMs as well as the corresponding cross-sectional images of a sample with 100 nm pore diameter in average and a sample with a pore diameter of about 30 nm.

A further possibility to influence the morphology of the resulting porous structure is the modification of the electrolyte concentration, where the current density is kept constant. The pore diameter decreases and the pore distance increases, resulting in lower porosity with an increase of the HF concentration [71,72].

Table 10.1 Relationship between pore diameters and distances between the pores as well as the associated porosities of the porous structure. Furthermore, the applied current densities are added. All morphologies have been prepared in an aqueous 10 wt% hydrofluoric acid solution and using an n$^+$ Si wafer

AVERAGE PORE DIAMETER (nm)	AVERAGE PORE DISTANCE (nm)	POROSITY (%)	APPLIED CURRENT DENSITY (mAcm^{-2})
30	60	10	50
50	55	20	70
60	50	40	80
80	40	70	100
100	30	80	125

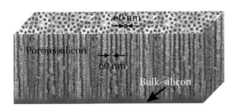

Figure 10.4 Composition of top view and cross-sectional SEM images demonstrating the appearance of a typical porous silicon sample with a quasi-regular pore arrangement. (From Rumpf, K., et al., *Phys. Status Solidi C*, 6, 1592–1595, 2009. With permission.)

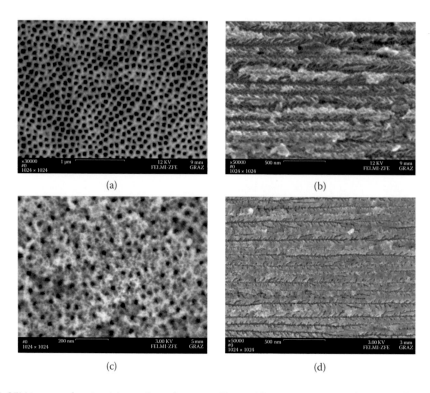

Figure 10.5 SEM images showing (a) top view of porous silicon with an average pore diameter of 100 nm, (b) the corresponding cross section offering a moderate dendritic pore growth, (c) top view of a morphology offering an average pore diameter of 30 nm, and (d) corresponding cross section with increased side pore growth.

By increasing the HF concentration from 10 wt% to 15 wt%, and by applying a current density of 75 mAcm^{-2} and using an n$^+$ silicon substrate, the pore diameter decreases from 55 to 25 nm. This behavior can be explained due to the fact that the HF concentration c is related to the critical current density J_{PS} which is present at the pore tips [70]:

$$J_{PS} = Cc^{3/2} \exp\left(-E_a/kT\right) \tag{10.1}$$

where C = 3300 A/cm^2

c ... in wt % HF

E_a = 0.345 eV (activation energy)

k ... Boltzmann's constant

T ... temperature.

Since the critical current density J_{PS} is indirectly proportional to the pore diameter

$$d = i\left(\frac{J}{J_{PS}}\right)^{1/2} \tag{10.2}$$

an increase in the HF concentration results in a decrease in the porosity.

To influence the dendritic pore growth of mesopores, an experimental sophisticated etching method has been employed by the group of Koshida [73], showing a decrease in the side pore growth. An external magnetic field up to 8 T has been applied during the anodization, which ensures an effective confinement of electronic holes at the pore tips. Due to the high magnetic field, a correspondence between the pore-tip radius and the cyclotron radius confining the motion of the holes is assumed [74], which improves the smoothness and thus the uniformity of the pores.

10.2.2 INCORPORATION OF METALS WITHIN THE PORES

The most common techniques to deposit a metal within the pores of porous silicon are electrodeposition, electroless deposition, atomic layer deposition (ALD), and evaporation. Electrodeposition under cathodic conditions generally is not accompanied by oxide formation of the silicon surface, but electroless metal deposition results in oxide formation on the pores [75]. Ogata et al. investigated the deposition process of different metals into porous silicon very accurately [29,76]; nevertheless, the mechanism of cathodic metal deposition within doped semiconductor materials is not fully understood so far and therefore is still under investigation. However, it can be said that because of the high field strength at the pore tips and a simultaneous dielectric breakdown of the oxide layer, the deposition ideally starts at the pore bottom [77]. The deposition of a metal forms a Schottky barrier between silicon and electrolyte in the case the work function of the metal is higher than that of silicon for n-type and smaller for p-type silicon. For the deposition of noble metals, the metal nucleation starts at the pore bottom due to the electron transfer via valence band. In the case of depositing less noble metals, the precipitation has to be supported by illumination, which enables the reduction at the pore walls resulting in the formation of metal tubes covering the pore walls [29]. It has also been shown that Cu can be deposited as a rod within p-type macroporous silicon, whereas Ni can be deposited at the pore walls. Zinc, platinum, and silver deposition into microporous silicon has been shown [78,79] to exhibit an enhanced electrodeposition compared to mesoporous silicon.

10.2.2.1 Electrodeposition of ferromagnetic metals

In depositing magnetic materials, the arising magnetic properties of the nanocomposite strongly depend not only on the size, shape, and mutual arrangement within the pores but also on the morphology of the porous template. Therefore, the tunability of the metal deposits by adjusting the electrochemical parameters is desired to enable the fabrication of systems with specific properties. It could be shown that the elongation of Ni structures within porous silicon mainly depends on the pulse frequency of the applied current density [80] and the filling ratio is dependent on the strength of the current density. A variation of the pulse frequency from 0.025 to 0.2 Hz results in an increase of the elongation from about 60 nm (spherical particles) to a few micrometers (wires) [80] and an increase of the current density from 15 to 40 mAcm^{-2} results in an enhancement of the filling fraction from moderate to densely packed (Figure 10.6).

The kinetics of Ni particle growth within mesoporous silicon of different porosities and various thicknesses of the porous layers (between 0.5 and 4 μm) has been investigated by Michelakaki et al. [81]. The aspect ratio was in maximum 100. Furthermore, direct and pulsed electrodeposition has been compared where the authors report on similar results gained by the two methods. With increasing aspect ratio, the filling of the pores becomes more challenging. In this case of using pulsed electrodeposition, the filling down to the pore tips is facilitated because an exchange of the ions is possible without agitation of the electrolyte. In using high aspect ratio pores (~1000), filling from pore tips to the surface could be achieved, but the filling with continuous wires along the entire pore was not possible so far. The deposition of Ni, Co, and NiCo has been investigated mostly, reporting that the best tunability could be achieved [82]. Figure 10.7 shows an example of Ni particles distributed more or less homogeneously over the entire porous layer of about 30 μm.

Besides Ni and Co, also Fe has been deposited within mesoporous silicon by Gauthier et al., where the dependence of deposition on the surface chemistry of the porous silicon has been investigated [83].

A further approach is the filling of the pores with two different magnetic materials to enhance the tunability of the magnetic properties of the nanocomposite systems. In anodized alumina templates, wires consisting of stacked layers of magnetic/nonmagnetic materials have been investigated regarding the interactions between the magnetic parts. The magnetic properties of Ni/Cu segmented nanowires have been tuned by varying the length of the Ni segments between 10 and 140 nm. By increasing the length of the magnetic segments, the dominant antiferromagnetic coupling between the Ni segments changed to ferromagnetic coupling [84]. For the combined incorporation of Ni and Co structures within porous silicon, two approaches have been used. First, a silicon wafer has been porosified on both sides (front and back sides) and subsequently on one side Ni has been deposited and on the other side Co has been deposited. Second, a typical single-sided sample has been filled with Ni and Co alternatingly from one electrolyte. The double-sided sample offers magnetic properties due to the two distinct magnetic materials exhibiting a clear kink, and in the case of the single-sided sample the two magnetic materials exchange couple and thus a smooth hysteresis is observed [85].

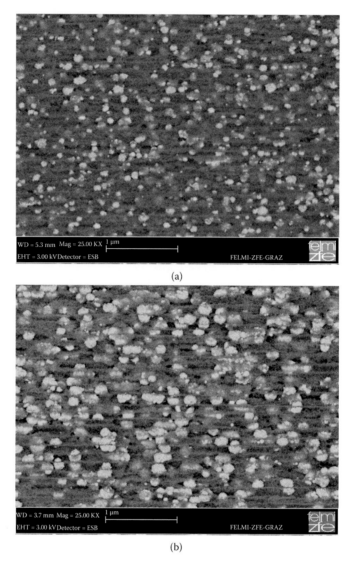

Figure 10.6 Cross-sectional SEM image of (a) a sample moderately filled with Ni particles. The applied current density j was 15 mAcm^{-2}. (b) Porous silicon with high filling of Ni particles (j = 40 mAcm^{-2}).

10.2.2.2 Infiltration of synthesized iron oxide nanoparticles

Besides the chemical deposition of magnetic materials into porous silicon, one further possibility is the infiltration of readily synthesized magnetic particles. The used particles are Fe_3O_4 particles in a size range between 4 and 10 nm, which are in hexane solution and all of them are coated with oleic acid to avoid agglomeration and for stabilization. The thickness of the coating is about 2 nm, which excludes magnetic exchange interactions between them. Figure 10.8 shows a transmission electron microscopy (TEM) image of 8 nm Fe_3O_4 particles coated with oleic acid.

The infiltration of the particles into the porous silicon depends on the pore to particle ratio and also on the concentration of the particle solution. The infiltration of bigger particles (8 and 10 nm) into pores between 50 and 60 nm has to be performed under the application of an external magnetic field of 1 T. Smaller particles (4 and 5 nm) can be infiltrated without the help of a magnetic field [86]. Considering the infiltration into the porous layer depends on the concentration of the particle solution in taking an infiltration time of 20 min, one sees that the smaller particles (4 and 5 nm) are loaded down to the pore tips within this time frame, whereas, with increasing particle size, the filling fraction of the pores decreases. Changing the concentration

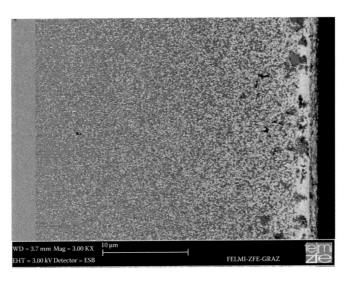

Figure 10.7 Cross-sectional SEM image showing Ni particles distributed over the whole porous layer of about 30 µm. Only a slight decrease of the packing density toward the pore tips (left) is present.

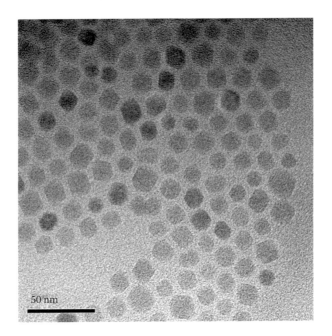

Figure 10.8 TEM image of Fe_3O_4 nanoparticles of 8 nm in size.

of the particle solution by diluting with hexane, an increase in the infiltration depth with decreasing concentration can be observed [86]. Figure 10.9 shows energy-dispersive X-ray spectra (EDX) at the pore tips comparing two samples infiltrated with different concentrations of the solution of 8 nm Fe_3O_4 nanoparticles.

10.3 MAGNETIC PROPERTIES OF THE NANOCOMPOSITE SYSTEMS

In this section, the magnetic behavior of the discussed nanocomposite systems will be reviewed. In general, the magnetic properties depend on the type of deposited metal, size and shape of the metal deposits, the filling fraction and also the mutual arrangement of the magnetic structures within the pores. A further

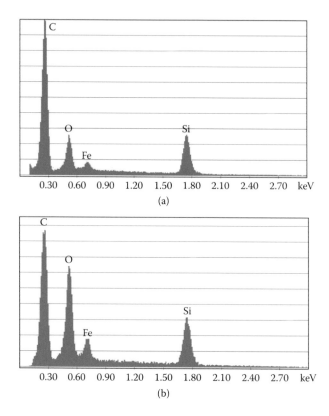

Figure 10.9 EDX spectra at the pore tips of porous silicon with infiltrated 8 nm Fe_3O_4 nanoparticles. (a) High concentration of particle solution (8 mL hexane) and (b) low concentration (2 mL hexane).

crucial parameter is the morphology of the porous silicon template which mainly influences the magnetic coupling between structures of adjacent pores. Therefore, an important issue for the magnetic behavior of the nanocomposite systems is a quasi-regular pore arrangement and to control the dendritic pore growth.

10.3.1 TUNING OF THE HYSTERESIS BY THE SHAPE OF THE ENCAPSULATED MAGNETIC STRUCTURES

To create nanocomposites with a specific hysteresis behavior, electrodeposition has been used to fill the pores selectively with Ni and Co nanostructures. Depending on the elongation of the metal deposits, the magnetic characteristics such as coercivity and remanence can be modified. The coercivity as well as the remanence of the nanocomposite decreases with increasing elongation of the deposited nanostructures [87]. Figure 10.10 shows the dependence of the coercivity on the elongation of the deposits. For these investigated samples, the average pore diameter is 80 nm and the mean distance between the pores is 40 nm, which means that the dipolar magnetic coupling between structures of adjacent pores has still to be taken into account.

Magnetic anisotropy between easy axis (magnetic field applied parallel to the pores) and hard axis (magnetic field applied perpendicular to the pores) magnetization could be observed for elongated deposits and also for densely packed particles because they magnetically couple within the pores [87]. Codeposition can be used to keep the particle size roughly constant within the pores, whereas a variation of the filling fraction can be used to modify the magnetic behavior of the nanocomposites [88]. In Figure 10.11, Codeposits of a mean size of 80 nm within porous silicon with different filling fractions can be seen. In the case of a low filling fraction, the magnetic coupling between nanoparticles within the pores can be neglected and the resulting coercivity offers a value of 810 Oe. An increase of the filling fraction from low to densely packed leads to a decrease of the coercivity due to increasing magnetic interactions between the particles to 580 Oe and further to 420 Oe. This behavior shows that a densely packed filling offers a magnetic behavior of needle-like structures.

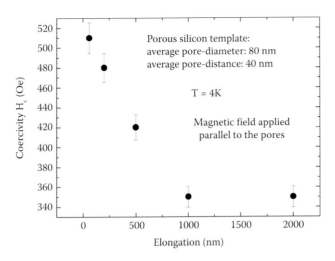

Figure 10.10 Dependence of coercivity of porous silicon samples with equal morphology filled with Ni nanostructures on the elongation of the Ni deposits. The coercivity increases with decreasing particle elongation. The filling factor of the porous silicon template is moderate, which allows to neglect the magnetic coupling between structures within one pore (mean distance between structures within an individual pore >200 nm).

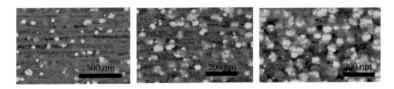

Figure 10.11 Different filling fractions of Co particles within porous silicon templates with the same morphology.

10.3.1.1 Influence of the metal filling factor of the pores

The filling fraction, packing density, and the spatial distribution of the deposits within the pores can be tuned by the electrochemical parameters, mainly the pulse frequency and the applied current density [89]. Mostly favorable is a more or less homogeneous distribution of the metal structures over the entire porous layer. In choosing the structure size by the applied pulse frequency and the packing density by the current density, the magnetic response of the nanocomposite can be tailored. In the case of Ni deposition, a pulse frequency of 0.025 Hz, a current density of 25 mAcm^{-2}, and a deposition time of 15 min sphere-like particles of 60 nm (corresponding to the pore diameter) with a moderate filling are achieved. The corresponding hysteresis offers a coercivity of 520 Oe for easy axis magnetization and 510 Oe for hard axis magnetization, which indicates that dipolar coupling between adjacent pores can be neglected. In varying the pulse frequency to 0.2 Hz and keeping the other parameters equal, wire-like structures of more than 1 μm are deposited. In this case, the coercivity is reduced to 350 Oe for easy axis magnetization and to 220 Oe for hard axis magnetization, which indicates shape anisotropy due to the wire-like structures. Dipolar coupling should be negligible because of the use of the same template morphology as for the spherical particles. By choosing a pulse frequency of 0.025 Hz and a current density of 40 mAcm^{-2} to achieve a high packing density of the spherical particles, the coercivity is reduced to 390 Oe for easy axis magnetization and to 280 Oe for hard axis magnetization, which shows that in this case the particles inside the pores magnetically couple and therefore the nanocomposite offers a wire-like behavior [82].

10.3.2 MAGNETIC INTERACTIONS BETWEEN DEPOSITED NANOSTRUCTURES

Magnetic coupling between deposited nanostructures can occur in two ways. First, metal deposits within the pores can interact and thus result in wire-like behavior. Second, metal deposits of adjacent pores can interact if the pore distance is small enough (in the range of the pore diameter), which mainly influences the magnetic anisotropy behavior between easy axis and hard axis magnetization.

10.3.2.1 Dipolar coupling of metal wires of adjacent pores

Considering wire-like Ni structures, with an aspect ratio greater than 50, magnetic coupling between structures of adjacent pores strongly influences these features in the case of small pore distances. If the distance between the pores is in the range of the pore diameters, dipolar interaction plays a significant role. The magnetic coupling between structures of neighboring pores also increases with increasing length of the side pores, since the metal deposits correspond to the surface morphology of the pores. As a result, the coercivity of the nanocomposite decreases in using templates with increasing dendritic growth. In Table 10.2, the morphological data and the corresponding coercivities are summarized. Due to this behavior, the modification of the magnetic coupling between nanostructures by different morphologies of the templates enables the tuning of the hysteresis curves of the nanocomposites. Weaker magnetic coupling occurs when the side pore length decreases because of an enhanced effective pore distance and leads to an increase of the coercivity of nanowires from 270 to 550 Oe [90]. Furthermore, the magnetic anisotropy between easy axis (magnetic field applied parallel to the pores) and hard axis (magnetic field perpendicular to the pores) increases with decreasing dendritic pore growth. With negligible magnetic crosstalk between the deposited magnetic nanowires, the magnetic behavior approximates to the one of single nanowires [91].

10.3.2.2 Dipolar coupling of nanoparticles within the pores

Magnetic nanoparticles arranged densely packed within the pores give rise to dipolar coupling between them. The coupling depends on the distance between the particles and also on the particle size. Considering readily synthesized iron oxide nanoparticles, they can also be used to tune the magnetic properties of the composite system by varying the nanoparticle size and the magnetic interactions between them within the pores, respectively. The iron oxide particles are coated with an organic shell of about 2 nm of thickness, and thus magnetic exchange coupling between the particles can be excluded and only dipolar particle interaction has to be taken into account in the case of a densely packed filling of the pores. Since the iron oxide particles are superparamagnetic [92] due to their size, magnetic coupling between them shifts the blocking temperature of the nanocomposite to higher temperatures. So called blocking temperature T_B indicates the transition temperature between superparamagnetic behavior and blocked state of the system. To figure out the blocking temperature of the system, zero-field-cooled (ZFC)/field-cooled (FC) measurements have been performed. For this reason, the samples have been heated up to 300 K and subsequently cooled down to 4 K in temperature steps of 5 K. Then a magnetic field of 500 Oe has been applied and the magnetization has been measured up to 300 K and down to 4 K in temperature steps of 2 K. The peak present in the ZFC curve indicates the blocking temperature. The blocking temperature can be shifted on the one hand due to the particle size [93] and on the other hand due to magnetic coupling between the particles. For smaller particles (4 and 5 nm), the blocking temperature is mainly determined by the particle size and magnetic coupling is negligible. The blocking temperature is 12 K for 4 nm and 15 K for 5 nm particles. For bigger particles of 8 and 10 nm size, T_B is shifted to higher temperatures of 160 and 170 K, respectively. These high temperatures are not only caused by the bigger particle size but also by dipolar coupling between the particles [94]. A further parameter to tune T_B is the variation of the filling fraction of the pores from low to densely packed, which results for

Table 10.2 **Summary of the average values of the morphological data (pore diameter, pore distance, and side-pore length) of the porous silicon templates and the corresponding coercivities. All samples are filled with wire-like Ni structures (aspect ratio > 40) and a moderate filling fraction**

PORE DIAMETER (nm)	PORE DISTANCE (nm)	SIDE-PORE LENGTH (nm)	COERCIVITY H_C (Oe)
100	30	15	550
80	40	20	350
60	50	30	320
50	55	40	285
30	60	50	270

8 nm particles in a shift of T_B from 50 to 160 K [95]. Dipolar coupling between particles does not occur in the case of the smaller particles because of their smaller magnetic moment and the same thickness of the coating (2 nm) as for the bigger ones. Due to the superparamagnetic behavior, the hysteresis curves of all samples show a negligible remanence at room temperature. The hysteresis measured at 4 K shows coercivities between 380 Oe for 8 nm particles, 200 Oe for 5 nm particles, and 120 Oe for 4 nm particles [94].

This nanocomposite system consisting of porous silicon and iron oxide nanoparticles, which offers biocompatibility, is used in the form of microparticles. The nanocomposite microparticles are prepared after the infiltration process of the particles into the pores. For this purpose, the porous layer which offers typically a thickness of about 35 μm is removed mechanically from the substrate and after milling and filtering particles in the range of 2 μm in size are achieved. The magnetic properties of samples on the substrate and samples removed from the substrate are the same.

10.4 SUMMARY

In this chapter, the versatility of porous silicon and its usability as template material, especially for the incorporation of magnetic materials, have been discussed. Briefly, the formation processes and the various morphologies are addressed. For the use as template material, the porous silicon is fabricated by self-organization; nevertheless, the porous structure and porosity are accurately tunable in a broad range. For metal filling, mesoporous and macroporous templates are mainly used. In the case of encapsulation of magnetic nanostructures, pore diameters between 30 and 100 nm are investigated due to the arising magnetic properties. Besides electrodeposition of metals, also the infiltration of synthesized iron oxide nanoparticles has been discussed. The magnetic properties of the nanocomposite systems can be adjusted, on the one hand, by the size, shape, and arrangement of the metal structures and, on the other hand, by the morphology of the porous silicon. In considering one template morphology, the coercivity depends on the elongation of the deposit. By varying the template morphology and keeping the shape of the deposits equal, it was observed that the magnetic properties are dependent on the distance between the pores and thus on the inter-pore magnetic coupling. The discussed systems enable the combination of semiconducting and ferromagnetic materials, which is of interest for on-chip applications with integration in microtechnological processes. In encapsulating superparamagnetic iron oxide nanoparticles, a biocompatible system arises, which is a promising candidate for biomedical applications.

ACKNOWLEDGMENTS

The authors thank the Institute for Electron Microscopy and Fine Structure Research, Graz University of Technology, Graz, Austria, especially Dr. P. Pölt for SEM and Dr. M. Albu for TEM investigations of the samples.

REFERENCES

1. Uhlir, A. Jr., Electrolytic shaping of germanium and silicon, *Bell Syst. Tech. J.*, 1956, 35, 333–347.
2. Turner, R., Electropolishing silicon in hydrofluoric acid solution, *J. Electrochem. Soc.*, 1958, 105, 402–408.
3. Canham, L.T., Silicon quantum wire array fabrication by electrochemical and chemical dissolution of wafers, *Appl. Phys. Lett.*, 1990, 57, 1046–1048.
4. Lehmann, V., Gösele, U., Porous silicon formation: A quantum wire effect, *Appl. Phys. Lett.*, 1991, 58, 856.
5. Koshida, N., Koyama, H., Visible electroluminescence from porous silicon, *Appl. Phys. Lett.*, 1992, 60, 347.
6. Jiang, D.T., Coulthard, I., Sham, T.K., Lorimer, J.W., Frigo, S.P., Feng, X.H., Rosenberg, R.A., Observations on the surface and bulk luminescence of porous silicon, *J. Appl. Phys.*, 1993, 74, 6335–6340.
7. Koch, F., Petrova-Koch, V., The surface state mechanism for light emission from porous silicon. In *Porous Silicon*, Feng, Z.C., Tsu, R., Eds. World Scientific, Singapore, 1994.
8. Tamura, H., Rückschloss, M., Wirschem, Th., Veprek, S., Origin of the green/blue luminescence from nanocrystalline silicon, *Appl. Phys. Lett.*, 1994, 65, 1537.
9. Lehmann, V., Grüning, U., The limits of macropore array fabrication, *Thin Solid Films*, 1997, 297, 13–17.
10. Lehmann, V., Föll, H., Formation mechanism and properties of electrochemically etched trenches in n-type silicon, *J. Electrochem. Soc.*, 1990, 137, 653–659.

11. Föll, H., Christophersen, M., Carstensen, J., Hasse, G., Formation and application of porous silicon, *Mater. Sci. Eng.*, 2002, R39, 93–141.

12. Kochergin, V., Föll, H., Eds., *Porous Semiconductors, Optical Properties and Applications*, Springer, London, 2009.

13. Wehrspohn, R.B., Linear and non-linear optical experiments based on macroporous silicon photonic crystals, In *Nanophotonic Materials, Photonic Crystals, Plasmonics and Metamaterials*, Wehrspohn, R.B., Kitzerow, H.-S., Busch, K., Eds. Wiley-VCH, Weinheim, Germany, 2008.

14. Perova, T.S., Tolmachev, V.A., Astrova, E.V., Zharova, Y.A., O'Neill, S.M. Tunable one-dimensional photonic crystal structures based on grooved Si infiltrated with liquid crystal E7. *Phys. Status Solidi C*, 2007, 4, 1961–1965.

15. Boarino, L., Baratto, C., Geobaldo, F., Amato, G., Comini, E., Rossi, A.M., Faglia, G., Lerondel, G., Sberveglieri, G., NO_2 monitoring at room temperature by a porous silicon gas sensor. *Mater. Sci. Eng. B*, 2000, 69–70, 210–214.

16. Levitsky, I.A., Porous silicon structures as optical gas sensors, *Sensors*, 2015, 15, 19963–19991.

17. Umann, K., Walter, J.G., Schepert, Th., Segal, E., Label-free optical biosensor based on aptamer-functionalized porous silicon scaffolds, *Anal. Chem.*, 2015, 87, 1999–2006.

18. Mariani, S., Strambini, L.M., Barillaro, G., Femtomole detection of proteins using a label-free nanostructured porous silicon interferometer for perspective ultrasensitive biosensing, *Anal. Chem.*, 2016, 88, 8502–8509.

19. Canham, L.T., Bioactive silicon structure fabrication through nanoetching techniques, *Adv. Mater.*, 1995, 7, 1033–1037.

20. Santos H.A., Porous silicon for biomedical applications, In *Biomaterials*, ed., H.A. Santos, Woodhead Publishing Series, Cambridge, UK, 2014.

21. Henstock, J.R., Canham, L.T., Anderson, S.I., Silicon: The evolution of its use in biomaterials, *Acta Biomater.*, 2015, 11, 17–26.

22. Alvarez, S.D., Derfus, A.M., Schwartz, M.P., Bhatia, S.N., Sailor, M.J., The compatibility of hepatocytes with chemically modified porous silicon with reference to in vitro biosensors, *Biomaterials*, 2008, 30, 26–34.

23. Johansson, F., Kanje, M., Linsmeier, C.E., Wallman, L., The influence of porous silicon on axonal outgrowth in vitro, *IEEE Trans. Biomed. Eng.*, 2008, 55, 1447–1449.

24. Whitehead, M.A., Fan, D., Mukherjee, P., Akkaraju, G.R., Canham, L.T., Coffer J.L., High-porosity poly(epsilon-caprolactone)/mesoporous silicon scaffolds: Calcium phosphate deposition and biological response to bone precursor cells, *Tissue Eng. Part A*, 2008, 14, 195–206.

25. Shabir, Q., Skaria, C., O'Brien, H., Loni, A., Barnett C., Canham, L.T., Taste and mouthfeel assessment of porous and non-porous silicon microparticles, *NRL*, 2012, 7, 407.

26. Buriak, J.M., High surface area silicon materials: Fundamentals and new technology, *Philos. Trans. Royal Soc. A Math. Phys. Eng. Sci.*, 2006, 15, 217–225.

27. Herino, R., Impregnation of porous silicon, In *Properties of Porous Silicon*, Canham, L., Ed. INSPEC, London, UK, 1997.

28. Ronkel, F., Schultze, J.W., Arens-Fischer, R., Electrical contact to porous silicon by electrodeposition of iron, *Thin Solid Films*, 1996, 276, 40–43.

29. Ogata, Y.H., Kobayashi, K., Motoyama, M., Electrochemical metal deposition on silicon. *Curr. Opin. Solid State Mater. Sci.*, 2006, 10, 163–172.

30. Koyama, A., Fukami, K., Sakka, T., Abe, T., Kitada, A., Murase, K., Kinoshita, M., Penetration of platinum complex anions into porous silicon: Anomalous behavior caused by surface-induced phase transition, *J. Phys. Chem. C*, 2015, 119, 19105–19116.

31. Darwich, W., Garron, A., Bockowski, P., Santini, C., Gaillard, F., Haumesser, P.H., Impact of surface chemistry on copper deposition in mesoporous silicon, *Langmuir*, 2016, 32, 7452–7458.

32. Zacharatos, F., Nassiopoulou, A.G., Copper-filled macroporous Si and cavity underneath for microchannel heat sink technology, *Phys. Status Solidi A Appl. Res.*, 2008, 205, 2513–2517.

33. Gusev, S.A., Korotkova, N.A., Rozenstein, D.B., Fraerman, A.A., Ferromagnetic filaments fabricated in porous Si matrix, *J. Appl. Phys.*, 1994, 76, 6671.

34. Dolgyi, A., Bandarenka, H., Prischepa, S., Yanushkevich, K., Nenzi, P., Balucani, M., Bondarenko, V., Electrochemical deposition of Ni into mesoporous silicon electrodeposition of metals on restricted substrates, *ECS Trans.*, 2012, 41, 111.

35. Aravamudhan, S., Luongo, K., Poddar, P., Srikanth, H., Bhansali, S., Porous silicon templates for electrodeposition of nanostructures, *Appl. Phys. A*, 2007, 87, 773.

36. Granitzer, P., Rumpf, K., Pölt, P., Reichmann, A., Krenn, H., Self-assembled mesoporous silicon in the crossover between irregular and regular arrangement applicable for Ni filling, *Phys. E*, 2007, 38, 205.

37. Granitzer, P., Rumpf, K., Pölt, P., Plank, H., Albu, M., Ferromagnetic nanostructure arrays self-assembled in mesoporous silicon, *J. Phys. Conf. Ser.*, 2010, 7, 72037–72042.

38. Zhang, X., Tu, K.N., Preparation of hierarchically porous nickel from macroporous silicon, *J. Am. Chem. Soc.*, 2006, 29, 15036–15037.

39. Park, J.-H., Derfus, A.M., Segal, E., Vecchio, K.S., Bhatia, S.N., Sailor, M.J., Local heating of descrete droplets using magnetic porous silicon-based photonic crystals, *J. Am. Chem. Soc.*, 2006, 128, 7938–7946.

40. Lundquist, C.M., Loo, C., Meraz, I.M., Cerda, J.D., Liu, X., Serda, R.E., Characterization of free and porous silicon-encapsulated superparamagnetic iron oxide nanoparticles as platforms for the development of theranostic vaccines, *Med. Sci.*, 2014, 2, 51–69.

41. Granitzer, P., Rumpf, K., Venkatesan, M., Roca, A.G., Cabrera, L., Morales, M.P., Poelt, P., Albu, M., Magnetic study of Fe3O4-nanoparticles within mesoporous silicon, *J. Electrochem. Soc.*, 2010, 157, K145.

42. Chazalviel, J.-N., The silicon/electrolyte interface, In *Porous Silicon Science and Technology*, Vial, J.-C., Derrien, J., Eds. Springer, New York, USA, 1995, Winter School Les Houches 8–12 February, Les Editions de Physique.

43. Tserepi, A., Tsamis, C., Gogolides, E., Nassiopoulou, A.G., Dry etching of porous silicon in high density plasmas, *Phys. Status Solidi A*, 2003, 197, 163–167.

44. Laiho, R., Pavlov, A., Preparation of porous silicon films by laser ablation, *Thin Solid Films*, 1995, 255, 9–11.

45. Lehmann, V., Stengl, R., Luigart, A., On the morphology and the electrochemical formation mechanism of mesoporous silicon, *Mater. Sci. Eng. B*, 2000, 69/70, 11–22.

46. Kleinmann, P., Badel, X., Linnros, J., Toward the formation of three-dimensional nanostructures by electrochemical etching of silicon, *Appl. Phys. Lett.*, 2005, 86, 183108.

47. Bisi, O., Ossicini, S., Pavesi, L., Porous silicon: A quantum sponge structure for silicon based optoelectronics, *Surf. Sci. Rep.*, 2000, 38, 1–126.

48. Lehmann, V., Grüning, U., The limits of macropore array fabrication, *Thin Solid Films*, 1997, 297, 13–17.

49. Ouyang, H., Christophersen, M., Fauchet, P.M., Enhanced control of porous silicon morphology from macropore to mesopore formation, *Phys. Status Solidi A*, 2005, 202, 1396–1401.

50. Parkhutik, V., Porous silicon—Mechanisms of growth and applications, *Solid State Electron.*, 1999, 43, 1121–1141.

51. Cozzi, C., Polito, G., Strambini, L.M., Barillaro, G., Electrochemical preparation of in-silicon hierarchical networks of regular out-of-plane macropores interconnected by secondary in-plane pores through controlled inhibition of breakdown effects, *Electrochim. Acta*, 2016, 187, 552–559.

52. Han, H., Huang, Z., Lee, W., Metal-assisted chemical etching of silicon and nanotechnology applications, *Nano Today*, 2014, 9, 271–304.

53. McSweeney, W., Geaney, H., O'Dwyer, C., Metal assisted chemical etching of silicon and the behavior of nanoscale silicon materials as Li-ion battery anodes, *Nano Res.*, 2015, 8, 1395–1442.

54. Cao, Y., Zhou, Y., Liu, F., Zhou, Y., Zhang, Y., Liu, Y., Guo, Y., Progress and mechanism of Cu assisted chemical etching of silicon in a low Cu^{2+} concentration region, *ECS J. Solid State Technol.*, 2015, 4, P331–P336.

55. Huang, Z., Geyer, N., Werner, P., de Boor, J., Gösele, U., Metal assisted chemical etching of silicon: a review, *Adv. Mater.*, 2011, 23, 285–308.

56. Fuller, C.S., Ditzenberger, J.A., Diffusion of donor and acceptor elements in silicon, *J. Appl. Phys.*, 1957, 27, 544–553.

57. Archer, R.J., Stain film on silicon, *J. Phys. Chem. Solids*, 1960, 14, 104–110.

58. Kolasinski, K.W., Barclay, W.B., The stoichiometry of Si electroless etching in V_2O_5 + HF solutions, *Angew. Chem. Int. Ed. Engl.*, 2013, 9, 6731–6734.

59. Kolasinski, K.W., Gogola, J.W., Barclay, W.B., A test of Marcus theory predictions for electroless etching of silicon, *J. Phys. Chem. C*, 2012, 116, 21472–21481.

60. Kolasinski, K.W., The mechanism of galvanic/metal-assisted etching of silicon, *Nanoscale Res. Lett.*, 2014, 9, 432.

61. Masuda, H., Fukuda, K., Ordered metal nanohole arrays made by a two-step replication of honeycomb structures of anodic alumina, *Science*, 1995, 268, 1466–1468.

62. Masuda, H., Hasegawa, F., Ono, S., Self-ordering of cell arrangement of anodic porous alumina formed in sulphuric acid solution, *J. Electrochem. Soc.*, 1997, 144, L127–L130.

63. Pirota, K.R., Navas, D., Hernandez-Velez, M., Nielsch, K., Vazquez, M., Novel magnetic materials prepared by electrodeposition techniques: Arrays of nanowires and multi-layered microwires, *J. Alloys Compd.*, 2004, 369, 18–26.

64. Reyes, D., Biziere, N., Warot-Fonrose, B., Wade, T., Gatel, C., Magnetic configurations in Co/Cu multilayered nanowires: Evidence of structural and magnetic interplay, *Nano Lett.*, 2016, 16, 1230–1236.

65. Gerngross, M.D., Carstensen, J., Föll, H., Electrochemical and galvanic fabrication of a magnetoelectric composite sensor based on InP, *Nanoscale Res. Lett.*, 2012, 7, 379.

66. Lantiat, D., Toudert, J., Babonneau, D., Camelio, S., Tromas, C., Simonot, L., Self-organization and optical response of silver nanoparticles dispersed in a dielectric matrix, *Rev. Adv. Mater. Sci.*, 2007, 15, 150–157.

67. Kompan, M.E., Gorodetski, A.E., Tarasova, I.L., Self formation of porous silicon structure: Primary microscopic mechanism of pore separation, *Solid State Phenom.*, 2004, 97–98, 181–184.

68. Granitzer, P., Rumpf, K., Poelt, P., Albu, M., Chernev, B., The interior interfaces of a semiconductor/metal nanocomposite and their influence on its physical properties, *Phys. Status Solidi C*, 2009, 6, 2222–2227.

69. Rumpf, K., Granitzer, P., Pölt, P., Krenn, H., Transition metals specifically electrodeposited into porous silicon, *Phys. Status Solidi C*, 2009, 6, 1592–1595.

70. Lehmann, V., *Electrochemistry of Silicon, Instrumentation, Science, Materials and Applications*, Wiley-VCH, Weinheim, Germany, 2002.

71. Halimaoui, A., Porous silicon: Material processing, properties and applications, In *Porous Silicon Science and Technology*, Vial, J.C., Derrien, J., Eds. Springer, NY, 1995, Winter School Les Houches, 8–12 February 1994, Les Editions de Physique.

72. Nakagawa, T., Sugiyama, H., Koshida, N., Fabrication of periodic Si nanostructure by controlled anodization, *Jpn. J. Appl. Phys.*, 1998, 37, 7186–7189.

73. Hippo, D., Nakamine, Y., Urakawa, K., Tsuchiya, Y., Mizuta, H., Koshida, N., Oda, S., Formation mechanism of 100-nm scale periodic structures in silicon using magnetic-field-assisted anodization, *Jpn. J. Appl. Phys.*, 2008, 47, 7398–7402.

74. Gelloz, B., Masunaga, M., Shirasawa, T., Mentek, R., Ohta, T., Koshida, N., Enhanced controllability of periodic silicon nanostructures by magnetic field anodization, *ECS Trans.*, 2008, 16, 195–200.

75. Jeske, M., Schultze, J.W., Thönissen, M., Münder, H., Electrodeposition of metals into porous silicon, *Thin Solid Films*, 1995, 255, 63–66.

76. Fukami, K., Kobayashi, K., Matsumoto, T., Kawamura, Y.L., Sakka, T., Ogata, Y.H., Electrodeposition of noble metals into ordered macropores in p-type silicon, *J. Electrochem. Soc.*, 2008, 155, D443–D448.

77. Fang, C., Foca, E., Xu, S., Carstensen, J., Föll, H., Deep silicon macropores filled with Copper by electrodeposition, *J. Electrochem. Soc.*, 2007, 154, D45–D49.

78. Koda, R., Fukami, K., Sakka, T., Ogata, Y.H., A physical mechanism for suppression of zinc dendrites caused by high efficiency of the electrodeposition within confined nanopores, *ECS Electrochem. Lett.*, 2013, 2, D9–D11.

79. Koda, R., Fukami, K., Sakka, T., Ogata, Y.H., Electrodeposition of platinum and silver into chemically modified microporous silicon electrodes, *Nanoscale Res. Lett.*, 2012, 7, 330.

80. Granitzer, P., Rumpf, K., Koshida, N., Poelt, P., Michor, H., Electrodeposited metal nanotube/nanowire arrays in mesoporous silicon and their morphology dependent magnetic properties, *ECS Trans.*, 2014, 58, 139–144.

81. Michelakaki, E., Valalki, K., Nassiopoulou, A.G., Mesoscopic Ni particles and nanowires by pulsed electrodeposition into porous Si, *J. Nanopart. Res.*, 2013, 15, 1499.

82. Rumpf, K., Granitzer, P., Albu, M., Pölt, P., Electrochemically fabricated silicon/metal hybrid nanosystem with tailored magnetic properties, *Electrochem. Solid State Lett.*, 2010, 13, K15.

83. Bardet, B., Defforge, T., Negulescu, B., Valente, D., Billoue, J., Poveda, P., Gautier, G., Shape-controlled electrochemical synthesis of mesoporous Si/Fe nanocomposites with tailored ferromagnetic properties, *Mater. Chem. Front.*, 2017, 1, 190–196. DOI: 10.1039/C6QM00040A.

84. Susano, M., Proenca, M.P., Moraes, S., Sousa, C.T., Araújo, J.P., Tuning the magnetic properties of multisegmented Ni/Cu electrodeposited nanowires with controllable Ni lengths, *Nanotechnology*, 2016, 27, 335301.

85. Rumpf, K., Granitzer, P., Michor, H., Porous silicon nanocomposites with combined hard and soft magnetic properties, *Nanoscale Res. Lett.*, 2016, 11, 398.

86. Granitzer, P., Rumpf, K., Coffer, J., Pölt, P., Reissner, M., Assessment of magnetic properties of nanostructured silicon loaded with superparamagnetic iron oxide nanoparticles, *ECS J. Solid State Sci. Technol.*, 2015, 4, N44–N46.

87. Rumpf, K., Granitzer, P., Hilscher, G., Albu, M., Poelt, P., Magnetically interacting low dimensional Ni-nanostructures within porous silicon, *Microelectron. Eng.*, 2012, 90, 83.

88. Rumpf, K., Granitzer, P., Pölt, P., Influence of the electrochemical process parameters on the magnetic behaviour of a silicon/metal nanocomposite, *ECS Trans.*, 2010, 25, 157.

89. Rumpf, K., Granitzer, P., Reissner, M., Pölt, P., Albu, M., Investigation of Ni and Co deposition into porous silicon and the influence of the electrochemical parameters on the physical properties, *ECS Trans.*, 2012, 41, 59.

90. Granitzer, P., Rumpf, K., Ohta, T., Koshida, N., Poelt, P., Reissner, M., Porous silicon/Ni composites of high coercivity due to magnetic field assisted etching, *Nanoscale Res. Lett.*, 2012, 7, 384.

91. Granitzer, P., Rumpf, K., Ohta, T., Koshida, N., Reissner, M., Poelt, P., Enhanced magnetic anisotropy of Ni nanowire arrays fabricated on nanostructured silicon templates, *Appl. Phys. Lett.*, 2012, 101, 033110.

92. Coey, J.M.D., *Magnetism and Magnetic Materials*, Cambridge University Press, Cambridge, 2010.

93. Granitzer, P., Rumpf, K., Tian, Y., Akkaraju, G., Coffer, J., Poelt, P., Reissner, M., Size-dependent assessment of Fe_3O_4-nanoparticles loaded into porous silicon, *ECS Trans.*, 2013, 50, 77.

94. Granitzer, P., Rumpf, K., Tian, Y., Coffer, J., Akkaraju, G., Poelt, P., Reissner, M., Assessment of magnetic properties of nanostructured silicon loaded with superparamagnetic iron oxide nanoparticles, *ECS Trans.*, 2015, 64, 1–7.

95. Granitzer, P., Rumpf, K., Tian, Y., Akkaraju, G., Coffer, J., Poelt, P., Reissner, M., Fe_3O_4 nanoparticles within porous silicon: Magnetic and cytotoxicity characterization, *Appl. Phys. Lett.*, 2013, 102, 193110.

Heat and mass transfer in silicon-based nanostructures

Hyung Hee Cho and Beom Seok Kim

Contents

11.1 INTRODUCTION: ADVANTAGES OF NANOSCALE INTERFACES IN HEAT TRANSFER

Silicon-based nano- and nano/micro-hierarchical structures synthesized by metal-assisted chemical etching have been widely adopted due to their functional potential in various application fields such as solar energy conversion, energy storage, chemical and biological sensing techniques, interfacial static/dynamic wetting functionalization, and heat/energy transfer techniques.

In this chapter, we present an overview of the interface functionalization methods based on the synthesis/fabrication of silicon-based nanostructures by metal-assisted chemical etching, and extend their impact on boiling heat transfer, which is a powerful energy transfer scheme. First, we discuss the essential aspects of nanoscale interface manipulations and their functionality for engineering applications including heat/energy transfer. We introduce the principal factors and their fabrication procedure: metal-assisted chemical etching for silicon-based nanostructures, which is one of the representative top-down methods for fabrication, as well as a possible alternative for hierarchical structures and bulk micromachining. Recently, the metal-assisted chemical etching method has attracted increasing attention due to its advantages in synthesis and applications. Considering the simplicity and feasibility of the manipulation of silicon, morphology manipulation and sequential wettability control techniques enable its application in the heat transfer field.

In addition, we highlight the effects of interfacial manipulations on boiling heat transfer and introduce promising breakthroughs done and to be done in heat transfer technologies. Nano/micro-hierarchical structures can be very favorable for performance enhancement in boiling heat transfer by extending the interfacial area of the heat transfer between a heat-dissipating solid and a heat-absorbing liquid coolant. Furthermore, they can catalyze the heterogeneous fluidic behavior of the convective phase change. Reviewing the heat transfer processes indicates possible ways to reinforce the feasibility and functionality of silicon-based nanostructures and expand their application to advanced heat and mass transfer with more favorable interfacial peculiarities.

11.2 PREPARATION AND SYNTHESIS OF SILICON-BASED NANOSTRUCTURES

We present an overview of the synthesis and fabrication of silicon-based nano- and nano/micro-hierarchical structures by metal-assisted chemical etching, which is one of the representative top-down methods for their fabrication. The metal-assisted chemical etching method has attracted increasing attention recently due to its advantages for synthesis and its potential expandable applications, since it involves a simple and low-cost procedure with chemicals and noble-metal catalysts and can be easily used for large-area synthesis without requiring specialized/expensive equipment. In addition, it is compatible with other conventional nano- and microstructure fabrication methods and thus can be used to develop advanced fabrication techniques and novel applications.

In this section, we explain essential factors in fundamental fabrication procedures of silicon-based nano- and nano/micro-hierarchical structures, with details on how to control their shape and characteristic lengths through mask-based shadowgraph methods, and the possible application of metal-assisted chemical etching as a substitutional technique in bulk micromachining.

11.2.1 SYNTHESIS OF TOP-DOWN SILICON NANOSTRUCTURES

11.2.1.1 Fundamentals in reactions of metal-assisted chemical etching

In the metal-assisted chemical etching process, a silicon substrate is partially or selectively covered by a noble-metal layer and then immersed in a mixture of hydrofluoric acid (HF) and hydrogen peroxide (H_2O_2).

In a completely wet chemical procedure, electroless metal deposition on a silicon substrate and subsequent top-down silicon etching are the fundamental techniques used to create rough morphologies via nanoscale structures. This method is based on galvanic displacement reactions, which enable the reduction of metal ions and the subsequent oxidation of silicon atoms just under the reduced metal layer. For example, in the case of a mixture of HF and $AgNO_3$, electroless deposition of Ag occurs first on a silicon substrate exposed to the solution, since the valence band edge of silicon is higher than the energy levels of the Ag^+/Ag reduction system, as shown in Figure 11.1a (Peng et al., 2006). A series of oxidation-reduction and electrochemical etching processes involves the galvanic displacement among the substrate materials, the catalytic noble metal, and the etching solution. Figure 11.1b schematically describes a series of these reactions and the consequential syntheses of nanostructures on a silicon substrate for a mixture of HF and $AgNO_3$ (Peng et al., 2008). A series of galvanic displacement reactions can be expressed by the following reactions (Peng et al., 2006; Huang et al., 2011):

$$Ag^+ + e_{VB}^- \rightarrow Ag^0(s) \tag{11.1}$$

$$Si(s) + 2H_2O \rightarrow SiO_2 + 4H^+ + 4e_{VB}^- \tag{11.2}$$

$$SiO_2(s) + 6HF \rightarrow H_2SiF_6 + 2H_2O \tag{11.3}$$

Simultaneous electrochemical reactions involving cathodic and anodic reactions occur on a silicon substrate when the substrate is immersed in the mixture containing fluoric acid and Ag ions. In the cathode reaction, Ag ions are first electroless-deposited on the substrate (Equation 11.1). This process is

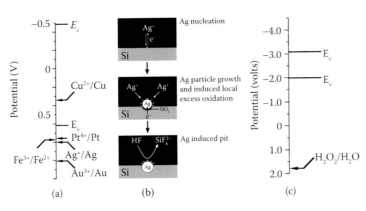

Figure 11.1 (a) Qualitative diagram of the comparison between the electrochemical electron energy levels of the Si band edges and various redox systems of metals. E_c and E_v are the conduction and valence bands, respectively. (b) Schematic of electroless Ag deposition on a silicon substrate and sequential etching of the substrate by HF/AgNO$_3$ solution. (Reproduced from Peng, K.Q., et al.: Fabrication of single-crystalline silicon nanowires by scratching a silicon surface with catalytic metal particles. *Advanced Functional Materials*. 2006. 16. 387–394. Copyright Wiley-VCH Verlag GmbH & Co. KGaA. With permission.) (c) Comparison of the silicon band edge energy and the H$_2$O$_2$/H$_2$O redox potential on the electrochemical energy scale, indicating that holes are injected deep into the valence band. (Reprinted with permission from Chattopadhyay, S., et al., (2002), In-plane control of morphology and tunable photoluminescence in porous silicon produced by metal-assisted electroless chemical etching. *Journal of Applied Physics*, 91, 6134–6140. Copyright 2002, American Institute of Physics.)

accompanied by hole injections from the solution into the silicon substrate through the reduced metal layer. In the anode reaction, silicon atoms are locally oxidized under the metal layer through electron supplies from the electron pool of the silicon substrate (Equation 11.2). The resulting oxidized portion of the silicon is then etched through the anode reaction with fluoric acid (Equation 11.3).

As an alternative to the electroless metal deposition through simultaneous reduction and deposition in a mixture of HF and a solution containing metallic ions, noble metals can be separately introduced onto the silicon substrate via a pre-process using either a solution containing metallic ions or a conventional metal deposition using the dry method. In these approaches, the HF/H$_2$O$_2$ mixture is widely used along with the equivalent etching reaction process for the top-down structuring explained above, and the reaction processes around the pre-deposited metallic layer or pattern are implemented as schematically described in Figure 11.2 (Huang et al., 2009; Peng et al., 2009; Huang et al., 2011). Herein, we can understand the practicable role of H$_2$O$_2$ in the reactions. The electrochemical potential of H$_2$O$_2$ is much more positive than the valence band of the silicon substrate and the oxidants used in silicon etching, for example,

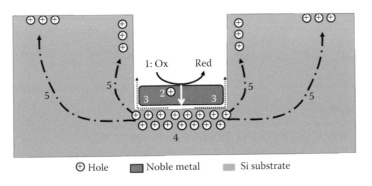

Figure 11.2 Scheme of processes involved in metal-assisted chemical etching. 1: preferential reduction of oxidant at the noble metal. 2: hole generation due to the reduction of the oxidant and sequential transfer to the Si in contact with the noble metal. 3: dissolve of Si at the Si/metal interface by HF. 4: concentration of holes at the interface. (Faster etch rate due to the noble metal rather than a bare Si surface without the metal.) 5: diffusion of holes through Si substrate. (Reproduced from Huang, Z., et al.: Metal-assisted chemical etching of silicon: A review. *Advanced Materials*. 2011: 23. 285–308. Copyright Wiley-VCH Verlag GmbH & Co. KGaA. With permission.)

HNO_3 and $FeNO_3$. Therefore, as shown in Figure 11.1c, holes in the solution (H_2O_2) can be readily injected into the silicon substrate (Chattopadhyay et al., 2002).

In light of the galvanic displacement and sequential chemical reactions, metal-assisted chemical etching of Si has been realized. As we discussed above, the electrochemical oxidation–reduction reactions occur preferentially near the noble metal, which acts as a catalytic layer for the galvanic displacement between silicon and the etching solution (Peng et al., 2002, 2003; Han et al., 2014). In detail, the noble-metal layer acts as a microscopic cathode on which the reduction of the oxidant occurs. The generated holes are then injected into the silicon substrate in contact with the noble metal. Accordingly, the silicon atoms just under the noble metal are oxidized due to the hole injection. The locally oxidized portion of the silicon is then etched away by hydrofluoric acid (HF) mixture in an etching solution. Other noble metals can be used as local electrodes in oxidation–reduction reactions by forming a catalytic layer, considering each metal's redox potential as indicated in Figure 11.1a.

11.2.1.2 Principal factors in synthesis

As metal-assisted chemical etching is a wet-solution-based method for inducing autocatalytic chemical reactions, the principal factors for the characterization of the structures' morphology are related to the reaction environments including solution temperature, concentration, etching time, and species of noble metal, which can be employed as the catalytic layer (Peng et al., 2002; Srivastava et al., 2010; Kim et al., 2011a).

Recent studies have revealed that the length of the vertically aligned silicon nanowires (SiNWs) via metal-assisted chemical etching is almost linearly correlated with etching time. In case of an HF/$AgNO_3$ solution, an approximately linear increase in the silicon nanowire's length occurs with etching time, as presented in Figure 11.3a (Cheng et al., 2008; Srivastava et al., 2010; Kim et al., 2011a). These researchers reported that the average length of the nanostructures depends purely on etching time, but the etch rate differs according to temperature (Figure 11.3b). This phenomenon is attributable to the

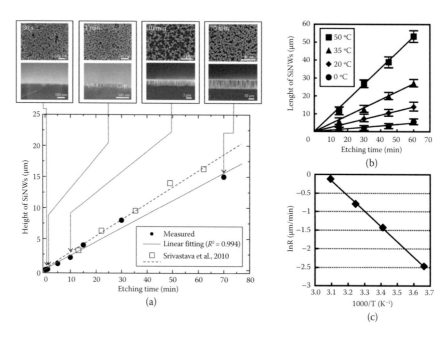

Figure 11.3 (a) Top and cross-sectional view of SEM images of silicon nanowires synthesized by electroless etching for various etching time, showing the height dependence of the silicon nanowires on etching time. (Black circles indicate average heights measured by SEM, and the red solid line represents a linear trend.) (Reprinted with permission from Kim, B.S., et al., (2011), Control of superhydrophilicity/superhydrophobicity using silicon nanowires via electroless etching method and fluorine carbon coatings, *Langmuir*, 27, 10148–10156. Copyright 2011 American Chemical Society.) (b) Length of silicon nanowires versus etching time curves at different etching temperatures. (c) Arrhenius plot of the formation rate versus reciprocal of absolute temperature. (From Cheng, S.L., et al., *J. Electrochem. Soc.*, 155, D711–D714, 2008. With permission.)

temperature-dependent diffusion of molecules in the etchant and leads to a more active circulation or a convective behavior of the etchant with temperature. Cheng et al. (2008) especially reported the apparent activation energy for the formation of silicon nanowires, given the temperature-dependent etch rate variations, which is presented in Figure 11.3c.

The role of noble metal was briefly discussed in the previous section. Without the noble metal, a series of oxidation–reduction reactions occurs, with very low silicon etch rate of up to 10 nm/h (Huang et al., 2011). For a feasible synthesis method, a noble metal should be used as a catalytic layer to expedite the chemical reaction, considering the metal's redox potential in comparison with that of silicon and the etchant solution. Due to the kinetic properties related to the reaction rate, the noble metals Ag, Au, Pt, Cu, and Pd have been extensively adopted as catalysts in the reactions between the etchant and the silicon substrate (Peng et al., 2002; Peng et al., 2005). However, the choice of noble-metal species slightly affects the morphology control and even the etch rate (Peng et al., 2005). For instance, it was reported that Au leads to a much slower etch rate than Pt (Peng et al., 2005; Peng et al., 2006). The selection of the noble metals can be made based on the electrochemical electron energy levels of the band edges of silicon and the metal's redox potential, as presented in Figure 11.1a. The difference in the catalytic activity of the noble metals is a key factor in determining the characteristics of the procedure. However, so far, the physical demonstrations regarding the difference in the etch rate and morphologies obtained by metal-assisted chemical etching have not been clearly suggested (Huang et al., 2011).

11.2.2 CONTROLLABLE NANO/MICROSCALE STRUCTURE SYNTHESIS

Nano/micro-hierarchical structures can be very favorable for performance enhancement in boiling heat transfer since they can catalyze the heterogeneous fluidic behavior of the phase change by adjusting their characteristic lengths to the effective cavity sizes (Li et al., 2008a; Chen et al., 2009; Kim et al., 2014b). In addition, hierarchical multiscale structures can expand the interfacial area of the heat transfer between a heat-dissipating solid and a heat-absorbing liquid. In order to achieve this in applications, various attempts have been made for more feasible and advanced control of the nanostructures' dimensions and shape. This section presents recent approaches on the nano- and nano/micro-hierarchical structure syntheses for various surface manipulations based on metal-assisted chemical etching. Based on their fundamental principles and parameters, we discuss their possibilities for substitutional bulk micromachining.

11.2.2.1 Micro/nano-hierarchical structures via the top-down method

Micro-nano hybrid structures (MNHSs) with hierarchical geometries can lead to the extension of the interface area much more effectively and the functionalization of the interfacial roughness and the resulting wettability characteristics for boiling heat transfer applications. An advantage of metal-assisted chemical etching is its compatibility with conventional fabrication techniques, which provides a way to realize MNHSs with reliability in the fabrication. As an example, Kim et al. (2011b) suggested combining metal-assisted chemical etching with silicon dry etching, which is a well-established fabrication technique for micropatterning, in order to extend its impact on MNHS fabrication without great expense. Furthermore, this approach would be efficient in maximizing the surface area with extremely high surface roughness and controllable surface wettability for possible applications in fluidics and related heat transfer. In their synthesis, Kim et al. (2011b) devised MNHSs consisting of micropillars or microcavities along with nanowires, which had a length-to-diameter ratio of about 100:1, perpendicular to the exposed surfaces of the micropatterns. MNHSs were fabricated by a two-step silicon etching process; first, dry etching of the silicon to produce the micropatterns and then metal-assisted chemical etching to synthesize the nanowires (Figure 11.4). The fabrication process is readily capable of producing MNHSs covering a wafer-scale area. By controlling the removal of polymeric passivation layers (C_4F_8) deposited on the etched surfaces during the silicon dry etching process (i.e., the Bosch process) (Ayon et al., 1999), they manipulated the geometries of the hierarchical structure with and without thin hydrophobic barriers; this technique can also be adopted for the manipulation of surface wettability. MNHSs without the barrier structures exhibited superhydrophilic behavior with a contact angle (CA) smaller than 10°, whereas MNHSs with the barrier structures preserved by the passivation layer displayed relatively hydrophobic characteristics with a CA near 60°.

The optimal MNHS usage can guarantee a conspicuously large heat transfer area and superhydrophilic characteristics. Introducing hydrophilic regions and removing hydrophobic sidewall structures, which are

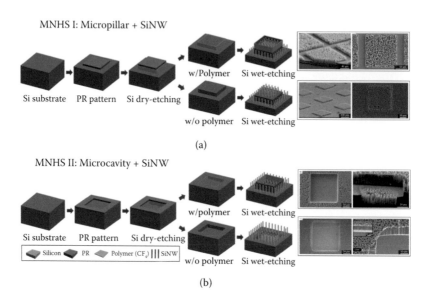

Figure 11.4 Schematics of the fabrication processes using acetone-based photoresist stripping and asher-mediated process for MNHS and SEM images. (a) MNHS with micropillar and nanowires. (b) MNHS with microcavities and nanowires. (With permission from Springer Science+Business Media: *Nanoscale Research Letters*, Micro-nano hybrid structures with manipulated wettability using a two-step silicon etching on a large area, 6, 2011, 333, Kim, B.S., et al., Copyright 2011.)

usually formed in the silicon dry etching process, improves the boiling performance over a broad heat-flux range. Several MNHSs with superhydrophilic characteristics functioning collaboratively can supply and refresh more water to the surface than MNHS with hydrophobic barrier structures. Thus, the burnout of the surface can be delayed by increasing the critical heat flux (CHF) limit and preventing film boiling. In view of these characteristics, the boiling performance can be improved by using MNHSs that are accompanied by the geometrical combination of micro- and nanoscale structures and their superhydrophilic surface wettability. A design study for optimal MNHSs and an experimental evaluation of the performance in boiling heat transfer are challenging topics for future research and are currently under investigation. More details on the requirements for boiling applications will be provided in the following sections relating to the fundamental physics of boiling heat transfer.

11.2.2.2 Nanosphere lithographs for nanoscale pillars

For the employment of regular and dimension-controlled nanostructures, shadow nanosphere lithography combined with metal-assisted chemical etching allowed us to precisely control the characteristic lengths of the structures without expensive equipment and processing (Kosiorek et al., 2004; Peng et al., 2007; Han et al., 2014). Based on this scheme, there have been successful approaches reporting the fabrication of hexagonally arranged nanopillar structures and manipulation of their characteristic lengths within the sub-micron range.

The overall fabrication process for the arranged nanopillar structures is schematically presented in Figure 11.5. First, the template consisting of a polymeric monolayer is allowed to self-assemble on a target silicon substrate (Huang et al., 2007). For a self-assembled monolayer synthesis, as a precoating process for the definition of the shape and dimensions of the target structures, the Langmuir–Blodgett method was adopted in applications (Dabbousi et al., 1994; Hsu et al., 2008). For convenience, polystyrene nanospheres are widely used for the coating of fine spherical structures. The polystyrene nanospheres with a given initial diameter form a hexagonally closed-packed monolayer on an air–water interface. The monolayer of polystyrene nanospheres is then applied on the silicon substrate. After drying to remove the remaining water, the nanospheres comprising the monolayer on the silicon substrate are transformed from closed-packed to nonclosed-packed by further dry etching with O_2

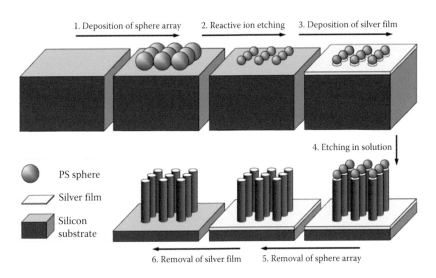

Figure 11.5 Schematic of the fabrication process for arranged nanopillar structures. (Reproduced from Huang, Z.P., et al.: Fabrication of silicon nanowire arrays with controlled diameter, length, and density. *Advanced Materials*. 2007. 19. 744–748. Copyright Wiley-VCH Verlag GmbH & Co. KGaA. With permission.)

plasma while maintaining their location and just shaving their external surfaces. A thin noble-metal layer, such as gold or silver, is then deposited. The diameter of the holes matches that of the diameter-reduced polystyrene spheres, and the exposed silicon portion acquired by the previous dry etching of the polystyrene monolayer is thus selectively covered with the noble metal. The Si substrate underneath the metal layer, which acts as a catalytic layer in the metal-assisted chemical etching of Si, is etched away in a mixture of HF and H_2O_2. This process allows the metal layer to sink vertically into the Si substrate, and to protrude as hexagonally arranged nanopillar structures. The pitch and the diameters of the pillars are equal to the initial diameter of the polystyrene nanospheres and that of the remaining polystyrene nanospheres, respectively. After the etching process, the metallic layer and the remaining polystyrene spheres have to be removed. By changing the initial diameter of the nanospheres and the size reduction degree of the coated nanospheres, the characteristic lengths of the pitch and diameters of the nanopillars can be controlled even on a large-area substrate. Figure 11.6 shows

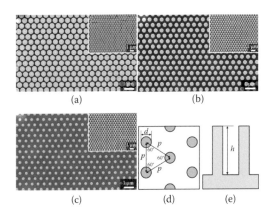

Figure 11.6 SEM images of the nanopillar structures with the same 610-nm pitch and 2-μm height. (a) Diameter of 500 nm. (b) Diameter of 360 nm. (c) Diameter of 280 nm. (d, e) Schematics describing nanopillars; *d*, *p*, and *h* represent the diameter, pitch, and height of nanopillars, respectively. Insets in (a) to (c) are SEM images in tilted view. (Reprinted with permission from Kim, B.S., et al., (2014), Interfacial wicking dynamics and its impact on critical heat flux of boiling heat transfer, *Applied Physics Letters*, 105, 191601. Copyright 2014, American Institute of Physics.)

various diameter-controlled nanopillar structures according to the size control of the nanospheres used as a masking monolayer (Kim et al., 2014a).

11.2.2.3 Possible bulk machining with conventional lithography

Typical bulk micromachining of silicon has been conducted mostly by the dry etching method, which is combined with photolithography, in order to define the shape and dimensions of the target structures using a photoresist material. Silicon anisotropic deep etching, such as deep reactive ion etching, consists of a series of cascading processes: sidewall passivation for the anisotropy of the target patterns, silicon substrate etching, and removal of residues (Ayon et al., 1999). However, this dry etching method allows neither the fabrication of anisotropic high-aspect-ratio structures nor large-area treatment, and is complex and costly. Therefore, metal-assisted chemical etching of silicon has been pursued as an alternative for silicon bulk micromachining. In a wet-etching process, metal-assisted chemical etching can facilitate the treatment of large substrate areas at a low cost. However, technical obstacles and some remaining questions must be resolved prior to its application to large pattern features, which have characteristic dimensions in the range of a few to hundreds of microns. By overcoming critical obstacles and optimizing related parameters, metal-assisted chemical etching can be developed into a simple, easy, and low-cost bulk micromachining method, a possible alternative to dry etching.

Regarding the success of bulk silicon micromachining through metal-assisted chemical etching, the most important issue is the optimization of the catalytic metal layer. The metal film, especially for the large patterns in width and depth, should have sufficient stability in the etchants and mutual cohesion during the etching (Kim and Khang 2014). For the easy access of etchants and the removal of byproducts over the whole area of large patterns, the catalytic metal layer must first guarantee its stability under the etchant, and permeability as a mesh-like layer. In these respects, the morphology and the stability of the metal layer are of importance and the optimization of the metal type for the catalytic layer and its morphology has been demonstrated as a key for the purpose.

A recent study on the role of the metal layer and its validity for bulk etching applications revealed that even with continuous film-like Ag and Au layers, etching was unsuccessful, producing tapered structures and unclear pattern boundaries with a porous medium between the patterns, respectively (Kim and Khang 2014). These aspects are presented in Figure 11.7a, b, d, e to illustrate the significance of the catalytic metal species and their morphology. Some researchers have pointed out that Ag,

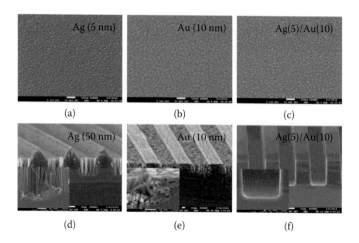

Figure 11.7 Effects of metal type and morphology on the bulk micromachining of Si. (a–c) Ag, Au, and Ag/Au metal film morphology and corresponding etching results in (d–f). Scale bars are 100 nm for (a–c), 10 μm for (d–f) and inset of (e), and 1 μm for the insets of (d) and (f), respectively. (Reproduced from Kim, S.-M., and Khang, D.-Y.: Bulk micromachining of Si by metal-assisted chemical etching. *Small.* 2014. 10. 37613766. Copyright Wiley-VCH Verlag GmbH & Co. KGaA. With permission.)

which is used for metal-assisted chemical etching, easily dissolves into the etchants and re-nucleates on any exposed silicon surface (Chen et al., 2008; Huang et al., 2009; Megouda et al., 2009; Huang et al., 2010). On the contrary, Au has poor cohesion on the thin layer form and is prone to disintegration, even though it exhibits good stability in etchants and produces a continuous film with at least 10 nm thickness (Kim and Khang 2014). The metal layer during the silicon etching process sinks, acts as an intermediary hole supplier from the etchant to the silicon, and catalyzes its oxidation just underneath the layer. However, this sinking of the entire metal mesh film layer cannot occur harmoniously and uniformly if the layer's morphology is not uniform and the stress level leads to cracks. Once the cracks open, they propagate and the area cannot be etched with a uniform etch rate under the remaining metal layer. To optimize the layer, an Ag/Au bilayer facilitates successful anisotropic etching of silicon with well-defined patterns and the results are presented in Figure 11.7c and f. The Ag, which is in direct contact with silicon and protected by the more stable Au, is an etching front and guides the morphology of the overlying Au layer. The Au layer protects the dissolution of Ag into etchants and provides mechanical cohesion for the underlying discontinuous Ag layer (Kim et al., 2011c; Kim and Khang 2014). According to the optimized catalytic metal combination and morphology, bulk micromachined silicon patterns can be well-delineated with various shapes and dimensions, as presented in Figure 11.8.

Another requirement for the application of this wet process in bulk silicon micromachining is the uniformity of the top surface of the target pattern feature against chemical attack during etching. In this wet process, the chemical attack on the top surface is caused by either the dissolved metal species or the diffusion of excess holes. Changing the etch bath and increasing the HF concentration have been suggested to resolve this issue (Kim and Khang 2014). Particularly, excess hole diffusion occurs when the hole consumption is not sufficient or is delayed (Chartier et al., 2008; Lianto et al., 2012; Kim and Khang 2014). In this case, extra holes will diffuse from the source below the metal layer to areas that are not in contact with it, inducing uncontrolled oxidation and subsequent etching. Thus, increasing the HF concentration, to expedite the etching/removal of the oxidized silicon layer, can effectively prevent excess hole diffusion and poor uniformity of pattern definitions. Figure 11.9 presents the effects of the dissolved metal and the top attack and the produced well-protected microscale patterns with a uniformly protected and etched surface as well as good pattern anisotropy (Kim and Khang 2014).

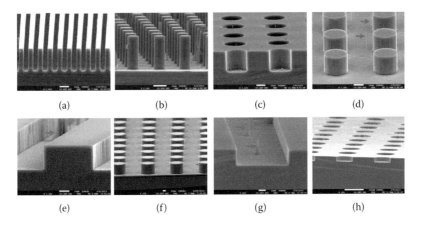

(a)　　　　(b)　　　　(c)　　　　(d)

(e)　　　　(f)　　　　(g)　　　　(h)

Figure 11.8 Bulk micromachining of silicon using optimized metal catalyst. The etching for etch depth of 30 µm was done with Ag (5 nm)/Au (10 nm) bilayer film. (a) 5 µm lines. (b) 5 µm posts. (c) 25 µm holes. (d) 25 µm posts. (e) 50 µ lines. (f) 50 µm posts. (g) 75 µm lines. (h) 100 µm holes. Scale bars are 10 µm for (a) to (g), and 100 µm for (h), respectively. (Reproduced from Kim, S.-M., and Khang, D.-Y.: Bulk micromachining of Si by metal-assisted chemical etching. *Small*. 2014. 10: 3761–3766. Copyright Wiley-VCH Verlag GmbH & Co. KGaA. With permission.)

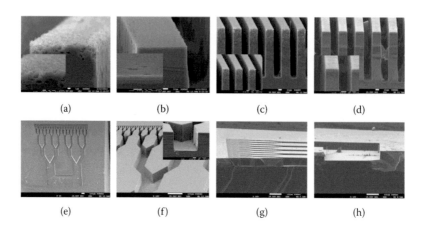

(a) (b) (c) (d)

(e) (f) (g) (h)

Figure 11.9 Reduction of top attack, and etch uniformity on sample having different feature sizes. (a, b) Effect of etch bath change to remove dissolved metals during the etching. (c, d) The reduction of top attack by increasing HF concentration. (e–h) Uniformity on sample having different feature size. Scale bars are 1 µm for (a), 10 and 1 µm for (b) and its inset, 10 µm for (c) and (d), 100 µm for (e) to (h), and 10 µm for inset of (f), respectively. (Reproduced from Kim, S.-M., and Khang, D.-Y.: Bulk micromachining of Si by metal-assisted chemical etching. *Small*. 2014. 10. 3761–3766. Copyright Wiley-VCH Verlag GmbH & Co. KGaA. With permission.)

11.3 BREAKTHROUGH IN HEAT TRANSFER VIA SILICON NANOSTRUCTURES

Heat and mass transfer is the convergence of conduction, convection, and radiation. In a separate or integrated manner in these transfer schemes, engineering applications, that is, effective cooling and thermal energy transfer technologies assuring stability of local/temporal thermal characteristics have been applied to guarantee the reliability of engineering devices by releasing concentrated thermal load and dissipating it efficiently. A promising convective thermal energy transfer scheme refers to boiling heat transfer, which is the most powerful heat/energy dissipating mechanism for efficient cooling and energy transfer. The heat/mass transfer and fluid dynamics scheme accompanies the heterogeneous phase change of a working fluid on a surface and the intricate hydrodynamics phenomena involving a series of cascading ebullition and convective behavior of the multiphase flow. Therefore, the interface definitely governs the heterogeneous phase change of a working fluid and thereby the convective fluidic behavior during boiling: boiling is an interface-dependent physical phenomenon.

The ultimate goals in boiling heat transfer are to maximize the allowable heat dissipation capacity (i.e., critical heat flux, CHF) and to increase the heat dissipation efficiency (heat transfer coefficient, HTC). There have been numerous studies proving that the surface morphology and wettability characteristics, which are interrelated, are two principal factors determining the boiling heat transfer outcome. Roughened surface, at first, offers an extension of heat-dissipating surface area on a solid–liquid interface. In addition, the roughed surface with cavity-like structures can facilitate ebullition by catalyzing the nucleation of the coolant. Moreover, the wettability characteristics are important in determining the liquid accessibility toward the boiling surface. Taking into account the static and dynamic wetting characteristics, extremely rough surface morphologies with nano/microstructures can also lead to strong capillary pumping due to morphologically induced hemi-wicking phenomenon. Morphology manipulations such as structural interface modifications using nano/microscale and combined hierarchical structures thus present a promising solution to take a major step forward in improving the capacity and efficiency of thermal energy dissipation as well as the stability of local/temporal heat transfer characteristics.

In this section, we highlight recent findings and attempts regarding the manipulation of silicon-based fine structures' interfacial wetting characteristics and then deal with their applications in boiling heat transfer. Novel designs of silicon-based interfacial structures and their functionalization offer the potential for great breakthroughs in developing heat/energy transfer systems through functionalized interfaces.

11.3.1 INTERFACIAL CHARACTERISTIC MANIPULATIONS VIA NANOSCALE MORPHOLOGY CONTROL

Practically, manipulation of surface wettability is important in diverse fields including physics, chemistry, biology, and engineering, and functionalized interfaces with favorable wettability characteristics have been widely applied in fluid mechanics and heat transfer (Choi et al., 2006; Tuteja et al., 2007). For instance, superhydrophobic surfaces have been used to enhance fluidic performance by decreasing surface friction–induced pressure drop, and to produce waterproof self-cleaning surfaces that mimic a lotus leaf (Shibuichi et al., 1996; Wang et al., 1999; Lau et al., 2003; Xiu et al., 2007). On the contrary, hydrophilic surfaces have been applied to antifogging technology and bio-molecular purifications, drug delivery, and low-temperature fuel cell systems to prevent their performance degradation by choking or flooding (Suzuki et al., 2010; Tsougeni et al., 2010). Particularly, hydrophilic and superhydrophilic surfaces have been employed to enhance the boiling heat transfer performance by catalyzing interfacial liquid refreshing (Li et al., 2008a; Chen et al., 2009; Patankar 2010; Kim et al., 2014a; Kim et al., 2014b).

Based on numerous theoretical and experimental approaches, principal factors determining the wettability characteristics of the intrinsic surface-free energy and the surface morphology have been discussed in depth. In this section, we focus on the fundamentals of how the interfacial morphology affects wettability. According to the fundamental physical rules, we extend the applicability of silicon materials for the efficient and feasible wettability manipulations.

11.3.1.1 Static wetting control

Understanding how the interfacial roughness defines the wetting characteristics is a key topic for the physical demonstration of wetting phenomena and their expression in engineering systems. Numerous morphology manipulation techniques for controlling surface wettability and their associated design guidelines have been proposed.

It is possible to transform the surface wettability of a silicon substrate from hydrophilic to hydrophobic and to control the surface energy control by polymer layer coating. One of the typical approaches is to use the top-down etched silicon nanowires to create superhydrophilic surfaces by realizing extremely rough and porous surfaces, since synthesizing vertically aligned silicon nanowires, with their morphological advantages of a high aspect ratio, is strategic and efficient to greatly enrich the surface roughness. The surface roughness permits hydrophilicity or hydrophobicity to be increased to the extreme, depending on the intrinsic wettability determined by the surface-free energy of the base substrate.

The thermodynamic stability of a liquid droplet on a solid surface can be given by Young's equation (Yoshimitsu et al., 2002), which determines the static CA θ (Figure 11.10):

$$\cos\theta = \frac{\gamma_{SV} - \gamma_{SL}}{\gamma_{LV}} \tag{11.4}$$

where γ_{SV}, γ_{SL}, and γ_{LV} are the interfacial free energies per unit area of the solid–vapor, solid–liquid, and liquid–vapor interfaces, respectively. Densely distributed silicon nanowires can be treated as uniformly distributed circular pillar structures: the liquid droplet behavior on the nanowire surface is illustrated in

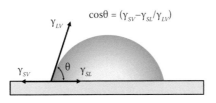

Figure 11.10 Thermodynamic stability of a liquid droplet on a solid surface and corresponding Young's equation.

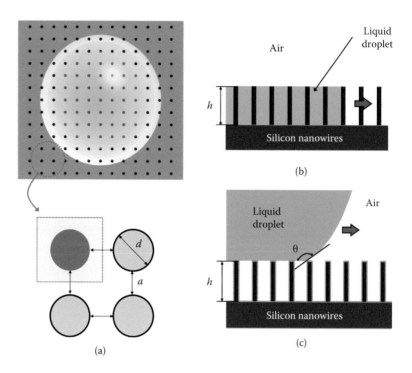

Figure 11.11 (a) Schematic of liquid droplet on an interface with vertically standing silicon nanowires (SiNWs). (b) Liquid spreading through SiNWs based on hemi-wicking. (c) Liquid droplet on SiNWs according to the Cassie–Baxter model.

Figure 11.11. When an intrinsic surface of the substrate has high surface-free energy (e.g., Si: 52.96 mJ/m²) and a sequential hydrophilic static CA less than 90° (Kim et al., 2011a), the roughness effect on the CA can be explained by Wenzel's model as follows (Wenzel 1936):

$$\cos\theta = r \cdot \cos\theta^* \quad (r \geq 1) \tag{11.5}$$

where θ, r, and θ^* are the apparent CA, roughness factor defined as the ratio of the actual to the projected surface area, and equilibrium CA on an ideal plain surface, respectively. This model indicates that the surface becomes more hydrophilic as the roughness increases in the hydrophilic regime (0< θ^*<90), and the apparent angles presented (as presented in Figure 11.12a) decreases gradually with increasing roughness, which is determined by the height of the silicon nanowires depending on the etching time in metal-assisted chemical etching (Kim et al., 2011a).

Surface roughening can also enhance intrinsic hydrophobicity. In the Cassie–Baxter state, a surface with rough cavity structures can retain air and prevent the invasion of liquid droplets from the surface (Figure 11.11c) under thermodynamic stability. Polymeric materials have lower surface-free energies than silicon. Therefore, easy wettability conversion, as schematically described in Figure 11.13, has been reported using materials such as Teflon-like fluorocarbon, octadecyltrichlorosilane (OTS), self-assembled monolayers (SAMs), fluorinated SAMs, and siloxane oligomers, which can lower the surface-free energy on a prepared roughened surface showing superhydrophilicity (Wei-Fan and Li-Jen 2009; Choi et al., 2010; Dawood et al., 2011; Kim et al., 2011a; Yoon and Khang 2012). Similarly, hydrophobicity can also be strengthened by increasing the surface roughness, and the static CA variations are presented in Figure 11.12b. Accordingly, superhydrophobic surfaces can be readily produced from superhydrophilic ones, which are created by silicon nanowires, in advance, by diminishing the surface-free energy of the superhydrophilic surface much lower than that of silicon using fluorine carbon coatings (Kim et al., 2011a). Figure 11.14 shows the results of the reversible wettability conversion process on silicon nanowire surfaces by applying contact printing of siloxane oligomers. Considering the variation in the surface-free energy, Yoon and Khang (2012) presented guideline

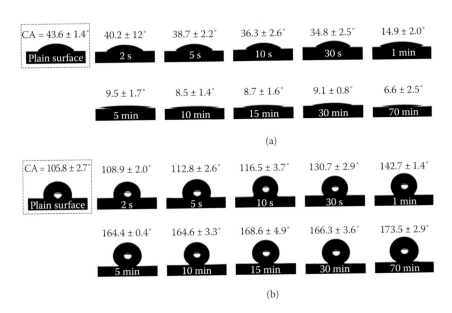

Figure 11.12 Apparent contact angles in relation to etching time for synthesizing SiNWs based on metal-assisted chemical etching and hydrophobic C_4F_8 layer coating. (a) Hydrophilic characteristics on SiNWs-decorated surface. (b) Hydrophobic characteristics on SiNWs-decorated surfaces with a deposited C_4F_8 layer. (Reprinted with permission from Kim, B.S., et al., (2011), Control of superhydrophilicity/superhydrophobicity using silicon nanowires via electroless etching method and fluorine carbon coatings, *Langmuir*, 27, 10148–10156. Copyright 2011 American Chemical Society.)

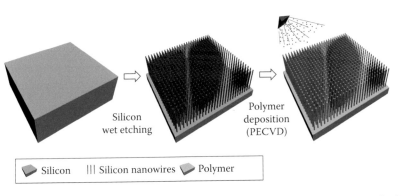

Figure 11.13 Schematic of the interface manipulation methodology using SiNWs for increasing hydrophilicity and converting hydrophilic surfaces into hydrophobic surfaces. (Reprinted with permission from Kim, B.S., et al., (2011), Control of superhydrophilicity/superhydrophobicity using silicon nanowires via electroless etching method and fluorine carbon coatings, *Langmuir*, 27, 10148–10156. Copyright 2011 American Chemical Society.)

functions, which can be expressed by geometrical variables of the surface morphology, for the feasible application of surface manipulation with nanoscale structures on macroscale systems.

Moreover, there are empirical and analytical parametric guidelines for morphology manipulations and critical dimension control. For example, the critical height of the nanowires should be greater than one micrometer and a few micrometers for superhydrophilicity and superhydrophobicity, respectively (Kim et al., 2011a). Silicon nanowires fabricated with a height of a few micrometers provide an appropriate physical ground for surface roughening and contribute to the creation of a superhydrophilic (<10°) and a superhydrophobic interface (>160°). Even though superhydrophilicity is a consequence of hemi-wicking, which is a peculiar hydrodynamic behavior causing liquid to spread on an interface, the static characteristics can be separately discussed, first according to the manipulation of surface morphology and then

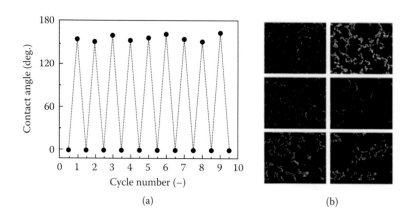

(a)

(b)

Figure 11.14 Switchable wettability of vertical silicon nanowire array, from superhydrophilic (H_2SO_4 dip) to superhydrophobic (PDMS contact printing) (a) Cycle test of wettability switching. (b) SEM images showing the changes in aggregation state of SiNWs upon cycling, as prepared (top row), after two cycles (middle), and after four cycles (bottom), for ~10 μm long NWs (left column) and ~70 μm long NWs (right column), respectively. Scale bars are 1 μm for all images. (Reproduced from Yoon, S.-S., and Khang, D.-Y., (2012), Switchable wettability of vertical Si nanowire array surface by simple contact-printing of siloxane oligomers and chemical washing, *Journal of Materials Chemistry*, 22, 10625–10630. By permission of The Royal Society of Chemistry.)

with regard to reduction in the intrinsic surface energy. Dynamic wetting, that is, hemi-wicking, will be discussed in the following section.

11.3.1.2 Dynamic wetting control

Solid surfaces with a rough morphological texture can be considered as porous surfaces that can absorb liquids. Imbibition occurs on the formed solid–liquid interfaces. This morphologically induced dispersion of the liquid is called hemi-wicking (Figure 11.15) (Bico et al., 2002; Quéré 2008). The capillary-driven hemi-wicking flow of the liquid has potential applications in biological and industrial processes, which include biomedical devices for chromatography, DNA electrophoresis and drug delivery, oil recovery, sensors, and thermal management in phase-changing boiling heat transfer schemes. Besides the static wetting characteristics on an interface, the dynamic factor of hemi-wicking has an effect on the boiling heat transfer, due to the concurrent liquid refreshment toward the surface and vaporization of the liquid.

Patterns on a hydrophilic solid at a scale much smaller than the capillary length (above which gravity dominates over surface tension) can induce superhydrophilicity. The Wenzel effect, according to which roughness enhances hydrophilicity, provides that the liquid is distributed in a pattern, leaving the rest of the solid dry as in the case of partial wetting. However, interfacial structures may also guide the liquid within the array they form, in a manner similar to wicking. The phenomenon that occurs here is not classical wicking but hemi-wicking: as the film progresses in the microstructures, it develops an interface with air, leaving (possibly) a few dry islands behind it. We discuss how this phenomenon affects the wetting laws and conclude with a few considerations regarding the dynamics of these films (Quéré 2008).

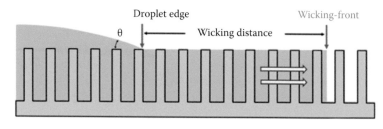

Figure 11.15 Schematic of hemi-wicking on a morphologically roughened interface. (Redrawn from Kim BS, et al., *Applied Physics Letters*, 105, 191601, 2014a.)

Considering Young's equation, the hydrodynamic behavior of a liquid should be governed by the morphological characteristics of the surface, such as roughness and porosity, as well as by the intrinsic characteristics of the substratum including the interfacial free energy. As discussed, hydrophilicity is enhanced by roughening the surface morphology of a silicon substrate according to Wenzel's theoretical model (Wenzel 1936). Especially, strong hydrophilicity can be realized when a geometrical prerequisite meets a theoretical criterion for the occurrence of the hemi-wicking phenomenon (Bico et al., 2002; Kim et al., 2011a). Hemi-wicking, which drives the propagation or attraction of liquids through a forest of interfacial structures (schematically described in Figure 11.15), is closely related to morphological factors such as dimension, shape, and configuration. According to Wenzel's model with Young's equation for the equilibrium of interfacial forces, the morphological criterion for hemi-wicking can be expressed as a function of geometric variables and should satisfy the following equation (Bico et al., 2002; Kim et al., 2011a):

$$\theta^* < \theta_c \left(\cos\theta_c = \frac{1-\varphi}{r-\varphi} \right) \tag{11.6}$$

where θ_c and φ represent the critical CA demanded as the prerequisite and solid fraction of the solid–liquid interface contacting the liquid droplet, respectively. With a dimensional approach, the characteristic dimensions of the interfacial structure are the principal factors determining r and φ. In particular, a few micron-height silicon nanopillars can feasibly secure the prerequisite condition of Equation 11.4 (Kim et al., 2011a). When we consider the hexagonally arranged pillars with an average diameter d, a pitch p, and a height h presented in Figure 11.6, the determinants of r and φ can be expressed as the function of the design variables as follows:

$$r \equiv 1 + \frac{\pi dh}{p^2 \sin 60}, \quad \varphi \equiv \frac{\pi d^2}{4p^2 \sin 60} \tag{11.7}$$

These deliver the analytical formulation and determine whether the morphological treatments and designs meet the hemi-wicking prerequisite of Equation 11.3. For instance, nanoscale structures (e.g., nanopillars) with the characteristics presented in Table 11.1 are efficient to satisfy the prerequisite with the critical CAs of θ_c over 80° in most cases, whereas this cannot be readily achieved by macro- or micropatterns. Since specific geometric variables fall in the domain for the prerequisite of hemi-wicking, water can be permeated immediately and continuously on the interface having Si-nanopillar structures. Figure 11.16 illustrates hemi-wicking phenomenon along (a) vertical and (b) planar directions on nano-inspired surfaces and qualitative comparison of the effect of their characteristic lengths.

Taking into account the morphologically induced phenomena, hemi-wicking can be further evaluated to reveal dynamic characteristics involving its velocity and maximum displacement of the liquid spreading against viscous friction caused by interfacial structure forests. Ishino et al. (2007) and Mai et al. (2012) presented hemi-wicking evaluations using silicon nanopillars and quantified the shape effects of the interfacial structures on wicking control. Because the hemi-wicking flow is laminar, the liquid supply to the hemi-wicking front can be linearly correlated to the hydraulic resistance for a fully developed flow. As wickability depends on fluidic resistance and capillary pressure, the wicking distance x can be expressed as a function of time and characteristic lengths of wicking passages based on Washburn's model (Washburn 1921; Tas et al., 2004):

$$x = \left(\frac{\gamma A t \cos\theta^*}{C\mu} \right)^{0.5} = W \cdot t^{0.5} \tag{11.8}$$

where A, μ, C, W, and t are the characteristic hydraulic length, viscosity of the liquid, a constant, wicking coefficient or wickability, and time, respectively. Here A and C determine the dimensions and shape of the wicking channel and refer to the height of the wicking passage and a constant, respectively (for rectangular cross-sectional channels, $C = 3$; Tas et al., 2004). Assuming that the volume of a liquid droplet is sufficient

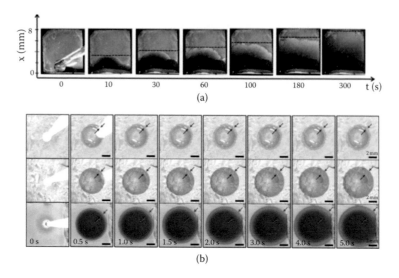

Figure 11.16 Hemi-wicking phenomena on nanostructure-employed Si interfaces. (a) Vertical hemi-wicking process of silicone oil on a surface with vertically standing Si nanopillars ($h = 4$ μm and $p = 1$ μm). The red dotted line marks the liquid front. (Reprinted with permission from Mai, T.T., et al., (2012), Dynamics of wicking in silicon nanopillars fabricated with interference lithography and metal-assisted chemical etching, *Langmuir*, 28, 11465–11471. Copyright 2012 American Chemical Society.) (b) Planar hemi-wicking by a DI-water droplet on Si-nanopillar surfaces with different diameter (silicon nanowires with $d = 500$ nm, $d = 360$ nm, and $d = 280$ nm which correspond to Figure 11.6a–c, respectively). (Reprinted with permission from Kim, B.S., et al., (2014), Interfacial wicking dynamics and its impact on critical heat flux of boiling heat transfer, *Applied Physics Letters*, 105, 191601. Copyright 2014, American Institute of Physics.)

Table 11.1 Geometrical variables and determinants for hemi-wicking with the manipulated nano- and micropillar structures

CASES	GEOMETRIC VARIABLES			DETERMINANTS FOR HEMI-WICKING			CAs
	d (nm)	p (nm)	h (nm)	r	φ	θ_C (°)	θ (°)
Nano_1	500	610	2000	10.8	0.609	87.8	10.5
Nano_2	360	610	2000	8.02	0.316	84.9	9.56
Nano_3	280	610	2000	6.46	0.191	82.6	6.97
Micro_1	6000	8000	2000	1.68	0.510	65.2	–
Micro_2	4000	8000	2000	1.45	0.227	50.9	–
Micro_3	2000	8000	2000	1.23	0.0567	36.3	–

Source: Kim, B.S., et al., *Appl. Phys. Lett.*, 105, 191601, 2014a.
Note: Experimental results of Nano_1, Nano_2, and Nano_3.

and the wicking passage created by the nanopillars has a rectangular cross-sectional shape for direct wicking propagation, the wicking coefficient can be characterized by controlling the dimensional hydraulic length, that is, the spacing between adjacent nanopillars. Figures 11.16b and 11.17 show that the nanopillars shown in Figure 11.6a–c, satisfying the hemi-wicking criterion, lead to interfacial hemi-wicking. The corresponding hemi-wicking coefficients in these three cases are 0.73, 0.98, and 1.28 mm/s$^{0.5}$, respectively, by expanding space between nanopillars with decreasing hydraulic resistance. These results explain that surface roughening using the nanopillars induces hemi-wicking, and then designing the characteristic lengths can allow us to control the hemi-wicking performance related to the hydrodynamic balance between the morphology-driven capillary and the counteracting hydraulic resistance. Even in the case of microscale structures, as presented in Figure 11.18, properly designing their geometry and dimensions can lead to interfacial hemi-wicking.

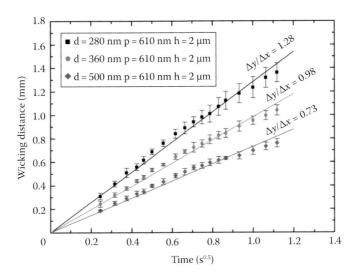

Figure 11.17 Wicking propagation plots of hemi-wicking performance as corresponding to silicon nanowires presented in Figure 11.6. The wicking coefficient, W, is defined by the average slope of wicking distance versus the square root of time. (Reprinted with permission from Kim, B.S., et al., (2014), Interfacial wicking dynamics and its impact on critical heat flux of boiling heat transfer, *Applied Physics Letters*, 105, 191601. Copyright 2014, American Institute of Physics.)

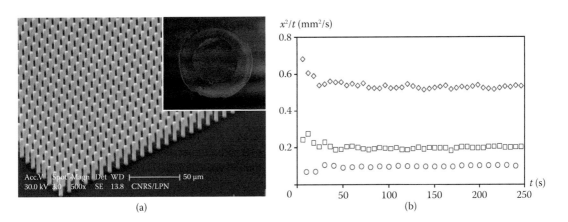

Figure 11.18 (a) Silicon surface decorated with a forest of micropillars of diameter $d = 1.3$ μm and height $h = 26$ μm regularly displayed on a square pattern of pitch $p = 10$. The bar indicates 50 μm. The inset presents an ethanol drop deposited on a surface. The drop acts as a lens, which deforms the colors arising from the presence of a regular texture. (b) Hemi-wicking performances on a micropillar decorated surface ($h = 6$ μm and $p = 10$ μm) according to variation of liquid viscosity of 19 mPa-s (diamonds), 48 mPa-s (squares), and 97 mPa-s (circles), respectively. (The source of the material Ishino, C., et al., Wicking within forests of micropillars, *Europhysics Letters*, 79, 56005, 2007 is acknowledged.)

11.3.2 EFFECTIVE HEAT DISSIPATION IN MULTIPHASE FLOW (BOILING HEAT TRANSFER)

There are three modes in heat transfer: conduction, radiation, and convection. Conduction is energy transfer through collision and diffusion of particles or quasi-particles such as phonons within a body or between contiguous bodies. Radiation is a physical phenomenon describing electromagnetic irradiation by thermal motion of charged particles in matter, and it is possible to transfer energy between noncontacting media through the electromagnetic wave. Finally, convection is based on the movement of groups or aggregates of molecules within a fluid. In everyday life, it is important to increase the transfer of thermal energy by enhancing the efficiency and feasibility of higher and more reliable dissipation schemes in various engineering cooling systems. Thus, systematic thermal designs must be accompanied by the integration of the three heat transfer schemes.

The heat transfer schemes are clearly governed by interface characteristics, which define and confine the related physical behaviors and performance. Particularly, convection, as one of the most pragmatic heat/energy transfer mechanisms, is definitely based on the interfacial interaction between solid surfaces and counteracting fluidic media transferring energy. Therefore, interface-inspired physics with respect to fluidic interactivity must be decisive factors for characterizing heat and energy transfer performance. Recent studies have focused on the peculiarities of nano-inspired functionalizations and their impact on interfaces, which can contribute greatly to the enhancement of convection performance. Interfacial functionalization is widely applied, especially to boiling heat transfer, which is the most powerful heat/mass transfer and fluid dynamic scheme. Boiling heat transfer involves heterogeneous phase changes of the working fluid on a solid–liquid interface. The intricate hydrodynamic phenomena involving ebullition and subsequent convective behavior of the multiphase flow can thus be definitely controlled and reinforced by the particular interface characteristics.

In this section, we briefly review the fundamentals of boiling heat transfer and the principal factors determining its performance. We then highlight recent attempts to reinforce the interface's peculiarities through nanoscale surface manipulations. As a representative controllable and feasible material, silicon has been widely used for interface functionalization. Recent findings and state-of-the-art demonstrations can be pragmatically applied to various engineering fields involving heat and mass transfer for powerful and reliable cooling/energy transferring systems.

11.3.2.1 Boiling heat transfer and solid–liquid interfaces

In convective heat transfer schemes, there are many specific methods. Macroscopically, heat transfer schemes can be classified according to the presence of pumping power into natural convection and forced convection. In addition, they can be distinguished by the type of transferring media (i.e., liquid or gas). Table 11.2 presents representative convection schemes and the corresponding HTCs, which refer to the efficiency of thermal energy transfer and dissipation. Among the various types of convective schemes, boiling heat transfer exhibits a remarkable heat transfer performance due to the phase change of the working fluid and the subsequent strong convection. Since boiling heat transfer can offer much higher heat dissipation efficiency (i.e., HTC) and heat dissipation capacity (i.e., allowable heat dissipation capacity or CHF), it has been widely adopted in high-temperature and heat-flux-generating systems as a powerful heat-transferring/dissipation scheme.

Fundamental boiling performance curves with their intrinsic characteristics are presented in Figure 11.19a (Chen et al., 2009). According to the thermal load of the wall superheat (i.e., the degree of overheating of the boiling surface, $T_w - T_{sat}$, where T_w and T_{sat} indicate the wall temperature and saturation temperature of the working fluid, respectively), we can simply define boiling domains from the nucleation region to the film boiling region. The principal performance indicators in boiling are CHF and HTC, which can be interpreted as the maximum peak for an allowable heat dissipation capacity and the local/overall inclination (i.e., $dq''/\Delta T$, where q'' is the heat flux transferred from a solid hot spot of the heat source to the environment) in the boiling curve, respectively. Effective cooling and thermal energy transferring technologies assuring stability of

Table 11.2 **Order of magnitude for heat transfer coefficients depending on cooling scheme and working fluid**

HEAT TRANSFER PROCESS		HEAT TRANSFER COEFFICIENT (W/m²K)
Convection without phase change		
Free convection	Gases	2–40
	Liquids	50–1200
Forced convection	Gases	25–250
	Liquids	100–20000
Convection with phase change		
Boiling heat transfer		2500–1,00000

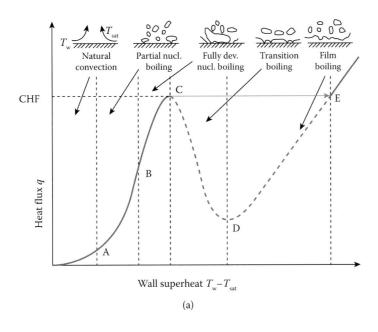

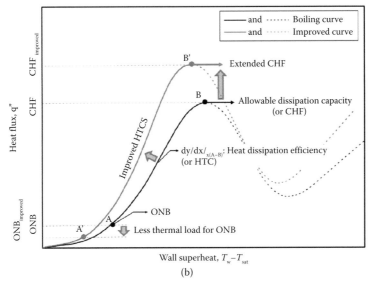

Figure 11.19 (a) A representative boiling curve qualitatively showing heat flux versus wall superheat T_w-T_{sat} with schematics describing characteristic nucleation behavior. (Reprinted with permission from Chen, R., et al., (2009), Nanowires for enhanced boiling heat transfer, *Nano Letters*, 9, 548–553. Copyright 2009 American Chemical Society.) (b) Boiling curves representing performance factors including allowable heat dissipation capacity, heat dissipation efficiency, and onset of nucleate boiling. CHF, critical heat flux; HC, heat transfer coefficients; ONB, onset of nucleate boiling. (Reprinted from *International Journal of Heat and Mass Transfer*, 70, Kim, B.S., et al., Stable and uniform heat dissipation by nucleate-catalytic nanowires for boiling heat transfer, 23–32, Copyright 2014, with permission from Elsevier.)

ocal/temporal thermal characteristics should be applied to guarantee the reliability of engineering devices by releasing concentrated thermal load and dissipating it efficiently. Therefore, in order to develop a promising heat/energy transferring scheme, the ultimate goals in boiling heat transfer are to maximize the allowable heat dissipation capacity (CHF) and increase the heat dissipation efficiency (HTC). Figure 11.19b shows boiling performance curves and the related research targets in performance improvements (Kim et al., 2014b).

The phase change of the fluid from liquid to gas during boiling involves a heterogeneous ebullition procedure: vaporized bubble nucleation, growth, detachment, and confluent two-phase convection on the solid–liquid interface (Dhir 1998; Pioro et al., 2004; Wu and Sundén 2014). A series of multiphase convection behaviors are specifically determined by the characteristics of the boiling surface. In particular, nucleative boiling depends on the interfacial hydrodynamics between the phase changing of the working fluids and counteracting re-wetting of the interface. Based on the associated thermo-physical phenomena on boiling interfaces, we can present two principal factors, which dominantly determine the boiling performance: surface roughness and wettability characteristics. First, roughness defines the contact area between the heat-dissipating solid surface and the liquid. Higher heat dissipation capacity can be attributed to an increase in surface roughness when the contact area between the solid and the counteracting working fluid is extended. Typical surface roughening, as a strategy for heat transfer enhancement, has been pragmatically adopted in various engineering devices in the form of ribs, pin-fin, cavities, pillars, dimples, and their conjugated structures on a heat transfer surface in single- and multiphase heat transfer schemes (Shoji and Takagi 2001; Yu et al., 2006). Moreover, regarding surface wettability, several studies have indicated that the hydrophilic characteristics of surfaces enhance their re-wetting characteristics and can lead to CHF extension (Dhir and Liaw 1989; Dhir 1998; Kandlikar 2001; Hsu and Chen 2012). The effects of surface roughness and wettability on boiling performance and CHF variations according to the wetting characteristics of the boiling surface are schematically illustrated in Figure 11.20a, b and c, b, respectively.

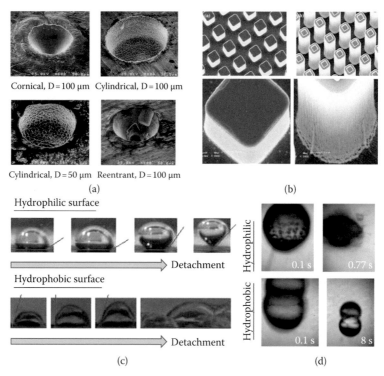

Figure 11.20 Effects of surface roughness and wettability on boiling behaviors. (a) Variation of artificial cavities. (Reprinted from *International Journal of Heat and Mass Transfer*, 44, Shoji, M., and Takagi Y., Bubbling features from a single artificial cavity, 2763–2776, Copyright 2001, with permission from Elsevier.) (b) SEM views of micro-pin-fins. (With permission from Springer Science+Business Media: *Heat and Mass Transfer*, Experimental study of boiling phenomena and heat transfer performances of FC-72 over micro-pin-finned silicon chips, 41, 2005, 744–755, Wei, J.J., et al., Copyright 2005.) (c) Contact angle change during growth time on hydrophilic and hydrophobic surfaces. (Reprinted from *International Journal of Heat and Mass Transfer*, 52, Phan, H.T., et al., Surface wettability control by nanocoating: The effects on pool boiling heat transfer and nucleation mechanism, 5459–5471, Copyright 2009, with permission from Elsevier.) (d) Evaporation of water drop on manipulated surfaces heated at 132°C exhibiting a hydrophilic network and hydrophobic network. (Reprinted with permission from Betz, A.R., et al., (2010), Do surfaces with mixed hydrophilic and hydrophobic areas enhance pool boiling?, *Applied Physics Letters*, 97, 141909. Copyright 2010, American Institute of Physics.)

11.3.2.2 Functionalized nano-interfaces and their impact on boiling performance

Both surface roughness and wettability, which are controlled by surface morphology, have been properly manipulated by emerging nanostructure synthesis techniques for boiling heat transfer enhancements. Novel establishments of nano/material technologies for nano-inspired functional interfaces have triggered a possible leap in recent boiling research with their peculiar characteristics. Extremely high surface roughness via nanoscale structures can extend the interfacial contact area for greater heat dissipation. For example, nano-inspired interfaces with vertically aligned nanowires (Chen et al., 2009; Kim et al., 2011a; Shin et al., 2011; Li et al., 2012) and micro/nano-hierarchical structures (Kim et al., 2011b; Chu et al., 2013) can increase the heat transfer area for more efficient heat dissipation than macroscale structures do. In addition, they accompany the intensification of hydrophilicity toward a superhydrophilic regime, which is clearly favorable to wetting or refreshing of the interface by liquid-phase working fluid against a surface dry-out.

Representative recent approaches by Chen et al. (2009) and Kim et al. (2014a, 2016a and 2016b) revealed that the peculiarities of nano-inspired interfaces can contribute greatly to the enhancement of boiling heat transfer. The peculiarities are derived from the intrinsic functionality of extremely rough morphologies and subsequent interfacial fluidic interactivity, which are favorable for the associated easy nucleation and interfacial liquid refreshing. In accordance with the aims in these boiling study, advanced techniques, such as the fabrication of novel and nano/micro-hierarchical structures (Kim et al., 2011b; Shin et al., 2012; Chu et al., 2013), the nanoscale fabrication technique including nanostructure patterning by conventional photolithography (Li et al., 2008c) or maskless method (Wu et al., 2008), the chemicophysical functionalization of interfaces using metal-assisted chemical etching (Chen et al., 2009; Lu et al., 2011; Li et al., 2012), template-assisted electrodeposition (Shin et al., 2012), and thermodynamic nanoparticle deposition (You et al., 2003; Kim et al., 2007), can be used to easily manipulate surface wettability (Verplanck et al., 2007; Jokinen et al., 2008) and have been applied to boiling. These techniques allow for the syntheses of finer structures for unlimited manipulation of interfacial structures with nanoscale functional peculiarities. The outcome structures secure the advantages of rough morphology and byproductive wetting for activating nucleation, areal pin-fin effects, and favorable surface re-wetting (Li et al., 2008a; Lu et al., 2011; Chu et al., 2012; Kim et al., 2014b). For instance, a 3D macro-porous metallic surface layer with nanoscale porous structures exhibited an improved HTC, especially at a low heat flux of 1 W/cm^2, by over 17 times, as compared with the plain surface (Li et al., 2008b). Moreover, top-down etched silicon nanowires and electrodeposited copper nanowires with an improved boiling performance by up to 100% as compared with a plain silicon surface were developed by increasing surface wettability at which the nanowires exhibited superhydrophilic behavior (Chen et al., 2009). With tilted copper nanorods synthesized by an electron-beam evaporator, pool-boiling heat transfer characteristics were also enhanced by the reinforced hydrophilicity and in the presence of nucleation sites that resulted from the intrinsic nature of the dense nanowires (Li et al., 2008a). Carbon nanotube coated surfaces were also adopted for improving boiling performance (Khanikar et al., 2009). It may be inferred from these references that nanoscale structures appear to greatly increase the surface area and wettability and lead to the enhancement of the boiling behavior by supplying adequate liquid to the boiling surface through secondary morphologically driven capillary pumping effects.

A promising interface for boiling applications is developed through the silicon-based interface modification technique, which is effective for interface manipulation and functionalization. Silicon nanowires synthesized by metal-assisted chemical etching show relatively low-conductive heat dissipation ability with low thermal conductivity of up to 10 W/m-K, which is much less with an order of magnitude than that of a bulk silicon substrate at 140 W/m-K (Hochbaum et al., 2008). Nevertheless, the boiling performance on an interface with silicon nanowires is greatly improved with profound HTC increase by more than 100% compared with an untreated surface as presented in Figure 11.21b (Kim et al., 2014b). This can be explained by the additional advantages produced by the morphological surface roughening for catalytic nucleation albeit its low-conductive effect. Taking into account this improvement, an additional merit can be discovered from nanostructures in that they can offer favorable nucleation sites such as cavity-like structures. Interfacial morphologies combined with nano/microscale cavity-like structures can act as a nucleation seed and be favorable for their easy nucleation for the vaporizing of the coolant. Figure 11.22 shows apparent cavity-like structures formed on a nanowire forest and an analytic prediction of the

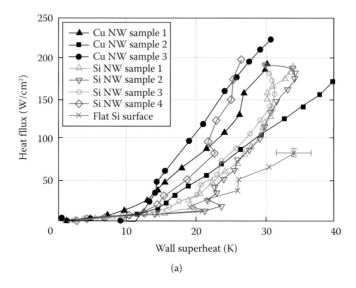

(a)

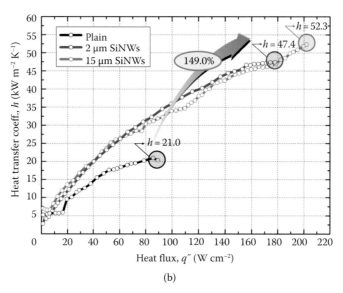

(b)

Figure 11.21 Pool-boiling heat transfer performance on nano-inspired surfaces. (a) Boiling curves for plain silicon surface, silicon nanowires, and Cu nanowires. (Reprinted with permission from Chen, R., et al., (2009), Nanowires for enhanced boiling heat transfer, *Nano Letters, 9,* 548–553. Copyright 2009 American Chemical Society.) (b) Local heat transfer coefficients on plain silicon, short (2 μm long) and long silicon nanowires (15 μm long). Insets are surface morphology with the short and long silicon nanowires with conglomerated silicon nanowires forming natural microscale cavities. (Reprinted from *International Journal of Heat and Mass Transfer,* 70, Kim, B.S., et al., Stable and uniform heat dissipation by nucleate-catalytic nanowires for boiling heat transfer, 23–32, Copyright 2014, with permission from Elsevier.)

effective cavity sizes (Li et al., 2008a; Chen et al., 2009). Especially for the nucleation procedure, cavity structures catalyze the initiation of bubble generation. Vapor nuclei are formed when the pressure becomes greater than that of the surrounding liquid. The pressure difference between the nuclei and surrounding liquid is explained by the Young–Laplace equation ($\Delta P = 2\sigma/r_b$, where r_b is the bubble radius), and the pressure elevation can be obtained by superheat, which is a primary thermal energy source for further growth. At the onset of nucleation, morphological cavities on a heterogeneous boiling surface can act as catalytic nucleation seeds by diminishing the required superheat for the pressure elevation. Hsu (1962) demonstrated the role of cavities and suggested a model describing an effective cavity size, which can be beneficial

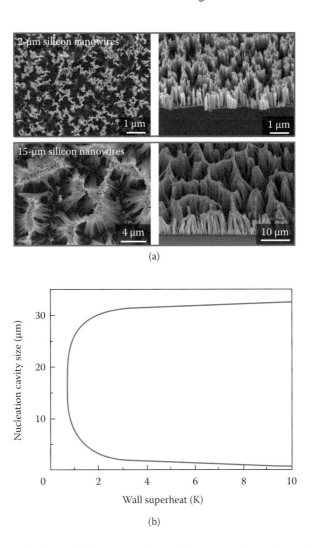

(a)

(b)

Figure 11.22 (a) Surface morphology with the short and long silicon nanowires with conglomerated silicon nanowires forming apparent cavity-like structures with microscale characteristic sizes. (Reprinted from *International Journal of Heat and Mass Transfer, 70*, Kim, B.S., et al., Stable and uniform heat dissipation by nucleate-catalytic nanowires for boiling heat transfer, 23–32, Copyright 2014, with permission from Elsevier.) (b) Predictions for the range of effective cavity size for the nucleation at a given wall superheat. (Reprinted with permission from Shin, S., et al., (2012), Double-templated electrodeposition: Simple fabrication of micro-nano hybrid structure by electrodeposition for efficient boiling heat transfer, *Applied Physics Letters, 101*, 251909. Copyright 2012, American Institute of Physics.)

for the initiation of nucleation. The presented analytic predictions in Figure 11.22b follow this model. Since byproductive cavity-like structures meet the analytic prerequisite in the region of a few micron, they enhance the catalytic nucleation behavior during boiling.

The advantageous impacts of nano-inspired extended surface area, easy re-wetting, and byproductive cavity-like structures via long silicon nanowires could lead to improvement of spatial heat transfer uniformity and temporal stability. Kim et al. (2014b) evaluated local and temporal heat transfer characteristics on a boiling surface with silicon nanowires and they are attributed to the byproductive cavity-like structures via the in situ nanostructures, which can facilitate nucleation behaviors. The catalyzed ebullition could make heat transfer more stable with less local and temporal temperature fluctuations by diminishing bubble size and accelerating its easy detachment.

In addition, the functionalized silicon micro/nanostructures can guarantee enhancement of both CHF and HTC by more than 100%, as compared with an untreated plain surface. In order to achieve static wetting for a promising boiling interface manipulation, hydrophilic interfaces, which can be presented via nano/microscale functionalized structures, facilitate the coolant access to the boiling surface. Besides roughness, static wetting characteristics in terms of CAs have been recognized as a principal factor in boiling. However, a question remains, as to whether the complexity of the CHF results from the independent effects of the factors: roughness and static wetting. According to the classical aspects of the demonstrations, there is unclear connection between the observed quantitative performance variations and roughness or static wetting. For example, smaller CAs, even though they are obtained by either lowering the surface tension or due to superhydrophilic nanostructures, would be favorable for higher CHF. However, this cannot clearly explain the remarkable enhancement of silicon nanowired surfaces (Hanley et al., 2013; Kim et al., 2014a). In fact, we cannot presume the exact answer why the static CA does not clearly quantify the absolute value of CHF not in relative comparison, either. This questions whether the principal factors used so far are indeed independent and whether other undiscovered factors can explain this phenomenon. In an extension of such inference, we can bring up an additional peculiarity from nano-inspired functionality, that is, the interfacial hydrodynamics regarding dynamic wetting by morphologically driven hemi-wicking.

11.3.2.3 Near-field hydrodynamics via nanostructures

Boiling performance including CHF has been remarkably improved in combination with novel nanomaterial technologies. However, it arouses unresolved issues regarding CHF: what are the direct grounds of the recent breakthroughs in extending CHF by developing nano-inspired surfaces? Can we predict the CHF as a physical function of the design variables of the interfacial surfaces? Especially for CHF predictions and demonstrations, previous models, which mostly considered the two principal factors of surface roughness and wettability, could not perfectly explain the drastic enhancement of CHF via nano-inspired interfaces. Indeed, these previous models were insufficient to present a plausible physical demonstration revealing peculiar contributions of nanoscale structures to boiling enhancement as briefly presented in Figure 11.23. In light of the peculiarities of nano-inspired interfaces as apparent reasons that led to performance improvements, the *near-field* hydrodynamic characteristics might provide a direct clue (Lu et al., 2011; Kim et al., 2016a).

Hemi-wicking, driven by morphological aspects determining the hydrodynamic conditions on the interfaces, can be one of the most important factors due to its direct liquid refreshing toward the boiling interface. Thus, recent studies have reported that the manipulation and improvement of hemi-wicking can contribute greatly to the enhancement of boiling performance. Liquid refreshing is required for a better boiling performance. In order to demonstrate the hemi-wicking process analytically, we first present a morphological requisite of the interface accompanying the wicking as $\theta^* < \theta_c$ expressed in (Equation 11.3), where $\cos\theta_c = (1 - \varphi)/(r - \varphi)$. In the presence of hemi-wicking, by using the proper nanostructure employments

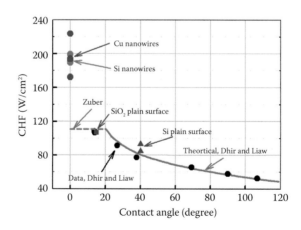

Figure 11.23 Experimental critical heat flux (CHF) results on various manipulated surfaces and analytic predictions of CHF considering static contact angle characteristics on a boiling surface (the analytic predictions are based on models suggested by Zuber (1959) and Dhir and Liaw (1989). (Reprinted with permission from Chen, R., et al., (2009), Nanowires for enhanced boiling heat transfer, *Nano Letters*, 9, 548–553. Copyright 2009 American Chemical Society.)

satisfying this prerequisite, the CHF prediction models can be further discussed. The maximum heat dissipation capacity should be balanced by the thermodynamically absorbed heat during a phase change of the refreshing liquids on a boiling surface. Then, it is possible to calculate the volumetric capacity of the porous interface for considering the solid fraction φ, which is already defined as the solid fraction of the solid–liquid interface contacting the liquid. Thus, $(1-\varphi)\cdot h$ corresponds to the volumetric capacity according to the height of the manipulated nanostructures on a unit area. Because the wicking coefficient W indicates the liquid propagation ability due to hemi-wicking (Washburn 1921; Tas et al., 2004; Ishino et al., 2007), the refreshing rate of the working fluid (liquid phase) can be approximated by W^2 (wicking speed per unit second) in the porous volume of the nano-inspired interface layer. Herein, the interface layer is confined to the solid–liquid interface with a thickness of h, which is the characteristic height of nanoscale structures leading to hemi-wicking. The thermodynamic equilibrium of CHF then can be explained considering the hydrodynamic refreshing of a liquid toward the interface. When a liquid perfectly propagates due to hemi-wicking through the interface (as schematically illustrated in Figure 11.24), the total amounts of refreshing liquid can be simply expressed as $(1-\varphi)\cdot h\cdot W^2$. Then, the total heat dissipation capacity on a unit area, that is, the amount of heat absorption for a phase change of the liquid, can be expressed as follows:

$$l_{lg}\rho_l h\left(1-\varphi\right)W^2 \tag{11.9}$$

where l_{lg} and ρ_l are the latent heat of a working fluid and its density in the liquid phase, respectively. In this physical demonstration, it is reasonable to assume that hemi-wicking occurs with a uniform velocity profile along a vertical direction from 0 (at the substrate) to h (at the top of the wicking-inducing structure) within the interface layer, and that refreshing is uniform in a unit area.

One of the limitations of this model is that it does not define the valid region pertaining to a spatial area. When there is no certain constraints, the model can cause misconception for defining the effective area of the boiling. On a confined boiling surface, the hydrodynamic stability would be maintained by balancing the up-flowing vapor columns and the down-flowing liquid (Zuber 1959; Lienhard and Dhir 1973). Based on the hydrodynamic instability model described in Figure 11.25, CHF occurs when the hydrodynamic stability is lost, inducing a large retarding force on the down-flowing liquid (Lu et al., 2011). Therefore, the impacts of hemi-wicking can be confined considering the hydrodynamic stability by adopting the Rayleigh–Taylor interfacial stability wavelength at the effective length scale. The Rayleigh–Taylor interfacial stability wavelength λ_{RT} is given by $2\pi(\sigma/(g(\rho_l-\rho_v)))^{1/2}$ in the most critical cases (Lamb 1959;

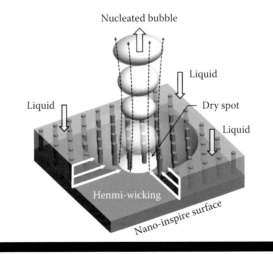

Figure 11.24 Schematic describing bubble nucleation in boiling heat transfer and counteracting interfacial hemi-wicking induced by nanostructures extending critical heat flux. (Reprinted with permission from Kim, B.S., et al., (2014), Interfacial wicking dynamics and its impact on critical heat flux of boiling heat transfer, *Applied Physics Letters*, 105, 191601. Copyright 2014 American Chemical Society.)

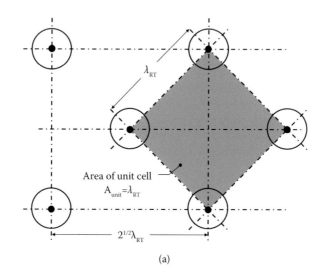

(a)

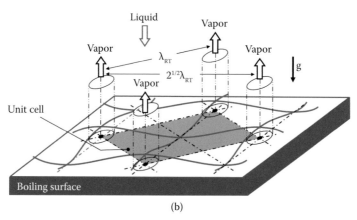

(b)

Figure 11.25 Conceptual schematics showing λ_{RT}. (a) Top view and (b) perspective view of the resulting λ_{RT} spaced array of locations of vapor rise and superpositioned liquid on an upward-facing horizontal surface. (Redrawn from Liter S., Pool-boiling enhancement and liquid choking limits within and above a modulated porous-layer coating, PhD Thesis, The University of Michigan, 2000.)

Liter 2000). The wavelength simply indicates the characteristic length of an unstable interface between two fluids of different densities (i.e., up-flowing vapor and down-flowing liquid). Therefore, the amount of heat dissipated by hemi-wicking can be revised using the concept of the valid heating area for hydrodynamic stability as a criterion (Kim et al., 2014a):

$$\frac{l_{lg}\,\rho_l h\left(1-\varphi\right)W^2}{\lambda_{RT}^2} \tag{11.10}$$

As the wicking-CHF model is not valid without the prerequisite of morphologically driven wicking, it cannot cover the CHF results on a plain surface, which do not lead to any wicking behavior. Thus, Kim et al. (2014a) suggested a compensating model using Kandlikar's model to demonstrate the approximation of CHFs on a plain surface as follows (Kandlikar 2001):

$$q''_{C,2} = l_{lg}\,\rho_v^{1/2}\left(\frac{1+\cos\theta}{16}\right)\left[\frac{4\sigma}{\lambda_{RT}} + \frac{\lambda_{RT}}{8}g\left(\rho_l - \rho_v\right)\left(1+\cos\theta\right)\right]^{1/2} \tag{11.11}$$

This approach is reasonable because the previous model was established considering the static wetting characteristics of the interface, and it is already validated compared with the experimental results, especially on a plain interface (Kandlikar 2001). The total amount of heat dissipation by interfacial re-wetting and subsequent phase changing can then be specified on the *near field*, and the critical amount of heat dissipation can be estimated by adopting the prerequisite stable boiling area with the critical characteristic length of λ_{RT} as follows (Kim et al., 2014a):

$$q''_{C,1} = \frac{C_1 l_{lg}\, \rho_l h\left(1-\varphi\right) W^2}{\lambda_{RT}^2}, \; q''_{CHF} = q''_{C,1} + q''_{C,2} \tag{11.12}$$

where the shape factor C_1 is a correlating coefficient with regard to h, defined as the specific wicking space. Figure 11.26 reported in a recent study shows that this converging approach of predicting CHFs on the hydrodynamics is in agreement with the experimental results (Kim et al., 2016a). The morphological design variable φ and the resulting capillary momentum of wickability over a manipulated interface can be the principal factors determining the CHF and can suggest a clue revealing the peculiarity of the nano-inspired interface on CHF. This model elucidates that the amount of critical heat dissipation can be incremented by reinforcing the capillary flow momentum through a structured forest as well as by decreasing the counteracting fluidic resistance with a lower φ (Washburn 1921; Tas et al., 2004; Quéré 2008; Kim et al., 2014a). Hemi-wicking is generated by an interfacial capillary pressure, and stronger interfacial refreshing leads to a greater liquid supply directly to a boiling surface.

We reviewed recent approaches regarding wickability, which is a quantitative indicator of hemi-wicking and reflects the confluent characteristics of both roughness and wettability. Strong hemi-wicking, through the interfacial capillary effect, can be important in boiling heat transfer performance by improving hydrodynamic stability against surface dry-out. Wickability can thus be a key factor for a reasonable model to account for the hydrodynamic criteria on CHF regarding interfacial liquid propagation and refreshment. Considering the effects of hemi-wicking on a boiling interface, we can conclude that the recent demonstrations and approaches can be used to put forward a physical model to predict CHF, which cannot be explained by classical models pertaining to just wettability or roughness independently (Hanley et al., 2013; Kim et al., 2016a).

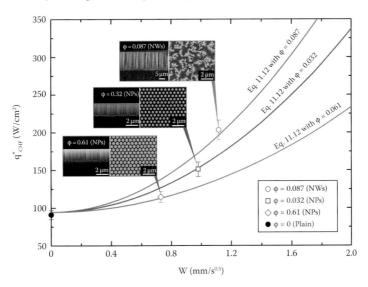

Figure 11.26 CHF estimations reflecting hydrodynamic criteria with regard to interfacial liquid refreshing based on hydrodynamic liquid accessibilities. As a parametric presentation, the solid fraction (φ) is demonstrated from each surface with SiNWs ($\varphi = 0.087$ with *davg* = 100 nm, *pavg* = 300 nm, and *havg* = 15 μm) and silicon nanopillars (SiNPs) ($\varphi = 0.32$ with *davg* = 360 nm, *pavg* = 610 nm, and *havg* = 2 μm and $\varphi = 0.61$ with *davg* = 500 nm, *pavg* = 610 nm, and *havg* = 2 μm). (Redrawn with permission from Kim et al., *Applied Physics Letters* 105, 191601, 2014.) Insets of SEM images present cross-sectional and top-view of the fabricated nanostructures. (Redrawn with permission from Kim et al., Scientific Reports 6:34348, 2016a).

11.3.3 POSSIBLE APPLICATIONS OF NANOSTRUCTURES IN HEAT TRANSFER

In light of previous efforts to enhance the boiling performance by employing nucleation sites and reinforcing surface wettability, micro-nano hybrid structures (MNHSs) (Vlad et al., 2008; Kuan and Chen 2009) may offer extraordinary boiling heat transfer performance. Specifically, hierarchical MNHSs can increase the boiling surface area with higher surface roughness significantly more than single-scale structures do while retaining the favorable wettability characteristics. However, there has been relatively little research on the fabrication and application of hierarchical MNHSs to further enhance the boiling heat transfer performance. Regarding the rarity of studies dealing with the applications of nanoscale surface manipulation in macroscale heat transfer performance improvement, recent approaches can be remarkable but still not known. The nano/microscale structures for functional interfaces will take their own validity by widening the application field of material synthesis, interface manipulations, and nanoscale researches, and exhibiting the possibility of performance improvements of macro-systems in the heat transfer field.

On the contrary, quite recent studies have aimed to uncover the underlying mechanism for the fluidic interactivity of nano-inspired structures, and extend its impact into the practical heat/energy transferring scheme of single-phase convection (Kim et al., 2017). The peculiarities of silicon-based nano-inspired interfaces come from their intrinsic functionality of extremely rough morphologies and subsequent fluidic interactivity. The interactivity of nanostructures with fluidic boundary layers can offer advantages of quasi-fin effects to dissipate thermal energy and diminish macroscale friction loss. Even though the friction loss and subsequent pressure drop are common trade-offs in heat transfer following implementation of macroscale surface roughening, strategic nanostructures can prevent these adverse effects by guaranteeing the stability of a specific fluidic boundary layer. In these strategies, achieving a balance in the characteristic lengths of the nanostructures can facilitate application in both fluidics and heat transfer. The recent findings and state-of-the-art demonstrations reveal the possibility of pragmatic applications in engineering fields related to heat and mass transfer for powerful and reliable cooling/energy transferring systems.

REFERENCES

Ayon AA, Braff R, Lin CC, Sawin HH, Schmidt MA. (1999). Characterization of a time multiplexed inductively coupled plasma etcher. *Journal of the Electrochemical Society* 146:339–349.

Betz AR, Xu J, Qiu HH, Attinger D. (2010). Do surfaces with mixed hydrophilic and hydrophobic areas enhance pool boiling? *Applied Physics Letters* 97:141909.

Bico J, Thiele U, Quéré D. (2002). Wetting of textured surfaces. *Colloids and Surfaces A-Physicochemical and Engineering Aspects* 206:41–46.

Chartier C, Bastide S, Lévy-Clément C. (2008). Metal-assisted chemical etching of silicon in HF-H_2O_2. *Electrochimica Acta* 53:5509-5516.

Chattopadhyay S, Li X, Bohn PW. (2002). In-plane control of morphology and tunable photoluminescence in porous silicon produced by metal-assisted electroless chemical etching. *Journal of Applied Physics* 91:6134–6140.

Chen C-Y, Wu C-S, Chou C-J, Yen T-J. (2008). Morphological control of single-crystalline silicon nanowire arrays near room temperature. *Advanced Materials* 20:3811–3815.

Chen R, et al. (2009). Nanowires for enhanced boiling heat transfer. *Nano Letters* 9:548–553.

Cheng SL, Chung CH, Lee HC. (2008). A study of the synthesis, characterization, and kinetics of vertical silicon nanowire arrays on (001) Si substrates. *Journal of The Electrochemical Society* 155:D711–D714.

Choi C, et al. (2010). Strongly superhydrophobic silicon nanowires by supercritical CO_2 drying. *Electronic Materials Letters* 6:59–64.

Choi CH, Ulmanella U, Kim J, Ho CM, Kim CJ. (2006). Effective slip and friction reduction in nanograted superhydrophobic microchannels. *Physics of Fluids* 18:087105.

Chu K-H, Enright R, Wang EN. (2012). Structured surfaces for enhanced pool boiling heat transfer. *Applied Physics Letters* 100:241603.

Chu K-H, Joung YS, Enright R, Buie CR, Wang EN. (2013). Hierarchically structured surfaces for boiling critical heat flux enhancement. *Applied Physics Letters* 102:151602.

Dabbousi BO, Murray CB, Rubner MF, Bawendi MG. (1994). Langmuir-Blodgett manipulation of siye-selected CdSe nanocrzstallites. *Chemistry of Materials* 6:216–219.

Dawood MK, et al. (2011). Mimicking both petal and lotus effects on a single silicon substrate by tuning the wettability of nanostructured surfaces. *Langmuir* 27:4126–4133.

Dhir VK. (1998). Boiling heat transfer. *Annual Review of Fluid Mechanics* 30:365–401

Dhir VK, Liaw SP. (1989). Framework for a unified model for nucleate and transition pool boiling. *Journal of Heat Transfer* 111:739–746.

Han H, Huang Z, Lee W. (2014). Metal-assisted chemical etching of silicon and nanotechnology applications. *Nano Today* 9:271–304.

Hanley H, et al. (2013). Separate effects of surface roughness, wettability, and porosity on the boiling critical heat flux. *Applied Physics Letters* 103:024102.

Hochbaum AI, et al. (2008). Enhanced thermoelectric performance of rough silicon nanowires. *Nature* 451:163–U165.

Hsu CC, Chen PH. (2012). Surface wettability effects on critical heat flux of boiling heat transfer using nanoparticle coatings. *International Journal of Heat and Mass Transfer* 55:3713–3719.

Hsu CM, Connor ST, Tang MX, Cui Y. (2008). Wafer-scale silicon nanopillars and nanocones by Langmuir-Blodgett assembly and etching. *Applied Physics Letters* 93:133109.

Hsu YY. (1962). On the size range of active nucleation cavities on a heating surface. *Journal of Heat Transfer-Transactions of the ASME* 84:207–213.

Huang Z, Geyer N, Werner P, de Boor J, Gösele U. (2011). Metal-assisted chemical etching of silicon: A review. *Advanced Materials* 23:285–308.

Huang Z, et al. (2010). Oxidation rate effect on the direction of metal-assisted chemical and electrochemical etching of silicon. *The Journal of Physical Chemistry C* 114:10683–10690.

Huang ZP, Fang H, Zhu J. (2007). Fabrication of silicon nanowire arrays with controlled diameter, length, and density. *Advanced Materials* 19:744–748.

Huang ZP, et al. (2009). Ordered arrays of vertically aligned 110 silicon nanowires by suppressing the crystallographically preferred etching directions. *Nano Letters* 9:2519–2525.

Ishino C, Reyssat M, Reyssat E, Okumura K, Quéré D. (2007). Wicking within forests of micropillars. *Europhysics Letters* 79:56005.

Jokinen V, Sainiemi L, Franssila S. (2008). Complex droplets on chemically modified silicon nanograss. *Advanced Materials* 20:3453–3456.

Kandlikar SG. (2001). A theoretical model to predict pool boiling CHF incorporating effects of contact angle and orientation. *Journal of Heat Transfer-Transactions of the ASME* 123:1071–1079.

Khanikar V, Mudawar I, Fisher T. (2009). Effects of carbon nanotube coating on flow boiling in a micro-channel. *International Journal of Heat and Mass Transfer* 52:3805–3817.

Kim BS, Choi G, Shin S, Gemming T, Cho HH. (2016a). Nano-inspired fluidic interactivity for boiling heat transfer: impact and criteria. Scientific Reports 6:34348.

Kim BS, Choi G, Shim DI, Kim KM, Cho HH (2016b). Surface roughening for hemi-wicking and its impact on convective boiling heat transfer. International Journal of Heat and Mass Transfer 102:1100–110.

Kim BS, Lee H, Shin S, Choi G, Cho HH. (2014a). Interfacial wicking dynamics and its impact on critical heat flux of boiling heat transfer. *Applied Physics Letters* 105:191601.

Kim BS, et al. (2014b). Stable and uniform heat dissipation by nucleate-catalytic nanowires for boiling heat transfer. *International Journal of Heat and Mass Transfer* 70:23–32.

Kim BS, Shin S, Shin SJ, Kim KM, Cho HH. (2011a). Control of superhydrophilicity/superhydrophobicity using silicon nanowires via electroless etching method and fluorine carbon coatings. *Langmuir* 27:10148–10156.

Kim BS, Shin S, Shin SJ, Kim KM, Cho HH. (2011b). Micro-nano hybrid structures with manipulated wettability using a two-step silicon etching on a large area. *Nanoscale Research Letters* 6:333.

Kim J, Kim YH, Choi S-H, Lee W. (2011c). Curved silicon nanowires with ribbon-like cross sections by metal-assisted chemical etching. *ACS Nano* 5:5242–5248.

Kim SJ, Bang IC, Buongiorno J, Hu LW. (2007). Surface wettability change during pool boiling of nanofluids and its effect on critical heat flux. *International Journal of Heat and Mass Transfer* 50:4105–4116.

Kim S-M, Khang D-Y. (2014). Bulk micromachining of Si by metal-assisted chemical etching. *Small* 10:3761–3766.

Kosiorek A, Kandulski W, Chudzinski P, Kempa K, Giersig M. (2004). Shadow nanosphere lithography: Simulation and experiment. *Nano Letters* 4:1359–1363.

Kuan WF, Chen LJ. (2009). The preparation of superhydrophobic surfaces of hierarchical silicon nanowire structures. *Nanotechnology* 20:035605.

Lamb H. (1959). *Hydrodynamics*. Cambridge: Cambridge University Press.

Lau KKS, et al. (2003). Superhydrophobic carbon nanotube forests. *Nano Letters* 3:1701–1705.

Li C, et al. (2008a). Nanostructured copper interfaces for enhanced boiling. *Small* 4:1084–1088.

Li D, et al. (2012). Enhancing flow boiling heat transfer in microchannels for thermal management with monolithically-integrated silicon nanowires. *Nano Letters* 12:3385–3390.

Li SH, Furberg R, Toprak MS, Palm B, Muhammed M. (2008b). Nature-inspired boiling enhancement by novel nanostructured macroporous surfaces. *Advanced Functional Materials* 18:2215–2220.

Li XR, et al. (2008c). Photolithographic approaches for fabricating highly ordered nanopatterned arrays. *Nanoscale Research Letters* 3:521–523.

Lianto P, Yu S, Wu J, Thompson CV, Choi WK. (2012). Vertical etching with isolated catalysts in metal-assisted chemical etching of silicon. *Nanoscale* 4:7532–7539.

Lienhard JH, Dhir VK. (1973). *Extended hydrodynamic theory of the peak and minimum pool boiling heat fluxes.* NASA Report (CR2270). Washington, DC: NASA.

Liter S. (2000). Pool-boiling enhancement and liquid choking limits within and above a modulated porous-layer coating. PhD Thesis, The University of Michigan.

Lu MC, Chen RK, Srinivasan V, Carey V, Majumdar A. (2011). Critical heat flux of pool boiling on Si nanowire array-coated surfaces. *International Journal of Heat and Mass Transfer* 54:5359–5367.

Mai TT, et al. (2012). Dynamics of wicking in silicon nanopillars fabricated with interference lithography and metal-assisted chemical etching. *Langmuir* 28:11465–11471.

Megouda N, Hadjersi T, Piret G, Boukherroub R, Elkechai O. (2009). Au-assisted electroless etching of silicon in aqueous HF/H_2O_2 solution. *Applied Surface Science* 255:6210–6216.

Patankar NA. (2010). Supernucleating surfaces for nucleate boiling and dropwise condensation heat transfer. *Soft Matter* 6:1613–1620.

Peng KQ, et al. (2006). Fabrication of single-crystalline silicon nanowires by scratching a silicon surface with catalytic metal particles. *Advanced Functional Materials* 16:387–394.

Peng KQ, Lu AJ, Zhang RQ, Lee ST. (2008). Motility of metal nanoparticles in silicon and induced anisotropic silicon etching. *Advanced Functional Materials* 18:3026–3035.

Peng KQ, Wang X, Wu XL, Lee ST. (2009). Fabrication and photovoltaic property of ordered macroporous silicon. *Applied Physics Letters* 95:143119.

Peng KQ, et al. (2005). Uniform, axial-orientation alignment of one-dimensional single-crystal silicon nanostructure arrays. *Angewandte Chemie-International Edition* 44:2737–2742.

Peng KQ, Yan YJ, Gao SP, Zhu J. (2002). Synthesis of large-area silicon nanowire arrays via self-assembling nanoelectrochemistry. *Advanced Materials* 14:1164–1167.

Peng KQ, Yan YJ, Gao SP, Zhu J. (2003). Dendrite-assisted growth of silicon nanowires in electroless metal deposition. *Advanced Functional Materials* 13:127–132.

Peng KQ, et al. (2007). Ordered silicon nanowire arrays via nanosphere lithography and metal-induced etching. *Applied Physics Letters* 90:163123.

Phan HT, Caney N, Marty P, Colasson S, Gavillet J. (2009). Surface wettability control by nanocoating: The effects on pool boiling heat transfer and nucleation mechanism. *International Journal of Heat and Mass Transfer* 52:5459–5471.

Pioro IL, Rohsenow W, Doerffer SS. (2004). Nucleate pool-boiling heat transfer. I: Review of parametric effects of boiling surface. *International Journal of Heat and Mass Transfer* 47:5033–5044.

Quéré D. (2008). Wetting and roughness. *Annual Review of Materials Research* 38:71–99.

Shibuichi S, Onda T, Satoh N, Tsujii K. (1996). Super water-repellent surfaces resulting from fractal structure. *Journal of Physical Chemistry* 100:19512–19517.

Shin S, Kim BS, Choi G, Lee H, Cho HH. (2012). Double-templated electrodeposition: Simple fabrication of micro-nano hybrid structure by electrodeposition for efficient boiling heat transfer. *Applied Physics Letters* 101:251909.

Shin S, et al. (2011). Tuning the morphology of copper nanowires by controlling the growth processes in electrodeposition. *Journal of Materials Chemistry* 21:17967–17971.

Shoji M, Takagi Y. (2001). Bubbling features from a single artificial cavity. *International Journal of Heat and Mass Transfer* 44:2763–2776.

Srivastava SK, et al. (2010). Excellent antireflection properties of vertical silicon nanowire arrays. *Solar Energy Materials and Solar Cells* 94:1506–1511.

Suzuki T, et al. (2010). Nonvolatile buffer coating of titanium to prevent its biological aging and for drug delivery. *Biomaterials* 31:4818–4828.

Tas NR, Haneveld J, Jansen HV, Elwenspoek M, van den Berg A. (2004). Capillary filling speed of water in nano-channels. *Applied Physics Letters* 85:3274–3276.

Tsougeni K, Papageorgiou D, Tserepi A, Gogolides E. (2010). "Smart" polymeric microfluidics fabricated by plasma processing: Controlled wetting, capillary filling and hydrophobic valving. *Lab on a Chip* 10:462–469.

Tuteja A, et al. (2007). Designing superoleophobic surfaces. *Science* 318:1618–1622.

Verplanck N, Coffinier Y, Thomy V, Boukherroub R. (2007). Wettability switching techniques on superhydrophobic surfaces. *Nanoscale Research Letters* 2:577–596.

Vlad A, et al. (2008). Nanowire-decorated microscale metallic electrodes. *Small* 4:557–560.

Wang R, Sakai N, Fujishima A, Watanabe T, Hashimoto K. (1999). Studies of surface wettability conversion on TiO_2 single-crystal surfaces. *Journal of Physical Chemistry B* 103:2188–2194.

Washburn EW. (1921). The dynamics of capillary flow. *Physical Review* 17:273–283.

Wei JJ, Guo LJ, Honda H. (2005). Experimental study of boiling phenomena and heat transfer performances of FC-72 over micro-pin-finned silicon chips. *Heat and Mass Transfer* 41:744–755.

Wei-Fan K, Li-Jen C. (2009). The preparation of superhydrophobic surfaces of hierarchical silicon nanowire structures. *Nanotechnology* 20:035605.

Wenzel RN. (1936). Resistance of solid surfaces to wetting by water. *Industrial and Engineering Chemistry* 28:988–994.

Wu W, Dey D, Memis OG, Katsnelson A, Mohseni H. (2008). A novel self-aligned and maskless process for formation of highly uniform arrays of nanoholes and nanopillars. *Nanoscale Research Letters* 3:123–127.

Wu Z, Sundén B. (2014). On further enhancement of single-phase and flow boiling heat transfer in micro/minichannels. *Renewable and Sustainable Energy Reviews* 40:11–27.

Xiu Y, Zhu L, Hess DW, Wong CP. (2007). Hierarchical silicon etched structures for controlled hydrophobicity/superhydrophobicity. *Nano Letters* 7:3388–3393.

Yoon S-S, Khang D-Y. (2012). Switchable wettability of vertical Si nanowire array surface by simple contact-printing of siloxane oligomers and chemical washing. *Journal of Materials Chemistry* 22:10625–10630.

Yoshimitsu Z, Nakajima A, Watanabe T, Hashimoto K. (2002). Effects of surface structure on the hydrophobicity and sliding behavior of water droplets. *Langmuir* 18:5818–5822.

You S, Kim J, Kim K. (2003). Effects of nanoparticles on critical heat flux of water in pool boiling heat transfer. *Applied Physics Letters* 83:3374–3376.

Yu CK, Lu DC, Cheng TC. (2006). Pool boiling heat transfer on artificial micro-cavity surfaces in dielectric fluid FC-72. *Journal of Micromechanics and Microengineering* 16:2092–2099.

Zuber N. (1959). *Hydrodynamic aspects of boiling heat transfer.* Ph.D. thesis, Research Laboratory, Los Angeles and Ramo-Wooldridge Corporation, University of California, Los Angeles, CA.

12 Electrodeposited silicon from ionic liquids

Frank Endres

Contents

12.1 INTRODUCTION

Electrodeposition or electrosynthesis of any material is dependent on the electrochemical window of the electrolyte, the electrode material, the bath temperature, additives, and the electrochemical parameters, to mention the most important pertinent properties. Let us have a look at an undergraduate experiment, which is quite easy to perform even in a high school chemical laboratory, namely copper electroplating. The teacher will use an aqueous solution containing a copper salt, for example, $CuSO_4$, mixed with an acidic electrolyte to ensure conductivity, usually H_2SO_4; a steel plate to be plated with copper; and a copper plate for a counterelectrode. If a standard AA primary battery with 1.5 V nominal voltage is connected to this simple cell (minus pole = cathode: steel; plus pole = anode: copper plate), one can see with the naked eye the steel plate being slowly covered by a dull, reddish copper layer, and the copper counterelectrode as such being roughened, manifested as either darkening or brightening of the electrode, depending on the exact conditions. The following reactions occur:

$$\text{Cathode:} \quad Cu^{2+} + 2e^- \rightarrow Cu$$
$$\text{Anode:} \quad Cu \rightarrow Cu^{2+} + 2e^-$$

This reaction will run until all Cu at the anode has been consumed and deposited on the steel plate, and the original $CuSO_4$ concentration in the electrolyte will remain constant. With this simple experiment, the steel plate is covered with Cu, but the Cu will neither adhere, nor will it be bright. Bright deposits require additives and—depending on the system—special electrochemical techniques. If Cu deposition is well performed, the Faraday efficiency of this reaction will be 100%, that is, all electricity will be used for the synthesis of Cu and no other reaction occurs. In principle, the H^+ ions from H_2SO_4 could be reduced to H_2, but the standard electrode potential of Cu/Cu^{2+} in aqueous solutions will be +350 mV versus NHE (normal hydrogen electrode), thus positive enough to avoid hydrogen formation during Cu plating. The electrochemical window of water is determined by the thermodynamics of the hydrogen combustion reaction:

$$2 H_2 + O_2 \rightarrow 2 H_2O$$

The heat of formation ΔG consist of ΔH and ΔS, giving

$$\Delta G = \Delta H - T * \Delta S \tag{12.1}$$

This equation, which is derived in all textbooks of physical chemistry, can be related to the electrode potential by:

$$\Delta G = -z * F * \Delta E \tag{12.2}$$

where z is the number of exchanged electrons, and F is the Faraday constant (96485 Coulomb/mol).

Although not an exact derivation, this formula can be easily remembered, if an electron in a vacuum between two differently charged plates is considered. In this case, the force on the electron in scalar notation is:

$$F = q * E \tag{12.3}$$

where q is the charge of the electron and E is the electric field.

After integration over x ($W = F*x$), it results in $U = E*x$:

$$W = q * U \tag{12.4}$$

The unit of ΔG and W is joule (or J/mol), and the unit of U and ΔE is volt. The charge (q or z*F) is thus the proportionality constant between energy and voltage. With the values for the hydrogen combustion reaction (see textbooks of physical chemistry) it results, that the thermodynamic electrochemical window of water is 1.23 V at 298 K. It is limited in the cathodic regime by hydrogen evolution ($2H^+ + 2e^- \rightarrow H_2$) and in the anodic regime by water oxidation ($H_2O \rightarrow 2H^+ + ½ O_2$). At pH = 0, the electrode potential for the hydrogen evolution is 0 volt (enabling Cu plating) if the activity of hydrogen is 1. Thus, under these conditions, the electrode potential for oxygen evolution is +1.23 volt. Both electrode potentials are pH dependent, but the difference remains constant at all pH values. Thus, at pH = 14, the difference is still 1.23 Volt, but hydrogen evolution occurs at −826 mV; with reference to NHE, oxygen evolution occurs at +404 mV versus NHE. If we have a look at the electrochemical potential scale, the electrode potential for the deposition of zinc $E_0(Zn/Zn^{2+})$ is found to be at about −800 mV versus NHE. Thus, although zinc is thermodynamically quite reactive, it is stable in concentrated basic solutions and can thus be deposited from a suitable basic aqueous electrolyte. This is the principle of industrial zinc plating. What would happen if we intended to deposit silicon from an aqueous solution? In a very naive approach, one would think about adding a silicon compound (e.g., $SiCl_4$) to an aqueous solution. One would—in this naive approach—expect that the following electrochemical reaction occurs:

$$SiCl_4 + 4 e^- \rightarrow Si + 4 Cl^-$$

In fact, the experiment would already fail when $SiCl_4$ is added to water due to an immediate and exothermic chemical reaction:

$$SiCl_4 + 4 H_2O \rightarrow Si(OH)_4 + 4 HCl$$

$Si(OH)_4$ will finally age to SiO_2. The electrode potential of $Si/SiCl_4$ (dependent on the electrolyte) is about −2 Volt versus NHE. The exothermic reaction with water leads, according to Equations 12.1 and (12.2), to thermodynamically more stable SiO_2, thus shifting the electrode potential for Si deposition from

SiO$_2$ to way more negative electrode potentials, making Si deposition from aqueous electrolytes simply impossible under all conditions. Just hydrogen would evolve. As a conclusion, all reactive elements must be deposited under water- and oxygen-free conditions, ideally in an inert gas glovebox. How to deposit elemental silicon? If water is not stable enough, one can consider an organic solvent as the basis of an electrolyte. Organic electrolytes like, for example, dimethylsulfoxide (DMSO), acetonitrile (AcN), or propylenecarbonate (PC) with a respective conductivity salt have much wider electrochemical windows than water. PC, as an example, allows the electrodeposition of lithium, enabling today's lithium ion batteries. Lithium has an electrode potential E_0(Li/Li$^+$) of roughly −3 Volt versus NHE, and both thermodynamic and kinetic reasons allow the deposition of Li. Thus, PC and other organic solvents are in principle stable enough to allow the electrodeposition of Si from SiCl$_4$,as an example. The reader is referred to references [1–3] for an insight into Si deposition from organic electrolytes. When doing the first experiments on Si deposition, we struggled a lot with organic solvents. On the one hand they are difficult to dry to water values below 10 ppm or to keep such low water contents under the conditions of the experiment even in an inertgas glovebox, on the other hand the deposition of Si also leads to decomposition products of the organic solvent that can strongly alter the quality of the deposit. On top, organic solvents can attack polymers, thus the synthesis of nanowires by the help of templates or of photonic crystals by the help of artificial polymer opal structures is quite difficult. For this reason, we decided to focus on ionic liquids.

12.2 IONIC LIQUIDS

Especially in the recent 10 years, there have been a lot of worldwide activities dealing with ionic liquids. Regarded as an exotic laboratory curiosity just 15 years ago, there were almost 10,000 papers published on this topic in 2015 alone. Thus, what makes ionic liquids that interesting that several thousands of papers are published every year? A soft definition suggests that an ionic liquid is a liquid, solely consisting of cations and anions with a melting point of <100°C. That is what can be found, for example, in Wikipedia or in many original and review articles. Having a look on an early paper from Paul Walden [4], who is more or less regarded as the "father of ionic liquids," one can find that he just stated that the alkylammonium compounds, that he synthesized, had melting points around 100°C. He stated that these compounds with a melting point of around 100°C behave like high-temperature molten salts—not more and not less. Nevertheless, the above-mentioned soft definition is widely accepted now, and most of the activities are concentrated on ionic liquids that are liquid below but not far above 20°C. Until today, roughly 1000 different ionic liquids have been published, at least 300 of them are commercially available in different qualities. Figure 12.1 shows a snapshot of often-used cations and anions, however, without completeness.

For an insight into the versatility of ionic liquids in synthesis, physical chemistry, and electrochemistry, the reader is referred to several books [5–8]. Here, only the most important properties for electrochemistry are summarized.

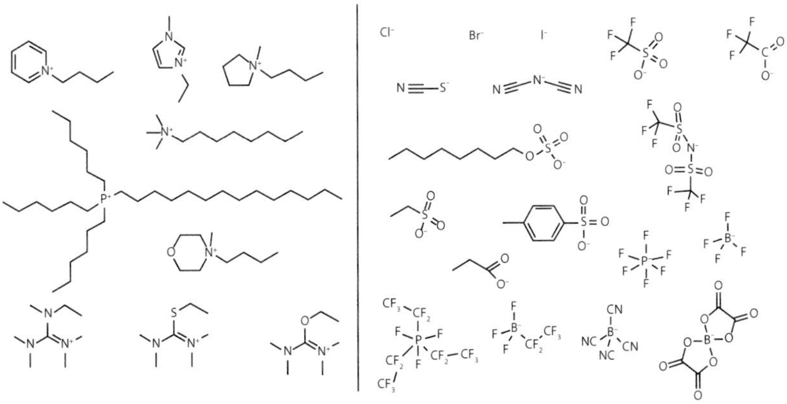

Figure 12.1 Often-reported cations and anions of ionic liquids.

Depending on the individual liquid, temperature, and impurities, ionic liquids have the following properties:

- Wide electrochemical windows of 3–6 V
- Wide thermal windows of up to 300 K; thus, they are stable between about –50°C and +250°C
- in most cases, a negligible vapor pressure of up to 100°C, thus enabling variable temperature electro-chemistry or plasma electrochemistry and a deep drying under vacuum
- Low interfacial tensions; thus, even polymers are quite well wetted by them

Furthermore, ionic liquids form on electrode surfaces so-called "solvation layers," which are altered by temperature and solutes and can influence electrochemical reactions [9]. They are presumably the reason why in one ionic liquid a metal is deposited as a microcrystalline material, in another one with another type of cation, however, it is deposited as a nanocrystalline material. So far, we cannot yet precisely predict which liquid leads to a nanomaterial, and which one leads to a micromaterial. The interactions between a solute and the respective ionic liquid, both of them with the interface, and the influence of temperature and electrode potential on these interactions are far from being deeply understood. In a recent paper [10], we showed from the example of Ta deposition from TaF_5 that there is a complex electrochemical behavior. In one ionic liquid, the Ta deposition is possible. In situ atomic force microscopy (AFM) measurements showed that TaF_5 strongly disturbs the solvation layers. Thus, TaF_5 approaches the electrode surface and is subsequently reduced to elemental Ta. In another liquid with the same cation but with different anion, there is no difference between the solvation layers with/without TaF_5; the electrochemical reduction currents are practically zero and consequently no Ta can be deposited. For details on this complicated behavior, the reader is referred to [10] and references therein. The properties of ionic liquids and their chemical versatility are, maybe, the reason why there are so many activities in various fields.

12.3 ELECTRODEPOSITION OF SILICON

12.3.1 THE ELECTROCHEMICAL WINDOW OF AN IONIC LIQUID

Figure 12.2 shows the cyclic voltammogram (CV) of 1-butyl-1-methylpyrrolidinium tris(pentafluoroethyl) trifluorophosphate ($[Py_{1,4}]FAP$) on Au(111) with a scan rate of 10 mV/s. This liquid is suitable to the electrodeposition of Si. Cyclic voltammetry [11] is one of the first experiments an electrochemist performs, as relatively easily a first insight into the electrochemical behavior of a system is obtained.

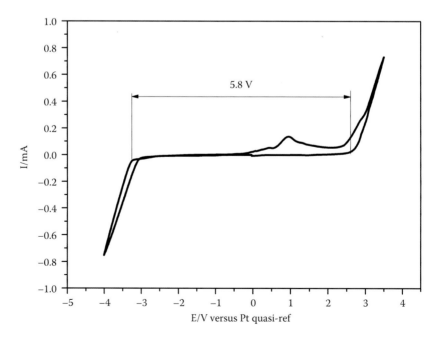

Figure 12.2 Electrochemical window of $[Py_{1,4}]FAP$ on Au(111), 22°C, v =10 mV/s.

Between the cathodic limit, where in this case the irreversible decomposition of the cation starts, and the anodic limit, where the oxidation of gold starts, there are only quite low currents. A fundamental view on the processes occurring between these limits shows that complicated surface processes occur that are due to gold reconstruction and ionic liquid (IL) adsorption, see, for example, Ref. [12]. Apart from these rather fundamental processes, one can see in the CV that the electrochemical window at around 20°C is almost 6 V, thus more than 4 × better than the one of aqueous electrolytes; it is strongly dependent on the electrode material and a little dependent on the temperature. Usually, the electrochemical window gets narrower when the temperature is increased, but at 150°C, $[Py_{1,4}]FAP$ has still an electrochemical window of more than 4 V, depending on the electrode material. There is no general rule for the width of IL electrochemical windows; furthermore, kinetic and thermodynamic contributions are difficult to distinguish. As Si is quite a reactive material, so far only liquids with pyrrolidinium or tetraalkylammonium ions are suitable to Si electrodeposition.

12.3.2 SOLVATION LAYERS

When we started with ionic liquids almost 20 years ago, it was a certain surprise that the electrochemical behavior is not only far away from the one of the aqueous or organic electrolytes, we also found that ionic liquids can influence electrochemical reactions. The electrodeposition of Al from three different ionic liquids having all the same anion showed that in two of them nanocrystalline Al was obtained, whereas in the other one, under all conditions, a microcrystalline Al was obtained [13]. At that time, we speculated that the adsorption of the IL to the electrode surface must somehow be different and lead to a grain refinement in some cases. Together with Rob Atkin (Australia), we investigated in a series of papers the interfacial behavior of different ionic liquids with the help of an AFM. In short, the cantilever is approached to the surface under electrochemical control. On the way down in the IL, the deflection is constant until the cantilever undergoes an interaction. If this is the case, the cantilever is bended, a signal is detected, and the approach curves can be transformed to force/distance curves by a simple algorithm [14]. Figure 12.3 shows the force curves of $[Py_{1,4}]TFSA$ at three different electrode potentials on Au(111) [15].

There are a few interesting observations. First, several nanometers above the electrode surface, there is a layer that the cantilever has to push through. Second, the force to push through such layers increases nearer the surface. Third, especially with more negative electrode potentials the forces increase, compared to the open-circuit potential, where no electrode potential was applied. The width of the steps can be correlated to the size of cations or to the size of an ion pair. For details, the reader is referred to our original papers. It should be mentioned that these solvation layers are not only dependent on the liquid itself and the electrode potential, the electrode material also plays a role as well as temperature and solute, for example, $SiCl_4$ concentration. Thus, it has to be expected that the electrodeposition of materials is altered by such layers and no one should expect that all ILs behave in the same way. The interaction of $SiCl_4$ or $SiBr_4$ with ionic liquids and both of them with the electrode surface are quite complicated and under study at TU Clausthal. No such layers, by the way, have been observed in aqueous or organic solutions. These solvation layers might explain why ILs have an excellent wetting behavior for solid substrates.

12.3.3 CYCLIC VOLTAMMETRY OF $SiCl_4$

Despite the complicated interfacial behavior, $[Py_{1,4}]TFSA$ is suited to the electrodeposition of Si. Figure 12.4 shows a CV of $[Py_{1,4}]TFSA/SiCl_4$ (saturated) at 10 mV/s.

The CV of Si deposition shows a prominent reduction current with a peak, which is correlated to the electrodeposition of Si. This has been proved later with an electrochemical quartz crystal microbalance (EQCM) by Komadina et al. [16] and ourselves. The EQCM data suggest that $SiCl_4$ is reduced in a four-electron step to elemental Si on the timescale of the electrochemical experiment. When presenting single CV's at a constant scan rate as here, often the question arises why the scan rate has not been varied to check for the mechanism of the reaction. In our experience, there are several reasons why results on the scan rate dependence should not be overinterpreted. In the case of Si deposition, the reaction is irreversible. This is due to the fact that under the conditions presented, the Si is not stripped upon applying a more positive potential. The failed re-oxidation has to do with complexation chemistry. Furthermore, often it occurs that the CVs change from cycle to cycle, which we can correlate to the varying interfacial layers. Still, unpublished results

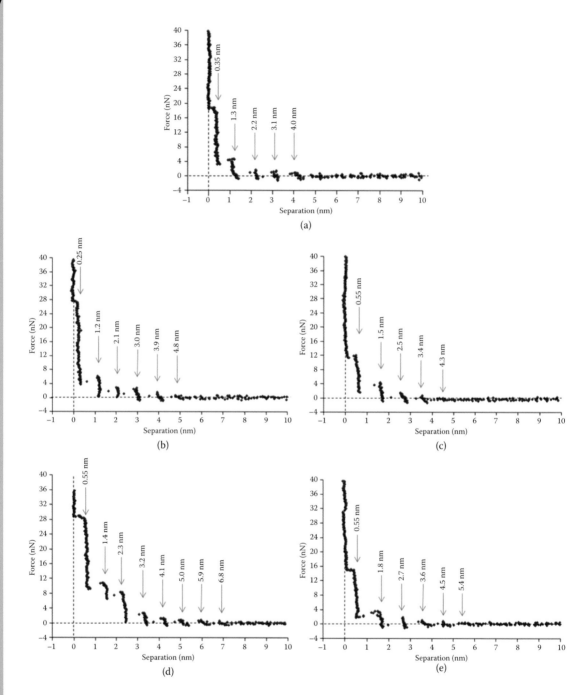

Figure 12.3 Typical force versus distance profiles for an AFM tip approaching a Au(111) surface in $[Py_{1,4}]FAP$ at (a) open-circuit potential (OCP, −0.16 V), (b) −1.0 V (vs. Pt), (c) +1.0 V (vs. Pt), (d) −2.0 V (vs. Pt), and (e) +2.0 V (vs. Pt). (Reprinted with permission from Hayes, R. et al., Double layer structure of ionic liquids at the Au(111) electrode interface: An atomic force microscopy investigation, *J. Phys. Chem. C*, 115, 6855–6863. Copyright 2011 American Chemical Society.)

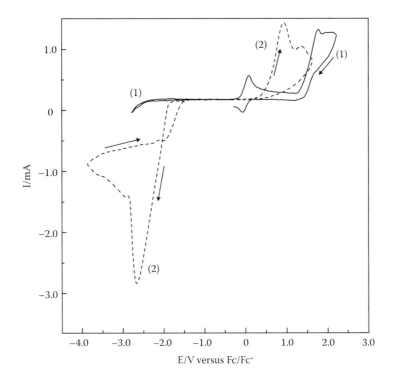

Figure 12.4 Electrochemical window of [Py$_{1,4}$]TFSA on HOPG with ferrocene/ferrocinium as internal reference (1). Cyclic voltammogram of [Py$_{1,4}$]TFSA saturated with SiCl$_4$. (From Zein El Abedin, S., et al., *Electrochem. Commun.*, 6, 510, 2004. With permission.)

on a relatively simple model system reveal that during deposition, the solvation layers are altered. If after CV the electrode potential is held constant for a moment, the layers re-arrange, and the interfacial behavior is no more the same as it was a short moment before. If this occurs, it is more or less meaningless to evaluate such data with models that focus solely on diffusion or electron exchange and that have been developed for aqueous electrolytes. This does not mean that such experiments are worthless, it strongly depends on the electrode material and on the individual ionic liquid. The Al deposition from AlCl$_3$-based ionic liquids on tungsten shows an almost reversible behavior, as it is known from aqueous electrolytes. We interpret it such that such liquids, maybe, do not form strongly adherent solvation layers, and thus, the electrochemical behavior is similar to that of aqueous solutions. There is no general rule, and there is no other way but to investigate in detail any IL-based electrolyte for its own. The minimum information from a CV is to find a suited electrode potential for the electrosynthesis of the material.

12.3.4 DEPOSITION OF Si FROM SiCl$_4$

From the point of view of the electrosynthesis, the most important information one gets from a CV is the electrode potential for deposition. Having a look on Figure 12.4 suggests that the electrode potential, where the reduction current has a peak, is a good starting point, in this case about –2.7 V versus the reference electrode [17]. If the Si deposition is done for a while on a gold substrate (the time can vary from a few seconds to minutes, depending on the IL), we can get deposits as shown in Figure 12.5.

The deposit shows a granular structure with a certain porosity. The interesting information is that the particles are built-up of smaller particles, and in that early experiment, we found smallest particle sizes of around 50 nm. The granular growth and the nonuniform deposition are the result of interfacial effects and of the low electronic conductivity of elemental Si. Usually, Si deposited at or near room temperature is amorphous, and no peaks are obtained in the X-ray diffraction pattern. Furthermore, if the ionic liquid is removed from the material, it can happen that the Si is washed-off (like with highly oriented pyrolytic graphite, HOPG, as the substrate) or it is oxidized under air on the way to the X-ray photoelectron

500 nm

Figure 12.5 SEM image of electrodeposited Si, made potentiostatically at −2,7 V versus Fc/Fc⁺. (From Zein El Abedin, S., et al., *Electrochem. Commun.*, 6, 510, 2004. With permission.)

spectroscopy (XPS) chamber. Thus, an analysis is not necessarily trivial. In our early approach, we performed tunneling spectroscopy. If Si deposition is performed on the nanoscale and probed with a scanning tunneling microscope (STM), the bandgap of the deposited material can be determined. Figure 12.6 shows one of the first tunneling spectra that we obtained on Si that has been deposited on HOPG [17].

Whereas HOPG has no band gap, a film with a thickness of about 100 nm shows a band gap of around 1 eV. In this case, the STM would probe the indirect band gap of Si. If this Si is rinsed with a suited solvent and kept even under the conditions of argon with water and oxygen below 1 ppm, a surface oxidation cannot be avoided. In Ref. [18], we investigated the surface oxidation and found that on this amorphous Si a several nanometer thick oxide layer forms quite rapidly even in an argon-filled glovebox. Thus, the electrodeposited Si is amorphous and subject to a rapid oxidation. In a recent master's thesis [19], we investigated the kinetics of Si oxidation under air. In short, Si was deposited in a glovebox from an ionic liquid in an EQCM cell. The substrate was as well as possible rinsed with an ionic liquid to remove the remaining electrolyte. The cell was closed with an adhesive tape and put under air. With the EQCM running, a small hole was cut into the tape so that air could enter the EQCM cell. After a short period, the oxidation of this Si began, which can be seen in a frequency shift of the resonance frequency and the damping of the quartz (Figure 12.7).

After just 20 min, the oxidation was complete. Thus, if amorphous nanoscale Si is made by electrodeposition in an ionic liquid, care has to be taken to avoid oxidation. This can be achieved by a surface passivation with organic or inorganic groups or by covering the Si inside of the glovebox with a paint. Unpublished results reveal that under certain conditions at elevated temperatures, crystalline Si is obtained by electrodeposition; furthermore, the amorphous Si can be transformed to crystalline one by annealing under high vacuum.

12.3.5 SiGe ELECTRODEPOSITION

The silicon–germanium alloy has been the focus of scientists for decades, as this material, where the composition can be varied over a wide range, is of high interest in semiconductor industry. We were simply curious if, based on our experience with the electrodeposition of Ge and Si, SiGe can be codeposited. This was a fundamental approach; we were not interested in any application. The experimental approach was quite simple. We selected a suitable ionic liquid, added SiCl₄ and GeCl₄, and ran a CV, as shown in Figure 12.8.

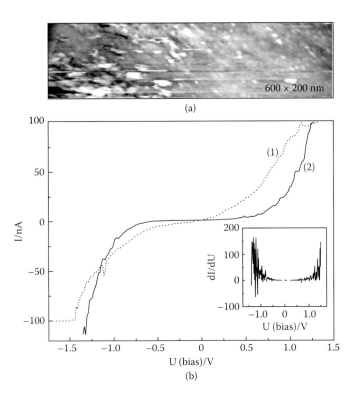

Figure 12.6 (a) STM image of a Si film with a thickness of about 100 nm, 600 nm × 200 nm. (b) In situ current voltage tunneling spectra of HOPG (curve 1) and of Si on HOPG. (From Zein El Abedin, S., et al., *Electrochem. Commun.*, 6, 510, 2004. With permission.)

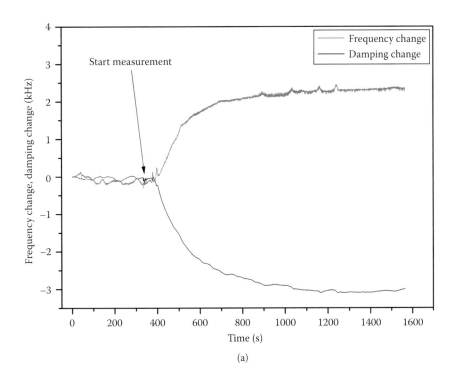

Figure 12.7 Oxidation of electrochemically deposited Si under air. (a) The curves show the frequency shift and the damping change with time.   *(Continued)*

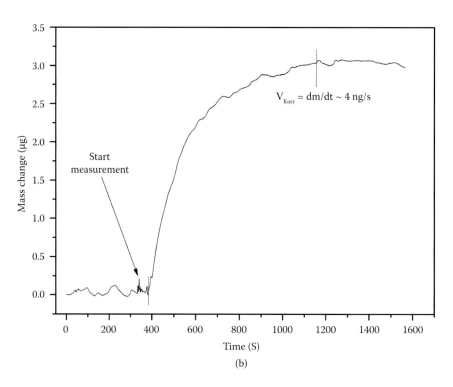

Figure 12.7 (Continued) (b) After a short induction time, the oxidation starts and is completed after less than 20 min, with a rate of about 4 ng/s in this experiment.

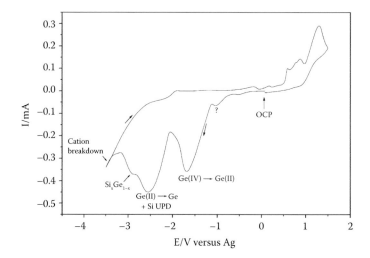

Figure 12.8 Cyclic voltammogram of SiCl$_4$:GeCl$_4$ (1:1 molar ratio) in (Py$_{1,4}$)TFSA on Au(111) at a scan rate of 10 mV/s. (Reproduced from Al-Salman, R., et al., Electrodeposition of Ge, Si and Si$_x$Ge$_{1-x}$ from an air- and water-stable ionic liquid, *Phys. Chem. Chem. Phys.*, 31, 4650–4657, 2008. With permission from the PCCP Owner Societies.)

The CV itself is not spectacular. We found two main reduction peaks, which can be related to the reactions Ge(IV) → Ge(II) and Ge(II) → Ge. As marked in Figure 12.8, in the regime of Ge deposition, there was a shoulder that we could allocate to Si$_x$Ge$_{1-x}$ deposition. As shown by Al-Salman et al. [20], we found that there was a change in color during deposition. Such a color change was not observed if Si and Ge were deposited alone, just if codeposited. Based on SEM results, we concluded that we made Si$_x$Ge$_{1-x}$ nanoparticles with a quantum confinement effect. Later studies [21,22] revealed that the obtained

Si_xGe_{1-x} is rather complex, both consisting of Si_xGe_{1-x} and of tiny Si and Ge nanoparticles that are responsible for the observed colors. These results show at a minimum that the electrodeposition of semiconductors from ionic liquids leads sometimes to surprising results, whose technical applications are currently not foreseeable.

12.3.6 INFLUENCE OF ANION ON Si DEPOSITION

In the recent maybe 5 years, we had many hints that not only the cation of an ionic liquid influences the electrodeposition of materials but also the anion can have an effect. For this purpose, we investigated Si deposition in three different ionic liquids, all having the same cation, that is, $[Py_{1,4}]^+$. The anions were FAP, TFSA, and trifluoromethylsulfonate (TfO). We added 0.15 mol/L $SiCl_4$ and performed—as first experiment—cyclic voltammetry at room temperature and at 100°C. As described in more detail in [23], we found that the CV's were similar, but not identical, and that the interface processes, probed at room temperature with the STM, are far from being identical in these three liquids. In $[Py_{1,4}]$FAP, we observed the underpotential deposition (upd) of Si on Au(111), in $[Py_{1,4}]$TFSA no upd was found, but just adsorption processes of the ionic liquid were observed. These experiments suggest that the surface chemistry is influenced by the anion, which consequently should also affect the final material of interest. Indeed, the structure of the deposits is slightly dependent on the anion of the liquid. It can currently not finally be commented, to which extent bulk properties (viscosity) and surface processes interact, at a minimum; however, we can conclude that the structure of the final material can be influenced by several parameters. For details, the reader is referred to Ref. [23].

12.3.7 MACROPOROUS Si

Together with Yao Li from the Harbin Institute of Technology (China), we synthesized inverse-opal macroporous structures of Ge [24]. Germanium has a high dielectric constant in the IR regime; thus, it is an interesting material to make photonic crystals. The procedure is relatively simple: polystyrene (PS) spheres with a diameter of roughly 500 nm are adsorbed from an alcoholic suspension onto an electrode surface. By self-assembling, a hexagonal or an fcc structure forms, resulting in an artificial opal structure. Figure 12.9 shows an example of PS spheres adsorbed on Au(111), the layer is about 10 monolayers of PS spheres high; thus, the thickness of the layer is in the regime of 4–5 μm.

The cracks result from inhomogenities and from the drying process. There have been many activities in this field, and Bartlett et al. have shown in a series of papers that in conventional electrolytes three-dimensional ordered macroporous (3DOM) structures of relatively noble elements and conducting polymers can be synthesized [25,26]. In the case of reactive elements, the low electrochemical window of water prevents the deposition of reactive elements on the one hand; on the other hand, organic solvents are not compatible with the relatively loose PS spheres. Although Ge can be deposited in aqueous solutions, the huge formation of hydrogen during deposition makes approaches to make 3DOM Ge in aqueous solutions more or less hopeless. We could show in Ref. [24] that the synthesis of 3DOM Ge is not too difficult. Ge grows, due to the help of the good wetting behavior of ILs, quite uniformly from the electrode surface at the bottom through the voids of the PS opal structure. When the PS spheres are removed by an organic solvent, a 3DOM Ge structure remains showing the typical properties of a photonic crystal [27]. What about 3DOM Si? 3DOM silicon would be quite an interesting material for current micro and optoelectronic technologies, due to its potential to enable, for example, low threshold lasers through control of spontaneous emission, see Ref. [28] and references therein. Apart from that, 3DOM Si is interesting as a catalyst but also as matrix for lithium in Li-ion batteries. A Li/Si anode has a much higher capacity than graphite, but the huge volume changes during charging/discharging require a nanostructure [29]. Apart from nanowires, 3DOM structures would be an option, too. In our joint work, it was attempted to adapt the procedure of 3DOM Ge deposition to the electrosynthesis of 3DOM Si. Our first approach [27] failed totally and instead of a 3DOM Si structure, the PS opal structure was completely disrupted upon Si deposition. Although $SiCl_4$ and $GeCl_4$ are chemically quite similar, from an electrochemical point of view, $SiCl_4$ reduction "only" requires a more negative electrode potential, but the final result is totally different. We have yet unpublished results that imply that $GeCl_4$ behaves in the same IL differently at the interface than $SiCl_4$ does. Furthermore, the considerably lower electronic conductivity of elemental Si versus the one of Ge

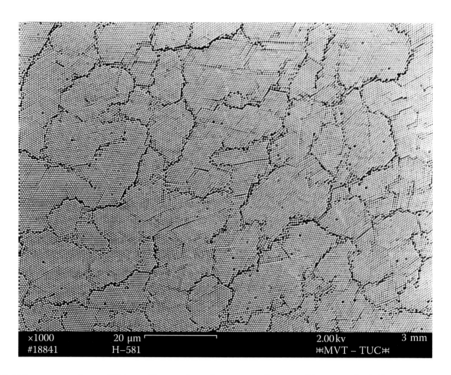

Figure 12.9 Polystyrene spheres (581 nm diameter) opal structure deposited from an alcoholic suspension on Au(111).

seems to play a role in the deposition. After a lot of detailed work in which the parameters for Si deposition (electrode potential, temperature, concentration) were adjusted, finally 3DOM Si could be made. Figure 12.10 shows one example, for details on the synthesis, the reader is referred to [30].

This 3DOM Si shows the typical behavior of a photonic crystal. Whereas 3DOM Ge is quite stable under air, the electrodeposited 3DOM Si is subject to a rapid oxidation under air; even in the glovebox, in quite a short time, the first few nanometers are oxidized by the remaining oxygen of the argon inertgas atmosphere. To protect the 3DOM Si from oxidation, it has to be covered with a type of a paint, which subsequently even allows its handling under air (Yao Li [Harbin Institute of Technology, Harbin, China], pers. comm.). It should be mentioned that the electrosynthesis of 3DOM Si_xGe_{1-x} is feasible as well; however, the complicated build-up of the deposit (see above) makes it quite difficult to make a material reproducible. In Ref. [31], we presented the peculiarities of this approach. In all cases published so far, the obtained Si, Ge and Si_xGe_{1-x} were XRD-amorphous. An interesting question is, whether it is possible to find parameters that enable the deposition of crystalline such semiconductors or whether the amorphous 3DOM structure can be transformed to a crystalline one by annealing at elevated temperatures.

12.3.8 TEMPLATE-FREE SYNTHESIS OF Si–Sn NANOWIRES

In a joint study with Jeremy Mallet [32], we were interested whether it is possible to use ionic liquids and templates to make Si nanowires. Here, only the principle is summarized. Commercially available track-etched polymer membranes with thicknesses of about 20 µm are made by ion bombardment and chemical etching such that they have from one side to the other one channels with—depending on the procedure to make them—variable diameters. If filled with a material, followed by dissolution of the membrane, nanowires are obtained. Si nanowires are interesting from a fundamental point of view and also act as anode host material for Li in Li-ion batteries [29]. The procedure is not too difficult: one side of the membrane is coated with a conductive layer and put on a metal plate. If put into an ionic liquid, it wets the electrode on the bottom and Si grows from the bottom of the electrode through the membrane. By a control of the reduction current versus time curve, a criterion can be found to decide whether the growth through the membrane is complete or not. Apart from Si, we have used this technique to synthesize several materials as nanowires. An interesting observation was that nanotubes are also feasible [33].

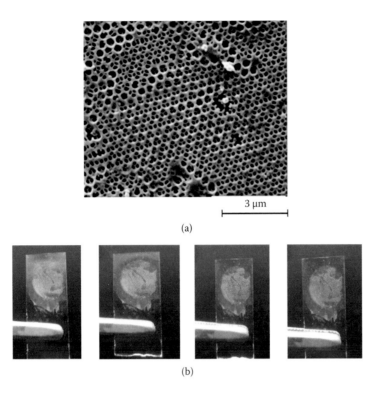

$3\ \mu m$

(a)

(b)

Figure 12.10 (a) SEM image of 3DOM silicon from PS template with 455 nm particles. (Reproduced from Liu, X., et al., Three-dimensionally ordered macroporous silicon films made by electrodeposition from an ionic liquid, *Phys. Chem. Chem. Phys.*, 14, 5100–5105, 2012. With permission from the PCCP Owner Societies.) (b) Optical behavior of 3DOM Si (sphere size 582 nm). (Reproduced from Liu, X., et al., Three-dimensionally ordered macroporous silicon films made by electrodeposition from an ionic liquid, *Phys. Chem. Chem. Phys.*, 14, 5100–5105, 2012. With permission from the PCCP Owner Societies.)

In many experiments done over the years, we found often that sometimes nanowires can be obtained even without a template. The problem has always been that the results were in most cases not reproducible. We found Zn nanowires, Al nanowires, and Ge nanowires without the template route, but—unfortunately—these were only "one shot" experiments; thus, the reproducibility was not given and consequently the results have not been published. Obviously, tiny experimental parameters decide about whether a thin layer or nanowires form. The first system where we found reproducible parameters for the template-free deposition of nanowires was the deposition of Sn–Si [34]. Although the topic of this book deals with Si nanostructures, the incorporation of even small amounts of Sn creates a strain in the silicon lattice, which changes its electronic and optical properties [35]. However, our main motivation was to try to investigate the ways to make Si anode host materials for Li-ion batteries. On the basis of our knowledge that the anion of an ionic liquid can influence material deposition and on restrictions on the solubility of metal salts in ILs, we selected $[Py_{1,4}]TfO$ as electrolyte and $SnCl_2$ and $SiCl_4$ as precursors in an equal concentration of 0.1 mol/L. The CVs of Si, Sn, and Si–Sn depositions are presented in Figure 12.11.

The CVs are no real surprise. Si is deposited as expected and the deposition of Sn occurs at more positive potentials. If Sn–Si is codeposited, the CV is not just an addition of the two single CVs; thus, an interaction occurs. If based on these electrode potentials a potential is selected for the codeposition of Sn–Si, usually granular deposits are obtained. The XRD analysis does not show crystalline Si, but the peaks for Sn can be well identified. If, however, the deposition is performed after a CV has been run before, we got—reproducibly—nanowires. Figure 12.12 shows an example from Ref. [34].

We interpreted this surprising result such that the CV disturbs the interfacial layers, and for still unknown reasons, a type of an intrinsic template forms that leads to the deposition of nanowires.

Functional materials

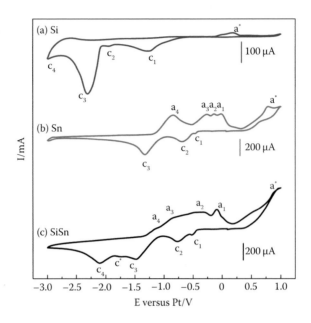

Figure 12.11 Cyclic voltammograms of (a) 0.1 M $SiCl_4$, (b) 0.1 M $SnCl_2$, and (c) 0.1 M $SnCl_2$ + 0.1 M $SiCl_4$ in $[Py_{1,4}]$ TfO on gold substrates. Scan rate: 10 mV s^{-1}. (From Elbasiony, A.M.A., et al., *ChemElectroChem*, 2, 1361–1365, 2015. With permission.)

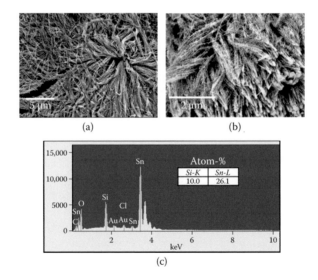

Figure 12.12 SEM-EDX of the SnSi nanowires obtained potentiostatically from 0.1 M $SnCl_2$ +0.1 M $SiCl_4$ in $[Py_{1,4}]$TfO at a potential of –2 V versus platinum for 1 h after running CV: (a) scale bar = 5 μm, (b) scale bar = 2 μm, and (c) EDX analysis of SnSi deposit. (From Elbasiony, A.M.A., et al., *ChemElectroChem*, 2, 1361–1365, 2015. With permission.)

It has to be mentioned that such nanowires are not obtained with any substrate; thus, if performed on a more technical substrate, the result might be different. Parallel to us, the group of Jürgen Janek found with another liquid and with another tin precursor also Sn-Si nanowires [36]. They gave a different interpretation and could obtain Sn–Si nanowires without running first a CV. These results show that there are a lot of parameters, and a deep understanding of the electrochemical processes at the electrode/IL interface is required to understand the processes and to find reproducible ways to make without any template nanowires. The author of this chapter is convinced that parameters can be found that allow the template-free electrodeposition of Si, Ge, and Si_xGe_{1-x} nanowires.

12.3.9 Si NANOPARTICLES BY PLASMA ELECTROCHEMISTRY

Till now, this chapter has shown that ionic liquids are suited to the electrodeposition of silicon. We have learnt that an amorphous deposit is obtained with nanoparticles in the regime of 50 nm and below. The ionic liquid itself plays a role in the deposition of Si, and by the help of templates, 3DOM materials or nanowires can be made. Under certain circumstances, even free-standing nanowires can be made without any template, shown at the example of Sn–Si nanowires. The question arises whether by any electrochemical method also, isolated Si nanoparticles can be made. Classic electrodeposition always leads to deposits, from which individual nanoparticles cannot be separated easily. However, plasma electrochemistry is a method.

The principle of plasma electrochemistry is rather simple and was shown by Gubkin more than 100 years ago [37]. An electrochemical cell with one electrode in the electrolyte and the other one in the gas phase is flooded with, for example, argon, subsequently a vacuum is applied. If a voltage of about 300–500 V is applied to the electrodes, the gas is ionized, and depending on the polarization, electrons move from the gas phase to the surface of the electrolyte. This way, Gubkin could show that the principle works, but the high vapor pressure of water makes a stable plasma quite difficult. Instead of having a stable plasma over all of the surface, rather a flash-like electron beam was obtained. Having this problem in mind, we wondered whether it was possible to use an ionic liquid, as most ionic liquids have around 20°C only negligible vapor pressures. The experimental setup is sketched in Figure 12.13.

The only main deviation from Gubkin's original experiment is that the aqueous solution is replaced by an ionic liquid, which has practically no vapor pressure; thus, the plasma should be stable over the whole of the surface, once ignited. In a joint paper with Jürgen Janek, we could show that silver nanoparticles can easily be made by this method. The particles in suspension were quite small with diameters of only a few ten nanometers [38]. In succeeding papers, we could show that tiny Cu nanoparticles can be made, with diameters of below 10 nm. The plasma electrochemical synthesis of germanium nanoparticles was also possible, but we had to select quite a special precursor, as $GeCl_4$, which in classic electrochemistry is quite a good educt for Ge deposition, has a too high vapor pressure and is simply pumped off. When trying to make Si nanoparticles with this method, we had the same problem with $SiCl_4$ and $SiBr_4$ as educts, both are simply pumped off. With SiI_4 as educt, the synthesis was possible. In contrast to liquid $SiCl_4$ and $SiBr_4$, SiI_4 is a solid material. Although we got a product, this product rather was a low valent SiI_x than Si nanoparticles [39]. To solve the problem of the educt being pumped off, we designed a new cell, which is presented in Figure 12.14.

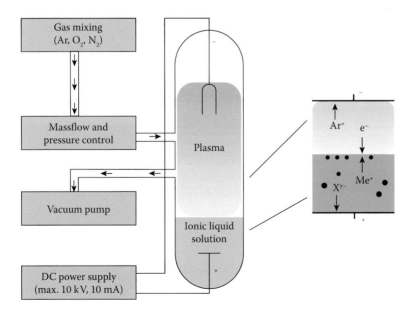

Figure 12.13 Sketch of the plasma electrochemistry setup. The cell is setup inside an argon-filled glovebox. (From Spitczok, N., et al., *J. Mol. Liq.*, 192, 59–66, 2014. With permission.)

Functional materials

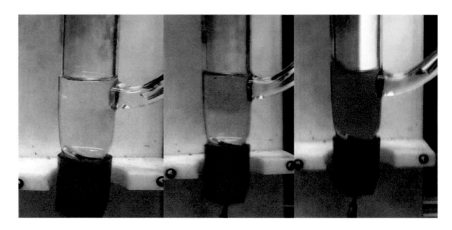

Figure 12.14 Plasma electrochemical cell with an inlet for SiCl$_4$. The three images show an ongoing experiment (30 min) where nanoparticles form in the ionic liquid Si.

This cell is fitted with an inlet for SiCl$_4$ inside of the liquid. Thus, even if the formerly SiCl$_4$ saturated liquid loses SiCl$_4$ during evacuation and the plasma electrochemical experiment, it is replaced by this inlet. The plasma electrochemical reaction takes place at the liquid/plasma interface, and as in the case of Ag, Cu and Si particles form now in the ionic liquid, shown in yellow in Figure 12.14. The analysis of the particle sizes is quite challenging. As with electrochemically made Si, the particles are subjected to a rapid oxidation once removed from the liquid. Consequently, SEM and TEM rather show SiO$_2$ than Si nanoparticles. To overcome this problem, we did an analysis with dynamic light scattering, in special with photon cross-correlation spectroscopy (PCCS). The data show that there are tiny particles but also obviously larger aggregates. On top, if the IL is diluted, for example, with an alcohol, to get better PCCS data, the particle size distribution changes. This is due to a chemical reaction of the Si nanoparticles with the cosolvent. Thus, to make single Si nanoparticles in suspension, both aggregation and a chemical reaction with solvent and cosolvent have to be avoided. However, the size of the obtained Si nanoparticles is in the range of 20 nm, and Figure 12.15, where the photoluminescence spectra of the pure liquid (1-ethyl-3-methylimidazolium = [EMIm] TFSA), of the plasma-treated pure liquid and of the SiCl$_4$ containing plasma-treated liquid are compared, show that the particles are optically active.

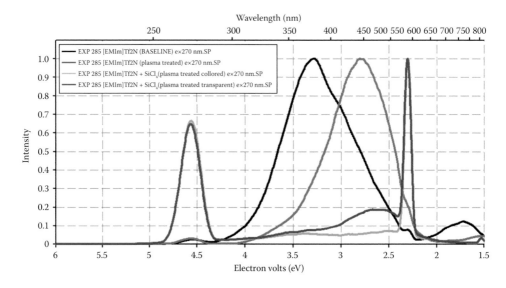

Figure 12.15 Photoluminescence spectra of pure [EMIm]TFSA (black), a plasma-treated (red), and a SiCl$_4$/[EMIm] TFSA solution (blue, green) after synthesis.

12.4 SUMMARY AND OUTLOOK

This chapter has given a first insight into the prospects of Si electrodeposition from ionic liquids. Due to its limited electrochemical window, water is not suitable to Si deposition at all, as pointed out in the introduction. Organic solvents can be used; however, it is difficult to remove water from them and to keep an electrolyte dry during the whole of the experiment. In the presence of even trace amounts of water in the organic solvent, $SiCl_4$ hydrolyzes to undesired SiO_2; furthermore, organic solvents tend to be decomposed as a side reaction to the desired Si deposition. Due to quite low vapor pressures, usually negligible even at 100°C, ionic liquids can be dried easily by putting them stirred under vacuum. Water concentrations below 10 ppm can be achieved in just a few hours. Ionic liquids have electrochemical windows of 3–6 V, and especially with pyrrolidinium cations, Si can be deposited. At bath temperatures below 100°C, usually amorphous Si is obtained. Apart from Si, Si-compounds such as Si–Ge or Sn–Si can also be obtained. Due to the excellent wetting behavior, ILs also wet polymer surfaces; thus, with the help of artificial opal structures or track-etched membranes, 3DOM materials (photonic crystals) and nanowires can be made. An observation that is far from being understood is that under certain conditions, free-standing nanowires can be made without any template. The few available data in literature suggest that under certain conditions, the liquid seems to act as a template. This interesting observation is worth of being studied in more detail in future. Last but not least it was shown that plasma electrochemistry allows the electrochemical synthesis of individual Si nanoparticles. There are many open questions in this approach such as, what is the best ionic liquid to avoid agglomeration and chemical corrosion of the particles and can Si precursors be found that well dissolve in the IL without being pumped off during the experiment.

REFERENCES

1. J. Gobet, H. Tannenberger, Electrodeposition of silicon from a nonaqueous solvent, *J. Electrochem. Soc.* 135 (1988), 109–112.
2. T. Munisamy, A. J. Bard, Electrodeposition of Si from organic solvents and studies related to initial stages of Si growth, *Electrochim. Acta* 55 (2010), 3797–3803.
3. M. Bechelany, J. Elias, P. Brodard, J. Michler, L. Philippe, Electrodeposition of amorphous silicon in non-oxygenated organic solvent, *Thin Solid Films* 520 (2012), 1895–1901.
4. P. Walden, Ueber die Molekulargroesse und Leitfaehigkeit einiger geschmolzenen Salze, *Bull. Acad. Imper. Sci. St. Petersburg* 8 (1914), 405–422.
5. P. Wasserscheid, T. Welton, Editors, *Ionic liquids in synthesis*, 2nd edition, Wiley-VCH Verlag GmbH & Co. KgaA, Weinheim, Germany, 2008.
6. M. Freemantle, *An introduction to ionic liquids*, Royal Society of Chemistry, Cambridge, 2010.
7. H. Ohno, *Electrochemical aspects of ionic liquids*, 2nd edition, Wiley, Hoboken, NJ, 2011.
8. F. Endres, A. Abbott, D. R. MacFarlane, *Electrodeposition from ionic liquids*, Wiley-VCH Verlag GmbH & Co. KgaA, Weinheim, Germany, 2008.
9. F. Endres, O. Höfft, N. Borisenko, L. H. S. Gasparotto, A. Prowald, R. Al-Salman, T. Carstens, R. Atkin, A. Bund, S. Zein El Abedin, Do solvation layers of ionic liquids influence electrochemical reaction? *Phys. Chem. Chem. Phys.* 12 (2010), 1724–1732.
10. T. Carstens, A. Ispas, N. Borisenko, R. Atkin, A. Bund, F. Endres, In situ scanning tunneling microscopy (STM), atomic force microscopy (AFM) and quartz crystal microbalance (EQCM) studies of the electrochemical deposition of tantalum in two different ionic liquids with the 1-butyl-1-methylpyrrolidinium cation, *Electrochim. Acta* 197 (2016), 374–387.
11. A. J. Bard, L. R. Faulkner, *Electrochemical methods: Fundamentals and applications*, 2nd edition, Wiley, New York, 2001.
12. R. Atkin, N. Borisenko, M. Druschler, S. Zein El Abedin, F. Endres, R. Hayes, B. Huber, B. Roling, An in situ STM/AFM and impedance spectroscopy study of the extremely pure 1-butyl-1-methylpyrrolidinium tris(pentafluoroethyl) trifluorophosphate/Au(111) interface: Potential dependent solvation layers and the herringbone reconstruction, *Phys. Chem. Chem. Phys.* 13 (2011), 6849–6857.
13. S. Zein El Abedin, E. M. Moustafa, R. Hempelmann, H. Natter, F. Endres, Electrodeposition of nano- and microcrystalline aluminium in some water and air stable ionic liquids, *ChemPhysChem.* 7 (2006), 1535–1543.
14. T. J. Senden, Force microscopy and surface interactions, *Curr. Opin. Colloid Interface Sci.* 6 (2001), 95–101.
15. R. Hayes, N. Borisenko, M. K. Tam, P. C. Howlett, F. Endres, R. Atkin, Double layer structure of ionic liquids at the Au(111) electrode interface: An atomic force microscopy investigation, *J. Phys. Chem. C* 115 (2011), 6855–6863.

16. J. Komadina, T. Akiyoshi, Y. Ishibashi, Y. Fukunaka, T. Homma, Electrochemical quartz crystal microbalance study of Si electrodeposition in ionic liquid, *Electrochim. Acta* 100 (2013), 236–241.

17. S. Zein El Abedin, N. Borissenko, F. Endres, Electrodeposition of nanoscale silicon in a room temperature ionic liquid, *Electrochem. Commun.* 6 (2004), 510.

18. F. Bebensee, N. Borissenko, M. Frerichs, O. Höfft, W. Maus-Friedrichs, S. Zein El Abedin, F. Endres, Surface analysis of nanoscale aluminium and silicon films made by electrodeposition in ionic liquids, *Z. Phys. Chem.* 222 (2008), 671–686.

19. N. Behrens, Aufbau einer elektrochemischen Quarzmikrowaage (EQCM) und Untersuchung des elektrochemischen Verhaltens von Silicium bei dessen Abscheidung aus Ionischen Flüssigkeiten. Master Thesis, TU Clausthal-Zellerfeld, 2016.

20. R. Al-Salman, S. Zein El Abedin, F. Endres, Electrodeposition of Ge, Si and Si_xGe_{1-x} from an air- and water-stable ionic liquid, *Phys. Chem. Chem. Phys.* 31 (2008), 4650–4657.

21. A. Lahiri, M. Olschewski, O. Hoefft, S. Z. El Abedin, F. Endres, Insight into the electrodeposition of Si_xGe_{1-x} thin films with variable compositions from a room temperature ionic liquid, *J. Phys. Chem. C* 117 (2013), 26070–26076.

22. A. Lahiri, M. Olschewski, O. Höfft, S. Z. El Abedin, F. Endres, In situ spectroelectrochemical investigation of Ge, Si, and Si_xGe_{1-x} electrodeposition from an ionic liquid, *J. Phys. Chem. C* 117 (2013), 1722–1727.

23. G. Pulletikurthi, A. Lahiri, T. Carstens, N. Borisenko, S. Z. El Abedin, F. Endres, Electrodeposition of silicon from three different ionic liquids: Possible influence of the anion on the deposition process, *J. Solid State Electrochem.* 17 (2013), 2823–2832.

24. X. Meng, R. Al-Salman, J. Zhao, N. Borissenko, Y. Li, F. Endres, Electrodeposition of 3D ordered macroporous germanium from ionic liquids: A feasible method to make photonic crystals with a high dielectric constant, *Angew. Chem. Int. Ed.* 48 (2009), 2703–2707.

25. P. N. Bartlett, P. R. Birkin, M. A. Ghanem, C.-S. Toh, Electrochemical synthesis of highly ordered macroporous conducting polymers grown around self-assembled colloidal templates, *J. Mater. Chem.* 11 (2001) 849–853.

26. J. Hu, M. Abdelsalam, P. Bartlett, R. Cole, Y. Sugawara, J. Baumberg, S. Mahajan, G. Denuault, Electrodeposition of highly ordered macroporous iridium oxide through self-assembled colloidal templates, *J. Mater. Chem.* 19 (2009), 3855–3858.

27. R. Al-Salman, X. Meng, J. Zhao, Y. Li, U. Kynast, M.M. Lezhnina, F. Endres, Semiconductor nanostructures via electrodeposition from ionic liquids, *Pure Appl. Chem.* 82 (2010), 1673–1689.

28. C. Ciminelli, F. Peluso, M. N. Armenise, Parametric analysis of 2D guided-wave photonic band gap structures, *Opt. Express* 13 (2005), 9729–9746.

29. V. Etacheri, O. Haik, Y. Goffer, G. A. Roberts, I. C. Stefan, R. Fasching, D. Aurbach, Effect of fluoroethylene carbonate (FEC) on the performance and surface chemistry of Si-nanowire Li-ion battery anodes, *Langmuir* 28 (2012), 965–976.

30. X. Liu, Y. Zhang, D. Ge, J. Zhao, Y. Li, F. Endres, Three-dimensionally ordered macroporous silicon films made by electrodeposition from an ionic liquid, *Phys. Chem. Chem. Phys.* 14 (2012), 5100–5105.

31. W. Xin, J. Zhao, D. Ge, Y. Ding, Y. Li, F. Endres, Two-dimensional SixGe1-x films with variable composition made via multilayer colloidal template-guided ionic liquid electrodeposition, *Phys. Chem. Chem. Phys.* 15 (2013), 2421–2426.

32. R. Al-Salman, J. Mallet, M. Molinari, P. Fricoteaux, F. Martineau, M. Troyon, S. Zein El Abedin, F. Endres, Template assisted electrodeposition of germanium and silicon nanowires in an ionic liquid, *Phys. Chem. Chem. Phys.* 10 (2008), 6233–6237.

33. A. Lahiri, A. Willert, S. Z. El Abedin, F. Endres, A simple and fast technique to grow free-standing germanium nanotubes and core-shell structures from room temperature ionic liquids, *Electrochim. Acta* 121 (2014), 154–158.

34. A. M. A. Elbasiony, M. Olschewski, S. Z. El Abedin, F. Endres, Template-free electrodeposition of SnSi nanowires from an ionic liquid, *ChemElectroChem* 2 (2015), 1361–1365.

35. E. Simoen, C. Claeys, V. Neimash, A. Kraitchinskii, N. Kras'ko, O. Puzenko, A. Blondeel, P. Clauws, Deep levels in high-energy proton-irradiated tin-doped n-type Czochralskii silicon, *Appl. Phys. Lett.* 76 (2000), 2838–2840.

36. R. Al-Salman, H. Sommer, T. Brezesinski, J. Janek, Template-free electrochemical synthesis of high aspect ratio Sn nanowires in ionic liquids: A general route to large-area metal and semimetal nanowire arrays? *Chem. Mater.* 27 (2015), 3830–3837.

37. J. Gubkin, Elektrolytische Metallabscheidung an der freien Oberfläche einer Salzlösung, *Ann. Phys.* 32 (1887), 114–115.

38. S. A. Meiß, M. Rohnke, L. Kienle, S. Zein El Abedin, F. Endres, J. Janek, Employing plasmas as gaseous electrodes at the free surface of ionic liquids: Deposition of nanocrystalline silver particles, *ChemPhysChem* 8 (2007), 50.

39. N. Spitczok, V. Brisinski, O. Hoefft, F. Endres, Plasma electrochemistry in ionic liquids: From silver to silicon nanoparticles, *J. Mol. Liq.* 192 (2014), 59–66.

13 Sonosensitizing properties of silicon nanoparticles

Liubov A. Osminkina and Andrew P. Sviridov

Contents

13.1 INTRODUCTION

Combination therapy of cancer is a rather new method, which involves the simultaneous use of various anticancer drugs and therapeutic strategies, and is becoming increasingly important for improving the quality of long-term cancer treatment (Lammers et al. 2010; Eldar-Boock et al. 2013). Typically, the combination therapy can modulate different signaling pathways in cancer cells in order to maximize the therapeutic effect compared with mono-drug therapy. One of the popular fields in the combination therapy is targeted delivery of antitumor drugs in various containers based on polymers (Rapoport 2007), carbon nanotubes (Liu et al. 2008), gold (You et al. 2010), magnetic nanoparticles (Purushotham et al. 2009), silica nanoparticles (Zou et al. 2013), and composite nanoparticles (Yagüe et al. 2010), and their release under external influence generated by infrared radiation (Gannon et al. 2007), alternating magnetic field (Bañobre-López et al. 2013) or radiofrequency radiation (Gannon et al. 2008). The main problem of this approach is the problem of drug container toxicity. Although the results of in vivo experiments on the suppression of tumor growth in mice give very promising results, the toxicity of materials during a prolonged treatment is questionable. For example, the intravenous administration of carbon nanotubes at a concentration of $100~\mu g \cdot mL^{-1}$ after 7 days leads to inflammation in tissues (Lam et al. 2004). Gold and silicon dioxide are inert materials, their nanoparticles do not obtain a pronounced toxicity (Lasagna-Reeves et al. 2010), but the circulation time of the latter is a big problem. In addition, the complexities of this approach include physical limitations of external exposure, such as small penetration depth (IR radiation), difficulties of focusing (magnetic field), and side effects (radiation emission).

From this point of view, ultrasonic irradiation (USI) seems to be an attractive one. As a result of years of research, two approaches emerged in the ultrasonic cancer therapy: low-invasive high-intensity focused ultrasound surgery (high-intensity focused ultrasound, HIFU) (Diederich and Hynynen 1999) and sonodynamic therapy (SDT) (Yumita et al. 2007). HIFU surgery requires high-intensity ultrasound $(0.1–10~kW \cdot cm^{-2})$ and expensive precision equipment for its execution (Diederich and Hynynen 1999). The method of SDT for the treatment of malignant tumors is much simpler both technically and operationally, and in some cases may act as an alternative to HIFU surgery. Since the intensity of ultrasound in the SDT lies in the range of $1–10~W \cdot cm^{-2}$, the desired therapeutic effect is implemented by the activator of USI, or sonosensitizer (Yumita et al. 2007). The selectivity of USI is determined by (i) selective accumulation of the sonosensitizer in the tumor, (ii) predominant effect of ultrasound on the tumor nidus, and (iii) preferential ability of healthy tissues to recover (Serpe et al. 2012; Nikolaev et al. 2015).

The beginning of SDT utilization can be associated with the negligible temperature elevation effect of noncytotoxic USI on organs and tissues detected in 1979. Then it was found that such hyperthermia can enhance the cytotoxicity of drugs, and ultrasound, with its focusing ability, was suggested as an attractive tool to achieve it (Kremkau 1979; Rosenthal et al. 2004). In 1989, Yumita et al. showed that hematoporphyrin increased the sensitivity of tumor cells to USI (Yumita et al. 1989). Then it was demonstrated that the combination of typical anticancer medicaments like porphyrin-based drugs as sonosensitizers and therapeutic ultrasound can initiate a specific sonochemical reaction to produce reactive oxygen species (ROS) with high toxicity (Rosenthal et al. 2004; Costley et al. 2015). However, traditional organic sonosensitizer molecules suffer from low bioavailability, fast elimination out of the body, and poor tumor accumulation (Qian et al. 2016).

Microbubble contrast agents, which are air bubbles stabilized by proteins, are often used as sonosensitizers. They have been used in clinical practice for more than two decades, which serves as an incentive for the development of new methods of contrast imaging and new contrast agents. For example, Optison® and Levovist® microbubbles are used in echocardiography (Podell et al. 1999; Raisinghani et al. 2004; Wei et al. 2008), while SonoVue® is used in radiology for the detection of lesions (Schneider 1999; Nicolau et al. 2004; von Herbay et al. 2004). Moreover, microbubbles may themselves act as drug carriers. For example, microbubbles coated with albumin or charged lipid bilayer are capable of directly capturing the genetic material (Porter et al. 1996; Miller 2000), while the administration of Optison® helped to enhance intercellular uptake of green calcein fluorescent molecules (Hallow et al. 2006).

However, microbubbles have a number of disadvantages. Although they lower the threshold of cavitation, the stable cavitation generation in vivo is a difficult task in its essence, because of its stochastic nature. Over the past years, the circulation time of ultrasound contrast agents in the bloodstream has significantly increased, but, for example, residence time of SonoVue is only 6 minutes (Ferrara et al. 2007). Even PEGylated microbubbles (Claudon et al. 2013) are characterized by low circulation time, since they can be absorbed by immune cells and are able to accumulate in liver or spleen. Another problem is the size of agents: microbubbles are generally too large to penetrate into the tumor or other target tissues; moreover, they cannot reach remote tissues, the blood vessel diameter in which is less than few microns.

Nowadays a lot of studies show that the presence of sonosensitizers, especially solid ones, can lead to the USI energy release via cavitation, which, in its turn, can (i) significantly increase the temperature of surrounding environment, (ii) initiate sonochemical reactions, (iii) lead to sonoluminescence, that is, light generation, upon irradiation of a solution with ultrasound (Costley et al. 2015; Qian et al. 2016). Recently developed nanobiotechnology and nanomedicine have provided solid evidence of micro/nanoparticles enhancing the therapeutic efficiency of various modalities (Qian et al. 2016). Thus, TiO_2, polyhydroxy fullerene, alumina, and platinum nanoparticles can respond to USI to generate ROS for the SDT (Tuziuti et al. 2005; Serpe et al. 2012; Qian et al. 2016). Graphene oxide nanosheet is of a great interest because of its high heat-conducting capability, which could be used to increase the local tissue temperature upon USI. Mesoporous silica nanoparticles promote the cavitation effect of ultrasound (Qian et al. 2016). Different nanocomposites and nanocontainers can be used to enhance synergistic effects of SDT (Qian et al. 2016).

The main problem of solid sonosensitizers is their toxicity (Yildirimer et al. 2011) and low biodegradability (Canham 2007). In this regard, biocompatible and biodegradable porous silicon nanoparticles (PSi NPs) seem to be very promising for various medical applications (Canham 2014a; Shabir 2014). In this review, the preparation techniques and results of investigation of structural and physical properties of PSi NPs are summarized and discussed in the frame of perspective of their application as sensitizers for the SDT.

13.2 PREPARATION, AND STRUCTURAL AND PHYSICAL PROPERTIES OF SILICON NANOPARTICLES

PSi NPs in the form of powders or suspensions are usually obtained by ultrasonic fragmentation or mechanical ball-grinding of porous silicon (PSi) or porous silicon nanowires (PSi NWs) (Heinrich and Curtis 1992; Canham 2007; Osminkina et al. 2014). Mesoporous (mPSi) and microporous (μPSi) silicon films are commonly formed from boron-doped c-Si substrates with specific resistivity of 1–50 mΩ cm and 10–20 Ω cm, respectively. Typical electrolyte composition and current density for etching are the following: HF(49%):C$_2$H$_5$OH=1:1 and 50–100 mA/cm^2, respectively. The prepared PSi films are lifted off by a short increase of the current up to 500–800 mA/cm^2 (Cullis et al. 1997). PSi NWs are formed by metal-assisted chemical etching (MACE) of c-Si wafer with specific resistivity of 1–5 mΩ·cm (Osminkina et al. 2014). To prevent fast dissolution, PSi NPs can be covered by different biopolymers, such as dextran (Park et al. 2009), polylactic-co-glycolic acid (PLGA), and polyvinyl alcohol (PVA) (Gongalsky et al. 2012). To enhance the specificity of PSi NPs treatment, nanoparticles can be functionalized with cancer cell–targeting antibodies, and pore of PSi NPs can be loaded with different anticancer drugs (Anglin et al. 2008; Santos et al. 2011; Secret et al. 2013; Alhmoud et al. 2015; Tamarov et al. 2016).

Figure 13.1a shows a cross-sectional SEM image of a PSi film, which consists of interconnected silicon nanocrystals (nc-Si) and pores (Osminkina et al. 2015). Inset in Figure 13.1a shows a photographic image of the PSi film grown on a 4″ c-Si substrate. Porosity of PSi was determined by the gravimetric method (Herino et al. 1987) and amounted 80 ± 5%. The powder of PSi NPs obtained by milling of PSi films consists of porous nanoparticles as shown in Figure 13.1b.

The brown-yellowish color of the prepared PSi film and powder (see insets in Figure 13.1a and b) indicates that their optical properties changed significantly in comparison with those of c-Si because of quantum confinement effect in nc-Si (Canham 2014).

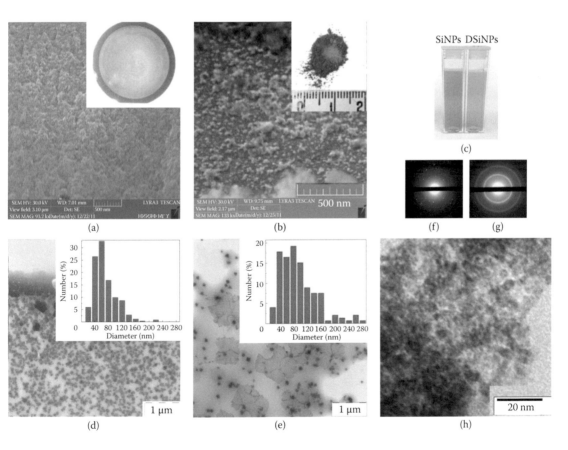

Figure 13.1 (a) Cross-sectional SEM image of a PSi film (the inset shows a photographic image of the PSi film grown on a 4″ c-Si substrate). (b) SEM image of a powder of PSi NPs obtained by the hand milling of PSi films (the inset shows a photo of the powder of PSi NPs). (c) Photo of a PSi NPs (left) and DPSi NPs (right) aqueous suspensions with the concentration of 1 mg/mL. (d) TEM image of PSi NPs (the inset shows the size distribution of PSi NPs). (e) TEM image of DPSi NPs (the inset shows the size distribution of DPSi NPs). (f) Pattern of the PSi NPs electron diffraction. (g) Pattern of the DPSi NPs electron diffraction. (h) TEM image of the porous structure of DPSi NPs. (From Osminkina, L.A., et al., *Microporous and Mesoporous Mater*, 210, 169–175, 2015. With permission.)

Figure 13.1c shows digital images of standard optical cells filled with aqueous suspensions of PSi NPs (left) and PSi NPs covered with dextran (DPSi NPs) (right). Such suspensions are stable at least for several weeks. The brown-yellowish color of the suspension is similar to that for PSi films. This fact demonstrates that optical properties of silicon nanocrystals did not change dramatically after their dispersion and storage in water. Note that the deeper color of the PSi NP suspension in comparison with the DPSi NP suspension indicates changes in the optical properties of the latter. It can be related to changes in the light scattering because of dextran coating of PSi NPs.

The size of the nanoparticles was analyzed using TEM of dried PSi NPs and DPSi NPs suspensions (see Figure 13.1d and e). The insets in Figure 13.1d and e show the size distribution of nanoparticles centered near 60 nm and 80 nm, respectively. The larger size of DPSi NPs can be related to the dextran layer on the nanoparticle's surface. Unbound dextran molecules are visible as gray surroundings of DPSi NPs, which are represented as black points in the TEM images (see Figure 13.1e).

Figure 13.1f and g show the corresponding electron diffraction pattern obtained in the "transmission" geometry for PSi NPs and DPSi NPs, respectively. The diffraction pattern indicates the crystalline structure of the nanoparticles. While the diffraction pattern of PSi NPs contains both the diffraction rings and bright spots, DPSi NPs are characterized by the diffraction rings only. This fact indicates larger degree of misorientation of nc-Si in DPSi NPs in comparison with PSi NPs. Indeed, 2–10 nm-sized nc-Si within individual DPSi NPs are assembled in a porous sponge-like structure (see Figure 13.1h).

The nanocrystalline structure of the prepared nanoparticles was confirmed by Raman spectroscopy (see Figure 13.2a). While the Raman spectrum of c-Si represents a sharp line at 520 cm^{-1}, PSi NPs and DPSi NPs are characterized by the broader spectra with peak position at 518 cm^{-1}. This low-frequency

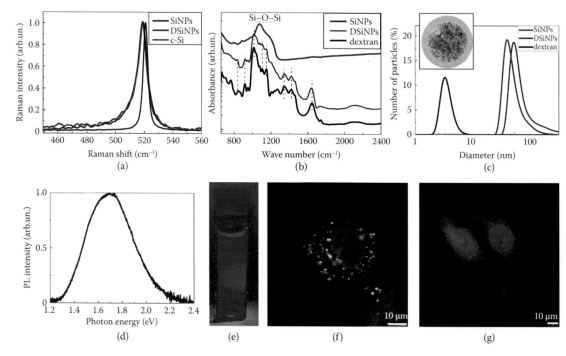

Figure 13.2 (a) Raman scattering spectra of PSi NPs (blue curve), DPSi NPs (red curve) and c-Si (black curve). (b) FTIR absorbance spectra of PSi NPs obtained by the mechanical grinding in water (blue curve), DPSi NPs (red curve), and a dried suspension of 10% dextran (30,000–40,000 mol wt) suspended with 0.9 % NaCl (black curve). The vertical dashed lines indicate vibration frequencies of dextran. (c) DLS spectra of aqueous suspensions of PSi NPs (blue curve), DPSi NPs (red curve), and a suspension of 10% saline dextran solution (black curve). Inset represents the schematic view of a DPSi NP. (d) PL spectrum of an aqueous suspension of DPSi NPs. (e) Optical image of the cuvette with DPSi NP suspension under UV excitation. (f) Luminescent image of living CF2Th cells with added 0.1 mg/mL DPSi NPs, and (g) the same for the reference group. Green, blue, and red colors in (f) and (g) correspond to the luminescence of cell cytoplasm, cell nuclei, and DPSi NPs, respectively. (From Osminkina, L.A., et al., *Microporous and Mesoporous Mater*, 210, 169–175, 2015. With permission.)

shift and broadening is usually attributed to the phonon confinement in nc-Si with sizes of 3–4 nm (Campbell and Fauchet 1986).

As-prepared PSi films have hydrophobic properties because of hydrogen surface coverage, as evidenced by the IR absorption peaks at 2082–2190 and 906 cm^{-1} both related to the Si–H surface bonds (Theiß 1997; Ogata 2014). During the milling process in water, hydrogen is desorbed from the surface of silicon nanocrystals and replaced by oxygen. The oxidation of PSi NPs surface is markedly revealed by the absorption peaks of Si–O–Si vibrations at 1070 cm^{-1} (see Figure 13.2c, blue curve). The oxide coverage of SiNP surface ensures the hydrophilic properties of the latter and determines the possibility to form stable aqueous suspensions. The coverage of SiNPs by dextran leads to a change of their FTIR spectra caused by different absorbance bands of dextran (Heyn 1974). The same absorbance bands can also be seen in the FTIR spectra of dried dextran, which was used for the DPSi NP formation. The FTIR data indicate that the dextran coating of PSi NPs corresponds mainly to physical adsorption (Suh et al. 2005) rather than formation of new chemical bonds between dextran molecules and nc-Si surfaces.

According to the Dynamic Light Scattering (DLS) data (see Figure 13.2c), the size (diameter) distributions of nanoparticles in suspensions are characterized by maxima at 70 nm and 96 nm for PSi NPs and DPSi NPs, respectively. The mean size of dextran molecules in suspension is about 4 nm. The larger size of DPSi NPs in comparison with PSi NPs can be explained by the adsorbed dextran layer on the nanoparticle's surface as it is schematically shown in the inset of Figure 13.2c.

The zeta potential (ZP) of PSi NPs in as-prepared aqueous suspension was negative and accounted –28 ± 5 mV. The latter value is typical for porous silicon nanoparticles and is related to the negative charge of hydroxyl groups on PSi NPs surfaces. It was found that the value of ZP for as-prepared DPSi NPs was –20 ± 3 mV. The small difference between the ZP values of PSi NPs and DPSi NPs indicates a weak interaction between dextran and nanoparticles. This interaction can be covered by the Van-der-Waals forces between the adsorbed dextran and oxidized surface of SiNPs.

The suspensions of PSi NPs exhibit bright photoluminescence (PL) with a broad spectrum centered at 1.7 eV (Figure 13.2d). This PL band can be explained by the radiative recombination of excitons confined in nc-Si with sizes of 3–5 nm (Canham 1990). Note that the PL emission of the suspension with a concentration of 0.1 mg/mL could be easily observed with the naked eye (see Figure 13.2e). PL properties of PSi NPs can be used for cells diagnostics (see details in the following sections).

13.3 SILICON NANOPARTICLES AS SONOSENSITIZERS

13.3.1 FUNDAMENTALS OF SONOSENSITIZING PROPERTIES OF SILICON NANOPARTICLES

As it was previously mentioned, ultrasound irradiation is widely used in modern medicine (Hill et al. 2002). This is due to a specific interaction between body tissues and USI: thus, effective absorption of USI leading to tissue ablation is used in HIFU therapy (Coussios et al. 2007) while differences of acoustic propagation in organs are exploited for USI diagnostics (Frinking et al. 2000). The obvious advantages of USI for medical purposes are its low invasiveness, relatively low cost, opportunity for real-time imaging, and deep penetration depth (Fenster et al. 2001; Hill et al. 2002). Among the most recent advances in therapeutic USI-based modalities are USI-triggered drug release (Frenkel 2008) and SDT (Chen et al. 2014). The last one is based on the usage of special substances that can enhance the effect of USI—sonosensitizers. Historically, the term "sonodynamic therapy" emerged from the name of its well-known alternative—photodynamic therapy, or PDT (Celli et al. 2010). This is so because the first substance used as a sonosensitizer was hematoporphyrin—a popular photosensitizer. In these early experiments done by Umemura and colleagues, it was revealed that being exposed to USI of megahertz frequency range, the above substance could generate reactive oxygen species (ROS) (Yumita et al. 1989). Among such species are hydroxyl, peroxide radicals, and even singlet oxygen (Wang et al. 2011), which lead to cell apoptosis and necrosis. There is no definite explanation of the ROS generation process, but commonly the effects of acoustic cavitation, sonoluminescence, pyrolysis, sonoporation, and sonochemistry are considered as the main mechanisms (Qian et al. 2016).

Functional materials

However, in comparison to traditional organic photosensitizers for PDT, sonosensitizer molecules have not found wide application in SDT because of their low availability, short circulation time, and poor accumulation in tumor. In this regard, the latest advances in nanobiotechnology have helped to find a good substitute for sonosensitizer molecules in the form of organic and inorganic micro- and nanoparticles, which can enhance the therapeutic effect of various modalities. For example, they can serve as drug-delivery carriers providing sustained chemotherapeutic release and protection against recognition by the immune cells (Allen and Cullis 2004; de las Heras Alarcón et al. 2005). Micro- and nanoparticles can also perform as agents enhancing the therapeutic effect of different physical treatments like radiotherapy (Butterworth et al. 2012) and ultrasound hyperthermia (Chen et al. 2015). As for phototherapy, specially designed nanoparticles can be used not only for photothermal energy conversion (in photo-ablation) (Hu et al. 2006) but also for singlet oxygen generation (in PDT) (Dolmans et al. 2003). The following considerations are valid regarding the usage of micro- and nanoparticles for the SDT. First of all, they can stand as carriers of traditional organic sonosensitizers improving the capability of intracellular delivery (Serpe et al. 2012). Second, nanoparticles can be functionalized for the targeted delivery of drugs to the tumor tissue, providing accumulation of the sonosensitizer in deep-seated tumors (Gao et al. 2015). Moreover, they can help to preserve the sonochemical properties of sonosensitizers (McEwan et al. 2014). Next, micro- and nanoparticles can act as energy transducers in the SDT, which manifests, for instance, in the local increase of USI absorption (Umemura et al. 1990). Finally, particles can carry nucleation sites for the generation of cavitation bubbles, thus decreasing the cavitation thresholds (Sazgarnia et al. 2011).

As one can see, there are two different approaches in the utilization of micro- and nanoparticles in the SDT and, as a consequence, their synthesis techniques. The first one is aimed at the loading of organic sonosensitizer molecules into the particles using their porous structure (Wang et al. 2012) or special coverings like PLGA, lipid layers, etc. (McEwan et al. 2014; Li et al. 2016) The second one is based on the usage of micro- and nanoparticles like silicon-based or TiO_2 nanoparticles as sonosensitizers themselves (Yamaguchi et al. 2011; Osminkina et al. 2015). Another example is the synthesis of so-called CO_2 "nanobombs" designed for the low-intensity (1 W·cm^{-2}) ultrasound-triggered cancer therapy (Zhang et al. 2015). Hollow mesoporous silica nanoparticles (HMSNPs) were loaded with L-arginine capable of absorbing and generating CO_2 bubbles in response to either pH or temperature changes. After intravenous administration of HMSNPs, they accumulated in tumor and led to cancer cell necrosis and local blood vessel destruction.

The above-mentioned results of application of silicon-based nanoparticles for the SDT can be explained mainly by the effects of hyperthermia and cavitation (Sviridov et al. 2013; Sviridov et al. 2015; Zhang et al. 2015), which can lead to the effective destruction of lesions in case the problem of targeting is solved. The propagation of USI in suspensions and soft media in the presence of solid micro- and nanoparticles is accompanied by additional attenuation associated with USI scattering and absorption. The absorption of USI waves is due to the Stokes friction exerted on the nanoparticles oscillating in the wave field. The absorption is significantly dependent on the nanoparticle properties (surface hydrophobicity/hydrophilicity and roughness), their size and shape, as well as concentration. The absorption of USI energy leads to the heating of the surrounding medium, while temperature increase may sufficiently exceed the values typical for pure liquids and soft media. This effect can be utilized for the purposes of ultrasound therapy of cancer (Hill et al. 2002), when silicon nanoparticles can accumulate, for example, due to the presence of micropores in tumor blood vessels.

Energy transferred by the USI beam is attenuated during its propagation through the viscous medium. If we set the intensity of traveling plane wave as I_0 at $x = 0$, it exponentially decreases down to $I(x)$ within the distance of x:

$$I(x) = I_0 e^{-2\alpha x},$$ (13.1)

where α is the amplitude attenuation coefficient.

The attenuation of USI has been investigated for a long time. The work of R. J. Urick (1948) is considered to be a classic work in this field of study. The attenuation coefficient of USI wave of frequency ω in a medium with spherical particles of radius a can be written as a sum of two terms, the first of which describes the process of energy scattering and is negligibly small for nano-sized particles, while the second one is associated directly with the absorption:

$$2\alpha = C \cdot \left[\frac{1}{6} k^4 a^3 + k(\sigma - 1)^2 \frac{s}{s^2 + (\sigma + \tau)^2} \right] \tag{13.2}$$

where the following notations are used:

$$s = \frac{9}{4\beta a}\left(1 + \frac{1}{\beta a}\right), \quad \tau = \frac{1}{2} + \frac{9}{4\beta a}, \quad \sigma = \frac{\rho_1}{\rho_0}, \quad \beta = \left(\frac{\omega}{2\mu}\right)^{1/2}, \quad C = \frac{4}{3}\pi a^3 n,$$

$k = 2\pi f/c$ is the wavenumber, c is the speed of sound in the medium, μ is the kinematic viscosity coefficient, ρ_0 is the medium density, ρ_1 is the nanoparticle density, and C is the volume concentration of nanoparticles.

The absorption of USI energy is connected with relative motion of solid nanoparticles in the USI wave. In a viscous medium, the Stokes force is exerted on a particle leading to inconvertible heat loses, which depend on the ratio of particle and medium densities, as well as medium viscosity. The change of temperature T of the medium in which the USI wave of intensity I propagates can be calculated using the heat-transfer equation:

$$\frac{\partial T}{\partial t} = \chi \Delta T + \frac{2\alpha_a I}{\rho_0 c_p}, \tag{13.3}$$

where $\chi = \kappa/\rho_0 c_p$ is the thermal conductivity, Δ is Laplacian operator, α_a is the USI absorption coefficient in the medium, ρ_0 and c_p are the medium density and specific heat capacity, respectively. In case of relatively fast heating, when the exposure time is much smaller than the characteristic time of heat transfer, the change of temperature within period t can be evaluated from Equation 13.3 using a simple formula:

$$\tilde{T} = \frac{2\alpha_a I}{\rho_0 c_p} t = \frac{2\alpha_a W}{S\rho_0 c_p} t, \tag{13.4}$$

where I is USI intensity, W is USI power, and S is the cross-sectional area of USI beam.

It should be noted that local hyperthermia effect of SiNPs has been studied both experimentally and theoretically (Sviridov et al. 2013). It has been revealed that efficient absorption of USI energy by SiNPs with the mean size of around 100 nm at a concentration of about 1 g·L^{-1} could efficiently increase the water temperature under exposure to USI at the frequencies of 1–2.5 MHz and intensities of 1–20 W·cm^{-2} (see Figure 13.3).

The efficiency of tissue heating with silicon-based nanoparticles can be significantly enhanced by the usage of acoustic cavitation effect, which is in the generation and activity of gas or vapor bubbles in the medium exposed to USI (Hill et al. 2002). Cavitation is normally used for the destruction of drug-loaded capsules to release medicine into the tumor under exposure to USI of sufficient amplitude. The process of acoustic cavitation is induced at a necessary level of pressure, which is characteristic for a certain medium known as the cavitation threshold. Many nanoparticles, including silicon nanoparticles, obtain porous surface, which enables their usage for the decrease of cavitation threshold in biological media and application of ultrasound of relatively low intensity for the drug delivery. The cavitation threshold depends on temperature, which changes both due to external conditions and absorption of a part of the USI energy (Brabec and Morstein 2007).

There are two distinct types of cavitation: inertial (collapsing or nonstable) and noninertial (stable) cavitation. Inertial cavitation starts when a cavity filled with gas is swelling during the rarefaction phase and rapidly collapses to the size several times smaller the initial one afterward. The collapse is accompanied by high temperatures and pressures, which results in energy transfer and destruction of neighboring

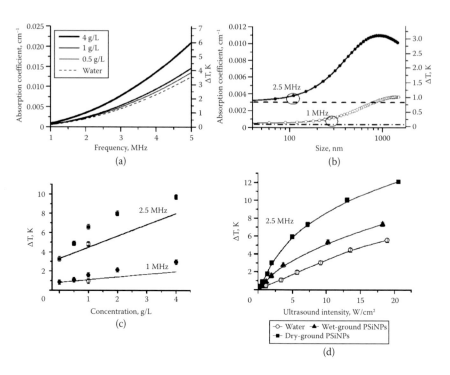

Figure 13.3 (a) Theoretically calculated dependence of the USI absorption coefficient of PSi NP suspensions on the USI frequency at different nanoparticle concentrations. (b) Theoretically calculated dependence of the USI absorption coefficient of PSi NP suspensions on the nanoparticle size at different frequencies. (c) Experimentally measured dependence of USI-induced heating in PSi NP suspensions on the nanoparticle concentration and the same theoretically calculated dependence. (d) Experimentally observed heating in PSi NP suspensions of two types and pure water under exposure to MHz-frequency ultrasound at various USI intensities. (From Sviridov, A.P., et al., *Appl. Phys. Lett.*, 103(19), 193110, 2013. With permission.)

elements of the medium. Noninertial cavitation describes the activity of different gaseous inclusions under exposure to USI. Such an activity may comprise their translational motion, surface distortion, size growth, as well as initiation of microflows. Small bubbles can grow due to a process called rectified diffusion. This phenomenon consists in the following: for the period of acoustic oscillation, gas alternately diffuses inside the bubble during the rarefaction phase and out of the bubble during the compression phase. Since the bubble surface area during the rarefaction phase exceeds the surface area during the compression phase, the total gas flow is directed inside the bubble, which results in the bubble growth. In order for the bubble to grow due to the rectified diffusion, amplitude of the acoustic pressure must exceed some threshold value. The threshold of rectified diffusion is considered to determine the cavitation threshold.

It was shown that the speed of gas flow directed inside the bubble during the rarefaction phase can be defined as:

$$\frac{dm}{dt} = \frac{8\pi}{3} DC_\infty R_0 \left(\frac{P_a}{P_0}\right)^2, \tag{13.5}$$

where D is the diffusion coefficient and C_∞ is the gas concentration in the liquid without the bubble. To take into account the effects of surface tension, the right part of Equation 13.5 should be multiplied by $(1 + 2\sigma / R_0 P_a)$. The speed of gas diffusion during the compression phase is equal to:

$$\frac{dm}{dt} = -4\pi DC_\infty R_0 \left(1 + \frac{2\sigma}{R_0 P_a} - \frac{C_\infty}{C_0}\right), \tag{13.6}$$

where C_0 is gas concentration of the saturated liquid. The pressure threshold of cavitation is reached when the gas flow in the rarefaction phase becomes equal to the flow in the compression phase. This condition is described by the equation:

$$\left(\frac{P_a}{P_0}\right)^2 = \frac{3}{2}\left[1 - \frac{C_\infty}{C_0}\left(1 + \frac{2\sigma}{R_0 P_a}\right)^{-1}\right].\tag{13.7}$$

When the threshold value of the acoustic field is exceeded, the bubbles will grow and finally reach the resonant size, while Equation 13.7 is applicable only for bubbles the size of which is much smaller than the resonant size.

More precisely, the threshold pressure value is described by the following equation (Neppiras 1969):

$$\left(\frac{P_a}{P_0}\right)^2 = \left(1 + \frac{2\sigma}{R_0 P_a} - \frac{C_\infty}{C_0}\right)\cdot\left(1 + \frac{2\sigma}{R_0 P_a}\right)^{-1}\cdot\left[\left(1-\beta^2\right)^2 + \delta^2\beta^2\right],\tag{13.8}$$

where $\beta^2 = (\omega/\omega_0)^2 = \rho R_0^2\omega^2/3\gamma P_0$, δ is the attenuation coefficient, ω is the cyclic frequency, ω_0 is the resonant bubble size, and γ is the gas isentropic volume exponent in the bubble. The resonance condition is determined by the formula:

$$\omega_0 = \frac{1}{R_0}\left\{\frac{3\gamma P_0}{\rho}\left[1 + \frac{2\sigma(3\gamma - 1)}{3\gamma R_0 P_0}\right]\right\}^{1/2}.\tag{13.9}$$

Recently, it has been experimentally shown that PSi NPs at the concentration of 0.3–0.5 g·L^{-1} induces cavitation in aqueous media and lowers the cavitation thresholds by more than two times as compared to pure water (see Figure 13.4). This effect can be explained by the presence of residual air bubbles inside the nanopores of PSi NPs, formation of gas (hydrogen) bubbles because of PSi dissolution, and formation of small voids filled with water vapor due to the relative motion of PSi NPs and water (Sviridov et al. 2015).

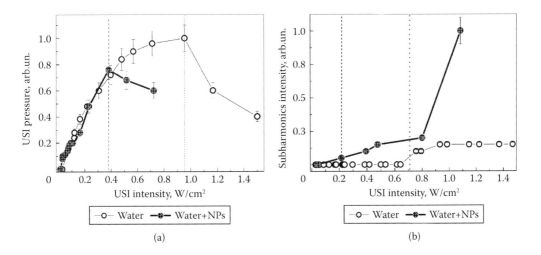

Figure 13.4 (a) Experimentally measured USI pressure transmitted through the PSi NP suspension and pure water as an indicator of acoustic cavitation initiation. (b) Experimentally measured subharmonic (half of the fundamental frequency) magnitude, which indicates the start of intense acoustic cavitation in the PSi NP suspension and pure water. (From Sviridov, A.P., et al., *Appl. Phys. Lett.*, 107(12), 123107, 2015. With permission.)

13.3.2 APPLICATION OF PSi NPs AS SONOSENSITIZERS FOR THE SDT: IN VITRO AND IN VIVO STUDIES

Because of high biocompatibility and biodegradability properties (Park et al. 2009; Canham 2014; Low and Voelcker 2014; Tolstik et al. 2016a), PSi NPs are very promising for different biomedical applications, in particular for the SDT. PSi NPs can passively penetrate into the living cells due to endocytosis or can also be functionalized by targeting antibodies (Park et al. 2009; Secret et al. 2013; Alhmoud et al. 2015). In both cases, nanoparticle sizes have to be limited to about 100 nm because of the cell and blood vessel morphology (Chithrani and Chan 2007; Peer et al. 2007; Perrault et al. 2009).

Silicon nanocrystals with sizes of 2–5 nm are known to exhibit room temperature PL in the visible and near-infrared (NIR) spectral regions (Canham 1990). The PL properties of SiNPs have been successfully used for bioimaging both in vitro (see Figure 13.2f) and in vivo (Park et al. 2009). Recently, several new approaches for visualization of PSi NPs in cancer cells are realized by means of the linear (high-resolution structured illumination microscopy [HR-SIM]; Raman spectroscopy) and nonlinear (coherent anti-Stokes Raman scattering [CARS]; two-photon excited fluorescence [TPEF]) optics in vitro (see Figure 13.5) (Tolstik et al. 2016b). Note that using these imaging techniques, a study of cancer tissue morphology with submicron resolution, as well as identification of tumor cell compositions with high specificity, becomes possible.

Sonosensitizing properties of PSi NPs were for the first time described by Osminkina et al. (2011a, 2011b). In this study, ultrasonic exposure was carried out in a standard ultrasonic bath UZV6-0.063/37 with working frequency of 37 kHz and maximum power of 120 W. The average power density of ultrasonic waves in the bath was less than 0.5 W·cm⁻². A distribution of the power density in different parts of the bath was assessed both for water surface in the bath and surface of the liquid in culture vessels placed in the bath (Marangopoulos et al. 1995). Then, the culture vessels were placed in zones with a maximum power of ultrasound (center of the bath, power density up to 5 W·cm⁻²) and in the zone of minimal USI intensity (edge of the bath, power density below 0.5 W·cm⁻²). Note that these power densities lay in the range of standard therapeutic ultrasound (Speed 2001). It is known that the USI with high intensity can lead to

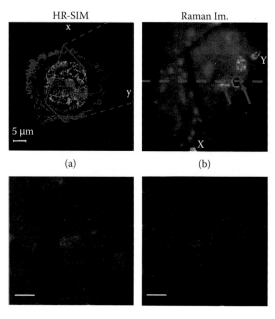

Figure 13.5 (a) Fluorescent HR-SIM image of MCF-7 breast cancer cells incubated with PSi NPs for 24 h. The cell nuclei were stained with Hoechst 34580 and the cytoplasm actin was stained with Alexa Fluor® 488 Phalloidin (colored in cyan and green, respectively). The PSi NPs are marked in red. (b) Raman spectroscopy images (xy-cross-section of the Raman image reconstructed with VCA) of MCF-7 cells incubated with PSi NPs for 24 h. The PSi NPs depicted in red and pointed with yellow arrows are located in the cell cytoplasm depicted in cyan; (c) merged images of CARS and TPEF of PSi NPs in MCF-7 cells. PSi NPs depicted in red within the cells depicted in green; (d) merged images of TPEF of PSi NPs in MCF-7 cells. (From Tolstik, E., et al., *Int. J. Mol. Sci.*, 17(9), 1536, 2016. With permission.)

heating of water and biological systems. In order to avoid the thermal effect of USI, a thermostat which allowed the researchers to maintain the temperature in the bath at 37°C was used.

A strong decrease of Hep2 cell population was detected in vitro after the treatment in ultrasound bath in the presence of PSi NPs. So, after the combined treatment with USI (37 kHz, 5 W·cm⁻²) and PSi NPs with concentration above 0.5 mg/mL, full destruction of cancer cells was observed (see Figure 13.6a). Moreover, the cell proliferation was not revealed even after further cultivation of cells for 3 days.

An interesting fact is that such USI treatment with minimal power (37 kHz, 5 W·cm⁻²) in the presence of PSi NPs did not destruct the cancer cells instantly, but they lost their proliferation properties and died within 20–80 h after the treatment (see Figure 13.6b). In this case, the measured phase composition of cells indicates that the cells died due to the apoptosis mechanism (Osminkina et al. 2011b).

In order to explain the observed effect, the possible mechanisms of cell death were proposed: (1) local heating (hyperthermia); (2) "nano-scalpel" effect of nanoparticles, which destroy the cancer cells mechanically; (3) inception of cavitation, which results in shock wave generation and additional dissipation of the ultrasound energy (Osminkina et al. 2011a). It is worth noting that the USI with frequency in the kHz range is usually accompanied by strong cavitation and the third mechanism should be dominating. However, the therapeutic USI is usually operated in the MHz-frequency range and all the mechanisms can be important for the sonosensitizing properties of PSi NPs.

The sonosensitization effect under the therapeutic USI with MHz frequency was soon observed both in vitro and in vivo. In in vitro experiments, USI (0.88 MHz, 1 W·cm⁻², pulse mode, modulation 2/20) generated by standard therapeutic equipment, UST-1.3.01 F 'MeDTeKo' was used. Degassed distilled water (at 37°C) was used as a contact medium between flat emitters with a radius of 2 cm and cuvette filled with the sample (see inset in Figure 13.7 [Osminkina et al. 2014]). The distance between the sample plane and emitter was adjusted so that the USI itself did not destroy the cells. In the control group, the cells without PSi NPs were investigated.

Figure 13.7 represents the living cell viability dependence on USI duration.

It has been shown in vitro that USI has essentially no effect on the cell viability within the time interval from 0 to 10 minutes. At the same time, the combined action of ultrasound and SiNPs leads to a 50% drop in the number of living cells as compared to the control (Osminkina et al. 2014). To achieve a full destruction of Hep2 cells, the prolongation of combined impact of PSi NPs and USI till 20 min can be used (Osminkina et al. 2015; Osminkina et al. 2016).

In in vivo experiments intravenous injection of DPSi NPs at a dose of 10 mg/kg in a volume of 0.2 mL was used. After the administration of nanoparticles, the tumor (melanoma B16) was affected simultaneously by two ultrasonic frequencies: 0.88 MHz and 2.64 MHz with an intensity of 1 and 2 W·cm⁻², respectively.

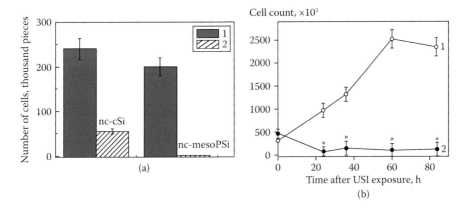

Figure 13.6 (a) Number of Hep2 cells after a 30-min ultrasound treatment (37 kHz, 5 W·cm⁻²) in the reference group (dark gray columns) and in the presence of nonporous silicon nanoparticles (nc-cSi) or porous PSi NPs (primed columns) (Osminkina et al. 2011a). (b) Changes in the amount of Hep2 cells after a 30-min ultrasound treatment (37 kHz, 0.5 W·cm⁻²) in the reference group (only USI, empty circles) and in the presence of PSi NPs (filled circles) (Osminkina et al. 2011b).

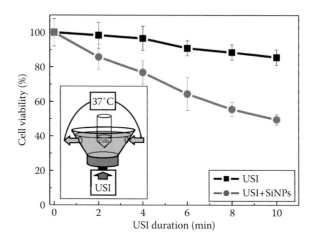

Figure 13.7 Dependence of in vitro cell viability on USI duration. The cells have been exposed to USI (black curve) or to the combined action of USI and SiNPs (red curve). Inset: schematic representation of the setup for in vitro USI experiments. (From Osminkina, L.A., et al., *Nanoscale Res. Lett.*, 9(1), 463, 2014. With permission.)

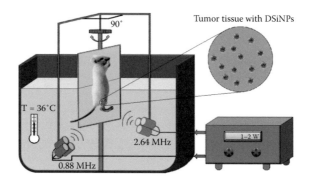

Figure 13.8 Schematic image of the experimental setup for USI experiments in vivo. (From Osminkina, L.A., et al., *Microporous Mesoporous Mater.*, 210, 169–175, 2015. With permission.)

In this case, flat emitters with a radius of 0.8 cm were placed at an angle of 90° relative to each other (see Figure 13.8). Degassed distilled water (at 36° C) was used as a contact medium. Exposure for 6 min was carried out in different time intervals and varied from 15 min to 24 h after the administration of DPSi NPs (Osminkina et al. 2015).

The efficacy of treatment was assessed by the dynamics of tumor growth in the control and test groups, respectively. Tumor volume doubling period «τ» was determined from the time dependence of V/V_0, where V_0 denotes the initial tumor volume. Antitumor enhancement coefficient «C» was calculated in the following way: $C = \tau/\tau 0$, where $\tau 0$ is the time of tumor volume doubling for the control group.

Figure 13.9a shows the dynamics of tumor growth for the group of mice. The time interval between the injection of DPSi NPs and USI treatment was 1 h. Note that DSiNPs themselves did not influence the tumor growth (C = 1). While the employed USI treatment itself was characterized by the antitumor enhancement coefficient C = 2, the combined action of USI and DSiNPs enhanced the therapeutic effect of ultrasound more than twofold (C = 4.5) (see Figure 13.9a) (Osminkina et al. 2015).

The inhibition of tumor growth for different time intervals between the DPSi NP injection and USI treatment can be analyzed using the following expression:

$$\text{Inhibition} = (1-V/V_c)\cdot100\% \tag{13.10}$$

where V_c is the tumor volume for the control group of mice, into which sterile water was injected. The higher value of inhibition indicates a more intense suppression of the tumor growth.

Functional materials

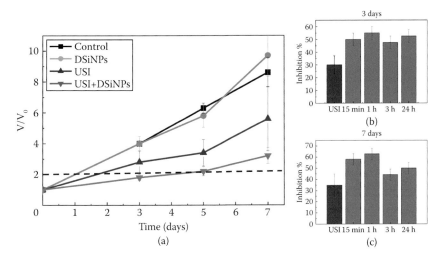

Figure 13.9 (a) Dynamics of tumor growth for mice with intravenously injected DPSi NPs followed after 1 h of USI treatment (red down triangles) for mice with intravenously injected DPSi NPs only (green circles), for mice with the USI treatment (blue up triangles), and for the control group of mice (black squares). The horizontal dashed line indicates the doubled tumor volume. (b) Inhibition of the tumor growth on the third day after the USI treatment performed for different time intervals between the injection of DPSi NPs and USI treatment (red bars). (c) Inhibition of the tumor growth on the seventh day after the USI treatment performed for different time intervals between the injection of DSiNPs and USI treatment (red bars). The effect of USI without DPSi NPs is shown by the blue bars. (From Osminkina, L.A., et al., *Microporous Mesoporous Mater.*, 210, 169–175, 2015. With permission).

Figure 13.9b and c shows the inhibition of tumor growth on the third (b) and on the seventh (c) days after the USI treatment (blue bars) and after the combined action of USI and DSiNPs (red bars). One can see that the strongest inhibition of the tumor growth was achieved in the case when the time interval separating the DPSi NPs administration and USI treatment was 1 h. Note that the inhibition effect was also observed for different time intervals between the administration of DPSi NPs and USI treatment. The obtained results indicate that the accumulation time of DPSiNPs in tumors is shorter than 15 min, and the removal time is longer than 24 h (Osminkina et al. 2015).

So, the observed strong suppression of the cancer tumor growth in vivo after the combined treatment of USI and SiNPs can be explained by the sonosensitizing properties of silicon nanoparticles.

13.4 CONCLUSIONS

In this review, the recent data about sonosensitizing properties of PSi NPs are discussed. Promising perspectives of usage of PSi NPs for application in the SDT of cancer are confirmed in several physical and biomedical experiments. According to the model physical experiments, the efficient absorption of USI energy by PSi NPs leads to local hyperthermia and cavitation effects. So, combination of USI and PSi NPs treatment causes the total destruction of cancer cells in vitro and the significant inhibition of tumor growth in vivo. The major advantage of SDT in cancer treatment is the increased penetration of USI through mammalian tissue up to more than 10 cm. At the same time, PSi NPs are characterized not only by low toxicity but also by the property of biodegradation, which is a significant advantage compared with other sonosensitizers.

Moreover, the sonosensitizing properties of PSi NPs can be combined with PSi-based cancer diagnostics, for example, by means of the optical linear and nonlinear bioimaging, which opens new possibilities for applications of PSi NPs in cancer theranostics.

ACKNOWLEDGMENTS

This work was financially supported by the Russian Science Foundation (Grant No. 16-13-10145).

Functional materials

REFERENCES

Alhmoud, H., Delalat, B., Elnathan, R., Cifuentes-Rius, A., Chaix, A. et al., Porous silicon nanodiscs for targeted drug delivery, *Advanced Functional Materials* 25, no. 7 (2015): 1137–1145.

Allen, T.M., Cullis, P.R., Drug delivery systems: Entering the mainstream, *Science* 303, no. 5665 (2004): 1818–1822.

Anglin, E.J., Cheng, L., Freeman, W.R., Sailor, M.J., Porous silicon in drug delivery devices and materials, *Advanced Drug Delivery Reviews* 60, no. 11 (2008): 1266–1277.

Bañobre-López, M., Teijeiro, A., Rivas, J., Magnetic nanoparticle-based hyperthermia for cancer treatment, *Reports of Practical Oncology & Radiotherapy* 18, no. 6 (2013): 397–400.

Brabec, K., Mornstein, V., Detection of ultrasonic cavitation based on low-frequency analysis of acoustic signal, *Open Life Sciences* 2, no. 2 (2007): 213–221.

Butterworth, K.T., McMahon, S.J., Currell, F.J., Prise, K.M., Physical basis and biological mechanisms of gold nanoparticle radiosensitization, *Nanoscale* 4, no. 16 (2012): 4830–4838.

Campbell, I.H., Fauchet, P.M., The effects of microcrystal size and shape on the one phonon Raman spectra of crystalline semiconductors, *Solid State Communications* 58, no. 10 (1986): 739–741.

Canham, L.T., Silicon quantum wire array fabrication by electrochemical and chemical dissolution of wafers, *Applied Physics Letters* 57, no. 10 (1990): 1046–1048.

Canham, L.T., Nanoscale semiconducting silicon as a nutritional food additive, *Nanotechnology* 18, no. 18 (2007): 185704.

Canham, L.T., Color of porous silicon, in *Handbook of Porous Silicon*, Ed. L.T. Canham, Cham, Switzerland: Springer International Publishing (2014a): 255–262.

Canham, L.T., Porous silicon for medical use: From conception to clinical use, in *Porous Silicon for Biomedical Applications*, Ed. H.A. Santos, Woodhead Publishing, Cambridge, UK (2014b): 3–20.

Celli, J.P., Spring, B.Q., Rizvi, I., Evans, C.L., Samkoe, K.S. et al., Imaging and photodynamic therapy: Mechanisms, monitoring, and optimization, *Chemical Reviews* 110, no. 5 (2010): 2795–2838.

Chen, H., Zhou, X., Gao, Y., Zheng, B., Tang, F., Huang, J., Recent progress in development of new sonosensitizers for sonodynamic cancer therapy, *Drug Discovery Today* 19, no. 4 (2014): 502–509.

Chen, Y., Chen, H., Shi, J., Nanobiotechnology promotes noninvasive high-intensity focused ultrasound cancer surgery, *Advanced Healthcare Materials* 4, no. 1 (2015): 158–165.

Chithrani, B.D., Chan, W.C., Elucidating the mechanism of cellular uptake and removal of protein-coated gold nanoparticles of different sizes and shapes, *Nano Letters* 7, no. 6 (2007): 1542–1550.

Claudon, M., Dietrich, C.F., Choi, B.I., Cosgrove, D.O., Kudo, M. et al., Guidelines and good clinical practice recommendations for contrast enhanced ultrasound (CEUS) in the liver–update 2012, *Ultraschall in der Medizin-European Journal of Ultrasound* 34, no. 1 (2013): 11–29.

Costley, D., Mc Ewan, C., Fowley, C., McHale, A.P., Atchison, J. et al., Treating cancer with sonodynamic therapy: A review, *International Journal of Hyperthermia* 31, no. 2 (2015): 107–117.

Coussios, C., Farny, C.H., Ter Haar, G., Roy, R.A. Role of acoustic cavitation in the delivery and monitoring of cancer treatment by high-intensity focused ultrasound (HIFU), *International Journal of Hyperthermia* 23, no. 2 (2007): 105–120.

Cullis, A.G., Canham, L.T., Calcott, P.D.J., The structural and luminescence properties of porous silicon, *Journal of Applied Physics* 82, no. 3 (1997): 909–965.

de las Heras Alarcón, C., Pennadam, S., Alexander, C., Stimuli responsive polymers for biomedical applications, *Chemical Society Reviews* 34, no. 3 (2005): 276–285.

Diederich, C.J., Hynynen, K., Ultrasound technology for hyperthermia, *Ultrasound in Medicine & Biology* 25, no. 6 (1999): 871–887.

Dolmans, D.E., Fukumura, D., Jain, R.K., Photodynamic therapy for cancer, *Nature Reviews Cancer* 3, no. 5 (2003): 380–387.

Eldar-Boock, A., Polyak, D., Scomparin, A., Satchi-Fainaro, R., Nano-sized polymers and liposomes designed to deliver combination therapy for cancer, *Current Opinion in Biotechnology* 24, no. 4 (2013): 682–689.

Fenster, A., Downey, D.B., Cardinal, H.N., Three-dimensional ultrasound imaging, *Physics in Medicine and Biology* 46, no. 5 (2001): R67.

Ferrara, K., Pollard, R., Borden, M., Ultrasound microbubble contrast agents: Fundamentals and application to gene and drug delivery, *Biomedical Engineering* 9 (2007): 415–447.

Frenkel, V., Ultrasound mediated delivery of drugs and genes to solid tumors, *Advanced Drug Delivery Reviews* 60, no. 10 (2008): 1193–1208.

Frinking, P.J., Bouakaz, A., Kirkhorn, J., Ten Cate, F.J., De Jong, N., Ultrasound contrast imaging: Current and new potential methods, *Ultrasound in Medicine & Biology* 26, no. 6 (2000): 965–975.

Gannon, C.J., Cherukuri, P., Yakobson, B.I., Cognet, L., Kanzius, J.S. et al., Carbon nanotube-enhanced thermal destruction of cancer cells in a noninvasive radiofrequency field, *Cancer* 110, no. 12 (2007): 2654–2665.

Gannon, C.J., Patra, C.R., Bhattacharya, R., Mukherjee, P., Curley, S.A., Intracellular gold nanoparticles enhance non-invasive radiofrequency thermal destruction of human gastrointestinal cancer cells, *Journal of Nanobiotechnology* 6, no. 1 (2008): 1.

Gao, Y., Li, Z., Wang, C., You, J., Jin, B. et al., Self-assembled chitosan/rose bengal derivative nanoparticles for targeted sonodynamic therapy: Preparation and tumor accumulation, *RSC Advances* 5, no. 23 (2015): 17915–17923.

Gongalsky, M.B., Kharin, A.Y., Osminkina, L.A., Timoshenko, V.Y., Jeong, J. et al., Enhanced photoluminescence of porous silicon nanoparticles coated by bioresorbable polymers, *Nanoscale Research Letters* 7, no. 1 (2012): 446.

Hallow, D.M., Mahajan, A.D., McCutchen, T.E., Prausnitz, M.R., Measurement and correlation of acoustic cavitation with cellular bioeffects, *Ultrasound in Medicine & Biology* 32, no. 7 (2006): 1111–1122.

Heinrich, J.L., Curtis, C.L., Luminescent colloidal silicon suspensions from porous silicon, *Science* 255, no. 5040 (1992): 66.

Herino, R., Bomchil, G., Barla, K., Bertrand, C., Ginoux, J.L., Porosity and pore size distributions of porous silicon layers, *Journal of the Electrochemical Society* 134, no. 8 (1987): 1994–2000.

Heyn, A.N.J., The infrared absorption spectrum of dextran and its bound water, *Biopolymers* 13, no. 3 (1974): 475–506.

Hill, C., Bamber, J., Haar, G.T., *Physical Principles of Medical Ultrasonics*, Somerset, NJ: Wiley (2002).

Hu, M., Chen, J., Li, Z.Y., Au, L., Hartland, G.V. et al., Gold nanostructures: Engineering their plasmonic properties for biomedical applications, *Chemical Society Reviews* 35, no. 11 (2006): 1084–1094.

Kremkau, F.W., Cancer therapy with ultrasound: A historical review, *Journal of Clinical Ultrasound* 7, no. 4 (1979): 287–300.

Lam, C.W., James, J.T., McCluskey, R., Hunter, R.L., Pulmonary toxicity of single-wall carbon nanotubes in mice 7 and 90 days after intratracheal instillation, *Toxicological Sciences* 77, no. 1 (2004): 126–134.

Lammers, T., Subr, V., Ulbrich, K., Hennink, W.E., Storm, G., Kiessling, F., Polymeric nanomedicines for image-guided drug delivery and tumor-targeted combination therapy, *Nano Today* 5, no. 3 (2010): 197–212.

Lasagna-Reeves, C., Gonzalez-Romero, D., Barria, M.A., Olmedo, I., Clos, A. et al., Bioaccumulation and toxicity of gold nanoparticles after repeated administration in mice, *Biochemical and Biophysical Research Communications* 393, no. 4 (2010): 649–655.

Li, W.P., Su, C.H., Chang, Y.C., Lin, Y.J., Yeh, C.S., Ultrasound-induced reactive oxygen species mediated therapy and imaging using a fenton reaction activable polymersome, *ACS Nano* 10, no. 2 (2016): 2017–2027.

Liu, Z., Chen, K., Davis, C., Sherlock, S., Cao, Q. et al., Drug delivery with carbon nanotubes for in vivo cancer treatment, *Cancer Research* 68, no. 16 (2008): 6652–6660.

Low, S.P., Voelcker, N.H., Biocompatibility of porous silicon, in *Handbook of Porous Silicon*, Ed. L.T. Canham, Cham, Switzerland: Springer International Publishing (2014): 381–393.

Marangopoulos, I.P., Martin, C.J., Hutchison, J.M.S., Measurement of field distributions in ultrasonic cleaning baths: Implications for cleaning efficiency, *Physics in Medicine and Biology* 40, no. 11 (1995): 1897.

McEwan, C., Fowley, C., Nomikou, N., McCaughan, B., McHale, A.P., Callan, J.F., Polymeric microbubbles as delivery vehicles for sensitizers in sonodynamic therapy, *Langmuir* 30, no. 49 (2014): 14926–14930.

Miller, M.W., Gene transfection and drug delivery, *Ultrasound in Medicine & Biology* 26, no. 1 (2000): S59–S62.

Neppiras, E.A., Subharmonic and other low-frequency emission from bubbles in sound-irradiated liquids, *The Journal of the Acoustical Society of America* 46, no. 3B (1969): 587–601.

Nicolau, C., Catalá, V., Vilana, R., Gilabert, R., Bianchi, L. et al., Evaluation of hepatocellular carcinoma using SonoVue, a second generation ultrasound contrast agent: Correlation with cellular differentiation, *European Radiology* 14, no. 6 (2004): 1092–1099.

Nikolaev, A.L., Gopin, A.V., Bozhevol'nov, V.E., Treshalina, H.M., Andronova, N.V. et al., Combined method of ultrasound therapy of oncological diseases, *Russian Journal of General Chemistry* 85, no. 1 (2015): 303–320.

Ogata, Y.H., Characterization of porous silicon by infrared spectroscopy, in *Handbook of Porous Silicon*, Ed. L.T. Canham, Cham, Switzerland: Springer International Publishing (2014): 473–480.

Osminkina, L.A., Gongalsky, M.B., Motuzuk, A.V., Timoshenko, V.Y., Kudryavtsev, A.A., Silicon nanocrystals as photo- and sono-sensitizers for biomedical applications, *Applied Physics B* 105, no. 3 (2011a): 665–668.

Osminkina, L.A., Kudryavtsev, A.A., Zinovyev, S.V., Sviridov, A.P., Kargina, Y.V. et al., Silicon nanoparticles as amplifiers of the ultrasonic effect in sonodynamic therapy, *Bulletin of Experimental Biology and Medicine* 161, no. 2 (2016): 296–299.

Osminkina, L.A., Luckyanova, E.N., Gongalsky, M.B., Kudryavtsev, A.A., Gaydarova, Akh. et al., Effects of nanostructurized silicon on proliferation of stem and cancer cell, *Bulletin of Experimental Biology and Medicine* 151, no. 1 (2011b): 79–83.

Osminkina, L.A., Nikolaev, A.L., Sviridov, A.P., Andronova, N.V., Tamarov, K.P. et al., Porous silicon nanoparticles as efficient sensitizers for sonodynamic therapy of cancer, *Microporous and Mesoporous Materials* 210 (2015): 169–175.

Osminkina, L.A., Sivakov, V.A., Mysov, G.A., Georgobiani, V.A., Natashina, U.A. et al., Nanoparticles prepared from porous silicon nanowires for bio-imaging and sonodynamic therapy, *Nanoscale Research Letters* 9, no. 1 (2014): 463.

Park, J.H., Gu, L., Von Maltzahn, G., Ruoslahti, E., Bhatia, S.N., et al., Biodegradable luminescent porous silicon nanoparticles for in vivo applications, *Nature Materials* 8, no. 4 (2009): 331–336.

Peer, D., Karp, J.M., Hong, S., Farokhzad, O.C., Margalit, R., Langer, R., Nanocarriers as an emerging platform for cancer therapy, *Nature Nanotechnology* 2, no. 12 (2007): 751–760.

Perrault, S.D., Walkey, C., Jennings, T., Fischer, H.C., Chan, W.C., Mediating tumor targeting efficiency of nanoparticles through design, *Nano Letters* 9, no. 5 (2009): 1909–1915.

Podell, S., Burrascano, C., Gaal, M., Golec, B., Maniquis, J., Mehlhaff, P., Physical and biochemical stability of Optison®, an injectable ultrasound contrast agent, *Biotechnology and Applied Biochemistry* 30, no. 3 (1999): 213–223.

Porter, T.R., Iversen, P.L., Li, S., Xie, F., Interaction of diagnostic ultrasound with synthetic oligonucleotide-labeled perfluorocarbon-exposed sonicated dextrose albumin microbubbles, *Journal of Ultrasound in Medicine* 15, no. 8 (1996): 577–584.

Purushotham, S., Chang, P.E.J., Rumpel, H., Kee, I.H.C., Ng, R.T.H. et al., Thermoresponsive core-shell magnetic nanoparticles for combined modalities of cancer therapy, *Nanotechnology* 20, no. 30 (2009): 305101.

Qian, X., Zheng, Y., Chen, Y., Micro/nanoparticle-augmented sonodynamic therapy (SDT): Breaking the depth shallow of photoactivation, *Advanced Materials* 28, no. 37 (2016): 8097–8129.

Raisinghani, A., Rafter, P., Phillips, P., Vannan, M.A., DeMaria, A.N., Microbubble contrast agents for echocardiography: Rationale, composition, ultrasound interactions, and safety, *Cardiology Clinics* 22, no. 2 (2004): 171–180.

Rapoport, N., Physical stimuli-responsive polymeric micelles for anti-cancer drug delivery, *Progress in Polymer Science* 32, no. 8 (2007): 962–990.

Rosenthal, I., Sostaric, J.Z., Riesz, P., Sonodynamic therapy—A review of the synergistic effects of drugs and ultrasound, *Ultrasonics Sonochemistry* 11, no. 6 (2004): 349–363.

Santos, H.A., Bimbo, L.M., Lehto, V.P., Airaksinen, A.J., Salonen, J., Hirvonen, J., Multifunctional porous silicon for therapeutic drug delivery and imaging, *Current Drug Discovery Technologies* 8, no. 3 (2011): 228–249.

Sazgarnia, A., Shanei, A., Meibodi, N.T., Eshghi, H., Nassirli, H., A novel nanosonosensitizer for sonodynamic therapy in vivo study on a colon tumor model, *Journal of Ultrasound in Medicine* 30, no. 10 (2011): 1321–1329.

Schneider, M., SonoVue, a new ultrasound contrast agent, *European Radiology* 9, no. 3 (1999): S347.

Secret, E., Smith, K., Dubljevic, V., Moore, E., Macardle, P. et al., Antibody-functionalized porous silicon nanoparticles for vectorization of hydrophobic drugs, *Advanced Healthcare Materials* 2, no. 5 (2013): 718–727.

Serpe, L., Foglietta, F., Canaparo, R., Nanosonotechnology: The next challenge in cancer sonodynamic therapy, *Nanotechnology Reviews* 1, no. 2 (2012): 173–182.

Shabir, Q., Biodegradability of porous silicon, in *Handbook of Porous Silicon*, Ed. L.T. Canham, Cham, Switzerland: Springer International Publishing (2014): 395–401.

Speed, C.A., Therapeutic ultrasound in soft tissue lesions, *Rheumatology* 40, no. 12 (2001): 1331–1336.

Suh, K.Y., Yang, J.M., Khademhosseini, A., Berry, D., Tran, T.N.T. et al., Characterization of chemisorbed hyaluronic acid directly immobilized on solid substrates, *Journal of Biomedical Materials Research Part B: Applied Biomaterials* 72, no. 2 (2005): 292–298.

Sviridov, A.P., Andreev, V.G., Ivanova, E.M., Osminkina, L.A., Tamarov, K.P., Timoshenko, V.Y., Porous silicon nanoparticles as sensitizers for ultrasonic hyperthermia, *Applied Physics Letters* 103, no. 19 (2013): 193110.

Sviridov, A.P., Osminkina, L.A., Nikolaev, A.L., Kudryavtsev, A.A., Vasiliev, A.N., Timoshenko, V.Y., Lowering of the cavitation threshold in aqueous suspensions of porous silicon nanoparticles for sonodynamic therapy applications, *Applied Physics Letters* 107, no. 12 (2015): 123107.

Tamarov, K., Xu, W., Osminkina, L., Zinovyev, S., Soininen, P. et al., Temperature responsive porous silicon nanoparticles for cancer therapy—Spatiotemporal triggering through infrared and radiofrequency electromagnetic heating, *Journal of Controlled Release* 241, no. 10 (2016): 220–228.

Theiß, W., Optical properties of porous silicon, *Surface Science Reports* 29, no. 3 (1997): 91–192.

Tolstik, E., Osminkina, L.A., Akimov, D., Gongalsky, M.B., Kudryavtsev, A.A. et al., Linear and non-linear optical imaging of cancer cells with silicon nanoparticles, *International Journal of Molecular Sciences* 17, no. 9 (2016): 1536.

Tolstik, E., Osminkina, L.A., Matthäus, C., Burkhardt, M., Tsurikov, K.E. et al., Studies of silicon nanoparticles uptake and biodegradation in cancer cells by Raman spectroscopy, *Nanomedicine* 12, no. 7 (2016):1931–1940.

Tuziuti, T., Yasui, K., Sivakumar, M., Iida, Y., Miyoshi, N., Correlation between acoustic cavitation noise and yield enhancement of sonochemical reaction by particle addition, *The Journal of Physical Chemistry A* 109, no. 21 (2005): 4869–4872.

Umemura, S.I., Yumita, N., Nishigaki, R., Umemura, K., Mechanism of cell damage by ultrasound in combination with hematoporphyrin, *Japanese Journal of Cancer Research* 81, no. 9 (1990): 962–966.

Urick, R.J., The absorption of sound in suspensions of irregular particles, *The Journal of the Acoustical Society of America* 20, no. 3 (1948): 283–289.

von Herbay, A., Vogt, C., Willers, R., Häussinger, D., Real-time imaging with the sonographic contrast agent SonoVue differentiation between benign and malignant hepatic lesions, *Journal of Ultrasound in Medicine* 23, no. 12 (2004): 1557–1568.

Wang, C., Cao, S., Tie, X., Qiu, B., Wu, A., Zheng, Z., Induction of cytotoxicity by photoexcitation of TiO2 can prolong survival in glioma-bearing mice, *Molecular Biology Reports* 38, no. 1 (2011): 523–530.

Wang, X., Chen, H., Chen, Y., Ma, M., Zhang, K. et al., Perfluorohexane-encapsulated mesoporous silica nanocapsules as enhancement agents for highly efficient high intensity focused ultrasound (HIFU), *Advanced Materials* 24, no. 6 (2012): 785–791.

Wei, K., Mulvagh, S.L., Carson, L., Davidoff, R., Gabriel, R. et al., The safety of definity and optison for ultrasound image enhancement: A retrospective analysis of 78,383 administered contrast doses, *Journal of the American Society of Echocardiography* 21, no. 11 (2008): 1202–1206.

Yagüe, C., Manuel, A., Santamaria, J., NIR-enhanced drug release from porous Au/SiO 2 nanoparticles, *Chemical Communications* 46, no. 40 (2010): 7513–7515.

Yamaguchi, S., Kobayashi, H., Narita, T., Kanehira, K., Sonezaki, S. et al., Sonodynamic therapy using water-dispersed TiO 2-polyethylene glycol compound on glioma cells: Comparison of cytotoxic mechanism with photodynamic therapy, *Ultrasonics Sonochemistry* 18, no. 5 (2011): 1197–1204.

Yildirimer, L., Thanh, N.T., Loizidou, M., Seifalian, A.M., Toxicology and clinical potential of nanoparticles, *Nano Today* 6, no. 6 (2011): 585–607.

You, J., Zhang, G., Li, C., Exceptionally high payload of doxorubicin in hollow gold nanospheres for near-infrared light-triggered drug release, *ACS Nano* 4, no. 2 (2010): 1033–1041.

Yumita, N., Nishigaki, R., Umemura, K., Umemura, S.I., Hematoporphyrin as a sensitizer of cell-damaging effect of ultrasound, *Japanese Journal of Cancer Research* 80, no. 3 (1989): 219–222.

Yumita, N., Okuyama, N., Sasaki, K., Umemura, S., Sonodynamic therapy on chemically induced mammary tumor: Pharmacokinetics, tissue distribution and sonodynamically induced antitumor effect of gallium–porphyrin complex ATX-70, *Cancer Chemotherapy and Pharmacology* 60, no. 6 (2007): 891–897.

Zhang, K., Xu, H., Chen, H., Jia, X., Zheng, S. et al., CO2 bubbling-based "Nanobomb" system for targetedly suppressing Panc-1 pancreatic tumor via low intensity ultrasound-activated inertial cavitation, *Theranostics* 5, no. 11 (2015): 1291–1302.

Zou, Z., He, D., He, X., Wang, K., Yang, X. et al., Natural gelatin capped mesoporous silica nanoparticles for intracellular acid-triggered drug delivery, *Langmuir* 29, no. 41 (2013): 12804–12810.

14 Silicon metamaterials for infrared applications

Xianglei Liu and Zhuomin M. Zhang

Contents

14.1 INTRODUCTION

The mid-infrared (IR) region from about 2.5 to 25 µm wavelengths (or wavenumbers from about 400 to 4000 cm^{-1}) (Zhang 2007) is the main platform for thermal emission from ambient objects and resonance absorption of many molecular species, thus playing a vital role in thermal management, thermal imaging, energy harvesting, chemical and biological sensing, and environmental monitoring. Besides, the atmosphere's transparent window (8–13 µm) falls in the mid-IR region and is important for radiative cooling and free-space optical communication. Noble metals have a low emissivity in the mid-IR due to the poor impedance mismatch with air, thus precluding the potential applications in efficient thermal management and thermal signature detection. The absolute value of the permittivity being much larger than the dielectrics obscures the application prospects in sensing due to the weak confinement of light and light–matter interaction. Doped silicon and its nanostructures or related metamaterials, as alternative engineered "metals" due to its free carriers, can relieve these problems via the much smaller plasma frequency lying in the mid-IR. Besides, the resonance frequency can also be tuned by chemical doping, as in stark contrast to the fixed plasma frequency of noble metals. Therefore, more flexibility and tunability can be achieved, enabling the manipulation of both the structure design and doping control for various applications.

The doping can be introduced by well-established ion implantation or thermal diffusion approaches after fabricating silicon nanostructures via current optical, nanoparticle, and e-beam lithography techniques or chemical vapor depositions. Alternatively, penetration of dopant atoms can be involved during the fabrication of silicon nanostructures via bottom-up technique. For example, doped silicon nanowires can be fabricated by employing the vapor–liquid–solid technique (Schmidt et al. 2010; Chou et al. 2012). The vapor, liquid, and solid denote gaseous Si precursors such as SiH$_4$ gas, metal–Si alloy liquid, and solid silicon nanowire, respectively. The metals such as Au act as the catalyst. Doped nanowires can be obtained by adding vapor-phase dopants such as AsCl$_2$ and PCl$_3$ (Schmidt et al. 2010). One of the advantages of this bottom-up technique is that the doping concentration or free carrier density can be controlled with more flexibility. For example, doped-Si nanowires with controlled doping-level distribution along the nanowire were fabricated by Chou and Filler (2013). Doped-Si nanostructures are compatible with standard silicon processing technologies and are relatively easier to integrate with nanoelectronics than metals, opening the possibility of realizing on-chip mid-IR devices for sensing, thermal management, and energy harvesting.

This chapter focuses on some unique mid-IR radiative properties enabled by doped-Si nanostructures and metamaterials for applications in plasmonic sensing, subwavelength thermal imaging, and thermal management. The arrangement of this chapter is as follows:

Section 14.2 presents the advantages of plasmonic sensing via the mid-IR wavelengths, where molecule's fingerprint resonances occur. Advantages of doped silicon acting as a better replacement of noble metals for excitation of mid-IR surface plasmons are discussed, such as the relatively strong confinement of surface modes to the subwavelength scale and the comparable magnitude of permittivities of doped silicon to common dielectrics such as air. Efforts on employing doped-Si metamaterials or particles in mid-IR plasmonic sensing are briefly reviewed.

Section 14.3 discusses low-loss all-angle negative refraction in the mid-IR based on doped-Si nanowire arrays. At oblique incidence, the figure of merit (FOM) counterintuitively increases with the material loss in the direction along the optical axis of the metamaterials. The underlying physical mechanisms accounting for the large FOM are elaborated based on the loss-enhanced transmission, impedance matching, and absence of resonances.

Section 14.4 briefly reviews the capability to achieve tunable omnidirectional perfect absorption of mid-IR electromagnetic waves via doped-Si metamaterials. Simple two-layer configuration and relatively complex nanowires are discussed. The predicted tunable perfect absorption can have a wide range of applications in noncontact temperature measurement, radiation cooling, free-space thermal management, and so on.

Section 14.5 shows the utilization of doped-Si-based metamaterials or metasurfaces in enhancing near-field thermal radiation, which has promising applications in energy harvesting, heat dissipation, nano-manufacturing, as well as noncontact thermal diodes and transistors. Various doped-Si nanostructures including nanowires, nanoholes, multilayers, gratings, graphene-covered nanowires, and metasurfaces are also discussed. The underlying mechanisms for the enhancement are summarized, including the hyperbolic modes, low-loss surface resonance modes, and hybridization of graphene plasmons with hyperbolic modes.

14.2 MID-IR PLASMONICS FOR SENSING AND DETECTION

Surface plasmon polaritons (SPPs) induced by the collective oscillations of electrons coupled with incident electromagnetic waves may be excited at the interface of two materials when their dielectric functions have opposite signs. SPPs are featured with the exponential decay of electric and magnetic fields when being away from the interface and can propagate along the interface, as shown in Figure 14.1. For nonmagnetic materials, only transverse magnetic (TM) waves (magnetic field perpendicular to the plane of incidence) can support the excitation of SPPs. The dispersion relation of SPPs can be obtained from the pole of Fresnel reflection coefficients at the metal–dielectric interface and is given as (Zhang 2007)

$$\beta = \beta' + j\beta'' = \frac{\omega}{c_0}\sqrt{\frac{\varepsilon_d \varepsilon_m}{\varepsilon_d + \varepsilon_m}} \tag{14.1}$$

where β is the tangential wavevector parallel to the interface, which is expressed in a real part β' and imaginary part β'', ω is the angular frequency, c_0 is the speed of the light in the vacuum, ε_d and ε_m are the dielectric functions of the dielectric and metal, respectively. The square of the wavevector is given as $k^2 = \varepsilon(\omega/c_0)^2 = k_x^2 + k_y^2 + k_z^2 = \beta^2 + k_z^2$, where ε is the dielectric function of the medium. While k and k_z are medium dependent, β is the same in both media. When Equation 14.1 is satisfied, the incident photon energy can be strongly absorbed by the collectively oscillating electrons, leading to a sharp change in the intensity, phase, or polarization of the reflected or transmitted signals. The spectral or spatial location (e.g., angle of incidence) of this sharp change heavily depends on the dielectric environment of the metal surface, such as optical constants, thickness, or concentration. This phenomenon lays down the foundation for plasmonic sensing of chemical and biomedical molecules (Liedberg et al. 1983). This revolutionary noninvasive sensing technique relieves the need of label and enables on-site analysis and, thus, has attracted much attention from both researchers and practitioners in the past few decades due to its wide applications in fundamental chemical and biological researches, industrial process control, clinical diagnosis, and environmental monitoring (Homola et al. 1999; Singh 2016).

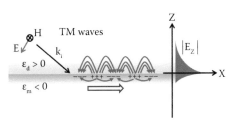

Figure 14.1 Schematic of propagating surface plasmon polaritons at the metal–dielectric interface.

Extensive research and commercialization efforts of plasmonic sensing have been mainly devoted to the visible and near-field spectral regions due to the readily available noble metals as the substrate supporting SPPs as well as light sources and detectors. Plasmonic sensing working in mid-IR wavelengths, however, has been recently demonstrated to exhibit some unique advantages. For example, the mid-IR region is safer to the living cells compared with possible photodamage and phototoxicity caused by the visible light (Ziblat et al. 2006). Besides, different chemical or biological analytes have distinct mid-IR absorption spectra, caused by their "fingerprint" resonant vibrations. Therefore, a large contrast of dielectric functions of different analytes is expected, leading to the improvement of both sensitivity and selectivity (Cleary et al. 2010; Mizaikoff 2013). Although noble metals can still exhibit negative permittivities and support SPPs in the mid-IR, their magnitudes of dielectric functions are much larger than those of dielectrics. As a result, the dispersion relation of SPPs is close to the light line, leading to weak light confinement and less sensitivity to the dielectric environment.

The optical response of Ag and doped silicon at different doping levels can be modeled using the Drude model, that is, $\varepsilon(\omega) = \varepsilon_\infty - \dfrac{\omega_p^2}{\omega(\omega + j\gamma)}$. For Ag, $\varepsilon_\infty = 1$, $\omega_p = 1.37 \times 10^{16}$ rad/s, and $\gamma = 2.73 \times 10^{13}$ rad/s (Zhang 2007). For heavily doped silicon (*n*-type) in the mid-IR, $\varepsilon_\infty = 11.7$ is a good approximation with $\omega_p = 1.08 \times 10^{15}$ or 2.43×10^{15}) rad/s and $\gamma = 9.34 \times 10^{13}$ or 2.00×10^{14} rad/s if the doping level is 1×10^{20} or 5×10^{20} cm³, respectively (Basu et al. 2010a). The dispersion of SPPs showing the relationship between $k_0 = \omega/c_0$ with β' is plotted in Figure 14.2a between the air and Ag or doped silicon. For the air–Ag interface, the dispersion overlaps with the light line since $|\varepsilon_{Ag}| \gg 1$, as shown by the black dotted line. For doped silicon, the free carrier concentration is reduced by two orders of magnitude compared with noble metals. The surface resonance frequency where $\varepsilon_\infty = -1$ shifts from the ultraviolet to the mid-IR. Subsequently, the huge mismatch of dielectric function between air and Ag is relieved to some extent by employing doped silicon. The dispersion of SPPs at the air and doped Si interface lies at the top side of the light line, indicating that the incident light from the air even at the grazing angle cannot excite SPPs. Nevertheless, the momentum difference can be compensated via the well-known Otto setup, Kretschmann configuration, or the coupling with nanostructures. Note that for doped silicon, the high-frequency regime where the permittivity becomes positive is not drawn since SPPs exist only when the dielectric and the metal substrate have opposite signs.

The confinement length of SPPs in the dielectric material is given as

$$\delta = \frac{1}{k_0 \, \mathrm{Im}\left(\sqrt{\dfrac{\varepsilon_d^2}{\varepsilon_d + \varepsilon_m}} \right)} \tag{14.2}$$

Figure 14.2b shows the SPP confinement length when air is the top dielectric medium and the substrate is either Ag or doped silicon. The field of SPPs with Ag substrate penetrates deeply into the air with a length more than one order of magnitude over the wavelength. This precludes strong field enhancement close to the interface, as a result, the mid-IR plasmonic sensing based on the Ag substrate is less sensitive to the existence of analytes especially when the thickness is not so large. Employing doped silicon as the substrate, on the contrary, enables the confinement width to be comparable and even smaller than the wavelength. Then, the sensitivity is improved prominently, making the detection of subwavelength-scale analytes possible. Another advantage of doped silicon–based mid-IR plasmonic sensing lies in the tunability, achieved

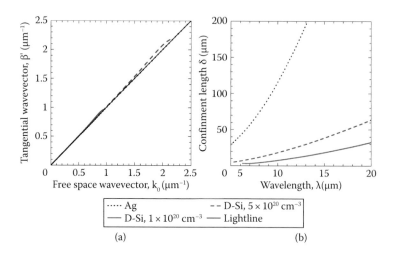

Figure 14.2 (a) Dispersion relation of SPPs at the interface of air with different materials; and (b) confinement length of SPPs versus incident wavelength.

via chemical doping. Mediating the doping level can adjust the plasmon frequency and scattering rate of free charge carriers. As a result, the operation wavelength can be tuned to maximize the sensing performance of different target analytes.

Chen (2009) first proposed a mid-IR plasmonic sensor based on doped-Si gratings and theoretically evaluated the performance. The schematic of the developed sensor system, including a mid-IR light source, a linear polarizer, and a detector, is shown in Figure 14.3a. Figure 14.3b presents the reflectance spectra for the doped silicon grating with the incident angle of 45°. The grating period and depth are chosen as 10 and 0.6 μm, respectively, and the filling ratio is set at 0.5. The culture fluid is assumed to be the free space. A reflectance dip is clearly observed when the resonance condition is satisfied to render energy transfer from the incident light to the collective charge movements. When the refraction index increases slightly, the reflectance dip redshifts accordingly. A high sensitivity of 17,200 nm/RIU was predicted. The light source and detector are inside the culture fluid. As a result, the light may be absorbed, causing the increase of noise-to-signal ratio. DiPippo et al. (2010) modified the design to address this issue and to relieve the fabrication challenges to some extent. The modified mid-IR plasmonic sensing system with a thin gold film covering on a doped silicon grating is shown in Figure 14.4a. The doped-Si grating provides the momentum difference required to excite SPPs at the Au–sample interface. Changing the doping concentration of the gratings can tune the operation wavelength as shown in Figure 14.4b. Nevertheless, experimental demonstrations are still missing though much needed for practical applications. Active tunability, which may be achievable by externally applying magnetic field or voltage, also deserves much attention.

Besides employing SPPs for mid-IR sensing, localized surface plasmon (LSP), confined to a metallic nanoparticle, has also been employed for sensing applications. When LSP is excited, the optical absorption will be maximized due to the strong light–substance interaction. The shift of the absorption peak wavelength with the dielectric environment surrounding the nanoparticle enables LSP to be used in sensing applications. The absorption and scattering properties of spherical particles can be obtained analytically from the well-known Mie theory. For arbitrary nonspherical particles, numerical methods such as discrete dipole approximations and finite-difference time-domain (FDTD) approaches can be readily employed to deal with their scattering properties. In the quasi-static limit (i.e., the incident wavelength λ is much larger than the particle diameter D), high-order multipole oscillations beyond dipoles can be neglected, and the extinction cross-section of spherical nanoparticles from the Mie theory can be simplified as (Petryayeva and Krull 2011)

$$C_{\text{ext}} = \frac{3\pi^2 D^3 \varepsilon_d^{3/2} N}{\lambda \ln(10)} \frac{\varepsilon_m''}{(\varepsilon_m' + 2\varepsilon_d)^2 + \varepsilon_m''^2} \tag{14.3}$$

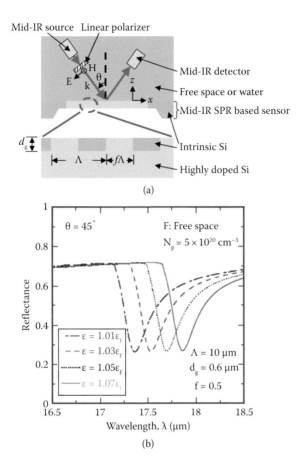

(a)

(b)

Figure 14.3 (a) Setup of mid-IR sensor based on doped silicon grating; and (b) reflectance spectra with varying dielectric environment. (Adapted from Chen, Y.B., (2009), Development of mid-infrared surface plasmon resonance-based sensors with highly-doped silicon for biomedical and chemical applications, *Opt. Express*, 17, 3130–3140. With permission of Optical Society of America.)

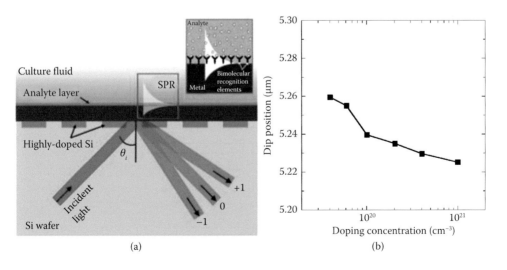

(a)

(b)

Figure 14.4 (a) Configuration of doped silicon biosensor platform; and (b) reflectance dip position in wavelength versus doping concentration. (Adapted from DiPippo, W., et al., (2010), Design analysis of doped-silicon surface plasmon resonance immunosensors in mid-infrared range, *Opt. Express*, 18, 19396–19406. With permission of Optical Society of America.)

Functional materials

where ε'_m and ε''_m are, respectively, the real and imaginary parts of the metallic nanoparticle's permittivity. Clearly, at localized surface plasmon resonance (LSPR) when $\varepsilon'_m = -2\varepsilon_d$, the extinction cross-section achieves the maximum value. Taking the air as the surrounding dielectric with the permittivity ε_d equal to 1, the LSPR wavelength of dipoles made of different materials, λ_{LSPR}, is given in Table 14.1. The permittivities of Ag and doped Si are from the above Drude models, and the dielectric functions for other materials, that is, Pd, Ti, Cu, and Au, are from Rakic et al. (1998). As shown in Table 14.1, the LSPR wavelength of common noble metals lies in the ultraviolet or visible range. There have been a great deal of theoretical and experimental researches on LSPR in the UV, visible, and even the near-IR regions (Haes and Van Duyne 2002; Link and El-Sayed 2003; Hutter and Fendler 2004; Endo et al. 2005; Sherry et al. 2005; Willets and Van Duyne 2007; Chan et al. 2008; Mayer and Hafner 2011; Teranishi et al. 2011; Xuan et al. 2014). Doped silicon, on contrary, can extend λ_{LSPR} to the mid-IR. Therefore, the possibility of mid-IR sensing via LSPR by using nanosize particles is opened.

Chou et al. (2012) investigated the absorption properties of phosphorus-doped Si nanowires fabricated by using the vapor–liquid–solid approach. The nanowire diameter and doping concentration are around 67 nm and on the order of 10^{19}–10^{20} cm^{-3}, respectively. The nanowire length is adjustable by changing the deposition time. The schematic of nanowires with varying lengths they fabricated is shown in Figure 14.5a. The resonance wavelength for a nanowire length of 135 nm was found to be 6.13 μm. When increasing the nanowire length, a redshift of the resonance wavelength was observed, as shown in Figure 14.5b. For example, when the nanowire length increases to 1160 nm, λ_{LSPR} redshifts to 13.5 μm. The agreement with Mie–Gans theory is good by varying the doping concentration. Rowe et al. (2013) studied the LSPR of phosphorus-doped Si nanocrystals synthesized based on a nonthermal plasma technique. The doping concentration is tuned by varying the fractional PH$_3$ flow rate. The reported LSPR wavelength ranged from 4 to 17 μm. Zhou et al. (2015) reported mid-IR LSPR of boron-doped Si nanocrystals and found that the LSPR wavelength of boron-doped nanocrystals is shorter than that for phosphorous-doped counterparts when the doping concentration is similar. The above very recent demonstrations of mid-IR LSPR supported by doped Si nanoparticles are exciting, and we expect employing them in direct sensing applications to come in the near future.

Table 14.1 Localized surface plasmon resonance wavelength for dipoles made of different materials

MATERIALS	Al	Pd	Ag	Ti	Cu	Au	DOPED-Si (5×10^{20} cm^{-3})	DOPED-Si (10^{20} cm^{-3})
λ_{LSPR}(μm)	0.1434	0.1537	0.2380	0.2447	0.3403	0.4761	3.014	6.819

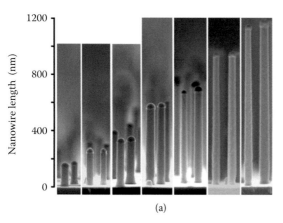

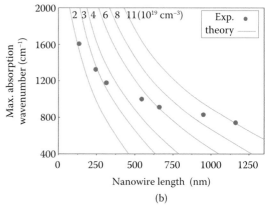

(a) (b)

Figure 14.5 (a) Phosphorous-doped silicon nanowires with lengths from 135 to 1160 nm; and (b) absorption peak position versus nanowire length. (Reprinted with permission from Chou, L.-W., et al. (2012), Tunable mid-infrared localized surface plasmon resonances in silicon nanowires, *J. Am. Chem. Soc.*, 134, 16155–16158. Copyright 2012 American Chemical Society.)

14.3 LOW-LOSS NEGATIVE REFRACTION OF MIDDLE IR WAVELENGTHS

Due to the potential applications for flat lens and sub-diffraction imaging (Shalaev 2007; Zhang 2007; Cai and Shalaev 2009), negative refraction has attracted much attention in recent years. Double-negative materials with both permeability and permittivity being negative in certain frequency range have been demonstrated to bend light in the other direction by using artificial nanostructured metamaterials, such as combinations of split-ring resonators and nanowires (Smith et al. 2000; Shelby et al. 2001), nanostrip pairs (Chettiar et al. 2006), and fishnet structures (Zhang et al. 2005; Dolling et al. 2007; García-Meca et al. 2011). Nevertheless, since the resonance is usually needed to obtain negative permeability, losses are usually too high for it to be used in practical optical devices. Besides, the performance sustains only for narrow ranges of wavelengths and incidence angles due to the conditions required to excite magnetic resonances. Single-negative material with only the permittivity being negative, such as silver films, could also support negative refraction for p-polarization (Pendry 2000; Fang et al. 2005; Zhang and Lee 2006), but the light penetration depth is very small because only evanescent waves can exist in these materials.

Hyperbolic metamaterials, for which the permittivities have different signs for orthogonal directions, on the contrary, can relieve the above limitations and thus were proposed as good candidates for all-angle negative refraction (Smith et al. 2004). Focusing and negative refraction have been experimentally demonstrated in the visible and microwave regions using either one-dimensional (1D) metal–dielectric multilayers (Shin and Fan 2006) or metallic nanowires (Liu et al. 2008; Yao et al. 2008). More recently, Hoffman et al. (2007) reported negative refraction for wide incident angles in the mid-IR based on $In_{0.53}Ga_{0.47}As/Al_{0.48}In_{0.52}As$ multilayers. Similarly, by employing Al:ZnO/ZnO multilayers, Naik et al. (2012) demonstrated negative refraction in the near-IR. Nevertheless, for multilayered metamaterials, the negative refraction performance is in a narrow band, and a Lorentzian resonance generally needs to be excited to obtain the negative permittivity for primary extraordinary waves. Later, Liu and Zhang (2013) proposed an alternative approach to achieve low-loss negative refraction in the mid-IR based on doped silicon nanowire (D-SiNW) arrays.

A schematic of the D-SiNW arrays is shown in Figure 14.6. The diameter and thickness of the nanowires are d and H, respectively. A plane electromagnetic wave is incident from air at an incidence angle of θ_0. The array period is a; thus, the volume filling ratio can be calculated by $f = 0.25\pi d^2/a^2$ considering a unit cell. D-SiNW arrays are a mixture of air and nanowires. However, when the characteristic dimension of nanowires d is considerably smaller than the wavelength of the incident light, the inhomogeneous medium can be treated as an effective medium with a homogeneous dielectric function. This has been verified by the FDTD method (Liu et al. 2013b) and finite-element method (Liu et al. 2008). Only TM waves are considered here. In this case, the anisotropic dielectric function of the D-SiNW array is a second-order tensor (Liu et al. 2013b; Wang et al. 2013):

$$\overset{=}{\varepsilon} = \begin{pmatrix} \varepsilon_x & 0 & 0 \\ 0 & \varepsilon_y & 0 \\ 0 & 0 & \varepsilon_z \end{pmatrix} = \begin{pmatrix} \varepsilon_O & 0 & 0 \\ 0 & \varepsilon_O & 0 \\ 0 & 0 & \varepsilon_E \end{pmatrix} \qquad (14.4)$$

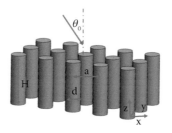

Figure 14.6 Schematics of D-SiNW arrays with *H* as the wire length, *a* as the period, and *d* as the wire diameter. Note that only TM waves are considered here.

where ε_O and ε_E are the principal dielectric functions for ordinary and primary extraordinary waves, respectively, and are given as

$$\varepsilon_O = \frac{\varepsilon_{\text{D-Si}} + 1 + (\varepsilon_{\text{D-Si}} - 1)f}{\varepsilon_{\text{D-Si}} + 1 - (\varepsilon_{\text{D-Si}} - 1)f} \tag{14.5}$$

$$\varepsilon_E = 1 + (\varepsilon_{\text{D-Si}} - 1)f \tag{14.6}$$

For ordinary waves, the effective permittivity ε_O follows Lorentz model since electrons in the nanowires are bounded by the surrounding air (Liu et al. 2013b). For primary extraordinary waves, the effective permittivity ε_E is essentially a diluted Drude model since it is the weighted average of the permittivities of doped silicon ($\varepsilon_{\text{D-Si}}$) and air ($\varepsilon_{\text{air}} = 1$). Since the optical axis is in the z direction and according to Equation 14.4, $\varepsilon_x = \varepsilon_y + \varepsilon_O$ and $\varepsilon_z = \varepsilon_E$.

The dispersion relation of the effective homogeneous medium is given as

$$\frac{k_x^2}{\varepsilon_z} + \frac{k_z^2}{\varepsilon_x} = \frac{\omega^2}{c_0^2} = k_0^2 \tag{14.7}$$

When both ε_x and ε_z are positive, the isofrequency contour shown in Equation 14.7 is elliptical. Note that Equation 14.7 is for the plane of incidence perpendicular to the y axis such that $k_y = 0$. A three-dimensional (3D) ellipsoidal shape can be formed by rotation of the elliptical surface around the z axis. If loss is considered, let $\varepsilon_x = \varepsilon_x' + i\varepsilon_x''$ and $\varepsilon_z = \varepsilon_z' + i\varepsilon_z''$, Equation 14.7 is usually based on the real part of the permittivities (Zhang and Zhang 2015). When $\varepsilon_x' > 0$ and $\varepsilon_z' > 0$, negative energy refraction will be supported (Hoffman et al. 2007; Liu et al. 2008). The wavelength-dependent ε_x' and ε_z' are shown in Figure 14.7a and b, respectively, for different doping levels when the volume filling ratio f is set as 0.1. Note that ε_x' is almost a constant near 1.2 across the whole wavelength range, although there is a Lorentz resonance featured with higher absorptance and a small bump of ε_x'. On the other hand, ε_z' decreases monotonously with increasing wavelength. The transition wavelength at which $\varepsilon_z' = 0$ is denoted as λ_T. It can be seen that $\lambda_T = 8.581$ and 2.767 μm for doping level N equal to 10^{20} and 10^{21} cm^{-3}, respectively. Therefore, increasing the doping level or the volume filling ratio will blueshift the transition toward shorter wavelengths. The values of ε_x'' and ε_z'' are shown in Figure 14.7c and d, respectively. It can be seen that ε_x'' is almost negligible even when Lorentz resonances occur. Note that ε_z'' is much greater than ε_x'', especially at higher doping levels due to the increased number of free carriers. At first glance, this seems to be an undesired feature as a large ε_z'' implies high loss for primary extraordinary waves. However, as demonstrated by Liu and Zhang (2013), higher ε_z'' actually can help enhance the transmission at oblique incidence, that is, reducing the loss in this type of hyperbolic metamaterial.

The tangent of the refraction angle for the wavevector is $\tan\theta_k = k_x/\text{Re}(k_z)$ and that for the Poynting vector can be expressed as

$$\tan\theta_s = \frac{S_x}{S_z} = \frac{\text{Re}(k_x/\varepsilon_z)}{\text{Re}(k_z/\varepsilon_x)} \tag{14.8}$$

where S_x and S_z are the components of the Poynting vector S, and k_z can be solved from Equation 14.7 as $k_z = k_0\sqrt{\varepsilon_x - (\varepsilon_x/\varepsilon_z)\sin^2\theta_0}$ since $k_x = k_0\sin\theta_0$. It should be noted that both the Poynting vector and wavevector will be positively refracted when the wavelength λ is shorter than the transition wavelength for the given doping level and filling ratio. The isofrequency contour in this case is elliptic since ε_x' and ε_z' are both positive. When $\lambda = 10$ μm, due to the hyperbolic dispersion, negative refraction of the Poynting vector occurs for both doping levels at any oblique incident angle, as shown in Figure 14.8a. For $N = 10^{21}$ cm^{-3}, the magnitude of θ_s becomes much smaller with a maximum of 1.5°, meaning that the waves will propagate nearly parallel to the nanowires inside the D-SiNW array for any incidence angles. This phenomenon

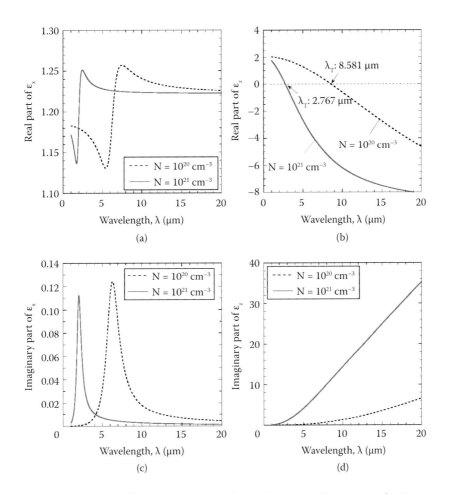

Figure 14.7 Effective permittivities of D-SiNW arrays with $f = 0.1$. Real part of permittivity for (a) primary ordinary wave and (b) primary extraordinary wave. Imaginary part of permittivity for (c) primary ordinary wave and (d) primary extraordinary wave. (From Liu, X.L., and Zhang, Z.M., *Appl. Phys. Lett.*, 103, 103101, 2013. With permission.)

makes the array act as an optical collimator, which could find potential applications in sensors and detectors to increase the acceptance angles. The collimation effect has recently been shown in epsilon-near-zero (ENZ) metamaterials (Feng 2012); however, it will be difficult to fabricate broadband ENZ materials.

The dimensionless hyperbolic isofrequency contour is given in Figure 14.8b for $N = 10^{20}$ cm^{-3} and $\lambda = 10$ μm to theoretically demonstrate the negative refraction of the Poynting vector, which is normal to the isofrequency surface. As the incident angle increases, $|\theta_s|$ increases monotonically and gradually reaches the maximum for $\theta_0 = 90°$. Note that the propagation loss, determined by Im(k_z) is very small, even though the values of ε_z'' as shown in Figure 14.7d are quite large. The mechanism will be explained later.

The field distribution can be obtained by FDTD for a Gaussian beam incident from air. Consider a D-SiNW array with thickness $H = 20$ μm and $N = 10^{20}$ cm^{-3}, the magnetic field distributions are shown in Figure 14.8c and d using the commercial FDTD software (Lumerical Solutions, Inc.) with a Gaussian beam with a width of 20 μm at $\lambda = 10$ μm and $\theta_0 = 20°$. The interface between air and the D-SiNW array is delineated as thin horizontal lines at $z = \pm 10$ μm. Figure 14.8c is for the actual 3D nanowires structures shown in Figure 14.6 with $a = 2$ μm and $f = 0.1$ (i.e., $d = 714$ nm). Some vertical fringes represent the boundary between individual D-SiNW and air. The field distribution obtained by FDTD using effective permittivity tensor is shown in Figure 14.4b and the result agrees well with that in Figure 14.8c. The beam is clearly shown to be bended negatively in the film containing the nanowire array, and the agreement between Figure 14.8c and d confirms the applicability of the anisotropic effective medium approach. Note that thinner nanowires will be necessary for EMT to be applicable at shorter wavelengths. The reflection loss is relatively small due to

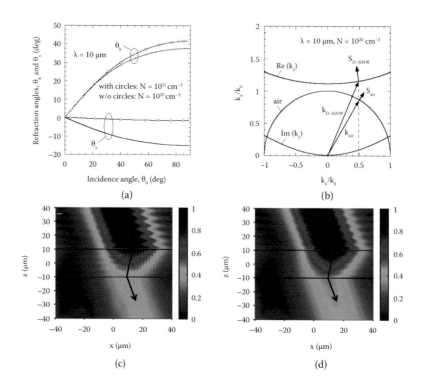

Figure 14.8 (a) Refraction angles for wavevector and Poynting vector versus incidence angle for $\lambda = 10$ μm; (b) normalized hyperbolic isofrequency contour for doping concentration of 10^{20} cm^{-3} and $\lambda = 10$ μm, where Re(k_z) and Im(k_z) are for D-SiNW array; and dimensionless magnetic field amplitude |H| obtained by FDTD for a Gaussian beam at $\theta_0 = 20°$ and $\lambda = 10$ μm for (c) the actual nanowire structure with $a = 2$ μm and (d) the effective anisotropic medium. (From Liu, X.L., and Zhang, Z.M., *Appl. Phys. Lett.*, 103, 103101, 2013. With permission.)

the good impedance matching between air and D-SiNW arrays. This can be understood by the fact that ε'_x is close to 1 and $\varepsilon''_x \ll 1$ in the hyperbolic dispersion region as shown in Figure 14.7.

The FOM given in the following was introduced by Hoffman et al. (2007) to describe the performance of a hyperbolic metamaterials in terms of light propagation:

$$\text{FOM} = \frac{\text{Re}(k_z)}{\text{Im}(k_z)} \tag{14.9}$$

A high FOM means that waves can propagate more freely with less loss inside the medium. This definition takes into account the dependence of loss on specific propagating direction. FOMs for different doping levels when θ_0 is at 20° and 60° are shown in Figure 14.9. Note that when $N = 10^{20}$ cm^{-3}, negative refraction starts from $\lambda_T = 8.581$ μm as marked by the square symbols. With $N = 10^{20}$ cm^{-3}, the FOM value at $\lambda = 10$ μm and $\theta_0 = 20°$ is 23.4, which is greater than those reported earlier (Hoffman et al. 2007; Naik et al. 2012). It can be seen that FOM increases monotonously with wavelength to more than 100 at $\lambda = 20$ μm. For hyperbolic materials made with nanowires, the Lorentz resonance is away from the wavelengths at which negative refraction occurs, implying that lower loss or less absorption can be achieved (Liu et al. 2008). Note that the FOM values are much higher for a higher doping concentration. This is counterintuitive since the loss is much higher (i.e., ε''_z is much greater) for $N = 10^{21}$ cm^{-3} than $N = 10^{20}$ cm^{-3}. For $N = 10^{21}$ cm^{-3}, the FOM at $\theta_0 = 20°$ and $\lambda = 10$ μm reaches 219, which is more than one order of magnitude greater than the previously reported values for multilayered hyperbolic metamaterials based on all semiconductors (Hoffman et al. 2007; Naik et al. 2012). The mechanism of loss-assisted FOM or transmission at oblique incidence is explained in the following based on the work of Liu and Zhang (2013).

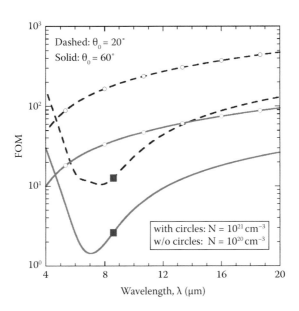

Figure 14.9 FOM as a function of wavelength at two incidence angles, $\theta_0 = 20°$ and $60°$, for different doping levels. The squares denote the transition wavelength for doping level of 10^{20} cm^{-3}. (From Liu, X.L., and Zhang, Z.M., *Appl. Phys. Lett.*, 103, 103101, 2013. With permission.)

From Figure 14.7a and c, when $\lambda \geq \lambda_T$, ε''_x could be one or even three orders of magnitude less than ε'_x especially for longer wavelengths. As a result, ε''_x can be neglected and k_z can be simplified as

$$k_z \approx k_0 \sqrt{\varepsilon'_x} \sqrt{1 - \sin^2 \theta_0 / \varepsilon_z} = k_0 \sqrt{\varepsilon'_x} \sqrt{1 - \frac{\varepsilon'_z}{\varepsilon'^2_z + \varepsilon''^2_z} \sin^2 \theta_0 + i \frac{\varepsilon''_z}{\varepsilon'^2_z + \varepsilon''^2_z} \sin^2 \theta_0} \qquad (14.10)$$

Figure 14.7b and d shows that ε'_z is much smaller than $\varepsilon'^2_z + \varepsilon''^2_z$; thus, the second term inside the square root is much less than 1. Therefore,

$$k_z \approx k_0 \sqrt{\varepsilon'_x} \sqrt{1 + i \frac{\sin^2 \theta_0}{\varepsilon''_z + \varepsilon'^2_z / \varepsilon''_z}} \qquad (14.11)$$

For $\lambda \geq \lambda_T$, $\dfrac{\sin^2 \theta_0}{\varepsilon''_z + \varepsilon'^2_z / \varepsilon''_z} \ll 1$ especially with high doping levels; therefore, the FOM can be approximated as follows when $\theta_0 \neq 0$

$$\text{FOM} \approx \frac{2(\varepsilon''_z + \varepsilon'^2_z / \varepsilon''_z)}{\sin^2 \theta_0} \qquad (14.12)$$

It can be shown that for given θ_0 and ε'_z, the FOM will increase with ε''_z as long as $\varepsilon''_z \geq |\varepsilon'_z|$, which is satisfied here. For $N = 10^{21}$ cm^{-3}, $|\varepsilon'_z| \ll \varepsilon''_z$ and thus FOM $\approx 2\varepsilon''_z / \sin^2 \theta_0$ at oblique incidence. In this case, the FOM will depend only on ε''_z at a given θ_0. Since $\varepsilon''_z = f\varepsilon''_{D\text{-Si}}$, higher doping concentrations, that is, larger $\varepsilon''_{D\text{-Si}}$ and ε''_z, will lead to an enhanced FOM. It can be shown that Equation 14.12 is a good approximation of the exact FOM. For example, compared with exact values of 23.3 and 219.1, the simplified formula predicts FOMs of 27.1 and 291.0 for doping levels of 10^{20} and 10^{21} cm^{-3}, respectively, when $\lambda = 10 \, \mu$m and $\theta_0 = 20°$. The agreement holds for other doping levels and incident angles as long as θ_0 is not too small. For near normal incidence, loss is dominated by ε''_x, that is, ordinary waves. While Equation 14.12 breaks down for very small θ_0, Equation 14.9 is applicable to normal incidence, at which FOM = n'/n'', where n' and n'' are the real and imaginary parts of the ordinary refractive index, respectively.

While simple, Equation 14.12 provides a physical insight into the loss-enhanced transmission and may be useful for estimation of FOM. This counterintuitive loss-assisted FOM or transmission has been recently shown for ENZ metamaterials (Feng 2012; Sun et al. 2012). Higher doping concentrations can increase both FOM and transmission for sufficiently large θ_0. Moreover, with increasing doping level, θ_s will be significantly reduced and highly efficient collimators can be obtained. The doping concentration may be adjusted to achieve a desired FOM and θ_s in a suitable wavelength region to meet the requirements for specific applications. In sum, metal-free all-angle negative refraction with low loss can be supported in the mid-IR region using D-SiNW arrays, as being verified by both EMT and FDTD approaches. Besides, the performance is broadband, and there is no upper limit for λ since the hyperbolic dispersion will hold as long as $\lambda > \lambda_T$.

14.4 TUNABLE BROADBAND OMNIDIRECTIONAL PERFECT ABSORBERS/EMITTERS

Controlling the spectral and directional absorption properties of surfaces has attracted considerable interest due to the wide applications, such as solar cells, thermophotovoltaic (TPV) systems, sensing, free space heat dissipation, and bolometers (Zhang and Ye 2012). Spectrally and spatially coherent absorbers or emitters can be achieved by exciting surface plasmon or phonon polaritons (Greffet et al. 2002), photonic crystals (Narayanaswamy and Chen 2004; Lee et al. 2008a), resonance microcavities (Maruyama et al. 2001; Dahan et al. 2008), and Fabry–Perot type resonances (Wang et al. 2012; Shu et al. 2013). The resonance emission or absorption peaks are usually sensitive to the incidence angle and polarization. Due to increasing energy needs and environmental concerns, many efforts have been taken to design broadband diffuse absorbers/emitters, which are important for enhancing TPV cell efficiency for harvesting waste heat as well as for improving thermal management in space.

Complex gratings achieved by packing different cells of different periods together were proposed to broaden the emission or absorption spectrum as well as to create peaks that are less sensitive to emission or incidence angles (Chen and Zhang 2007, 2008; Cui et al. 2011; Hendrickson et al., 2012; Wang and Wang 2013; Devarapu and Foteinopoulou 2014). Nevertheless, different resonance peaks supported by each cell may interact with each other and thus reduce the absorption. Metamaterials that can support magnetic (plasmon) polaritons, the strong coupling between the magnetic resonance inside nanostructures and the external electromagnetic waves (Lee et al. 2008b; Wang and Zhang 2012; Zhao et al. 2013; Feng et al. 2014; Xuan and Zhang 2014; Zhao and Zhang 2014; Liu et al. 2015b; Song et al. 2015b), are another type of candidate for diffuse absorbers. The wavelength band featured with perfect absorption is not wide since resonance conditions are required. Carbon nanotubes exhibit high absorptance over a broad range of wavelengths (Yang et al. 2008; Mizuno et al. 2009; Lehman et al. 2010; Wang et al. 2010; Ye et al. 2012). However, the absorption by carbon nanotubes tends to cover an overly broad spectral range that is not tunable. Therefore, realizing broadband omnidirectional perfect absorbers with good tunabilities still needs more efforts.

Due to the tunable free electron densities via changing chemical doping, relieved optical impedance mismatch with the air, easy integration with optoelectronic devices, and compatible with current semiconductor fabrication techniques, doped silicon and nanostructure counterparts were proposed as promising candidates for perfect absorbers/emitters to meet different applications. Liu et al. (2013b) theoretically demonstrated broadband omnidirectional perfect absorption in the mid-IR based on D-SiNW arrays. Both the effective medium theory (EMT) combined with anisotropic thin-film optics and the FDTD methods are used to predict the radiative properties, and the agreement with each other is very good for different polarizations and incident angles when the period of nanowire arrays is much smaller than the interested mid-IR regimes.

The absorptance of a free-standing D-SiNW array with the thickness of 100 μm and filling ratio of 0.15 is compared with that of a doped-Si film with the same thickness as shown in Figure 14.10a, with a doping concentration of 10^{20} cm^{-3}, at normal incidence. At such a high doping level, a 100-μm doped-Si film is essentially opaque. Thus, the absorptance of the doped-Si film is the same as that of bulk doped silicon. The reflectance of Si has a dip between 5 and 6 μm close to the valley of the refractive index of doped silicon, resulting in a peak in the absorptance spectrum. On the other hand, the reflectance of D-SiNW array

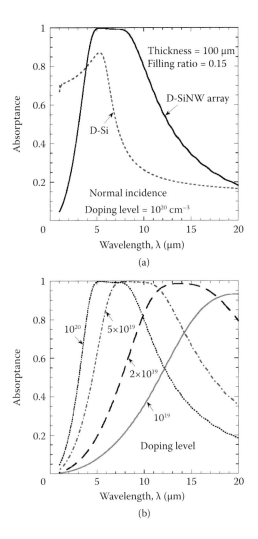

Figure 14.10 (a) Absorptance at normal incidence for a free-standing D-SiNW array (f = 0.15) compared to a doped-Si film with the same thickness (100 μm) and doping concentration of 10^{20} cm^{-3}. (b) Effects of doping concentration on the absorption. (From Liu, X.L., et al., *J. Heat Transfer*, 135, 061602, 2013. With permission.)

is suppressed to be less than 2% (not shown); subsequently, the absorptance A is related to the transmittance T by $A = 1 - T$. The near-zero reflectance can be explained by the index match between air and the D-SiNW array whose (ordinary) refraction index (n_O) is close to unity (Liu et al. 2013b). In the short wavelength region (1 μm < λ < 4 μm), the absorptance of the D-SiNW array is smaller than that of doped silicon due to the smaller extinction coefficient, that is, the lack of absorption. There is a broadband near-unity absorptance region ranging from 5 to 8 μm, due to the reflection suppression and small penetration depth. At even longer wavelength, the absorptance of the D-SiNW film drops again, which is consistent with the reduction of the extinction coefficient and the increase of transmission toward longer wavelengths. Figure 14.10b illustrates the tunable absorptance of free-standing D-SiNW arrays via mediating the doping level. The absorptance plateau shifts to shorter wavelengths with increasing doping level, since a higher doping concentration results in a high concentration of free electrons and thus a smaller plasmon resonance wavelength. Therefore, the location and bandwidth of the absorptance plateau can be tuned by controlling the doping level to meet different applications.

To investigate the effects of incidence angle and polarization, the absorptance contours for TE and TM waves are plotted in Figure 14.11a and b, respectively. The geometry parameters are the same as those used in Figure 14.10a. The absorptance is close to unity for both polarizations at wavelengths between 5 and 8 μm and

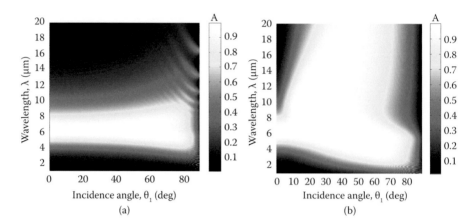

Figure 14.11 Contour plots of the absorptance as a function of wavelength and incidence angle for (a) TE waves and (b) TM waves. (From Liu, X.L., et al., *J. Heat Transfer*, 135, 061602, 2013. With permission.)

is insensitive to the incidence angle, indicating that the SiNW array is a quasi-diffuse perfect absorber or emitter in this wavelength region. The absorption bandwidth is broader for TM waves than that for TE waves at oblique angles. The high absorption for TM waves is because the extinction coefficient of a uniaxial medium is higher for extraordinary waves when the electric field is parallel to the wires or optical axis. The contour plot also shows some interference fringes for TE waves. These fringes are not seen in the absorption contour for TM waves since the waves cannot reach the other interface due to the high absorption in the D-SiNW array. The omnidirectional broadband absorbers with tunable operating wavelengths demonstrated here based on D-SiNW array may have important applications in energy harvesting devices, IR filters, and radiation detectors.

Besides the patterned doped silicon structures, simple two-layer configurations may also be able to support strong absorption properties. Streyer et al. (2013) both experimentally and numerically demonstrated a perfect absorptance in the mid-IR based on thin germanium film on doped-Si substrate, as shown by the inset of Figure 14.12. The absorptance was shown to be insensitive to either polarization states or incidence angles.

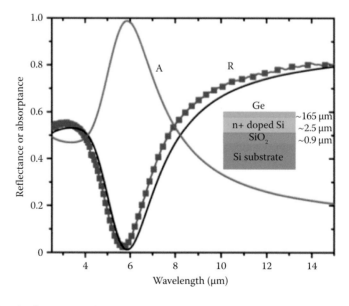

Figure 14.12 Calculated reflectance and absorptance (solid lines) and measured reflectance (with square markers) of the considered configuration shown as the inset. (Adapted from Streyer, W., et al., Strong absorption and selective emission from engineered metals with dielectric coatings, *Opt. Express*, 21, 9113–9122, 2013. With permission of Optical Society of America.)

The distribution of doping concentration in the direction normal to the surface was estimated and incorporated into the transfer matrix method to obtain the reflectance and absorptance. Their simulation results agree well with the experiments, as shown in Figure 14.12. The high absorption band was shown to be adjustable via the control of Ge film thickness. Rather than employing doped silicon as the substrate, Cleary et al. (2013), on the other hand, investigated doped-Si film on a sapphire substrate. An absorptance peak of as high as 96% was theoretically predicted at a wavelength of 12.1 μm based on a doped-Si film with a thickness of 600 nm and a doping level of 2×10^{19} cm^{-3}. Thus, doped silicon has been proposed and demonstrated to be a promising candidate for achieving perfect absorption in the mid-IR region. The high absorption band can be mediated by adjusting the doping concentration, which is a unique property not possessed by noble metals. The tunable perfect absorption properties combined with the easy integration with optoelectronic devices enable doped silicon to be used in wide applications such as radiation cooling, noncontact temperature measurement, free-space thermal management, and energy harvesting.

14.5 SUPER-PLANCKIAN THERMAL RADIATION OF DOPED SILICON METAMATERIALS AND METASURFACES

The mid-IR contains the major thermal radiation energy of most objects ranging from 100 to 1000 K. The absorption properties of objects with time-reverse symmetry and at thermal equilibrium are related to their thermal emission by the well-known Kirchhoff's law saying that the spectral directional absorptance is equal to the spectral directional emissivity. The heat dissipation through thermal radiation is maximum when the absorptance or emissivity is equal to unity for any polarization states, wavelengths, and angle of incidence. This upper limit can be readily obtained from the integration of Planck's distribution or the Stefan–Boltzmann law. However, only propagating waves, for which the tangential wavevector is smaller than the free-space wavevector, are considered in this scenario. When the distance between objects is comparable or smaller than the characteristic wavelength (10 μm around room temperature), evanescent waves can tunnel through and even play a dominant role especially for deep submicron gap spacing (Rytov et al. 1989). Recent experimental and theoretical studies have demonstrated super-Planckian near-field radiation beyond the Stefan–Boltzmann limit by orders of magnitude when resonance-based surface plasmon (or phonon) polaritons or hyperbolic modes are excited (Joulain et al. 2005; Zhang 2007; Shen 2013; Xuan 2014; Liu et al. 2015a; Song et al. 2015a). As a result, new applications of thermal radiation in efficient energy harvesting, subwavelength thermal imaging, nanofabrication, and contactless thermal management such as modulation, rectification, and amplification are opened via near-field thermal radiation.

Noble metals, such as gold and silver, are known to support surface plasmon resonance in the ultraviolet or visible range. However, the Bose–Einstein distribution magnitude exponentially decays with the frequency and thus is too low to thermally excite SPPs for objects around room temperature. This mismatch can be alleviated by highly doped silicon, for which the surface plasmon wavelength lies in the mid-IR due to the much lower density of free electrons compared with noble metals. Rousseau et al. (2010) presented asymptotic expressions to predict near-field heat transfer between doped silicon substrates. They found that the dominant contribution came from frustrated total internal reflection of propagating waves and surface mode coupling for low and high doping levels, respectively. van Zwol et al. (2012) experimentally confirmed the role of surface plasmon of doped silicon in enhancing the near-field radiation between silicon oxide sphere and doped silicon film on silicon oxide substrate. Shi et al. (2013) measured the near-field radiative heat transfer between doped silicon substrates and a silicon oxide sphere. The tunability was observed by changing the doping concentration. The desire of more efficient radiative energy transfer for further improving heat dissipation and energy harvesting motivates people to consider doped silicon metamaterials. Basu and Wang (2013) investigated the near-field thermal radiation between D-SiNW arrays via the fluctuation dissipation theory combined with Bruggeman effective medium approximation. Enhanced radiative heat flux over the bulk doped silicon was predicted. Similar results were theoretically demonstrated by Liu et al. (2014a) for both nanowire and nanohole arrays.

Liu et al. (2014c) conducted a thorough investigation of different doped silicon metamaterials including nanowires, nanoholes, multilayers, and 1D gratings in terms of their potentials for enhancing near-field radiative heat transfer at ambient temperature. The schematics of different metamaterials considered are

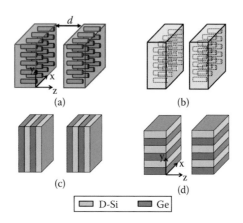

Figure 14.13 Schematics of two semi-infinite nanostructured metamaterials separated by a vacuum gap at a distance d: (a) D-SiNW arrays; (b) D-SiNH arrays; (c) multilayers composed of doped silicon and Ge, which is modeled as a dielectric; and (d) 1D gratings composed of doped silicon and Ge. (From Liu, X.L., et al., *Int. J. Heat Mass Transfer*, 73, 389–398, 2014. With permission.)

shown in Figure 14.13. Both the multilayer and 1D grating structures are composed of doped silicon and germanium (Ge), whereas nanowire and nanohole configurations are surrounded by vacuum. The dielectric function of Ge is largely independent of wavelength in the IR region and can be approximated as a constant with $\varepsilon_{Ge} = 16$. EMT is used to obtain the anisotropic dielectric function and is combined with fluctuational electrodynamics to calculate the near-field radiative heat transfer coefficient due to its simplicity and low computational demand. The aforementioned nanostructures are treated as homogeneous uniaxial materials; this assumption is valid in the far field as long as the characteristic thermal wavelength is much greater than the nanostructure period. In the near field, as pointed out by some investigators, nonlocal effects would arise at the surface resonance frequency or very large wavevectors (Orlov et al. 2011; Kidwai et al. 2012; Tschikin et al. 2013). Nevertheless, when the gap distance is much greater than the period of nanostructures to some extent, EMT should be applicable in the near field as well (Liu et al. 2014d).

The heat transfer coefficient at temperature T between two anisotropic planar media separated by a vacuum gap d can be calculated by (Biehs et al. 2011)

$$h = \frac{1}{8\pi^3} \int_0^\infty g(\omega, T)\, d\omega \int_0^{2\pi} \int_0^\infty \xi(\omega, \beta, \phi)\beta\, d\beta\, d\phi \tag{14.13}$$

where

$$g(\omega, T) = \frac{\partial}{\partial T}\left(\frac{\hbar\omega}{(e^{\hbar\omega/k_B T} - 1)} \right) = \frac{(\hbar\omega)^2 e^{\hbar\omega/k_B T}}{k_B T^2 (e^{\hbar\omega/k_B T} - 1)^2} \tag{14.14}$$

and $\xi(\omega, \beta, \phi)$ is called the energy transmission coefficient (to distinguish from Fresnel's transmission coefficient which is based on the field). The energy transmission coefficient can be expressed in a matrix formulation as [36]

$$\xi(\omega, \beta, \phi) = \begin{cases} \mathrm{Tr}\left[(I - R_2^* R_2) D (I - R_1 R_1^*) D^* \right], & \beta < k_0 \\ \mathrm{Tr}\left[(R_2^* - R_2) D (R_1 - R_1^*) D^* \right] e^{-2|k_z|d}, & \beta > k_0 \end{cases} \tag{14.15}$$

where β is the transverse wavevector, and $k_{z0} = \sqrt{k_0^2 - \beta^2}$ is the wavevector component in the z direction in vacuum. Note that the symbol "*" denotes the conjugate transpose (Hermitian transpose), and Tr stands for trace. $R_{1,2}$ is the reflection coefficient matrix considering both s- and p-polarization states and the coupling of them if existing.

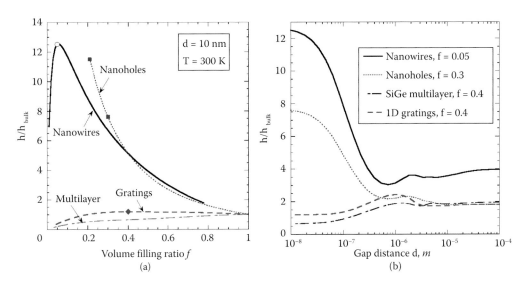

Figure 14.14 (a) Ratio of the heat transfer coefficient of the nanostructures to bulk doped silicon at a gap distance of 10 nm; and (b) enhancement of heat transfer coefficients versus gap distance. (From Liu, X.L., et al., *Int. J. Heat Mass Transfer*, 73, 389–398, 2014. With permission.)

The radiative heat transfer coefficient versus the filling ratio of doped silicon for the four nanostructures at a distance d = 10 nm is shown in Figure 14.14a. The radiative heat transfer coefficient is normalized to that of bulk doped silicon to show the enhancement or reduction of heat transfer by the nanostructures. Note that the heat transfer coefficient for bulk doped silicon at this distance with the same doping level is 5022 W/m²-K, which is 820 times the value between two blackbodies (i.e., $4\sigma T^3$, where σ is the Stefan–Boltzmann constant). The enhancement of heavily doped Si is due to surface waves (Rousseau et al. 2010; Basu et al. 2010b). Figure 14.2 suggests that D-SiNW array with f = 0.05 can achieve an enhancement factor of 12.5 over bulk doped silicon. With the nanowires, the predicted heat transfer coefficient exceeds 60,000 W/m²-K at d = 10 nm, making D-SiNW array very attractive for applications ranging from high-efficiency near-field radiative cooling to local heating. The enhancement over bulk by nanoholes increases with decreasing f and reaches the maximum of 11.3 times when f = 1 − π/4. For 1D gratings, the calculation is based on Δ=0° unless otherwise specified since the alignment case yields the maximum heat transfer. For 1D gratings, there is a slight enhancement when f > 0.2, and the maximum enhancement is around f = 0.4 with a ratio h/h_{bulk} = 1.19. However, the near-field radiative transfer coefficient between two multilayers is always smaller than that of the bulk and increases monotonously with f. Therefore, multilayer structures are not as effective in terms of enhancing near-field radiative transfer at ambient temperature at deep submicron gap distances. Similar trends can be obtained for d = 100 nm with reduced heat flux, although the results are not shown here.

To see whether the performance achieved with these nanostructures will still hold at different gap distances, relative heat transfer coefficient is plotted in Figure 14.14b. The filling ratios for nanowires and 1D gratings are taken as the optimized value at d = 10 nm, that is, f = 0.05 and 0.4, respectively. For clearer comparison between multilayers and 1D gratings, the filling ratio is also chosen as 0.4 for multilayers. The filling ratio of nanoholes is taken as 0.3 based on practical consideration to stay away from the physical limitation. It is interesting to note that in the far field, multilayers give slightly higher heat transfer coefficient than nanoholes or 1D gratings. Among all the structures, D-SiNW arrays result in the largest heat transfer coefficient at any gap distance.

Nanowires and nanoholes could achieve an enhancement of about one order of magnitude as shown in Figure 14.14a and b; their energy transport mechanisms are more interesting and thus are summarized here. For the near-field thermal radiation of multilayers and gratings, one can refer to the original study of Liu et al. (2014c) for more details. The energy transmission coefficient for p-polarization waves (s-polarization waves cannot support surface resonance modes nor hyperbolic modes for doped-Si nanostructures, and thus not considered here) is shown for nanowires and nanoholes in Figure 14.15a and b, respectively. For D-SiNW arrays, when the electric field is perpendicular to the nanowires (ordinary waves), the effective optical properties of the nanowire medium is essentially a dielectric. On the other hand, the effective dielectric function of the nanowires for extraordinary waves is described by the dilute Drude model and is therefore metallic. As a result, the dielectric functions in orthogonal directions may have opposite signs, leading to a hyperbolic dispersion relation rather than conventional elliptic or spherical dispersions. One of the unique properties of hyperbolic dispersion lies the supporting of propagating waves no matter how large the tangential wavevector is. High-wavevector waves are propagating inside the metamaterials but evanescent in the vacuum gap, the frustrated total internal reflection occurs and contributes dominantly to the super-Planckian thermal radiation between hyperbolic metamaterials (Biehs et al. 2012; Guo et al. 2012; Liu et al. 2013a; Liu and Shen 2013; Chang et al. 2015). For D-SiNW array, as shown in Figure 14.15a, there is a broad hyperbolic region below 1.72×10^{14} rad/s, as delineated by the arrows between the white line and the horizontal axis. In this hyperbolic band, the energy transmission coefficient is high due to the efficient photon tunneling of high β propagating waves. That's why, the heat transfer coefficient between D-SiNW arrays at $d = 10$ nm reaches 12.5 times that between bulk doped silicon.

There is also a hyperbolic band though being narrow (2.44×10^{14} rad/s $< \omega < 2.74 \times 10^{14}$ rad/s) for doped silicon nanohole (D-SiNH) arrays, as marked in Figure 14.15b. It contributes 10.8% to the total heat transfer coefficient between D-SiNH arrays. Below 2.24×10^{14} rad/s, the anisotropic dielectric functions are negative for both in-plane and out-plane directions, enabling the excitation of SPPs. The dispersion relation of SPPs derived as shown by the black curve agrees well with the contour plot. These surface modes are largely responsible for the near-field heat transfer enhancement in D-SiNH arrays. While SPP modes can also exist between doped silicon, due to the high loss in the bulk material, high $\xi_p(\omega,\beta)$ occurs at relatively small β values. The cut-off wavevector increases with reducing material loss (Liu et al. 2014d). At the resonance frequency (2.88×10^{14} rad/s), the imaginary part of the dielectric function of doped silicon is nearly twice of that for D-SiNH arrays at resonance frequency of 2.24×10^{14} rad/s. Therefore, the number of contributing modes of low-loss D-SiNH arrays is much greater than that of doped silicon, resulting in a much greater enhancement of near-field thermal radiation.

Hyperbolic modes have been demonstrated to support super-Planckian radiation in a broadband. Nevertheless, since hyperbolic modes are resonance free, the energy transmission coefficient unavoidably decays with increasing k. By combining graphene plasmons and hyperbolic modes, Liu et al. (2014b) demonstrated perfect energy transmission coefficient close to the theoretical limit across a broad frequency range and large k-space based on graphene-covered D-SiNW array. The schematic considered is shown in Figure 14.16a. As demonstrated in Figure 14.16b, ξ_p approaches the theoretical limit of 1 across a broad frequency range up to 1.5×10^{14} rad/s and a large k-space up to $20k_0$. The exponential decay feature of hyperbolic modes and the narrow-band nature of graphene plasmons are overcome via the hybridization of the two modes. All the photons emitted in this regime will be absorbed, which is the blackbody behavior in the near field. As a result, the heat transfer coefficient of this hybrid structure could achieve 614.7 W/m²-K, much larger than that for plain D-SiNW array or suspended graphene sheets alone.

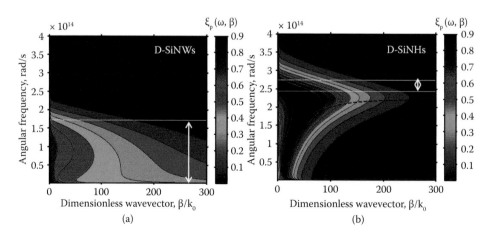

Figure 14.15 Contour plots of the energy transmission factor for p-polarization $\xi_p(\omega,\beta)$ of (a) D-SiNW arrays and (b) D-SiNH arrays for $d = 10$ nm. (From Liu, X.L., et al., *Int. J. Heat Mass Transfer*, 73, 389–398, 2014. With permission.)

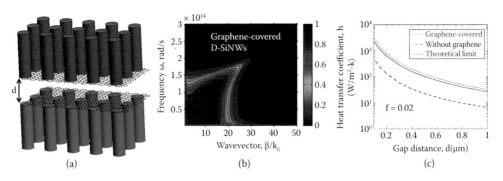

Figure 14.16 (a) Schematic of near-field radiative heat transfer between graphene-covered semi-infinite doped silicon nanowires separated by a vacuum gap of distance d; (b) energy transmission coefficient of p-polarization at $d = 200$ nm; (c) heat transfer coefficient varying with gap distances for D-SiNW arrays, graphene-covered D-SiNW arrays with optimal chemical potential, and the theoretical limit. (From Liu, X.L., et al., *ACS Photon*, 1, 785–789, 2014. With permission.)

Figure 14.16c plots the heat transfer coefficient of D-SiNW array and graphene-covered nanowires optimized by adjusting the chemical potential, considering only p-polarized evanescent modes. The minimum d used here is 100 nm to ensure that EMT is valid, given that practical nanowire diameter cannot be infinitesimal. By optimizing the chemical potential, the inclusion of graphene can improve the near-field heat transfer between D-SiNW array. Heat transfer coefficient for graphene-covered D-SiNW array lies between that for D-SiNW array and the theoretical limit of hyperbolic materials given as $k_B^2 T \ln(2)/12\hbar d^2$ (Biehs et al. 2012). With increasing gap distance, the heat transfer coefficient gets closer to the near-field limit. Note that the theoretical limit from Biehs et al. (2012) is only for hyperbolic materials rather than a physical upper limit for all materials. Nevertheless, hybridization of graphene plasmons and hyperbolic modes is demonstrated to achieve a near-perfect energy transmission coefficient and efficient heat transfer that is close to the theoretical limit.

As mentioned earlier, the near-field radiative heat transfer between doped silicon nanostructures has been investigated by several groups. However, metasurfaces, planar metamaterials with subwavelength thicknesses, received little attention in the nanoscale thermal radiation field although they have been extensively investigated for far-field manipulation of light propagation, polarization states, and absorption in an unprecedented way. Besides, they have some peculiar advantages over conventional metamaterials, such as less volumetric propagation loss, relative easy fabrication, and compatible integration with other nanodevices. Based on exact formalisms, Liu and Zhang (2015) theoretically demonstrated that perforating or patterning doped silicon into metasurfaces can enhance the radiative heat flux. Both 1D and two-dimensional (2D) periodically patterned metasurfaces are considered, as shown in Figure 14.17a and b, respectively. The temperatures of the top and bottom metasurfaces, which have identical geometry and are separated by a vacuum gap of d, are $T_1 = 310$ K and $T_2 = 290$ K, respectively. The period, width, and thickness of the patterned metasurfaces are denoted by P, W, and h, respectively. The scattering theory and Green's function method are used for 1D and

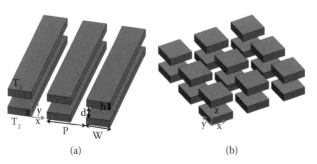

Figure 14.17 Schematic of near-field radiation between (a) 1D and (b) 2D metasurfaces separated by a vacuum gap of d with temperatures as T_1 and T_2, respectively. Note that h is the thickness, P the period, and W the width of the 1D or 2D patterns. (From Liu, X.L., and Zhang, Z.M., *ACS Photonics*, 2, 1320, 2015. With permission.)

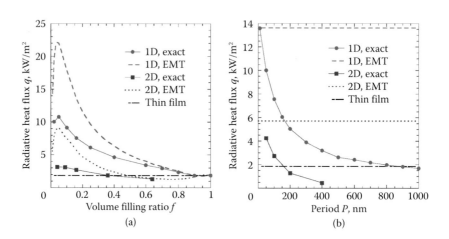

Figure 14.18 (a) Radiative heat flux as a function of the volume filling ratio for P = 100 nm; and (b) effects of period for f = 0.16, d = 100 nm, and h = 400 nm as default. (From Liu, X.L., and Zhang, Z.M., *ACS Photonics*, 2, 1320, 2015. With permission.)

2D metasurfaces, respectively. In the long-wavelength limit, 1D and 2D metasurfaces can be homogenized as uniaxial materials. Therefore, besides numerical solutions based on the exact formulation, EMT is also be employed to show the heat flux in the limit when the period approaches zero.

Figure 14.18a gives the radiative heat flux for both configurations with varying volume filling ratios. The volume filling ratio f is defined as W/P and W^2/P^2 for 1D and 2D metasurfaces, respectively. The geometric parameters are taken as P = 100 nm, d = 100 nm, and h = 400 nm. These values are used as default in this work unless otherwise specified. The radiative heat flux between thin films (f = 1) of the same thickness (h = 400 nm) is denoted by the dash-dotted line and is 1860 W/m². This value exceeds that between bulk doped silicon of 1629 W/m² and is more than 15 times that between blackbodies. The underlying mechanism for the enhancement can be attributed to the coupling of SPPs inside the thin film, as discussed in Francoeur et al. (2008), although different materials were used. Patterning the film into 1D metasurface can enhance thermal radiation for all practical volume filling ratios. Interestingly, while the 2D metasurface yields a radiative heat flux higher than that of thin films at moderate filling ratios, it does not support a heat flux as high as that of the 1D metasurface. Beyond f = 0.36, 2D patterning will deteriorate the radiative transfer as shown in Figure 14.18a. The EMT approximation, which is valid only when P is sufficiently small, overpredicts the heat flux for both 1D and 2D configurations. The disagreement between the exact method and EMT is expected to be diminished when reducing the period, as was demonstrated for multilayered metamaterials (Liu et al. 2014d). The optimal heat flux supported by 1D and 2D metasurfaces reaches 10.79 and 3.17 kW/m², respectively. Reducing P is expected to further increase the radiative flux. As indicated by the dashed and dotted curves predicted by EMT, the maximum possible heat flux of 1D and 2D configurations is 22.11 kW/m² and 9.16, respectively. These values are about 12 and 5 times as large as that between two films with the same thickness. Therefore, patterned thin metamaterials may increase the heat flux over the counterpart thin films and bulks.

The effects of period are shown in Figure 14.18b for f = 0.16, with other geometric parameters remaining the same. With decreasing period, the radiative heat flux for both configurations based on exact methods approaches the values predicted by the corresponding EMT as expected. As an example, the heat flux of 2D metasurface at P = 50 nm is 4.26 kW/m², which is 25% less than that predicted by EMT. This discrepancy further decreases at small P. At P = 10 nm, the heat flux for 1D metasurface is 13.6 kW/m², which is essentially the same as that from EMT. With increasing P, the heat flux decreases and will eventually approach to the limit governed by the proximity approximation based on pair-wise addition, that is, the heat flux of thin films multiplied by a factor of f. Metasurfaces support a higher radiative heat flux than thin films unless the period exceeds 900 and 150 nm for 1D and 2D configurations, respectively. The reason why 1D metasurface has a higher radiative heat flux than the 2D counterpart lies in the support of surface plasmon resonance modes besides hyperbolic modes (Liu and Zhang 2015).

Besides enhancing the near-field radiative energy transport, doped-Si metamaterials consisting of doped-Si and Ge multilayers have also been demonstrated to support extremely large later shifts of the streamlines of emitted energy (Bright et al. 2014). It was predicted to be several thousand times of the vacuum gap distance. This may promote the spreading of the heat, enabling doped-Si hyperbolic metamaterials as good heat spreaders, especially when locally cooling down hot spots via noncontact near-field radiation technique.

14.6 CONCLUSIONS AND OUTLOOK

This chapter surveys some exotic radiative properties of doped silicon in the mid-IR, including plasmonic sensing, negative refraction, tunable omnidirectional absorption, and super-Planckian thermal radiation. The advantages of doped silicon over noble metals for the aforementioned applications lie in three aspects. First of all, the magnitude of the permittivity of doped silicon is comparable with that of dielectrics, enabling excitation of mid-IR propagating and localized surface resonances and strong confinement of light. Besides, the plasma frequency can be adjusted through doping, enabling great flexibilities in device designs and functionalities. In addition, fabrication of doped-Si devices is compatible with standard silicon processing techniques and can easily integrate with silicon-based optoelectronic systems.

Mid-IR plasmonic sensing based on doped silicon and corresponding metamaterials is more sensitive to the existence of analytes due to the strong fundamental molecular absorption in the mid-IR. Detecting ultrasmall amount targets is further enhanced due to the small confinement length of SPPs propagating along the interface of analytes and doped silicon. Doped silicon metamaterials and metasurfaces with delicate designs may further improve the sensitivity and enable the achievement of deep subwavelength plasmonic sensing so that the sensing instrument can easily integrate with nano-electronic systems to obtain lap-on-chip diagnostic devices.

Metal-free all-angle negative refraction in the mid-IR region with low loss may be achieved via D-SiNW arrays. The performance holds for very broad wavelength regimes, where the dispersion is hyperbolic. The mechanism of loss-enhanced transmission is elucidated using an approximate expression of FOM. Experimental demonstrations are demanding, and the extension to the design of low-loss mid-IR flat lens is necessary.

Wavelength-tunable omnidirectional IR absorbers with perfect absorptance are possible via simple two-layer configuration and nanowires made of doped silicon. The insensitivity of the absorption to the incidence angle nor the polarization status is demonstrated. A good tunability can be achieved by varying the doping concentration or volume filling ratio to satisfy different specified needs. Wide applications in radiant energy harvesting, bolometers, and radiation cooling are expected.

Different practically achievable nanostructures based on doped silicon are discussed in terms of their potentials in enhancing near-field radiative heat transfer around ambient temperature. D-SiNW and D-SiNH array can provide an enhancement over bulks by more than one order of magnitude in the deep submicron gap region, due to broadband hyperbolic modes and low-loss surface modes, respectively. Perfect energy transmission coefficient across a broad frequency region and over a large k-space is possible via the hybridization of graphene plasmons and hyperbolic modes. The near-field radiative heat transfer coefficient can achieve as high as 80% of a theoretical limit of hyperbolic materials. Patterning thin doped silicon film into 1D and 2D metasurfaces can help to improve increase the radiative heat flux. Promising applications in noncontact thermal management, nanofabrication, heat-assisted magnetic recording, and TPV are anticipated in the near future.

Overall, doped-Si nanostructures and related metamaterials provide a unique platform for lots of exciting applications related to the mid-IR, such as sensing, thermal imaging, heat dissipation, and energy harvesting. The easy integration with current electronic and photovoltaic system helps to promote the development of doped-Si devices. We hope this chapter will help to remind the importance of doped silicon in mid-IR applications and contribute to the development.

ACKNOWLEDGMENTS

This work was supported by the Department of Energy, Office of Science, Basic Energy Sciences (DE-FG02-06ER46343), the National Science Foundation (CBET-1603761), and a startup fund (90YAH16057) from the Nanjing University of Aeronautics and Astronautics for X.L. Liu.

Functional materials

REFERENCES

Basu S, Lee BJ, Zhang ZM. (2010a). Infrared radiative properties of heavily doped silicon at room temperature. *J Heat Transfer* 132:023301.

Basu S, Lee BJ, Zhang ZM. (2010b). Near-field radiation calculated with an improved dielectric function model for doped silicon. *J Heat Transfer* 132:023302.

Basu S, Wang L. (2013). Near-field radiative heat transfer between doped silicon nanowire arrays. *Appl Phys Lett* 102:053101.

Biehs S-A, Rosa FSS, Ben-Abdallah P. (2011). Modulation of near-field heat transfer between two gratings. *Appl Phys Lett* 98:243102.

Biehs SA, Tschikin M, Ben-Abdallah P. (2012). Hyperbolic metamaterials as an analog of a blackbody in the near field. *Phys Rev Lett* 109:104301.

Bright TJ, Liu XL, Zhang ZM. (2014). Energy streamlines in near-field radiative heat transfer between hyperbolic metamaterials. *Opt Express* 22:A1112–A1127.

Cai WS, Shalaev VM. (2009). *Optical Metamaterials: Fundamentals and Applications.* New York: Springer.

Chan GH, Zhao J, Schatz GC, Duyne RPV. (2008). Localized surface plasmon resonance spectroscopy of triangular aluminum nanoparticles. *J Phys Chem C* 112:13958–13963.

Chang J-Y, Basu S, Wang L. (2015). Indium tin oxide nanowires as hyperbolic metamaterials for near-field radiative heat transfer. *J Appl Phys* 117:054309.

Chen YB. (2009). Development of mid-infrared surface plasmon resonance-based sensors with highly-doped silicon for biomedical and chemical applications. *Opt Express* 17:3130–3140.

Chen YB, Zhang ZM. (2007). Design of tungsten complex gratings for thermophotovoltaic radiators. *Opt Commun* 269:411–417.

Chen YB, Zhang ZM. (2008). Heavily doped silicon complex gratings as wavelength-selective absorbing surfaces. *J Phys D-Appl Phys* 41:095409.

Chettiar UK, Kildishev AV, Klar TA, Shalaev VM. (2006). Negative index metamaterial combining magnetic resonators with metal films. *Opt Express* 14:7872–7877.

Chou LW, Filler MA. (2013). Engineering multimodal localized surface plasmon resonances in silicon nanowires. *Angew Chem Int Ed* 52:8079–8083.

Chou L-W, Shin N, Sivaram SV, Filler MA. (2012). Tunable mid-infrared localized surface plasmon resonances in silicon nanowires. *J Am Chem Soc* 134:16155–16158.

Cleary JW, Medhi G, Peale RE, Buchwald WR, Edwards O, Oladeji I. (2010). Infrared surface plasmon resonance biosensor. *Proc SPIE*:767306, vol. 7673, 767306.

Cleary JW, Soref R, Hendrickson JR. (2013). Long-wave infrared tunable thin-film perfect absorber utilizing highly doped silicon-on-sapphire. *Opt Express* 21:19363–19374.

Cui Y, Xu J, Fung KH, Jin Y, Kumar A, He S, Fang NX. (2011). A thin film broadband absorber based on multi-sized nanoantennas. *Appl Phys Lett* 99:253101.

Dahan N, Niv A, Biener G, Gorodetski Y, Kleiner V, Hasman E. (2008). Extraordinary coherent thermal emission from SiC due to coupled resonant cavities. *J Heat Transfer* 130:112401.

Devarapu GCR, Foteinopoulou S. (2014). Broadband Mid-IR superabsorption with aperiodic polaritonic photonic crystals. *J Eur Opt Soc* 9:14012.

DiPippo W, Lee BJ, Park K. (2010). Design analysis of doped-silicon surface plasmon resonance immunosensors in mid-infrared range. *Opt Express* 18:19396–19406.

Dolling G, Wegener M, Soukoulis CM, Linden S. (2007). Negative-index metamaterial at 780 nm wavelength. *Opt Lett* 32:53–55.

Endo T, Kerman K, Nagatani N, Takamura Y, Tamiya E. (2005). Label-free detection of peptide nucleic acid-DNA hybridization using localized surface plasmon resonance based optical biosensor. *Anal Chem* 77:6976–6984.

Fang N, Lee H, Sun C, Zhang X. (2005). Sub-diffraction-limited optical imaging with a silver superlens. *Science* 308:534–537.

Feng R, Qiu J, Cao Y, Liu L, Ding W, Chen L. (2014). Omnidirectional and polarization insensitive nearly perfect absorber in one dimensional meta-structure. *Appl Phys Lett* 105:181102.

Feng S. (2012). Loss-induced omnidirectional bending to the normal in ϵ-near-zero metamaterials. *Phys Rev Lett* 108:193904.

Francoeur M, Mengüç MP, Vaillon R. (2008). Near-field radiative heat transfer enhancement via surface phonon polaritons coupling in thin films. *Appl Phys Lett* 93:043109.

García-Meca C, Hurtado J, Martí J, Martínez A, Dickson W, Zayats AV. (2011). Low-loss multilayered metamaterial exhibiting a negative index of refraction at visible wavelengths. *Phys Rev Lett* 106:067402.

Greffet J-J, Carminati R, Joulain K, Mulet J-P, Mainguy S, Chen Y. (2002). Coherent emission of light by thermal sources. *Nature* 416:61–64.

Guo Y, Cortes CL, Molesky S, Jacob Z. (2012). Broadband super-Planckian thermal emission from hyperbolic metamaterials. *Appl Phys Lett* 101:131106.

Haes AJ, Van Duyne RP. (2002). A nanoscale optical biosensor: Sensitivity and selectivity of an approach based on the localized surface plasmon resonance spectroscopy of triangular silver nanoparticles. *J Am Chem Soc* 124:10596–10604.

Hendrickson J, Guo J, Zhang B, Buchwald W, Soref R. (2012). Wideband perfect light absorber at midwave infrared using multiplexed metal structures. *Opt Lett* 37:371–373.

Hoffman AJ, Alekseyev L, Howard SS, Franz KJ, Wasserman D, Podolskiy VA, Narimanov EE, Sivco DL, Gmachl C. (2007). Negative refraction in semiconductor metamaterials. *Nat Mater* 6:946–950.

Homola J, Yee SS, Gauglitz G. (1999). Surface plasmon resonance sensors: Review. *Sens Actuators B: Chem* 54:3–15.

Hutter E, Fendler JH. (2004). Exploitation of localized surface plasmon resonance. *Adv Mater* 16:1685–1706.

Joulain K, Mulet JP, Marquier F, Carminati R, Greffet JJ. (2005). Surface electromagnetic waves thermally excited: Radiative heat transfer, coherence properties and Casimir forces revisited in the near field. *Surf Sci Rep* 57:59–112.

Kidwai O, Zhukovsky SV, Sipe JE. (2012). Effective-medium approach to planar multilayer hyperbolic metamaterials: Strengths and limitations. *Phys Rev A* 85:053842.

Lee B, Chen Y-B, Zhang ZM. (2008a). Surface waves between metallic films and truncated photonic crystals observed with reflectance spectroscopy. *Opt Lett* 33:204–206.

Lee BJ, Wang LP, Zhang ZM. (2008b). Coherent thermal emission by excitation of magnetic polaritons between periodic strips and a metallic film. *Opt Express* 16:11328–11336.

Lehman J, Sanders A, Hanssen L, Wilthan B, Zeng JA, Jensen C. (2010). Very black infrared detector from vertically aligned carbon nanotubes and electric-field poling of lithium tantalate. *Nano Lett* 10:3261–3266.

Liedberg B, Nylander C, Lunström I. (1983). Surface plasmon resonance for gas detection and biosensing. *Sens Actuators* 4:299–304.

Link S, El-Sayed MA. (2003). Optical properties and ultrafast dynamics of metallic nanocrystals. *Annu Rev Phys Chem* 54:331–366.

Liu B, Shen S. (2013). Broadband near-field radiative thermal emitter/absorber based on hyperbolic metamaterials: Direct numerical simulation by the Wiener chaos expansion method. *Phys Rev B* 87:115403.

Liu B, Shi J, Liew K, Shen S. (2014a). Near-field radiative heat transfer for Si based metamaterials. *Opt Commun* 314:57–65.

Liu XL, Bright TJ, Zhang ZM. (2014d). Application conditions of effective medium theory in near-field radiative heat transfer between multilayered metamaterials. *J Heat Transfer* 136:092703.

Liu XL, Wang LP, Zhang ZM. (2013b). Wideband tunable omnidirectional infrared absorbers based on doped-silicon nanowire arrays. *J Heat Transfer* 135:061602.

Liu XL, Wang LP, Zhang ZM. (2015a). Near-field thermal radiation: Recent progress and outlook. *Nanoscale Microscale Thermophys Eng* 19:98–126.

Liu XL, Zhang ZM. (2013). Metal-free low-loss negative refraction in the mid-infrared region. *Appl Phys Lett* 103:103101.

Liu XL, Zhang ZM. (2015). Near-field thermal radiation between metasurfaces. *ACS Photonics* 2:1320.

Liu XL, Zhang RZ, Zhang ZM. (2013a). Near-field thermal radiation between hyperbolic metamaterials: Graphite and carbon nanotubes. *Appl Phys Lett* 103:213102.

Liu XL, Zhang RZ, Zhang ZM. (2014b). Near-perfect photon tunneling by hybridizing graphene plasmons and hyperbolic modes. *ACS Photon* 1:785–789.

Liu XL, Zhang RZ, Zhang ZM. (2014c). Near-field radiative heat transfer with doped-silicon nanostructured metamaterials. *Int J Heat Mass Transfer* 73:389–398.

Liu XL, Zhao B, Zhang ZM. (2015b). Blocking-assisted infrared transmission of subwavelength metallic gratings by graphene. *J Opt* 17:035004.

Liu YM, Bartal G, Zhang X. (2008). All-angle negative refraction and imaging in a bulk medium made of metallic nanowires in the visible region. *Opt Express* 16:15439–15448.

Maruyama S, Kashiwa T, Yugami H, Esashi M. (2001). Thermal radiation from two-dimensionally confined modes in microcavities. *Appl Phys Lett* 79:1393–1395.

Mayer KM, Hafner JH. (2011). Localized surface plasmon resonance sensors. *Chem Rev* 111:3828–3857.

Mizaikoff B. (2013). Waveguide-enhanced mid-infrared chem/bio sensors. *Chem Soc Rev* 42:8683–8699.

Mizuno K, Ishii J, Kishida H, Hayamizu Y, Yasuda S, Futaba DN, Yumura M, Hata K. (2009). A black body absorber from vertically aligned single-walled carbon nanotubes. *Proc Natl Acad Sci U S A* 106:6044–6047.

Naik GV, Liu J, Kildishev AV, Shalaev VM, Boltasseva A. (2012). Demonstration of Al:ZnO as a plasmonic component for near-infrared metamaterials. *Proc Nat Acad Sci U S A* 109:8834.

Narayanaswamy A, Chen G. (2004). Thermal emission control with one-dimensional metallodielectric photonic crystals. *Phys Rev B* 70:125101.

Orlov AA, Voroshilov PM, Belov PA, Kivshar YS. (2011). Engineered optical nonlocality in nanostructured metamaterials. *Phys Rev B* 84:045424.

Pendry JB. (2000). Negative refraction makes a perfect lens. *Phys Rev Lett* 85:3966–3969.

Petryayeva E, Krull UJ. (2011). Localized surface plasmon resonance: Nanostructures, bioassays and biosensing—A review. *Anal Chim Acta* 706:8–24.

Rakic AD, Djurisic AB, Elazar JM, Majewski ML. (1998). Optical properties of metallic films for vertical-cavity opto-electronic devices. *Appl Opt* 37:5271–5283.

Rousseau E, Laroche M, Greffet J-J. (2010). Radiative heat transfer at nanoscale: Closed-form expression for silicon at different doping levels. *J Quant Spectrosc Radiat Transfer* 111:1005–1014.

Rowe DJ, Jeong JS, Mkhoyan KA, Kortshagen UR. (2013). Phosphorus-doped silicon nanocrystals exhibiting mid-infrared localized surface plasmon resonance. *Nano Lett* 13:1317–1322.

Rytov SM, Kravtsov YA, Tatarskii VI. (1989). *Principles of Statistical Radiophysics.* New York: Springer.

Schmidt V, Wittemann J, Gosele U. (2010). Growth, thermodynamics, and electrical properties of silicon nanowires. *Chem Rev* 110:361–388.

Shalaev VM. (2007). Optical negative-index metamaterials. *Nature Photon* 1:41–48.

Shelby RA, Smith DR, Schultz S. (2001). Experimental verification of a negative index of refraction. *Science* 292:77–79.

Shen S. (2013). Experimental studies of radiative heat transfer between bodies at small separations. *Annu Rev Heat Transfer* 16:327–343.

Sherry LJ, Chang S-H, Schatz GC, Van Duyne RP, Wiley BJ, Xia Y. (2005). Localized surface plasmon resonance spectroscopy of single silver nanocubes. *Nano Lett* 5:2034–2038.

Shi J, Li P, Liu B, Shen S. (2013). Tuning near field radiation by doped silicon. *Appl Phys Lett* 102:183114.

Shin H, Fan S. (2006). All-angle negative refraction and evanescent wave amplification using one-dimensional metal-lodielectric photonic crystals. *Appl Phys Lett* 89:151102.

Shu S, Li Z, Li YY. (2013). Triple-layer Fabry-Perot absorber with near-perfect absorption in visible and near-infrared regime. *Opt Express* 21:25307–25315.

Singh P. (2016). SPR biosensors: Historical perspectives and current challenges. *Sens Actuators B Chem* 229:110–130.

Smith DR, Kolinko P, Schurig D. (2004). Negative refraction in indefinite media. *J Opt Soc Am B* 21:1032–1043.

Smith DR, Padilla WJ, Vier DC, Nemat-Nasser SC, Schultz S. (2000). Composite medium with simultaneously negative permeability and permittivity. *Phys Rev Lett* 84:4184–4187.

Song B, Fiorino A, Meyhofer E, Reddy P. (2015a). Near-field radiative thermal transport: From theory to experiment. *AIP Adv* 5:053503.

Song J, Wu H, Cheng Q, Zhao J. (2015b). 1D trilayer films grating with W/SiO2/W structure as a wavelength-selective emitter for thermophotovoltaic applications. *J Quant Spectrosc Radiat Transfer* 158:136–144.

Streyer W, Law S, Rooney G, Jacobs T, Wasserman D. (2013). Strong absorption and selective emission from engineered metals with dielectric coatings. *Opt Express* 21:9113–9122.

Sun L, Feng S, Yang X. (2012). Loss enhanced transmission and collimation in anisotropic epsilon-near-zero metamaterials. *Appl Phys Lett* 101:241101.

Teranishi T, Eguchi M, Kanehara M, Gwo S. (2011). Controlled localized surface plasmon resonance wavelength for conductive nanoparticles over the ultraviolet to near-infrared region. *J Mater Chem* 21:10238–10242.

Tschikin M, Biehs SA, Messina R, Ben-Abdallah P. (2013). On the limits of the effective description of hyperbolic materials in the presence of surface waves. *J Opt* 15:105101.

van Zwol PJ, Thiele S, Berger C, de Heer WA, Chevrier J. (2012). Nanoscale radiative heat flow due to surface plasmons in graphene and doped silicon. *Phys Rev Lett* 109:264301.

Wang H, Liu XL, Wang LP, Zhang ZM. (2013). Anisotropic optical properties of silicon nanowire arrays based on effective medium calculation. *Int J Therm Sci* 65:62.

Wang H, Wang L. (2013). Perfect selective metamaterial solar absorbers. *Opt Express* 21:A1078–A1093.

Wang L, Basu S, Zhang ZM. (2012). Direct measurement of thermal emission from a Fabry–Perot cavity resonator. *J Heat Transfer* 134:072701.

Wang LP, Zhang ZM. (2012). Wavelength-selective and diffuse emitter enhanced by magnetic polaritons for thermophotovoltaics. *Appl Phys Lett* 100:063902.

Wang XJ, Wang LP, Adewuyi OS, Cola BA, Zhang ZM. (2010). Highly specular carbon nanotube absorbers. *Appl Phys Lett* 97:163116.

Willets KA, Van Duyne RP. (2007). Localized surface plasmon resonance spectroscopy and sensing. *Annu Rev Phys Chem* 58:267–297.

Xuan Y. (2014). An overview of micro/nanoscaled thermal radiation and its applications. *Photon Nanostruct Fundam Appl* 12:93–113.

Xuan Y, Duan H, Li Q. (2014). Enhancement of solar energy absorption using a plasmonic nanofluid based on TiO 2/Ag composite nanoparticles. *RSC Adv* 4:16206–16213.

Xuan Y, Zhang Y. (2014). Investigation on the physical mechanism of magnetic plasmons polaritons. *J Quant Spectrosc Radiat Transf* 132:43–51.

Yang Z-P, Ci L, Bur JA, Lin S-Y, Ajayan PM. (2008). Experimental observation of an extremely dark material made by a low-density nanotube array. *Nano Lett* 8:446–451.

Yao J, Liu ZW, Liu YM, Wang Y, Sun C, Bartal G, Stacy AM, Zhang X. (2008). Optical negative refraction in bulk metamaterials of nanowires. *Science* 321:930.

Ye H, Wang XJ, Lin W, Wong CP, Zhang ZM. (2012). Infrared absorption coefficients of vertically aligned carbon nanotube films. *Appl Phys Lett* 101:141909.

Zhang RZ, Zhang ZM. (2015). Negative refraction and self-collimation in the far infrared with aligned carbon nanotube films. *J Quant Spectrosc Radiat Transfer* 158:91–100.

Zhang S, Fan WJ, Panoiu NC, Malloy KJ, Osgood RM, Brueck SRJ. (2005). Experimental demonstration of near-infrared negative-index metamaterials. *Phys Rev Lett* 95:137404.

Zhang ZM. (2007). *Nano/Microscale Heat Transfer*. New York: McGraw-Hill.

Zhang ZM, Lee BJ. (2006). Lateral shift in photon tunneling studied by the energy streamline method. *Opt Express* 14:9963–9970.

Zhang ZM, Ye H. (2012). Measurements of radiative properties of engineered micro/nanostructures. *Annu Rev Heat Transfer* 16:345–396.

Zhao B, Wang L, Shuai Y, Zhang ZM. (2013). Thermophotovoltaic emitters based on a two-dimensional grating/thin-film nanostructure. *Int J Heat Mass Transf* 67:637–645.

Zhao B, Zhang ZM. (2014). Study of magnetic polaritons in deep gratings for thermal emission control. *J Quant Spectrosc Radiat Transf* 135:81–89.

Zhou S, Pi X, Ni Z, Ding Y, Jiang Y, Jin C, Delerue C, Yang D, Nozaki T. (2015). Comparative study on the localized surface plasmon resonance of boron-and phosphorus-doped silicon nanocrystals. *ACS Nano* 9:378–386.

Ziblat R, Lirtsman V, Davidov D, Aroeti B. (2006). Infrared surface plasmon resonance: A novel tool for real time sensing of variations in living cells. *Biophys J* 90:2592–2599.

Antireflective silicon nanostructures

Young J. Yoo, Eun K. Kang, Yong T. Lee, and Young M. Song

Contents

15.1 INTRODUCTION

In optoelectronic devices or optical components, the suppression of surface reflections caused by discontinuity of the refractive index (RI) between two different optical media is crucial. For instance, the efficiency of a photovoltaic device is primarily limited by the Fresnel reflection loss on the surface of the device. Such a reflection loss is especially highlighted in materials with a high RI such as Si, GaAs, and GaP. Other optoelectronic devices, including light-emitting diodes (LEDs), laser diodes, and photodetectors, have similar problems [1,2]. A "ghost image" in glasses and other transparent materials is also generated by the Fresnel reflection loss [3]. Thin-film technology is commonly used for mass production of antireflection coatings (ARCs) with quarter wavelength stacks. However, it shows antireflection (AR) properties only in specific wavelength ranges and for limited incidence angles. Additionally, thin-film multilayers have problems with material appropriateness, thermal mismatch, and instability of the stacks [4].

Nowadays, the ideas of bionics have intruded into many technological fields. In the field of optical science and technology, a variety of biomimetic concepts, including photonic crystals in opals, vivid colors in butterfly wings, and deformable eye lenses in birds and mammals, have been developed [5–7]. A few decades ago, Bernhard and Miller discovered that the outer surface of facet lenses in the eyes of a moth consists of an array of cuticular protuberances termed "corneal nipples" [8]. A set of facet lenses of the insect's eye, the cornea, is approximately a hemisphere (Figure 15.1a). The convex outer surface of the facet lenses consists of the corneal nipples, which are locally arranged in a highly regular hexagonal lattice (Figure 15.1b and c) [9]. The optical action of the corneal nipple array is a severe reduction in the reflectance of the facet lens surface [10]. The operation of the moth-eye surface may be understood most easily in terms of a surface layer, in which the RI varies gradually from unity to that of the bulk material [11]. The insight that nipple arrays can significantly reduce the surface reflectance has been widely applied [12].

Figure 15.1 Scanning electron micrographs of (a) the Attacus atlas moth eye showing the compound eye structure. (b) The nipple array in one facet lens. (c) The local arrangement of domains with highly ordered nipple arrays. (Reproduced from Ko, D.-H., et al., Biomimetic microlens array with antireflective "moth-eye" surface, *Soft Matter* 7 (2011), 6404, By permission of The Royal Society of Chemistry.) (d) A hawkmoth, *Cephonodes hylas*, male. The letters under the animal are visible due to the transparent part of the wing. (e) The SEM image of the oblique view of the rough wing with protuberances. (Reproduced from Yoshida, A., et al., Antireflective nanoprotuberance array in the transparent wing of a hawkmoth, *Cephonodes Hylas*, *Zool. Sci.* 14 (1997), 737–41, By permission of The Zoological Society of Japan.)

Closely packed nipple arrays can also be found in the wings of some moth species. The wings of adult *Cephonodes* are transparent except for small parts, that is, the wing margin and veins (Figure 15.1d). This transparent part has regular and hexagonal protuberances with no scales. The center-to-center distance between neighboring structures is about 200 nm, as shown in Figure 15.1e. Protuberances are dome- or nipple-shaped and have a constriction around their middle height [13]. Similar nipple arrays have also been observed on the wings of cicadas, which can be used for producing superhydrophobic surfaces for water-repellent applications [14,15].

To overcome the limitation of multilayer structures, the biomimetic optical concept of the moth eye and effective media has been transferred into the technological world within the past decades. The interest in tapered subwavelength gratings was also driven by the new possibilities of numerical simulation and of micro and nanofabrication. Moreover, highly efficient optical devices with such moth-eye structures were reported. In this chapter, we begin by explaining the basic principles of the optic behavior in the moth-eye structures and other graded index media. Several fabrication techniques, including top-down, bottom-up, and soft molding, for creating antireflective nanostructures are discussed. Examples of silicon-based optoelectronic devices illustrate the power of these concepts. We also discuss design guidelines and parametric studies for specific optoelectronic devices.

15.2 THEORY OF ANTIREFLECTIVE NANOSTRUCTURES

For traditional layered ARCs widely used in many optical and optoelectronic devices, the basic principle of a single-layer dielectric thin film with a low RI (n) on a substrate with a different RI (n_s), where $n_s > n$, follows the film interference law. Two interfaces are generated in the thin-film configuration, which produces two reflected waves, and destructive interference occurs when these two waves are out of phase. The minimum reflection loss can be achieved for the optimized thickness and RI of an ARC, which are dependent on the wavelength, angle, and polarization of incident light. Therefore, single-layer ARCs can only obtain a high AR performance toward incident light under specific conditions, that is, wavelength, incidence angle, and polarization [16–18]. ARCs based on micro and nanostructure arrays follow an alternative way of reflectance reduction. Depending on the characteristic scale of the structures, two different ways of the interaction between the arrayed structures and incident light exist [17–19]. If the size of an individual unit is much larger than the wavelength, namely, a macrostructure unit, incident light is reflected normally and scattered after being absorbed partly. If the depth and the space between individual structural units are on the same scale as the light wavelength, light rays are trapped in the gaps leading to multiple internal reflections [20,21]. Thus, incident radiation can be absorbed reducing the reflection in the visible range to a very low level. The amount of the reflection is strongly dependent on the geometry of the structures.

On the contrary, when the AR structures have dimensions less than the wavelength, that is, located in the subwavelength scale or nanoscale, an alternative means is employed. If the AR surface has a gradient RI, light is insensitive to the AR structures and tends to bend progressively [22,23]. Even though the angle of incidence is changed, the coating still exhibits a relatively smooth change in the RI toward the incident direction of light, suppressing the reflection of light for a broad range of the wavelength [20,22]. Moreover, natural light always shows some degree of polarization, including s- and p-polarizations, which have the electric field perpendicular and parallel to the incidence plane, respectively. For the subwavelength-scale or nanoscale arrays with a smoothly graded RI from air to a substrate, the reflection of light with either s- or p-polarization can be suppressed to a very low level because the transmission of light with different polarizations is insensitive to media with extremely low disparity of the RI. Therefore, this type of ARCs based on nanostructure arrays with a gradient RI can realize broadband, omnidirectional, and polarization-insensitive AR performance, which is superior to that of layered ARCs.

Generally, different nanostructure arrays may have different RI profile curves, such as linear, parabolic, cubic, quintic, exponential, and exponential–sinusoidal curves, exhibiting different AR performances [21,24]. Theoretical computation is important in developing and optimizing high-performance AR surfaces. In the view of the effective medium theory (EMT), which is an essential concept for many computational models in the area of antireflectivity, the surface RI depends on the topology and the composition and can be calculated as a function of the volume fraction of inclusion for a material mixture [17,18,25–27]. As for a nanostructure array film, the effective RI can be determined by considering the surface consisting of layers of homogeneous mixtures of nanostructured materials and the air in the interspaces. The rigorous coupled-wave analysis (RCWA), first proposed by Moharam and Gaylord in 1981, is widely used in the theoretical calculations to optimize the ARC design [28]. This analytical model is a relatively straightforward technique to exactly solve Maxwell's equations to accurately analyze the diffraction of electromagnetic waves. In the RCWA, the cross-section of the structures is treated as consisting of a large number of thin layers parallel to the surface. A particular formulation without any approximations has been developed to analyze both transmission and reflection from planar and surface-relief structures accurately and efficiently. Using this model, the performance of AR structures can be predicted and the structure optimization of ARCs can be conducted.

As illustrated in Figure 15.2, ARCs can be classified into two basic types: those based on inhomogeneous layers and those that consist of a homogeneous layer [29,30]. A single inhomogeneous layer with RI n_m on a substrate with RI n_s at a wavelength λ has an optical thickness equal to $\lambda/4$ and the RIs satisfy the relation $n = \sqrt{(n_s n_m)}$ (Figure 15.2a). However, these conditions can be satisfied only at a specific wavelength because of material dispersion. Double-inhomogeneous layers have a medium with a lower RI on the bottom coating layer (Figure 15.2b). Although the two-step change of the RI from air to the substrate can reduce the reflection, discontinuity of the RI at each layer remains. Finally, multi-inhomogeneous layers have a graded index, indicating a smooth change in the RI of each layer. By the use of multiple thin-film layers, zero reflectance can be obtained at one or more wavelengths even if the RI relationship given above is not satisfied (Figure 15.2c). A lower effective index n of the homogeneous layer is achieved when the layer is patterned or structured. A homogeneous porous layer acts like a single inhomogeneous layer because a porous pattern without tapering means invariance of the effective index along the depth of the patterned layer (Figure 15.2d). A tapered porous layer has a graded index profile; however, the effective index of the layer changes abruptly at the end of the structure, causing a reflection loss at the surface (Figure 15.2e).

Functional materials

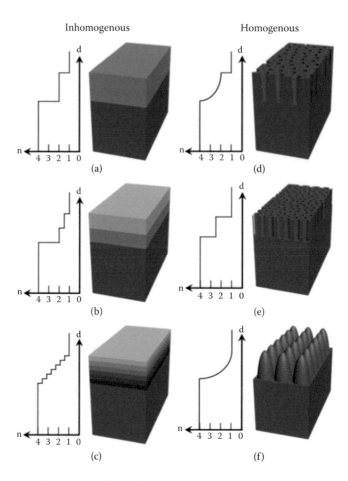

Figure 15.2 Structure and effective RI profiles of various types of ARCs. (a)–(c) Inhomogeneous single-layer, multilayer, and graded-index ARCs and (d)–(f) homogeneous porous, tapered porous, and moth-eye ARCs.

A smooth change in the effective RI can be achieved by introducing artificial moth-eye structures, provided that its thickness exceeds at least one or two waves and that the lateral dimensions of the patterns are less than the light wavelength (Figure 15.2f). When these conditions are satisfied, the EMT can be applied and the material–air structure can be represented by a series of thin films with RIs that vary gradually from unity (air) to n_s, the substrate index.

A perfect ARC should meet the requirements of excellent AR properties, namely, broadband, omnidirectional, and polarization-insensitive antireflectivity [17,18]. The fact that there may be a difference in RI matching or optical impedance matching required for different wavelength regions, such as visible, ultraviolet (UV), and near-infrared regions, impairs the broadband AR performance of ARCs. The incidence angles have a significant impact on the reflectance. For example, most glass and plastics with an RI of ~1.5 exhibit a reflectance of 4% at normal incidence, but a much higher reflectance, even a reflectance of 100%, can be reached as the incidence angles are increased [17,18]. This causes difficulties in the case of solar cells that should be mechanically oriented to face the sun throughout the day, which needs additional control devices and energy consumption. Therefore, omnidirectional antireflectivity is important for the practical applications of ARCs in photovoltaic modules. Moreover, ARCs have to be insensitive to the light polarization because at smaller angles, p-polarized light is maximally reflected. Therefore, a perfect antireflective coating should exhibit broadband, omnidirectional, and polarization-insensitive antireflectivity. A traditional layered AR film has difficulties with satisfying all the requirements because of the fundamental interference destructive principles. There has been considerable progress toward perfect ARCs based on nanostructure arrays in recent years.

15.3 ARTIFICIALLY ENGINEERED MOTH-EYE STRUCTURES

15.3.1 QUARTER-WAVELENGTH ARC AND MULTILAYER ARC

The AR performance of traditional quarter-wavelength ARCs depends on the coating thickness and the material RI. Careful control of both factors would result in lowering the amount of the reflection from the surface [31,32]. As mentioned in Chapter 2, optical reflection can be efficiently suppressed if the RI of the coating material is equal to the geometric mean of the RIs of the two media at the interface [33]. A quarter-wavelength coating will allow light reflected from the surrounding medium/ARC interface and the ARC/substrate interface to interfere destructively, eliminating the reflection. The amount of the reflection also depends on the angle of the incident light [24,34]. Numerous methods of quarter-wavelength ARC production have been developed, for example, vacuum-based deposition processes and layer-by-layer deposition of polyelectrolyte and/or nanoparticle films [31].

A limited number of materials with an adequate RI (generally for low-RI substrates) are one of the major limitations of single-layer coating. The use of composite single-layer ARCs, whose RI can be tuned by changing the ratio (filling factor) of the constituents of the composite, is one of the solutions to this problem. As shown in Figure 15.3a, mesoporous silica nanoparticles are used for fabricating antireflective coatings on glass substrates [35]. The combination of mesoporous silica nanoparticles in conjunction with a suitable binder material allows mechanically robust single-layer coatings with a reflectance of <0.1% to be produced by using simple wet-processing techniques (Figure 15.3b). Further advantages of these films are that their structure results in antireflective properties with a minimum reflection that can be tuned between 400 and 1900 nm. The ratio of the binder material to mesoporous nanoparticles allows the RI control [35]. Double-layer ARCs are also widely used for reflection reduction. In the case of double-layer ARCs, the upper film facing air usually has the lowest RI and another layer is made successively based on the ascending order of their RIs. Similar to single-layer coating, the interference conditions should be fulfilled to destructively cancel bouncing back waves off the substance surface. Hence, the thickness of each layer is usually a quarter or half of the wavelength. Similar to single-layer ARC, each layer of a double-layer configuration can be made of a composite material with a tunable RI to provide more design flexibility [36]. The AR in a wide wavelength range originates from a reasonable RI gradient from air to the substrate produced by mesopores in the bottom-layer silica coating and particle-packed pores in the top-layer silica coating (Figure 15.3c and e). The maximum transmittance of the broadband ARC was approximated to 100.0% at the peak value and above 99.6% in the visible region from 400 to 800 nm. Meanwhile, the average transmittance of the coating was more than 99.0% over the range of 360–920 nm (Figure 15.3d). In addition, the double-layer broadband silica ARC showed a considerable mechanical performance and a high environmental stability [37].

15.3.2 GRADED-INDEX ARC

A gradual change in the film RI from the substrate RI (n_s) to the air RI (n_{air}) is an alternative way of reflectance reduction. Interference effects in the stacking layers of a dielectric rely on multipass light circulation inside the optical media formed by the films [36,38]. This means that the reflection of a multilayer coating strongly correlates with the thickness and RI of each layer. Nanostructured multilayers for graded-index ARC are achieved using oblique-angle deposition [39,40]. Oblique-angle deposition provides the control on the thin-film porosity and thickness, resulting in consistency of the ARC design parameters with actual parameters [41,42].

Figure 15.4 shows several examples of multilayered gradient media for broadband AR characteristics [43–45]. For instance, a three-layer graded-index ARC to Si is composed of the following layers: the first layer of TiO_2 ($n = 2.66$ at 550 nm), the second layer of SiO_2 ($n = 1.47$ at 550 nm), and the third layer of low-n SiO_2 ($n = 1.07$ at 550 nm) (Figure 15.4a). The first layer of TiO_2 and the second layer of SiO_2 are deposited by means of reactive sputtering. The third layer of porous SiO_2 is deposited using the oblique-angle e-beam evaporation technique (Figure 15.4b). The desired low RI is achieved by mounting the sample such that the substrate normal is at 85° to the incoming flux. By averaging over the wavelength range from 400 to 1100 nm and the incidence angle range of 0°–90°, it is found that polished Si reflects ~37% of

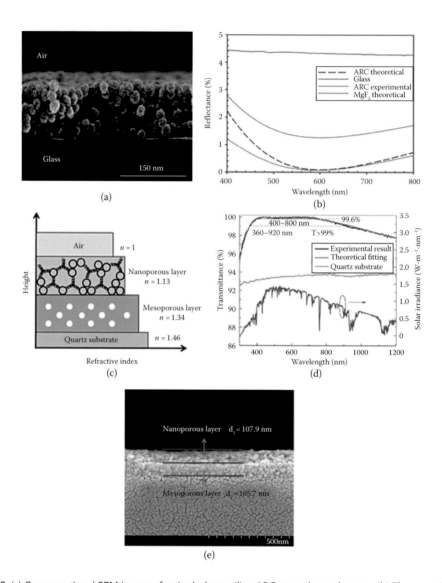

Figure 15.3 (a) Cross-sectional SEM image of a single-layer silica ARC on a glass substrate. (b) The experimental reflection spectra of glass and the ARC and the theoretically calculated reflection spectra of the single-layer silica ARC and MgF$_2$. (Reprinted with permission from Moghal, J., et al., High-performance, single-layer antireflective optical coatings comprising mesoporous silica nanoparticles, *ACS App. Mater. Interfaces* 4, 2 (2012), 854–859. Copyright 2012 American Chemical Society.) (c) The schematic illustration of the double-layer broadband silica ARC. A reasonable RI gradient from air to the substrate was produced with ordered mesopores in the bottom layer and particle-packed pores in the top layer. (d) The transmittance spectra of the experimental and theoretical fitting double-layer silica ARCs and the quartz substrate, overlaid with the solar spectral irradiance at air mass (AM) 1.5. The prepared broadband coating showed consistency of the optical transmission of the double-layer broadband ARC with the solar spectrum. (e) The cross-sectional SEM image of the double-layer broadband silica ARC, from which the thickness of each layer was clearly observed. The broadband ARC was coated on a silicon wafer, and the cross-section of the coating was sprayed with Au nanoparticles prior to SEM imaging. (Reproduced from Sun, J., et al., A broadband antireflective coating based on a double-layer system containing mesoporous silica and nanoporous silica, *J. Mater. Chem. C*, 3 (2015), 7187–7194. By permission of The Royal Society of Chemistry.)

incident radiation. The reflection losses are reduced to only 5.9% by applying a three-layer graded-index ARC to Si (Figure 15.4c) [43]. A four-layer tailored- and low-RI ARC on an inverted metamorphic (IMM) triple-junction solar cell device is demonstrated (Figure 15.4d). By utilizing the oblique-angle e-beam deposition and physical vapor deposition (sputtering) methods, the four-layer ARC was fabricated on an IMM solar cell device (Figure 15.4e). By incorporating tailored- and low-RI materials into the ARC design

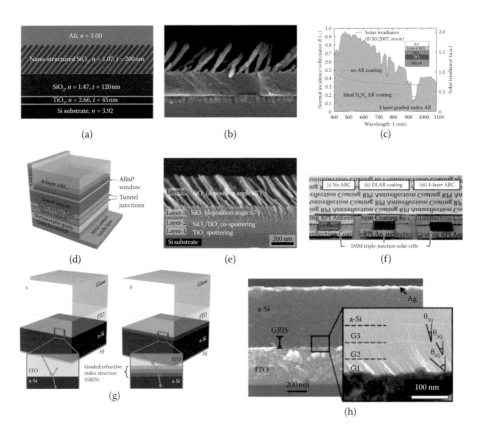

Figure 15.4 (a, b) Schematic and SEM image of three-layer graded-index AR coating; the refractive-index values are measured at 550 nm. (c) Solar spectrum and reflectance of silicon substrate with no AR coating, ideal Si3N4 AR coating, and three-layer graded-index AR coating. (Reprinted with permission from Chhajed, S., et al., Nanostructured multilayer graded-index antireflection coating for Si solar cells with broadband and omnidirectional characteristics, *Appl. Phys. Lett.*, 93 (2008), 251108. Copyright 2008 American Institute of Physics.) (d) Schematic layer sequence of an inverted metamorphic (IMM) triple-junction solar cell with four-layer AR coating. (e) SEM cross-sectional image of the four-layer AR coating deposited on a silicon substrate. (f) Photograph of three IMM solar cells with (i) no AR coating, (ii) double-layered AR coating, and (iii) four-layer AR coating. (Reproduced from Yan, X., et al.: Enhanced omnidirectional photovoltaic performance of solar cells using multiple-discrete-layer tailored- and low-refractive index anti-reflection coatings. *Adv. Funct. Mater.* 2013. 23. 583–590. Copyright Wiley-VCH Verlag GmbH & Co. KGaA. With permission.) (g) Schematic diagrams of typical superstrate-type thin-film solar cell structure (Sample A) and graded-index structure (Sample B). The magnified images in red squares express the light propagation through the interface between ITO and a-Si layer with and without the graded-index structure. (h) SEM image of the structure with the graded-index structure. The inset is a TEM image of the graded-index structure. (Reproduced from Jang, S.J., et al., Antireflective property of thin film a-Si solar cell structures with graded refractive index structure, *Opt. Express*, 19 (2011), 108–117. With permission of Optical Society of America.)

and fabrication, significant progress in terms of the optical efficiency was achieved in the implementation of quasi-continuously graded RI ARCs with so-called quintic or modified-quintic profiles. ARCs with these RI profiles are able to achieve improved broadband and omnidirectional AR characteristics in comparison to conventional double-layer ARCs (Figure 15.4f) [44].

The graded RI structure can also be applied to a thin-film amorphous-Si (a-Si) solar cell (Figure 15.4g and h). The graded RI structure fabricated using oblique angle deposition suppresses optical reflection in wide wavelength and incidence angle ranges compared to the conventional structure without graded index layers. Such graded index media are embedded inside the device structures, which are not available in conventional micro or nanostructures. The average reflectance of the thin-film a-Si solar cell structure with the graded RI structure is suppressed by 54% at normal incidence owing to effective RI matching between indium tin oxide and a-Si [45].

15.3.3 POROUS ARC

As mentioned in Chapter 2, subwavelength-scale porous materials could have AR properties. Usually, such porous materials are generated using chemical etching of a substrate. Among the fabrication methods of Si nanostructures, metal-assisted chemical etching has attracted increasing attention in recent years owing to simplicity and low operating costs [46,47]. A nanoscale porous structure is the typical morphology of Si etched using the metal-assisted chemical etching method. In a typical metal-assisted chemical etching procedure, a Si substrate partly covered by a noble metal is subjected to an etchant composed of HF and an oxidative agent. Typically, the Si beneath the noble metal is etched much faster than Si without noble metal coverage. As a result, the noble metal sinks into the Si substrate, generating pores in the Si substrate or, additionally, Si wires [47].

Such porous silicon structures show a prominent AR property. Porous Si fabricated by etching a polycrystalline Si substrate in a Ag particle–loaded Si substrate in a H_2O_2/HF solution revealed reduced reflectances in the wavelength range of 300–800 nm (Figure 15.5a and b) [48]. Wafer-scale silicon nano and microwires are also fabricated by using metal-assisted electroless etching for solar cell applications (Figure 15.5c–g). The optical reflectance spectra of the co-integrated wire samples are varied as a function of KOH etching in the wavelength range of 300–1000 nm (Figure 15.5h). The highest reflectance of the agglomerated nanowires (KOH 0 s) is not greater than ~2.5%, with an average reflectance of ~2.0% in the range of 300–1000 nm [49].

15.3.4 MOTH-EYE ARC

The biomimetic moth-eye structure can be understood easily in terms of a thin film, in which the RI changes gradually from the structure top to the bulk materials [50]. In case of a structured film with a gradient RI, we can regard the reflectance of the moth-eye surface as a resultant of an infinite series of reflections at each incremental change in the index [51]. To fabricate a high-performance artificial moth-eye structure, three structural features are generally required: the height and period of the arrays and the distance between the arrays. For the moth-eye surface, the period of the structures should be sufficiently small so that the array cannot be resolved by incident light. If this condition is fulfilled, we assume that at any depth, the effective RI is the mean of that of air and the bulk materials, weight in proportion to the volume of the materials. The condition that the moth-eye array should not be resolved by light is that the direction of the first diffracted order is over the horizon. For an ideal case of the moth-eye structure, it should show a tapered profile, the period should be as fine as possible, and the depth should be as great as possible to provide the widest bandwidth and almost omnidirectional antireflective properties [52].

Many techniques based on top-down lithography, such as electron-beam lithography [53,54], focused ion beam [55], interference lithography [56–58], and nanoimprint lithography [59], have been applied to fabricated AR structured surfaces. To avoid scattering from the optical interface, its structure dimension has to be smaller than the wavelength of the incident light [11,60]. For UV and visible light applications, the feature size should always be below 200 nm. In such a small size range, conventional top-down lithographic technologies (electron-beam etching and fast atom beam) require sophisticated equipment and are time-consuming and expensive for large-area fabrication for practical applications. To overcome the limitations of electron-beam etching, nanoimprint lithography is introduced to fabricate various functional polymer nanopillar arrays with a high throughput [61,62]. Moreover, by using such polymer arrays as masks, many functional material nanopillar arrays can be prepared by utilizing reactive ion etching (RIE).

Colloidal lithography is a low-cost and relatively high-throughput technique for patterning nanostructures [41]. The advantage of colloidal lithography in nanofabrication is that large-area self-assembly of colloids having well-ordered structures can be performed without expensive equipment [63,64]. Colloidal lithography is a simple, cost-effective, time-effective, and reproductive method to fabricate moth-eye structures on many materials [65–69]. Two-dimensional (2D) colloidal crystals are used as etch masks on a substrate [64,70]. Then, substrates are etched by implementing RIE [71]. In the process of RIE, the sphere masks can also be etched by reactive ions, and with an increase in the etching duration, the sphere masks were etched away gradually. Therefore, as the traverse diameter of a sphere was decreased, the etched area of an underlying substrate increased gradually, whereas the top diameter of the obtained structure was nearly the same as the traverse diameter of the sphere above, which led to the shape transformation of the

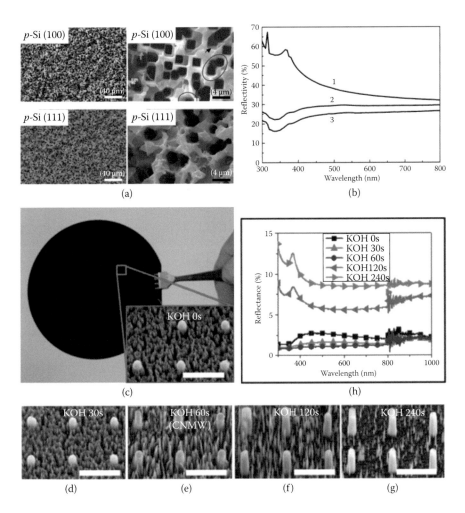

Figure 15.5 (a) SEM images of p-Si wafers subjected to MacEtch in 1:1:1 (v:v:v) HF (49%):H$_2$O$_2$ (30%):ethanol solution for 12 h at 300 K by using catalytic Ag particles deposited on Si via silver-mirror reaction. The top macroporous layers were removed by dipping in NaOH (1%) for 15 min. The used wafers are p-Si (100) and p-Si (111). The images are displayed at different magnifications and orientations. (b) The reflectivity of the samples before and after MacEtch. Curves 1, 2, and 3 represent the unetched p-Si (100), etched p-Si (111), and etched p-Si (100) samples, respectively. (Reproduced with kind permission from Springer Science+Business Media: *J. Electron. Mater.*, Metal-assisted chemical etching using Tollen's reagent to deposit silver nanoparticle catalysts for fabrication of quasi-ordered silicon micro/nanostructures, 40, 2011, 2480–2485, Geng, et al.) (c) The photograph showing the wafer scale (4-inch) fabrication of a co-integrated, tapered nano and microwire (CNMW) structure. The effect of the KOH etching time after the formation of a CNMW structure is shown using 30°-tilted view SEM images: (c) 0, (d) 30, (e) 60, (f) 120, and (g) 240 s with the same scale bars of 10 µm. (h) The optical reflectance spectra of the CNMW samples as a function of the KOH etching time. (The source of the material Jung, J.-Y., et al., A waferscale Si wire solar cell using radial and bulk p–n junctions, *Nanotechnology*, 21, 445303, 2010 is acknowledged.)

obtained structure from cylinder to frustum of cone. When the sphere disappeared, the obtained textured profile started to change from the frustum of a cone to a cone because the top region of the etched substrate was nearer to the plasma, and thus it was etched more rapidly than narrow bottom areas between the frustum of the cones. Using colloidal lithography, subwavelength pyramidal and honeycomb structures are fabricated and optimized in solar cells. Subwavelength-scale monolayer and bilayer polystyrene spheres are combined using the one-step reactive ion etching process to fabricate optimized pyramid- and honeycomb shaped AR structures, respectively. A close-packed monolayer of 350 nm-diameter polystyrene spheres was coated on a silicon substrate using a dip coater (Figure 15.6a). The period of the pyramidal structure remained 350 nm, and the height was 480 nm (Figure 15.6b). Etched silicon has a height of 310 nm and a

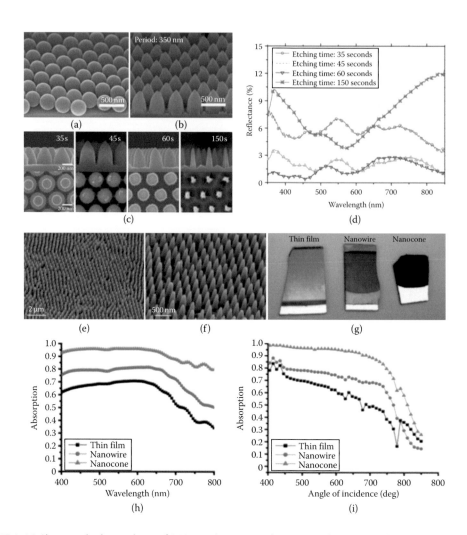

Figure 15.6 (a) Close-packed monolayer of 350-nm-diameter polystyrene spheres coated on a silicon substrate. (b) The SEM image of pyramidal structures (period: 350 nm) transferred to silicon. (c) The top-view and cross-sectional images of the textured profiles obtained after etching for 30, 60, 100, and 150 s. (d) The measured reflectance spectra of the textured silicon samples prepared using various etching durations. (Reproduced from Chen, H.L., et al., Using colloidal lithography to fabricate and optimize sub-wavelength pyramidal and honeycomb structures in solar cells, *Opt. Express*, 15 (2007), 14793–14803. With permission of Optical Society of America.) (e) SEM images in a large area of a monolayer of a-Si:H nanocone arrays. (f) Zoom-in SEM images of a-Si:H nano-cone arrays. (g) Photographs of a-Si:H thin film (left), nanowire arrays (middle), and nanocone arrays (right). (h-i) Measured value of absorption on samples with a-Si:H thin film, nanowire arrays, and nanocone arrays as top layer over (h) a large range of wavelengths at normal incidence and (i) different angles of incidence (at wavelength λ = 488 nm). (Reprinted with permission from Zhu, J., et al., Optical absorption enhancement in amorphous silicon nanowire and nanocone arrays, *Nano Lett.*, 9 (2009), 279–282. Copyright 2009 American Chemical Society.)

small flat roof appeared when the etching was performed for 35 s. Upon increasing the etching duration, the top of etched silicon became sharper and the width narrower. The height of etched silicon increased to 480 nm (60 s) then decreased to 380 nm (150 s) (Figure 15.6c). The reflectance within the wavelength regime from 350 nm to 850 nm was less than 2.5% for the sample obtained at an etching duration of 60 s (Figure 15.6d) [68].

Nanocone arrays using a wafer-scale Langmuir–Blodgett assembly and etching technique are fabricated for enhanced broadband AR in solar cell applications (Figure 15.6e–f). The Langmuir–Blodgett method was used to assemble silica nanoparticles into a close-packed monolayer on top of an a-Si:H thin film. These silica nanoparticles were then used as an etch mask during a chlorine-based RIE process. It is believed that the conical shape is due to the gradual shrinkage of the size of the silica nanoparticle. After RIE, the silica nanoparticles were so small that they were no longer observable on top of nanocones. The sample with nanocone arrays looks black, exhibiting enhanced absorption due to suppression of reflection from the front surface (Figure 15.6g). From Figure 15.6, it is obvious that, under identical conditions, the sample with nanocone arrays absorbed the most light, whereas the thin-film sample reflected the most light. The nano-cone arrays provided excellent impedance matching between a-Si:H and air through a gradual reduction in the effective RI away from the surface and, therefore, exhibited enhanced absorption due to superior AR properties over a large range of wavelengths and angles of incidence (Figure 15.6h–i) [72].

15.3.5 IDEAL GEOMETRY OF THE MOTH-EYE ARC

By the observation of corneal nipple arrays of various butterfly species and reflectance calculations of three different types of subwavelength structures (SWSs) using a thin-film multilayer model, Stavenga et al. showed that the parabola shape provides better AR properties than the cone and Gaussian-bell shapes [10]. In the EMT, the parabola shape yields a nearly linear RI gradient, which is efficient to reduce the surface reflection. To obtain SWSs with a conical profile, many fabrication methods, such as electron-beam/interfer-ence lithography [58,73–78], nanoimprint lithography [74,79,80], nanosphere or colloid formation [81–84], metal nanoparticles [85,86], and Langmuir–Blodgett assembly [72,87], have been proposed. However, it is difficult for these techniques to guarantee the formation of the parabola shape because the shape of the SWSs depends on complicated process control. However, the combination of interference lithography, thermal reflow, and subsequent pattern transfer is used to obtain the ideal shape of the moth-eye structure (Figure 15.7a). The fabricated SWSs consist of parabolic grating patterns, resulting in a linearly graded index profile. An increase in the process pressure improves the etch selectivity of PR, which leads to a taller height.

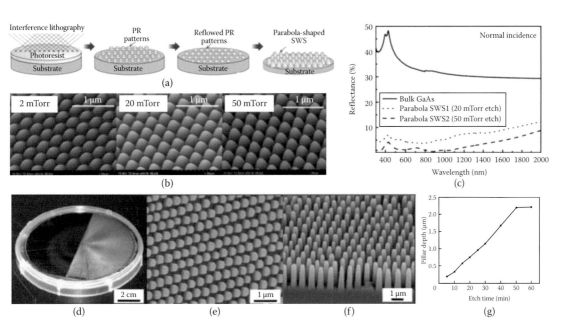

Figure 15.7 (a) Schematic illustration of the fabrication procedure of the parabola-shaped SWSs on the GaAs substrate used in this experiment. The scale bar in the SEM images is 500 nm. (b) The SEM images of the parabola-shaped SWSs fabricated at process pressures of 2, 20, and 50 mTorr. (c) The measured reflectance as a function of the wavelength for the fabricated parabola-shaped SWSs on the GaAs substrate at process pressures of 20 mTorr and 50 mTorr. The measured reflectance of bulk GaAs is shown as a reference. Templated silicon pillar arrays by using a nonclose-packed colloidal monolayer. (Reproduced from Song, Y.M., et al.: Bioinspired parabola subwavelength structures for improved broadband antireflection. *Small*. 2010. 6. 984–7. Copyright Wiley-VCH Verlag GmbH & Co. KGaA. With permission.) (d) The photograph of a 4-inch silicon wafer with the right half cov-ered by subwavelength pillars and the left half unetched. The sample is illuminated by white light. (e) The silicon pillars after 10 min RIE. (f) The silicon pillars after 50 min RIE. (g) The pillar depth dependence on the RIE dura-tion. (Reproduced from Min, W.-L., et al.: Bioinspired self-cleaning antireflection coatings. *Adv. Mater.* 2008. 20. 3914–3918. Copyright Wiley-VCH Verlag GmbH & Co. KGaA. With permission.)

The etch selectivity is varied from 1 to 3 simply by increasing the process pressure from 2 to 50 mTorr. This means that the aspect ratio, that is, the height under the same period, can be controlled easily by adjusting the process pressure during the ICP etching procedure without the use of a complex gas mixture or additional process steps. The morphology of the etched surface through the lens-like PR mask is smooth, and these grating patterns are uniform (Figure 15.7b). Two SWS samples on the GaAs substrate suppress drastically the Fresnel reflection compared to that of the flat GaAs surface (Figure 15.7c) [56]. Based on a simple and scalable spin-coating technique that enables wafer-scale production of colloidal crystals with ncp structures, colloidal lithography is also used to obtain uniform moth-eye structures with a high aspect ratio. Broadband silicon moth-eye structures have been fabricated by using a 2D ncp colloidal crystal as the etching mask during an SF_6 RIE process (Figure 15.7d–g). The nonwetting properties are also obtained to ultimately realize self-cleaning broadband ARCs on silicon substrates [88].

15.3.6 LARGE-SCALE FABRICATION OF ARCs

For high-performance broadband antireflective properties of the moth-eye structure over large areas, high-aspect ratio silicon random nanotip arrays are fabricated by using high-density electron cyclotron resonance plasma etching (Figure 15.8a–c). The geometric features of the silicon nanotip arrays were characterized by the apex diameter in the range of 3–5 nm, a base diameter of 200 nm, and lengths from 1 to 16 μm. Such AR structured surfaces can suppress light reflection in the wavelength range from UV, through the visible part of the spectrum, to the terahertz region. Reflection is suppressed in a wide range of the incidence angles and for both s- and p-polarized light. Excellent antireflective properties of such an antireflective structure are close to those of ideal antireflective surfaces (Figure 15.8d–g) [89].

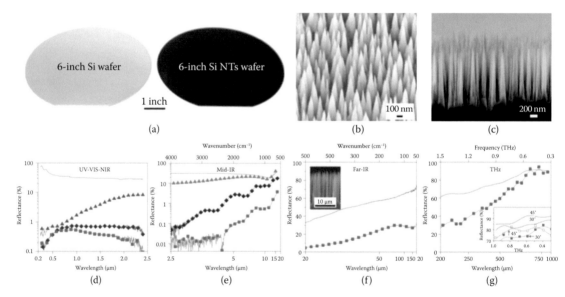

Figure 15.8 (a) Photographic images showing the 6 inch polished silicon wafer (left) and the wafer coated with Si nanotip structures (right). The SEM images showing the tilted top view (b) and the cross-sectional view (c) of Si nanotip structures with a length of 1,600 nm. (d) The hemispherical reflectance (using an integrating sphere) as a function of the wavelength for a planar Si wafer (solid line, black) and Si nanotip structures (symbols) for L = 1.6 (green), 5.5 (blue), and 16 μm (red) at the UV, visible, and near-infrared wavelengths. (e) The specular reflectance (without an integrating sphere) as a function of the wavelength in the mid-infrared region for an incidence angle of 30°. (f–g) Comparison of the specular reflectance as a function of the wavelength for a planar silicon wafer (solid line, black) and Si nanotip structures with L = 16 μm (red) in the far-infrared (f) and terahertz (g) regions for an incidence angle of 30°. The inset in (f) shows the cross-sectional SEM image of the Si nanotip structures with L = 16 μm. The inset in (g) compares the reflectance in planar silicon (solid line, black) and Si nanotip structures (symbols, red) with unpolarized light and an incidence angle of 30° (filled squares) and 45° (open squares). The solid red lines in the inset of (f) are guides to the eye. (Reprinted by permission from Macmillan Publishers Ltd: *Nature Nanotechnology* [89], copyright © 2007.)

15.3.7 COMPOUND-EYE ARC—HIERARCHICAL MICRO AND NANOSTRUCTURES

An antireflective compound-eye surface structure was generated by applying the moth-eye structure to another structure. By employing KOH etching and silver catalytic etching, pyramidal hierarchical structures were generated on the crystalline silicon wafer (Figure 15.9a–c). The hierarchical structures exhibited strong AR and superhydrophobic properties after fluorination (Figure 15.9d). Furthermore, a flexible superhydrophobic substrate was fabricated by transferring a hierarchical Si structure to NOA 63 film using UV-assisted imprint lithography. This method is of potential application in optical, optoelectronic, and wettability control devices [90].

The details of the AR technologies discussed in this chapter, including AR mechanisms, AR materials, fabrication methods, AR structure, and final reflection, are summarized in Table 15.1.

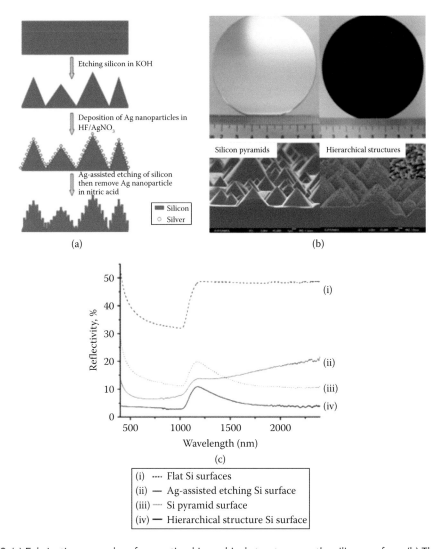

(a)   (b)

(c)

(i) ····	Flat Si surfaces
(ii) —	Ag-assisted etching Si surface
(iii) ····	Si pyramid surface
(iv) —	Hierarchical structure Si surface

Figure 15.9 (a) Fabrication procedure for creating hierarchical structures on the silicon surface. (b) The photographs of the polished silicon wafer (left) and the hierarchically structured silicon wafer (right). (c) The cross-sectional SEM photographs of silicon pyramids created using KOH etching and hierarchical structures generated by utilizing Ag-assisted etching. The inset is the magnified SEM image. (d) The hemispherical reflectance spectra of flat silicon (i), nanohole textured silicon surface (ii), pyramid textured silicon surface (iii), and hierarchically structured silicon (iv). (Reprinted with permission from Qi, D., et al., Simple approach to wafer-scale self-cleaning antireflective silicon surfaces, *Langmuir*, 25, 14 (2009), 7769–7772. Copyright 2009 American Chemical Society.)

Table 15.1 Summary of technical details of the AR technologies discussed in this chapter

TYPE	FABRICATION METHODS	MATERIALS	REFLECTION, %	REFERENCE
Quarter-wavelength	Seeding and growth method	ZnO nanostructures	~6.6	[91]
	Convective assembly	SiO_2	~11	[92]
	Spin coating	TiO_2	~8	[93]
Multilayer	Remote PECVD	Porous SiO_2/SiN	<1	[94]
	Electron cyclotron resonance PECVD	SiO_2/TiO_2	<0.5	[95]
	PECVD	SiN	<3	[96,97]
	Immersion in carbonate-based solutions followed by stain etching	Carbonate-based solutions, HNO_3/HF and SiN_x	3.16	[98]
Graded index	Oblique-angle deposition	TiO_2/SiO_2/ nanostructured SiO_2	5.9	[43]
	Electron beam lithography followed by SF6 etching	EB-positive resist	0.5	[54]
	Spin coating, etching	SiO_2	<2	[88,99]
	Reactive ion etching	–	<3	[100]
Porous ARC	Electrochemical etching	Ethanol-based solution of 33.3 wt% HF	8.93	[101]
	Interference lithography and then metal-assisted chemical etching	Si photonic crystals	~3	[102]
	Electochemical etching, HF:ethanol, (1:4 ratio)	HF:ethanol, (1:4 ratio)	~1	[103]
	Spin coating, RIE, templating	SiO_2 and polymer posts, gold nanoholes	<0.5	[104]
Moth eye	Spin coating and then anisotropic wet etching	SiO_2	~2	[82]
	Nanoimprint technique	UV-sensitive resist	~1	[105]
	Microreactor-assisted nanomaterial deposition	Solution-processed Ag nanoparticles, ZnO	3.4	[106]
	Reactive ion etching	Polystyrene nanobeads	~3.8	[107]
	UV-nanoimprint lithography	Anodized aluminum oxide	4	[108]
	Spin coating, RIE	SiO_2	<2.5	[83]

15.4 OPTICAL DEVICE APPLICATIONS

Thin-film crystalline silicon (c-Si) solar cells are one of the promising candidates for low-cost photovoltaic applications because of commercially compatible mass-production processes. However, a relatively thin absorption region tends to degrade the cell efficiency, which is the main drawback of thin-film solar cells [109]. To improve light absorption in thin-film solar cells, the Fresnel reflection at the air/silicon interface in the range of the entire solar spectrum should be minimized. For silicon solar cell applications, broadband AR is needed to cover the whole absorption range of silicon in the solar spectrum. Moreover, to achieve a high absorption efficiency for the entire day, the angle-independent AR property is required [20]. Conventional thin-film ARCs exhibit reflection reduced by their interference principle; however, this can only work in a limited wavelength range. Although multilayer ARC is commonly used for broadband AR, it has problems related to material selection, thermal mismatch, and instability of thin-film stacks [4]. As an alternate to thin-film coatings, submicrometer grating (SMG) structures with a tapered feature, originally inspired by the excellent antireflective capability of the corneal of night-active insects, have been focused on as a more practical method for ultra broadband and omnidirectional ARs (Figure 15.10a). In comparison to single-layer and double-layer ARCs, the calculated reflectance of SMG structures on the top surface of c-Si has broader AR regions (Figure 15.10b). Tapered structures with a taller height and a shorter period are desirable for broadband AR properties; however, a taller structure requires complex process steps, which increase the fabrication cost. Therefore, the optimum geometry with an appropriate height and period is needed in practical solar cell applications. SMG structures with a period of 400 nm and a height of 400 nm exhibit a higher cell efficiency in a reasonable process range (Figure 15.10c). The cell efficiency of c-Si solar cells with a flat surface drops rapidly as the incidence angle increases because of the increased reflection loss. However, in SMG-integrated solar cells, the cell efficiency is stable at an incidence angle greater than 60°, and it is degraded by only 8.3% at 70° compared to that at 0° (Figure 15.10d) [20].

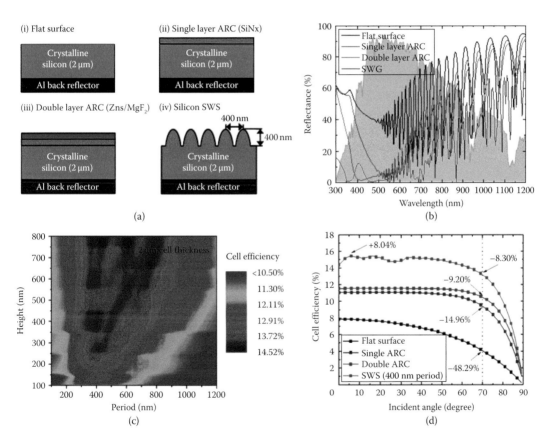

Figure 15.10 (a) Thin-film c-Si solar cells with four different surface structures: (i) flat surface, (ii) single-layer ARC, (iii) double-layer ARC, and (iv) antireflective SWSs. (b) The calculated reflectance spectra of thin-film c-Si solar cells with an Al metallic back reflector for four different surface structures. AM 1.5 solar spectrum is also shown as a reference. (c) The contour map for the cell efficiency of the SWS integrated thin-film c-Si solar cells as a function of the cone period and height. (d) The cell efficiencies of the 2-μm-thick c-Si solar cells with four different surface structures as a function of incidence angle. The period and height of the SWSs are 400 nm. (Reproduced form Song, Y.M., et al., Antireflective submicrometer gratings on thin-film silicon solar cells for light-absorption enhancement, *Opt. Lett.*, 35, 3 (2010), 276–278. With permission of Optical Society of America.)

The device exhibits a higher performance at a slightly tilted incidence angle compared to the normal incidence angle. This efficiency enhancement is attributed to the extended optical path length [110].

Silicon nanostructure arrays have the potential to increase the power conversion efficiency of photovoltaic devices. Nevertheless, so far, photovoltaic cells based on nanostructured silicon exhibit lower power conversion efficiencies than conventional cells owing to enhanced photocarrier recombination associated with the nanostructures [111]. Increased photocarrier recombination at the dramatically increased surface area of nanostructured silicon decreases the cell efficiency by reducing the device short-circuit current J_{sc} and open-circuit voltage V_{oc}. For example, in antireflective nanostructured silicon solar cells made by gold-nanoparticle catalyzed etching, increased recombination was observed, and it dramatically reduced the collection of photocarriers generated by blue and green (i.e., 350–600 nm) photons [112]. However, it was only possible to model the measured quantum efficiency of these nanostructured cells by including a dead layer extending ~500 nm beneath the front surface, in which the minority carrier lifetime was extremely low. This dead layer was roughly as thick as the nanoporous layer itself, and modeling with increased planar surface recombination alone could not be performed. It has recently been reported that the blue quantum efficiency could be increased by decreasing the nanopore depth, but this inevitably compromised the beneficial effects of the nanostructure for photon management [111,113].

Small cone-like subwavelength structures for solar cells were fabricated uniformly on the Si surface, although the periodicity was not perfect (Figure 15.11a). The period and height for the SWS were around 100 and 300 nm, respectively. Test solar cells were fabricated to demonstrate the wideangle AR effect of the SWS in solar cells. A Si SWS with a high aspect ratio was fabricated on p-type Czochralski-grown (CZ) Si wafers by dry etching using anodic porous alumina masks. After the formation of the SWS, phosphorus diffusion was carried out using a standard quartz tube furnace with $POCl_3$ at 850°C to fabricate a p–n junction. A Ag-based front grid and Al rear contact were formed by vacuum evaporation and successive firing. The finger grid pattern on the front side with a covered area of about 11% was fabricated by photolithography. The spectral reflectivity of the SWS measured using diffuse reflection optics shows a very low reflectivity over a wide range of wavelengths (Figure 15.11b). The SWS cell clearly showed higher short-circuit current density, J_{sc}, and slightly higher open-circuit voltage, V_{oc}, than those of the flat cell. These improvements are attributable to the reduction of surface reflection loss by the SWS. However, the obtained J_{sc} for the SWS cell was lower than expected from its very low reflectivity (R ~ 1%). This is mainly due to the carrier recombination at the surface and highly doped emitter region. The SWS cell shows lower internal quantum efficiency (IQE) at shorter wavelengths than the flat cell, which causes the relatively low J_{sc} (Figure 15.11c) [114]. A sub-100 nm surface-oxidized silicon nanocone forest structure is created and integrated onto the existing texturization microstructures on a photovoltaic device surface by a one-step high-throughput plasma-enhanced texturization method (Figure 15.11d). The surface of silicon can be nanotexturized with the simultaneous plasma-enhanced reactive ion etching and synthesis (SPERISE) process carried out in a plasma etcher with O_2 and HBr gas mixture. The bromine ion plays the role of etching and texturizing while the oxygen ion plays the role of nanosynthesis and oxidizing passivation. According to the EMT, the gradually varying RI of the cone forest structure can dramatically reduce the reflection and thus make the silicon surface black. The nanocone forest structure dramatically reduces the optical reflection by 70.25% (Figure 15.11e). The I–V characteristics of the solar cell measured under one sun indicates that our nanotexturization method can improve the open-circuit voltage by a little, short-circuit current by 7.09%, fill factor by 7.0%, and conversion efficiency by 14.66%. The quantum efficiency is also increased by 14.31% (Figure 15.11f). Although the reflectance of the solar cell surface is suppressed by the nanotexturization in the whole wavelength range from UV to IR, the EQE only improved in most of the visible and near-IR region. In terms of the conversion efficiency, there must be some competition between the increase by absorption enhancement and the decrease by surface recombination. That is, the absorption enhancement dominates in the visible and near-IR region, whereas the surface recombination dominates in other regions [115].

Despite great efforts to directly produce antireflective nanostructures on Si solar cells in various ways, several problems are still difficult to be considered. For so-called superstrate-type thin-film solar cells, where active cells are deposited onto transparent glass covered by a transparent conducting oxide, the

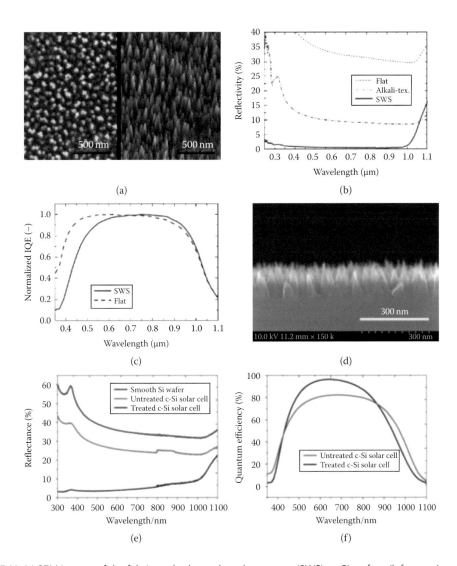

Figure 15.11 (a) SEM images of the fabricated subwavelength structure (SWS) on Si surface (left: top view, right: oblique view). (b) Measured reflectivity of the fabricated Si SWS at normal incidence. (c) Normalized IQE of the test solar cells with SWS and flat surface. (Reprinted from Sai, H., et al., Wide-angle antireflection effect of subwavelength structures for solar cells, *Jpn. J. Appl. Phys.* 46 (2007): 3333–6. Copyright 2007 The Japan Society of Applied Physics. With permission.) (d) Cross-sectional SEM image of sub-100 nm surface-oxidized silicon nanocone forest structure. (e) Diffusive reflection spectra of smooth Si wafer (blue), untreated c-Si solar cell (red), and nanotexturized c-Si solar cell (black). (f) EQE spectra of commercial solar cell before (red curve) and after (black curve) nanotexturization treatment. (Reproduced from Xu, Z., et al., Lithography-free sub-100 nm nanocone array antireflection layer for low-cost silicon solar cell, *Appl. Opt.*, 51 (2012): 4430–4435. With permission of Optical Society of America.)

reflection at the interface between air and glass should be minimized. An alternative way to apply antireflective nanostructures onto a thin-film solar cell is the fabrication of disordered submicron structures (d-SMSs) on the top glass substrate of superstrate-type thin-film hydrogenated amorphous silicon (a-Si:H) solar cells to improve the light absorption (Figure 15.12a). The d-SMSs with a tapered shape were fabricated on the back side of an SnO_2:F covered glass substrate by using plasma etching of thermally dewetted silver nanoparticles without any lithography process (Figure 15.12b). The glass substrates with the d-SMSs showed a very low reflectance compared to that of the glass substrates with the flat surface in a wide specular and angular range (Figure 15.12c). Thin-film a-Si:H solar cells were prepared on the opposite side of the d-SMS integrated glass substrates, and the devices exhibited an increased short-circuit current density (J_{sc})

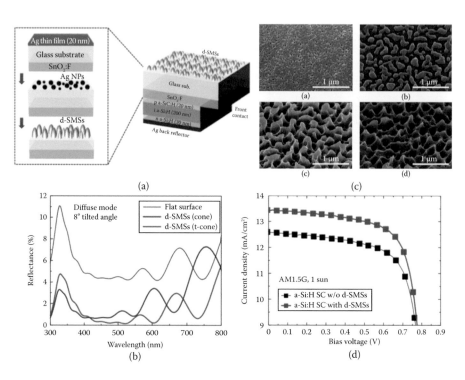

Figure 15.12 (a) Schematic illustration of thin-film amorphous silicon (a-Si) solar cells with d-SMSs for broadband AR. The inset on the left shows the fabrication procedure of the d-SMSs by using thermally dewetted silver nanoparticles. (b) The SEM images of (i) as a Ag thin film deposited on a glass substrate, (ii) Ag nanoparticles thermally dewetted at 500°C for 1 min. The fabricated d-SMSs using dry etching of Ag nanoparticles in ICP-RIE for (iii) 7 min and (iv) 9 min, respectively. (c) The measured reflectances of three different types of glass substrates (flat surface, cone shaped d-SMSs, and truncated cone-shaped d-SMSs) in the diffuse mode. (d) The J–V characteristics of thin-film a-Si:H solar cells with and without d-SMSs. (Reprinted from *Solar Energy Mater. Solar Cells*, 101, Song, Y.M., et al., Disordered submicron structures integrated on glass substrate for broadband absorption enhancement of thin-film solar cells, 73–78, Copyright 2012, with permission from Elsevier.)

by 6.84% compared to the reference cells with the flat surface without detrimental changes in the open-circuit voltages (V_{oc}) and the fill factor (Figure 15.12d) [116].

Besides silicon materials, antireflective nanostructures composed of many other materials were explored as ARCs, in spite of relatively less reports compared to those of silicon nanostructures. Antireflective nanostructures can be applied to various optoelectronic devices with AR surfaces constructed by nonsilicon antireflective nanostructures, mainly including group III–V compounds, and polymers. Group III–V semiconductors are widely used in optoelectronics, such as solar cells, LEDs, and lasers, because of their high carrier mobility and direct energy gaps. However, there are some challenges in the fabrication of broadband ARCs of most group III–V semiconductors, owing to their bandwidth disparity. Nanostructure arrays of group III–V materials with a gradient RI directly grown on the same substrates may be able to address the problems. Nanostructure arrays of group III–V semiconductors were usually fabricated through the vapor deposition growth or RIE methods, rather than solution methods owing to their intrinsic physicochemical properties. Antireflective polymer films were investigated intensively because of their advantageous characteristics compared to inorganic materials, such as easily controllable morphology and porosity, adherence to a flexible substrate, and ease of large-area processing. For preparing polymer ARCs based on nanostructure arrays, the most widely used technique is template imprinting [17]. Parabola-shaped ARNSs were fabricated using simple process steps based on the combination of laser-interference lithography, thermal reflow, and subsequent pattern transfer [56]. The use of the additional thermal-reflow process makes lens-shaped photoresist patterns, which enable pattern transfer to realize the parabola shape. Parabola-shaped ARNSs on a GaAs substrate fabricated by the lens-like shape-transfer method resulted in a linearly graded index profile. The morphology of the etched surface is smooth, and the grating patterns are uniform.

Functional materials

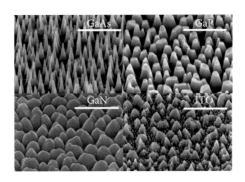

Figure 15.13 SEM images of ARNSs on GaAs, gallium phosphide, gallium nitride, and indium tin oxide substrates fabricated using dry etching of thermally dewetted silver nanoparticles. Scale bars: μm. (Reproduced from Song, Y.M., and Lee, Y.T., Antireflective nanostructures for high-efficiency optical devices, *SPIE Newsroom*, November 30, 2010. With permission of SPIE.)

The fabricated sample shows AR properties in a broader wavelength range than that of the flat surface or even conventional cone-shaped ARNSs. Another approach for ARNSs is the overall dry-etching process using thermally dewetted silver nanoparticles. In this method, the thermal dewetting process of thin metal films deposited by implementing electron-beam evaporation provides nanoscale etch-mask patterns without lithography, enabling cost-effective fabrication. Moreover, the average nanoparticle size and separation can be controlled by the film thickness at a given annealing temperature. Because the fabrication process is not limited to certain materials, ARNSs can be fabricated on various substrates (Figure 15.13) [117].

15.5 SUMMARY

The principle of AR layers/structures was discovered a few decades ago, and a number of manufacturing methods have been proposed for various materials, including silicon, germanium, III–V compound semiconductors, and transparent glasses/polymers. Nowadays, a promising application field of AR structures concerns optoelectronic devices/systems. In this review, we discussed several ARCs (i.e., single- or double-layer ARCs and gradient-index ARCs) and AR structures (i.e., tapered and nontapered porous structures and moth-eye structures). We also showed various fabrication methods of the above-mentioned layers/structures. In addition, optoelectronic device applications and design guidelines of ARCs for specific devices were discussed. We believe that our review will contribute to giving insights not only in developing simpler fabrication methods of ARCs on silicon or other semiconductor/dielectric materials but also in clear understanding of the reflection behavior of optoelectronic devices and the optimal geometry of ARCs.

REFERENCES

1. Karen Forberich, Gilles Dennler, Markus C. Scharber, Kurt Hingerl, Thomas Fromherz, and Christoph J. Brabec, Performance Improvement of Organic Solar Cells with Moth Eye Anti-Reflection Coating, *Thin Solid Films* 516, 20 (2008): 7167–70, doi: 10.1016/j.tsf.2007.12.088.
2. Jia Zhu, Ching-Mei Hsu, Zongfu Yu, Shanhui Fan, and Yi Cui, Nanodome Solar Cells with Efficient Light Management and Self-Cleaning, *Nano Lett.* 10, 6 (2010): 1979–84, doi: 10.1021/nl9034237.
3. Jeri'Ann Hiller, Jonas D. Mendelsohn, and Michael F. Rubner, Reversibly Erasable Nanoporous Anti-Reflection Coatings from Polyelectrolyte Multilayers, *Nat. Mater.* 1 (2002): 59–63, doi: 10.1038/nmat719.
4. Philippe Lalanne and G. Michael Morris, Design, Fabrication, and Characterization of Subwavelength Periodic Structures for Semiconductor Antireflection Coating in the Visible Domain, *Proc. SPIE* 2776 (1996): 300–8, doi: 10.1117/12.246835.
5. Hyun Myung Kim, Sang Hyeok Kim, Gil Ju Lee, Kyujung Kim, and Young Min Song, Parametric Studies on Artificial Morpho Butterfly Wing Scales for Optical Device Applications, *J. Nanomater.* 2015, 451834 (2015): 7, doi:10.1155/2015/451834.
6. Rick C. Schroden, Mohammed Al-Daous, Christopher F. Blanford, and Andreas Stein, Optical Properties of Inverse Opal Photonic Crystals, *Chem. Mater.* 14, 8 (2002): 3305–15, doi: 10.1021/cm020100z.

7. Inhwa Jung, Jianliang Xiao, Viktor Malyarchuk, Chaofeng Lu, Ming Li, Zhuangjian Liu, and Jongseung Yoon, Dynamically Tunable Hemispherical Electronic Eye Camera System with Adjustable Zoom Capability, *Proc. Natl. Acad. Sci. U S A.* 108, 5 (2011): 1788–93, doi: 10.1073/pnas.1015440108.

8. C. G. Bernhard and William H. Miller, A Corneal Nipple Pattern in Insect Compound Eyes, *Acta Physiol. Scand.* 56 (1962): 385–6, doi: 10.1111/j.1748-1716.1962.tb02515.x.

9. Doo-Hyun Ko, John R. Tumbleston, Kevin J. Henderson, Larken E. Euliss, Joseph M. DeSimone, Rene Lopez, and Edward T. Samulski, Biomimetic Microlens Array with Antireflective "Moth-Eye" Surface, *Soft Matter* 7 (2011): 6404, doi: 10.1039/c1sm05302g.

10. Doekele Stavenga, S. Foletti, G. Palasantzas, and Kentaro Arikawa, Light on the Moth-Eye Corneal Nipple Array of Butterflies, *Proc. R. Soc. B* 273 (2006): 661–7, doi: 10.1098/rspb.2005.3369.

11. S. J. Wilson and M. C. Hutley, The Optical Properties of 'Mote Eye' Antireflection Surfaces, *Optica Acta* 29, 7 (1982): 993–1009, doi: 10.1080/713821334.

12. G. Palasantzas, J. T. M. de Hosson, K. F. L. Michielsenand, and D. O. Stavenga, Optical Properties and Wettability of Nanostructured Biomaterials: Moth-Eyes, Lotus Leaves, and Insect Wings, in *Handbook of Nanostructured Biomaterials and Their Applications in Biotechnology, Vol. 1: Biomaterials*, ed. Hari Singh Nalwa (California: American Scientific Publishers, 2005), 273–301.

13. Akihiro Yoshida, Mayumi Motoyama, Akinori Kosaku, and Kiyoshi Miyamoto, Antireflective Nanoprotuberance Array in the Transparent Wing of a Hawkmoth, *Cephonodes Hylas, Zool. Sci.* 14 (1997): 737–41, doi: 10.2108/zsj.14.737.

14. Taolei Sun, Lin Feng, Xuefeng Gao, and Lei Jiang, Bioinspired Surfaces with Special Wettability, *Acc. Chem. Res.* 38, 8 (2005): 644–52, doi: 10.1021/ar040224c.

15. Hongbo Xu, Nan Lu, Gang Shi, Dianpeng Qi, Bingjie Yang, Haibo Li, Weiqing Xu, and Lifeng Chi, Biomimetic Antireflective Hierarchical Arrays, *Langmuir* 27, 8 (2011): 4963–7, doi: 10.1021/la1040739.

16. Surojit Chattopadhyay, Yi-Fan Huang, Y. J. Jen, Abhijit Ganguly, K. H. Chen, and L. C. Chen, Anti-Reflecting and Photonic Nanostructures, *Materials Science and Engineering R* 69 (2010): 1–35, doi: 10.1016/j.mser.2010.04.001.

17. Jinguang Cai and Limin Qi, Recent Advances in Antireflective Surfaces Based on Nanostructure Arrays, *Mater. Horiz.* 2 (2015): 37–53, doi: 10.1039/C4MH00140K.

18. Hemant Kumar Raut, V. Anand Ganesh, A. Sreekumaran Nair, and Seeram Ramakrishna, Anti-Reflective Coatings: A Critical, in-Depth Review, *Energ. Environ. Sci.* 4 (2011): 3779–804, doi: 10.1039/c1ee01297e.

19. Eugene Hecht, *Optics*, 4th Edition (San Francisco, CA: Addison-Wesley, 2002), 3.

20. Young Min Song, Jae Su Yu, and Yong Tak Lee, Antireflective Submicrometer Gratings on Thin-Film Silicon Solar Cells for Light-Absorption Enhancement, *Opt. Lett.* 35, 3 (2010): 276–8, doi: 10.1364/OL.35.000276.

21. Won Il Nam, Young Jin Yoo, and Young Min Song, Geometrical Shape Design of Nanophotonic Surfaces for Thin Film Solar Cells, *Opt. Express* 24, 14 (2016): A1033, doi: 10.1364/OE.24.0A1033.

22. Young Min Song, Hee Ju Choi, Jae Su Yu, and Yong Tak Lee, Design of Highly Transparent Glasses with Broadband Antireflective Subwavelength Structures, *Opt. Express* 18, 12 (2010): 73–8, doi: 10.1364/OE.18.013063.

23. Young Jin Yoo, Ki Soo Chang, Suck Won Hong, and Young Min Song. Design of ZnS Antireflective Microstructures for Mid- and Far-Infrared Applications, *Opt. Quant. Electron.* 47 (2015): 1503–8, doi: 10.1007/s11082-015-0143-0.

24. J.-Q. Xi, Martin F. Schubert, Jong Kyu Kim, Ef Schubert, Minfeng Chen, S. H. Lin, Wen Liu, and Joseph A. Smart, Optical Thin-Film Materials with Low Refractive Index for Broadband Elimination of Fresnel Reflection, *Nature Photonics* 1 (2007): 176–9, doi: 10.1038/nphoton.2007.26.

25. J. C. Maxwell Garnett, Colours in Metal Glasses, in Metallic Films, and in Metallic Solutions. II, *Phil. Trans. R. Soc. Lond. A* 205 (1906): 237–88, doi: 10.1098/rsta.1906.0007.

26. J. C. Maxwell Garnett, Colours in Metal Glasses, in Metallic Films, and in Metallic Solutions. II, *Phil. Trans. R. Soc. Lond. A* 203 (1904): 385–420, doi: 10.1098/rsta.1904.0024.

27. Dirk Anton George Bruggeman, Berechnung Verschiedener Physikalischer Konstanten von Heterogenen Substanzen. I. Dielektrizitätskonstanten Und Leitfähigkeiten Der Mischkörper Aus Isotropen Substanzen, *Annalen Der Physik* 416 (1935): 636–64, doi:10.1002/andp.19354160802.

28. M. G. Moharam and T. K. Gaylord, Rigorous Coupled-Wave Analysis of Planar-Grating Diffraction, *J. Opt. Soc. Am.* 71, 7 (1981): 811–18, doi: 10.1364/JOSA.71.000811.

29. Jerzy Adam Dobrowolski, Daniel Poitras, Penghui Ma, Himanshu Vakil, and Michael Acree, Toward Perfect Antireflection Coatings: Numerical Investigation, *Appl. Opt.* 41 (2002): 3075–83, doi: 10.1364/AO.41.003075.

30. Jerzy Adam Dobrowolski, Optical Properties of Films and Coatings, in *Handbook of Optics: Fundamentals, Techniques, and Design* (New York: McGraw-Hill, 1994, pp. 41).

Functional materials

31. Khalid Askar, Blayne M. Phillips, Yin Fang, Baeck Choi, Numan Gozubenli, Peng Jiang, and Bin Jiang, Self-Assembled Self-Cleaning Broadband Anti-Reflection Coatings, *Colloids and Surfaces A: Physicochem. Eng. Aspects* 439 (2013): 84–100, doi: 10.1016/j.colsurfa.2013.03.004.

32. Kevin T. Cook, Kwadwo E. Tettey, Robert M. Bunch, Daeyeon Lee, and Adam J. Nolte, One-Step Index-Tunable Antireflection Coatings from Aggregated Silica Nanoparticles, *ACS App. Mater. Interfaces* 4 (2012): 6426–31, doi: 10.1021/am3020586.

33. William H. Southwell, Gradient-Index Antireflection Coatings, *Opt. Lett.* 8, 11 (1983): 584–6, doi: 10.1364/OL.8.000584.

34. Grant R. Fowles, *Introduction to Modern Optics,* 2nd Edition (New York: Dover Publications, 1975).

35. Jonathan Moghal, Johannes Kobler, Jürgen Sauer, James Best, Martin Gardener, Andrew A. R. Watt, and Gareth Wakefield, High-Performance, Single-Layer Antireflective Optical Coatings Comprising Mesoporous Silica Nanoparticles, *ACS App. Mater. Interfaces* 4, 2 (2012): 854–9, doi: 10.1021/am201494m.

36. Mehdi Keshavarz Hedayati and Mady Elbahri, Antireflective Coatings: Conventional Stacking Layers and Ultrathin Plasmonic Metasurfaces, A Mini-Review, *Materials* 9, 6 (2016): 497, doi: 10.3390/ma9060497.

37. Jinghua Sun, Xinmin Cui, Ce Zhang, Cong Zhang, Ruimin Ding, and Yao Xu, A Broadband Antireflective Coating Based on A Double-Layer System Containing Mesoporous Silica and Nanoporous Silica, *J. Mater. Chem. C* 3 (2015): 7187–94, doi: 10.1039/C5TC00986C.

38. Mikhail A. Kats, Romain Blanchard, Patrice Genevet, and Federico Capasso, Nanometre Optical Coatings Based on Strong Interference Effects in Highly Absorbing Media, *Nature Materials* 12, 1 (2012): 20–4, doi: 10.1038/nmat3443.

39. Chan Il Yeo, Hee Ju Choi, Young Min Song, Seok Jin Kang, and Yong Tak Lee, A Single-Material Graded Refractive Index Layer for Improving the Efficiency of III–V Triple-Junction Solar Cells, *J. Mater. Chem. A* 3 (2015): 7235–40, doi: 10.1039/C4TA06111J.

40. Sung Jun Jang, Young Min Song, Jae Su Yu, Chan Il Yeo, and Yong Tak Lee, Antireflective Properties of Porous Si Nanocolumnar Structures with Graded Refractive Index Layers, *Opt. Lett.* 36, 2 (2011): 253–5, doi: 10.1364/OL.36.000253.

41. Matthew M. Hawkeye and Michael J. Brett, Glancing Angle Deposition: Fabrication, Properties, and Applications of Micro- and Nanostructured Thin Films, *J. Vac. Sci. Technol. A* 25 (2007): 1317–35, doi: 10.1116/1.2764082.

42. Scott R. Kennedy and Michael J. Brett, Porous Broadband Antireflection Coating by Glancing Angle Deposition, *App. Opt.* 42, 22 (2003): 4573–9, doi: 10.1364/AO.42.004573.

43. Sameer Chhajed, Martin F. Schubert, Jong Kyu Kim, and E. Fred Schubert, Nanostructured Multilayer Graded-Index Antireflection Coating for Si Solar Cells with Broadband and Omnidirectional Characteristics, *Appl. Phys. Lett.* 93 (2008): 251108, doi: 10.1063/1.3050463.

44. Xing Yan, David J. Poxson, Jaehee Cho, Roger E. Welser, Ashok K. Sood, Jong Kyu Kim, and E. Fred Schubert, Enhanced Omnidirectional Photovoltaic Performance of Solar Cells Using Multiple-Discrete-Layer Tailored- and Low-Refractive Index Anti-Reflection Coatings, *Adv. Funct. Mater.* 23 (2013): 583–90, doi: 10.1002/adfm.201201032.

45. Sung Jun Jang, Young Min Song, Chan Il Yeo, Chang Young Park, and Jae Su Yu, Antireflective Property of Thin Film a-Si Solar Cell Structures with Graded Refractive Index Structure, *Opt. Express* 19 (2011): 108–17, doi: 10.1016/j.surfcoat.2010.08.131.

46. Volker Schmidt, Stephan Senz, and Ulrich Gösele, Diameter-Dependent Growth Direction of Epitaxial Silicon Nanowires, *Nano Lett.* 5 (2005): 931–5, doi: 10.1021/nl050462g.

47. Zhipeng Huang, Nadine Geyer, Peter Werner, Johannes De Boor, and Ulrich Gösele, Metal-Assisted Chemical Etching of Silicon: A Review, *Adv. Mater.* 23 (2011): 285–308, doi: 10.1002/adma.201001784.

48. Xuewen Geng, Meicheng Li, Liancheng Zhao, and Paul W. Bohn, Metal-Assisted Chemical Etching Using Tollen's Reagent to Deposit Silver Nanoparticle Catalysts for Fabrication of Quasi-Ordered Silicon Micro/Nanostructures, *J. Electron. Mater.* 40 (2011): 2480–5, doi: 10.1007/s11664-011-1771-1.

49. Jin-Young Jung, Zhongyi Guo, Sang-Won Jee, Han-Don Um, Kwang-Tae Park, Moon Seop Hyun, Jun Mo Yang, and Jung-Ho Lee, A Waferscale Si Wire Solar Cell Using Radial and Bulk p–n Junctions, *Nanotechnology* 21 (2010): 445303, doi: 10.1088/0957-4484/21/44/445303.

50. Jyh-Yuan Chen, Wui-Lee Chang, Chao-Kai Huang, and Kien Wen Sun, Biomimetic Nanostructured Antireflection Coating and Its Application on Crystalline Silicon Solar Cells, *Opt. Express* 19 (2011): 14411–19, doi: 10.1364/OE.19.014411.

51. Taiji Zhang, Yurong Ma, and Limin Qi, Bioinspired Colloidal Materials with Special Optical, Mechanical, and Cell-Mimetic Functions, *J. Mater. Chem. B* 1 (2013): 251–64, doi: 10.1039/c2tb00175f.

52. Yunfeng Li, Junhu Zhang, and Bai Yang, Antireflective Surfaces Based on Biomimetic Nanopillared Arrays, *Nano Today* 5 (2010): 117–27, doi: 10.1016/j.nantod.2010.03.001.

53. Hiroshi Toyota, Koji Takahara, Masato Okano, Tsutom Yotsuya, and Hisao Kikuta, Fabrication of Microcone Array for Antireflection Structured Surface Using Metal Dotted Pattern, *Jpn. J. Appl. Phys.* 40 (2001): L747–9, doi: 10.1143/JJAP.40.L747.

54. Yoshiaki Kanamori, Minoru Sasaki, and Kazuhiro Hane, Broadband Antireflection Gratings Fabricated upon Silicon Substrates, *Opt. Lett.* 24 (1999): 1422–4, doi: 10.1364/OL.24.001422.

55. Keiichiro Watanabe, Takayuki Hoshino, Kazuhiro Kanda, Yuichi Haruyama, and Shinji Matsui, Brilliant Blue Observation from a Morpho-Butterfly-Scale Quasi-Structure, *Jpn. J. Appl. Phys.* 44 (2005): L48–50, doi: 10.1143/JJAP.44.L48.

56. Young Min Song, Sung Jun Jang, Jae Su Yu, and Yong Tak Lee, Bioinspired Parabola Subwavelength Structures for Improved Broadband Antireflection, *Small* 6, 9 (2010): 984–7, doi: 10.1002/smll.201000079.

57. P. B. Clapham and M. C. Hutley, Reduction of Lens Reflexion by the "Moth Eye" Principle, *Nature* 244 (1974): 281–2, doi: 10.1038/244281a0.

58. K. Hadobás, S. Kirsch, A. Carl, M. Acet, and E. F. Wassermann, Reflection Properties of Nanostructure-Arrayed Silicon Surfaces, *Nanotechnology* 11 (2000): 161–4.

59. Zhaoning Yu, He Gao, Wei Wu, Haixiong Ge, and Stephen Y. Chou, Fabrication of Large Area Subwavelength Antireflection Structures on Si Using Trilayer Resist Nanoimprint Lithography and Liftoff, *J. Vac. Sci. Technol. B* 21 (2003): 2874–7, doi: 10.1116/1.1619958.

60. William H. Southwell, Pyramid-Array Surface-Relief Structures Producing Antireflection Index Matching on Optical Surfaces, *J. Opt. Soc. Am. A* 8 (1991): 549–53, doi: 10.1364/JOSAA.8.000549.

61. Matthias Geissler and Younan Xia, Patterning: Principles and Some New Developments, *Adv. Mater.* 16 (2004): 1249–69, doi: 10.1002/adma.200400835.

62. L. Jay Guo, Nanoimprint Lithography: Methods and Material Requirements, *Adv. Mater.* 19 (2007): 495–513, doi: 10.1002/adma.200600882.

63. Jun Hyuk Moon, Se Gyu Jang, Jong Min Lim, and Seung Man Yang, Multiscale Nanopatterns Templated from Two-Dimensional Assemblies of Photoresist Particles, *Adv. Mater.* 17 (2005): 2559–62, doi: 10.1002/adma.200501167.

64. Seung Man Yang, Se Gyu Jang, Dae Geun Choi, Sarah Kim, and Hyung Kyun Yu, Nanomachining by Colloidal Lithography, *Small* 2 (2006): 458–75, doi: 10.1002/smll.200500390.

65. Chih Hung Sun, Peng Jiang, and Bin Jiang, Broadband Moth-Eye Antireflection Coatings on Silicon, *Appl. Phys. Lett.* 92 (2008): 06112, doi: 10.1063/1.2870080.

66. Wei Lun Min, Amaury P. Betancourt, Peng Jiang, and Bin Jiang, Bioinspired Broadband Antireflection Coatings on GaSb, *Appl. Phys. Lett.* 92 (2008): 141109, doi: 10.1063/1.2908221.

67. Tsutomu Nakanishi, Toshiro Hiraoka, Akira Fujimoto, Satoshi Saito, and Koji Asakawa, Nano-Patterning Using an Embedded Particle Monolayer as an Etch Mask, *Microelectronic Engineering* 83 (2006): 1503–8, doi: 10.1016/j.mee.2006.01.193.

68. H. L. Chen, S. Y. Chuang, Chun-Hung Lin, and Y. H. Lin, Using Colloidal Lithography to Fabricate and Optimize Sub-Wavelength Pyramidal and Honeycomb Structures in Solar Cells, *Opt. Express* 15 (2007): 14793–803, doi: 10.1364/OE.15.014793.

69. Yunfeng Li, Junhu Zhang, Shoujun Zhu, and Bai Yang, Biomimetic Surfaces for High-Performance Optics, *Adv. Mater.* 21 (2009): 4731–4, doi: 10.1002/adma.200901335.

70. Junhu Zhang, Yunfeng Li, Xuemin Zhang, and Bai Yang, Colloidal Self-Assembly Meets Nanofabrication: From Two-Dimensional Colloidal Crystals to Nanostructure Arrays, *Adv. Mater.* 22 (2010): 4249–69, doi: 10.1002/adma.201000755.

71. Stephen J. Pearton and David P. Norton, Dry Etching of Electronic Oxides, Polymers, and Semiconductors, *Plasma Process. Polym.* 2 (2005): 16–37, doi: 10.1002/ppap.200400035.

72. Jia Zhu, Zongfu Yu, George F. Burkhard, Ching-Mei Hsu, Stephen T. Connor, Yueqin Xu, Qi Wang, Michael McGehee, Shanhui Fan, and Yi Cui, Optical Absorption Enhancement in Amorphous Silicon Nanowire and Nanocone Arrays, *Nano Lett.* 9 (2009): 279–82, doi: 10.1021/nl802886y.

73. Stuart Boden and D. M. Bagnall, Tunable Reflection Minima of Nanostructured Antireflective Surfaces, *Appl. Phys. Lett.* 93 (2008): 133108, doi: 10.1063/1.2993231.

74. Xudi Wang, Yanlin Liao, Bin Liu, Liangjin Ge, Guanghua Li, Shaojun Fu, Yifang Chen, and Zheng Cui, Free-Standing SU-8 Subwavelength Gratings Fabricated by UV Curing Imprint, *Microelectronic Engineering* 85 (2008): 910–13, doi: 10.1016/j.mee.2007.12.060.

75. Yoshiaki Kanamori, M. Ishimori, and Kazuhiro Hane, High Efficient Light-Emitting Diodes with Antireflection Subwavelength Gratings, *IEEE Photonics Technology Letters* 14 (2002): 1064–6, doi: 10.1109/LPT.2002.1021970.

76. Yoshiaki Kanamori, Hisao Kikuta, and Kazuhiro Hane, Broadband Antireflection Gratings for Glass Substrates Fabricated by Fast Atom Beam Etching, *Jpn. J. Appl. Phys.* 39 (2000): L735–7.

77. Philippe Lalanne and G. Michael Morris, Antireflection Behavior of Silicon Subwavelength Periodic Structures for Visible Light, *Nanotechnology* 8 (1999): 53–6, doi: 10.1088/0957-4484/8/2/002.

78. Young Min Song, Si Young Bae, Jae Su Yu, and Yong Tak Lee, Closely Packed and Aspect-Ratio-Controlled Antireflection Subwavelength Gratings on GaAs Using a Lenslike Shape Transfer, *Opt. Lett.* 34 (2009): 1702–4, doi: 10.1364/OL.34.001702.

79. Guoyong Xie, Guoming Zhang, Feng Lin, Jin Zhang, Zhongfan Liu, and Shichen Mu, The Fabrication of Subwavelength Anti-Reflective Nanostructures Using a Bio-Template, *Nanotechnology* 19 (2008): 095605, doi: 10.1088/0957-4484/19/9/095605.

80. Guoming Zhang, Jin Zhang, Guoyong Xie, Zhongfan Liu, and Huibo Shao, Cicada Wings: A Stamp from Nature for Nanoimprint Lithography, *Small* 2 (2006): 1440–3, doi: 10.1002/smll.200600255.

81. Chih-Hung Sun, Brian J. Ho, Bin Jiang, and Peng Jiang, Biomimetic Subwavelength Antireflective Gratings on GaAs, *Opt. Lett.* 33 (2008): 2224–6, doi: 10.1364/OL.33.002224.

82. Chih Hung Sun, Wei Lun Min, Nicholas C. Linn, Peng Jiang, and Bin Jiang, Templated Fabrication of Large Area Subwavelength Antireflection Gratings on Silicon, *Appl. Phys. Lett.* 91 (2007): 231105, doi: 10.1063/1.2821833.

83. Wei-Lun Min, Peng Jiang, and Bin Jiang, Large-Scale Assembly of Colloidal Nanoparticles and Fabrication of Periodic Subwavelength Structures, *Nanotechnology* 19 (2008): 475604, doi: 10.1088/0957-4484/19/47/475604.

84. Hongbo Xu, Nan Lu, Dianpeng Qi, Juanyuan Hao, Liguo Gao, Bo Zhang, and Lifeng Chi, Biomimetic Antireflective Si Nanopillar Arrays, *Small* 4 (2008): 1972–5, doi: 10.1002/smll.200800282.

85. Sen Wang, Xiao Zheng Yu, and Hong Tao Fan, Simple Lithographic Approach for Subwavelength Structure Antireflection, *Appl. Phys. Lett.* 91 (2007): 061105, doi: 10.1063/1.2767990.

86. Yasuhiko Kojima and Takahisa Kato, Nanoparticle Formation in Au Thin Films by Electron-Beam-Induced Dewetting, *Nanotechnology* 19 (2008): 255605, doi: 10.1088/0957-4484/19/25/255605.

87. Ching-Mei Hsu, Stephen T. Connor, Mary X. Tang, and Yi Cui, Wafer-Scale Silicon Nanopillars and Nanocones by Langmuir–Blodgett Assembly and Etching, *Appl. Phys. Lett.* 93 (2008): 133109, doi: 10.1063/1.2988893.

88. Wei-Lun Min, Bin Jiang, and Peng Jiang, Bioinspired Self-Cleaning Antireflection Coatings, *Adv. Mater.* 20 (2008): 3914–18, doi: 10.1002/adma.200800791.

89. Yi-Fan Huang, Surojit Chattopadhyay, Yi-Jun Jen, Cheng-Yu Peng, Tze-An Liu, Yu-Kuei Hsu, Ci-Ling Pan, et al., Improved Broadband and Quasi-Omnidirectional Anti-Reflection Properties with Biomimetic Silicon Nanostructures, *Nat. Nanotechnol.* 2, 12 (2007): 770–4, doi: 10.1038/nnano.2007.389.

90. Dianpeng Qi, Nan Lu, Hongbo Xu, Bingjie Yang, Chunyu Huang, Miaojun Xu, Liguo Gao, Zhouxiang Wang, and Lifeng Chi, Simple Approach to Wafer-Scale Self-Cleaning Antireflective Silicon Surfaces, *Langmuir* 25, 14 (2009): 7769–72, doi: 10.1021/la9013009.

91. Yun-Ju Lee, Douglas S. Ruby, David W. Peters, Bonnie B. McKenzie, and Julia W. P. Hsu, ZnO Nanostructures as Efficient Antireflection Layers in Solar Cells, *Nano Lett.* 8, 5 (2008): 1501–5, doi: 10.1021/nl080659j.

92. Brian G. Prevo, Emily W. Hon, and Orlin D. Velev, Assembly and Characterization of Colloid-Based Antireflective Coatings on Multicrystalline Silicon Solar Cells, *J. Mater. Chem.* 17 (2007): 791–9. doi: 10.1039/B612734G.

93. Srinivasan Venkatesh, Peng Jiang, and Bin Jiang, Generalized Fabrication of Two-Dimensional Non-Close-Packed Colloidal Crystals, *Langmuir* 23 (2007): 8231–5, doi: 10.1021/la7006809.

94. Henning Nagel, Armin G. Aberle, and Rudolf Hezel, Optimised Antireflection Coatings for Planar Silicon Solar Cells Using Remote PECVD Silicon Nitride and Porous Silicon Dioxide, *Prog. Photovolt. Res. Appl.* 7 (1999): 245–60, doi: 10.1002/(SICI)1099-159X(199907/08)7:4<245::AID-PIP255>3.0.CO;2-3

95. Charles Martinet, Victor Paillard, Aline Gagnaire, and J. Joseph, Deposition of SiO_2 and TiO_2 Thin Films by Plasma Enhanced Chemical Vapor Deposition for Antireflection Coating, *Journal of Non-Crystalline Solids* 216 (1997): 77–82, doi: 10.1016/S0022-3093(97)00175-0.

96. Parag Doshi, Gerald E. Jellison, Jr., and Ajeet Rohatgi, Characterization and Optimization of Absorbing Plasma-Enhanced Chemical Vapor Deposited Antireflection Coatings for Silicon Photovoltaics, *Appl. Opt.* 36, 30 (1997): 7826–37, doi: 10.1364/AO.36.007826.

97. C. C. Johnson, T. Wydeven, and K. Donohoe, Plasma-Enhanced CVD Silicon Nitride Antireflection Coatings for Solar Cells, *Sol. Energ.* 31 (1983): 355–8, doi: 10.1016/0038-092X(83)90133-0.

98. N. Marrero, R. Guerrero-lemus, B. González-díaz, and D. Borchert, Effect of Porous Silicon Stain Etched on Large Area Alkaline Textured Crystalline Silicon Solar Cells, *Thin Solid Films* 517 (2009): 2648–50, doi: 10.1016/j.tsf.2008.09.070.

99. Chih-Hung Sun, Wei-Lun Min, Nicholas C. Linn, Peng Jiang, and Bin Jiang, Large-Scale Assembly of Periodic Nanostructures with Metastable Square Lattices, *J. Vac. Sci. Technol. B* 27 (2009): 1043–7, doi: 10.1116/1.3117347.

100. Saleem H. Zaidi, Douglas S. Ruby, and James M. Gee, Characterization of Random Reactive Ion Etched-Textured Silicon Solar Cells, *IEEE Transactions on Electron Devices* 48 (2001): 1200–6, doi: 10.1109/16.925248.

101. Bum Sung Kim, Don Hee Lee, Sun Hee Kim, Guk Hwan An, Kun Jae Lee, Nosang V. Myung, and Yong Ho Choa, Silicon Solar Cell with Nanoporous Structure Formed on a Textured Surface, *J. Am. Ceram. Soc.* 92 (2009): 2415–17, doi: 10.1111/j.1551-2916.2009.03210.x.

102. Mohammad Malekmohammad, Mohammad Soltanolkotabi, Reza Asadi, M. H. Naderi, Alireza Erfanian, Mohammad Zahedinejad, Shahin Bagheri, and Mahdi Khaje, Combining Micro- and Nano-Texture to Fabricate an Antireflective Layer, *J. Micro/Nanolith. MEMS MOEMS* 11 (2012): 013011, doi: 10.1117/1.JMM.11.1.013011.

103. Asmiet Ramizy, Z. Hassan, Khalid Omar, Y. Al-Douri, and M. A. Mahdi, New Optical Features to Enhance Solar Cell Performance Based on Porous Silicon Surfaces, *Appl. Surf. Sci.* 257 (2011): 6112–17, doi: 10.1016/j.apsusc.2011.02.013.

104. Wei Han Huang, Chih Hung Sun, Wei Lun Min, Peng Jiang, and Bin Jiang, Templated Fabrication of Periodic Binary Nanostructures, *J. Phys. Chem. C* 112 (2008): 17586–91, doi: 10.1021/jp807290u.

105. Qin Chen, G. Hubbard, Philip Shields, Chaowang Liu, D. W. E. Allsopp, Wang Nang Wang, and S. Abbott, Broadband Moth-Eye Antireflection Coatings Fabricated by Low-Cost Nanoimprinting, *Appl. Phys. Lett.* 94 (2009), doi: 10.1063/1.3171930.

106. Seung-Yeol Han, Brian K. Paul, and Chih-Hung Chang, Nanostructured ZnO as Biomimetic Anti-Reflective Coatings on Textured Silicon Using a Continuous Solution Process, *J. Mater. Chem.* 22 (2012): 22906–12, doi: 10.1039/C2JM33462C.

107. Haesung Park, Dongheok Shin, Gumin Kang, Seunghwa Baek, Kyoungsik Kim, and Willie J. Padilla, Broadband Optical Antireflection Enhancement by Integrating Antireflective Nanoislands with Silicon Nanoconical-Frustum Arrays, *Adv. Mater.* 23 (2011): 5796–800, doi: 10.1002/adma.201103399.

108. Fang Jiao, Qiyu Huang, Wangchun Ren, Wei Zhou, Fangyi Qi, Yizhou Zheng, and Jing Xie, Enhanced Performance for Solar Cells with Moth-Eye Structure Fabricated by UV Nanoimprint Lithography, *Microelectronic Engineering* 103 (2013): 126–30, doi: 10.1016/j.mee.2012.10.012.

109. Martin A. Green, Keith Emery, Yoshihiro Hishikawa, and Wilhelm Warta, Solar Cell Efficiency Tables (Version 31), *Prog. Photovolt: Res. Appl.* 16 (2008): 61–7, doi: 10.1002/pip.808.

110. Ning-Ning Feng, Jurgen Michel, Lirong Zeng, Jifeng Liu, Ching-Yin Hong, Lionel C. Kimerling, and Xiaoman Duan, Design of Highly Efficient Light-Trapping Structures for Thin-Film Crystalline Silicon Solar Cells, *IEEE Trans. Electron Devices* 54 (2007): 1926–33, doi: 10.1109/TED.2007.900976.

111. Jihun Oh, Hao-Chih Yuan, and Howard M. Branz, An 18.2%-Efficient Black-Silicon Solar Cell Achieved through Control of Carrier Recombination in Nanostructures, *Nat. Nanotechnol.* 7 (2012): 743–8, doi: 10.1038/nnano.2012.166.

112. Hao Chih Yuan, Vernon E. Yost, Matthew R. Page, Paul Stradins, Daniel L. Meier, and Howard M. Branzm, Efficient Black Silicon Solar Cell with a Density-Graded Nanoporous Surface: Optical Properties, Performance Limitations, and Design Rules, *Appl. Phys. Lett.* 95 (2009): 123501, doi: 10.1063/1.3231438.

113. Fatima Toor, Howard M. Branz, Matthew R. Page, Kim M. Jones, and Hao Chih Yuan, Multi-Scale Surface Texture to Improve Blue Response of Nanoporous Black Silicon Solar Cells, *Appl. Phys. Lett.* 99 (2011): 103501, doi: 10.1063/1.3636105.

114. Hitoshi Sai, Homare Fujii, Koji Arafune, Yoshio Ohshita, Yoshiaki Kanamori, Hiroo Yugami, and Masafumi Yamaguchi, Wide-Angle Antireflection Effect of Subwavelength Structures for Solar Cells, *Jpn. J. Appl. Phys.* 46 (2007): 3333–6, doi: 10.1143/JJAP.46.3333.

115. Zhida Xu, Jing Jiang, and Gang Logan Liu, Lithography-Free Sub-100 nm Nanocone Array Antireflection Layer for Low-Cost Silicon Solar Cell, *Appl. Opt.* 51 (2012): 4430–5, doi: 10.1364/AO.51.004430.

116. Young Min Song, Ji Hoon Jang, Jeong Chul Lee, Eun Kyu Kang, and Yong Tak Lee, Disordered Submicron Structures Integrated on Glass Substrate for Broadband Absorption Enhancement of Thin-Film Solar Cells, *Solar Energy Materials and Solar Cells* 101 (2012): 73–8, doi: 10.1016/j.solmat.2012.02.013.

117. Young Min Song and Yong Tak Lee, Antireflective Nanostructures for High-Efficiency Optical Sevices, *SPIE Newsroom*, November 30, 2010.

16 Black silicon antireflection nanostructures

Martin Steglich and Oliver Puffky

Contents

16.1 INTRODUCTION

Silicon is mankind's most important and viable semiconductor, not only enabling powerful analog and digital electronics, but also cost-efficient optoelectronic light sources and sensors. Besides that, silicon is by far the most-used material for the fabrication of solar cells, being responsible for the tremendous learning curve in industrial photovoltaic (PV) production and accounting for more than 90% market share in the PV market (ITRPV 2016). Moreover, silicon is a widespread optical material in the field of infrared optics. Here, it is most frequently used in the short- to mid-wavelength infrared range (SWIR to MIR, 2–6 µm wavelength) since it offers low cost, weight, and dispersion, as well as excellent structural quality and chemical stability (Rogalski 2002).

Unfortunately, all the named application fields suffer strongly from the exceptionally high refractive index of silicon (Figure 16.1a), which leads, according to Fresnel's equations, to severe reflection losses. As an example, considering normal incidence (0° angle of incidence) from air and a wavelength λ of 1000 nm ($n_{Si} \approx 3.6$, $k_{Si} \approx 0$), the reflection R from a perfect plane silicon surface amounts to

$$R = \frac{(n_1 - n_2)^2}{(n_1 + n_2)^2} = \frac{(n_{Si} - 1)^2}{(n_{Si} + 1)^2} \approx 32\% \qquad (16.1)$$

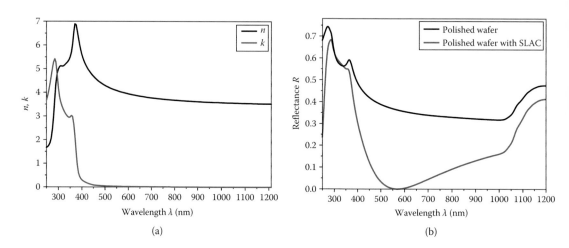

Figure 16.1 Complex refractive index $\tilde{n} = n + i \cdot k$ of silicon (a) and calculated reflectance R of double-side polished silicon wafer with and without SLAC (70 nm Si_3N_4) on one side (b).

Especially in infrared optical systems including multiple silicon elements (like lenses, windows), this high degree of reflection makes the application of pure silicon (without antireflection coating or structure) unimaginable as reflection losses quickly add up for multiple air—silicon surfaces. As an example, for three silicon optical elements (two surfaces per element) the optical transmission already diminishes to $T = (1 - R)^n = (1 - R)^6 \approx 12\%$. The same problem holds in principle for the field of optoelectronics, especially for PV, where reflection causes an undesired reduction in the achievable photon conversion efficiency.

The classical approach toward reflection suppression is to make use of antireflection coatings (ARC). In the simplest case, a good antireflection (AR) effect can be accomplished with a single-layer AR coating (SLAC), represented by a thin film of transparent material on top of the silicon, whose refractive index is between the indices of the incident medium (typically air, $n = 1$) and silicon (Stenzel 2005). Ideally, both the amplitude condition (d_{SLAC} ... film thickness)

$$d_{SLAC} = \frac{\lambda}{4 \cdot n_{SLAC}}$$ (16.2)

and the phase condition (n_{inc} ... refractive index of incidence medium)

$$n_{SLAC} = \sqrt{n_{inc} \cdot n_{Si}}$$ (16.2)

are met, since then the reflectance vanishes entirely at the design wavelength λ (Figure 16.1b). In the real world, however, the phase condition might not be sufficiently fulfilled as a material and the optimum refractive index might not be available. In order to suppress the silicon reflectance from air optimally, silicon nitride Si_3N_4 with a well suitable index of $n = 2.0$–2.2 is generally used, yielding a minimum reflectance of less than 1%. Beyond that, silicon dioxide SiO_2 ($n \approx 1.45$) is occasionally applied due to its extensive dissemination in semiconductor technology.

Of course, for better AR performance more complex optical layer structures can be used. Double layer AR coatings already allow for an increased AR bandwidth (Zhao and Green 1991; Lee et al. 2000; Wright et al. 2005). Multilayer optical film systems, which can easily be calculated numerically, may further boost the AR bandwidth and/or decrease the reflection dependence on the angle of incidence (AOI) (Stenzel 2005).

Another useful approach is gradient index systems (Southwell 1983). These implement a smooth transition from the ambient medium to the silicon refractive index, ideally in infinitely small index changes, and thus lower the overall reflectance drastically. If the index matching occurs sufficiently slowly, the interface reflectance almost vanishes over broad wavelength ranges. On silicon, satisfactory gradient index systems can be established with silicon oxynitride layers SiO_xN_y where x lowers gradually (and y increases) toward the silicon substrate (Snyder et al. 1992; Callard 1997).

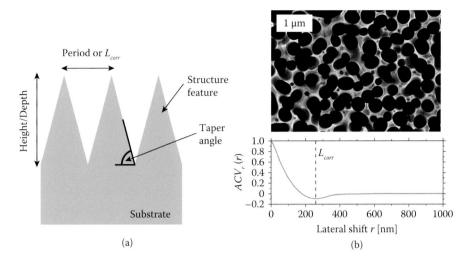

Figure 16.2 (a) Cross section of an ARS. (b) While in a periodical structure $L_{corr} = p$, statistical ARS demand a different L_{corr} definition. Shown is the example of a statistical black silicon surface (top view) obtained by reactive ion etching. Here, it is reasonable to calculate a radially averaged autocovariance function and use the characteristic minimum as L_{corr}, also representing a mean etch pore diameter.

This chapter, however, deals with another approach which might be considered as being closely related to gradient index systems: silicon antireflection structures (SiARS) or black silicon.[*] Generally, these terms bundle all monolithic silicon structures that exhibit a tapered geometrical profile—regardless of the specific structure dimensions or arrangements (Figure 16.2a).[†] As will be shown in the following, such ARS are not only able to provide reasonably low reflectances with an inherent broadband behavior and AOI coverage, but are sometimes even markedly easy to produce. Furthermore, black silicon may also yield a beneficial scattering for the transmitted part of incident light which is beneficial for a series of light-harvesting applications. Lastly, due to their monolithic character, ARS exhibit higher laser damage thresholds (Hobbs and MacLeod 2007; Schulze et al. 2012) compared to their optical thin-film counterparts, and do not suffer from thermal or film stress (Hobbs and Macleod 2005).

To give a comprehensive and instructional overview of the current state-of-the-art of SiARS, we will begin with a brief introduction into the optics of ARS, covering the relevant theories that apply to ARS of different physical dimensions with respect to the wavelength. We will then continue by explaining the most common fabrication technologies for SiARS—first, the required techniques to manufacture deterministic ARS and second, the most widespread self-organization approaches will be covered.

16.2 PHYSICS OF ANTIREFLECTION NANOSTRUCTURES

Understanding the optics of silicon ARS requires some insight into basic physical models in micro- and nanooptics. These models can be relatively clearly distinguished according to the dimensions (with respect to the wavelength of light λ) and the nature (periodic or stochastical) of the ARS under consideration. Thus, a correct choice of the appropriate optical model can be usually made by answering the questions "How big are the structure's characteristic geometrical features?" and "Is the structure periodically ordered or (partially) stochastically arranged?"

While the degree of lateral (dis-)order of the ARS is primarily important to understand the light-scattering behavior both in reflection and transmission, the characteristic feature sizes determine whether the ARS has to be treated rigorously (i.e., direct solution of Maxwell's equations with numerical methods), or if the approximations of geometrical optics or effective medium theory can be applied to considerably facilitate the theoretical treatment (Figure 16.3). For the latter, it is important to have an eye for the ratio of the characteristic

---

[*] We will use these terms synonymously in the following.

[†] In fact, even arbitrary binary profiles always decrease the overall interface reflectance (given that they are monolithic).

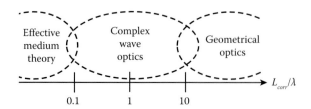

Figure 16.3 Rough classification of suitable optical approaches as a function of L_{corr}/λ.

lateral structure dimensions—which is, for the sake of universality, denoted as lateral autocorrelation length L_{corr} in the following—over the wavelength of light: L_{corr}/λ. For periodic structures L_{corr} is simply given by the period p (which may differ in x and y directions). In statistically arranged structures which, obviously, exhibit no distinct period, another L_{coor} definition is required (Figure 16.2b). For example, a reasonable approach is to use a characteristic site (x', y') of the structure's surface autocovariance function $(x', y') = \iint h(x, y) \cdot h(x + x', y + y')dxdy$, where $h(x, y)$ represents the structure topology (Schröder et al. 2011; Steglich et al. 2015). The so-defined characteristic point (x', y') can then serve as a quantitative measure of the lateral structure dimension and may, in a simplified fashion, be interpreted as an "averaged pseudoperiod."

16.2.1 RAY-OPTICAL WAVELENGTH REGIME

Whenever the characteristic lateral sizes of an ARS are large compared to the wavelength of light under consideration, (i.e., $L_{corr} \geq 10\cdot\lambda$) the well-known approximations of geometrical optics can be used to greatly facilitate the calculation of the structure's optical behavior. Then, the wavefront of the incident light may be represented by rays undergoing partial (or total) reflection and transmission, as well as refraction, while interacting with the ARS' optical interfaces. Modeling these effects simply requires finding the local AOI of the ray and then calculating the angle of refraction according to Snell's law, as well as the partial transmittance according to Fresnel's equations. By tracing both the reflected and the transmitted rays in such a way until they are completely absorbed inside the structured material, or leaving the calculation volume (through a boundary plane in reflection or transmission direction), the propagation of light can be accurately modeled. Macroscopic results on the optical sample performance can be calculated by repeating this ray-tracing procedure for a great number of homogenously distributed rays having the desired global AOI with respect to the sample normal. It should be noted that in contrast to the most common ray-tracing techniques used in the engineering of optical instruments, a so-called nonsequential ray-tracing has to be performed (Gross 2015). Here, each ray impinging on a surface is either split into two daughter rays (with accordingly lowered intensities) to account for the partial reflection and transmission, or it is statistically decided which unweakened ray to trace—either the reflected or the transmitted one (Monte Carlo ray-tracing).

Within the ray-optical regime, the suppression of interface reflectance by tapered silicon structures can be easily understood. A ray hitting the tapered surface is partially reflected (Figure 16.4), depending on the local AOI on the structure surface. The reflected part, however, is not directed into reflection but rather toward a neighboring structure feature, getting another opportunity to couple into the material. As this process can, depending on the taper angle, structure depth, and neighbor distance, repeat many times, a clearly increased part of the incident light will enter the silicon material. Hence, the overall macroscopic sample reflectance is greatly reduced to values significantly lower than for unstructured silicon. As the transmitted fraction of light undergoes refraction at the silicon–air interface, it changes its angle of propagation upon launching into the material. This gives rise to so-called light trapping effect due to the increased degree of total internal reflection (TIR) not only inside the structure features, but also within the whole structured sample (e.g., the substrate rear side). Especially in silicon, pronounced light trapping occurs as the high refractive index yields a rather low critical angle of total internal reflectance of only 16°. As a result, light can travel back and forth in the sample, being (on average) strongly reflected both at the sample rear and the structured front surface. Since light trapping greatly enhances the effective optical pathway within the silicon, it is of major importance for photon-harvesting applications such as solar thermal absorbers or solar cells (Füchsel et al. 2015).

The light trapping effect, in conjunction with the structure-related AR effect and the possibility to combine it with classical AR coatings, makes ray-optical silicon ARS ideal for photovoltaic applications. There is

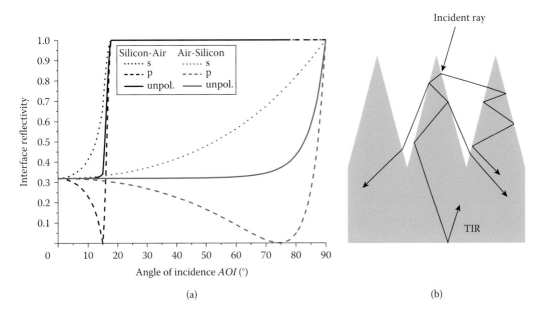

(a)   (b)

Figure 16.4 (a) Interface reflectivities for s-, p-, and unpolarized light incident on silicon from air and on air from silicon, calculated using Fresnel's equations. (b) Ray-tracing principle at huge feature ARS. Only some rays are traced exemplarily. Ray deflection though refraction gives rise to TIR both at the feature interfaces and the substrate rear, leading to light trapping.

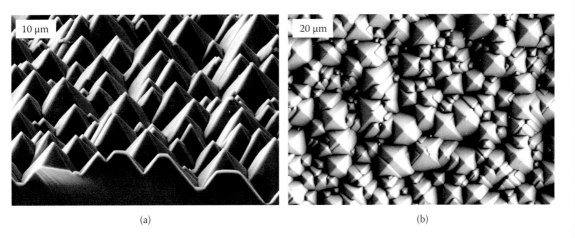

(a)   (b)

Figure 16.5 Random texture obtained by KOH etching of (100) silicon in cross section (a) and top view (b).

a long history of AR surface textures in photovoltaics dating back to the COMSAT solar cell established in 1975, representing a huge breakthrough in the development of high-efficiency silicon solar cells (Allison et al. 1975; Arndt et al. 1975; Sze and Ng 2006). The COMSAT cell (and basically all monocrystalline silicon solar cells commercially available nowadays) utilizes a random pyramidal surface texture that can be fabricated by etching (100) oriented silicon wafers in potassium hydroxide (KOH) containing etch solutions (Figure 16.5) (Füchsel et al. 2015). As KOH etching is inherently slow on (111) silicon planes, this procedure gives a random distribution of upright standing silicon pyramids with 54.7° taper angle (Köhler 1999). Combined with a standard Si_3N_4 single-layer antireflection coating, such structured surfaces can easily accomplish low surface reflectances of less than 3% in the wavelength range between 450 and 1000 nm and a strong light trapping effect which is crucial for absorptance maximizing (Duttagupta et al. 2012). KOH etching can also be done in predefined areas, through lithographically processed etch masks to give regular arrangements of inverted pyramids or V-grooves on (100) silicon (Köhler 1999). A popular example is the PERT solar cell which combined this optimized ray-optical AR and light trapping structure with a double layer AR coating to result in a notable conversion efficiency of 24.7% (Zhao and Green 1991; Zhao et al. 1999).

Another example for ray-optical silicon ARS are laser-textured black silicon surfaces which have gained a lot of attention lately. In contrast to KOH-etched surfaces, these exhibit conical instead of pyramidal structure features that evolve during repeated bombardment of silicon surfaces with short laser pulses (Her et al. 1998). Here, although the structure dimensions may differ greatly depending on the processing conditions (number of irradiating laser pulses, pulse energy, wavelength), most of the fabricated structures are rather large with dimensions in the range of 10 μm—which can thus be widely understood in terms of geometrical optics principles if the considered wavelengths are not too high.

16.2.2 EFFECTIVE MEDIUM WAVELENGTH REGIME

With respect to the ray-optical regime, we find the effective medium wavelength regime on the far opposite of the L_{corr}/λ-scale. Here, the structures are considerably smaller than the wavelength of the incident light: $L_{corr} \ll \lambda$. In that case, the optical wave is not able to resolve the structure any more but rather "sees" a homogenous optical medium. The complex refractive index of this medium may then depend on the structure geometry (shape birefringence) and is composed of the single refractive indices of the structure medium and surrounding medium, respectively. This effect allows a simplified mathematical treatment of small optical structures as stacks of optical thin films, and is widely known as the effective medium approximation (EMA).

The condition $L_{corr} \ll \lambda$ for the validity of the EMA can be found quite often in the literature although it can be formulated by far less strict means for periodic structures. In periodic surface relief structures, L_{corr} can be replaced by the one-dimensional (1D) or two-dimensional (2D) period p, and the grating equation

$$\sin \varphi_m = \sin \varphi_{in} + m \frac{\lambda}{n \cdot p} \tag{16.1}$$

can be used to calculate the diffraction angle φ_m of the m-the diffraction order. Here, φ_{in} is the angle of incidence, λ is the vacuum wavelength, n the substrate refractive index and m an integer (0, ±1, ±2, ±3, …). The grating equation only gives reasonable results if the right side in Equation 16.1 is smaller than 1, allowing to calculate the number of diffraction orders that can propagate behind the grating. Thus, for normal incidence $\varphi_{in} = 0$, we can find the so-called zero-order condition telling us at which point even the ±1st diffraction orders vanish ($m = 1$):

$$p < \frac{\lambda}{n}. \tag{16.2}$$

This means, as a periodic surface structure only allows for the 0th order whenever its period p is smaller than λ/n, it may be regarded as an effective medium[*]. Nevertheless, the rather strict (and less distinct) effective medium condition $L_{corr} \ll \lambda$ retains its meaning for statistical structures since there is clearly no specific lateral structure period that could be assigned.

Given that the EMA is valid for an existent sub-wavelength structure, the appropriate method to calculate the effective index depends on the structure's nature. For periodic, 1D gratings with fill factor f, the effective permittivity ε_{eff} is polarization dependent (TE polarization: field vector $\vec{E}$ is perpendicular to grating lines, TM polarization: $\vec{E}$ is parallel to grating lines) and can be deduced from Grann et al. (1994):

$$\varepsilon_{eff,TE} = f \cdot \varepsilon_1 + (1-f) \cdot \varepsilon_2$$

$$\varepsilon_{eff,TM}^{-1} = f \cdot \varepsilon_1^{-1} + (1-f) \cdot \varepsilon_2^{-1}. \tag{16.3}$$

---

[*] It should be noted that this criterion becomes more strict for oblique incidence on a high index medium, as then Equation 16.2 becomes $p < \dfrac{\lambda}{n + n_{in} \cdot \sin \varphi_{in}}$, where n_{in} is the index of the incidence medium.

Even more important, in particular for a consideration of ARS is the extension to 2D, quadratic grating ($p_x = p_y = p$) structures with linear fill factor $f_x = f_y = f$ which are intrinsically polarization independent (Grann et al. 1994):

$$\varepsilon_{\textit{eff}} = f^2 \cdot \varepsilon_1 + (1 - f^2) \cdot \varepsilon_2. \tag{16.4}$$

For nonperiodic, statistically distributed structures, the situation is more complex. Calculation of refractive indices is then often performed by means of the Maxwell-Garnett model

$$\varepsilon_{\textit{eff}} = \varepsilon_2 \frac{2 f_V (\varepsilon_1 - \varepsilon_2) + \varepsilon_1 + 2\varepsilon_2}{2\varepsilon_2 + \varepsilon_1 + f_V (\varepsilon_2 - \varepsilon_1)}, \tag{16.5}$$

which holds for small inclusions of the material with ε_1 being entirely surrounded by the host material with ε_2. Here, f_V denotes the volume fraction of the inclusion. Other effective medium models, such as the Bruggeman model, are also possible for permittivity calculation (Stenzel 2005).

The outcome of the EMA for an ARS can be easily understood (Figure 16.6). If we cut the tapered structure profile into multiple horizontal slices and assign an effective complex refractive index to each of the slices, we directly obtain a stack of optical layers whose refractive indices monotonically increase from the top to the bottom. The structure thus forms an effective medium gradient index layer stack that results in a decreasing reflection—just like true gradient index layer systems. Theoretically, there is an optimum graded index profile. For example, Southwell compared linear, cubic, and quantic index profiles and found the latter to be best for maximum antireflection effect (Southwell 1983). Such profile considerations are, however, not very meaningful from the practical point of view, as the required structure profile control in true effective medium ARS is hardly feasible. Luckily, also linear physical structure tapers, yielding quadratic effective indices, can result in markedly low reflectances.

More important than the physical structure profile and the associated gradient index is the depth of the ARS if a broadband, low-reflection behavior is desired. This is due to the obvious fact that the gradual index change has to be sufficiently smooth to effectively suppress reflection. As the required thickness scales linearly with wavelength, it depends linearly on the upper limit of the targeted AR bandwidth (in wavelength units). As a rule of thumb, the structure thicknesses should amount to at least $\lambda/2$ for a satisfyingly low residual reflection (less than 1% even for silicon with its high refractive index of about 3.6); higher thicknesses are always beneficial (Southwell 1991).

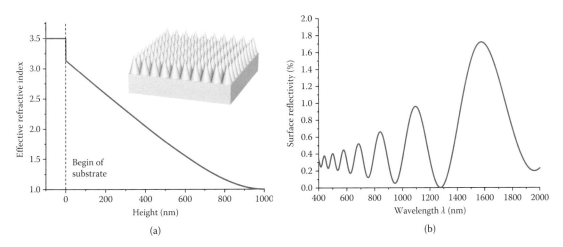

(a) (b)

Figure 16.6 (a) Calculated effective refractive index for 1000 nm deep, conical SiARS arranged in a square lattice. For simplicity, a constant refractive index of 3.5 for silicon was assumed. (b) Resulting surface reflectivity of silicon with such ARS (rear surface reflections are disregarded). Calculations carried out by Olaf Stenzel (Fraunhofer IOF Jena, Germany) using the method described in Tikhonravov et al. (2006).

While the required structure thickness is widely independent on the substrate index, the required maximum lateral structure correlation L_{corr} for the validity of the EMA decreases hyperbolically with n. This makes the realization of nonscattering or nondiffracting (in transmission) effective medium ARS rather difficult for silicon—compared to low-index materials like glasses. For example, consider a periodic structure capable of establishing an AR in the short- to mid-wavelength infrared (1 to 6 µm): while the required maximum acceptable period can be derived from the lower wavelength boundary to about $p \approx \lambda_{min}/n \approx 280$ nm, the upper limit requires at least a structure depth of approximately $h \approx \lambda/2 \approx 3000$ nm. The ARS to be fabricated thus has both a low period and a high aspect ratio $h/p \approx 10$. Manufacturing such structures is a very challenging, demanding task—even with the mature manufacturing methods available nowadays.

16.2.3 COMPLEX WAVE-OPTICAL WAVELENGTH REGIME

In between the ray-optical and the effective medium wavelength regime, there is a wide range of L_{corr}/λ ratios that need to be treated rigorously by directly solving Maxwell's equations for these structures. This applies especially whenever silicon ARS for optical wavelengths in the visual spectral range (Vis) are regarded. Here, structures with sufficient AR performance typically exhibit features in the submicrometer range such that a treatment with geometrical optics principles fails. On the other hand, in order to be handled in terms of EMA in the Vis, the structures should have geometric features smaller than approximately 100 nm. For example, for a periodic silicon ARS at a wavelength of 500 nm, the period has to be less than 110 nm to fulfill the zero-order condition for normal incidence. As such small periodic structures are extremely difficult to fabricate and since for statistical ARS the EMA criterion $L_{corr} \ll \lambda$ is even much harder to meet, most of the practical silicon ARS have to be treated wave-optically in the Vis to fully understand their behavior.

Nevertheless, even without rigorous calculations, some important conclusions can be drawn by discussing wave-optical ARS in the context of the limiting cases of geometrical optics (higher L_{corr}/λ) and effective medium theory (lower L_{corr}/λ). First of all, it is obvious that the optical behavior of structures being located in the wave-optical regime can be well anticipated by the consistent findings of geometrical optics and EMA. Therefore, it is intuitively clear that a structure's reflectivity will decrease with increasing height and decreasing L_{corr} for a given wavelength. It is also clear that such a structure will cause diffraction (periodic ARS) or scattering (statistical ARS) since the effective medium criterion is not met. For the former, the grating equation (Equation 16.1) can be used without restrictions to calculate the respective diffraction angles while the diffraction efficiency cannot be calculated analytically. Regarding the scattering in statistical ARS, a simple conclusion can be drawn by comparing with the EMA. As EMA, being valid at very small L_{corr}, does not allow for scattering, it is only reasonable that both the mean scattering angle and the scattering magnitude decreases monotonically with decreasing L_{corr} in the wave-optical regime. Nevertheless and similar to stated before, more precise statements on the optical behavior of ARS within this regime require numerical wave-optical treatments.

Widespread numerical methods for rigorous solutions to Maxwell's equations are the rigorously coupled wave analysis (RCWA), also known as Fourier modal method (FMM), and the finite differences in time domain method (FDTD). Both numerical approaches have their specific strengths and drawbacks which will be discussed shortly in the following in the context of (dielectric) ARS simulation.

RCWA assumes periodic structures and uses the Floquet–Bloch theorem to develop the electromagnetic waves laterally along the x and y directions (see Figure 16.7a). Within each unit cell of the periodic structure arrangement, both the fields and the physical structure are developed in Fourier series, thus enabling a solution of the vectorized Maxwell's equations in Fourier space (k_x, k_y). Nonbinary structures have to be split up in a vertical series of homogenous films where each film is separately Fourier expanded ("slicing"). The overall solution can be obtained by exploiting the boundary conditions between the slices and solving the final linear system of equations. This can be done quite fast if the number of used Fourier orders (i.e., the length of the Fourier sums), as well as the number of structure slices is not too high. Thus, the finiteness of the latter represents a source of some kind of numerical discretization error. The inherent strength of RCWA is the solution of dielectric, periodic structures. Therefore, it is ideally suited to calculate the reflection and transmission (as a function of Fourier order or angle, respectively) of periodic ARS.[*]

---

[*] Although it might not be necessary in view of the good results obtainable with common EMA models, the suitability of RCWA also applies to dielectric structures in the effective medium regime.

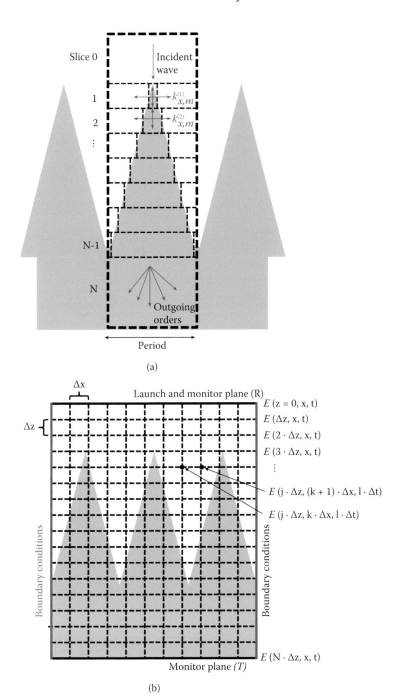

Figure 16.7 Scheme of RCWA (a) and FDTD (b). See text for description.

Compared to RCWA, FDTD is a rather direct approach to solve Maxwell's equations (Figure 16.7b). Here, the differential equations are solved in a time domain on a spatial grid by discretizing the differential operators in time and space. The light is then "turned on" by simply launching the optical excitation on the desired grid points. As the equations are then solved in small time steps subsequently, the simulated time has to be long enough to reach a steady-state—until the transient oscillation has ended. FDTD can indeed be regarded as a numerical experiment that directly simulates the propagation of electromagnetic waves in (almost) arbitrary geometries. Its numerical effort is, however, immense.

The decision whether to use RCWA or FDTD for rigorous calculations depends mainly on the nature of the silicon ARS under consideration. The reason for this is the different numerical scaling of the methods. FDTD scales with the amount of grid points in simulation volume ($V = L^3$) such that the required computing capacity increases with L^3. On the contrary, RCWA scales with the number of Fourier orders to the power of 6, where the required number for an accurate calculation increases linearly with L. As RCWA assumes periodic boundary conditions and since it is much faster than FDTD on small simulation domains (with respect to the wavelength λ), it is ideal for the simulation of periodic ARS. Statistical structures, on the other hand, in order to account for their statistical nature properly should be simulated on simulation domains as large as possible. Here, because of the better numerical scaling of FDTD versus RCWA (L^3 vs. L^6), FDTD is the better choice for accurate modeling.

16.2.4 STRUCTURE REQUIREMENTS FOR MAXIMUM ANTIREFLECTION

In view of the multitude of different silicon ARS available, the ideal choice of a distinct structure depends strongly on the target application. From that perspective it is reasonable to distinguish between two important application scenarios that make different demands on the applied silicon ARS.

The first scenario is the application of an ARS as a direct substitute for AR coatings, such as on silicon optical elements in infrared (imaging) systems. Here, scattering or diffraction into higher orders, especially in transmission, is generally undesired as it ruins the optical function of the device. The applied structure should therefore behave effectively in the target spectral range—which is best obtained with periodic ARS. For a maximum AR effect, the structure should moreover be as high as possible—where heights of $\lambda/2$ already yield reasonably low reflectances.

Another typical application is the maximization of light absorptance in optoelectronic devices like solar cells or photodiodes. For that, not only should the reflection be minimized, but also the magnitude of scattering or refraction for the transmitted part has to be maximized. Then, the injected light has a maximum optical pathway inside the absorbing medium due to oblique propagation and light trapping. Especially the latter plays an important role since scattering initiates TIR at the outer absorber interfaces, thus allowing the light to perform multiple roundtrips in the device. These optical requirements are best met by application of statistical ARS in the wave- or ray-optical regime. In particular, it was demonstrated that statistical black silicon ARS with $L_{corr} \approx \lambda$ provide superior scattering and light trapping behavior (Otto et al. 2014), being close to the performance of surfaces with perfect Lambertian scattering properties (also known as the Yablonovitch limit).

A further precondition for reaching an optimum AR effect is to keep the fraction of horizontal, non-tapered area as low as possible. This means, ideally the entire sample surface is covered with the tapered, three-dimensional ARS with no plane areas between the single structure features. For periodic ARS, which often exhibit conical structure features with circle shaped bases (due to technological reasons), this implies that the feature should rather be arranged in a hexagonal lattice than in a square lattice, in order to achieve a maximum areal packing density.

16.3 FABRICATION METHODS

As shown in the previous section, well-performing ARS have to fulfill challenging geometrical requirements simultaneously: low L_{corr}, high structure depth, and continuously tapered profile. In principle, there are two routes to fabricate such structures. First, a deterministic, lithographic patterning, followed by pattern transfer through dry etching can be followed up; primarily whenever periodic ARS are desired. Second, to reduce efforts and costs, several self-organization principles can be pursued to manufacture (mostly) statistical ARS.

16.3.1 DETERMINISTIC PATTERNING

Deterministic structure fabrication typically relies on lithographically defining a resist mask on the substrate and subsequent transfer of this pattern into the substrate by deep etching. Here, besides the challenges involved in deep submicrometer pattern generation, the applied etching procedure has to provide for the favored taper profile and high structure depth. Luckily, the nowadays mature reactive ion etching (RIE) technology for silicon is able to achieve the latter.

16.3.1.1 Electron beam lithography

Electron beam lithography (EBL) is a serial direct writing technique with the ability to structure arbitrary 2D nanometer scale patterns. Using a highly focused beam of electrons, complex nanoscale patterns can be written into an electron-sensitive resist by deflecting the beam locally and moving the substrate at the same time (Pease 1981). Due to serial writing and submicrometer spot size, the expectable throughput is very low. To overcome this serious limitation, advanced electron beam tools were developed, which use an unfocused flow of electrons to expose a larger area at the same time (see Figure 16.8). Here, apertures with variable size and shape (variable-shaped beam) or even lattice structures (character projection) are used for definition of desired exposure patterns (Pfeiffer 1978; Berger and Gibson 1990; Nakayama 1990).

16.3.1.2 Interference lithography

In contrast to EBL, interference lithography (IL) is a highly parallel technique which allows the fast creation of periodic resist patterns. In IL, two coherent waves are interfered to form a standing wave with alternating areas of high and low intensity (Figure 16.9). This intensity pattern is being recorded by placing a substrate coated with photosensitive resist into the region of wave intersection. The major advantages of IL are the comparable low equipment costs, the high throughput, and the capability to pattern large areas at once (van Wolferen and Abelman 2011).

By using two interfering beams, each with the angle of incidence θ to the substrate normal, a 1D grating with period

$$p = \frac{\lambda}{2\sin\theta} \tag{16.6}$$

can be recorded. Rotating the sample by 90° after a first half-exposure and exposing it again will result in a 2D grating structure of equivalent periods in x and y. More sophisticated exposure setups using three- and four-beam interference can be used to create hexagonal and rectangular lattice structures, respectively (Baker 1999; van Wolferen and Abelman 2011).

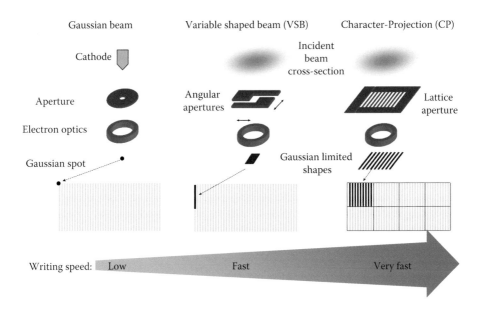

Figure 16.8 Comparison of different electron beam writing strategies with drastically improving writing speeds from Gaussian over variable-shaped beam to character projection lithography. Since the latter can be used to print whole arrays of shapes at once, it is particularly useful for microoptical applications.

Functional materials

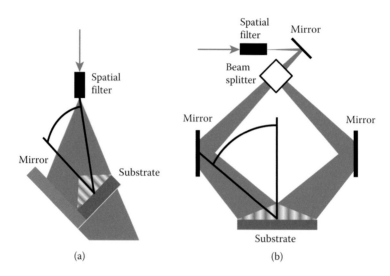

Figure 16.9 Basic interference lithography: (a) Lloyd's mirror configuration and (b) two-beam setup using beam splitter.

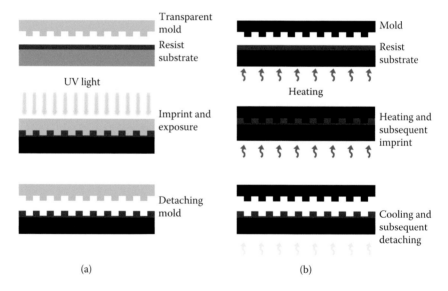

Figure 16.10 Nanoimprint lithography using UV light (a) and using thermal heating (b).

16.3.1.3 Nanoimprint lithography

Nanoimprint lithography (NIL) is a nonoptical patterning technique, which relies on direct mechanical contact (see Figure 16.10). Hence, resolution is not limited by light diffraction or beam scattering. Basically the surface nanopattern of a mold/stamp is replicated into a thin polymer coating deposited on the substrate by means of mechanical contact. During contact the polymer is pushed into the voids of the stamp, thus giving a negative of the stamp nanopattern. Good void filling is achieved with rather liquid, low viscosity polymers. Yet, after demolding the polymer should be hard to keep the pattern. This difference may be achieved by heating the stamp and substrate during imprint and cooling it down prior to demolding (Chou 1996; Chou et al. 1996). Alternatively, at room temperature the polymer can be hardened by UV light curing during contact (Resnick et al. 2003).

Finally, the achieved polymer structure can either be used directly as functional layer, or as mask layer to structure the underlying substrate.

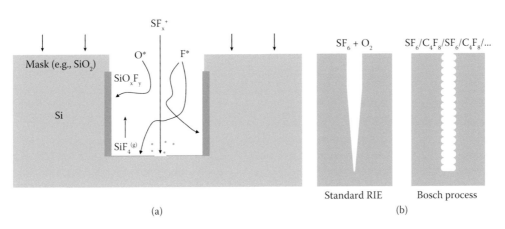

Figure 16.11 (a) Chemical reactions in deep RIE of silicon with a lithographically defined etch mask. Due to diffusion of F^* and O^* toward the etch hole sidewalls, a passivating SiO_xF_y film is formed here. The continuous bombardment by high-energetic ions (e.g., SF_x^+) removes this passivation at the hole bottom. Thus, etching occurs preferably in the vertical direction, leading to anisotropic etch profiles. (b) Typical etch profiles using standard RIE, exhibiting an "RIE lag" (left) and using a Bosch process, showing scallops (right).

Since NIL allows patterning in the nanoscale with high throughput and low cost, it is a promising lithography method for the production of periodic ARS on large areas. Yu et al. used NIL to create a 2D SiARS with conical profile, 200 nm period and 520 nm depth on an area of 16 cm², reaching an average reflectance in the Vis of less than 3% (Yu et al. 2003). Aiming at photovoltaic applications, Nakanishi et al. demonstrated a pattern transfer to 8″ substrates, where the resulting ARS achieved a reflectance of about 1% (Nakanishi et al. 2010).

16.3.1.4 Pattern transfer by RIE

After lithographic resist pattern definition, the pattern has to be transferred into the substrate by etching. For ARS on silicon, RIE is by far the best choice since it allows for deep anisotropic (i.e., etching occurs only in the vertical, but not in the horizontal directions) etch profiles. By that, structures with enormous aspect ratios >20 can be easily realized (Köhler 1999).

As a dry etching technique, RIE utilizes gaseous etchants that are generated in a plasma by dissociation of precursor gases. In the case of silicon, fluorine radicals F^* are the preferred etchants formed by dissociating fluorine-containing gases like SF_6 or CF_4 (Jansen et al. 1996). Since chemical etching by F^* alone results in isotropic etch profiles, further passivating precursors like O_2, C_4F_8, or CHF_3 have to be introduced into the reaction chamber to achieve anisotropic profiles (Figure 16.11). During etching, fragments from these passivation gases form solid films (silicon oxides or carbon halogens) on the vertical silicon surfaces that serve as protection from the chemical etch attack by F^*. On the sample's horizontal surface, however, such films do not occur because of the continuous and directional sputtering effect (physical etching) by ions impinging from the plasma. Here, the self-bias that evolves between the plasma and the substrate cathode accelerates the ions toward the substrate such that the amount of physical etching can be controlled by self-bias adjustment.

As the achievable etch depth is limited by diffusion of etching species to—as well as reaction products from—the etch pit bottom (so-called RIE lag), more cyclic RIE techniques were developed to allow for even deeper etch pits and higher aspect ratios. In these "Bosch processes" the passivation and etching reactions are performed temporarily separated from each other—and repeated many times (Laermer et al. 1993). In the simplest case, for instance, an etch step is performed with SF_6, followed by a passivation step with C_4F_8, and the whole procedure is periodically repeated (Abdolvand and Ayazi 2008; Rhee et al. 2008). Due to the isotropic etching behavior during the etch step, characteristic "scallops" evolve in Bosch processes.

16.3.2 SELF-ORGANIZATION APPROACHES

The inherent disadvantage of deterministic ARS fabrication is the huge effort, both in terms of cost and time, being incorporated with lithographic pattern definition and mask transfer by dry etching.

Therefore, a vast of scientific work has been devoted to finding methods to produce black silicon ARS by self-organization techniques during the last decades. These promise a cost-efficient yet constructive pathway toward high-performance black silicon ARS, suitable for industrial utilization.

In the following, an attempt is made to sketch the fabrication method and the specific (dis-) advantages of the most relevant self-organization approaches for black silicon ARS.

16.3.2.1 Random micromasking in RIE

Creation of black silicon ARS by self-organized RIE processes has a long history dating back to 1979, when Schwartz and Schaible first reported on the observation of black silicon surfaces in the course of silicon etching in Cl_2 plasmas (Schwartz and Schaible 1979). In their contribution, they also introduced the descriptive and very appropriate term black silicon in the scientific community. Shortly afterward, IBM patented a similar processing technology for black silicon by etching in Cl_2–SF_6 plasmas (Hansen et al. 1980). Black silicon fabricated in Cl_2 RIE can be seen as a random superposition of inverted, pyramidal etch pits whose base edges are oriented along the <110> crystal directions (Figure 16.12a). This morphology ensures both a strong AR effect and forward scattering of light, and furthermore indicates that structure formation is due to a crystal anisotropy of the etching process (Figure 16.12b). Yet, strangely enough, there have been no further studies with regard to the underlying physical principles in the past decades.

This lack of investigations regarding black silicon formation in Cl_2 plasmas most certainly is owed to the rapid dissemination of fluorine-based chemistries, in particular SF_6, for silicon etching. Here, recipes that produce needle-like black silicon surfaces were quickly found, too. Jansen et al. first published a simple step-by-step manual on how to find an etching regime for black silicon: silicon is etched several times in a mixture of SF_6 and O_2, where after each step the amount of O_2 is slightly increased (Jansen et al. 1995). With this procedure, a black silicon recipe is readily found at a certain, suitable gas composition of SF_6 and O_2.[*]

Black silicon formation during RIE of silicon in SF_6–O_2 chemistry relies on two chemical reactions, as well as on physical etching by impinging ions (Steglich et al. 2014). The first reaction is the etching reaction, mainly by fluorine radicals F^* that form upon SF_6 dissociation in the plasma:

$$Si + 4 \cdot F^* \rightarrow SiF_4^{(g)}. \tag{16.7}$$

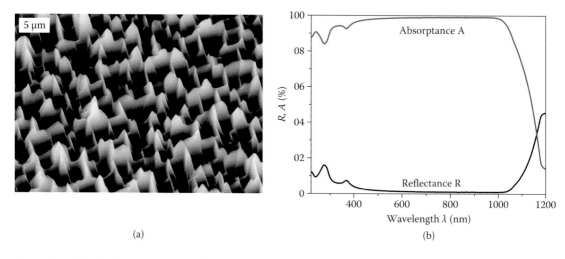

(a)

(b)

Figure 16.12 Black silicon ARS obtained by RIE in Cl_2 (a) and optical spectra (reflectance and absorptance) of such structure (b).

---

[*] The intention of Jansen's publication was to give a general method to find suitable RIE conditions, in particular gas compositions, that can be capitalized for anisotropic silicon deep etching. Finding the black silicon regime is only one intermediate step in this method.

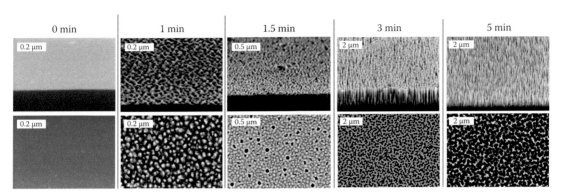

| 0 min | 1 min | 1.5 min | 3 min | 5 min |

Figure 16.13 Evolution of black silicon ARS during RIE in SF$_6$–O$_2$ plasma. After initial surface roughening, first etch pores develop that continuously grow afterward until black silicon is finally formed.

Silicon is thus etched by F^* where the produced, gaseous $SiF_4^{(g)}$ is pumped out of the reaction chamber. The second reaction is the passivation reaction with F^* and oxygen radicals O^*, formed by O$_2$ dissociation:

$$Si + x \cdot O^* + y \cdot F^* \rightarrow SiO_xF_y. \tag{16.8}$$

Here, O^*—possibly together with some F^* radicals—develops a thin (≤ 1 nm) solid film on the exposed silicon surface that is very resistant against the etch attack as given by Equation 16.1. This film hence protects, or passivates, the silicon surface from the etching reaction.

As soon as the etching process in SF$_6$ and O$_2$ is started, a uniform passivating layer forms on the silicon surface (even if the preexistent native silicon oxide was removed before, such as by dipping in diluted HF) that initially protects the surface from being etched (Figure 16.13). Afterward, in the initial phase of structure evolution, a considerable surface roughening is observed. Numerical calculations by Abi-Saab et al. strongly suggest that this increasing roughness can be explained solely by an increased physical ion etching (in other words: sputtering) around the heights of the finite starting roughness of the substrate—even for polished silicon surfaces with extremely low rms roughnesses of 0.1 nm or even less (Abi-Saab et al. 2014). As this roughness increase continues, the increased physical material removal in the valleys beside the roughness hills likewise rises. After a while, the protecting SiO$_x$F$_y$ passivation will be entirely removed in some valley points. Now the much faster chemical etching (Equation 16.7) sets in here, and quickly forms circular etch holes that are randomly distributed over the sample. Due to the directionality of the ion bombardment, rather stable SiO$_x$F$_y$ passivation layers build up on the vertical sidewalls of these holes. Thus, they will grow fast in the vertical direction, but—due to the finite etch stopping capabilities of SiO$_x$F$_y$—also slowly in the lateral directions. In the course of pore growth, finally black silicon evolves when the pores start to interconnect with each other. Black silicon fabricated by RIE in SF$_6$ and O$_2$ can thus be regarded as a random distribution of laterally connected, circular to elliptical etch holes of relatively uniform depth, with needle-like structure features at the pores' cutting points.

Black silicon ARS typically exhibit L_{corr} values in the range 100–500 nm and structure depths between 500 and 3000 nm. Due to their statistical nature they have to be treated rigorously, ideally by means of FDTD, to calculate their optical behavior, in particular their scattering properties. Reflectances below 0.5% over a broad wavelength range from 500 to 1500 nm and low dependency on the angle of incidence (e.g., less than 2% up to 50° angle of incidence at 633 nm) are readily reached. In addition, thanks to the distinct amorphous character of lateral structure distribution, these structures are very strong scatterers, getting close to ideal Lambertian scattering behavior (Otto et al. 2014).

16.3.2.2 Metal-assisted chemical etching

Metal-assisted chemical etching (MACE) is a comparably simple and cost-efficient method for black silicon fabrication and has thus attracted much attention lately, especially in the photovoltaic community. The low-cost character of this pure (electroless) wet-etching method results primarily from the inexpensiveness of the required equipment and etch media, as well as the feasibility of batch production.

In general, MACE relies on the SiO$_2$ etching capabilities of hydrofluoric acid (HF), and the introduction of an additional, oxidizing species. For example, pure silicon etching can be easily done in mixtures of nitric acid (HNO$_3$) and HF, where the former acts as oxidizer and the latter as oxide etchant. With this chemical composition, advances toward a self-organized structure formation were made by Dimova-Malinovska et al. in 1997 by performing the aforementioned etching procedure on silicon coated with thin aluminum films (Dimova-Malinovska et al. 1997). They found a drastically increased etch speed for the formation of porous silicon which could only be ascribed to a catalytic behavior of the incorporated Al film.

It was discovered later on that basically any noble metal (e.g., Au, Pt, Ag, Pd) gives rise to such catalytic etching behavior underneath the metal or metal particle, respectively (Li and Bohn 2000). Thus, needle-like black silicon ARS can be obtained by preparation of a noncoalescent (i.e., an only partly covered) noble metal film, such as by sputtering or evaporation, on the silicon surface and etching it afterward in solutions of HF and an oxidizing agent. For the latter, typically H$_2$O$_2$ is used since it only slowly oxidizes silicon in the uncoated areas, in contrast for example to HNO$_3$ which also readily attacks bare Si in conjunction with HF. When the partly metal-coated substrate is brought into the HF/H$_2$O$_2$ solution, the H$_2$O$_2$ is quickly reduced at the metal surface, thereby producing positively charged holes (Huang et al. 2011). These diffuse through the metal and reach the silicon, which is first oxidized by the holes and then dissolved by the HF. As the hole concentration is maximal directly underneath (or in close vicinity to) the metal particles, so is the effective etching rate here. By balancing the HF/H$_2$O$_2$ concentration, a steady-state can be reached in which the production of holes at the metal surface (through reduction of H$_2$O$_2$) equals the consumption due to Si oxidation and etching by HF. The catalytic etching process then occurs solely underneath the metal-coated areas while the uncoated parts remain (nearly) unetched. Of course, the described procedure can also be combined with a deterministic patterning of the deposited metal film, enabling regular arrangements of etch holes, trenches, etc.

Besides coating the substrate in a previous step, it is also possible to form the noncoalescent metal film directly during the etching procedure by applying noble metal plating solutions as oxidizers (Huang et al. 2011). Here, the metal ions are reduced, acting as an oxidizer, and simultaneously form metal nuclei on the silicon surface. The most popular solutions for this facilitated, self-organization process are HF/AgNO$_3$ and HF/KAuCl$_4$.

Generally, in MACE the etch depth increases with solution temperature (following an Arrhenius dependency) and etching time. Vertical etching is always observed for (100) substrates. However, (111) and (110) substrates are sometimes etched in the (100) directions, thus nonvertically, which can be explained by the crystal plane dependent back-bond strength and the different connected etch speeds.

Black silicon fabricated by MACE may have very small (~10 nm) or rather large (~ few μm) lateral feature sizes, as well as large structure depths up to several tens of μm (see Figure 16.14). Consequently, low surface reflectances in the range of 1% to 5% can be realized.

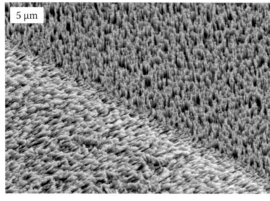 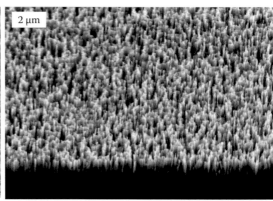

Figure 16.14 Black silicon ARS fabricated by MACE in HF/AgNO$_3$. Partial nonvertical etching can be observed on polycrystalline substrates (left), ideally vertical etching on single-crystalline (100) Si. (Images with kind permission from Guobin Jia, Leibniz Institute of Photonic Technology.)

16.3.2.3 Laser structuring

The possibility to create black silicon by repeated irradiation with short laser pulses is known since the 1980s, when first reliable, high-power laser systems became available (Rothenberg and Kelly 1984).

The structure formation can be understood as a result of laser-induced periodic surface structures (LIPSS) forming when the silicon surface is melted and quenched by incident high-energy laser pulses: periodic surface undulations are frozen on the silicon surface due to interference of the incident and reflected laser pulse (Her et al. 2000). When these sinusoidal corrugations are then repeatedly hit by more and more laser pulses, material in the valleys of the surface structure is removed more quickly than on the elevated areas. Thus, the structure is deepened more and more until a fairly regular arrangement of conical microneedles or bubbles is formed (Figure 16.15).

The mean physical dimensions and distances of the structure features increase with increasing laser wavelength (since it is directly related to the period of the LIPSS), pulse energy, and number of irradiating pulses. Furthermore, distinct dependencies on the ambient atmosphere during laser structuring were observed. Since the molten silicon is quite volatile, it might react with Cl_2 or SF_6, resulting in an etching effect and yielding conical microstructures with markedly sharp tips. On the other hand, structure fabrication in O_2 or N_2 results in rather bubble-like structures and cause partial oxidation or nitridation, respectively. Another side effect of laser structuring in nonvacuum atmospheres is the strong contamination involved (Winkler et al. 2011). Gas molecules from the atmosphere (or a pre-deposited solid source) may enter the silicon melt and freeze inside the silicon lattice during rapid cooling. By that, high doping densities, far above the limit of solid solubility can be achieved. This so-called hyperdoping might have an undesired, detrimental effect on the electronic properties of the silicon or—alternatively—can be seen as a novel method to produce artificial silicon alloys with new optical and electronic properties (Huang et al. 2006).

Laser-structured black silicon typically exhibits rather large features with distances of a few µm and depths of 10 µm or even more. Reflectances can be lower than 3% where structures with sharp features achieve the lowest values. At the same time, however, as these sharp structures have to be realized in an SF_6 or Cl_2 atmosphere, the related hyperdoping with S or Cl results in an (possibly unwanted) additional absorption mechanism that extends far into the infrared (i.e., in the sub-band gap region of silicon) (Wu et al. 2001).

16.3.2.4 Self-alignment of nanoparticles

Further possibilities to fabricate black silicon ARS arise from various self-alignment processes of molecules or clusters that can afterward be used as an etch mask for RIE pattern transfer.

Colloidal particles like polystyrene (Haynes and Van Duyne 2001; Frommhold et al. 2012) or silica nanospheres (Gonzalez et al. 2014) have the ability to form ordered structures on surfaces by self-organization due to electrostatic particle–particle interactions. The particles which cover the surface in

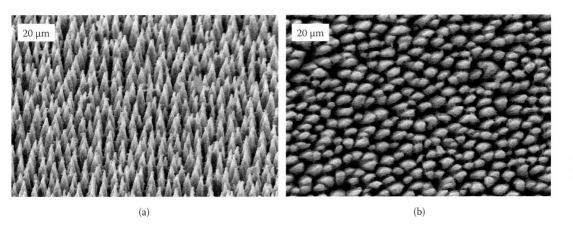

(a)   (b)

Figure 16.15 Laser-structured black silicon ARS under a viewing angle of 30° (a) and in top view (b). Typical structures have dimensions in the µm range.

Functional materials

well-arranged hexagonal arrays can then be used, either directly as etch mask or as lift-off material, to create SiARS with periods in the 100 nm range (Sun et al. 2008).

Furthermore, certain block copolymers like polystyrene-polymethyl methacrylate (PS-PMMA) arrange highly regular on plane substrates if treated correctly (Lohmüller et al. 2008). Here, since the ordering appears on the geometrical scale of single polymer chain blocks, markedly small periods in the range 10–20 nm are feasible (Chang et al. 2013). The geometry of the block copolymer arrangements—1D or 2D periodicity and filling factor—depend heavily on the used polymer and the applied processing conditions (Fredrickson and Bates 1996; Fukunaga et al. 2000; Darling 2007).

REFERENCES

Abdolvand, R., and F. Ayazi. 2008. An Advanced Reactive Ion Etching Process for Very High Aspect-Ratio Sub-Micron Wide Trenches in Silicon. *Sensors and Actuators A* 144 (1): 109–16. doi:10.1016/j.sna.2007.12.026.

Abi-Saab, D., P. Basset, M. J. Pierotti, M. Trawick, and D. E. Angelescu. 2014. Static and Dynamic Aspects of Black Silicon Formation. *Physical Review Letters* 113 (26): 265502. doi:10.1103/PhysRevLett.113.265502.

Allison, J. F., R. Arndt, and A. Meulenberg. 1975. A Comparison of the COMSAT Violet and Non-Reflective Solar Cell. *COMSAT Technology Review* 5: 211–24.

Arndt, R. A., J. F. Allison, J. G. Haynos, and A. Meulenberg. 1975. Optical Properties of the COMSAT Non-Reflective Cell. In *Conference Record of the Eleventh IEEE Photovoltaic Specialists Conference*, 40, Scottsdale, AZ.

Baker, K. M. 1999. Highly Corrected Close-Packed Microlens Arrays and Moth-Eye Structuring on Curved Surfaces. *Applied Optics* 38 (2): 352–6. doi:10.1364/AO.38.000352.

Berger, S. D., and J. M. Gibson. 1990. New Approach to Projection-Electron Lithography with Demonstrated 0.1 µm Linewidth. *Applied Physics Letters* 57 (2): 153. doi:10.1063/1.103969.

Callard, S. 1997. Fabrication and Characterization of Graded Refractive Index Silicon Oxynitride Thin Films. *Journal of Vacuum Science & Technology A: Vacuum, Surfaces, and Films* 15 (4): 2088. doi:10.1116/1.580614.

Chang, C.-C., D. Botez, L. Wan, P. F. Nealey, S. Ruder, T. F. Kuech, C. Chun-Chieh, W. Lei, C.-C. Chang, and L. Wan. 2013. Fabrication of Large-Area, High-Density Ni Nanopillar Arrays on GaAs Substrates Using Diblock Copolymer Lithography and Electrodeposition. *Journal of Vacuum Science & Technology B* 31 (3): 031801. doi:10.1116/1.4798464.

Chou, S. Y. 1996. Nanoimprint Lithography. *Journal of Vacuum Science & Technology B: Microelectronics and Nanometer Structures* 14 (6): 4129. doi:10.1116/1.588605.

Chou, S. Y., P. R. Krauss, and P. J. Renstrom. 1996. Imprint Lithography with 25-Nanometer Resolution. *Science* 272 (5258): 85–7. doi:10.1126/science.272.5258.85.

Darling, S. B. 2007. Directing the Self-Assembly of Block Copolymers. *Progress in Polymer Science* 32 (10): 1152–204. doi:10.1016/j.progpolymsci.2007.05.004.

Dimova-Malinovska, D., M. Sendova-Vassileva, N. Tzenov, and M. Kamenova. 1997. Preparation of Thin Porous Silicon Layers by Stain Etching. *Thin Solid Films* 297 (12): 9–12. doi:10.1016/S0040-6090(96)09434-5.

Duttagupta, S., F. Ma, B. Hoex, T. Mueller, and A. G. Aberle. 2012. Optimised Antireflection Coatings Using Silicon Nitride on Textured Silicon Surfaces Based on Measurements and Multidimensional Modelling. *Energy Procedia* 15 (2011): 78–83. doi:10.1016/j.egypro.2012.02.009.

Fredrickson, G. H., and F. S. Bates. 1996. Dynamics of Block Copolymers: Theory and Experiment. *Annual Review of Materials Science* 26 (1): 501–50. doi:10.1146/annurev.ms.26.080196.002441.

Frommhold, A., A. P. G. Robinson, and E. Tarte. 2012. High Aspect Ratio Silicon and Polyimide Nanopillars by Combination of Nanosphere Lithography and Intermediate Mask Pattern Transfer. *Microelectronic Engineering* 99: 43–9. doi:10.1016/j.mee.2012.06.008.

Füchsel, K., M. Kroll, M. Otto, M. Steglich, A. Bingel, T. Käsebier, R. B. Wehrspohn, E.-B. Kley, T. Pertsch, and A. Tünnermann. 2015. Black Silicon Photovoltaics. In *Photonmanagement in Solar Cells*, edited by R. B. Wehrspohn, U. Rau, and A. Gombert, 117–51. Weinheim, Germany: Wiley-VCH.

Fukunaga, K., H. Elbs, R. Magerle, and G. Krausch. 2000. Large-Scale Alignment of ABC Block Copolymer Microdomains via Solvent Vapor Treatment. *Macromolecules* 33 (3): 947–53. doi:10.1021/ma9910639.

Gonzalez, F. L., L. Chan, A. Berry, D. E. Morse, and M. J. Gordon. 2014. Simple Colloidal Lithography Method to Fabricate Large-Area, Moth-Eye Antireflective Structures on Si, Ge, and GaAs for IR Applications. *Journal of Vacuum Science & Technology B, Nanotechnology and Microelectronics: Materials, Processing, Measurement, and Phenomena* 32 (5): 051213. doi:10.1116/1.4895966.

Grann, E. B., M. G. Moharam, and D. A. Pommet. 1994. Artificial Uniaxial and Biaxial Dielectrics with Use of Two-Dimensional Subwavelength Binary Gratings. *Journal of the Optical Society of America A* 11 (10): 2695. doi:10.1364/JOSAA.11.002695.

Gross, H. 2015. Raytracing. In *Handbook of Optical Systems*, edited by H. Gross, 173–228. Weinheim, Germany: Wiley-VCH Verlag GmbH & Co. KGaA.

Hansen, T. A., C. Johnson, and R. R. Wilbarg. 1980. *Method for Fabricating Non-Reflective Semiconductor Surfaces by Anisotropic Reactive Ion Etching*. US4229233.

Haynes, C. L., and R. P. Van Duyne. 2001. Nanosphere Lithography: A Versatile Nanofabrication Tool for Studies of Size-Dependent Nanoparticle Optics. *The Journal of Physical Chemistry B* 105: 5599–611.

Her, T., R. J. Finlay, C. Wu, and E. Mazur. 2000. Femtosecond Laser-Induced Formation of Spikes on Silicon. *Applied Physics A Materials Science & Processing* 70: 383–5. doi:10.1007/s003390000406.

Her, T.-H., R. J. Finlay, C. Wu, S. Deliwala, and E. Mazur. 1998. Microstructuring of Silicon with Femtosecond Laser Pulses. *Applied Physics Letters* 73 (12): 1673. doi:10.1063/1.122241.

Hobbs, D. S., and B. D. Macleod. 2005. Design, Fabrication, and Measured Performance of Anti-Reflecting Surface Textures in Infrared Transmitting Materials. *Proceedings of SPIE* 5786: 578640.

Hobbs, D. S., and B. D. MacLeod. 2007. High Laser Damage Threshold Surface Relief Micro-Structures for Anti-Reflection Applications. edited by G. J. Exarhos, A. H. Guenther, K. L. Lewis, D. Ristau, M. J. Soileau, and C. J. Stolz. *Proceedings of SPIE* 6720: 67200L. doi:10.1117/12.754223.

Huang, Z., J. E. Carey, M. Liu, X. Guo, E. Mazur, and J. C. Campbell. 2006. Microstructured Silicon Photodetector. *Applied Physics Letters* 89 (3): 033506. doi:10.1063/1.2227629.

Huang, Z., N. Geyer, P. Werner, J. de Boor, and U. Gösele. 2011. Metal-Assisted Chemical Etching of Silicon: A Review. *Advanced Materials* 23 (2): 285–308. doi:10.1002/adma.201001784.

ITRPV. 2016. *International Technology Roadmap for Photovoltaic*. Vol. 7. German Mechanical Engineering Industry Association, Frankfurt, Germany.

Jansen, H., M. de Boer, R. Legtenberg, and M. Elwenspoek. 1995. The Black Silicon Method: A Universal Method for Determining the Parameter Setting of a Fluorine-Based Reactive Ion Etcher in Deep Silicon Trench Etching with Profile Control. *Journal of Micromechanics and Microengineering* 5: 115–20.

Jansen, H., H. Gardeniers, M. de Boer, M. Elwenspoek, and J. Fluitman. 1996. A Survey on the Reactive Ion Etching of Silicon in Microtechnology. *Journal of Micromechanical and Microengineering* 6: 14–28. doi:10.1088/0960-1317/6/1/002.

Köhler, M. 1999. *Etching in Microsystem Technology*. 1st ed. Weinheim: Wiley-VCH.

Laermer, F., A. Schilp, and Robert Bosch GmbH. 1993. *Method of Anisotropically Etching Silicon*. US5501893A.

Lee, S. E., S. W. Choi, and J. Yi. 2000. Double-Layer Anti-Reflection Coating Using MgF_2 and CeO_2 Films on a Crystalline Silicon Substrate. *Thin Solid Films* 376 (1–2): 208–13. doi:10.1016/S0040-6090(00)01205-0.

Li, X., and P. W. Bohn. 2000. Metal-Assisted Chemical Etching in HF/H_2O_2 Produces Porous Silicon. *Applied Physics Letters* 77 (16): 2572. doi:10.1063/1.1319191.

Lohmüller, T., M. Helgert, M. Sundermann, R. Brunner, and J. P. Spatz. 2008. Biomimetic Interfaces for High-Performance Optics in the Deep-UV Light Range. *Nano Letters* 8 (5): 1429–33. doi:10.1021/nl080330y.

Nakanishi, T., T. Hiraoka, A. Fujimoto, T. Okino, S. Sugimura, T. Shimada, and K. Asakawa. 2010. Large Area Fabrication of Moth-Eye Antireflection Structures Using Self-Assembled Nanoparticles in Combination with Nanoimprinting. *Japanese Journal of Applied Physics* 49 (7): 075001. doi:10.1143/JJAP.49.075001.

Nakayama, Y. 1990. Electron-Beam Cell Projection Lithography: A New High-Throughput Electron-Beam Direct-Writing Technology Using a Specially Tailored Si Aperture. *Journal of Vacuum Science & Technology B: Microelectronics and Nanometer Structures* 8 (6): 1836. doi:10.1116/1.585169.

Otto, M., M. Algasinger, H. Branz, B. Gesemann, T. Gimpel, K. Füchsel, T. Käsebier, et al. 2014. Black Silicon Photovoltaics. *Advanced Optical Materials* 3 (2): 147–64. doi:10.1002/adom.201400395.

Pease, R. F. W. 1981. Electron Beam Lithography. *Contemporary Physics* 22 (3): 265–90. doi:10.1080/00107518108231531.

Pfeiffer, H. C. 1978. Variable Spot Shaping for Electron-Beam Lithography. *Journal of Vacuum Science and Technology* 15 (3): 887. doi:10.1116/1.569621.

Resnick, D. J., W. J. Dauksher, D. Mancini, K. J. Nordquist, T. C. Bailey, S. Johnson, N. Stacey, et al. 2003. Imprint Lithography for Integrated Circuit Fabrication. *Journal of Vacuum Science & Technology B: Microelectronics and Nanometer Structures* 21 (6): 2624. doi:10.1116/1.1618238.

Rhee, H., H. Kwon, C.-K. Kim, H. J. Kim, J. Yoo, and Y. W. Kim. 2008. Comparison of Deep Silicon Etching Using SF_6/C_4F_8 and SF_6/C_4F_6 Plasmas in the Bosch Process. *Journal of Vacuum Science & Technology B* 26 (2): 576. doi:10.1116/1.2884763.

Rogalski, A. 2002. Infrared Detectors: An Overview. *Infrared Physics and Technology* 43 (3–5): 187–210. doi:10.1016/S1350-4495(02)00140-8.

Rothenberg, J. E., and R. Kelly. 1984. Laser Sputtering. Part II. The Mechanism of the Sputtering of Al_2O_3. *Nuclear Instruments and Methods in Physics Research B* 1 (2–3): 291–300. doi:10.1016/0168-583X(84)90083-1.

Schröder, S., A. Duparré, L. Coriand, A. Tünnermann, D. H. Penalver, and J. E. Harvey. 2011. Modeling of Light Scattering in Different Regimes of Surface Roughness. *Optics Express* 19 (10): 9820–35. doi:10.1364/OE.19.009820.

Schulze, M., M. Damm, M. Helgert, E.-B. Kley, S. Nolte, and A. Tünnermann. 2012. Durability of Stochastic Antireflective Structures—Analyses on Damage Thresholds and Adsorbate Elimination. *Optics Express* 20 (16): 1422–4.

Schwartz, G. C., and P. M. Schaible. 1979. Reactive Ion Etching of Silicon. *Journal of Vacuum Science and Technology* 16 (2): 410. doi:10.1116/1.569962.

Snyder, P. G., Y.-M. Xiong, J. A. Woollam, G. A. Al-Jumaily, and F. J. Gagliardi. 1992. Graded Refractive Index Silicon Oxynitride Thin Film Characterized by Spectroscopic Ellipsometry. *Journal of Vacuum Science & Technology A: Vacuum, Surfaces, and Films* 10 (4): 1462. doi:10.1116/1.578266.

Southwell, W. H. 1983. Gradient-Index Antireflection Coatings. *Optics Letters* 8 (11): 584. doi:10.1364/OL.8.000584.

Southwell, W. H. 1991. Pyramid-Array Surface-Relief Structures Producing Antireflection Index Matching on Optical Surfaces. *Journal of the Optical Society of America A* 8 (3): 549. doi:10.1364/JOSAA.8.000549.

Steglich, M., T. Käsebier, F. Schrempel, E.-B. Kley, and A. Tünnermann. 2015. Self-Organized, Effective Medium Black Silicon for Infrared Antireflection. *Infrared Physics & Technology* 69: 218–21. doi:10.1016/j.infrared.2015.01.033.

Steglich, M., T. Käsebier, M. Zilk, T. Pertsch, E.-B. Kley, and A. Tünnermann. 2014. The Structural and Optical Properties of Black Silicon by Inductively Coupled Plasma Reactive Ion Etching. *Journal of Applied Physics* 116 (17): 173503. doi:10.1063/1.4900996.

Stenzel, O. 2005. *The Physics of Thin Film Spectra: An Introduction.* 1st ed. Berlin: Springer.

Sun, C. H., P. Jiang, and B. Jiang. 2008. Broadband Moth-Eye Antireflection Coatings on Silicon. *Applied Physics Letters* 92 (6): 061112. doi:10.1063/1.2870080.

Sze, S. M., and K. K. Ng. 2006. *Physics of Semiconductor Devices.* 3rd ed. Hoboken, NJ: Wiley.

Tikhonravov, A. V., M. K. Trubetskov, T. V. Amotchkina, M. A. Kokarev, N. Kaiser, O. Stenzel, S. Wilbrandt, and D. Gäbler. 2006. New Optimization Algorithm for the Synthesis of Rugate Optical Coatings. *Applied Optics* 45 (7): 1515. doi:10.1364/AO.45.001515.

van Wolferen, H., and L. Abelman. 2011. Laser Interference Lithography. In *Lithography: Principles, Processes and Materials*, edited by T. C. Hennessy, 133–48. Hauppauge, NY: Nova Science Publishers.

Winkler, M. T., D. Recht, M. J. Sher, A. J. Said, E. Mazur, and M. J. Aziz. 2011. Insulator-to-Metal Transition in Sulfur-Doped Silicon. *Physical Review Letters* 106 (17): 178701. doi:10.1103/PhysRevLett.106.178701.

Wright, D. N., E. S. Marstein, and A. Holt. 2005. Double Layer Anti-Reflective Coatings for Silicon Solar Cells. In *Conference Record of the Thirty-First IEEE Photovoltaic Specialists Conference, 2005*, 1237–40. IEEE.

Wu, C., C. H. Crouch, L. Zhao, J. E. Carey, R. Younkin, J. A. Levinson, E. Mazur, R. M. Farrell, P. Gothoskar, and A. Karger. 2001. Near-Unity below-Band-Gap Absorption by Microstructured Silicon. *Applied Physics Letters* 78 (13): 1850–2. doi:10.1063/1.1358846.

Yu, Z., H. Gao, W. Wu, H. Ge, and S. Y. Chou. 2003. Fabrication of Large Area Subwavelength Antireflection Structures on Si Using Trilayer Resist Nanoimprint Lithography and Liftoff. *Journal of Vacuum Science & Technology B: Microelectronics and Nanometer Structures* 21 (6): 2874. doi:10.1116/1.1619958.

Zhao, J., and M. A. Green. 1991. Optimized Antireflection Coatings for High Efficiency Silicon Solar Cells. *IEEE Transactions on Electron Devices* 38 (8): 1925–34. doi:10.1109/16.119035.

Zhao, J., A. Wang, and M. A. Green. 1999. 24.5% Efficiency Silicon PERT Cells on MCZ Substrates and 24.7% Efficiency PERL Cells on FZ Substrates. *Progress in Photovoltaics: Research and Applications* 7: 471–4. doi:10.1002/(SICI)1099-159X(199911/12)7:6<471::AID-PIP298>3.0.CO;2-7.

17 Silicon nanowires in biomedicine

Alp Özgün and Bora Garipcan

Contents

17.1 INTRODUCTION TO SILICON NANOSTRUCTURES

Silicon (Si) nanostructures such as Si quantum dots (SiQD), Si nanoparticles (SiNP), Si nanohybrids (SiNH), Si nanocones (SiNCs) and Si nanowires (SiNWs) have received an increasing interest due to their unique physical and chemical properties in electronic and optic fields (Cao et al. 2008; Cao et al. 2006a; Cao et al. 2006b; O'Farrell et al. 2006), (Li et al. 2015), (Wang et al. 2014). In addition to that, SiNWs as one-dimensional (1D) semiconductive nanomaterials exhibit unique biological properties (nontoxic degradation products and biocompatibility), smaller size than cells and a comparable size within biomolecules. These promising 1D nanostructures have found useful applications in the biomedical field. Applications in cell assays, gene and drug delivery, and cell adhesion behavior have seen them emerge as a promising material in recent years.

This chapter comprehensively outlines the synthesis, fabrication, modification, and properties (physical, chemical, and biological) of SiNWs and their applications in biomedicine such as biosensors (sensing and diagnostics), biomaterials, tissue engineering, and drug delivery. Future challenges of SiNWs are also discussed at the end of the chapter. We do believe this chapter will be very useful not only for the life science community but also for engineers and physicists to broaden their research areas.

17.2 SILICON NANOWIRES

17.2.1 SILICON NANOWIRE SYNTHESIS

These methods include different bottom-up approaches where SiNWs are grown vertically from a substrate surface or dispersed on a collector surface, randomly. Each technique is discussed in the following subsections.

17.2.1.1 Vapor–liquid–solid growth

Growing Si whiskers from Si precursors by utilizing metal impurities was first discovered in the early 1960s when Si and iodine (I) were reacted in a vacuum tube in the presence of a small amount of metal (Wagner and Ellis 1964a). The obtained whiskers had diameters ranging from subnanometer scales to tens of micrometers with lengths up to several centimeters. At the time the role of impurities was not entirely understood but it was shown that presence of copper (Cu), gadolinium (Gd), gold (Au), magnesium (Mg), nickel (Ni), osmium (Os), and palladium (Pd) resulted in formation of whiskers while impurities like carbon (C), manganese (Mn), zinc (Zn), and absence of impurities resulted in deposition of films and nodular structures (Wagner and Ellis 1964b) (Cui et al. 2001a). The growth mechanism was explained the same year and referred to as the vapor–liquid–solid (VLS) mechanism, where liquid droplets act as catalytic sites for crystal deposition. Impurities which can form eutectic mixtures with Si create liquid droplets on the substrate surface upon reaching the melting point of the eutectic mixture. When an Si precursor gas such as silicon tetrachloride ($SiCl_4$) or silane (SiH_4) interacts with the droplets, precursor molecules decompose to allow formation of a liquid alloy with elemental Si. The droplets quickly become supersaturated with Si and precipitation starts to occur on the substrate surface. In a self-perpetuating process, crystallized precipitates start growing from underneath the droplet, elevating the droplet from the substrate surface and forming a nanowire with the metal catalyst still on the tip. This growth mechanism means that the size of liquid droplets determines the final diameters of nanowires while the temperature and type of Si precursor are additional controlling factors (Wagner and Ellis 1964; Cui et al. 2001a). Although high process temperatures limit substrates to heat-resistant materials, with plasma-assisted modes of this method one can achieve SiNW growth at lower temperatures that can be compatible with some organic materials (Tian et al. 2016). As nanowires grow, metal catalyst atoms become incorporated in the structure and the growth stops once the catalyst runs out. This means that SiNWs synthesized with this method contain metal impurities which can have detrimental effects for some applications since this will introduce new energy levels and change the band gap properties of Si (Wagner and Ellis 1964). However, recently catalyst-free epitaxial growth alternatives have been developed that eliminate this problem (Ishiyama et al. 2016). This problem can also be turned into an advantage by using catalyst droplets as a mixing bowl for precursors of other nanomaterials which results in nanocrystals embedded in SiNWs (Panciera et al. 2015). VLS growth is a low-yield method but may offer opportunities for precise control over the process. It was shown that SiNW growth direction and diameters can be manipulated by applying an electric field during growth (Panciera et al. 2016). Moreover, when Au particle colloids are used as catalysts they enable tuning of the spatial distribution of nanowires as well as the diameters. Monodispersed colloids result in a narrow diameter distribution and nanowire density is controllable by colloid concentration. Moreover, these colloids can be deposited on substrate surfaces in patterns by using soft lithography methods to obtain patterned SiNW arrays (Hochbaum et al. 2005).

Composition of nanowires can also be altered with this method. Using pulsed precursor vapor sources enables block-by-block synthesis of nanowires. Laser pulses were used to evaporate material from a Germanium (Ge) target within the reaction chamber which resulted in Si- and Ge- rich domains within nanowires whose lengths can be controlled by changing pulse durations (Wu et al. 2002).

The VLS method is also convenient in terms of allowing additional fabrication steps to be performed especially in device designs such as nanowires which are directly grown on low-cost substrates. It was shown that SiNWs grown on glass substrates with the VLS method can be coated with additional layers to obtain core–shell structured wires which can greatly enhance broadband absorption in solar cell devices (Adachi et al. 2013). The resemblance of SiNWs grown with the VLS method to fibrous proteins in the extracellular matrix (ECM) was utilized for obtaining bioinspired surfaces for cell culture studies (Yang et al. 2016).

17.2.1.2 Oxide-assisted growth method

In the late 1990s efforts for modifying the VLS growth method led to the discovery of a catalyst-free SiNW growth mechanism (Zhang et al. 1998). It was shown that laser ablation of Si–silicon dioxide (SiO$_2$) targets as a precursor source resulted in formation of SiNWs on the substrate without the need for metal impurities or high temperature while pure Si or SiO$_2$ did not produce the same results. The proposed growth mechanism involves Si suboxide clusters in the vapor favoring the formation of Si–Si bonds and chemically attaching to the Si substrate surface. These clusters then act as nucleation sites that promote bonding of more suboxide clusters and oxygen (O) atoms are expelled toward the edges as nanowire formation occurs which form an inert oxide sheath, preventing lateral growth. Even when a metal catalyst is used, the reaction follows the same mechanism and resulting nanowires lack solidified catalyst tips obtained with the VLS method (Zhang et al. 2003). This method allows synthesis of long, uniformly aligned SiNWs in the direction of the carrier gas flow (Shi et al. 2000). Process temperature and pressure affects growth yield and lower pressures yield thinner, smoother SiNWs (Yao et al. 2005) (Fan et al. 2001). Besides producing an abundance of smooth nanowires, this method also creates a small number of spring-shaped and fishbone-shaped nanowires (Tang et al. 1999). Using particulate substrates can allow limited control over nanowire morphology by changing particle shapes (Zhang et al. 2001). A recent relatively simple method that involves oxidation and subsequent reduction of Si wafer surfaces to grow SiNWs utilizes this mechanism in an economical and up-scalable process (Behura et al. 2014).

17.2.1.3 Molecular beam epitaxy

Molecular beam epitaxy is a thin layer growth method used in nanofabrication processes which takes place in ultrahigh vacuum and allows formation of uniform films of different compositions controllable down to a single-atom thickness in a very slow process (Cho and Arthur, 1975). When this method is applied to a Si substrate with metal droplets it results in SiNW growth from underneath the droplets very much like the VLS method. However, the growth mechanism is entirely different in this case. Deposited vapor is not a Si precursor gas but a uniform vapor flux from an ultrahigh purity Si target. This flux of Si atoms supersaturate the droplets and start crystallizing while also being deposited on bare substrate surface around the droplets. However, interaction of Au with substrate creates an excess of elastic surface energy, which makes it thermodynamically favorable for Si atoms around the droplet to migrate toward the droplets. This phenomenon causes faster deposition under the droplets while creating pits around the growing pillars (Schubert et al. 2013). Growth occurs under droplets whose diameters are within a certain range which limits minimum nanowire diameter to about 40 nm (Zakharov et al. 2006) since nanowire diameter is determined by droplet diameter. Unlike the VLS method this mode of growth causes smaller droplets to grow faster than larger ones. Average growth rate is substantially slower than other methods and maximum achievable nanowire length is in the submicron scale (Werner et al. 2006) but resulting SiNWs show vertical epitaxial alignment and lack metal impurities (Nguyen et al. 2016).

17.2.1.4 Wet synthesis

Solution based synthesis allows formation of extremely thin, monodisperse, and pure crystalline nanowires. The method involves supersaturating an organic solvent with a Si precursor such as diphenylsilane and adding metal nanocrystals to seed nanowire growth. Reaction takes place under pressure and above the eutectic temperature of metal-Si complex (e.g., 363°C for Si-Au). At these temperatures Si precursor molecules decompose and generated Si atoms dissolve in metal nanocrystals. After saturation SiNW growth occurs at nanocrystal surface. Obtained nanowires are single crystalline and crystal orientations are highly controllable by adjusting the reaction pressure. Nanowire diameters are determined by nanocrystal size. It is possible to obtain down to tens of angstroms (Å) thick nanowires with this method (Holmes, 2000). Besides Au (Lu et al. 2003), Ni (Tuan et al. 2005) and Cu (Tuan et al. 2008) nanocrytals can also be used for seeding. This mechanism was also shown to work under atmospheric pressure in high boiling point solvents (Heitsch et al. 2008). Increasing Au nanocrystal reactivity with an Au etchant such as potassium iodide (KI) allows this mechanism to take place in aqueous solutions under atmospheric pressure (Park and Choi, 2014).

17.2.1.5 Laser ablation

This method involves material evaporation from a Si–SiO$_2$ (Wang et al. 1998)or Si-metal (Zhang et al. 1998) target with a pulsed laser beam in an inert carrier gas environment. SiNWs get deposited on reactor walls in the vicinity of a cold finger. While the exact growth mechanism is unknown the termination of nanowires in nanoclusters suggests the process involves VLS mechanism (Morales and Lieber, 1998). Resulting SiNWs differ from conventional VLS growth method for having smooth curves and a lack of impurities. Growth rate and nanowire yield are also substantially higher than VLS mode of growth (Wang et al. 1998). Smallest obtained nanowire diameters are just below 10 nm ranging up to several hundred nanometers. Thickness distribution can be adjusted to an extent by changing the type of carrier gas (Zhang et al. 1999).

17.2.2 SILICON NANOWIRE FABRICATION

SiNW fabrication techniques are top-down methods that involve selective etching of bulk Si to leave behind 1D structures. Besides lithographic methods that generally create horizontal SiNWs, metal-assisted chemical etching is a popular top-down method for vertical SiNW fabrication. This process involves etching of Si substrates in metal containing hydrofluoric acid (HF) solutions near room temperature (RT). Metal atoms get deposited onto Si surface and form nuclei that act as cathodes in a micro-electrochemical process which etches away the Si atoms underneath it leaving behind pillars of metal-free zones (Peng et al. 2003). This simple and low-cost method can create different SiNW morphologies depending on temperature, solution concentration, immersion time and metal particles spacing (Han et al. 2014). While random deposition of metal nuclei creates chaotic and noncontrollable morphologies (Peng et al. 2006) if lithographic techniques such as nanosphere lithography are used highly ordered vertical SiNW arrays can be obtained (Jia et al. 2016). Additionally, in this method laser ablation of nanospheres gives control over spatial distribution of ordered SiNWs (Brodoceanu et al. 2016). While lithographic techniques are especially useful for device-integrated horizontal nanowires (Singh, et al. 2006), it is also possible to fabricate vertical SiNWs (Nakamura et al. 2013).

17.2.3 SILICON NANOWIRE MODIFICATION

Depending on the area of application, modifying surface chemistry of SiNWs becomes a vital requirement especially in sensor applications where surface functional groups exclusively determine interactions with analyte molecules (Patolsky et al. 2006a). Clean Si surfaces oxidize spontaneously upon exposure to ambient conditions forming a layer of silicon oxide (SiO$_x$) called native oxide layer (Morita et al. 1990). This layer, also present on SiNW surfaces, becomes useful for surface modification as they readily react with alkoxysilane groups after a brief cleaning procedure and form self-assembled monolayers as illustrated in Figure 17.1 (Aswal et al. 2006). Several alkoxysilanesare available designed for functionalizing surfaces with desired chemical groups were summarized in Table 17.1. For example amine (-NH$_2$) groups on SiNWs are especially useful for further coupling of other molecules due to their reactivity toward aldehyde (CHO), carboxylic acid (-COOH) and epoxy (COC) groups (Hermanson, 2013). 3-aminopropyltriethoxysilane (APTES) (Susarrey-Arce et al. 2016), (3-aminopropyl)-dimethylethoxysilane (APDMES) (Lichtenstein et al. 2014) and N-(2-aminoethyl)-3-aminopropyltrimethoxysilane (AEAPS) (Bi et al. 2008) are the most commonly used reagents for this purpose. Amine functionalization is often followed glutaraldehyde linking to form aldehyde groups on the surface which can bind to other amine moieties (Yen et al. 2016). Highly reactive aldehyde groups can also be obtained directly on SiNW surface with the same principle using triethoxysilane aldehyde (TEA) (Bi et al. 2009). This route results in a thinner organic layer around SiNWs but brings the drawback of aldehyde oxidation therefore requires execution of following modification steps immediately (Zheng et al. 2005). Thiol groups can also be created on SiNW surface for ion detection or thiol based conjugation strategies using (3-mercaptopropyl)trimethoxysilane (MPTMS) (Guo et al. 2016). Noncovalent routes can also be used for SiNW modification. Since SiO$_x$ surfaces are negatively charged at neutral pH values, polycations are deposited on SiNW surfaces through electrostatic interactions between oxide surface and polymer backbone. Polymer chains can be decorated

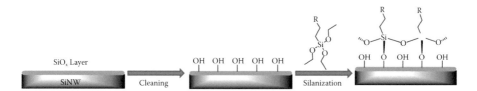

Figure 17.1 Modification of SiNW native oxide layer using alkoxysilane functionalized molecules.

Table 17.1 Modification routes of SiNWs with different surface groups

BOND TYPE	MODIFICATION	SURFACE GROUP	TARGET SURFACE	REFERENCE
Covalent	APTES	Amine (-NH$_2$)	SiO$_x$	Susarrey-Arce et al. 2016
	APDMES	Amine (-NH$_2$)	SiO$_x$	Lichtenstein et al. 2014
	AEAPS	Amine (-NH$_2$)	SiO$_x$	Bi et al. 2008
	TEA	Aldehyde (-CHO)	SiO$_x$	Bi et al. 2009
	MPTMS	Thiol (SH)	SiO$_x$	Guo et al. 2016
	Terminal alkyne	Fluorine	Si-H	Minh et al. 2016
	Terminal alkene	Amine (-NH$_2$)	Si-H	Bunimovich et al. 2006
Noncovalent	Polyelectrolytes	OEG/biotin	SiO$_x$	Duan et al. 2015
	Lipid bilayers	Lipid membrane	SiO$_x$	Römhildt et al. 2013

with selectively reactive groups for sensing applications or oligo-ethylene-glycol (OEG) groups can be attached to prevent protein adsorption (Duan et al. 2015). In a bioinspired approach, SiNWs can also be coated with fluid lipid bilayers akin to cell membranes which can become useful in studying interactions of molecules with cell membranes and membrane dynamics (Römhildt et al. 2013). Modifying oxide-free SiNWs is also possible by using terminal alkene or alkyne functionalities. Terminal unsaturated C–C groups react with Si–H groups on SiNW surface after oxide removal from SiNW surfaces with HF etching (Minh et al. 2016). This modification route creates true monolayers that are more stable than oxide layer modifications and result in improved performance for sensing applications (Bunimovich et al. 2006).

17.2.4 PROPERTIES OF SILICON NANOWIRES

17.2.4.1 Physical and chemical properties

SiNWs behave as semiconductors and can be doped to exhibit p-type or n-type behavior similar to bulk Si (Cui et al. 2000). The small sizes of SiNWs give them extraordinary properties compared to bulk Si where growth direction, size, morphology, and surface groups dictate their final electrical properties (Leao et al. 2007). It was calculated that reducing nanowire dimensions increases the band gap of bulk Si and the indirect band gap of bulk Si becomes direct in the case of SiNWs due to quantum confinement (Nolan et al. 2007). This effect is observed to a less extent when an oxide layer is present (Sacconi et al. 2007). The direct and tunable band gap of SiNWs make them optically active materials unlike bulk Si. Their photoluminescence in the visible and near-infrared region blue-shifts as diameter decreases due to increasing band gap (Guichard et al. 2006).

SiNWs are readily oxide terminated after synthesis, with oxide layer thickness depending on synthesis mode and parameters. Removal of the oxide layer with HF treatment yields hydrogen (H)-terminated SiNWs which are stable in air at room temperature but H desorption occurs at higher temperatures or in aqueous environments (Sun et al. 2003). Alkylation of H-terminated SiNW surfaces increases their oxidation resistance depending on alkyl chain length and molecular coverage (Bonds et al. 2008). H-terminated SiNWs and their noble metal modified derivatives show excellent catalytic activity exhibited by degradation of rhodamine B and oxidation of benzoyl alcohol to benzoic acid. Electron-deficient hydrides on SiNW surfaces likely act as electron sinks to promote catalytic activity (Shao et al. 2009). Decoration of H-terminated SiNWs with metal nanoparticles such as Ag and Cu enables them to catalyze reduction

Functional materials

reactions (Amdouni et al. 2016; Fellahi et al. 2015). Reduction of metal atom clusters can be utilized to grow metal clusters on H-terminated SiNW surfaces. Growing clusters oxidize H-terminated surfaces and detach from SiNWs after reaching a critical size (Sun et al. 2002).

17.2.4.2 Biological properties

Freestanding SiNWs are not particularly cytotoxic at concentrations used for biological applications such as drug delivery (Lv et al. 2010; Peng et al. 2013; Peng et al. 2014). Cell culture studies performed for demonstrating cytotoxic effects of freestanding SiNWs show reduced cell viability at high SiNW concentrations, with the effect becoming significant starting from 50 ug/mL (Alexander Jr et al. 2012) and 190 ug/mL (Adili et al. 2008) for different epithelial cell lines. On inspection of cytotoxicity mechanisms, it was discovered that apoptotic pathways are not involved but the effect is likely to be solely from the mechanical disturbances exerted on the cells by the overwhelming numbers of SiNWs, as differential studies conducted with Si nanoparticles lack this cytotoxicity (Adili et al. 2008). Surface-bound vertically aligned SiNW arrays have varying effects on different cell lines. They enhance cell adhesion and restrict cell spreading in studies conducted on epithelial HepG2 cells by activation of mechanosensitive pathways to upregulate adhesion-specific proteins (Qi et al. 2009) while suppressing adhesion of mouse osteoblast MC3T3-E1 cell line (Brammer et al. 2009). Mouse retinal cells on the other hand undergo phenotypic changes on these surfaces such as loss of neurites and reduction of retinal markers resulting in a viability decrease (Piret et al. 2014). Cell adhesion on SiNW arrays can be discouraged by a superhydrophobic chemical treatment such as octadecyltrichlorosilane (ODTS) and patterned cell attachment can be achieved via simple lithography techniques during chemical treatment (Piret et al. 2011). The ability of vertically aligned SiNWs to alter intracellular pathways culminates in influences on stem cell fate. Mesenchymal stem cells (MSCs) cultured on these surfaces show spontaneous differentiation into osteocytes and chondrocytes (Liu et al. 2013) but this effect weakens with SiNW lengths over 10 um (Kuo et al. 2012). Integrin, TGF-β/BMP, Akt, MAPK, Insulin, and Wnt pathways are shown to be involved in this process (Liu et al. 2014). In terms of biocompatibility, Si is known to be a relatively inert material (Stensaas and Stensaas 1978) and poor hemocompatibility of bare Si can be resolved with the presence of an oxide layer since platelet adhesion on SiO_2 is significantly lower than on bare Si (Weisenberg and Mooradian 2002). This effect is also visible in nanoscale counterparts as oxide-terminated SiNWs were shown to exhibit decreased fibrinogen adsorption and extended coagulation times (Garipcan et al. 2011). In vivo studies suggest a temporary inflammatory response to pulmonary exposure to freestanding SiNWs resulting in fibrous tissue formation as a defense mechanism and clearance by alveolar macrophages. SiNW clearance from lungs was found to be faster than carbon nanotube (CNT) clearance (Roberts et al. 2012). Similar results were obtained on implantation of SiO_x-coated gallium phosphide (GaP) nanowires into a rat brain. A temporary inflammatory period characterized by astrocyte activation resulting in nanowire phagocytosis by microglia and clearance was observed (Linsmeier et al. 2009). Injection of SiNWs into bloodstream gives clues about clearance mechanisms where their presence was detected in liver and spleen but not in kidneys, suggesting opsonization by the mononuclear phagocyte system and clearance by the liver (Jung et al. 2009). Despite the relatively bioinert qualities of Si, porous Si shows surprising bioactivity including biodegradation (Canham 1995). Porous Si slowly dissolves in simulated body fluids and releases orthosilicic acid $Si(OH)_4$, a naturally occurring Si derivative which is involved in bone and collagen formation (Anderson et al. 2003). This feature of porous Si allows formation of biodegradable SiNWs as drug and gene delivery vehicles (Chiappini et al. 2015). For more information on porous SiNWs we refer to Qu et al. (2011).

17.2.5 SILICON NANOWIRE APPLICATIONS IN BIOMEDICINE

17.2.5.1 Biosensors

SiNW-based biosensors are constructed in the form of field effect transistors (FET). In a typical FET device metal source and drain electrodes are connected through a positively (p-type) or negatively (n-type) doped Si region whose conductance is controlled via a gate electrode. The gate electrode is not physically in contact with the semiconductor but the voltage applied to it creates an electric field, which depletes or accumulates charge carriers within the semiconductor depending on dopant type as seen in Figure 17.2a (Sze and Ng 2007). This concept enables the amount of current that can pass from source to drain to be

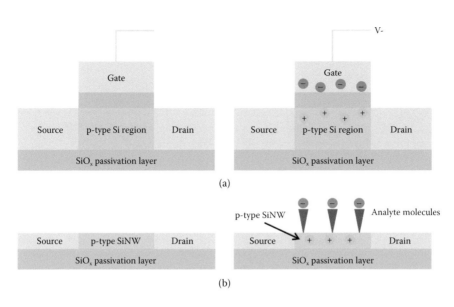

Figure 17.2 (a) Conventional p-type FET device. Applying negative voltage to gate electrode accumulates charge carriers, increasing the conductivity of the p-type region. (b) SiNW-based FET sensor. Binding of negatively charged molecules on an SiNW surface creates the same effect as negative gate voltage, except charge carriers accumulate in the bulk of the SiNW. (Cohen-Karni, T., and Lieber, C.M., *Pure Appl. Chem.*, 85, 883–901, 2003.)

determined by the polarity and density of charges at the gate electrode, which enables FET devices to be used as sensors. If instead of applying a voltage, charged molecules were to be attached to the gate surface, their electric field will also have an effect on the current (Turner and Wilson 1989; Domanský et al. 1998). Using this principle to sense molecule interactions with gate surfaces gained profound attention but suffered limited sensitivity due to charge carrier accumulation or depletion occurring in a superficial surface layer of the semiconductor (Janata 1994).

When the semiconductor region becomes 1D, such as in an n-type or p-type doped SiNW, on binding the charged species to the SiNW, carrier depletion or accumulation takes place in the bulk of the region as shown in Figure 17.2b, leading to a greater effect on conductance (Patolsky et al. 2006b). Increased surface-to-volume ratio of the 1D structure plays a principle role in this phenomenon as nonlinear sensitivity enhancement was observed with SiNW diameters under 100nm (Li et al. 2005). Calculations show that sensitivity also depends on nanowire length, dopant type, and doping density (Nair and Alam 2007). The first application of charged species detection was a pH sensor that was sensitive to protonation and deprotonation of NH_2 and oxide (SiOH) species bound on SiNW surfaces. Furthermore, biotin- and antigen-modified SiNWs were shown to selectively detect streptavidin and complementary antibodies, respectively (Cui et al. 2001b). Selective sensing of analytes requires immobilization of recognizing molecules on the SiNW surface using the aforementioned chemical modification techniques. After desired groups are created on the SiNW surface, bioconjugation techniques may be used for biomolecule immobilization (Nuzaihan et al. 2016). Nonselective modification of FET devices creates a challenge where recognizing molecules are immobilized on all modified surfaces including nonsensing parts. Recognizing molecules attached to nonsensing regions capture analyte molecules and increase the minimum amount of analyte needed for detection of diminishing sensor sensitivity. Selective bioconjugation on SiNW surfaces through the joule heating method overcomes this problem (Yun et al. 2013). SiNW FETs enable detection of clinically relevant markers with excellent sensitivity, making them promising candidates for point-of-care (POC) early disease diagnosis devices. Prostate-specific antigen (PSA), a blood-borne prostate cancer marker, can be detected in concentrations down to 1 femtogram (fg)/mL using SiNW FET biosensors modified with anti-PSA antibodies (Kim et al. 2007a). Fabrication of sensors on inexpensive single-use chips that are used with a readout platform can enable POC detection of cancer markers and rapid patient evaluation (Mohd Azmi et al. 2014). Acute diseases such as myocardial infarction can also be detected as an early alarm system through extremely low concentration sensing of cardiac troponin I using SiNW FETs (Kim et al. 2016). Such early alarm systems may involve implantation of the device for constant monitoring which requires ultra-thin metal oxide passivation of SiNWs for extended stability (Zhou et al. 2014). Decorating SiNW surfaces with antibodies against specific antigens of viruses enables selective detection of a desired virus type, making them relevant tools for not only disease diagnosis but also prevention of biological warfare and terrorism as conventional virus detection tools involve tedious sample preparation and long testing times. Using this concept, bottom-up synthesized SiNWs were fabricated into FET arrays and modified by antibodies to detect influenza A virus with a sensitivity high enough to detect binding of a single virus (Patolsky et al. 2004). Biosensing applications also include detection of specific nucleic acid sequences. Immobilization of a peptide nucleic acid sequence on an SiNW surface can detect the presence of a complementary sequence in tens of femtomolar (fM) concentrations while generating distinguishably different signals for mismatched sequences (Hahm and Lieber 2004). Recent advances in fabrication methods allow detection of single stranded DNA (ssDNA) sequences using ssDNA-modified SiNWs (Adam and Hashim 2015) with 1 fM and single-base-mismatch sensitivity (Lu et al. 2013). Nucleic acid-modified SiNWs are also extremely useful in studying biochemical pathways. DNA sequences immobilized on SiNW surfaces can shed light on protein DNA interactions where some proteins selectively bind to different sequences with slightly different affinities (Zhang et al. 2011). MicroRNA (miRNA) sequences can also be detected with this principle for studying post-transcriptional regulation mechanisms (Chen et al. 2015). High density arrays of SiNW FET devices can be interfaced with axons and dendrites of mammalian neurons for stimulation, inhibition, and detection of neuronal activity which enables in vitro investigation of neuronal networks (Patolsky 2006). Integration of these sensors with microfluidic flow systems provides precisely controlled sample injection volumes and flow rates, eliminating time-consuming manual sample switching and rinsing steps (De et al. 2013), paving the way for real-time highly sensitive, high-throughput, rapid drug screening systems (Wu 2014).

17.2.5.2 Biomaterials and tissue engineering

Since lost tissues and organs cannot always be replaced by mechanical devices, tissue engineering unites engineering principles with life sciences to offer alternative biological substitutes (Langer and Vacanti 1993). This often involves utilizing biomaterials seeded with stem cells and differentiation of stem cells to desired tissue type either by added biochemical factors (Mathews et al. 2012), or biochemical cues provided by the support material itself (Wang et al. 2012). SiNW diameters are comparable to biomolecules which enable them to interfere with biochemical processes and provide biophysical cues for stem cell differentiation as shown with osteogenic and chondrogenic differentiation of MSCs (Kuo et al. 2012; Liu et al. 2013). Using different supplements including chemicals, proteins, and cytokines MSCs can be enabled to transdifferentiate into cells of different tissue types such as skin, liver, and brain. Recently, human MSCs were transdifferentiated into neuron-like cells by culturing on sparsely distributed long SiNW arrays without any supplements. SiNWs with different lengths were tried and longer SiNWs were shown to be more effective at neuron-like transdifferentiation indicated by neurite extensions and increased expression levels of neuronal markers TUBB3, NEGR1, NEUROD, and NES (Kim et al. 2015). Vertically grown SiNWs also affect mature neuron growth by completely changing neurite development process. Hippocampal neurons plated on SiNW arrays show a much faster major neurite growth as well as a higher number of minor neurites. The absence of growth cones indicates a neuron development process different than conventional in vitro neuron development (Kang et al. 2016). Electrical properties of SiNWs also come in handy in tissue engineering applications. Induced pluripotent stem cells can be differentiated into cardiomyocytes to form contracting three-dimensional (3D) heart tissues. However, nonsynchronized individual contractions of cardiomyocytes prevent clinical translations of this concept. Incorporation of freestanding conductive SiNWs into 3D cardiomyocyte culture forms a network within the tissue, synchronizing the electrical activities of individual cells resulting in synchronized and enhanced contractions (Tan et al. 2015). Using SiNW-based FET biosensor principles, SiNW incorporated 3D cultures also enable noninvasive probing of cellular parameters deep within the structure such as local electric activity, drug response, and pH. This concept involves solution casting of freestanding SiNWs onto a sacrificial layer prior to FET nanofabrication. Removal of a sacrificial layer yields a freestanding network which is then incorporated into a biomaterial such as Matrigel to obtain 3D scaffolds (Tian et al. 2012). This approach was used to construct a 3D cardiac muscle tissue and record its extracellular action potentials using the embedded FET array to map tissue-wide electric activity (Dai et al. 2016).

17.2.5.3 Drug delivery

Nanoscale drug delivery strategies are especially promising in oncology applications since nanocarriers can diminish debilitating side effects of chemotherapeutics by encapsulating drug molecules and delivering them to the tumor site using targeting molecules or convoluted selective stimulus sensitive pathways (Wicki et al. 2015). Porous Si is one of the ideal candidate materials for nanocarrier fabrication since it can be formed into particles with varying size and shapes that can entrap drug molecules within the pores. Moreover, these carriers dissolve in physiologic conditions to form harmless products after payload delivery (Savage et al. 2013). Porous SiNWs were shown to have extremely high drug-loading capacities in experiments conducted with anti-cancer agent doxorubicin and loaded SiNWs outperformed free doxorubicin on tumor-bearing nude mice (Peng et al. 2013). Bioadhesion approaches are explored for retention of drug carriers around the target site for extended periods where nanocarrier surfaces are functionalized with groups that adhere to mucous constituents or cell membranes. Attaching SiNWs to carrier surfaces can become useful in bioadhesion applications by providing physical attachment to cell membranes and making them more permeable to drug payloads (Fischer et al. 2009; Uskoković et al. 2012). Penetration of SiNWs into cell membranes is also important for efficient delivery. Coating freestanding SiNWs with cell-penetrating peptides was shown to enhance cell internalization of SiNWs into hippocampal neurons (Lee et al. 2016). Vertical SiNWs can also be useful for intracellular delivery. Cells plated on vertical SiNW arrays coated with green fluorescent protein (GFP) show GFP expression, although with very low efficiency, by piercing the cell membrane and injecting the gene into the cytosol (Kim et al. 2007b). Membrane penetration efficiency of SiNW arrays can be enhanced by applying an electrical stimulus to create an electroporation effect (Xie et al. 2012), or coating SiNWs with surfactants to locally dissolve lipid membranes (Peer et al. 2012). SiNWs can also be coated with lipid bilayers which then merge with cell membranes and provide cytosolic access (Tian et al. 2010). These methods were proved useful for not only in vitro cytosolic delivery of DNA, RNA and peptide sequences, proteins, and small molecules (vertical silicon nanowires as a universal platform for delivering biomolecules into living cells), but also local in vivo delivery of growth factors (Chiappini et al. 2015).

17.2.6 THE FUTURE OF SILICON NANOWIRES IN BIOMEDICINE

SiNWs have already proved their invaluableness in the field of biomedicine. Their physical and chemical versatility will only increase the number of areas where they are useful in the future. FET biosensor applications already provide unmatched sensitivity which we predict to increase as SiNW fabrication and synthesis methods are coupled with new advanced nanofabrication methods. Multiplexing of these FET devices may lead to extremely high-throughput drug discovery and gene screening platforms. SiNW FET devices enable miniaturization of biosensor platforms which may pave the way for implantable chips for constant monitoring of vital patient parameters or disease markers in the blood if strategies for extended SiNW stability can be developed. Hybridization of biosensor and tissue engineering applications of SiNWs has already conceived exciting concepts like 3D tissue graft monitoring and tissue-electronic hybrids. These applications may outperform or even completely eliminate the need for animals for disease models in the future as FET arrays provide real-time tissue-wide information during testing. The challenge is integrating these sensor arrays into more complex organoids rather than conventional 3D cultures. Tissue-electronic hybrids may also become a standard in regenerative medicine, where tissue grafts are constantly monitored after implantation and required stimulations to the grafts can be administered immediately.

REFERENCES

Adachi MM, Anantram MP, Karim KS (2013). Core-shell silicon nanowire solar cells. *Sci Rep* 3:1546.
Adam T, Hashim U (2015). Highly sensitive silicon nanowire biosensor with novel liquid gate control for detection of specific single-stranded DNA molecules. *Biosens Bioelectron* 67:656–661.
Adili A, Crowe S, Beaux MF, Cantrell T, Shapiro PJ, McIlroy DN, Gustin KE (2008). Differential cytotoxicity exhibited by silica nanowires and nanoparticles. *Nanotoxicology* 2:1–8.

Alexander FA Jr, Huey EG, Price DT, Bhansali S (2012). Real-time impedance analysis of silica nanowire toxicity on epithelial breast cancer cells. *Analyst* 137:5823–5828.

Amdouni S, Coffinier Y, Szunerits S, Zaïbi MA, Oueslati M, Boukherroub R (2016). Catalytic activity of silicon nanowires decorated with silver and copper nanoparticles. *Semicond Sci Technol* 31:14011.

Anderson SHC, Elliott H, Wallis DJ, Canham LT, Powell JJ (2003). Dissolution of different forms of partially porous silicon wafers under simulated physiological conditions. *Phys Status Solidi (A) Appl Res* 197:331–335.

Aswal DK, Lenfant S, Guerin D, Yakhmi JV, Vuillaume D (2006). Self assembled monolayers on silicon for molecular electronics. *Anal Chim Acta* 568:84–108.

Behura SK, Yang Q, Hirose A, Jani O, Mukhopadhyay I (2014). Catalyst-free synthesis of silicon nanowires by oxidation and reduction process. *J Mater Sci* 49:3592–3597.

Bi X, Agarwal A, Yang K-L (2009). Oligopeptide-modified silicon nanowire arrays as multichannel metal ion sensors. pdf. *Biosens Bioelectron* 24:3248–3251.

Bi X, Wong WL, Ji W, Agarwal A, Balasubramanian N, Yang KL (2008). Development of electrochemical calcium sensors by using silicon nanowires modified with phosphotyrosine. *Biosens Bioelectron* 23:1442–1448.

Bonds SC, Bashouti MY, Stelzner T, Berger A, Christiansen S, Haick H (2008). Chemical passivation of silicon nanowires with C 1 - C 6 alkyl chains through covalent. *J Phys Chem C* 2:19168–19172.

Brammer KS, Choi C, Oh S, Cobb CJ, Connelly LS, Loya M, Kong SD, Jin S (2009). Antibiofouling, sustained antibiotic release by Si nanowire templates. *Nano Lett* 9:3570–3574.

Brodoceanu D, Alhmoud HZ, Elnathan R, Delalat B, Voelcker NH, Kraus T (2016). Fabrication of silicon nanowire arrays by near-field laser ablation and metal-assisted chemical etching. *Nanotechnology* 27:75301.

Bunimovich YL, Shin YS, Yeo W, Amori M, Kwong G, Heath JR (2006). Quantitative real-time measurements of DNA hybridization with alkylated nonoxidized silicon nanowires in electrolyte solution quantitative real-time measurements of DNA hybridization with alkylated nonoxidized silicon nanowires in electrolyte solution. *Biotechniques* 128:16323–16331.

Canham LT (1995). Bioactive silicon structure fabrication through nanoetching techniques. *Adv Mater* 7:1033–1037.

Cao L, Garipcan B, Atchison JS, Ni C, Nabet B, Spanier JE (2006a). Instability and transport of metal catalyst in the growth of tapered silicon nanowires. *Nano Lett* 6:1852–1857.

Cao L, Garipcan B, Gallo EM, Nonnenmann SS, Nabet B, Spanier JE (2008). Excitation of local field enhancement on silicon nanowires. *Nano Lett* 8:601–605.

Cao L, Nabet B, Spanier JE (2006b). Enhanced Raman scattering from individual semiconductor nanocones and nanowires. *Phys Rev Lett* 96:157402-157406.

Chen K-I, Pan C-Y, Li K-H, Huang Y-C, Lu C-W, Tang C-Y, Su Y-W, Tseng L-W, Tseng K-C, et al. (2015). Isolation and identification of post-transcriptional gene silencing-related micro-RNAs by functionalized silicon nanowire field-effect transistor. *Sci Rep* 5:17375.

Chiappini C, De Rosa E, Martinez JO, Liu X, Steele J, Stevens MM, Tasciotti E (2015). Biodegradable silicon nanoneedles delivering nucleic acids intracellularly induce localized in vivo neovascularization. *Nat Mater* 14:6–13.

Cho AY, Arthur JR (1975). Molecular beam epitaxy. *Prog Solid State Chem* 10:157–191.

Cohen-Karni T, Lieber CM (2003). Nanowire nanoelectronics: Building interfaces with tissue and cells at the natural scale of biology. *Pure Appl Chem* 85:883–901.

Cui Y, Duan X, Hu J, Lieber CM (2000). Doping and electrical transport in silicon nanowires. *J Phys Chem B* 104:5213–5216.

Cui Y, Lauhon LJ, Gudiksen MS, Wang JF, Lieber CM (2001a). Diameter-controlled synthesis of single-crystal silicon nanowires. *Appl Phys Lett* 78:2214–2216.

Cui Y, Wei Q, Park H, Lieber CM (2001b). Nanowire nanosensors for highly sensitive and selective detection of biological and chemical species. *Science* 293:1289–1292.

Dai X, Zhou W, Gao T, Liu J, Lieber CM (2016). Three-dimensional mapping and regulation of action potential propagation in nanoelectronics-innervated tissues. *Nat Nanotechnol.*11:776–782

De A, van Nieuwkasteele J, Carlen ET, van den Berg A (2013). Integrated label-free silicon nanowire sensor arrays for (bio)chemical analysis. *Analyst* 138:3221–3229.

Domanský K, Baldwin DL, Grate JW, Hall TB, Li J, Josowicz M, Janata J (1998). Development and calibration of field-effect transistor-based sensor array for measurement of hydrogen and ammonia gas mixtures in humid air. *Anal Chem* 70:473–481.

Duan X, Mu L, Sawtelle SD, Rajan NK, Han Z, Wang Y, Qu H, Reed MA (2015). Functionalized polyelectrolytes assembling on nano-BioFETs for biosensing applications. *Adv Funct Mater* 25:2279–2286.

Fan XH, Xu L, Li CP, Zheng YF, Lee CS, Lee ST (2001). E€ects of ambient pressure on silicon nanowire growth. *Chem Phys Lett* 334:229–232.

Fellahi O, Barras A, Pan G-H, Coffinier Y, Hadjersi T, Maamache M, Szunerits S, Boukherroub R (2015). Reduction of Cr(VI) to Cr(III) using silicon nanowire arrays under visible light irradiation. *J Hazard Mater* 304:441–447.

Fischer KE, Alemán BJ, Tao SL, Daniels RH, Li EM, Bünger MD, Nagaraj G, Singh P, Zettl A, Desai TA (2009). Biomimetic nanowire coatings for next generation adhesive drug delivery systems. *Nano Lett* 9:716–720.

Garipcan B, Odabas S, Demirel G, Burger J, Nonnenmann SS, Coster MT, Gallo EM, Nabet B, Spanier JE, Piskin E (2011). In vitro biocompatibility of n-type and undoped silicon nanowires. *Adv Eng Mater* 13:B3–B9.

Guichard AR, Barsic DN, Sharma S, Kamins TI, Brongersma ML (2006). Tunable light emission from quantum-confined excitons in TiSi2-Catalyzed silicon nanowires. *Nano Lett* 6:2140–2144.

Guo Z, Seol M-L, Gao C, Kim M-S, Ahn J-H, Choi Y-K, Huang X-J (2016). Functionalized porous Si nanowires for selective and simultaneous electrochemical detection of Cd(II) and Pb(II) ions. *Electrochim Acta* 211:998–1005.

Hahm J, Lieber CM (2004). Direct ultrasensitive electrical detection of DNA and DNA sequence variations using nanowire nanosensors. *Nano Lett* 4:51–54.

Han H, Huang Z, Lee W (2014). Metal-assisted chemical etching of silicon and nanotechnology applications. *Nanotoday* 9:271–304.

Heitsch AT, Fanfair DD, Tuan HY, Korgel BA (2008). Solution-liquid-solid (SLS) growth of silicon nanowires. *J Am Chem Soc* 130:5436–5437.

Hermanson GT (2013). Functional targets for bioconjugation. In: *Bioconjugate techniques* (Third edition), pp 127–228. Boston, MA: Academic Press.

Hochbaum AI, Fan R, He RR, Yang PD (2005). Controlled growth of Si nanowire arrays for device integration. *Nano Lett* 5:457–460.

Holmes JD (2000). Control of thickness and orientation of solution-grown silicon nanowires. *Science* 287:1471–1473.

Ishiyama T, Nakagawa S, Wakamatsu T (2016). Growth of epitaxial silicon nanowires on a Si substrate by a metal-catalyst-free process. *Sci Rep* 6:30608.

Janata J (1994). 20 years of ion-selective field-effect transistors. *Analyst* 119:2275–2278.

Jia G, Westphalen J, Drexler J, Plentz J, Dellith J, Dellith A, Andrä G, Falk F (2016). Ordered silicon nanowire arrays prepared by an improved nanospheres self-assembly in combination with Ag-assisted wet chemical etching. *Photonics Nanostructures - Fundam Appl* 19:64–70.

Jung Y, Tong L, Tanaudommongkon A, Cheng J-X, Yang C (2009). In vitro and in vivo nonlinear optical imaging of silicon nanowires. *Nano Lett* 9:2440–2444.

Kang K, Park YS, Park M, Jang MJ, Kim SM, Lee J, Choi JY, Jung DH, Chang YT, et al. (2016). Axon-first neuritogenesis on vertical nanowires. *Nano Lett* 16:675–680.

Kim A, Ah CS, Yu HY, Yang JH, Baek IB, Ahn CG, Park CW, Jun MS, Lee S (2007a). Ultrasensitive, label-free, and real-time immunodetection using silicon field-effect transistors. *Appl Phys Lett* 91:103901-103905.

Kim H, Kim I, Choi H-J, Kim SY, Yang EG (2015). Neuron-like differentiation of mesenchymal stem cells on silicon nanowires. *Nanoscale* 7:17131–17138.

Kim K, Park C, Kwon D, Kim D, Meyyappan M, Jeon S, Lee JS (2016). Silicon nanowire biosensors for detection of cardiac troponin I (cTnI) with high sensitivity. *Biosens Bioelectron* 77:695–701.

Kim W, Ng JK, Kunitake ME, Conklin BR, Yang P (2007b). Interfacing silicon nanowires with mammalian cells. *J Am Chem Soc* 129:7228–7229.

Kuo SW, Lin HI, Hui-Chun Ho J, Shih YR V, Chen HF, Yen TJ, Lee OK (2012). Regulation of the fate of human mesenchymal stem cells by mechanical and stereo-topographical cues provided by silicon nanowires. *Biomaterials* 33:5013–5022.

Langer R, Vacanti JP (1993). Tissue engineering. *Science* 260:920–926.

Leao CR, Fazzio A, Da Silva AJR (2007). Si nanowires as sensors: Choosing the right surface. *Nano Lett* 7:1172–1177.

Lee JH, Zhang A, You SS, Lieber CM (2016). Spontaneous internalization of cell penetrating peptide-modified nanowires into primary neurons. *Nano Lett* 16:1509–1513.

Li Y, Li M, Fu P, Li R, Song D, Shen C, Zhao Y (2015). A comparison of light-harvesting performance of silicon nanocones and nanowires for radial-junction solar cells. *Sci Rep* 5:11532.

Li Z, Rajendran B, Kamins TI, Li X, Chen Y, Williams RS (2005). Silicon nanowires for sequence-specific DNA sensing: Device fabrication and simulation. *Appl Phys A Mater Sci Process* 80:1257–1263.

Lichtenstein A, Havivi E, Shacham R, Hahamy E, Leibovich R, Pevzner A, Krivitsky V, Davivi G, Presman I, et al. (2014). Supersensitive fingerprinting of explosives by chemically modified nanosensors arrays. *Nat Commun* 5:4195.

Linsmeier CE, Prinz CN, Pettersson LME, Caroff P, Samuelson L, Schouenborg J, Montelius L, Danielsen N (2009). Nanowire biocompatibility in the brain—Looking for a needle in a 3D stack. *Nano Lett* 9:4184–4190.

Liu D, Yi C, Fong C-C, Jin Q, Wang Z, Yu W-K, Sun D, Zhao J, Yang M (2014). Activation of multiple signaling pathways during the differentiation of mesenchymal stem cells cultured in a silicon nanowire microenvironment. *Nanomedicine* 10:1153–1163.

Liu D, Yi C, Wang K, Fong C, Wang Z, Lo PK, Sun D, Yang M (2013). Reorganization of cytoskeleton and transient activation of Ca 2 + channels in mesenchymal stem cells cultured on silicon nanowire arrays. *ACS Appl Mater Interfaces* 5:13295–13304.

Lu N, Gao A, Dai P, Li T, Wang Y, Gao X, Song S, Fan C, Wang Y (2013). Ultra-sensitive nucleic acids detection with electrical nanosensors based on CMOS-compatible silicon nanowire field-effect transistors. *Methods* 63:212–218.

Lu X, Lu X, Hanrath T, Hanrath T, Johnston KP, Johnston KP, Korgel BA, Korgel BA (2003). Growth of single crystal silicon nanowires in supercritical solution from tethered gold particles on a silicon substrate. *Nano* 3:93–99.

Lv M, Su S, He Y, Huang Q, Hu W, Li D, Lee ST (2010). Long-term antimicrobial effect of silicon nanowires decorated with silver nanoparticles. *Adv Mater* 22:5463–5467.

Mathews S, Bhonde R, Gupta PK, Totey S (2012). Extracellular matrix protein mediated regulation of the osteoblast differentiation of bone marrow derived human mesenchymal stem cells. *Differentiation* 84:185–192.

Minh QN, Pujari SP, Wang B, Wang Z, Haick H, Zuilhof H, Rijn CJ (2016). Fluorinated alkyne-derived monolayers on oxide-free silicon nanowires via one-step hydrosilylation. *Appl Surf Sci* 387:1202–1210.

Mohd Azmi MA, Tehrani Z, Lewis RP, Walker KAD, Jones DR, Daniels DR, Doak SH, Guy OJ (2014). Highly sensitive covalently functionalised integrated silicon nanowire biosensor devices for detection of cancer risk biomarker. *Biosens Bioelectron* 52:216–224.

Morales AM, Lieber CM (1998). A laser ablation method for the synthesis of crystalline semiconductor nanowires. *Science* 279:208–211.

Morita M, Ohmi T, Hasegawa E, Kawakami M, Ohwada M (1990). Growth of native oxide on a silicon surface. *J Appl Phys J Appl Phys* 68:1272.

Nair PR, Alam MA (2007). Design considerations of silicon nanowire biosensors. *IEEE Trans Electron Dev* 54:3400–3408.

Nakamura J, Higuchi K, Maenaka K (2013). Vertical Si nanowire with ultra-high-aspect-ratio by combined top-down processing technique. *Microsyst Technol* 19:433–438.

Nguyen VH, Kato S, Usami N (2016). Evidence for efficient passivation of vertical silicon nanowires by anodic aluminum oxide. *Sol Energy Mater Sol Cells* 157:393–398.

Nolan M, O'Callaghan S, Fagas G, Greer JC, Frauenheim T (2007). Silicon nanowire band gap modification. *Nano Lett* 7:34–38.

Nuzaihan MMN, Hashim U, Md Arshad MK, Kasjoo SR, Rahman SFA, Ruslinda AR, Fathil MFM, Adzhri R, Shahimin MM (2016). Electrical detection of dengue virus (DENV) DNA oligomer using silicon nanowire biosensor with novel molecular gate control. *Biosens Bioelectron* 83:106–114.

O'Farrell N, Houlton A, Horrocks BR (2006). Silicon nanoparticles: Applications in cell biology and medicine. *Int J Nanomedicine* 1:451–472.

Panciera F, Chou Y-C, Reuter MC, Zakharov D, Stach EA, Hofmann S, Ross FM (2015). Synthesis of nanostructures in nanowires using sequential catalyst reactions. *Nat Mater* 14:820–825.

Panciera F, Norton MM, Alam SB, Hofmann S, Mølhave K, Ross FM (2016). Controlling nanowire growth through electric field-induced deformation of the catalyst droplet. *Nat Commun* 7:12271.

Park N-M, Choi C-J (2014). Growth of silicon nanowires in aqueous solution under atmospheric pressure. *Nano Res* 7:898–902.

Patolsky F, Timko BP, Yu G, Fang Y, Greytak AB, Zheng G, Lieber CM (2006). Detection, stimulation, and inhibition of neuronal signals with high-density nanowire transistor arrays. *Science* 313:1100–1104.

Patolsky F, Zheng G, Hayden O, Lakadamyali M, Zhuang X, Lieber CM (2004). Electrical detection of single viruses. *Proc Natl Acad Sci U S A* 101:14017–14022.

Patolsky F, Zheng G, Lieber CM (2006a). Nanowire-based biosensors. *Anal Chem* 78:4260–4269.

Patolsky F, Zheng G, Lieber CM (2006b). Fabrication of silicon nanowire devices for ultrasensitive, label-free, real-time detection of biological and chemical species. *Nat Protoc* 1:1711–1724.

Peer E, Artzy-Schnirman A, Gepstein L, Sivan U (2012). Hollow nanoneedle array and its utilization for repeated administration of biomolecules to the same cells. *ACS Nano* 6:4940–4946.

Peng F, Su Y, Ji X, Zhong Y, Wei X, He Y (2014). Doxorubicin-loaded silicon nanowires for the treatment of drug-resistant cancer cells. *Biomaterials* 35:5188–5195.

Peng F, Su Y, Wei X, Lu Y, Zhou Y, Zhong Y, Lee ST, He Y (2013). Silicon-nanowire-based nanocarriers with ultrahigh drug-loading capacity for in vitro and in vivo cancer therapy. *Angew Chemie—Int Ed* 52:1457–1461.

Peng K, Hu J, Yan Y, Wu Y, Fang H, Xu Y, Lee S, Zhu J (2006). Fabrication of single-crystalline silicon nanowires by scratching a silicon surface with catalytic metal particles. *Adv Funct Mater* 16:387–394.

Peng K, Yan Y, Gao S, Zhu J (2003). Dendrite-assisted growth of silicon nanowires in electroless metal deposition. *Adv Funct Mater* 13:127–132.

Piret G, Galopin E, Coffinier Y, Boukherroub R, Legrand D, Slomianny C (2011). Culture of mammalian cells on patterned superhydrophilic/superhydrophobic silicon nanowire arrays. *Soft Matter* 7:8642.

Piret G, Perez M-T, Prinz CN (2014). Substrate porosity induces phenotypic alterations in retinal cells cultured on silicon nanowires. *RSC Adv* 4:27888.

Qi S, Yi C, Ji S, Fong C-C, Yang M (2009). Cell adhesion and spreading behavior on vertically aligned silicon nanowire arrays. *ACS Appl Mater Interfaces* 1:30–34.

Qu Y, Zhou H, Duan X (2011). Porous silicon nanowires. *Nanoscale* 3:4060–4068.

Roberts JR, Mercer RR, Chapman RS, Cohen GM, Bangsaruntip S, Schwegler-Berry D, Scabilloni JF, Castranova V, Antonini JM, Leonard SS (2012). Pulmonary toxicity, distribution, and clearance of intratracheally instilled silicon nanowires in rats. *J Nanomater* 2012:1–17.

Römhildt L, Gang A, Baraban L, Opitz J, Cuniberti G (2013). High yield formation of lipid bilayer shells around silicon nanowires in aqueous solution. *Nanotechnology* 24:355601.

Sacconi F, Persson MP, Povolotskyi M, Latessa L, Pecchia A, Gagliardi A, Balint A, Fraunheim T, Di Carlo A (2007). Electronic and transport properties of silicon nanowires. *J Comput Electron* 6:329–333.

Savage DJ, Liu X, Curley SA, Ferrari M, Serda RE (2013). Porous silicon advances in drug delivery and immunotherapy. *Curr Opin Pharmacol* 13:834–841.

Schubert L, Werner P, Zakharov ND, Gerth G, Kolb FM, Long L, Gösele U, Tan TY (2013). Silicon nanowhiskers grown on ⟨111⟩ Si substrates by molecular-beam epitaxy Silicon nanowhiskers grown on Š 111 ‹ Si substrates by molecular-beam epitaxy. *Appl Phys Lett* 4968:111–114.

Shao M, Cheng L, Zhang X, Ma DDD, Lee ST (2009). Excellent photocatalysis of HF-treated silicon nanowires. *J Am Chem Soc* 131:17738–17739.

Shi W-S, Peng H-Y, Zheng Y-F, Wang N, Shang N-G, Pan Z-W, Lee C-S, Lee S-T (2000). Synthesis of large areas of highly oriented, very long silicon nanowires. *Adv Mater* 12:1343–1345.

Stensaas SS, Stensaas LJ (1978). Histopathological evaluation of materials implanted in the cerebral cortex. *Acta Neuropathol* 41:145–155.

Sun XH, Li CP, Wong NB, Lee CS, Lee ST, Teo BK (2002). Reductive growth of nanosized ligated metal clusters on silicon nanowires. *Inorg Chem* 41:4331–4336.

Sun XH, Wang SD, Wong NB, D Ma DD, Lee ST, Teo BK (2003). FTIR spectroscopic studies of the stabilities and reactivities of hydrogen-terminated surfaces of silicon nanowires. *Inorg Chem* 42:2398–2404.

Susarrey-Arce A, Sorzabal-Bellido I, Oknianska A, McBride F, Beckett AJ, Gardeniers JGE, Raval R, Tiggelaar RM, Diaz Fernandez YA (2016). Bacterial viability on chemically modified silicon nanowire arrays. *J Mater Chem B* 4:3104–3112.

Sze SM, Ng KK (2007). Physics of semiconductor devices. In *Bipolar transistors in physics of semiconductor devices*. New York: Wiley, pp. 241–292.

Tan Y, Richards D, Xu R, Stewart-Clark S, Mani SK, Borg TK, Menick DR, Tian B, Mei Y (2015). Silicon nanowire-induced maturation of cardiomyocytes derived from human induced pluripotent stem cells. *Nano Lett* 15:2765–2772.

Tang YH, Zhang YF, Wang N, Lee CS, Han XD, Bello I, Lee ST (1999). Morphology of Si nanowires synthesized by high-temperature laser ablation. *J Appl Phys* 85:7981–7983.

Tian B, Cohen-Karni T, Qing Q, Duan X, Xie P, Lieber CM (2010). Three-dimensional, flexible nanoscale field-effect transistors as localized bioprobes. *Science* 329:830–834.

Tian B, Liu J, Dvir T, Jin L, Tsui JH, Qing Q, Suo Z, Langer R, Kohane DS, Lieber CM (2012). Macroporous nanowire nanoelectronic scaffolds for synthetic tissues. *Nat Mater* 11:986–994.

Tian (田琳) L, Di Mario L, Minotti A, Tiburzi G, Mendis BG, Zeze DA, Martelli F (2016). Direct growth of Si nanowires on flexible organic substrates. *Nanotechnology* 27:225601.

Tuan H-Y, Ghezelbash A, Korgel BA (2008). Silicon nanowires and silica nanoitubes seeded by copper nanoparticles in an organic solvent. *Chem Mater* 20:2306–2313.

Tuan HY, Lee DC, Hanrath T, Korgel BA (2005). Catalytic solid-phase seeding of silicon nanowires by nickel nanocrystals in organic solvents. *Nano Lett* 5:681–684.

Turner, APF, Wilson GS (1989). *Biosensors Fundamentals and Applications*. London, UK: Oxford Science Publications.

Uskoković V, Lee PP, Walsh LA, Fischer KE, Desai TA (2012). PEGylated silicon nanowire coated silica microparticles for drug delivery across intestinal epithelium. *Biomaterials* 33:1663–1672.

Wagner RS, Ellis WC (1964). Study of hte filamentary growth of silicon crystals from the vapor. *J Appl Phys* 35:2993–3000.

Wagner RS, Ellis WC (1964). Vapor-liquid-sold mechanism of single crystal growth. *Appl Phys Lett* 4:89–90.

Wang H, Jiang X, Lee ST, He Y (2014). Silicon nanohybrid-based surface-enhanced raman scattering sensors. *Small* 10:4455–4468.

Wang N, Zhang YF, Tang YH, Lee CS, Lee ST (1998). SiO[sub 2]-enhanced synthesis of Si nanowires by laser ablation. *Appl Phys Lett* 73:3902.

Wang PY, Tsai WB, Voelcker NH (2012). Screening of rat mesenchymal stem cell behaviour on polydimethylsiloxane stiffness gradients. *Acta Biomater* 8:519–530.

Weisenberg BA, Mooradian DL (2002). Hemocompatibility of materials used in microelectromechanical systems: Platelet adhesion and morphology in vitro. *J Biomed Mater Res* 60:283–291.

Werner P, Zakharov ND, Gerth G, Schubert L, Göele U (2006). On the formation of Si nanowires by molecular beam epitaxy. *Int J Mater Res* 97:1008–1015.

Wicki A, Witzigmann D, Balasubramanian V, Huwyler J (2015). Nanomedicine in cancer therapy: Challenges, opportunities, and clinical applications. *J Control Release* 200:138–157.

Wu J (2014). Semiconducting silicon nanowire array fabrication for high throughput screening in the biosciences. In *Semiconducting silicon nanowires for biomedical applications*. S. L. Coffer (editor). pp. 171–191. Cambridge, UK: Woodhead Publishing Limited.

Wu Y, Fan R, Yang P (2002). Block-by-block growth of single-crystalline Si / SiGe superlattice nanowires. *Nano Lett* 2:83–86.

Xie C, Lin Z, Hanson L, Cui Y, Cui B (2012). Intracellular recording of action potentials by nanopillar electroporation. *Nat Nanotechnol* 7:185–190.

Yang S-P, Wen H-S, Lee T-M, Lui T-S (2016). Cell response on biomimetic scaffold of silicon nano- and micro-topography. *J Mater Chem B* 4:1891–1897.

Yao Y, Li F, Lee ST (2005). Oriented silicon nanowires on silicon substrates from oxide-assisted growth and gold catalysts. *Chem Phys Lett* 406:381–385.

Yen L-C, Pan T-M, Lee C-H, Chao T-S (2016). Label-free and real-time detection of ferritin using a horn-like polycrystalline-silicon nanowire field-effect transistor biosensor. *Sensors Actuators B Chem* 230:398–404.

Yun J, Jin CY, Ahn J-H, Jeon S, Park I (2013). A self-heated silicon nanowire array: selective surface modification with catalytic nanoparticles by nanoscale Joule heating and its gas sensing applications. *Nanoscale* 5:6851–6856.

Zakharov ND, Werner P, Gerth G, Schubert L, Sokolov L, Go\textasciidieresissele U (2006). Growth phenomena of Si and Si/Ge nanowires on Si (1 1 1) by molecular beam epitaxy. *J Cryst Growth* 290:6–10.

Zhang GJ, Huang MJ, Ang JJ, Liu ET, Desai KV (2011). Self-assembled monolayer-assisted silicon nanowire biosensor for detection of protein-DNA interactions in nuclear extracts from breast cancer cell. *Biosens Bioelectron* 26:3233–3239.

Zhang RQ, Lifshitz Y, Lee ST (2003). Oxide-assisted growth of semiconducting nanowires. *Adv Mater* 15:635–640.

Zhang YF, Tang YH, Peng HY, Wang N, Lee CS, Bello I, Lee ST (1999). Diameter modification of silicon nanowires by ambient gas. *Appl Phys Lett* 75:1842.

Zhang YF, Tang YH, Wang N, Yu DP, Lee CS, Bello I, Lee ST (1998). Silicon nanowires prepared by laser ablation at high temperature. *Appl Phys Lett* 72:1835–1837.

Zhang Z, Fan XH, Xu L, Lee CS, Lee ST (2001). Morphology and growth mechanism study of self-assembled silicon nanowires synthesized by thermal evaporation. *Chem Phys Lett* 337:18–24.

Zheng G, Patolsky F, Cui Y, Wang WU, Lieber CM (2005). Multiplexed electrical detection of cancer markers with nanowire sensor arrays. *Nat Biotechnol* 23:1294–1301.

Zhou W, Dai X, Fu TM, Xie C, Liu J, Lieber CM (2014). Long term stability of nanowire nanoelectronics in physiological environments. *Nano Lett* 14:1614–1619.

18 Silicon dots in radiotherapy

María L. Dell'Arciprete, Mónica C. Gonzalez, Roxana M. Gorojod, and Mónica L. Kotler

Contents

18.1 INTRODUCTION

Radiation therapy is an important treatment modality for many types of human cancer. Although this therapeutic approach is effective in many cases, normal tissue toxicity limits the total radiation dose a patient can receive. Advances in instrumentation and technology lead to more precise X-ray targeting which reduces collateral normal tissue damage. In particular, nanomaterials used as radiosensitizers are known to increase the efficiency of X-rays causing more localized damage to DNA and targeted organelles of tumor cells. The small size of the particles, their versatile surface modification with specific recognition molecules, and the capability of nanoparticles (NPs) to produce reactive radicals upon irradiation are key properties in the enhancement strategies of cytostatic and cytolytic activities in tumor cells by high-energy (MeV) ionizing radiation.

Silicon semiconductor nanoparticles or silicon dots (SiDs) have received great attention as they combine a size-dependent photoluminescence, a rich surface chemistry, and the capacity to photosensitize singlet oxygen ($^1O_2$) and to photoreduce O_2, methyl viologen and metal ions such as Au^{3+} and Ag^+. Remarkable properties of SiDs when compared to other materials are their biocompatibility, biodegradability, and tunable surface derivatization for drug delivery. In vivo assays indicate that SiDs are metabolized and eliminated from the body in relatively short times. In this regard, there is an increased interest in new applications of SiDs. In this chapter, we discuss the progress in SiDs studies as therapeutic agents enhancing the effects of radiation therapy.

18.2 IONIZING RADIATION IN CANCER THERAPY

Cancer is the second leading cause of morbidity and mortality worldwide, with approximately 14.9 million new cases and 8.2 million cancer-related deaths in 2013 (Torre et al. 2015). New cases are expected to rise by about 70% over the next two decades.

In the last few years, significant progress has been made toward the development of new cancer therapies. The deeper understanding of both the tumor biology and the molecular mechanisms involved in tumor growth and spread is rapidly expanding. However, considering its increasing incidence, the clinical management of this disease continues to be a challenge for the 21st century. Cancer treatment modalities comprise surgery, chemotherapy, radiation therapy (also called radiotherapy), immunotherapy, and hormonal therapy. The types of treatment that a patient receives will depend on both the type of cancer and the degree of advance of the disease.

Radiation therapy with high intensity ionizing radiation (including X-rays, gamma rays, and high-energy particles) is one of the main modalities for cancer treatment extensively employed for almost all types of solid tumors. Approximately 50% of all cancer patients receive radiotherapy during the course of the illness. Radiotherapy contributes toward 40% of curative treatment for cancer (Baskar et al. 2012). The accelerated development in this field is supported by the advances in imaging techniques, computerized treatment planning systems, radiation machines, and the improved understanding of the radiobiology of ionizing radiation (Baskar et al. 2012).

Despite the great advances made in focusing and dosing regulation of the ionizing radiation, some major issues still need to be developed. Unfortunately, ionizing radiation does not discriminate between cancerous and normal cells thus leading to a high morbidity of the healthy tissue. However, since healthy cells are able to repair themselves more efficiently and to maintain normal functionality better than cancer cells, tumor cells are generally more sensitive to the radiation treatment damage resulting in differential cancer cell killing (Begg et al. 2011). Of concern is the fact that malignant cells neighboring the radiation site might receive low radiation doses with the consequent development of cell resistance to radiation. Therefore, elevated radiation doses which might eventually also lead to death of the healthy tissue are needed to assure therapy effectiveness.

The delivery of a curative dose of radiation to a tumor while sparing normal tissues is still a great challenge in radiation therapy. The ongoing advances will endeavour to increase the survival and reduce treatment side effects for patients. Among the major approaches developed to improve the clinical outcome of radiotherapy, the enhancement of tumor tissues susceptibility to injury by radiation exposure has emerged as a persistent hotspot in radiation oncology. For this purpose, either therapeutic or otherwise inert agents are used. The National Cancer Institute (NCI 2016) defines the terms "radiosensitizer" and "radiosensitizing agent" as any substance that makes tumor cells easier to kill with radiation therapy. In this sense, application of tumor-targeted nanoparticles as radiosensitizers, has been an issue of considerable interest over the last few years.

18.2.1 IONIZING RADIATION MECHANISMS

The main goal of radiation therapy is to deprive cancer cells of their cell division potential and eventually produce cell death (Jackson and Bartek 2009). High-energy ionizing radiations such as gamma or X-rays are mainly employed to ionize cellular components and/or water. Particulate radiations such as alpha or beta particles, electron, proton, or neutron beams are also used in certain specific cases to target the cancer tissue (Kwatra et al. 2013). Their mechanisms might differ from that described here for high-energy X-ray radiation.

Since water is the major component of the cells, ionizing radiation causes damage mainly through water radiolysis resulting in the generation of charged species and radicals. Hydrated electrons (e^-_{aq}), hydroxyl radicals ($OH^•$), and H atoms are produced in the physical and chemical stages taking place in the picosecond time window after irradiation, reactions (R1) to (R3) in Scheme 18.1. In the presence of O_2, reactive e^-_{aq} and H atoms lead to the formation of superoxide ($O_2^{•-}$) in fast acid-base equilibrium with hydroperoxide ($HO_2^•$)—reactions (R4), (R5), and equilibrium (R6), respectively. Recombination of $O_2^{•-}$ and $HO_2^•$, leads ultimately to H_2O_2, reaction (R7). For a detailed discussion of all the reactions taking place during water radiolysis see the reviews by Le Caër (2011) and Gonzalez et al. (2004). Superoxide, $HO^•$ radicals, and H_2O_2

Scheme 18.1 Most important reactions taking place during liquid water radiolysis in the presence of molecular oxygen

$H_2O(l) + \text{ionizing radiation} \rightarrow [e^-, H_2O^+]$		(R1)
$[e^-, H_2O^+] + (H_2O) \rightarrow e^-_{aq} + HO^\bullet + H_3O^+$		(R2)
$e^-_{aq} + H_3O^+ \rightarrow H + H_2O$	$k_{R3} = 2.2{\times}10^{10}$ M^{-1}s^{-1}	(R3)
$e^-_{aq} + O_2 \rightarrow O_2^{\bullet-}$	$k_{R4} = 2{\times}10^{10}$ M^{-1}s^{-1}	(R4)
$H + O_2 \rightarrow HO_2^\bullet$	$k_{R5} = 1{\times}10^{10}$ M^{-1}s^{-1}	(R5)
$HO_2^\bullet \leftrightharpoons O_2^{\bullet-} + H_3O^+$	pK = 4.8	(R6)
$HO_2^\bullet + O_2^{\bullet-} + H_3O^+ \rightarrow H_2O_2 + H_2O + O_2$	$k_{R7} = 5{\times}10^7$ M^{-1}s^{-1}	(R7)

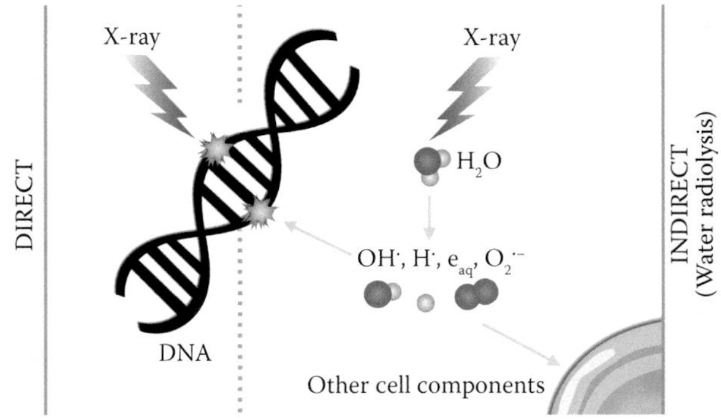

Figure 18.1 Effect of ionizing radiation on DNA. The joint action of direct and indirect X-ray interactions induce double strand DNA breaks leading to cell cycle dysfunction and death.

are among the reactive oxygen species known as ROS. DNA is arguably the most important molecular target involved in radiation-induced cell death along with many other cellular components which are also damaged (Kwatra et al. 2013; Panganiban et al. 2013). Accordingly, much attention has been focused on the comprehension of radiation-induced DNA damage.

Ionizing radiation interacts with DNA in two main ways: 1) by its direct interaction with DNA strands (direct effect) and 2) by generating ROS from water radiolysis which then attacks the biological molecules and structures (indirect effect). Figure 18.1 describes both mechanisms. Interaction of free radicals with the membrane structures also causes structural damages resulting in cell death induction (i.e., apoptosis). Particularly, the OH$^\bullet$ radical has been reported in multiple studies to be a major source of cellular damage and it is known to induce lipid peroxidation (Takeshita et al. 2004).

18.2.2 PHYSICAL MECHANISMS OF PHOTON BEAM ATTENUATION

To enhance the effects of radiation therapy, several basic topics on the interaction between X-rays and matter should be emphasized. In this regard, different behavior is expected depending on the X-ray energy and the chemical nature of the material as will be discussed below. The intensity $I(x)$ of a narrow monoenergetic photon beam attenuated by a material of thickness x, is given as: $I(x) = I(0)e^{-\mu(h\nu, Z)x}$ where $I(0)$ is the original intensity of the unattenuated beam; $\mu(h\nu, Z)$ is the linear attenuation coefficient, which depends on photon energy $h\nu$ and on the attenuator atomic number Z.

Ionizing radiation interacts with matter by well characterized mechanisms:

1. The photoelectric interaction, in which a photon with energy $h\nu$ interacts with a K-shell electron and the electron is emitted from the atom with energy $h\nu-E_K$, E_K being the binding energy of the electron in the K shell. The mass attenuation coefficient for the photoelectric effect is $\tau_m = Z^3/(h\nu)^3$.

As the atom is left in an ionized state, either a characteristic X-ray or an Auger-electron emission follows the ejection of a photoelectron. The probability that X-ray fluorescence takes place is w and that for the

Auger effect is $1 - w$. Radiation emission mechanisms are of importance for high-Z materials while the Auger effect particularly concerns materials of low Z and is dominant for $Z < 15$, and of no significance for materials with $Z > 60$ (Retif et al. 2015).

Certain selection rules prohibit photon emission and phonon emission occurs. In phonon emission, the excitation energy flows into the host lattice as low grade heat. This last quenching process is observed upon high-energy excitation in gold, attributed to a dominant photon–phonon transition in this material (Nelson and Reilly 1991; Mesbahi 2010).

2. The coherent Rayleigh scattering involves the elastic interaction between the photon and a bound orbital electron where the photon does not lose its energy but is scattered through a small angle. Rayleigh scattering plays no role in the energy transfer coefficient though it contributes to the attenuation coefficient with a mass attenuation coefficient $\sigma_R/\rho = Z/(h\nu)^2$.

3. The incoherent Compton scattering involves the interaction of the incident photon with energy $h\nu$ and the outermost (and hence loosely bound) valence electron at the atomic level. As a consequence of this interaction, the electron is ejected from the atom with a recoil angle ϕ and kinetic energy E_K while the photon is scattered at an angle θ and energy $h\nu'$. The higher the incident photon energy, the smaller the recoil electron angle ϕ. The mass Compton photon attenuation coefficient, σ_C/ρ, is independent of Z.

4. Pair production interaction in the nuclear Coulomb field leads to the formation of an electron–positron pair with a combined kinetic energy $h\nu - 2m_ec^2$. Since mass is produced out of photon energy in the form of an electron–positron pair, the minimum photon energy required for the pair production is of $2m_ec^2 = 1.02$ MeV. On the other hand, pair production occurring in the field of an orbital electron is referred to as triplet production because three particles share the available energy (an electron–positron pair and the orbital electron). The radiation energy threshold for this effect is $4m_ec^2$. The mass attenuation coefficient for pair production κ/ρ vary approximately as Z.

Pair and triplet production are followed by the annihilation of the positron with a "free" and stationary electron, producing two annihilation quanta, most commonly with energies of 0.511 MeV each and emitted at 180° from each other to satisfy the conservation of charge, momentum, and energy.

For an incident photon energy $h\nu$ and a material of atomic number Z, the coefficients for total mass attenuation (μ), energy transfer (μ_{tr}), and mass energy absorption (μ_{en}) are given as the sum of coefficients for the individual photon interactions as follows:

$$\mu/\rho = (\tau + \sigma_R + \sigma_C + \kappa)/\rho$$

$$\mu_{tr}/\rho = (\tau_{tr} + \sigma_{Ctr} + \kappa_{tr})/\rho$$

$\mu_{en}/\rho = \mu_{tr}\times(1-g)/\rho$ with g the average fraction of the kinetic energy of secondary charged particles produced in all types of interactions that is lost in photon emission.

Values of μ/ρ and μ_{en}/ρ for different materials as a function of incident X-ray energy may be found in the National Institute of Standards and Technology's (NIST) database (Hubbell and Seltzer 2004).

The emission of secondary electrons (also called delta rays) from Compton, photoelectron- and Auger-processes are believed to lead to cell damage. Ejected photoelectrons from low-energy photons might not carry a sufficient energy to cause subsequent ionizations.

18.3 THE ROLE OF NPs IN ENHANCING THE EFFECTS OF IONIZING RADIATION

Classical research in radio therapeutic sciences has been focused on interfering with the DNA repair mechanisms. Other strategies include targeting tumor hypoxia (which is protective toward radiation) or to interfere with genes and proteins that are needed specifically for the tumor survival and expansion. However, apart from the recognition that certain chemotherapeutic agents can be combined with radiation therapy to achieve at least additive if not synergistic effects on many solid tumors, little of this research has yet been effectively translated into significant improvements in therapy (Begg et al. 2011; Giaccia 2014).

Nanomaterial products represent an opportunity to achieve sophisticated targeting strategies and multifunctionality (Wicki et al. 2015). In this concern, recent developments of tumor-selective nanoparticles are expected to revolutionize therapeutic strategies against cancer. The goal of the targeted NPs delivery is to obtain high tumoral concentration of NPs together with low exposure of healthy cells and tissues. In passive targeting, NPs carrying the anticancer drug accumulate in a desired tissue thanks to their size and surface properties as well as those of the tissue (enhanced permeability and retention effect, EPR) (Kwatra et al. 2013). In active targeting, specific ligands are attached to the NP surface in order to be recognized by receptors expressed mainly by the target cell surfaces (Hirsjarvi et al. 2011). Therefore, radiation enhanced sensitivity using NPs will selectively affect cancer cells and protect healthy tissue.

Literature publications distinguish NPs with different mechanisms of action. A good and general discussion on the effect of diverse nanoparticles in radiation therapy may be found in the review by Retif and collaborators (Retif et al. 2015).

NPs with high atomic number. Densely packed high-Z nanoparticles can selectively scatter and/or absorb energy gamma/X-ray radiations. These NPs enhance the photoelectric and Compton effects and thus the subsequent emissions of secondary electrons (short range photoelectrons, Auger electrons) and ROS generation.

In fact, the most studied nanomaterials concern high atomic number NPs and in particular gold-based NPs (AuNPs). Since the early studies by Hainfeld and coworkers (Hainfeld et al. 2006), many biological studies have shown a significant gain in tumor growth control when low-energy X-rays are delivered in the presence of AuNPs (Rahman et al. 2009; Mesbahi 2010; Sicard-Roselli et al. 2014; Tsai and Hsiao 2016). Despite the highest achievable physical enhancement being predicted to occur for irradiation energies in the order of dozens of keV, low-energy photons are not applied in radiation therapy as they are associated with problems of high skin dose (the maximum dose is received on the skin) (Deeley 2013), the rapid drop off of absorbed dose with depth, and the development of cell resistance to radiation, vide supra. In fact, a high-energy photon beam (MeV photons) is the most common form of radiation used for cancer treatment. Such photon beams affect the cells along their path as they go through the body to the tumor, pass through the cancer tissue, and then exit the body. Despite the ratio of absorption between a high-Z material and water becomes theoretically negligible with MeV radiation energies, a radiosensitizing effect of the material is still reported as will be discussed further in the text.

Drug-releasing NPs triggered by X-rays. Polymeric nanoparticles formulated using chemotherapeutic agents such as paclitaxel, etanidazole, doxorubicin, and genexol-PM have been shown to serve as drug-releasing agents inside targeted tissues as a consequence of X-ray irradiation. The release of the chemotherapeutic drug allowed the enhancement of radiation effects due to different mechanisms in hypoxic cells, which are generally more resistant to radiation-induced injury (Kwatra et al. 2013).

Self-lighting photodynamic NPs. These particles are usually made of a lanthanide-doped high-Z core (Retif et al. 2015). Once irradiated with X-rays, the scintillator core emits a visible light and activates a photosensitizer that generates singlet oxygen ($^1O_2$) for tumor destruction. These NPs combine both photodynamic therapy that generates ROS and enhanced radiation therapy (high-Z core).

18.3.1 MECHANISMS LEADING TO THE ADSORPTION OF RADIATION ENERGY BY NPs

The term "physical enhancement" has been applied to the physical processes leading to an increased X-ray absorption. However, it must be remembered that the main idea behind the physical aspects of the enhancement of radiation is not able to explain all the effects of radiation therapy on cell lines and animal models (Mesbahi 2010).

In the presence of a nanomaterial, two mechanisms are involved. One is independent of the NPs location and is hence called average or remote physical enhancement. It is caused by energetic electrons that can travel microns in water and is uniformly spread throughout the whole sample volume. This mechanism is the most effective way to directly enhance the damage to a biological target at a distance at least tens to hundreds of nanometers away. The second mechanism involves low-energy electrons emitted from the NPs after absorbing X-ray photons. This mechanism is confined to tens of nanometers from the surface region of the nanomaterial (Davidson and Guo 2014).

A complete analysis of the action of ionizing radiation and NPs is not based only on the primary energy transfer mechanisms, but in secondary physical effects and also chemical and biological properties. Computational studies by McMahon et al. evaluated the macro- and microscale radiation dose enhancement following X-ray irradiation with both imaging and therapeutic energies on 20 nm diameter nanoparticles consisting of stable elements heavier than silicon (McMahon et al. 2016). The authors found that the mass energy absorption coefficient for monochromatic X-ray exposures in the 1–5 MeV range shows significantly less material dependence, as these physical effects are dominated by Compton interactions which are largely independent of atomic number. Consequently, Auger electrons are significantly less important, as interactions are dominated by Compton and electron scattering events, which are primarily outer-shell interactions. Secondary electrons are generated by the beam interacting with water molecules on the surface of the particle. In fact, low Z particles such as Fullerene C60 are also reported to enhance membrane damage and induce apoptotic cell death in combination with γ-radiation (Kwatra et al. 2013).

Additionally, the NPs size and shape show significant impact on the low-energy portion of the secondary spectra. Small and shaped nanoparticles (e.g., nanorods) with high surface-to-volume ratios show significantly less internal absorption, meaning electrons will have to get through less material before escaping the NPs. Material density is also an important factor, as less dense particle preparations (composites of atoms of a high-Z material in a crystal or organic molecule with lighter elements) may drive superior radiosensitization than a similar mass of pure high-Z material. Radiation-produced low-energy electrons in the latter situation have a decreased probability to escape from the material (McMahon et al. 2016). Low-energy secondary species could lead to a great energy deposition in the vicinity of NPs as they may be adsorbed by the particles surroundings leading to cell damage. The nature of the surface coating plays an important role in targeting purposes, biological compatibility, and might also have an impact on radiosensitization combining surface groups with the capability of enhancing ROS generation. Therefore, nanoparticles optimized designs would need to consider not only the core material, but the size, shape, density, and the surface coating.

To probe the effect of surface groups on radio sensitization enhancement, the mechanism of HO$^\bullet$ radical production by AuNPs under X-ray irradiation was evaluated (Sicard-Roselli et al. 2014). In such a system, increased HO$^\bullet$ production may be mainly achieved if the energy deposited in the NP is transferred into water through the emission of electrons, holes, or lower energy photons so that it becomes available to break H–OH bonds. The water layer can be structured in the nanoparticle vicinity as there is sufficient charge to align the water dipoles. These bonds could be already stressed, and the injection of energy by X-ray is likely to break them, leading to the production of HO$^\bullet$. Therefore, water radiolysis could be more efficient in the vicinity of nanoparticles.

18.3.2 METHODS MEASURING THE NPs-MEDIATED IMPROVEMENT IN RADIOTHERAPY

In radiotherapy, clinically relevant doses of radiation to generate DNA damage could lead to early cell death but rather result in cell death after one or more cell divisions. Thus, a cell is considered radiobiologically dead when its multiplicative cycle is lost. The standard to evaluate the cytotoxicity of ionizing radiation is the clonogenic assay or colony-forming unit. This method takes into account the cell reproductive death because it tests every cell in the population for its ability to undergo unlimited division.

Three approaches to assess the enhancement of radiation therapy efficacy by NPs have been described: (i) the determination of the dose modifying factor (DMF) based on survival curves, (ii) the evaluation of the nanoparticle-mediated enhancement ratio (NER) after a single radiation dose, and (iii) the variation of the ROS production upon irradiation (Retif et al. 2015). A detailed discussion of each of these approaches may be found elsewhere in the literature (Podgorsak 2005; Retif et al. 2015).

Although the role of ROS in the treatment of cancer is controversial, the high sensitivity and simplicity in data collection makes the evaluation of their production a frequently used assay for determining the radiosensitizing power of NPs. To assess the radiosensitizing power of NPs, experiments were performed in solution or cultured cells employing probes, some of which become fluorescent in the presence of ROS. Although most commonly used probes for cell culture (e.g., dihydrodichlorofluorescein diacetate (DCFH-DA), dihydrorhodamine (DHR), hydroethidine-dihydroethidium (DHE) are not specific for any one oxidant (Winterbourn 2014), they have value in providing information on changes to the redox environment of the cell (Takahashi and Misawa 2007; Gara et al. 2012; Klein et al. 2013).

18.4 SEMICONDUCTOR NANOPARTICLES

Semiconductor NPs have attracted wide attention from the scientific research community owing to their unique and highly desirable characteristics that only manifest themselves when bulk materials are reduced to nanometer scales. The discovery of the quantum confinement effect upon the density of states in several semiconductor NPs has considerably increased the number of applications of such NPs, often called quantum dots (QDs) (Vaverka 2008). Quantum confinement results in size-dependent band gaps with highly efficient photoluminescent quantum yields across the visible region with the ability to select the emission wavelength by selective control of the NP size (O'Farrell et al. 2006). These properties are potentially suitable for their use in a wide and ever-expanding range of applications (Caregnato et al. 2017).

Aqueous suspensions of semiconductor nanoparticles such as TiO_2, $ZnS:Ag$, CeF_3, and CdSe QDs showed a proportional increase in ROS generation with increasing doses of low-energy X-rays (20 to 170 keV) (Takahashi and Misawa 2007). The survival fraction of cells treated with CdSe was less than those without the QDs for all doses, suggesting that the colony-forming ability is damaged by the internalized QDs (Takahashi and Misawa 2007). CdSe QDs were also investigated in several cell lines and animal models with varying irradiation conditions (Retif et al. 2015). Moreover, excitation of QDs with megavoltage X-rays have been shown to transfer energy to the photosensitizer Photofrin, which in the presence of molecular oxygen led to the generation of cytotoxic singlet oxygen, thus enhancing H460 human lung carcinoma cell killing (Yang et al. 2008). Because of these properties, there is an increased interest in the application of semiconductor nanoparticles as radiosensitizers (Juzenas et al. 2008). However, chronic exposure to heavy metals such as cadmium, selenium, or tellurium is well known to pose a severe risk to human health (Derfus et al. 2004; Juzenas et al. 2008). As a result NPs derived from such metals are unusable in living systems. Silicon-based nanoparticles have shown promise to overcome these drawbacks since they display lower toxicity compared with transition metal QDs (O'Farrell et al. 2006).

18.4.1 SILICON DOTS

Silicon nanoparticles of <5 nm-size are known as Si quantum dots (SiQDs) due to the quantum confinement effects leading to a size-dependent emission (Dhara and Giri 2011). A variety of methods have been developed for SiQD synthesis, which have been shown to exhibit distinct optical properties with respect to one another (Huan and Shu-Qing 2014). The discrepancies in the data may be attributed to the lack of consistent characterization beyond size determination under conditions where surface termination is an important factor. Thus far, SiQDs have been conjugated successfully with different biomolecules and molecules with the aim of improving their versatility in biological environments (Juzenas et al. 2008; Rosso-Vasic et al. 2008; He et al. 2009; Bhattacharjee et al. 2010; He et al. 2011; Lillo et al. 2015). An important body of research has focused upon the use of SiQDs as fluorescent biological markers by exploiting the advantages that these QDs possess over conventional luminescent dyes within in vitro and in vivo systems (He et al. 2011; Shao et al. 2011; Nishimura et al. 2013). QDs used in these approaches are specifically engineered to closely interact with the target cells as it is crucial to ensure that they cause no undesirable effects within the healthy cells and tissues.

Values of μ/ρ and μ_{en}/ρ expected for silicon (Z = 14) upon irradiation with 1 to 4 MeV energy photons, of the order of the energy delivered in radiation therapy, are in the range of 6.4×10^{-2} to 3.24×10^{-2} and 2.7×10^{-2} to 2.0×10^{-2} cm^2g^{-1}, respectively. The photoelectric effect is relevant up to energies of 60 keV while the Compton scattering is the dominant process from 80 keV up to a few MeV. From 1.022 MeV on, pair production increases rapidly (Leroy and Rancoita 2012). In fact, Si semiconductor materials are known to create hole–electron pairs upon high-energy X-ray irradiation (Knoll 2000). It is worth mentioning that MeV X-rays also interact with soft tissue (Z_{eff} = 7.5) predominantly through the Compton scattering.

Passivated 2 nm-size SiDs interaction with low-energy X-rays (ca. 140 eV at the Si $L_{2,3}$ edge) are reported to show clear signatures of core and valence band exciton formation promoted by the spatial confinement of electrons and holes within the nanocrystals (Siller et al. 2009). However, to our knowledge, no physical interaction studies are reported in the literature on SiDs irradiation with high-energy X-rays.

18.4.1.1 Effect of radiation on SiDs surface

The FT-IR spectra of high-energy X-ray-irradiated SiDs show the increased intensity of Si–OH and Si–O–Si vibration modes (at 3400, 1630–1650, 800, and 1050–1100 cm^{-1}) with increasing irradiation doses. Also, corresponding X-ray photoemission spectroscopy (XPS) spectra of the SiDs show Si2p signals which strongly depend on the irradiation dose. An increased contribution of peaks at 101.4 eV assigned to Si$(-O-)_4$ silicon environments is observed at the lower irradiation doses while silicon oxidation to SiO$_2$ (peak at 103.6 eV) is the net process at the larger doses. The generation of Si$(-O-)_4$ centers may be related to the increased hydroxylation and siloxane bond formation in the particles surface. Interestingly, neither the stretching modes due to CH$_3$ groups nor Si-C contribution to the Si2p signal in the XPS spectra were significantly modified by X-ray irradiation up to 3 Gy. Altogether, these observations suggest the preferential oxidation of silicon to sylanol and siloxane functionalities rather than that of surface-attached organic groups.

18.4.1.2 ROS generation upon X-ray irradiation of aqueous suspensions of SiDs

As discussed before, HO$^\bullet$, O$_2^{\bullet-}$/HO$_2^\bullet$, and H$_2$O$_2$ are the main reactive species generated upon irradiation of air-saturated water with high-energy X-rays. Therefore, these species are available to initiate the depletion of added scavengers in air-saturated solutions. Since SiDs are known to produce singlet molecular oxygen, ^{1}O$_2$, formation of these species was also determined. To that purpose, X-ray irradiation assays were conducted in the presence of phenol, furfuryl alcohol, and histidine as scavengers of HO$^\bullet$, ^{1}O$_2$, and O$_2^{\bullet-}$, respectively (Kohn and Nelson 2007; Bosio et al. 2008). The concentration of ROS per 4 MeV dose achieved in experiments with 0.5 g/L SiDs aqueous suspensions are in the order of 10 µM, while fractions of µM concentrations were observed in experiments in the absence of SiDs but otherwise identical conditions. Singlet oxygen formation was unequivocally confirmed with the probe Singlet Oxygen Sensor Green, SOSG, which SOSG–^{1}O$_2$ fluorescent adduct was only observed in experiments with SiDs. Higher SOSG–^{1}O$_2$ fluorescence intensities were observed for the higher irradiation dose (Gara et al. 2012). Despite the particle surface becoming concomitantly oxidized to SiO$_2$ (vide supra), ROS formation per Gy is, within the experimental error, insensitive to these changes up to the 3 Gy irradiation tested.

As discussed above, the mass absorbed energy coefficient, μ_{en}/ρ, of 4 MeV X-rays reported for silicon and water are 0.0196 and 0.0207 cm^2 g^{-1}, respectively (Hubbell and Seltzer 2004). Therefore, SiDs absorbed energy of 4 MeV is expected to be of the same order or even lower than that of the same mass of water and cannot account for the 10-fold ROS enhancement observed upon irradiation of SiDs suspensions compared to water. As already discussed, Compton scattering is almost the most important process at 4 MeV generating photoelectrons and a scattering photon, both in water and silicon materials, as pair production accounts for less than 10% the total number of interactions (Khan 2003). The energy thus deposited is redistributed by subsequent interaction of the secondary electrons and photons within the material (Hubbell 1999); though, as discussed before, internal absorption is prevented by the small size of the particles. Electrons escaping the NPs may be captured by acceptors in the particles surface, such as adsorbed O$_2$ leading to O$_2^{\bullet-}$ generation. In addition, various decay processes may be associated with the dissipation of energy from the valence hole formed. In this context, the surface oxidation of the particles to Si$(-O-)_4$ and finally to SiO$_2$ at the expense of Si–Si functionalities may be related to the decay mechanisms of valence holes. Moreover, generation of ^{1}O$_2$ is supported by a mechanism whereby the formation of electron-hole excitons able to decay through an energy transfer mechanism with adsorbed O$_2$ molecules takes place (Gara et al. 2012), in agreement with reported processes for low-energy X-rays interactions with SiDs, vide supra.

18.4.1.3 Generation of ROS in cells

Employing rat glioma C6 cells our group has demonstrated that X-ray irradiation markedly increases the ROS generation induced by SiDs (Gara et al. 2012). C6 cells exposed to 50 µg/mL of SiDs for 6 h showed a 36% decrease in cell viability; a further 15% diminution was observed when irradiated with 1 Gy dose of 4-MeV X-rays, lower than the standard dose delivered to the cancer patients (2 Gy). Because the uptake of SiDs and X-ray irradiation have an impact on cell viability, C6 cells exposed to 50 µg/mL of SiDs for 6 h were used for ROS generation studies. Highly surface oxidized SiDs induce ROS generation in cells (c.a. 300%), even without radiation. However, ROS generation increases with the ionizing radiation dose, reaching 700% at 3 Gy.

Further studies with ultrasmall uncapped and amino silanized oxidized SiDs as radiosensitizers were performed by internalizing these nanoparticles into human breast cancer cell line MCF-7 and murine fibroblasts NIH/3T3 that were exposed to X-rays at a single dose of 3Gy (Klein et al. 2013). While surface oxidized SiDs did not increase the production of ROS in X-ray treated cells, the NH_2-capped SiDs significantly enhanced ROS formation. The enhanced ROS concentration by X-ray exposure of NH_2-coated SiDs internalized in MCF-7 and NIH/3T3 cells was of 180% and 120%, respectively. These observations were assigned to the amino functionality providing a positive surface charge to the particles thus favoring penetration into the mitochondrial membrane, where they provoked oxidative stress. NH_2-coated SiDs-induced ROS production was confirmed by the determination of an increased malondialdehyde level as a measure for the extent of membrane lipid peroxidation. Complementary cytotoxicity studies demonstrate the low cytotoxicity of these silicon nanoparticles (30%, 72 hs) for MCF-7 and (5%, 72 hs) for NIH/3T3 cells. The authors concluded that NH_2-coated SiDs are suitable as radiosensitizers for X-radiation in tumor cells.

18.5 CONCLUSIONS

Reviewed research has proved the capacity of different NPs to increase the efficacy of radiotherapy. Particularly, the most exhaustively investigated NPs, AuNPs, have been developed for their use as radiosensitizers in cancer radiotherapy (Babaei and Ganjalikhani 2014). Probable mechanisms involved in radiosensitizing effects of gold NPs are cell cycle changes and elevated ROS generation (Roa et al. 2009; Geng et al. 2011). Authors attributed the radiosensitivity to the increased localized absorption of X-rays, release of low-energy electrons from AuNPs and efficient deposition of energy in the form of radicals and electrons on the neighboring tissue (Babaei and Ganjalikhani 2014).

Considering the similar high-energy X-ray mass attenuation coefficients estimated for different tissues and water (Hubbell and Seltzer 2004) and that expected for a pure silicon network, $\mu/\rho \sim 0.050$ and $0.045 \text{ cm}^2 \text{ g}^{-1}$, respectively, ROS enhancement by SiDs in cells and in aqueous suspensions cannot rely on a preferential X-ray attenuation by silicon materials. Several radiation-induced physicochemical processes contribute to the overall ROS generation, as discussed before. Among them, the energy deposition in the SiDs surface as a consequence of the small size of the particles, the SiDs response to secondary particles and scattered photons promoted by the surrounding environment, and the formation of electron-hole excitons as strongly supported by the generation of cytotoxic $^1O_2$ only upon X-ray irradiation of SiDs.

On the other hand, in C6 cells, ionizing radiation resulted in a marked increase in SiDs-induced ROS generation. It is well known that induction of excessive ROS generation in cells promotes mitochondrial dysfunction, mitochondrial membrane permeability, and respiratory chain dysfunction triggering different types of cell death (Orrenius 2007). Mitochondrial dysfunction induces the release of electrons from the electron transport chain which in turn causes a second wave of ROS with complex I and complex III being sites of $O_2^{\bullet-}$ generation (St-Pierre et al. 2002). Interestingly, under our experimental conditions, ionizing radiation is not able to appreciably increase ROS generation in C6 cells in the absence of SiDs. In this regard, gliomas are highly resistant to ionizing radiation, due to their inability to generate ROS, and/or because of efficient ROS depletion by the cellular antioxidant defense system. Thus, our data employing SiDs represent a relevant contribution to the development of future radiotherapy strategies for the treatment of gliomas and other radio-resistant tumors.

As already mentioned, the subcellular location of NPs is an important factor for increasing the radiation cytotoxicity (Kong et al. 2008; Babaei and Ganjalikhani 2014). It has been reported that the radiation-sensitization effects induced in cancer cells is not only due to a physical interaction between NPs and radiation but also involves different cellular mechanisms (Kong et al. 2008) such as cell cycle arrest (Turner et al. 2005) and mitochondrial and lysosomal dysfunction (Nalika and Parvez 2015). In conclusion, alternative mechanisms other than ROS generation could be involved in the radiotoxic enhancement by SiDs.

The preliminary results discussed here are an overwhelming evidence of the SiDs potential capacity as radiosensitizers which are worth being investigated further.

REFERENCES

Babaei, M. & Ganjalikhani, M., 2014. The potential effectiveness of nanoparticles as radio sensitizers for radiotherapy. *BioImpacts : BI*, 4(1), pp. 15–20.

Baskar, R. et al., 2012. Cancer and radiation therapy: Current advances and future directions. *International Journal of Medical Sciences*, 9(3), pp. 193–199.

Begg, A.C., Stewart, F.A. & Vens, C., 2011. Strategies to improve radiotherapy with targeted drugs. *Nature Reviews Cancer*, 11(4), pp. 239–253.

Bhattacharjee, S. et al., 2010. Role of surface charge and oxidative stress in cytotoxicity of organic monolayer-coated silicon nanoparticles towards macrophage NR8383 cells. *Particle and fibre toxicology*, 7, p. 25.

Bosio, G.N. et al., 2008. Photodegradation of soil organic matter and its effect on gram-negative bacterial growth. *Photochemistry and Photobiology*, 84(5), pp. 1126–1132.

Caregnato, P. et al., 2017. Versatile silicon nanoparticles with potential uses as photoluminiscent sensors and photosensitizers. In A. Albini & E. Fasani, eds. *Photochemistry. SPR—Photochemistry.* London, UK: The Royal Society of Chemistry, pp. 322–345.

Davidson, R.A. & Guo, T., 2014. Average physical enhancement by nanomaterials under X-ray irradiation. *The Journal of Physical Chemistry C*, 118(51), pp. 30221–30228.

Deeley, T.J., 2013. *Principles of Radiation Therapy.* London: Butterworth-Heinemann.

Derfus, A.M., Chan, W.C.W. & Bhatia, S.N., 2004. Probing the cytotoxicity of semiconductor quantum dots. *Nano Letters*, 4(1), pp. 11–18.

Dhara, S. & Giri, P.K., 2011. Size-dependent visible absorption and fast photoluminescence decay dynamics from freestanding strained silicon nanocrystals. *Nanoscale Research Letters*, 6(1), p. 320.

Gara, P.M.D. et al., 2012. ROS enhancement by silicon nanoparticles in X-ray irradiated aqueous suspensions and in glioma C6 cells. *Journal of Nanoparticle Research*, 14(3), p. 741.

Geng, F. et al., 2011. Thio-glucose bound gold nanoparticles enhance radio-cytotoxic targeting of ovarian cancer. *Nanotechnology*, 22(28), p. 285101.

Giaccia, A.J., 2014. Molecular radiobiology: The state of the art. *Journal of Clinical Oncology*, 32(26), pp. 2871–2878.

Gonzalez, M.G. et al., 2004. Vacuum-ultraviolet photolysis of aqueous reaction systems. *Journal of Photochemistry and Photobiology C: Photochemistry Reviews*, 5(3), pp. 225–246.

Hainfeld, J.F. et al., 2006. Gold nanoparticles: A new X-ray contrast agent. *The British Journal of Radiology*, 79(939), pp. 248–253.

He, Y. et al., 2009. Photo and pH stable, highly-luminescent silicon nanospheres and their bioconjugates for immuno-fluorescent cell imaging. *Journal of the American Chemical Society*, 131(12), pp. 4434–4438.

He, Y. et al., 2011. One-Pot microwave synthesis of water-dispersible, ultraphoto- and pH-stable, and highly fluorescent silicon quantum dots. *Journal of the American Chemical Society*, 133(36), pp. 14192–14195.

Hirsjarvi, S., Passirani, C. & Benoit, J.-P., 2011. Passive and active tumour targeting with nanocarriers. *Current Drug Discovery Technologies*, 8(3), pp. 188–196.

Huan, C. & Shu-Qing, S., 2014. Silicon nanoparticles: Preparation, properties, and applications. *Chinese Physics B*, 23(8), p. 88102.

Hubbell, J.H., 1999. Review of photon interaction cross section data in the medical and biological context. *Physics in Medicine and Biology*, 44(1), p. R1.

Hubbell, J.H. & Seltzer, S.M., 2004. Tables of X-Ray mass attenuation coefficients and mass energy-absorption coefficients from 1 keV to 20 MeV for elements Z = 1 to 92 and 48 additional substances of dosimetric interest. *NIST Standard Reference Database 126.* Available at: http://www.nist.gov/pml/data/xraycoef/ [Accessed August 12, 2016].

Jackson, S.P. & Bartek, J., 2009. The DNA-damage response in human biology and disease. *Nature*, 461(7267), pp. 1071–1078.

Juzenas, P. et al., 2008. Quantum dots and nanoparticles for photodynamic and radiation therapies of cancer. *Advanced Drug Delivery Reviews*, 60(15), pp. 1600–1614.

Khan, F.M., 2003. *Physics of Radiation Therapy Third Edition* 3rd edition., T. Boyce, ed., Philadelphia, PA: Lippincott Williams & Wilkins.

Klein, S. et al., 2013. Oxidized silicon nanoparticles for radiosensitization of cancer and tissue cells. *Biochemical and Biophysical Research Communications*, 434(2), pp. 217–222.

Knoll, G.F., 2000. *Radiation Detection and Measurement.* Fourth Edition. New York: John Wiley & Sons Inc.

Kohn, T. & Nelson, K.L., 2007. Sunlight-mediated inactivation of MS2 coliphage via exogenous singlet oxygen produced by sensitizers in natural waters. *Environmental Science & Technology*, 41(1), pp. 192–197.

Kong, T. et al., 2008. Enhancement of radiation cytotoxicity in breast-cancer cells by localized attachment of gold nanoparticles. *Small*, 4(9), pp. 1537–1543.

Kwatra, D., Venugopal, A. & Anant, S., 2013. Nanoparticles in radiation therapy: A summary of various approaches to enhance radiosensitization in cancer. *Translational Cancer Research*, 2(4), pp. 330–342.

Le Caër, S., 2011. Water radiolysis: Influence of oxide surfaces on H2 production under ionizing radiation. *Water*, 3(1), p. 235.

Leroy, C. & Rancoita, P.-G., 2012. *Silicon Solid State Devices and Radiation Detection,* 1st edition. Singapore: World Scientific.

Lillo, C.R. et al., 2015. Organic coating of 1–2-nm-size silicon nanoparticles: Effect on particle properties. *Nano Research*, 8(6), pp. 2047–2062.

McMahon, S.J., Paganetti, H. & Prise, K.M., 2016. Optimising element choice for nanoparticle radiosensitisers. *Nanoscale*, 8(1), pp. 581–589.

Mesbahi, A., 2010. A review on gold nanoparticles radiosensitization effect in radiation therapy of cancer. *Reports of Practical Oncology and Radiotherapy*, 15(6), pp. 176–180.

Nalika, N. & Parvez, S., 2015. Mitochondrial dysfunction in titanium dioxide nanoparticle-induced neurotoxicity. *Toxicology Mechanisms and Methods*, 25(5), pp. 355–363.

NCI, 2016. NCI Dictionary of Cancer Terms. *NCI Dictionary of Cancer Terms*. https://www.cancer.gov/publications/dictionaries/cancer-terms.

Nelson, G. & Reilly, D., 1991. Gamma-ray interactions with matter. In D. Reilly, N. Ensslin, H. Smith, Jr, & S Kreiner, eds. *Passive Nondestructive Analysis of Nuclear Materials.* Los Alamos, NM: Los Alamos National Laboratory, pp. 27–42.

Nishimura, H. et al., 2013. Biocompatible fluorescent silicon nanocrystals for single-molecule tracking and fluorescence imaging. *The Journal of Cell Biology*, 202(6), pp. 967–983.

O'Farrell, N., Houlton, A. & Horrocks, B.R., 2006. Silicon nanoparticles: Applications in cell biology and medicine. *International Journal of Nanomedicine*, 1(4), pp. 451–472.

Orrenius, S., 2007. Reactive oxygen species in mitochondria-mediated cell death. *Drug Metabolism Reviews*, 39(2–3), pp. 443–455.

Panganiban, R.-A.M., Snow, A.L. & Day, R.M., 2013. Mechanisms of radiation toxicity in transformed and non-transformed cells. *International Journal of Molecular Sciences*, 14(8), pp. 15931–15958.

Podgorsak, E.B., 2005. *Radiation Oncology Physics*. E. B. Podgorsak, ed., Vienna: International Atomic Energy Agency.

Rahman, W.N. et al., 2009. Enhancement of radiation effects by gold nanoparticles for superficial radiation therapy. *Nanomedicine: Nanotechnology, Biology and Medicine*, 5(2), pp. 136–142.

Retif, P. et al., 2015. Nanoparticles for radiation therapy enhancement: The key parameters. *Theranostics*, 5(9), pp. 1030–1044.

Roa, W. et al., 2009. Gold nanoparticle sensitize radiotherapy of prostate cancer cells by regulation of the cell cycle. *Nanotechnology*, 20(37), p. 375101.

Rosso-Vasic, M. et al., 2008. Alkyl-functionalized oxide-free silicon nanoparticles: Synthesis and optical properties. *Small*, 4(10), pp. 1835–1841.

Shao, L., Gao, Y. & Yan, F., 2011. Semiconductor quantum dots for biomedicial applications. *Sensors*, 11(12), p. 11736.

Sicard-Roselli, C. et al., 2014. A new mechanism for hydroxyl radical production in irradiated nanoparticle solutions. *Small*, 10(16), pp. 3338–3346.

Siller, L. et al., 2009. Core and valence exciton formation in x-ray absorption, x-ray emission and x-ray excited optical luminescence from passivated Si nanocrystals at the Si L(2,3) edge. *Journal of Physics. Condensed Matter: An Institute of Physics Journal*, 21(9), p. 95005.

St-Pierre, J. et al., 2002. Topology of superoxide production from different sites in the mitochondrial electron transport chain. *Journal of Biological Chemistry*, 277(47), pp. 44784–44790.

Takahashi, J. & Misawa, M., 2007. Analysis of potential radiosensitizing materials for x-ray-induced photodynamic therapy. *Nanobiotechnology*, 3(2), pp. 116–126.

Takeshita, K. et al., 2004. In vivo monitoring of hydroxyl radical generation caused by x-ray irradiation of rats using the spin trapping/epr technique. *Free Radical Biology and Medicine*, 36(9), pp. 1134–1143.

Torre, L.A. et al., 2015. Global cancer statistics, 2012. *CA: A Cancer Journal for Clinicians*, 65(2), pp. 87–108.

Tsai, C.S. & Hsiao, J.K., 2016. Study on the radiosensitizing applications of gold nanoparticles. In *Advanced Materials, Structures and Mechanical Engineering II. Applied Mechanics and Materials*. Zurich, Switzerland: Trans Tech Publications, pp. 26–30.

Turner, J. et al., 2005. Tachpyridine, a metal chelator, induces G(2) cell-cycle arrest, activates checkpoint kinases, and sensitizes cells to ionizing radiation. *Blood*, 106(9), pp. 3191–3199.

Vaverka, A.S.M., 2008. *Characterization of the Electronic Structure of Silicon Nanoparticles Using X-ray Absorption and Emission*. Davis, CA: University of California.

Wicki, A. et al., 2015. Nanomedicine in cancer therapy: Challenges, opportunities, and clinical applications. *Journal of Controlled Release*, 200, pp. 138–157.

Winterbourn, C.C., 2014. The challenges of using fluorescent probes to detect and quantify specific reactive oxygen species in living cells. *Biochimica et Biophysica Acta (BBA)—General Subjects*, 1840(2), pp. 730–738.

Yang, W. et al., 2008. Semiconductor nanoparticles as energy mediators for photosensitizer-enhanced radiotherapy. *International Journal of Radiation Oncology Biology Physics*, 72(3), pp. 633–635.

Functional materials

19 Silicon-based anode materials for lithium-ion batteries

Gyeong S. Hwang and Chia-Yun Chou

Contents

19.1 INTRODUCTION

In the past few decades, lithium-ion batteries (LIBs) have attracted considerable attention as power sources for portable electronics and electrical vehicles (EVs). They have several advantages over other secondary battery technologies such as nickel-cadmium (NiCd) and nickel metal hydride (NiMH), including high operating voltage, high energy density, light weight, long cycle life, and zero-to-low memory effect. LIBs are currently one of the most commonly used rechargeable batteries for small portable electronic devices. However, in order to extend their effective use as large-scale energy storage systems for EVs and renewable energy, there is an urgent need to further increase the energy density, power density, and cycle life while retaining safety and cost at an affordable range. This presents knowledge and materials challenges. The materials challenge requires development of a firm understanding of the electrode and electrolyte materials, and the structure and chemistry at their interfaces that will allow us to identify and assess alternative energy storage strategies, in addition to improvements of existing technologies.

At present, graphite-based materials are commonly used as anode materials in commercial LIBs due to their long cycle life, low cost, and abundance, but exhibit relatively low gravimetric and volumetric specific capacity. The theoretical capacity of graphite is 372 mAh/g for LiC_6. While the capacity limitations of carbon-based anodes for use in advanced LIBs need to be overcome, silicon (Si) has emerged as one of the most promising anode materials because of its abundance, nontoxicity, desirable electrochemical potential to be coupled with high-voltage cathode materials, and more importantly the highest known specific Li storage capacity (close to 4000 mAh/g for Li_xSi, $x \leq 4.4$) which is one order of magnitude higher than that of graphite.

However, the practical use of Si as an anode material is hampered by its large volume expansion (~400%) causing pulverization, loss of electrical contact, and consequently early capacity fading; for graphite anodes, the only structural change is a ~12% isotropic expansion. Considerable efforts have been made to overcome these problems, for instance through structural modifications such as amorphous phases and nanostructured Si, combining Si with carbonaceous materials (Si-C), alloying Si with active/ inactive elements, and Si suboxides (SiO_x, $x < 2$). However, many fundamental aspects regarding the lithiation processes and properties remain unclear since direct characterization of the complex electro-chemical systems is rather difficult.

Recently, with advances in computing power and computational methodology, first-principles based computer simulations have been applied to evaluate the properties and performance of potential candidate materials for electrodes. Computer simulations support and complement experimental studies, and vice versa. They can provide explanations for experimental observations and guidelines for synthesizing and characteriz-ing new materials and developing improved materials systems, while experimental data are used for validating calculation results and testing theoretical predictions. The basic understanding gained from theoretical studies combined with experiments will be crucial for realizing next-generation electrical energy storage systems with long life at an affordable cost. In this chapter, we present the lithiation mechanisms and electrochemical prop-erties of Si-based anode materials, mostly based on the results of recent first-principles studies.

19.2 SINGLE Li ATOMS IN SILICON: STRUCTURE, BONDING, AND DIFFUSION

There have been extensive theoretical and experimental studies to determine minimum-energy configu-rations for interstitial atoms in crystalline Si. Experimental measurements showed that Li atoms intro-duced to crystalline Si tend to remain at an interstitial site with T_d symmetry and behave similarly to group-V shallow donors (Aggarwal et al. 1965). This observation has been supported by a series of first-principles studies (Patrocinio et al. 2010; Chou et al., 2011; Kim et al. 2011).

The interstitials, Si, or impurities, may be located at highly symmetric sites including tetrahedral, hexagonal, and bond-centered interstitial sites in the diamond lattice. Among them, the tetrahedral site is identified to be energetically the most favorable site for Li insertion (Patrocinio et al. 2010). The incorpora-tion of Li imposes a compressive strain on the surrounding host lattice and leads to a slight outward relax-ation of the neighboring Si atoms. The hexagonal state that is another important local minimum for a Si self-interstitial is found to be a saddle point for Li migration. The <110>-split state in which Li and Si atoms are aligned in the <110> direction while sharing a lattice site is also likely to be unstable. On the other hand, the <110>-split dumbbell structure is identified as energetically the most favorable for an Si self-interstitial. In addition, the <111> bond-centered state in which Li is located between two lattice Si atoms while the lattice Si atoms are substantially displaced outward in the <111> direction, is highly unlikely.

The first donor (+/0) level of an impurity in crystalline Si (c-Si) can be estimated using first-principles periodic supercell calculations. The ionization level (μ_i) is approximated in terms of the total energies of the positive and neutral states of the Li-containing supercell (E^+ and E^0) and the position of the valence band maximum in supercell E^+ (E_v^+)

$$E^+ + (E_v^+ + \mu_i) = E^0$$

In the periodic approach, a homogeneous background charge is included to maintain the overall charge neutrality of a charged supercell, which requires a careful monopole correction to account for the electro-static interaction with the background charge. Density-functional theory (DFT) calculations (Patrocinio et al. 2010) predicted the donor level of Li at the tetrahedral site to be located close to the conduction band minimum, indicating that Li interstitials act like a shallow donor in c-Si. This is apparently due to the fact that Li with one valence electron ([He]2s^1) is easily ionized via electron donation to the host Si matrix. Note that Li has a much lower electronegativity (~1.0) as compared to Si (~1.9).

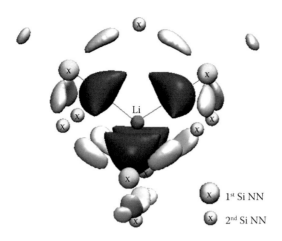

Figure 19.1 Valence charge density difference plot for Li insertion. The charge density difference ($\Delta\rho$) is calculated by subtracting the charge densities of an isolated Li and the Si matrix (with no Li) from the total charge density of the Li^0/Si matrix with no atomic displacement, i.e., $\Delta\rho = \rho(Li/Si) - \rho(Li) - \rho(Si)$. The positions of the Li and its first and second Si nearest neighbors are indicated. The dark gray and light gray isosurfaces represent the regions of charge gain ($+0.019$ e/Å³) and loss (-0.012 e/Å³), respectively.

Upon Li addition, the Fermi level shifts above the conduction band minimum of Si (Patrocinio et al. 2010). This implies that the charge transferred from Li fills the antibonding sp^3 state of neighboring Si atoms, which may in turn weaken the corresponding Si–Si bonds. The difference of the electron densities before and after the Li insertion at the tetrahedral site is shown in Figure 19.1. There is charge accumulation in the region between the Li and neighboring Si atoms with a noticeable shift toward the Si atoms. This suggests that the Li–Si bond can be considered polar covalent. In addition, electron densities are noticeably depleted in the nearby Si–Si bonding regions, which is indicative of the weakening of the Si–Si covalent bonds. This electronic structure analysis also shows that the excess electrons are largely localized within the first-nearest Si neighbors, thereby effectively screening the cationic Li. The partially ionized Li interstitials may prefer to remain isolated and well dispersed in the Si matrix, rather than aggregated.

A Li interstitial may undergo migration by jumping between adjacent tetrahedral sites via the hexagonal site which is identified to be a saddle point. The temperature-dependent diffusivity of Li in c-Si can be predicted using the Arrhenius equation,

$$D = D_0 \exp\left(-\frac{E_a}{k_B T}\right)$$

Within harmonic transition state theory, the prefactor can be derived by

$$D_0 = (1/2\alpha)\,\lambda l^2\,\nu_0$$

Using the Vineyard equation, the attempt frequency ν_0 can be approximated as

$$\nu_0 = \prod_{i=1}^{3N} \nu_i^* \Big/ \prod_{i=1}^{3N-1} \nu_i^{**}$$

The harmonic vibrational frequencies, ν_i^* and ν_i^{**} at the minimum and saddle points, respectively, can be obtained by diagonalizing a Hessian matrix obtained from numerical differentiation of forces. At the saddle point, there is one imaginary frequency for Li. DFT calculations (Milman et al., 1993; Patrocinio et al. 2010) predicted ν_0 and D_0 to be $8.6 - 10.1$ THz and $3.2 - 3.7 \times 10^{-3}$ cm²/sec, giving $D = 1 - 6 \times 10^{-13}$ cm²/sec at 298 K, while an experimental value of 2×10^{-14} cm²/sec was reported (Canham 1988).

19.3 LITHIATION OF SILICON: MECHANISM, STRUCTURE, AND PROPERTIES

19.3.1 ENERGETICS AND STRUCTURAL EVOLUTION

Experiments showed the formation of various crystalline Li-Si phases during high-temperature lithiation. However, room-temperature lithiation may lead to amorphous Li-Si phases that often show different properties from their crystalline counterparts.

The relative stabilities of the amorphous and crystalline structures were evaluated using first-principles DFT calculations (Kim et al. 2011; Chou et al. 2011). The variations in the mixing enthalpy per atom (ΔE_{mix}) for crystalline and amorphous Li–Si alloys as a function of Li atomic fraction y, with respect to the energies of c-Si (E_{Si}) and body-centered cubic Li (bcc-Li, E_{Li}), are shown in Figure 19.2.

$$\Delta E_{mix} = E_{Li_y Si_{1-y}} - yE_{Li} - (1-y)E_{Si}$$

For a-Li$_y$Si$_{1-y}$ at low Li contents, ΔE_{mix} appears to be positive in value and peaks around 25 at.% Li. As the Li content increases, ΔE_{mix} drops and changes from positive to negative at 40 at.% Li. Above this Li content, ΔE_{mix} continues to decrease and falls to a valley plateau between 60 - 80 at.% Li. The positive value of ΔE_{mix} at Li contents < 40 at.% may indicate the presence of an initial barrier for Li incorporation into the Si matrix, especially in the crystalline phase. Contrarily, the negative ΔE_{mix} at Li contents >40 at.% suggests favorable alloy formation. According to the trend, a-Li$_x$Si$_{1-x}$ alloys with 60–80 at.% Li are most stable with an energy gain of 0.16–0.18 eV/atom with respect to c-Si and bcc-Li. Experiments also reported the formation of a-Li$_{2.1}$Si ($\approx$68 at.% Li) with an energy gain of 0.12 eV (Limthongkul et al. 2003). For crystalline phases, a distinct ΔE_{mix} minimum occurs at 71 at.% Li, and on average the predicted total energies are ~0.1 eV/atom lower than their amorphous counterparts. This implies that the amorphous alloys would undergo recrystallization at elevated temperatures.

A host Si matrix may undergo a significant structural change during lithiation. A set of a-Li$_x$Si structures from first-principles computer simulations (Chou et al. 2011; Kim et al. 2011), together with c-Li$_x$Si structures for comparison, are shown in Figure 19.3. An amorphous structure can be characterized using pair-distribution function (PDF, $g(r)$), which is defined as

$$g(r) = \frac{V}{N}\frac{n(r)}{4\pi r^2 \Delta r}$$

The predicted PDFs for selected a-Li$_x$Si structures are also shown in Figure 19.3.

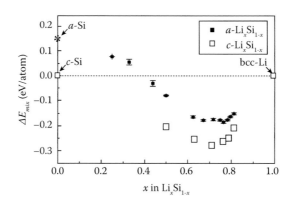

Figure 19.2 Variations in the predicted mixing enthalpies of amorphous and crystalline Li–Si alloys as a function of Li content.

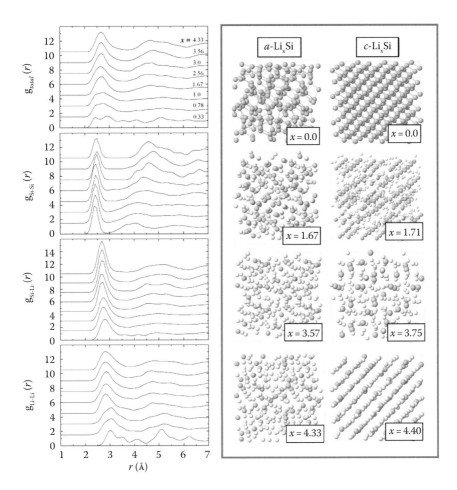

Figure 19.3 (Right panel) Atomic structures of amorphous and crystalline Li-Si alloys (Li$_y$Si where y indicates the Li content per Si). The yellow and white balls represent Si and Li atoms, respectively. (Left panel) Total and partial pair-distribution functions for selected a-Li$_x$Si alloys and the atomic structures of a-Li$_x$Si and c-Li$_x$Si, as indicated.

In contrast to the typical sharp lines found in crystalline materials, the PDF peaks are smoother and more broadened while no long-range order can be identified, indicating an amorphous nature. At low Li content, the first two distinct peaks are attributed to Si–Si pairs and a combination of Si–Li and Li–Li pairs, respectively. As the Li content x increases, the Si–Li and Li–Li peaks become stronger while the Si–Si peak dwindles. Moreover, with increasing Li content, the Si–Si peak position shifts to a larger pair distance while the opposite trend is found for the Li–Si and Li–Li peaks. The increased Si–Si pair distance is indicative of the weakened Si–Si bonds. Note that the charge transferred from Li fills up the antibonding sp^3 states of Si and thereby weakens the Si–Si bonds as mentioned earlier.

For both crystalline and amorphous phases, the volume increases nearly linearly with Li content, and the opposite trend is true for the density values. As expected, the crystalline phase is slightly denser than the amorphous phase of corresponding composition. The volume expansion of Si reaches up to 400% at full lithiation.

The tetrahedrally-bonded network of Si disintegrates into smaller Si fragments with increasing Li content. For the crystalline structures, the Si lattice breaks up into small clusters in various shapes. That is, c-Li–Si has a threefold-coordinated Si network that consists of interconnected chains and puckered eight-membered rings, c-Li$_{12}$Si$_7$ has two types of clusters (Si$_5$ rings and Si$_4$ stars), c-Li$_7$Si$_3$ has Si$_2$ dumbbells,

$Li_{13}Si_4$ has a mixture of Si_2 dumbbells and Si atoms, and c-$Li_{15}Si_4$ and c-$Li_{22}Si_5$ have only single Si atoms. Likewise, for the amorphous structures, the disintegration of Si into low-connectivity Si clusters also occurs as the Li content increases. It is worthwhile to note that roughly 30%–40% of Si atoms still form Si–Si pairs even in the highly lithiated amorphous structures such as a-$Li_{3.57}Si$ and a-$Li_{4.33}Si$.

19.3.2 COMPOSITION DEPENDENCE OF Li DIFFUSIVITY

Recent research showed that the mobility of Li atoms in Li–Si alloys is very sensitive to the Li:Si composition ratio, and the Li diffusivity is enhanced by orders of magnitude in highly lithiated stages as compared to the early stages of lithiation.

The diffusivities of Li and Si in lithiated Si can be estimated using ab initio molecular dynamics simulations (AIMD) (Chou et al. 2011; Kim et al. 2011). Once the mean-square displacements (MSD) of Li and Si atoms are linearly proportional to simulation time (t), their diffusivities (D) can be predicted based on the Einstein relation

$$D = D_0 \exp\left(\frac{E_a}{k_B T}\right) = \frac{MSD^2}{6t} = \frac{|R_i(t) - R_i(0)|^2}{6t}$$

From an Arrhenius plot of $\ln(D)$ versus $1/T$, the prefactor D_0 and the activation energy E_a can be extracted. That is, E_a and D_0 are obtained from the slope of the Arrhenius plot and by extrapolating D to infinite T, respectively.

For highly lithiated a-$Li_{3.57}Si$, the predicted E_a and D_0 are about 0.2 eV and 5×10^{-4} cm^2/s from AIMD simulations (Chou et al. 2011). Within the harmonic approximation to transition state theory, D_0 would be on the order of 10^{-3} cm^2/s. Taking the predicted E_a and D_0 values, the D_{Li} is estimated to be around 2×10^{-7} cm^2/s in a-$Li_{3.57}Si$ at room temperature. Note that the D_{Li} is several orders of magnitude greater than 2×10^{-13} cm^2/s as estimated for a single Li atom in c-Si with $E_a \approx 0.6$ eV (see Section 19.2). MD simulations (Chou et al. 2011; Cui et al. 2012) consistently predicted that the Li diffusivity may vary by orders of magnitude depending on the stages of lithiation.

It is also interesting to notice that there are discrepancies between the predicted and experimentally measured D_{Li}. First, the predicted values ($\sim10^{-7}$ cm^2/s) in lithiated Si at room temperature are orders of magnitude higher than the experimental values of $\sim0^{-12}$ cm^2/s (Ding et al. 2009). Second, while the D_{Li} is predicted to increase monotonically with increasing Li content in a-$Li_x Si$ alloys, experiments reported a W-shaped dependence with two minima around $y = 2.1 \pm 0.2$ and 3.2 ± 0.2 (coincides with c-Li_7Si_3 and c-$Li_{13}Si_4$) (Ding et al. 2009). One possible explanation is that unlike the simulated amorphous structures, experimentally lithiated samples could exhibit a certain degree of variety in terms of Li concentration and coexistence of crystalline and amorphous phases. As pointed out, the D_{Li} is highly sensitive to the alloy composition and atomic environment, so the above-mentioned factors are expected to influence Li diffusion strongly and thus contribute to the discrepancies. In addition, as experimental measurements are generally subject to sample-to-sample variation and differences in test conditions, the widely scattered D values (ranging from 10^{-14} to 10^{-8} cm^2/s) make it difficult to directly compare with simulation results. Nevertheless, theoretical efforts can provide valuable insights into understanding the atomistic-level factors and their effects on Li diffusion in lithiated Si.

19.3.3 ELECTRONIC AND MECHANICAL PROPERTIES

It is generally understood that Li–Si alloys are Zintl-like phases (Kauzlarich 1996) with the consideration that only a partial charge is transferred from Li to Si. Si is not electronegative enough to completely strip off the outer shell (2s) electron from Li. If every Li atom donates its 2s electron to Si, the charge transfer in $Li_x Si$ is represented as $(Li^+)_x Si^{x-}$, and in the case of $x < 4$, $(4 - x)$ Si–Si bonds per Si atom are formed to satisfy the $(8 - N)$ octet rule. For instance, in the Li–Si Zintl phase ($x = 1$), Si atoms are threefold coordinated ($4 - 1 = 3$), and crystalline $Li_{15}Si_4$ is the first Zintl phase where all the Si atoms are isolated and surrounded by Li atoms ($x = 3.75$). With the increasing net charge transfer in the higher lithiated phases, the filling of the Si antibonding p states lead to Si–Si bond weakening and consequently the fragmentation/disintegration of the host Si lattice.

As the Si–Si covalent bonds are replaced with more ionic Li–Si bonds, Li–Si alloys of higher Li contents can be expected to be softer. The effect of Li content on the mechanical properties of lithiated Si have been evaluated using first-principles calculations (Chou et al. 2011; Kim et al. 2011). Elastic constants, C_{ij} can be obtained by computing the energies of deformed unit cells. The deformation strain tensor, e_{ij} with six independent components is represented as

$$e_{ij} = \begin{pmatrix} e_1 & e_6/2 & e_5/2 \\ e_6/2 & e_2 & e_4/2 \\ e_5/2 & e_4/2 & e_3 \end{pmatrix}$$

For cubic phases (e.g., $Li_{15}Si_4$), orthorhombic, isotropic, and monoclinic distortions are applied to obtain three independent elastic constants, C_{11}, C_{12}, and C_{44} (expressed using Voigt notation). For tetragonal phases (e.g., Li-Si), six independent deformation modes are applied to calculate C_{11}, C_{12}, C_{13}, C_{33}, C_{44}, and C_{66}. Self-consistent relaxation is allowed in all strained unit cells, and the total energy change with respect to the strain tensor gives

$$E\left(e_{ij}\right) = E_0 - P(V)\Delta V + \frac{V}{2}\sum_{ij} C_{ij}e_{ij} + O\left[e_{ij}^3\right]$$

The calculation of the elastic constant requires a high degree of precision because the energy variation involved is very small, hence the geometry optimization needs to be performed with a tight force tolerance.

Once C_{ij} values are known, mechanical quantities such as Poisson's ratio, Young's modulus, shear modulus, and bulk modulus can be estimated. For instance, the bulk modulus B is given by $B = 1/9(C_{11} + C_{22} + C_{33} + 2C_{12} + 2C_{13} + 2C_{23})$. First-principles calculations (Shenoy et al. 2010; Chou et al. 2011; Kim et al. 2011) predicted $B = 53–65$ and $30–32$ GPa for crystalline Li-Si and Li_4Si_{15}, respectively. Clearly, the B decreases with increasing Li content leading to significant elastic softening, as also demonstrated by analysis of Li-Si bonding nature in Section 19.2.

19.4 LITHIATION OF SILICON NANOSTRUCTURES AND ALLOYS

19.4.1 NANOWIRES AND THIN FILMS

Si nanostructures can accommodate larger strain and provide better mechanical integrity because their dimensions would limit the size and propagation of cracks, which typically initiate the fracture process. Experiments showed that Si nanowires (SiNWs) and thin films exhibit excellent capacity retention and rate capability. Chan et al. reported SiNWs exhibiting 3,193 mAh/g discharge capacity after 10 cycles at C/20 rate (discharge in 20 hours) and stable capacity (~3,500 mAh/g) up to 20 cycles at C/5 rate (Chan et al. 2008). The lengths of SiNWs increased and their volume change appeared to be about 400% after lithiation, but NWs remained continuous without fractures. Likewise, Si thin films have consistently realized capacities above 2000 mAh/g. Maranchi et al. presented amorphous 250 nm-thick Si films with reversible capacities of about 3500 mAh/g at C/2.5 rate for 30 cycles with no obvious signs of failure (Maranchi et al. 2003). While nanostructured Si exhibits many beneficial properties as an anode material, the large surface area may also cause more significant solid-electrolyte interphase (SEI) formation and consequently larger capacity loss.

19.4.1.1 Surface effects

There has been an increasing amount of theoretical efforts to elucidate the lithiation mechanisms of Si nanostructures. To study Li behavior at the onset of lithiation, researchers looked at single Li insertion into SiNWs with different axis orientations and sizes (Chan and Chelikowsky 2010; Zhang et al. 2010). It was shown that surface sites are energetically the most favorable insertion positions, and the diffusion barrier is smaller compared to that in the bulk. Later, Chou and Hwang (Chou and Hwang 2013a) performed rigorous AIMD simulations to study the surface effects on the lithiation behavior of Si, through analysis of the near-surface composition, structural evolution, energetics, and Li diffusion properties of Li–Si alloys.

AIMD simulations predicted a slight surface enrichment of Li atoms when the Li content is sufficiently low, which may contribute to stabilizing the alloy surface. With increasing Li content, it was also found that the Coulomb repulsion between cationic Li atoms tends to become more pronounced and results in well-dispersed Li distribution in the bulk as well as near-surface regions. Detailed structural analysis showed that the near-surface and bulk Li–Si alloys share very similar structural features. The presence of surfaces tends to only affect the outermost surface layer with the reduced Si–Si connectivity, and below this layer the bulk-like connectivity is likely restored.

Indicative of the shallow surface effect, the predicted surface energy (E_{surf}) converges rapidly when increasing the thickness of thin alloy layers. With the introduction of Li ($x = 0.42$), E_{surf} decreases substantially from 0.07 eV/Å^2 for a-Si to less than 0.02 eV/Å^2, marking the significant surface stabilization due to the presence of Li atoms. With increasing Li content, the Li–Li repulsion grows more apparent and results in higher E_{surf}, which reaches a plateau around 0.03 eV/Å^2 between $x = 1.00$ and 3.57.

While the mobility of Li in lithiated Si strongly depends on the Li:Si composition ratio and local atomic environment as discussed in Section 19.3.2, AIMD simulations showed that the presence of surfaces significantly facilitate Li diffusion. The Li diffusivity in the near-surface region was predicted to be about twice larger than that in the bulk. There are also experimental evidences for the enhanced Li mobility under the influence of surface effects, in which the lithiation of SiNWs appears to proceed more rapidly in the surface region as compared to the center region, leading to a V-shaped lithiation front (Liu et al. 2011).

19.4.1.2 Ultrafast chemical lithiation of silicon nanowires

An affirmative description of lithiation kinetics in SiNWs is still limited. The reported experimental D values of Li are scattered by six orders of magnitude (from 10^{-14} to 10^{-8} cm^2/s) depending on sample and test conditions. From conventional ex situ electrochemical measurements, it would be very difficult to characterize Li diffusion along the axial direction of an individual NW and differentiate it from that through the electrolyte and/or the electrode-electrolyte interface layer.

Researchers characterized the concentration-gradient-driven lithiation process in individual SiNWs using an in situ scanning electron microscope (SEM) in the absence of an applied electric field (Seo et al. 2015). The structural evolution of an SiNW, which was brought into direct contact with Li metal at room temperature in vacuum is shown in Figure 19.4. The SiNW lithiation is purely driven by the Li concentration gradient, unlike conventional electrochemical approaches. At $t = 0.00$ s, the pristine SiNW was straight with a uniform diameter around 78.9 nm. Once the NW contacted the Li metal surface, it exhibited significant morphological changes captured in the time-lapse series of SEM images up to $t = 1.02$ s. Li atoms near the vicinity of the Li–SiNW contacting area diffused into the SiNW, and volume expansion occurred instantaneously as the lithiation front (indicated by the yellow triangle) propagated rapidly from the NW tip toward the other end. At $t = 1.02$, the entire SiNW swelled uniformly in the radial direction without noticeable elongation or bending. The cross section becomes circular after lithiation and the total

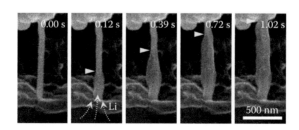

Figure 19.4 Microstructural evolution of an intrinsic Si nanowire in direct contact with Li metal. Volume expansion occurred instantaneously as the reaction front (marked by the yellow triangle) propagated from the nanowire tip toward the other end.

volume expansion is estimated around 330%. The lithiated SiNW was found to be completely amorphous from the scanning transmission electron microscope (STEM) and electron energy loss spectroscopy (EELS) analyses.

By tracking the lithiation front, the propagation speed was reported to be around 1×10^3 nm/s, which is at least two orders of magnitude higher than those reported from electrochemical measurements. In conventional electrochemical lithiation tests, four kinetic processes may need to be considered in determining the rate of lithiation: (1) Li diffusion in the electrolyte, (2) the redox reaction at the electrolyte-Si interface, (3) the formation of lithiated Si at the lithiation front, and (4) propagation of the lithiation phase boundary between amorphous lithiated Si and unlithiated crystalline Si. Li cations at the electrolyte-Si anode interface must combine with electrons (redox reaction) before the neutralized Li atoms start diffusing into the Si anode. In the previously presented technique, the first two factors can be ruled out while it may provide a perfect controlled environment to examine the latter two contributions.

If the Li_xSi alloy formation were the dominant rate-limiting factor, the corresponding length of the lithiation front propagation (h) would be linearly proportional to the lithiation time (t). Otherwise, if the propagation of the Li_xSi/Si lithiation front were the dominant factor, the relation between h and t would follow the one-dimensional (1D) random walk diffusion model, $h^2 = nDt$; h^2 was found to be linearly proportional to t, confirming that the propagation of lithiation front is the dominant rate-limiting step. Given 1D diffusion between a pair of semi-infinite solids, the Li diffusivity was estimated to be around 2.6×10^{-9} cm²/s.

A schematic illustration of the chemical lithiation process is shown in Figure 19.5. The mechanisms are based on the following rationales. Firstly, for an SiNW, the lithiation process may proceed faster along the outer surfaces than the center region, as discussed in the previous section. Secondly, the lithiation front may move much faster in the <110> direction, as compared to the <111> direction (Chan et al. 2012 Liu et al. 2012). Moreover, with increasing Li content, the Li diffusivity in Si will increase by orders of magnitude, as presented in Section 19.3.2.

The observed lithiation mechanisms were also cross-examined using first-principles calculations, where the rate-limiting step was modeled by the propagation of the $Li_xSi/Si(110)$ phase boundary. The predicted room-temperature Li diffusivity of $\approx 10^{-9}$ cm²/s was in excellent agreement with the experimentally estimated value. This suggests that the lithiation speed would be primarily controlled by Li diffusion across the lithiated and unlithiated Si phase boundary.

19.4.2 GOLD-COATED SILICON NANOWIRES

One common approach to grow SiNWs is via the vapor–liquid–solid (VLS) process with gold (Au) catalysts. In the VLS growth, a significant amount of Au was found to spread on the NW sidewalls. The presence of Au may influence the physical properties of SiNWs and also modify the NW morphology

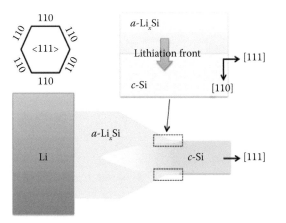

Figure 19.5 Schematic illustration of the chemical lithiation process of a SiNW.

for the following reasons: (1) Au is also reactive toward Li with a capacity of 451 mAh g^{-1} (based on the lithiated intermetallic compound of Li$_{15}$Au$_4$), (2) Au is mechanically ductile and electrically conductive, hence being able to maintain both the structural stability and conductive path, and (3) it has been shown that Au–Si alloys are likely to form at the Au/Si interface, which may in turn introduce intricate interfacial effects that influence the lithiation behavior.

A combined in situ characterization and first-principles study (Chou et al. 2015) demonstrated an intriguing stagewise lithiation behavior in Au-coated SiNWs (Au–SiNWs), as illustrated in Figure 19.6.

- Stages I and II: Axial lithiation of the Au shell with progressive and slow expansion. Li atoms are predicted to be incorporated preferentially in the Au shell while the Au–Si interface layer (provided it is sufficiently thick) may serve as a facile diffusion path at the early stage of lithiation. As the lithiation front proceeds in the axial direction through the entire Au shell, the lithiated region exhibits a gradual radial expansion till the full lithiation capacity of Au ($\approx a$-Li$_4$Au) is nearly reached.
- Stage III: Radial lithiation of the Si core with uniform and rapid expansion. Once the Au shell is close to fully lithiated in the Au-coated SiNW, Li atoms may start diffusing into the Si core, as a-Li$_x$Au alloys are far more stable than a-Li$_x$Si alloys. The radial diffusion of Li atoms into the Si core leads to the spontaneous formation of a-Li$_x$Si alloys. Since the diffusion distance in the radial direction is significantly shorter than in the axial direction, the corresponding radial expansion appears to occur much more rapidly and simultaneously over the entire NW length.

This proposed stagewise lithiation mechanism is in line with the experimental observation, in which an Au–SiNW was brought into direct contact with Li metal inside a SEM in the absence of an applied electric field. Upon lithiation, the structural evolution was monitored in situ and shown in Figure 19.7. At $t = 0.00$ s, the Au–SiNW was straight with a uniform diameter. Once the Au–SiNW contacted bulk Li, the morphological changes are captured in the time-lapse series of SEM images up to $t = 15.57$ s, and the transformation of the lithiated Au-SiNW appears to occur in distinct stages (Stages I, II, and III). In Stage I, from $t = 0$ to 7.13 s, Li atoms near the vicinity of the contacting area underwent diffusion into the Au–SiNW accompanied by small radial volume expansion as the effective reaction front (marked by the yellow triangle) propagated from the nanowire tip toward the other end, along the axial direction. In Stage II, the lithiation front had propagated through the entire length while the continuing lithiation process lead to slow and uniform volume expansion in the radial direction around $t = 9.17$ to 11.22 s. Finally, in Stage III, the entire Au–SiNW appeared to swell uniformly at once in the radial direction and reached the fully lithiated phase (corresponding to ~417% volume expansion) within approximately one second ($t = 13.26$ to 15.57 s).

These experimental observations and theoretical analysis highlight the importance of surface modifications and a quantitative understanding and description of the surface effects in explaining the lithiation behavior of SiNWs and designing nanostructured Si-based anodes with desired performance.

19.4.3 SILICON–CARBON COMPOSITES

By combining Si with carbonaceous materials (Si–C), the cycling stability can be further improved with the enhanced buffering effect and electrical conductivity from carbon. Several structurally superior designs have been demonstrated by researchers, including (1) making nano-sized Si–C composites in particular

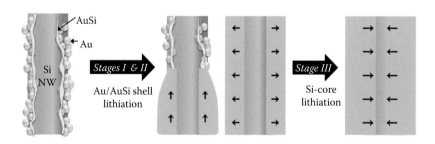

Figure 19.6 Proposed stagewise lithiation of an individual Au-SiNW. Stages I & II: Axial lithiation of the Au shell with progressive and slow expansion while the Au–Si interface layer may serve as a facile diffusion path. Stage III: radial lithiation of the Si core with uniform and rapid expansion.

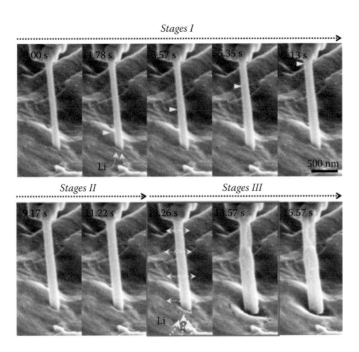

Figure 19.7 Temporal microstructure evolution of an Au–SiNW in direct contact with Li metal, which clearly demonstrates a stagewise lithiation behavior. The elapsed time during lithiation is indicated in second (s). Dotted red line indicates the bulk Li surface in the vicinity of the contacting area drastically sunk down due to the rapid Li diffusion into the Au–SiNW.

forms such as core–shell heterostructures; (2) depositing Si–C multilayer films; (3) making silicon-graphene composites by coating Si on graphene or encapsulating Si nanoparticles with crumpled graphene; and (4) putting C coating on SiNWs or coating tubular-forms of C with Si. As the result, excellent capacity retention was realized by composites made of Si and carbonaceous materials. Nano-Si-coated graphene granules have been reported with capacity exceeding 2000 mAh/g at the current density of 140 mA/g and excellent stability for 150 cycles, and near theoretical capacity (of Si) could be achieved (3890 mAgh/g) for more than 100 cycles in Si–C thin films, which is thought to be attributed to the buffering effect of intercalated C layers. Furthermore, nanostructured Si–C anodes have also been utilized to enhance the high-rate capabilities. Capacity as high as 2000 mAh/g was reported at 5C charging rate (full lithiation in 1/5 hour) for Si decorated carbon nanotubes, and capacity in excess of 1500 mAh/g was demonstrated at 8C rate using porous Si–C composites.

As a promising anode material for LIBs, Si–C composites exhibit many unique synergistic benefits, such as remarkable capacity retention and rate capabilities. However, in comparison to pure Si or C anodes, it is more challenging to investigate the lithiation behavior in Si-C composite anodes, especially concerning the complexities and intricate reactions at Si/C interfaces.

19.4.3.1 Interface effects

Chou and Hwang performed first-principles calculations to examine the atomic arrangement, bonding mechanism, Li diffusion, and lithiation energetics of Si–Gr composites, especially the Si–Gr interface effects (Chou and Hwang 2013b). A set of atomic structures from their simulations for selected Li_xSi-Gr systems (x = 1.00, 1.67 and 3.57) are shown in Figure 19.8.

There is a distinct Li enrichment at the Li_xSi/Gr interface, but away from the graphene layer, Li and Si atoms tend to be well mixed similar to bulk Li_xSi alloys. As partitioned by the vertical dashed lines, away from graphene ($|z|$ = 0.0 Å), the first layer ($|z| \approx 1.7$ to 3.0 Å) comprises only Li atoms with the average Li–Gr distance around 2.3 Å, the second layer ($|z| \approx 3.0$ to 4.5 Å) is mainly composed of Si atoms while the third layer consists of a blend of Li and Si atoms but slightly richer in Li. The distinct layering near

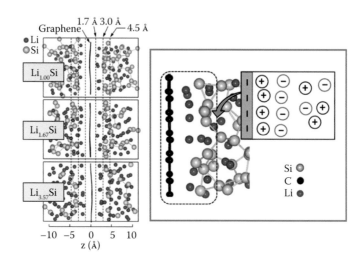

Figure 19.8 Atomic structures from AIMD simulations. The graphene position is set to zero in z-dimension. The illustration shows how Li cations and Si anions are distributed near the negatively charged graphene.

graphene tends to be pertinent within the first two to three atomic layers. Further away from graphene, the bulk-like structure is restored where Li and Si atoms are well mixed without segregation

From Bader charge analysis, it was found that the C atoms in Gr are more negatively charged with increasing Li content, while the Li charge state remains nearly constant, and the Si charge state varies significantly depending on the Si–Si connectivity. As such, it is apparent that the alternative layering of Li and Si is due to the electrostatic interaction between them. That is, the excess negative charges on Gr create an electric field, which attracts Li+ counterions while repelling Si− coions to screen the field, as illustrated in Figure 19.8.

From AIMD simulations, the Li diffusivity near Gr was predicted to be several times larger than that in the corresponding bulk Li–Si alloy. Such significantly enhanced Li mobility might contribute toward high-performance anodes with fast charge/discharge rates. However, the first-principles study showed that there is no significant interface effect on the predicted voltage profile and theoretical capacity, as Li atoms are mainly incorporated in the Si matrix rather than at the Li_xSi/Gr interface during lithiation.

19.4.4 SILICON SUBOXIDES (SiO_x, $X < 2$)

Si-rich oxides (SiO_x, $x < 2$) have been considered as a potential LIB anode. It was suggested that Si suboxide may form Si nanocrystallites dispersed in the a-SiO_2 matrix, leading to active-inactive structures that help buffer the strain during cycling and hence much improved reversible capacities. Over recent years, researchers have been toying with the idea of controlled oxidation, in which a-SiO_x of different O contents (x mainly lower than 0.5) have been tested to evaluate the effect of oxidation on the anode performance. Kim et al. recommended that the O concentration should be reduced below 18 at.% in order to increase the initial capacity (Kim et al. 2010), and Abel et al. later demonstrated that nanostructured Si thin films with homogeneous O incorporation ($\approx$ 13 at.% O) in combination with surface oxidation were able to deliver an excellent capacity ($\approx$ 2200 mAh/g) with nearly no capacity loss for the first 120 cycles, and 80% of the initial reversible capacity was retained after 300 cycles (Abel et al. 2012). Despite these encouraging improvements, the fundamental understanding regarding the nature and properties of lithiated a-SiO_x of relatively low O contents ($x \leq 0.5$) is still limited.

A first-principles DFT study (Chou and Hwang, 2013c) showed the structural evolution and voltage profile of lithiated a-$SiO_{1/3}$, along with discussion regarding the bonding mechanism of unique O complex formation during lithiation. From AIMD simulations, with increasing Li content x, it was found that (1) the Si-O network gradually disintegrates into smaller fragments of lower connectivity and (2) the charge states of Li and O are predicted to remain nearly constant around +0.8 to +0.9 and −1.6 to −1.8, respectively, while the Si charge state varies from +0.6 to −3.5 depending on the numbers of neighboring

Li and/or O atoms. Due to the electrostatic interaction between the cationic Li and anionic Si and O atoms are surrounded by Li atoms.

With lithiation, the Si–Li coordination number tends to monotonically increase up to $CN_{Si-Li} \approx 10$ when fully lithiated, whereas CN_{O-Li} becomes saturated at 6 far before full lithiation. This difference is partly due to the atomic size effect on packing efficiency, considering Si is much larger in size compared to O and hence able to neighbor with more Li atoms. Furthermore, electronic structure analysis revealed the tetrahedral arrangement of the four sp^3 hybrid orbitals of an isolated O^{2-} anion to minimize electron repulsion. The surrounding Li cations sit over the edges of the tetrahedron, forming a sixfold-coordinated $[Li_6O]^{4+}$ octahedron with O_h symmetry, as illustrated in Figure 19.9.

It is expected that the $[Li_6O]^{4+}$ formation is rather sensitive to the Si:O atomic ratio as well as the O spatial distribution in the host (suboxide) matrix. That is, the amount of Si anions should be sufficient to stabilize $[Li_6O]^{4+}$ polycations, otherwise Li_2O and/or various Li-silicates may form as commonly observed during lithiation of Si suboxides with high O contents ($x > 1$ in SiO_x). A further investigation is necessary to determine the critical O concentration which marks the transition from dispersed $[Li_6O]^{4+}$ complex to Li_2O and/or Li-silicate formation.

Prior to lithiation, the a-$SiO_{1/3}$ alloy ($B \approx 56$ GPa) is predicted to be considerably softer than a-Si ($B \approx 75$ GPa), which is mainly due to the flexible nature of Si−O−Si units. Compared to the pure Si case, the softer a-$SiO_{1/3}$ matrix may contribute to easier Li incorporation and better strain accommodation, especially during the early stages of lithiation. With increasing Li content x, the B values decrease monotonically in both alloy systems. The softening effect is attributed to the disintegration of the host matrix as well as the increasing metallic character. The comparison of predicted E_f between a-$Li_xSiO_{1/3}$ and a-Li_xSi is shown in Figure 19.10a. For both alloys, the E_f values decrease monotonically with increasing x, and approach the minimum-energy plateau as they are fully lithiated around $x = 4$. The predicted Li storage capacity for a-$SiO_{1/3}$ is close to that of pure Si, which is in line with experimental results which showed comparative first-cycle Li insertion capacity between a-Si films under controlled oxidation (~17 at.% O) and pure Si (Abel et al. 2012). Notice also that the E_f profile of a-$Li_xSiO_{1/3}$ is considerably lower in value compared to the a-Li_xSi case, and exhibits a much steeper descending trend. These differences indicate that Li incorporation is energetically more favorable in a-$SiO_{1/3}$ compared to a-Si, which is reasonable considering the stronger Li-O interaction relative to the Li-Si interaction, and also the relatively easier Li accommodation in the softer a-$SiO_{1/3}$ matrix.

The voltage-composition (V-x) curves for a-$Li_xSiO_{1/3}$ and a-Li_xSi are shown in Figure 19.10b. The lithiation voltage for a-$SiO_{1/3}$ is predicted to be around 0.2–0.8 V, which is within the desirable range for LIB anode application, but slightly higher than that of pure a-Si (0.1–0.5 V), especially during early stages of lithiation. The upshift in V reflects the more energetically favorable incorporation of Li in a-$SiO_{1/3}$ (relative to a-Si), as evidenced by experiments that showed a shift toward higher Li insertion potentials for Si suboxide-based anodes with increasing O contents (Abel et al. 2012).

At very high O:Si ratios (approaching 2 for instance), the formation of stable Li_2O and/or various Li-silicates during lithiation may cause a significant upshift in V. As a consequence, the Li extraction in the

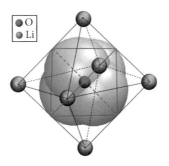

Figure 19.9 The dense close-packing arrangement of the $[Li_6O]^{4+}$ octahedron. The four sp^3 hybrid orbitals directed to the corners of a tetrahedron (in red) are surrounded by six Li cations each sits over an edge, forming a six-coordinated octahedron (in blue).

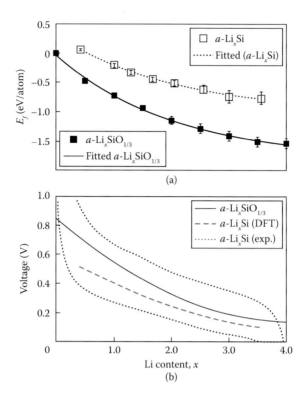

Figure 19.10 (a) Predicted formation energies (E_f) for a-Li$_x$SiO$_{1/3}$ ($0 \leq x \leq 4$) and a-Li$_x$Si ($0 \leq x \leq 3.57$). (b) Voltage-composition (V-x) curve for lithiated a-SiO$_{1/3}$ in comparison to that of pure Si.

following discharge cycle would be difficult, and those possibly trapped Li atoms could be partly responsible for the irreversible capacity loss and thus compromise the benefits of using Si suboxides as the anode material. Given this, the capacity and cycleability can be sensitive to the local atomic arrangement of the suboxide host matrix. Therefore, fine-tuning of the concentration and spatial distribution of O atoms will be essential in order to maximize the performance of Si suboxide-based anodes.

19.5 WHAT MAKES THE DISTINCT DIFFERENCE IN LITHIATION BETWEEN SILICON AND GERMANIUM?

Being in the same column in the periodic table, Si and Ge have many similarities. Si and Ge both exist in the tetrahedrally-bonded diamond structure with the lattice parameters differing only by 4%. Both are "alloy-type" anodes, which form a-Li$_x$Si/Li$_x$Ge with Li upon room-temperature lithiation accompanied by large structural/volume changes. Because of these similarities, one might expect Si and Ge to have very similar lithiation/delithiation behavior, but as highlighted by in situ characterizations, Si and Ge appear to have distinctively different responses to electrochemical lithiation/delithiation (Liu et al. 2012). For instance, lithiation of c-Si exhibits a strong orientation dependence (anisotropic), while c-Ge is lithiated isotropically. Moreover, in comparison to Si, Ge of comparable nano-architecture is able to withstand much faster charging rates with noticeably less crack formation. First-principles computational investigations (Chou and Hwang, 2014a; Chou and Hwang 2014b) showed that the distinctive difference in the lithiation mechanism and lithiation propagation speed between the two materials can be explained by the elastic responses of their lattices upon lithiation as well as the role of their lattice dynamics.

The structural evolution of the a-Li$_4$Si/Si(110) interface in comparison to that of the a-Li$_4$Ge/Ge(110) interface is shown in Figure 19.11. Upon lithiation, c-Si exhibits a sharp amorphous-crystalline interface (ACI), which propagates rather slowly and anisotropically in a layer-by-layer fashion, and the weakened Si–Si

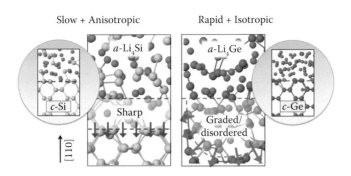

Figure 19.11 Schematic illustration showing the difference in the interface evolution between a-Li$_4$Si/c-Si and a-Li$_4$Ge/c-Ge from AIMD simulations.

bonds at the ACI break off mostly as dimers and dissolve into the lithiated a-Li$_4$Si phase. However, lithiated c-Ge exhibits distinctively different lithiation features. As Li atoms diffuse into the c-Ge region, the crystallinity is destroyed rapidly and Ge atoms are found to mostly break off as monomers instead of dimers in the Si case. The lattice distortion is farther extended beyond the interface layer, resulting in a graded fast-advancing lithiation front. This suggests that the lithiation process proceeds isotropically at a much higher speed in c-Ge compared to c-Si, as evidenced by experiments (Liu et al. 2012).

The underlying reasons behind these differences may relate to the elastic responses of the c-Si and c-Ge lattices upon lithiation as well as the role of their lattice dynamics. Firstly, in comparison to Si, the Ge lattice is more flexible and less resistive to distortion. Upon lithiation, the Li-induced lattice softening is more prominent in Ge as compared to Si. With 4 at.% and 9 at.% Li incorporation, the B of c-Si is reduced from 88.4 GPa by 1.5% and 7.1%, respectively, while more significant reductions are predicted for lithiated c-Ge as B decreases from 58.0 GPa by 8.2% and 20%, respectively. Secondly, the dynamic behavior of the c-Si and c-Ge lattices upon lithiation is also significantly different from one another. In comparison to the rather motionless c-Si matrix, the atomic rearrangements in the Ge lattice tend to extend far beyond the initial adjacent neighbors through a series of bond breaking, bond switching, and new bond forming events. For example, with a small degree of Li alloying, the rigid Si lattice may respond with less significant weakening as compared to the Ge lattice, and thereby well retains the crystallographic properties of unlithiated planes, leading to a sharp ACI and strong orientation dependence of lithiation.

AIMD simulations (Chou and Hwang 2014a) predicted that Li diffusivity is around $\times 10^{-13}$ cm^2 s^{-1} in c-Si and $\times 10^{-11}$ cm^2 s^{-1} in c-Ge. Upon lithiation, D_{Li} in a-Li$_x$Si alloys tends to rise by orders of magnitude from $\times 10^{-12}$ cm^2 s^{-1} ($x = 0.14$) to $\times 10^{-7}$ cm^2 s^{-1} ($x = 3.57$) with increasing Li content, whereas D_{Li} in a-Li$_x$Ge alloys exhibits virtually no concentration dependence and remains relatively constant around $\times 10^{-7}$. It turns out that this distinct difference of D_{Li} is attributed to the easier Li incorporation and facile Li diffusion in Ge (as compared to Si) at the onset of lithiation, and it is also attributed to the dynamic behaviors of the host Si and Ge matrices upon lithiation. In particular, when lightly lithiated Si and Ge alloys (a-Li$_{0.14}$Ge and a-Li$_{0.14}$Si) were annealed at a slightly elevated temperature (800 K) for 8 ps to facilitate sufficient atomic movements for assessing their lattice dynamic behaviors, in comparison to the rather motionless a-Si matrix, the atomic rearrangements of a-Ge can extend far beyond the initial adjacent neighbors through a series of bond breaking, bond switching, and new bond forming events, which in turn leads to facile self-diffusion and consequently the larger D_{Li}.

AIMD snapshots of the a-Li$_{0.14}$Si and a-Li$_{0.14}$Ge systems annealed at 800 K are shown in Figure 19.12. In each alloy, a randomly selected host atom (labeled as Si$_A$ and Ge$_A$ in red color) and its bonded neighbors (marked as Si$_B$ and Ge$_B$ in blue color) were tracked. For a time interval of 8 ps, in a-Li$_{0.14}$Si, Si$_A$ tends to remain bonded to the same four Si$_B$ neighbors, suggesting that the a-Si host network is nearly stationary, and barely rearranges its configuration in response to moving Li atoms. Contrarily, Ge$_A$ in a-Li$_{0.14}$Ge is found to undergo migration through a series of bond breaking, bond switching, and new bond forming events. The atomic rearrangements associated with Ge$_A$ extends far beyond the initial adjacent neighbors,

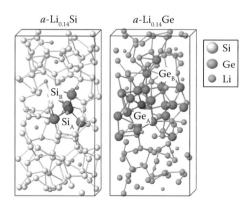

Figure 19.12 Comparison of the displacements of host Si and Ge atoms. The host atom of interest is labeled as Si_A and Ge_A in red color, and the host atoms involved in atomic rearrangement during a time interval of AIMD are marked as Si_B and Ge_B in blue color.

indicating a significant weakening of the Ge lattice even at the small degree of Li incorporation, which in turn allows facile self-diffusion.

As a result of the facile atomic rearrangements, D_{Ge} is orders of magnitude larger than D_{Si} and consequently leads to faster Li diffusion. As such, the dissimilar dynamic behaviors discussed here may help explain the different mechanical responses in Si and Ge NW lithiation/delithiation experiments. According to these results, with a small degree of Li alloying, the rigid Si lattice may respond with less significant weakening as compared to the Ge lattice, thereby well retaining the crystallographic properties of unlithiated planes, leading to a strong orientation dependence of lithiation. Contrarily, Ge shows pronounced lattice weakening even at low Li concentrations, such that upon lithiation, the original crystallographic characteristic is overshadowed by the sufficiently fast host atom rearrangements, thereby resulting in the isotropic lithiation. Likewise, since Ge is able to easily undergo atomic rearrangements, the matrix is more flexibly adjusted to the large strain variation during Li insertion/extraction and could subsequently reduce crack formation.

19.6 CONCLUSIONS AND OUTLOOK

Lithium-ion batteries are currently a principal power source for small portable electronics. However, in order to extend their effective use as large-scale energy storage systems for electric vehicles and renewable energy, there is an imminent need to further increase the energy density, power density, and cycle life while retaining safety and cost at an affordable range. This fundamentally represents a knowledge and materials challenge that needs to develop a deeper understanding of electrode and electrolyte materials as well as their interfaces. Over the last decade, there has been significant progress in first-principles modeling of anode materials and anode/electrolyte interfaces by virtue of rapid advances in computing power and computational methodology. First-principles simulations have been extensively used to explore the lithiation mechanisms of Si-based nanomaterials for LIB anodes. Despite recent progress, there is still considerable lack of fundamental investigations on the lithiation behavior of Si-based nanocomposites such as carbon scaffold Si nanoparticles and Si/metal oxide composites, which have been recognized as promising anode materials. Moreover, the formation and growth of SEI layers on Si-based anodes still remain largely unexplored despite the critical role of SEI in governing battery performance. Especially, there is a large disconnect between the decomposition of electrolyte and the formation of a SEI layer which needs to be further investigated. First-principles based computer simulations will continuously play a key role in the detailed mechanistic study of electrode materials and electrode/electrolyte interfaces. The improved understanding gained from computational studies combined with experiments will be crucial for realizing next-generation electrical energy storage systems.

REFERENCES

Aggarwal, R. L., P. Fisher, V. Mourzine, and A. K. Ramdas. 1965. Excitation Spectra of Lithium Donors in Silicon and Germanium. *Physical Review* 138 (3A): A882–93. doi:10.1103/PhysRev.138.A882.

Abel, P. R., Y.-M. Lin, H. Celio, A. Heller, and C. B. Mullins. 2012. Improving the Stability of Nanostructured Silicon Thin Film Lithium-Ion Battery Anodes through Their Controlled Oxidation. *ACS Nano* 6 (3): 2506–16. doi:10.1021/nn204896n.

Canham, L.T. 1988. *Properties of Silicon, Electronic Materials Information Service (EMIS),* Dataviews Series No. 4. Edited by Ravi, K.V., Hecking, N., Fengwei, W., Xiangqin, and Alexsandrev, Z. London: INSPEC, pp. 455.

Chan, C. K., H. Peng, G. Liu, K. McIlwrath, X. F. Zhang, R. A. Huggins, and Y. Cui. 2008. High-Performance Lithium Battery Anodes Using Silicon Nanowires. *Nature Nanotechnology* 3 (1): 31–35. doi:10.1038/nnano.2007.411.

Chan, M. K. Y., C. Wolverton, and J. P. Greeley. 2012. First Principles Simulations of the Electrochemical Lithiation and Delithiation of Faceted Crystalline Silicon. *Journal of the American Chemical Society* 134 (35): 14362–74. doi:10.1021/ja301766z.

Chan, T.-L., and J. R. Chelikowsky. 2010. Controlling Diffusion of Lithium in Silicon Nanostructures. *Nano Letters* 10 (3): 821–25. doi:10.1021/nl903183n.

Chou, C.-Y., and G. S. Hwang. 2013a. Surface Effects on the Structure and Lithium Behavior in Lithiated Silicon: A First Principles Study. *Surface Science* 612 (June). Elsevier B.V.: 16–23. doi:10.1016/j.susc.2013.02.004.

Chou, C.-y., and G. S Hwang. 2013b. Role of Interface in the Lithiation of Silicon-Graphene Composites: A First Principles Study. *The Journal of Physical Chemistry C* 117 (19): 9598–9604. doi:10.1021/jp402368k.

Chou, C.-Y., and G. S. Hwang. 2013c. Lithiation Behavior of Silicon-Rich Oxide (SiO 1/3): A First-Principles Study. *Chemistry of Materials* 25 (17): 3435–40. doi:10.1021/cm401303n.

Chou, C.-Y., and G. S. Hwang. 2014a. On the Origin of the Significant Difference in Lithiation Behavior between Silicon and Germanium. *Journal of Power Sources* 263 (October). Elsevier B.V: 252–58. doi:10.1016/j.jpowsour.2014.04.011.

Chou, C.-Y., and G. S. Hwang. 2014b. On the Origin of Anisotropic Lithiation in Crystalline Silicon over Germanium: A First Principles Study. *Applied Surface Science* 323 (December). Elsevier B.V.: 78–81. doi:10.1016/j.apsusc.2014.08.134.

Chou, C.-Y., H. Kim, and G. S. Hwang. 2011. A Comparative First-Principles Study of the Structure, Energetics, and Properties of Li–M (M = Si, Ge, Sn) Alloys. *The Journal of Physical Chemistry C* 115 (40): 20018–26. doi:10.1021/jp205484v.

Chou, C.-Y., J.-H. Seo, Y.-H. Tsai, J.-P. Ahn, E. Paek, M.-H. Cho, I.-S. Choi, and G.S. Hwang. 2015. Anomalous Stagewise Lithiation of Gold-Coated Silicon Nanowires: A Combined In Situ Characterization and First-Principles Study. *ACS Applied Materials & Interfaces* 7 (31): 16976–83. doi:10.1021/acsami.5b01930.

Cui, Z., F. Gao, Z. Cui, and J. Qu. 2012. A Second Nearest-Neighbor Embedded Atom Method Interatomic Potential for Li–Si Alloys. *Journal of Power Sources* 207 (June). Elsevier B.V.: 150–59. doi:10.1016/j.jpowsour.2012.01.145.

Ding, N., J. Xu, Y.X. Yao, G. Wegner, X. Fang, C.H. Chen, and I. Lieberwirth. 2009. Determination of the Diffusion Coefficient of Lithium Ions in Nano-Si. *Solid State Ionics* 180 (2–3). Elsevier B.V.: 222–25. doi:10.1016/j.ssi.2008.12.015.

Kauzlarich, S. M. 1996. *Chemistry, Structure, and Bonding of Zintl Phases and Ions.* New York: Wiley-VCH.

Kim, H., C.-Y. Chou, J. G. Ekerdt, and G. S. Hwang. 2011. Structure and Properties of Li–Si Alloys: A First-Principles Study. *The Journal of Physical Chemistry C* 115 (5): 2514–21. doi:10.1021/jp1083899.

Kim, K., J.-H. Park, S.-G. Doo, and T. Kim. 2010. Effect of Oxidation on Li-Ion Secondary Battery with Non-Stoichiometric Silicon Oxide (SiOx) Nanoparticles Generated in Cold Plasma. *Thin Solid Films* 518 (22). Elsevier B.V.: 6547–49. doi:10.1016/j.tsf.2010.03.176.

Limthongkul, P., Y.-I. Jang, N. J. Dudney, and Y.-M. Chiang. 2003. Electrochemically-Driven Solid-State Amorphization in Lithium-Silicon Alloys and Implications for Lithium Storage. *Acta Materialia* 51 (4): 1103–13. doi:10.1016/S1359-6454(02)00514-1.

Liu, X. H., H. Zheng, L. Zhong, S. Huang, K. Karki, L. Q. Zhang, Y. Liu, et al. 2011. Anisotropic Swelling and Fracture of Silicon Nanowires during Lithiation. *Nano Letters* 11 (8): 3312–18. doi:10.1021/nl201684d.

Liu, X. H., Y. Liu, A. Kushima, S. Zhang, T. Zhu, J. Li, and J. Y. Huang. 2012. In Situ TEM Experiments of Electrochemical Lithiation and Delithiation of Individual Nanostructures. *Advanced Energy Materials* 2 (7): 722–41. doi:10.1002/aenm.201200024.

Maranchi, J. P., A. F. Hepp, and P. N. Kumta. 2003. High Capacity, Reversible Silicon Thin-Film Anodes for Lithium-Ion Batteries. *Electrochemical and Solid-State Letters* 6 (9): A198. doi:10.1149/1.1596918.

Milman, V., M. C. Payne, V. Heine, R. J. Needs, J. S. Lin, and M. H. Lee. 1993. Free Energy and Entropy of Diffusion by Ab Initio Molecular Dynamics: Alkali Ions in Silicon. *Physical Review Letters* 70 (19): 2928–31. doi:10.1103/PhysRevLett.70.2928.

Patrocinio, A. O. T., L. G. Paterno, and N. Y. Murakami Iha. 2010. Role of Polyelectrolyte for Layer-by-Layer Compact TiO 2 Films in Efficiency Enhanced Dye-Sensitized Solar Cells. *The Journal of Physical Chemistry C* 114 (41): 17954–59. doi:10.1021/jp104751g.

Seo, J.-H., C.-Y. Chou, Y.-H. Tsai, Y. Cho, T.-Y. Seong, W.-J. Lee, M.-H. Cho, J.-P. Ahn, G. S. Hwang, and I.-S. Choi. 2015. Ultrafast Chemical Lithiation of Single Crystalline Silicon Nanowires: In Situ Characterization and First Principles Modeling. *Royal Society of Chemistry Advances* 5 (23): 17438–43. doi:10.1039/C4RA14953J.

Shenoy, V.B., P. Johari, and Y. Qi. 2010. Elastic Softening of Amorphous and Crystalline Li–Si Phases with Increasing Li Concentration: A First-Principles Study. *Journal of Power Sources* 195 (19). Elsevier B.V.: 6825–30. doi:10.1016/j.jpowsour.2010.04.044.

Zhang, Q., W. Zhang, W. Wan, Y. Cui, and E. Wang. 2010. Lithium Insertion In Silicon Nanowires: An Ab Initio Study. *Nano Letters* 10 (9): 3243–49. doi:10.1021/nl904132v.

Functional materials

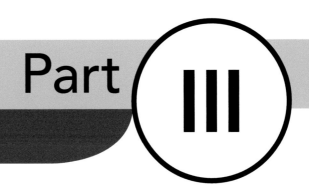

Part III

Industrial nanosilicon

Silicon nanopowder synthesis by inductively coupled plasma as anode for high-energy Li-ion batteries

20

Dominic Leblanc, Richard Dolbec, Abdelbast Guerfi, Jiayin Guo, Pierre Hovington, Maher Boulos, and Karim Zaghib

Contents

20.1 INTRODUCTION TO THERMAL PLASMAS

People are very familiar with the solid, liquid, and gaseous states of matter, but much less with plasma, the fourth state of matter, which is reached at sufficiently high temperatures and energy densities. The plasma state (Figure 20.1) is an ionized gas comprising molecules, atoms, ions (in their ground or in various excited states), electrons, and photons. It is electrically conductive since there are free electrons and ions present, and is in local electrical neutrality, since the numbers of free electrons and ions are equal.

The high energy content of plasmas compared to that of ordinary gases or even the highest temperature combustion flames offers unlimited potential for use in a number of significant modern industrial applications. In summary, plasmas offer:

- High temperature environment (5,000 to 10,000 K)
- High thermal conductivity
- High purity environment (no combustion involved)

20.1.1 GENERATION OF THERMAL PLASMA

Plasmas are generated by passing an electric current through a gas. Since gases at room temperature are excellent insulators, a sufficient number of charge carriers must be generated to make the gas electrically conducting. This process is known as electrical breakdown, and there are many possible ways to accomplish this breakdown. Breakdown of the originally nonconducting gas establishes a conducting path between a pair of

Figure 20.1 Thermal plasma in a quartz tube.

Table 20.1 Ionization and dissociation energies of the main plasma forming gases and surrounding atmosphere

SPECIES	Ar	He	H	N	O	H_2	N_2	O_2
Ionization energy (eV)	15.755	24.481	13.659	14.534	13.614	15.426	15.58	12.06
Dissociation energy (eV)	–	–	–	–	–	4.588	9.756	5.08

electrodes, as it would be the case in a direct current (DC) plasma. The passage of an electrical current trough the ionized gas leads to an array of phenomena known as gaseous discharges. Such gaseous discharges are the most common, but not the only, means for producing plasmas. For various applications, plasmas can also be produced by electrodeless radio frequency (RF) discharge, microwaves, shock waves, and laser or high-energy particle beams. Finally, plasmas can also be produced by heating gases (vapors) in a high-temperature furnace.

In thermal plasmas, external energy is supplied to the plasma forming gas. If this energy is sufficiently high, it results in the dissociation of molecules (in the case of molecular gases) and then in ionization of atoms. Table 20.1 summarizes the dissociation ($X_2 \rightarrow 2X$) and ionization energies ($X \rightarrow X^+ + e^-$) of the main plasma forming gases. The first observation is that the ionization energies of all plasma gases, except He, are rather close together (less than 2.1 eV difference). The ionization of molecular gases is far more energy demanding than most monatomic gases because of the need to supply both dissociation and ionization energies.

In the applications of induction plasma to material processing, the thermodynamic and transport properties of the plasma itself play a critical role in determining the efficiency of the processes.

20.1.2 PLASMA ENTHALPY

The plasma enthalpy depends strongly upon the dissociation and ionization phenomena. Figure 20.2 represents the variation with temperature of the specific enthalpy (MJ kg^{-1}) of the different plasma gases (Ar, He, N_2, O_2, and H_2). The steep variations of enthalpy are due to the heats of reaction (dissociation and ionization).

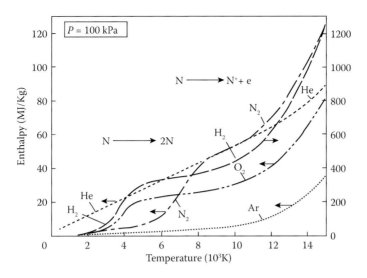

Figure 20.2 Temperature dependence of the specific enthalpy of Ar, He, N$_2$, O$_2$, and H$_2$ at atmospheric pressure (Boulos et al. *Thermal Plasmas: Fundamentals and Applications*, 1994).

The very high enthalpy of pure hydrogen (approximately a factor of 10 higher with respect to the other gases shown) is due to its low mass. This figure illustrates the fact that energy supply is independent of the gas, and the temperature is not determined by the chemical reactions (as in flames). Specifically, if an oxygen-fuel flame at 3000 K is used to heat a particle up to 2500 K, only 20% of its energy is used (Fauchais et al. 2014). If nitrogen plasma at 10,000 K is used for the same purpose, it is possible to recover almost 95% of the available energy in the gas. Adding secondary gases such as helium or hydrogen to the primary ones, argon or nitrogen, increases the enthalpy of the mixture.

For nanopowder synthesis applications, the enthalpy and thermal conductivity of the plasma are particularly important due to the fact that the residence time of particles in plasma is limited. In order to increase the heating rates of the particles injected into the plasma, the enthalpy and thermal conductivity should be maximized. The addition of hydrogen into a pure argon plasma can significantly increase its thermal conductivity, which is why an argon–hydrogen mixture plasma, rather than pure argon plasma, is always favored during metallic nanopowder synthesis.

20.1.3 RF INDUCTION PLASMA

The RF induction plasma is generated through an inductive coupling mechanism. When an alternating current of RF and high voltage are imposed on a spiral coil, the conductor placed in the center of the coil will be heated up under the alternative electromagnetic field. If a continuous gas flow is introduced into the coil, the gas is ionized and heated to form a plasma, allowing the conversion of electrical energy into thermal energy. The plasma is called an inductively-coupled plasma (ICP). Figure 20.3 shows the induction plasma generated by RF discharge.

The ICP is a heat source free of contamination, which is particularly suitable for high purity processes. The ICP has a relatively large volume with a high temperature zone and relatively low gas velocity, making it an ideal tool for high-temperature material processing, where melting or evaporation of material is required. In ICP, virtually any kind of gas (e.g., H$_2$, O$_2$, N$_2$, Ar, NH$_3$, CH$_4$, or various mixtures) can be used as the plasma working gas, making the induction plasma not only a heat source but a chemical reaction source as well.

Over the years, the induction plasma technology was developed in different laboratories and industrial units, including Tekna. The patent-protected plasma torches (shown in Figure 20.4) are widely used in both laboratories and industries. The torch consists essentially of a plasma confinement ceramic tube with its outer surface cooled using a high-velocity thin-film water stream. A water-cooled induction copper coil surrounds the ceramic tube creating an alternating magnetic field in the discharge cavity. On ignition, a conductive plasma load is produced within the discharge cavity, in which the energy is transferred through

Figure 20.3 Photograph of the inductively coupled plasma (Guo et al., 2010).

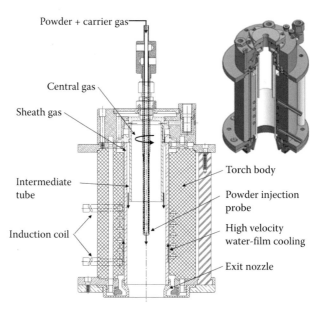

Powder + carrier gas

Central gas

Sheath gas

Intermediate tube

Induction coil

Torch body

Powder injection probe

High velocity water-film cooling

Exit nozzle

Figure 20.4 A proprietary plasma torch design (Boulos and Jurewicz, *High performance induction plasma torch with a water-cooled ceramic confinement tube*, 1993).

electromagnetic coupling. An internal gas flow introduced through the gas distributor head insures the shielding of the internal surface of the ceramic tube from the intense heat to which it is exposed. The hot plasma gases exit the torch cavity through the downstream flange-mounted nozzle.

Because of the close interdependence between the physical dimension of the induction plasma torch, its nominal design power rating and operating frequency, the induction plasma torch tends to have a relatively narrow dynamic range of operating conditions. For this reason, Tekna has developed a complete line of induction plasma torches with different internal diameters of the plasma confinement tube for operation at different power levels. Torches optimized for different power levels from 15 up to 200 kW (Figure 20.5) are commercially available.

20.1.4 PLASMA NANOPOWDER SYNTHESIS UNITS

The induction plasma torches developed by Tekna are integrated in turn-key units for various advanced materials-related applications, including nanopowder synthesis, powder processing (spheroidization) and coatings. A typical induction plasma system used for the synthesis of nanopowders is schematically represented in Figure 20.6. Such system is composed of the following sections: (a) plasma torch, (b) quench reactor, (c) a powder separation system (cyclone), (d) filter baghouse, and (e) nanopowder collection canister.

Figure 20.5 Tekna's induction plasma torches (40 to 200 kW).

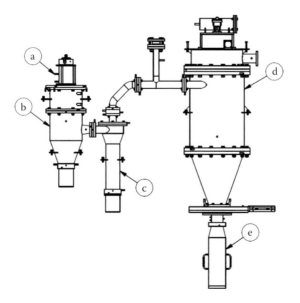

Figure 20.6 Setup developed for nanopowder synthesis using RF induction plasma technology. (a) plasma torch, (b) quench reactor, (c) a powder separation system (cyclone), (d) filter baghouse, and (e) nanopowder collection canister (Mamak et al., *Journal of Materials Chemistry* 20(44):9855–9857, 2010).

Industrial nanosilicon

Figure 20.7 Laboratory induction plasma torch system, 15 kW (Tekna, TekNano-15).

Plasma units are availaible in different power levels, depending on the end-use of the system. As an example of commercially available plasma unit, Figure 20.7 is showing a compact 15 kW induction plasma system used for developing nanopowders in laboratory scale quantities.

20.2 SILICON NANOPOWDER SYNTHESIS

Silicon nanopowder can be derived from silane gas or from vaporized micro-sized silicon powder by using the induction plasma synthesis process.

20.2.1 GENERAL SYNTHESIS PROCESS

As shown in Figure 20.8, the process starts with vaporization of precursor materials by the high enthalpy of the ICP, and subsequently the material vapor is transported to the tail or fringe of the plasma where the temperature decreases drastically. This temperature gradient allows the formation of a highly supersaturated vapor, which results in rapid production of numerous nanoparticles via homogeneous nucleation, heterogeneous condensation, and coalescence. Moreover, thermal plasmas are generated with any gas or gas mixture present, allowing as many chemical reactions to occur.

In thermal plasma synthesis, the precursors could be gases, liquids, or solids before injection into the plasma. The availability of gas-phase precursors for pure metals is severely limited, so the most commonly used reactants for plasma synthesis are solids material compositions that are the same as the nanopowder.

From Figure 20.8, it is clear that in-flight vaporization of the feed material is a prerequisite for the homogeneous nucleation of the nanoparticles and for the growth of uniform-sized nanopowders (i.e. narrow particle size distribution [PSD]). An analysis on the heating history of a particle from room temperature to vaporization in the plasma indicates that the particle gets energy from the plasma through convection heat transfer and internal conduction, while it loses energy to the environment through radiation. In order to ensure the vaporization of the particle, the total net energy it absorbs must be greater than the sum of the energy it needs to heat up to the vaporization point and its latent energy of melting and vaporization. Figure 20.9 illustrates such an energy balance analysis for in-flight plasma vaporization of powders.

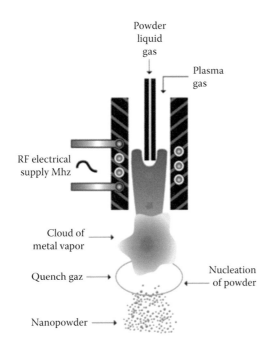

Figure 20.8 Schematic of nanopowder plasma synthesis process.

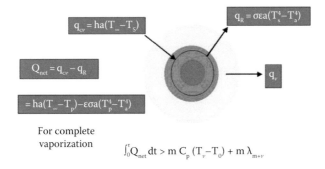

Figure 20.9 Analysis of thermal balance for in-flight plasma vaporization of fed powders (Guo et al., *Plasma Science and Technology* 12(2):188, 2010).

20.2.2 CONTROL OF PARTICLE SIZE

The PSD of nanopowders can be controlled during the plasma process through a number of means. The most important one is the cooling rate of the gaseous product. When the degree of supersaturation of the vapor in a gas mixture (S_v), defined as the ratio of the real vapor pressure over the saturation vapor pressure at the given temperature, exceeds a certain value, $S_v \geq 10$ for instance, vapor condenses by homogeneous nucleation to form liquid droplets. The diameter of the droplets is determined by the Kelvin equation:

$$d_p = \frac{4\sigma_o}{\rho_p (k/m) T \ln S_v} \tag{20.1}$$

where σ_o and ρ_p are the surface tension and density of the liquid, respectively; T is the temperature; k and m are the Boltzmann constant and the mass of the vapor molecule, respectively. Obviously, the size of the particle is predominantly determined by the degree of supersaturation S_v. The particle size is inversely proportional to S_v and the temperature T. Therefore, to obtain ultrafine nanometric particles, the cooling rate of the vapor has to be extremely fast, typically in the range of 10^4 to 10^6 K/s. Such a high quench rate

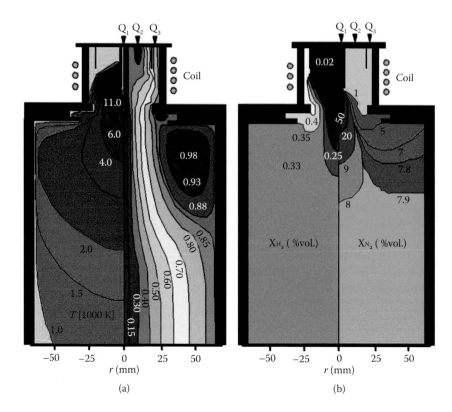

Figure 20.10 Flow and temperature fields (a) and concentration fields (b) for an Ar/H_2 plasma (71 slm Ar + 4 slm H_2) at 20 kW with a nitrogen gas injected through the probe with its tip at the center of the induction coil, pressure = 250 Torr (Rahmane et al., *International Journal of Heat and Mass Transfer* 37(14):2035–2046, 1994).

is usually achieved through proper design of the reactor and injection of a large amount of cooling gas. In addition, the plasma operational conditions (pressure, power level, etc.), to a certain extent, also have an effect on the size and distribution of the powders that are produced. Figure 20.10 shows the numerical modeling results of the temperature, flow, and concentration fields of a 20 kW plasma torch.

As mentioned earlier, the induction plasma offers a high-temperature environment, and the synthesis of nanopowders relies on in-flight vaporization of the precursor particles in the plasma. Due to the limited residence time, it is important to ensure that all of the particles are exposed to the plasma. The nanopowder production rate, to a great extent, depends on the evaporation rate of the particles, which is closely related to the dispersion of the particles in the plasma zone under the given conditions. Poorly dispersed particles are not fully exposed to high temperatures and thus are not uniformly evaporated when they pass through the plasma flame. Consequently, the design of the particle injector and the particle injection conditions become very important in determining the efficiency of the nanopowder synthesis process and its productivity.

20.2.3 SYNTHESIS PROCESS TYPE

In general, nanometric material synthesis in induction plasma is accomplished through one of two approaches: the first one is by a physical process (i.e., evaporation–condensation) and the second is by a chemical process (i.e., evaporation–reaction–condensation).

20.2.3.1 Physical process

The precursor material consists of micrometer-sized solid silicon particles which are prepared through mechanical milling and sizing (Leblanc et al. 2015a). In this process, the plasma is merely the heat source for evaporating materials. Rapid quenching of the vapor is necessary in order to nucleate particles before the vapor impinges on the cooled walls of the plasma reactor. The supersaturation of the vapor species due to high quench rates provides the driving force for particle nucleation. High quench rates lead to the production of

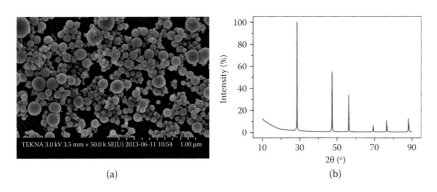

(a) (b)

Figure 20.11 (a) SEM micrograph of nanometer-sized Si spheres. (b) XRD diffractogram (Cu source) of particles shown in (a).

ultrafine particles (down to nm size) by homogeneous nucleation. The typical size of the nanoparticles ranges from 20 to 200 nm, depending on the operating conditions. Figure 20.11a shows silicon nanospheres produced by RF induction plasma by the evaporation–condensation approach. Silicon nanoparticles produced by evaporation of the solid precursor demonstrate a spherical morphology, narrow PSD, and high surface purity. Figure 20.11b presents the diffractogram measured by X-ray diffraction (XRD) spectrometry and shows the diamond cubic crystalline structure of the crystalline silicon powder.

The XRD peaks of the product at diffraction angles 2θ of 28.4°, 47.3°, 56.1°, 69.1°, and 76.4° correspond, respectively, to the (111), (220), (311), (400), and (331) planes of crystalline silicon. The silicon nanopowder observed using a transmission electron microscope (TEM) shows that the powder is mainly composed of monocrystalline spherical silicon particles (Figure 20.12).

As mentioned earlier, since the solid silicon precursor has a high boiling point (3265 °C) and a high specific enthalpy of vaporization (Figure 20.13), the precursor must be fed to an appropriate speed in order to evaporate it completely. The high power and the ability of feeding solid precursor through the thermal plasma zone yield a throughput of approximately 360 g of silicon per hour for a 98 kW system (Kambara et al. 2014).

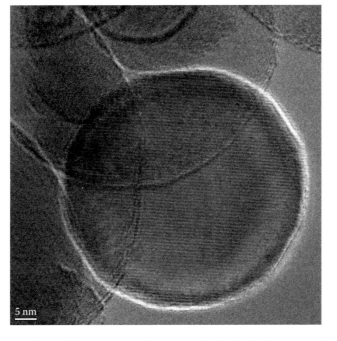

Figure 20.12 TEM micrograph of a silicon nanoparticles synthesized by induction plasma (Leblanc et al. 2015b).

Industrial nanosilicon

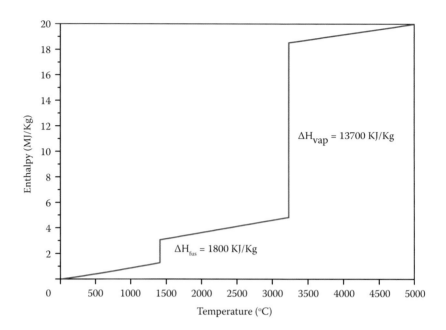

Figure 20.13 Enthalpy of silicon from 25°C to 5000°C (Chase, *NIST-JANAF Thermochemical Tables*, 1998).

The chemical composition of nanopowders produced by a physical process is directly related to the starting materials. However, depending on the difference in the boiling points of impurities in the starting material, the final nanopowder products may be purer than the starting powder due to the refining effect caused by volatility differences (Boulos 1983).

20.2.3.2 Chemical process

In the chemical process, one or more gases react with the vapor generated from the precursor material introduced into the plasma reactor, and the resultant product condenses to form nanoparticles. Any material exposed to thermal plasmas at these high temperatures ($T \geq 10^4$ K) will be decomposed, possibly into its elemental constituents. The gaseous reactant can be a component in the plasma or separately injected at appropriate locations (into the plasma zone or in the reactor).

The composition of nanopowders produced by a chemical process may be complicated due to unexpected intermediate reactions in the plasma. The operational conditions must be carefully controlled to ensure the completeness of the desired reaction/synthesis. It should be pointed out that in the case of evaporation–condensation with chemical reaction involved, attention should be paid to controlling the quench rate. Too fast a quench rate may bring about negative effects as some chemical reactions are lowered by the sudden drop in temperature.

Silicon nanopowders are prepared by the decomposition of monosilane (SiH_4) gas in a RF thermal plasma. In contrast to the chloride precursors ($SiCl_4$, $SiHCl_3$), monosilane does not produce corrosive chlorine gas in the plasma. The thermal plasma offers a high-temperature and contamination-free environment to produce pure silicon powder from SiH_4:

$$SiH_4(g) \xrightarrow{\Delta T} Si(s) + 2H_2(g) \tag{20.2}$$

with the evolution of hydrogen gas as a byproduct. Because the gas-phase precursor feed rate is easier to control precisely, a more homogeneous and uniform nucleation of the nanoparticles is expected to occur, leading to nanoparticles with a narrower size distribution. By properly controlling the coalescence of the nanoparticles in the quench stage, the monosilane gas also offers the possibility to produce particles with a smaller mean diameter as compared to the evaporation–condensation process. However, nanosilicon powder synthesized by silane decomposition in RF induction plasma appears to exhibit the same structure and morphology as the evaporation–condensation process (So et al. 2014). Introducing a secondary gas in the plasma provides a significant technical advantage to produce core–shell nanosilicon composites. As an example, addition of CH_4 to the plasma process produced a carbon coating on the primary Si particles (Kambara et al. 2014).

20.2.4 POWDER SURFACE PACIFICATION

As fine metal powders are very active, handling and storage of such powders are of great concern to users. This reactivity is sometimes detrimental because of unwanted surface-chemical modifications such as uncontrolled oxidation of bare silicon nanopowder surfaces in air, ligand attachments during solution processing or irreversible passive layer formation in some electrochemical reactions.

In the case of silicon nanopowders, reconstruction of the surface during nucleation leaves dangling bonds, which aggressively react with the surrounding environment. As a result of increased surface-to-volume ratio, the chemical state of the surface plays a critical role in the overall chemical reactivity of the powder. Controlled passivation of the particle surface is thus a very important concern during the preparation of silicon nanopowders; the passivation can be done inline during the synthesis process, in wet conditions such as immersion in liquid or ex situ in dry conditions such as controlled oxidation with air. It is also possible to introduce a broad range of organic functionalities at the silicon surface to change its reactivity with its environment (Buriak 2002).

20.3 CASE STUDY

Silicon nanoparticles exhibit novel physical and chemical properties, such as improved optical properties, dielectric constants, charge storage capacities, and catalytic activities, which reveals the potential in application areas of silicon nanoparticles in future technologies (Doğan and van de Sanden 2016). In the field of energy storage, silicon nanopowders are widely considered for future high-energy anodes in Li-ion batteries (LIBs).

20.3.1 ENERGY STORAGE: Li-ION BATTERIES

An LIB is a rechargeable battery in which lithium ions move reversibly from a positive host to a negative host during charge and back when discharging (Figure 20.14). LIBs use intercalation compounds as active electrode material. An intercalation compound is a rigid structure that is the host structure for intercalating ions (Li^+) which, together, produce a redox process. Lithium insertion and disinsertion may induce (or not) changes to the host material structure. An aprotic electrolyte containing a salt, usually $LiPF_6$, dissolved in a mixture of organic carbonates is required to avoid degrading the very reactive electrodes.

Current commercial LIBs anodes are presently based on carbonaceous materials (graphite), and different cathodes chemistries are available like $LiCoO_2$, $LiMn_2O_4$, and $LiFePO_4$ (Thackeray 2002).

Negative electrode: $$6C + Li^+ + e^- \leftrightarrow LiC_6 \qquad E = 0.2 \text{ V vs. Li/Li}^+ \qquad (20.3)$$

Positive electrode: $$LiFePO_4 \leftrightarrow FePO_4 + Li^+ + e^- \qquad E = 3.45 \text{ V vs. Li/Li}^+ \qquad (20.4)$$

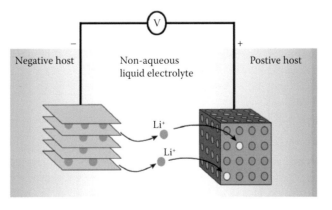

Figure 20.14 Schematic representation and operating principle of a rechargeable Li-ion battery (Adapted from Tarascon and Armand, *Nature* 414(6861):359–367, 2001).

According to Faraday's law, the theoretical gravimetric capacity density (C_{th}) of the graphite anode (equation 20.5) is:

$$C_{th} = \frac{nF}{3.6 \times M_w} = \frac{(1)(96485\,C.mol^{-1})}{(3.6\,C.mAh^{-1})(6 \times 12.011\,g.mol^{-1})} = 372\,mAh.g^{-1} \tag{20.5}$$

where n is the number of electrons participating in the reaction, F is the Faraday constant (96,485 C mol^{-1}), and M_w is the molar mass of the active material. According to Equation 20.5, a graphite anode has a maximum theoretical gravimetric capacity density of 372 mAh g^{-1}. This capacity is considered insufficient for applications requiring high-energy density over an extensive period of time; increasing energy density is the key requirement to realize a wide range of future applications. Silicon is an attractive alternative material due to its high theoretical gravimetric capacity (Kasavajjula et al. 2007). During lithiation of silicon, different intermetallic species are formed at room temperature, as depicted in the binary phase diagram Li–Si (Figure 20.15).

The Li–Si phase diagram shows the thermodynamically stable intermetallic phases when introducing lithium in the silicon lattice: Li–Si, $Li_{12}Si_7$, Li_7Si_3, $Li_{13}Si_4$ and, finally, $Li_{22}Si_5$. The alloying reaction of silicon with lithium ions to form the $Li_{22}Si_5$ phase is described by the redox reaction:

$$5Si + 22Li^+ + 22e^- \longleftrightarrow Li_{22}Si_5 \tag{20.6}$$

Using Faraday's law (and Equation 20.6), we can calculate the theoretical gravimetric capacity density (C_{th}) when the $Li_{22}Si_5$ phase is formed:

$$C_{th} = \frac{(22)(96485\,C.mol^{-1})}{(3.6\,C.mAh^{-1})(5 \times 28.0855\,g.mol^{-1})} \approx 4200\,mAh.g^{-1} \tag{20.7}$$

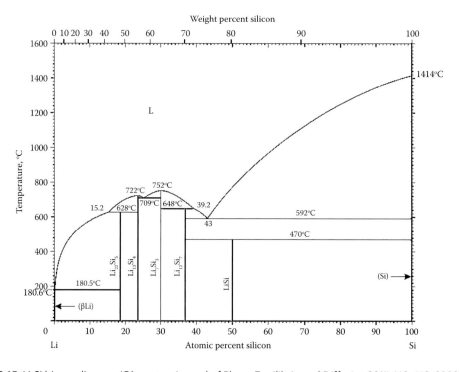

Figure 20.15 Li-Si binary diagram (Okamoto, *Journal of Phase Equilibria and Diffusion* 30(1):118–119, 2009).

With a theoretical gravimetric capacity of 4200 mAh g^{-1}, silicon is one of the promising elements for the negative electrode: its capacity is more than 10 times greater than graphitic carbon. However, silicon experiences high volume expansion during lithiation. The volume expansion (ξ) for an alloy anode is defined as:

$$\xi = \frac{v - v_o}{v_o} \times 100\% \qquad (20.8)$$

where v_o is the unlithiated molar volume of the alloy and v is the molar volume of the alloy anode (calculated per mole of host alloy atoms) (Obrovac et al. 2007). To evaluate the volume expansion of silicon during lithiation, the alloy molar volume as a function of the lithium content must be known. We can obtain the data from a crystallographic database of the Li–Si binary system (Table 20.2).

Figure 20.16 shows the volume expansion of the intermetallic compounds corresponding to the state-of-charge of the silicon anode. For each compound, the molar volume during lithiation falls on a linear function of the lithium content. The intermetallic compound (Li$_{22}$Si$_5$) has a total volume expansion of 320% of the corresponding initial silicon lattice volume. Since silicon is a brittle metalloid material, it is expected that internal mechanical stresses will build up during silicon lithiation and relax by cracking.

Table 20.2 Crystalline structure of the Li–Si binary system

PHASE	COMPOSITION (AT.% Si)	PEARSON SYMBOL	VOLUME PER Si (Å³)	EXPANSION (%)	CAPACITY (mAh/g)
(βLi)	0.0	cI2	- - -	- - -	- - -
Li$_{22}$Si$_5$	18.5	cF432	82.4	320	4200
Li$_{13}$Si$_4$	23.5	oP24	67.3	243	3100
Li$_7$Si$_3$	30.0	hR7	51.5	163	2230
Li$_{12}$Si$_7$	36.8	oP152	43.5	122	1640
LiSi	50.0	tI	31.4	60	950
(Si)	100.0	cF8	19.6	0	0

Source: Calculated from ICDD, PDF-4+ / Version 4.1504 (Database), International Centre for Diffraction Data, 2015.

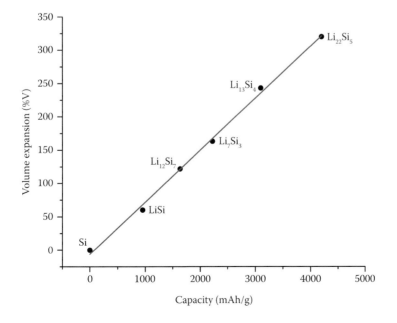

Figure 20.16 Volume expansion of silicon during lithiation (from data of Table 20.2).

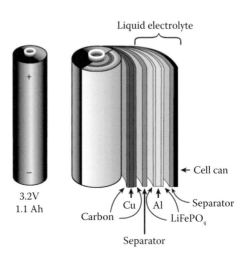

Liquid electrolyte

3.2V
1.1 Ah

Cell can

Cu | Al

Carbon

LiFePO$_4$

Separator

Separator

Figure 20.17 Schematic representation showing the components of a commercial 18650 format cylindrical Li-ion battery. (Adapted from Tarascon and Armand, 2001.)

The construction design of a commercial LIB is shown in Figure 20.17. The figure shows a cylindrical format composed of a roll of electrodes and separator inserted in a stainless steel can. The positive electrode (or cathode) is an aluminum foil on which a composite electrode mixture (active cathode material, carbon black, and polymeric binder) is applied on both faces. The negative electrode (or anode) is usually a copper foil on which a composite electrode mixture (graphite and polymeric binder) is also applied on both faces. To maintain a constant space between electrodes and to provide enough free space for liquid electrolyte, a porous polymeric separator is placed between the two electrodes.

The "real" volumetric energy density (E) of the complete device (Figure 20.17) is calculated for an 18,650 cylindrical cell (18 mm diameter and 65 mm length) by the equation:

$$E = \frac{(3.2\,\text{V})(1.1\,\text{Ah})}{(65\,\text{mm})\left(\frac{18\,\text{mm}}{2}\right)^2 \pi} = \frac{3.52\,\text{Wh}}{0.0165\,\text{L}} = 213\,\text{Wh.L}^{-1} \text{ or } 766\,\text{J/cm}^3 \tag{20.9}$$

The energy density is enhanced by increasing the voltage of the cell by selecting higher voltage cathode materials and/or increasing the quantity of lithium ions that is stored. A high-voltage cathode material such as LiMn$_{1.5}$Ni$_{0.5}$O$_4$ (LMNO) has a 4.7 V potential versus Li/Li$^+$ and is expected to be combined with a high-energy silicon anode to produce a high-energy cell that provides longer range for applications like electric vehicles.

As mentioned earlier, the large volume change associated with formation of Li–Si alloy phases during the battery charge/discharge cycles is responsible for the fracture of the silicon material and eventually the loss of effective electric paths within the electrode. As a result, the storage capacity of lithium ions decreases with each cycle and fades out to nearly zero in less than 10 charge/discharge cycles (Guerfi et al. 2011). Thus, measures to maintain the structural integrity of the electrodes are key to making the best use of silicon in LIBs (Kasavajjula et al. 2007). To reduce the effects of mechanical damage, the use of nanostructured silicon was proposed.

Nano-sized silicon structures are known to overcome the deteriorating effects of volumetric expansion as a result of their ability to relax the mechanical stresses. Various silicon nanostructures with high cycle life were demonstrated in the literature (silicon nanotubes, nanowires, nanoparticles). In general, as the size of silicon nanopowders decreases, their stability in cycling increases (Abel et al. 2012). In situ transmission electron microscopy (TEM) observations of the lithiation process suggested that no fracturing occurs when the Si particles are smaller than a threshold diameter of 150 nm (Liu et al. 2012; McDowell et al. 2013) (Figure 20.18).

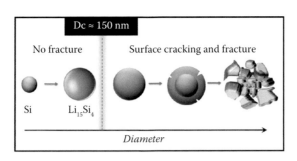

Figure 20.18 Critical particle diameter of $D_c \sim 150$ nm, below which the particles neither crack nor fracture upon first lithiation, and above which the particles initially form surface cracks and then fracture due to lithiation-induced swelling. (From Liu, X.H., et al., *ACS Nano*, 6, 1522–1531, 2012. With permission.)

In order to produce a silicon composite anode, a spherical silicon nanopowder was synthesized by induction plasma using a 3 MHz, 60 kW RF plasma torch (Leblanc et al. 2015b). The micrometric 99.999 wt% pure (5N) silicon [20–50 micrometer] was used in a pure evaporation–condensation process to produce the nanopowder. More specifically, micrometric silicon powder was heated and evaporated in the plasma torch, and the resultant vapor was subsequently quenched very rapidly and homogeneous nucleation led to the formation of a fine nanopowder aerosol. To prevent surface contamination, the nanosilicon powder and its packaging were handled in metal-plastic bags in an argon-filled glove box. Figure 20.11a shows a scanning electron microscope (SEM) micrograph of the morphology and typical size of silicon nanoparticles with primary spherical particles with diameters ranging from 50 to 200 nm. The average particle size (d_{50}) determined by measuring specific surface with the *Brunauer, Emmett* et *Teller method* (BET) was 85 nm (30 m^2 g^{-1}). Figure 20.12 is a TEM micrograph of a silicon nanosphere showing a crystalline lattice with a contamination-free atomic surface. Figure 20.19a presents the XRD diffractogram of the nanopowder with clear identification of crystalline silicon with a diamond cubic lattice ($a = 5.43$ Å) and no significant secondary impurity phase. Figure 20.19b shows the PSD by laser diffraction and confirms that the primary particles have a typical particle size between 50 and 200 nm with a few larger agglomerates.

The volume change of a single silicon particle was investigated by in situ TEM to understand the lithiation mechanism of a silicon nanosphere. Figure 20.20a shows a TEM micrograph of a typical 200 nm pristine spherical silicon particle that was synthesized by induction plasma. The same particle is contacted by Li/Li$_2$O in Figure 20.20b. At this point, a rapid reaction from the interior of the particle was observed, forming a core–shell structure. The crystalline core (dark gray contrast) was gradually transformed to an amorphous Li$_x$Si alloy (light gray). The thickness of the amorphous phase was not uniform around the particle and resulted in a anisotropic lithiation (Liu et al. 2011b). Crack formation was not observed; however, we were not able to fully lithiate the larger particles (200 nm) in our experiment. Since the particles were clustered together and no conductive binder was used, the conductivity

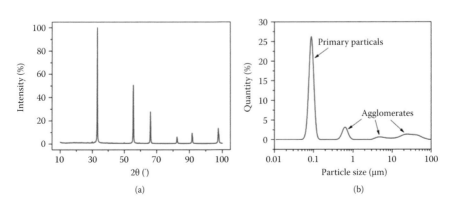

Figure 20.19 (a) XRD diffractogram (Co source). (b) PSD of the silicon nanopowder.

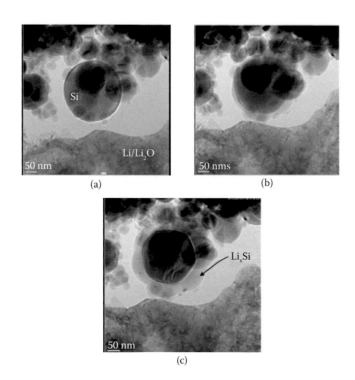

Figure 20.20 *In situ* TEM lithiation of a silicon nanosphere (a) before contact, (b) rapid lithiation at contact, and (c) end of lithiation.

might have been too low to achieve full lithiation. But, some smaller particles (~50 nm in Figure 20.20c) did fully react with lithium without fracturing. In a recent study, McDowell et al. also observed that the crystalline Si particles undergo anisotropic lithiation and volume expansion leading to a faceted Si core (McDowell et al. 2013).

The influence of silicon particle size on anode performances was investigated. Two composite electrodes were fabricated using micro-Si powder prepared by dry mechanical milling (Figure 20.21a) and nano-Si powder prepared in a plasma (Figure 20.21b). The two silicon powders were mixed with acetylene carbon black and sodium alginate to produce an electrode that was assembled with a separator and lithium foil anode in a button cell (Leblanc et al. 2015b). The electrolyte was composed of 1 M $LiPF_6$ in a mixture of ethylene carbonate (EC) and diethyl carbonate (DEC) (7:3 by volume) with the addition of 2 V% of vinylene carbonate (VC). The cells were galvanostatically charged and discharged at 25°C using a potentiostat at a $C/24$ rate

Figure 20.21 SEM micrographs showing particles of the powders used to make the two silicon anodes: (a) micro-Si particles prepared by mechanical milling and (b) nano-Si particles prepared by plasma.

Industrial nanosilicon

for formation cycles over the voltage range of 0.005–1.0 V versus Li/Li⁺. The theoretical maximum capacity (C) of the button cell was calculated from the active material loading in the electrodes (2.3 mg cm⁻²):

$$C = \left(2.3 \frac{\text{mg}}{\text{cm}^2}\right)\left(4200 \frac{\text{mAh}}{\text{g}}\right)\left(\frac{16\,\text{mm}}{2}\right)^2 \pi = 19\,\text{mAh} \qquad (20.10)$$

During galvanostatic cycling of batteries, the charge and discharge currents are often expressed as a C-rate, calculated from the battery maximum capacity. For example, a C-rate of $1C$ means that an appropriate current is applied to completely charge or discharge the cell in 1 hour. We calculate the current value for a C-rate corresponding to C/24 for the button cell formation cycling, as follows:

$$C/24 = \frac{19\,\text{mAh}}{24\,\text{h}} = 0.81\,\text{mA} \qquad (20.11)$$

Figure 20.22 shows the first two cycles (formation cycles) of the half-cells with a C/24 rate. The first discharge curve (insertion of lithium) illustrates outstanding specific capacity of silicon, which includes the irreversible capacity loss (Q_{irrev}) typically related to formation of the solid electrolyte interphase (SEI) on the electrode surface (Peled 1979; Verma et al. 2010). The micro-Si electrode (Figure 20.22a) shows a higher specific capacity (3940 vs. 3100 mAh g⁻¹). The irreversible capacity loss of the micro-Si electrode is rapidly increasing at the second charging/discharging cycle. This behavior is due to the loss of electrical contacts because of electrode cracking and the delamination of the composite electrode from the current collector. For the nano-Si electrode (Figure 20.22b), we observe a phenomenon referred to as "electrode slippage" by Dahn and coworkers (Burns et al. 2011), which is caused by the parasitic reaction of lithium with electrolyte at the silicon surface to form an ever-thickening SEI. This causes the measured lithiation capacity of the electrode to be larger than the measured delithiation capacity and hence the voltage-capacity curve continually slips to the right. However, even though lithium is being consumed irreversibly, there is more lithium available from the lithium foil counter/reference electrode so that the silicon electrode is fully lithiated during every cycle, and the capacity loss during cycling is almost nil. Finally, we notice a large hysteresis voltage loop for both electrodes. This phenomenon is typical of silicon and related to mechanical strain energy accumulated during volume expansion of the lithiated electrode (Huggins 2009).

In this experiment, both silicon anodes showed outstanding initial capacities compared to graphite, however, such high values are not required to improve actual commercial cells. A noticeable improvement of the LIB is achieved if the graphite anode was replaced with an electrode having a capacity on the order of 1000 mAh g⁻¹ (Yoshio et al. 2005). Therefore, the combination of a high-voltage cathodes and composite silicon anodes could boost the energy density of LIBs to levels comparable to the present-day primary lithium cells. Presently, the limiting issue for silicon anodes is not the initial capacity but rather the severe capacity fade during stability cycling.

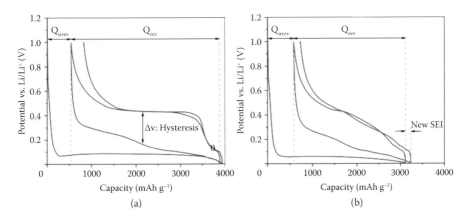

Figure 20.22 Formation cycles of Si/Li coin cells (2.3 mg/cm²; C/24): (a) micro-Si electrode and (b) nano-Si electrode.

In a full Li-ion cell, both electrodes must be balanced to optimize battery performance: the anode capacity (in mAh cm^{-2}) must be matched with the cathode capacity. Thus, performance degradation of both electrodes affects the useful life of the full cell that is normally limited by its weakest electrode. The stability of an electrode is measured by galvanostatic cycling until the cell fails, with the useful life of the electrode usually considered to be 80% of its initial capacity. The stability is reported by the graph of the reversible capacity (Q_{rev}) as a function of cycle number at a rate of $C/6$ (Figure 20.23):

Figure 20.23 clearly demonstrates that the nano-Si electrode has better cycling stability than a micro-Si electrode. The micro-Si electrode loses more than 20% of its initial capacity after the first cycle and has lower capacity than graphite (372 mAh g^{-1}) after only 9 cycles. The nano-Si electrode does approximately 10 cycles before reaching 80% of its initial capacity but stabilizes at a value around 1000 mAh g^{-1} after the 100th cycles. This clearly shows the improved ability of nano-sized silicon to relax the mechanical stress from volumetric expansion and contraction.

Post mortem analyses conducted on the nano-Si electrode after the 1st cycle, 10th cycles, and 50th cycles are shown in Figures 20.24 and 20.25. The deformation of the composite electrode appears even after the 1st cycle. Significant fractures are developed from the 1st cycle and getting worse with the number of cycles.

We also notice an increasing electrode thickness and deformation with cycle number (Figure 20.25). At cell failure, the electrode is partly detached from the current collector. The post mortem analysis clearly demonstrates the poor binding action of the polymer in the composite electrode and also the poor adhesion of the composite electrode with the current collector.

It is known that the capacity fade is severe when the silicon electrode is cycled at high depth of discharge (DoD), which results in the particles experiencing high mechanical stresses which induces cracks (Guerfi et al. 2011). Since volume expansion of the silicon electrode is affected by the DoD, the amplitude of the cyclic deformation can be reduced below the fatigue limit of the electrode (amplitude of cyclic stress that can be applied to the material without causing fatigue failure). In the case of the nano-Si electrode, limiting the DoD to 40% would increase significantly the life of the electrode in real operation.

Cycling stability is presently the main issue limiting commercialisation of silicon anodes. The use of nanosilicon powders synthesized by ICP has significantly improved the performances of silicon anodes; however, the DoD must be limited to prevent mechanical fatigue from producing rapid failure of the cell. New binder systems that provide higher electrode toughness and adhesion properties are needed for silicon anodes; the more promising candidates that maintain electronic conductivity and structural integrity are presently carbonized binders (Wilkes et al. 2016), conductive polymers (Liu et al. 2011a), and self-healing polymers (Wang et al. 2013).

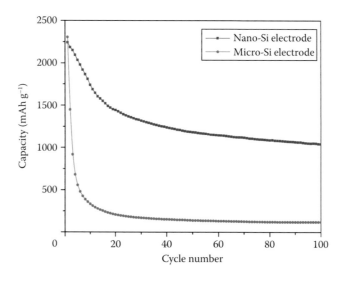

Figure 20.23 Stability cycles of the Si/Li coin cell (2.3 mg/cm^2; C/6): micro-Si electrode and nano-Si electrode.

Industrial nanosilicon

Figure 20.24 SEM micrograph showing top view of the nano-Si electrode: (a) fresh, (b) after 1st cycle, (c) after 10th cycles and (d) after 50th cycles.

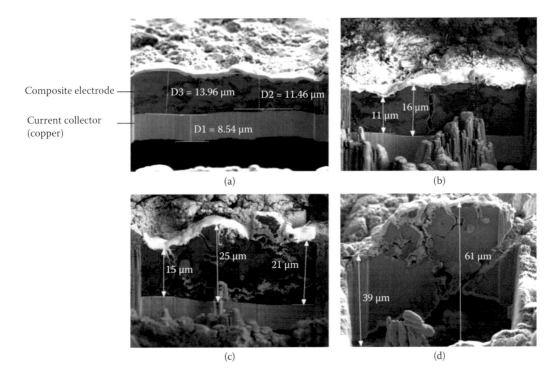

Figure 20.25 Dual beam (focused ion beam (FIB)/SEM) micrograph showing cross section of the nano-Si electrode (a) fresh, (b) after 1st cycle, (c) after 10th cycle, and (d) after 50th cycle.

Industrial nanosilicon

ACKNOWLEDGMENTS

We would like to acknowledge the contributions of the members of the IREQ SCE department: Myunghun Cho, Francis Barray, Daniel Clement, and Catherine Gagnon.

REFERENCES

Abel PR, Lin Y-M, Celio H, Heller A, Mullins CB (2012). Improving the stability of nanostructured silicon thin film lithium-ion battery anodes through their controlled oxidation. *ACS Nano* 6:2506–2516.

Boulos M (1983). Purification of metallurgical grade silicon, patent US 4,379,777.

Boulos MI, et al. (1994). *Thermal Plasmas: Fundamentals and Applications*, New York: Springer.

Boulos, M, Jurewicz J (1993). *High performance induction plasma torch with a water-cooled ceramic confinement tube*, patent US 5,200,595, Université de Sherbrooke.

Brunauer S, Emmett PH, Teller E (1938), Adsorption of Gases in Multimolecular Layers. *Journal of the American Chemical Society.* 60(2):309–319.

Buriak JM (2002). Organometallic chemistry on silicon and germanium surfaces. *Chemical Reviews* 102:1271–1308.

Burns JC, Krause LJ, Le D-B, Jensen LD, Smith AJ, Xiong D, Dahn JR (2011). Introducing symmetric Li-ion cells as a tool to study cell degradation mechanisms. *Journal of The Electrochemical Society* 158:A1417–A1422.

Chase, MWJ (1998). *NIST-JANAF Thermochemical Tables*, American Inst. of Physics.

Doğan İ, van de Sanden MCM (2016). Gas-phase plasma synthesis of free-standing silicon nanoparticles for future energy applications. *Plasma Processes and Polymers* 13:19–53.

Fauchais PL, Heberlein JVR, Boulos M (2014). Thermal spray fundamentals: From powder to part: Springer. http://www.springer.com/us/book/9780387283197 [Accessed March, 2016].

Guerfi A, Charest P, Dontigny M, Trottier J, Lagacé M, Hovington P, Vijh A, Zaghib K (2011). SiOx–graphite as negative for high energy Li-ion batteries. *Journal of Power Sources* 196:5667–5673.

Guo J, et al. (2010). Development of Nanopowder Synthesis Using Induction Plasma. *Plasma Science and Technology* 12(2):188.

Huggins RA (2009). *Advanced batteries: Materials science aspects*. Springer, London, Limited. http://www.springer.com/us/book/9780387764238 [Accessed March, 2016].

ICDD (2015). *PDF-4+ / Version 4.1504 (Database)*. Newtown Square, PA: International Centre for Diffraction Data.

Kambara M, Kitayama A, Homma K, Hideshima T, Kaga M, Sheem K-Y, Ishida S, Yoshida T (2014). Nano-composite Si particle formation by plasma spraying for negative electrode of Li-ion batteries. *Journal of Applied Physics* 115:143302.

Kasavajjula U, Wang C, Appleby AJ (2007). Nano- and bulk-silicon-based insertion anodes for lithium-ion secondary cells. *Journal of Power Sources* 163:1003–1039.

Leblanc D, Hovington P, Kim C, Guerfi A, Bélanger D, Zaghib K (2015a). Silicon as anode for high-energy lithium ion batteries: From molten ingot to nanoparticles. *Journal of Power Sources* 299:529–536.

Leblanc D, Wang C, He Y, Bélanger D, Zaghib K (2015b). In situ transmission electron microscopy observations of lithiation of spherical silicon nanopowder produced by induced plasma atomization. *Journal of Power Sources* 279:522–527.

Liu G, Xun S, Vukmirovic N, Song X, Olalde-Velasco P, Zheng H, Battaglia VS, Wang L, Yang W (2011a). Polymers with tailored electronic structure for high capacity lithium battery electrodes. *Advanced Materials* 23:4679–4683.

Liu XH, Zheng H, Zhong L, Huang S, Karki K, Zhang LQ, Liu Y, Kushima A, Liang WT, Wang JW, et al. (2011b). Anisotropic swelling and fracture of silicon nanowires during lithiation. *Nano Letters* 11:3312–3318.

Liu XH, Zhong L, Huang S, Mao SX, Zhu T, Huang JY (2012). Size-dependent fracture of silicon nanoparticles during lithiation. *ACS Nano* 6:1522–1531.

Mamak M, et al. (2010) Thermal plasma synthesis of tungsten bronze nanoparticles for near infra-red absorption applications. *Journal of Materials Chemistry.* 20(44):9855–9857.

McDowell MT, Lee SW, Harris JT, Korgel BA, Wang C, Nix WD, Cui Y (2013). In situ TEM of two-phase lithiation of amorphous silicon nanospheres. *Nano Letters* 13:758–764.

Obrovac MN, Christensen L, Le DB, Dahn JR (2007). Alloy design for lithium-ion battery anodes. *Journal of The Electrochemical Society* 154:A849–A855.

Okamoto H (2009). Li-Si (Lithium-Silicon). *Journal of Phase Equilibria and Diffusion* 30(1):118–119.

Peled E (1979). The electrochemical behavior of alkali and alkaline earth metals in nonaqueous battery systems—the solid electrolyte interphase model. *Journal of The Electrochemical Society* 126:2047–2051.

Rahmane M, et al. (1994). Mass transfer in induction plasma reactors. *International Journal of Heat and Mass Transfer* 37(14):2035–2046.

So K-S, Lee H, Kim T-H, Choi S, Park D-W (2014). Synthesis of silicon nanopowder from silane gas by RF thermal plasma. *Physica Status Solidi (a)* 211:310–315.

Tarascon JM, (2001). Issues and challenges facing rechargeable lithium batteries. *Nature* 414(6861):359–367.

Thackeray M (2002). Lithium-ion batteries: An unexpected conductor. *Nature Materials* 1:81–82.

Verma P, Maire P, Novák P (2010). A review of the features and analyses of the solid electrolyte interphase in Li-ion batteries. *Electrochimica Acta* 55:6332–6341.

Wang C, Wu H, Chen Z, McDowell MT, Cui Y, Bao Z (2013). Self-healing chemistry enables the stable operation of silicon microparticle anodes for high-energy lithium-ion batteries. *Nature Chemistry* 5:1042–1048.

Wilkes BN, Brown ZL, Krause LJ, Triemert M, Obrovac MN (2016). The electrochemical behavior of polyimide binders in Li and Na cells. *Journal of The Electrochemical Society* 163:A364–A372.

Yoshio M, Tsumura T, Dimov N (2005). Electrochemical behaviors of silicon based anode material. *Journal of Power Sources* 146:10–14.

21 Nanophotonics silicon solar cells

Boyuan Cai and Baohua Jia

Contents

21.1 INTRODUCTION

21.1.1 SILICON-BASED SOLAR CELLS

High-speed economic development in a modern society requires enormous energy consumption, in which electricity plays an important role. By the year 2040, the world energy consumption is predicted to boost to about 23 TW, which is mainly from the nonrenewable resources such as coal, oil, and gas, producing carbon dioxide (CO_2) contributing detrimentally to climate change through the greenhouse effect [1]. One of the most promising solutions is to utilize solar photovoltaics as a candidate for the resource, which comes from the sun. The annual solar energy on earth is 1.76×10^5 TW and the current world usage is estimated to be only 15 TW [2] which means the photovoltaic (PV) power holds the great potential to meet all the energy needs for the entire world.

Currently, the PV industry is majorly dominated by crystalline silicon (c-Si) solar cells with the thickness of 200–300 µm [3–5], the most developed solar PV devices, which are composed of monocrystalline or multicrystalline silicon [6–9]. The maximum efficiency of a single-junction c-Si solar cell under the AM1.5G solar spectrum is estimated to be around 30% from the Shockley method, and the highest efficiency experimentally achieved now for a c-Si solar cell is 25%. The efficiency gap between the theoretical limitation and the experiment results is mainly from the nonradiative recombination processes intrinsic to Si, such as Auger recombination. In terms of the cost of the c-Si solar cells, the cost of this kind of solar cell is quite high due to the involved complicated production process and the expensive material (around 50% of the cost is attributed to the raw materials themselves, with the reminder from the production and module installation). To make the PV energy competitive with the traditional energy solutions, it is an urgent need to investigate low-cost PV devices, for example, the thin-film solar cells with the thickness less than 2 µm, which are promising candidates for reducing the cost.

In the current mainstream market, thin-film technologies are cadmium telluride (CdTe) from First Solar with the highest laboratory efficiency at 17.3% [10,11]; copper indium gallium selenide (CIGS) from Zentrum fuer Sonnenenergie and Wasserstoff-Forschung, Baden-Wuerttemberg (ZSW) with an efficiency of around 20% [12]; and amorphous Si (a-Si) thin-film solar cells from Oerlikon with a stable efficiency 10.1% for single junction and 12% for tandem geometry [13,14]. Among all these thin-film technologies, the Si-based thin-film technology holds great potential for commercialization due to (a) low-cost and comparatively simple fabrication procedure, and the non-toxic material involved and (b) the utilization of the toxic and scarce elements for the CIGS and CdTe solar cells. The tendency of the development in Si-based solar cells is to make the solar cell active layer as thin as possible while maintaining the light absorption in the solar cells. There are two benefits for thinning the active layer. First, due to the reduced amount of raw materials required, the cost of the solar cells can be significantly decreased making the PV energy more competitive than traditional energy. Second, making the solar cell thinner could increase the operating voltage and provide a potentially higher efficiency compared with the normal thick ones, considering no absorption loss is involved. Thus, the key question now is, can the thinner solar cells absorb enough sunlight power to provide competitive efficiency with the normal thick cells? Traditionally, for normal thick c-Si solar cells, the absorption can be enhanced by introducing pyramid surface textures on the front surface to scatter light into the solar cell over a large angular range, thereby increasing the effective path length in the cell. However, this geometry is not suitable for thinner solar cells as the surface roughness would exceed the film thickness. In addition, the large surface textured area will also increase the carrier recombination on the surface leading to degraded solar cell performance. Motivated by solving this issue, many researchers devote themselves to investigating novel light trapping strategies, including metal/dielectric nanoparticles, nano-grating, and nanowire structures. In the following sections, a brief discussion of the nanoparticle and nanostructure light trapping methods will be presented with more details introduced in Sections 21.2 and 21.3.

21.1.2 LIGHT TRAPPING STRATEGY VIA NANOPARTICLES AND NANOSTRUCTURES

Recently, nanoparticle-based light management has attracted intensive attention in Si-based solar cells including metallic and dielectric nanoparticles. Metallic nanoparticles can support localised surface plasmons (LSPs), which result in great enhancement of the light-path length inside solar cells [15–23]. The resonances of noble metals are mostly in the visible or the infrared region of the spectrum, which is in the range of interest for Si-based solar cells. The surface plasmon resonance is mainly affected by the size, shape, and the dielectric properties of the surrounding media. By tailoring the geometry of the metallic nanoparticles, the surface plasmon resonance can be tuned, based on the requirements of the applications. There are two main mechanisms that account for enhancing the performance of photovoltaic devices: (a) the scattering from the metal particles (far-field effect); and (b) the near-field enhancement from small nanoparticles, which directly enhance the absorbance of the semiconductor in the close vicinity of the nanoparticles due to the enhanced field [24–27], as shown in Figure 21.1. To utilize the scattering effects, these nanoparticles are either on the front or the back surfaces of an absorbing layer, which are shown in Figure 21.1a and c. In terms of the near-field concentration effects, nanoparticles which are embedded into the absorbing layer are applied as shown in Figure 21.1b. This can lead to higher short-circuit current

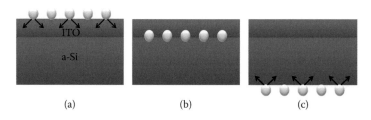

Figure 21.1 Nanoparticle light trapping geometries for thin-film solar cells. (a) Light trapping from metallic nanoparticles on top of solar cells. (b) Light absorption enhancement in the semiconductor by embedded nanoparticles due to the enhanced electric field around the nanoparticles. (c) Light trapping from metallic nanoparticles at the back of solar cells.

densities, in both relatively thick cells and thin solar cells on index-matched substrates, due to the LSP effect from the metallic nanoparticles.

Despite the effective light trapping by metallic nanoparticles, the obvious disadvantage is that it is challenging to avoid the absorption loss of the particle which prevents the solar cell performance from further improving. Therefore, lossless dielectric nanoparticles are proposed to be integrated into the solar cells for light management. For instance, TiO_2 nanoparticles have been demonstrated on top of the c-Si solar cells to act as antireflective coatings, thereby increasing the longer wavelength light-path length. In addition, closely packed SiO_2 nanoparticles have been applied on top of solar cells to generate whispering gallery modes (WGMs) to enhance the light absorption [28–32]. Dielectric nanoparticles can successfully avoid the shadowing problem and the parasitic absorption existing in the metal nanoparticles. High-index nanoparticles, Si for example, can support Mie resonance modes which yield almost zero reflection over the entire spectrum range when placed on an Si substrate. For nanostructure integrated solar cells, novel nano-cone and nanowire structures have been demonstrated for efficient light coupling to guided modes from the top surface of the solar cells. The above mentioned dielectric nanoparticles and nanostructures can provide effective light trapping schemes for thin-film solar cells. Detail discussions will be given in the following sections.

21.2 PLASMONIC LIGHT TRAPPING FOR SOLAR CELLS

21.2.1 METALLIC NANOPARTICLES FOR LIGHT TRAPPING

Subwavelength metallic nanoparticles have recently been widely investigated to couple light into different angles to increase the light path in solar cells, or localise the electric energy in the close vicinity of the nanoparticles, thereby increasing the absorbance in the active layer. This absorption enhancement is mainly attributed to the LSP effects of the nanoparticles. LSPs are nonpropagating excitations of electron collective oscillations within a metallic nanoparticle, which are illuminated by external light; enhanced near-field intensity and scattering can typically be observed at the resonance [33]. When the nanoparticles are placed on top or at the rear side of the absorber, the light can be scattered from the larger nanoparticles into the active layer accounting for the absorption enhancement. These nanoparticle arrays prefer scattering light into larger permittivity materials such as the Si when they are placed close to the absorber [34]. Another case is that when nanoparticles are integrated in the solar cell absorber, the enhanced near-field intensity around the particles related to the LSP of small metallic nanoparticles can improve the absorption of the Si layer. By tuning the sizes and shapes of nanoparticles, a particular frequency of optical excitation will lead to strongly enhanced fields in the close vicinity of nanoparticles.

To get a better understanding of the theory of the nanoparticles plasmonics, the formalism for obtaining the quasi static approximation for the resonant frequency of a spherical particle in a uniform external electric field is presented [35]. The scattering and absorption cross sections of the sphere can be expressed as follows, respectively:

$$C_{sca} = \frac{k^4}{6\pi}|\alpha|^2$$

$$C_{abs} = k\text{Im}(\alpha)$$

$$\alpha = 4\pi a^3 \frac{\varepsilon - \varepsilon_d}{\varepsilon + 2\varepsilon_d}$$

where K is the magnitude of the wave vector of the incident light, α is the polarisability, and a is radius of the sphere. The dielectric constants of the sphere and the dielectric medium are presented as ε and ε_d, respectively [36]. When $\varepsilon = -2\varepsilon_d$, the polarisability is at a maximum and the sphere exhibits a dipole surface plasmon resonance. If the medium surrounding the sphere is an ordinary dielectric material, ε_d is positive and the real part of ε must be negative to satisfy this condition. Obviously, it cannot be fulfilled with a dielectric sphere in a dielectric medium. However, in terms of metal materials, the real parts of dielectric constants $Re[\varepsilon_d]$ can be negative at specific wavelengths. At these wavelengths, resonant scattering cross section and absorption cross section can be achieved. This is regarded as the localised plasmon resonance.

21.2.1.1 Plasmonic light-scattering effect for solar cells

Plasmonic incoupling of scattered light from an external layer of metallic nanoparticles into the absorbing layer has been discussed for different kinds of solar cells. In this section, we focus on the discussion of metallic nanoparticles applied in Si-based solar cells. As introduced in Section 21.1, the nanoparticles are usually integrated either on top or at the rear side of the solar cells to scatter more light into the absorber as shown in Figure 21.1a and c.

In terms of the top surface integrated nanoparticles, metallic nanoparticles including the materials (Ag, Al), shapes, and coverage density have been systematically explored for Si-based solar cells. For example, Ag nanosphere arrays have been demonstrated to be capable of coupling light into the Si substrate over a broadband wavelength range [37] as shown in Figure 21.2a. Apart from the light-scattering effects, the Ag nanoparticles can also achieve better impendence matching the light propagating in air and in the substrate. Through the optimization of the size and pitch of Ag nanosphere arrays together with the thickness of SiN, the particle array demonstrates 50% enhanced incoupling compared with bare Si, and 8% higher than the cell with an optimized SiN antireflection coating layer. The enhancement of the absorption occurs mostly in the wavelength range near the band gap of Si as shown in Figure 21.2b.

Despite the high performance of the solar cell with integrated Ag nanoparticle arrays, the absorption loss of Ag nanoparticles can hardly be ignored, due to the plasmonic resonance in the visible wavelength range. Researchers are begining to investigate low-absorption and low-cost nanoparticles, Al for example, to further improve the solar cell performance and potentially decrease the fabrication cost. Compared with traditional metallic nanoparticles such as Ag and Au, the Al nanoparticles can provide much better light trapping due to the plasmonic resonance lying in the ultraviolet range, which would lead to less negative influence on the solar cell by the Fano resonance. As shown in Figure 21.3a, the absorption of the Si layer with Ag and Au nanoparticles in the plasmonic resonance range (350–430 nm for Ag and below 570 nm for Au) significantly reduces, which results from the absorption loss of the metallic nanoparticles. However for Al nanoparticles, the absorption over the entire spectrum is enhanced [38]. In addition, by applying the Al nanoparticles with an SiN antireflection layer, the overall performance of the Si solar cells is better than that with only an SiN layer as shown in Figure 21.3b. Besides, the Al nanoparticles have also been demonstrated experimentally that through combination of the graphene as a novel light trapping layer, the efficiency of the solar cell can be increased from 18.24% to 19.54% [39].

In terms of the rear surface integrated nanoparticles in solar cells, the case is quite different as the light can only interact with the nanoparticles after traveling through the solar absorber first. In that case, the preferable plasmonic material is not Al but Ag, since the absorption loss of an Ag nanoparticle can be neglected as the short wavelength light can be absorbed first by the absorber, before reaching the nanoparticles. Ag nanoparticles can provide a larger scattering cross section than Al with the same size and shape.

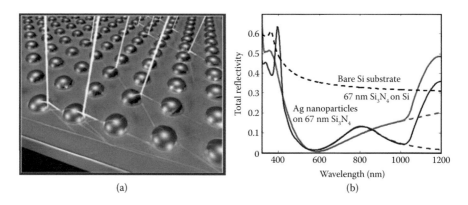

(a)   (b)

Figure 21.2 (a) The geometry of solar cells with Ag nanosphere arrays on top. (b) The comparison of reflectivity with optimized SiN and Ag+SiN on Si solar cells.

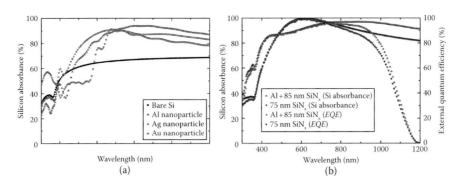

Figure 21.3 (a) Si absorption with optimized Ag, Au, and Al nanoparticles. (b) Comparison of the Si absorption with Al+SiN and SiN. (From Zhang, Y., et al., *Appl. Phys. Lett.*, 100, 151101, 2012. With permission.)

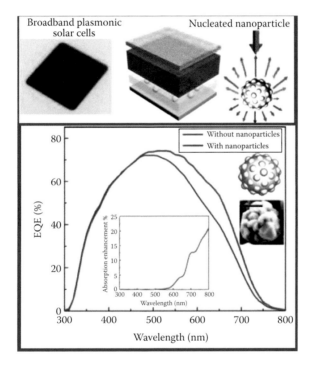

Figure 21.4 The nucleated Ag nanoparticle-incorporated solar cells. (From Chen, X., et al., *Nano Lett.*, 12, 2187–2192, 2012. With permission.)

Recently, for rear surface nanoparticles in solar cells, an interesting Ag nanoparticle structure (nucleated lumpy structure) has been proposed with a large Ag core particle surrounded by smaller Ag particles on the outer surface as shown in Figure 21.4. Compared with the normal sphere Ag nanoparticles, the back-scattering capability and the broadband-scattering property of nucleated nanoparticles can be adjusted by optimizing the size ratio of the outer surface small particles to the core large particles [20]. The a-Si solar cell performance has been experimentally demonstrated to be enhanced by 23% compared with reference cells. The significant improvement of the solar cell performance is mainly ascribed to the broadband light scattering, resulting from the optimized Ag nucleated nanoparticle structure. Since both the top and rear surface incorporated nanoparticles in solar cells can significantly boost the solar cell absorption, researchers proposed a concept for improving the absorption of the subcell in a tandem solar cell (a-Si/μ-Si) geometry via simultaneously applying tailored metallic nanoparticles both on the top and at the rear surfaces of the solar cells [40]. Light absorption enhancement as much as 56% can be obtained in the μ-Si bottom subcell.

Industrial nanosilicon

More importantly, the thickness of the μ-Si layer can be reduced by 57% without compromising the overall optical performance of the tandem solar cell.

21.2.1.2 Plasmonic near-field effect for solar cells

In comparison to the forward or backward plasmonic scattering from the comparative large particles to enhance the Si absorbance, small metallic particles can be applied in close vicinity of the active layer as shown in Figure 21.1b. This makes use of the LSP near-field enhancement of the electromagnetic field, which can give rise to the absorption enhancement in the active layer around the nanoparticles. Numerous publications have reported in recent years regarding the enhanced absorption arising from the embedded metallic nanoparticles in the active layer for thin-film solar cells, especially in organic solar cells [41–45]. High efficiency plasmonic tandem solar cells have also been reported, achieving 20% enhancement in the efficiency from 5.22% to 6.24% due to the LSP near-field absorption enhancement.

Theoretically, embedding metallic nanoparticles in the active layer of thin-film solar cells can lead to significant absorption enhancement. For example, it was predicated an ideal conversion efficiency of 18% can be achieved by combining an Ag/a-Si nanocomposite layer with an only 20-nm-thick active layer as shown in Figure 21.5 [46]. Ultrasmall Au nanoparticles have also been demonstrated to be capable of enhancing the a-Si solar cell efficiency via the plasmonic near-field concentration effects when incorporated into the active layer [47]. By tuning the evaporated Au film with subnanometer thickness, variation sizes, and coverage density Au nanoparticles can be achieved, thereby adjusting the plasmonic resonance. Through optimizing the Au nanoparticles, a dramatically enhancement in Jsc of 14% is achieved.

21.2.2 METALLIC NANOSTRUCTURES FOR LIGHT TRAPPING IN SOLAR CELLS

In a typical solar cell geometry, a metallic back reflector has been playing an important role for redirecting light back into the absorber, and as an electrode for electron collection. Recently, the designed nanostructure for the metallic back reflector has been regarded as a naturally integrated candidate for light management in solar cells as shown in Figure 21.6.

The nanostructures for the plasmonic back metal reflector can couple light into surface plasmon polaritons (SPPs) modes or photonic waveguide modes at the metal/semiconductor interface if the nanostructures are periodic. However, the SPP modes can be absorbed both in the metal and in the semiconductor, as shown in Figure 21.6b. Since the plasmon absorption loss cannot be ignored, the coupling to photonic modes is preferred. A large amount of research effort during the past few years has proved that a relative high absorption enhancement can be achieved by incorporating a nanostructured plasmonic back reflector into solar cells, such as nanocone structures and nanovoid structures.

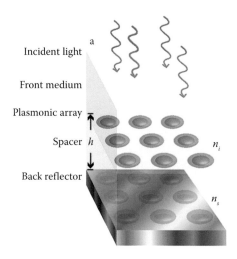

Figure 21.5 An array of metal core/semiconductor shell nanoparticles represents a close to ideal plasmonic near-field absorber. (From C. Hägglund and S. Apell, *J. Phys. Chem. Lett.*, 3, 1275-1285, 2012. With permission.)

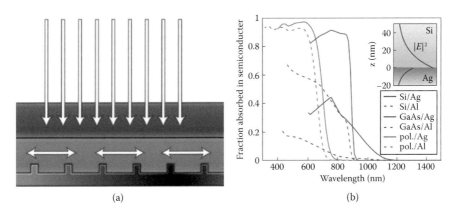

(a) (b)

Figure 21.6 The concept of a plasmonic back reflector for light management in solar cells (a) through surface plasmon polariton or photonic modes at the interface between the absorber and reflector (b) Fraction of the absorbed light in the semiconductor due to SPP. (From Atwater, H., and Polman, A., *Nat. Mater.*, 9, 205–213, 2010. With permission.)

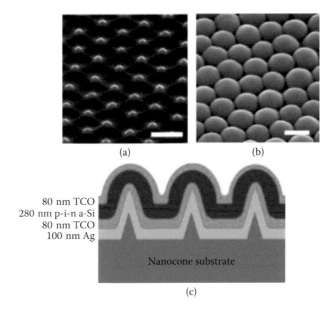

(a) (b)

80 nm TCO
280 nm p-i-n a-Si
80 nm TCO
100 nm Ag

Nanocone substrate

(c)

Figure 21.7 (a) The nanocone quartz substrate and (b) the a-Si nanocone solar cell after deposition multilayers of materials on nanocones. Scale bar: 500 nm. (c) Schematic of solar cells. (From Zhu, J., et al., *Nano Lett.*, 10, 1979–1984, 2010. With permission.)

Periodic nanocone structures have been designed for a-Si solar cells in the bottom Ag substrate as shown in Figure 21.7 [35,48]. This cone-structured solar cell could absorb 94% of the light from 400 nm to 800 nm with an only a 280-nm-thick a-Si layer compared to the flat solar cell which can only absorb 65% of the light. The unique nanocone geometry of the solar cells can provide an effective antireflection surface structure due to the gradually varied effective refractive index, resulting in a better impendence matching with air. In addition, the modes coupled into the a-Si layer and the scattered light along the inplane direction by the Ag nanocone structure can also increase the light traveling path, providing an additional plasmonic light trapping mechanism.

Different from the Ag cone-scattering light out of the structure, the plasmonic nanovoid structure prefers coupling the light inside the void, thereby exciting the hybrid void modes to improve the solar cell absorption as shown in Figure 21.8a–d [50]. The hybrid modes are excited by mixing the LSP modes with the Fabry–Pérot cavity modes and the rim dipole modes associated with the charge buildup at the void rims. The efficiency of the solar cell with the nanostructured plasmonic back reflector can be improved

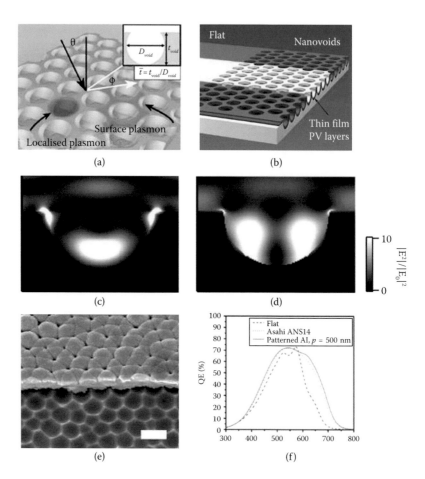

Figure 21.8 (a) Schematic of different modes excited in the void plasmonic structure. (b) Solar cell geometry fabricated on the plasmonic void substrate. (c) Distribution of the plasmonic electric field intensity at 375 nm. (d) Distribution of the plasmonic electric field intensity at 496 nm. (e) SEM image of the void structured a-Si solar cell. (f) EQE comparison of solar cells with and without the void structures. (From Huang, H., et al., *Energy Environ. Sci.*, 6, 2965–2971, 2013; Lal, N.N., et al., *Opt. Express*, 19, 11256–11263, 2011. With permission.)

from 5.6% to 7.1%, mainly due to the Jsc enhancement arising from the unique enhanced light coupling by the void structure as shown in Figure 21.8f [49,50].

21.3 DIELECTRIC NANOPARTICLE AND NANOSTRUCTURE FOR LIGHT TRAPPING

21.3.1 DIELECTRIC NANOPARTICLES FOR LIGHT TRAPPING

The intrinsic absorption loss of metal associated with metallic nanoparticles and nanostructures can hardly be avoided, despite optimizing the particle shape and size. Therefore, dielectric nanoparticles which benefit from low absorption loss provide another option for light management in solar cells. Even though the scattering cross section of dielectric nanoparticles is no larger than that of plasmonic nanoparticles, less absorption can also lead to comparable performance enhancement in the solar cells. Many works have been reported regarding solar cells with dielectric nanoparticles incorporated on the top surface. SiO_2 nanoparticles with a size of 150 nm were utilized in quantum well solar cells leading to an increased Jsc of 12.9% and efficiency enhancement of 17%, which mainly arises from the coupling of the incident light into lateral optical propagation [51]. Recently, a new concept for light management in a-Si thin-film solar cells by WGM generated by closely packed SiO_2 nanoparticles has been proposed to enhance the absorption and

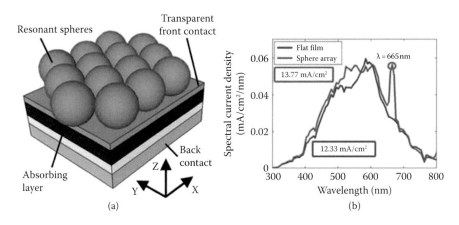

Figure 21.9 (a) The solar cell geometry with the presence of closely packed SiO$_2$ nanospheres. (b) Calculated current density in a 100 nm-thick a-Si layer with and without the presence of nanospheres. (From Grandidier, J., et al., *Adv. Mater.*, 23, 1272–1276, 2011; Grandidier, J., et al., *Phys. Status Solidi A*, 210, 255–260, 2013. With permission.)

the photocurrent [30,31]. These closely packed SiO$_2$ dielectric nanospheres can diffractively couple light from air and support confined resonant modes inside the nanospheres. In addition, the periodic distribution of the nanospheres can result in light coupling between the adjacent spheres, leading to mode splitting. The coupling modes can be leaked inside the a-Si solar cells, thereby enhancing the absorption near the resonance spectrum as shown in Figure 21.9.

As shown in Figure 21.9b, it can be found that the absorption enhancement mainly lies in a quite narrow spectrum range arising from the WGM coupling. Thus, to further broaden the enhanced absorption spectrum in solar cells, closely packed SiO$_2$ nanosphere arrays have been proposed to be partially embedded into the solar cells from the top surface, generating a hemisphere void nanostructure. The advantages of this structure are that both the dielectric particle light coupling effects and the plasmonic void structure of the Ag back reflector can be combined together to maximum the solar cell performance as shown in Figure 21.10. From the quantum efficiency (QE) measurement, the Jsc can be improved by 43.9% compared to the reference solar cell [52]. Apart from SiO$_2$ nanoparticles, closely packed TiO$_2$ nanoparticles have also been proved to be capable of improving the solar cell absorption as an effective antireflection layer on top of the solar cells [29].

Apart from the light management from the low refractive index nanoparticles such as SiO$_2$ and TiO$_2$, high-index nanoparticles [53], for example Si, are also utilized to improve the absorption by the light coupling to a variety of resonant modes in a solar cell. Commonly, these resonant modes can be identified by the electric field intensity distribution inside the absorber, as shown in Figure 21.11. Figure 21.11a–d represents a mixture of a Mie resonance with a guided resonance (a), excitation of a Faby-Pérot resonance (b), a guided resonance (c), and a diffracted resonance (d) by a lateral propagation light. These resonance modes excited inside the absorber via high-index nanoparticles or nanostructures can trap most of the light inside the solar cells in the specific wavelength ranges. For example, Si nanocylinder particles have been experimentally integrated on an Si wafer to provide an effective pathway to couple light into the substrate. Through designing the shape and the size of the Si nanoparticles, the Mie resonance induces a forward light coupling into the absorber due to the higher mode intensity.

21.3.2 ABSORBING NANOSTRUCTURES FOR LIGHT MANAGEMENT IN SOLAR CELLS

Introducing nanostructures in the absorber directly has a significant influence on both the solar cell optical properties. For instance, an InP nanowire array thin-film solar cell with p–i–n junction is a good example of this concept. As shown in Figure 21.12a and b, enhanced light absorption exceeding the ray optics limit was demonstrated, leading to a 13.8% efficiency [54]. For c-Si solar cells, systematic simulations have been performed on the possible benefits from nanostructuring both sides of the ultrathin c-Si solar cell as

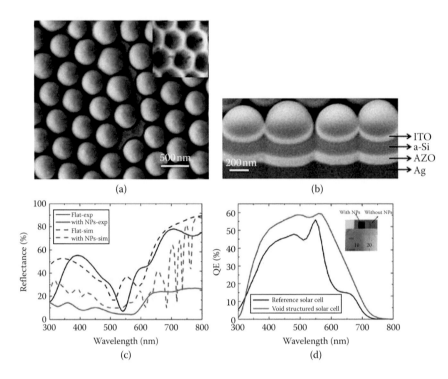

(a)   (b)

(c)   (d)

Figure 21.10 SEM images of 400 nm nanoparticle imprinted solar cells: (a) Top view of the solar cell. Insert figure is the a-Si nanostructures with SiO₂ nanoparticles peeled off. Scale bar is 500 nm. (b) Cross section of the solar cell. (c) Measured and calculated reflectance of 150-nm-thick a-Si solar cell with 400-nm-SiO₂ nanoparticles embedded, compared to the flat reference solar cell. (d) Comparison of the QE response of solar cells with the nanoparticles. (From Cai, B., et al., *J. Appl. Phys.*, 117, 223102, 2015. With permission.)

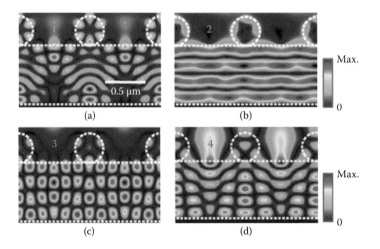

(a)   (b)

(c)   (d)

Figure 21.11 Electric field intensity distributions of various types of resonance excited: (a) Mie resonance, (b) Fabry-Pérot resonance, (c) guided resonance, and (d) diffracted resonance. (From Brongersma, M.L., et al., *Nat. Mater.*, 13, 451–460, 2014. With permission.)

shown in Figure 21.12d and e [55]. The high aspect ratio dense nanocone arrays at the front surface serve as an antireflection structure while low aspect ratio sparse nanocone arrays at the back surface function as a grating to couple light into the guided resonances in the absorber, leading to a photocurrent close to the Yablonovitch limit with a 2 μm-thick absorber. Similar nanocone array structure has also been applied into a-Si solar cells as shown in Figure 21.12c; broadband absorption enhancement in the solar cell through the

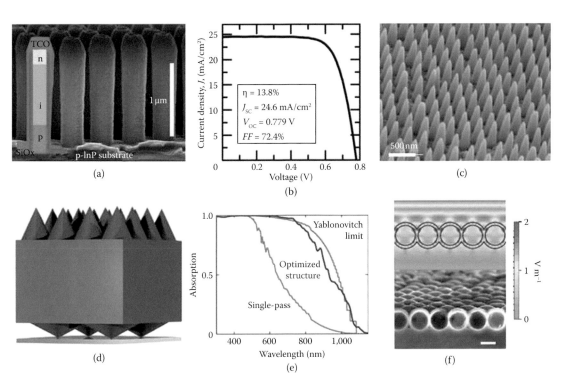

Figure 21.12 (a) SEM image of the InP nanowire solar cells. (b) Current-voltage (I-V) measurement for the nanowire solar cell shown in (a). (c) SEM of nanocone array structures of the a-Si solar cell. (d) Schematic of the ultrathin c-Si solar cell with double-sided nanocone structure design. (e) Calculated absorption curves of the optimized structures shown in (d). (f) Bottom: SEM cross section image of a monolayer of spherical nc-Si nanoshell arrays on a quartz substrate. Scale bar, 300 nm. Top: Full-field electromagnetic simulation of the electric field distribution inside the nanoshell showing the importance of the excitation of WGM to enhance the absorption of light. (From Wallentin, J., et al., *Science*, 339, 1057–1060, 2013; Wang, K.X., et al., *Nano Lett.*, 12, 1616–1619, 2012; Zhu, J., *Nano Lett.*, 10, 1979–1984, 2010. With permission.)

whole wavelength range was achieved due to the dramatically reduced reflection from the cone structure which provides an impedance match from the air to the a-Si substrate [48].

Recently, a novel spherical nanoshell nanocrystalline (nc-Si) structure has been proposed by Y. Yao to improve the broadband absorbance dramatically due to the excited low-Q WGMs as shown in Figure 21.12f [56]. The geometry of the nc-Si solar cell was designed into nanoshell structures resulting in a 20-fold enhancement in the absorber due to the coupling of light into the WGM modes. This nanoshell nc-Si solar cell could couple incident light into the WGM modes inside the spherical shells and induce a profitable recirculation of light inside the shell-shaped absorbers, leading to the increase of light path in the active layer and a reduction of the amount of material demanded for comparable light absorbance. By theoretical calculation of the solar cell efficiency with this structure, an 80-nm-thick nc-Si shell solar cell with an outer radius of 225 nm can provide an efficiency of 8.1% which is yielded with an active layer thickness of 1.5 μm in a flat cell.

21.4 OUTLOOK

Light management via metallic or dielectric nanoparticles and nanostructures have recently been widely reported in papers and proved to be an effective path to improve solar cell performance, especially with a much thinner absorber. However, the individual drawbacks of metallic or dielectric materials for light trapping hold back the further applications and development in solar cells. For instance, the intrinsic loss of the metal and the uncompetitive scattering cross section of the dielectric nanoparticles can hardly be avoided when integrated into the solar cells. Thus, there is a great demand to design a hybrid nanoparticle structure, which can utilize the benefits of both metallic and dielectric material together. Besides, through the combination of nanoparticle light trapping and solar cell nanostructure light coupling strategies, it holds great potential to improve solar cell performance. One further point we would like to mention is that even though the nanophotonic design of solar cells could boost the efficiency, the time consuming and high cost of nanofabrication of these structures on a large scale are not suitable for the manufacturing line. It is highly desired to develop cost-effective nanofabrication technology which can match the current solar panel production line.

For single-junction-based Si solar cells, there exists an efficiency limitation due to the band gap of semiconductor. This limitation cannot be overcome by the conventional light trapping strategies. However, based on the investigation of light management in single-junction solar cells, double junction or triple junction Si-based solar cells with matchable band gap material can be developed for a higher achievable efficiency through nanophotonic design.

REFERENCES

1. CO_2 Emissions from Fuel Combustion, International Energy Agency (2012).
2. N. S. Lewis. Powering the planet. *MRS Bull.* **32**, 808–820 (2007).
3. M. A. Green, High efficiency silicon solar cells, Proceedings of the International Conference, held at Sevilla, Spain, 27–31 October (1986).
4. J. Zhao, A. Wang, M. A. Green, and F. Ferrazza, 19.8% efficient "honeycomb" textured multicrystalline and 24.4% monocrystalline silicon solar cells, *Appl. Phys. Lett.* **73**, 1991 (1998).
5. M.A. Green, J. Zhao, A. Wang, and S. R. Wenham, Progress and outlook for high-efficiency crystalline silicon solar cells, *Sol. Energy Mater. Sol. Cells* **65**, 9–16 (2001).
6. O.Schultz, S. W. Glunz, and G. P. Willeke, Multicrystalline silicon solar cells exceeding 20% efficiency, *Prog. Photovolt. Res. Appl.* **12**, 553–558 (2004).
7. A. Rohatgi and S. Narasimha, Design, fabrication, and analysis of greater than 18% efficient multicrystalline silicon solar cells, *Sol. Energy Mater. Sol. Cells* **48**, 187–197 (1997).
8. P. Campbell and M. A. Green, High performance light trapping textures for monocrystalline silicon solar cells, *Sol. Energy Mater. Sol. Cells* **65**, 369–375 (2001).
9. T. M. Bruton, General trends about photovoltaics based on crystalline silicon, *Sol. Energy Mater. Sol. Cells* **72**, 3–10 (2002).
10. A. Morales, Towards 19% efficient CdTe solar cells, 27th European Photovoltaic Solar Energy Conference and Exhibition, Frankfurt, Germany, 2812–2814 (2013).
11. N. Romeo, A. Bosio, V. Canevari, and A. Podesta, Recent progress on CdTe/CdS thin film solar cells, *Sol. Energy Mater. Sol. Cells* **77**, 795–801 (2002).
12. P. Jackson, D. Hariskos, E. Lotter, S. Paetel, R. Wuerz, R. Menner, W. Wischmann, M. and Powalla, New world record efficiency for $Cu(In,Ga)Se_2$ thin-film solar cells beyond 20%, *Progress in Photovoltaics: Research and Applications* **19**, 894–897 (2011).
13. S. Benagli, D. Borrello, E. Vallat-Sauvain, J. Meier, U. Kroll, J. Hötzel, J. Spitznagel, J. Steinhauser, L. Castens, and Y. Djeridane, High-efficiency amorphous silicon devices on LPCVD-ZnO TCO prepared in industrial KAI-M R&D reactor, 24th European Photovoltaic Solar Energy Conference and Exhibition, 2293, Hamburg, Germany, 21–25 September (2009).
14. U. Kroll, J. Meier, L. Fesquet, J. Steinhauser, S. Benagli, J. B. Orhan, B. Wolf, et al., Recent developments of high-efficiency micromorph tandem solar cells in KAI-M Plasmabox PECVD reactors, 26th European Photovoltaic Solar Energy Conference and Exhibition, 2340, Hamburg, Germany, 05–09 September (2011).
15. D. Derkacs, S. H. Lim, P. Matheu, W. Mar, and E. T. Yu, Improved performance of amorphous silicon solar cells via scattering from surface plasmon polaritons in nearby metallic nanoparticles, *Appl. Phys. Lett.* **89**, 093103 (2006).
16. K. Nakayama, K. Tanabe, and H. A. Atwater. Plasmonic nanoparticle enhanced light absorption in GaAs solar cells, *Appl. Phys. Lett.* **93**, 121904 (2008).
17. S. Pillai, K. R. Catchpole, T. Trupke, and M. A. Green. Surface plasmon enhanced silicon solar cells. *J. Appl. Phys.* **101**, 093105 (2007).
18. F. J. Beck, A. Polman, and K. R. Catchpole, Tunable light trapping for solar cells using localized surface plasmons, *J. Appl. Phys.* **105**, 114310 (2009).
19. S. S. Kim, S. I. Na, J. Jo, D. Y. Kim, and Y. C. Nah, Plasmon enhanced performance of organic solar cells using electrodeposited Ag nanoparticles, *Appl. Phys. Lett.* **93**, 073307 (2008).

20. X. Chen, B. Jia, J. Saha, B. Cai, N. Stokes, Q. Qiao, Y. Wang, Z. Shi, and M. Gu, Broadband enhancement in thin-film amorphous silicon solar cells enabled by nucleated silver nanoparticles, *Nano Lett.* **12**, 2187–2192 (2012).

21. Y. A. Akimov and W. S. Koh, Resonant and nonresonant plasmonic nanoparticle enhancement for thin-film silicon solar cells, *Nanotechnology* **21**, 235201 (2010).

22. H. Tan, R. Santbergen, A. H. M. Smets, and M. Zeman, Plasmonic light trapping in thin-film silicon solar cells with improved self-assembled silver nanoparticles, *Nano Lett.* **12**, 4070–4076 (2012).

23. Z. Sun, X. Zuo, and Y. Yang, Role of surface metal nanoparticles on the absorption in solar cells, *Opt. Lett.* **37**, 641–643 (2012).

24. V. E. Ferry, L. A. Sweatlock, D. Pacifici, and H. A. Atwater. Plasmonic nanostructure design for efficient light coupling into solar cells, *Nano Lett.* **8**, 4391–4397 (2008).

25. P. Berini, Plasmon–polariton waves guided by thin lossy metal films of finite width: Bound modes of asymmetric structures, *Phys. Rev. B.* **63**, 125417 (2001).

26. V. Giannini, Y. Zhang, M. Forcales, and J. Gómez Rivas, Long-range surface plasmon polaritons in ultra-thin films of silicon, *Opt. Express* **16**, 19674–19685 (2008).

27. F. J. Haug, T. Söderström, O. Cubero, V. Terrazzoni-Daudrix, and C. Ballif, Plasmonic absorption in textured silver back reflectors of thin film solar cells, *J.Appl. Phys.* **104**, 064509 (2008).

28. P. Matheu, S. H. Lim, D. Derkacs, C. McPheeters, and E. T. Yu, Metal and dielectric nanoparticle scattering for improved optical absorption in photovoltaic devices, *Appl. Phys. Lett.* **93**, 113108 (2008).

29. D. Wan, H. Chen, T. Tseng, C. Fang, Y. Lai, and F. Yeh, Antireflective nanoparticle arrays enhance the efficiency of silicon solar cells, *Adv. Funct. Mater.* **20**, 3064–3075 (2010).

30. J. Grandidier, D. Callahan, J. Munday, and H. Atwater, Light absorption enhancement in thin-film solar cells using whispering gallery modes in dielectric nanospheres, *Adv. Mater.* **23**, 1272–1276 (2011).

31. J. Grandidier, R. Weitekamp, M. Deceglie, D. Callahan, C. Battaglia, C. Bukowsky, C. Ballif, R. Grubbs, and H. Atwater, Solar cell efficiency enhancement via light trapping in printable resonant dielectric nanosphere arrays, *Phys. Status Solidi A* 210, 255–260 (2013).

32. Y. A. Akimov, W. S. Koh, S. Y. Sian, and S. Ren, Nanoparticle-enhanced thin film solar cells: Metallic or dielectric nanoparticles, *Appl. Phys. Lett.* **96**, 073111 (2010).

33. C. F. Bohren and D. R. Huffman, *Absorption and Scattering of Light by Small Particles*, New York: Wiley (2008).

34. Y. A. Akimov and W. S. Koh, Design of plasmonic nanoparticles for efficient subwavelength light trapping in thin-film solar cells, *Plasmonics* **6**, 155–161 (2011).

35. H. Atwater and A. Polman, Plasmonics for improved photovoltaic devices, *Nat. Mater.* **9**, 205–213 (2010).

36. S. A. Maier, *Plasmonics: Fundamentals and Applications*, Heidelberg, Germany: Springer (2007).

37. P. Spinelli, M. Hebbink, R. de Waele, L. Black, F. Lenzmann, and A. Polman, Optical impedance m,atching using coupled plasmonic nanoparticle arrays, *Nano Lett.* **11**, 1760–1765 (2011).

38. Y. Zhang, Z. Ouyang, N. Stokes, B. Jia, Z. Shi, and M. Gu, Low cost and high performance Al nanoparticles for broadband light trapping in Si wafer solar cells, *Appl. Phys. Lett.* **100**, 151101 (2012).

39. X. Chen, B. Jia, Y. Zhang, and M. Gu, Exceeding the limit of plasmonic light trapping in textured screen-printed solar cells using Al nanoparticles and wrinkle-like graphene sheets, *Light Sci. Appl.* **2**, e92 (2013).

40. B. Cai, X. Li, Y. Zhang, and B. Jia, Significant light absorption enhancement in silicon thin film tandem solar cells with metallic nanoparticles, *Nanotechnology* **27**, 195401 (2016).

41. M. Kirkengena, J. Bergli, and Y. M. Galperin, Direct generation of charge carriers in c-Si solar cells due to embedded nanoparticles, *J. Appl. Phys.* **102**, 093713 (2007).

42. J. Yang, J. You, C. Chen, W. Hsu, H. Tan, X. Zhang, Z. Hong, and Y. Yang, Plasmonic polymer tandem solar cell, *ACS Nano* **5**, 6210–6217 (2011).

43. C. Hägglund and S. Apell, Plasmonic near-field absorbers for ultrathin solar cells, *J. Phys. Chem. Lett.* **3**, 1275–1285 (2012).

44. J. Zhu, M. Xue, H. Shen, Z. Wu, S. Kim, J. Ho, A. Hassani-Afshar, B. Zeng, and K. L. Wang, Plasmonic effects for light concentration in organic photovoltaic thin films induced by hexagonal periodic metallic nanospheres, *Appl. Phys. Lett.* **98**, 151110 (2011).

45. V. Gusak, B. Kasemo, and C. Hagglund, Thickness dependence of plasmonic charge carrier generation in ultrathin a-Si:H layers for solar cells, *ACS Nano* **5**, 6218–6225 (2011).

46. C. Hägglund and S. Apell, Plasmonic Near-Field Absorbers for Ultrathin Solar Cells, *J. Phys. Chem. Lett.* 3, 1275–1285 (2012).

47. B. Cai, B. Jia, Z. Shi, and M. Gu, Near-field light concentration of ultra-small metallic nanoparticles for absorption enhancement in a-Si solar cells, *Appl. Phys. Lett.* **102**, 093107 (2013).

48. J. Zhu, C. M. Hsu, Z. F. Yu, S. H. Fan, and Y. Cui, Optical absorption enhancement in amorphous silicon nanowire and nanocone arrays, *Nano Lett.* **10**, 1979–1984 (2010).

49. H. Huang, L. Lu, J. Wang, J. Yang, S. Leung, Y.Q. Wang, D. Chen, et al., Performance enhancement of thin-film amorphous silicon solar cells with low cost nanodent plasmonic substrates, *Energy Environ. Sci.* **6**, 2965–2971 (2013).

50. N. N. Lal, B. F. Soares, J. K. Sinha, F. Huang, S. Mahajan, P. N. Bartlett, N. C. Greenham, and J. J. Baumberg, Enhancing solar cells with localized plasmons in nanovoids, *Opt. Express* **19**, 11256–11263 (2011).

51. D. Derkacs, W. V. Chen, P. M. Matheu, S. H. Lim, P. K. L. Yu, and E. T. Yu, Nanoparticle-induced light scattering for improved performance of quantum-well solar cells, *Appl. Phys. Lett.* **93**, 091107 (2008).

52. B. Cai, B. Jia, J. Fang, G. Hou, X. Zhang, Y. Zhao, and M. Gu, Entire band absorption enhancement in double-side textured ultrathin solar cells by nanoparticle imprinting, *J. Appl. Phys.* **117**, 223102 (2015).

53. M. L. Brongersma, Y. Cui, and S. Fan, Light management for photovoltaics using high-index nanostructures, *Nat. Mater.* **13**, 451–460 (2014).

54. J. Wallentin, N. Anttu, D. Asoli, M. Huffman, I. Aberg, M. H. Magnusson, G. Siefer, et al., InP nanowire array solar cells achieving 13.8% efficiency by exceeding the ray optics limit, *Science* **339**, 1057–1060 (2013).

55. K. X. Wang, Z. Yu, V. Liu, Y. Cui, and S. Fan, Absorption enhancement in ultrathin crystalline silicon solar cells with antireflection and light-trapping nanocone gratings, *Nano Lett.* **12**, 1616–1619 (2012).

56. Y. Yao, J. Yao, V. K. narasimhan, Z. Ruan, C. Xie, S. Fan, and Y. Cui, Broadband light management using low-Q whispering gallery modes in spherical nanoshells, *Nat. Commun.* **3**, 664–671 (2012).

22 Photovoltaic structures based on porous silicon

Igor B. Olenych, Liubomyr S. Monastyrskii, and Olena I. Aksimentyeva

Contents

22.1 INTRODUCTION

Since the discovery of photoelectric transducers (PTs) based on monocrystalline silicon more than half a century ago, these devices have become key materials for the state-of-the-art photoelectric industry due to their high reliability and considerable energy conversion efficiency. However, the cost of reclaimed grade silicon wafers for solar energy conversion applications slows down the widespread use of silicon PT. In order to find a less expensive solution, much effort was put into the development of cheaper thin-film PT technology, which uses amorphous and polycrystalline silicon, cadmium and indium telluride, and gallium selenide sensitized dyes based on polymer materials. In addition, new designs of PT were suggested that exploit nontraditional approaches and allow a decrease in cost while increasing the efficiency of the solar cells (Alferov et al. 2004; Goetzberger et al. 2003; Goswami et al. 2004; Green 2003a).

Among the new strategies, of particular interest are the projects that utilize nanostructured materials such as nanocrystals (quantum dots), nanotubes, and nanorods (quantum wires) (Ahmed et al. 2013; Green 2003b; Peng and Lee 2001). For example, using quantum wires helps in solving the problem that arises in the case of the traditional PT fabrication technique and is related to the processes of solar radiation absorption in the bulk. Thick PT ensures complete absorption of incoming solar energy whereas in thinner PT, charge carriers are easier to collect. Long quantum wires create directional charge carriers flow from the bulk and by optimizing the base region one may achieve highly efficient conversion of solar radiation into electric energy.

A different route of enhancing the crucial characteristics and lowering the cost of the market-ready element is to use some of the porous semiconducting materials. Nanomaterials and heterostructures based on silicon, including porous silicon (PS), show great promise for PT and photodetector applications due to cost efficiency and optimal functionality (Bilyalov et al. 2001; Ünal et al. 2001: Wei et al. 1994; Yae et al. 2003). The material started to draw considerable attention after Canham discovered intensive red and orange photoluminescence of PS (Canham 1990), which is related to the quantum confinement effects in PS nanocrystals. PS exhibits specific structural features that ensure a large absorbing surface and a lower (as compared to bulk silicon) reflection coefficient, thus making it a good candidate for enhancing photodetector efficiency

(Bisi et al. 2000; Cullis et al. 1997; Föll et al. 2002). It is known that the texturized surface of a silicon PT that is formed by pyramid-shaped microstructures of several micrometers in height allows an increase in the energy conversion efficiency by several percent (Yamaguchi et al. 2006).

Moreover, any semiconductor photoelectric structure has losses related to the impossibility of absorbing photons with energies lower than the band gap of the material and the thermalization of photons with energies exceeding the band gap. In order to minimize these losses, a cascade strategy is employed in new generation PT: multilayered structures, where each layer has a different band gap, are used (Bailey and Flood 1998). Such elements are called multijunction cells or tandem cells (in case of two layers).

Since these elements work with a considerably larger portion of the solar spectrum they have much better conversion efficiency. Typically, multijunction solar cells are designed in a way that allows light to hit the element with the widest band gap first, so the higher-energy photons are absorbed first. Photons that were not absorbed in the first layer penetrate into the next element that has a lower band gap and so on. Materials used in cascade cells design are GaAs, GaInP$_2$, CuInSe$_2$, amorphous silicon, and related alloys (e.g., a-Si$_{1-x}$C$_x$:H, a-Si$_{1-x}$Ge$_x$:H).

On the other hand, changing the band energy structure of silicon when moving from bulk to nanocrystals due to quantum confinement allows to significantly widen the absorption spectra of PT and increase the dosage of absorbed solar energy. Laboratory experiments indicate that using multilayered photosensitive structures based on nanocrystalline materials achieve a notable improvement in sunlight-to-energy conversion efficiency (Peng and Lee 2001; Wu and Li 2015; Yamaguchi and Luque 1999).

The variability of PS band gap opens a possibility to optimize the solar energy absorption: a wide band gap layer of PS can serve as a window for multilayered elements (Osorio et al. 2011). Besides, PS is a good substrate for growing high-quality epitaxial films of GaAs and ZnO that are used in photosensitive elements (Hasegawa et al. 1989; Hsu et al. 2005; Lin et al. 1987). Photoluminescent properties of PS can be utilized to transform UV radiation into long-wavelength radiation, which, in turn, can be more efficiently converted into photoelectric energy (Bisi et al. 2000; Canham 1990; Cullis and Canham 1991).

Thus, the small size of PS nanocrystals, their large surface area, and increased band gap as compared to bulk silicon govern the prospective of using this material to enhance the efficiency of PTs and to design the new generation photodetectors (Kim et al. 2007; Martin-Palma et al. 2000; Menna et al. 1995; Selj et al. 2010; Strehlke et al. 2000). That said, it should be noted that the magnitude and sign of the photovoltaic signal observed in PS-based structures were found to depend not only on light intensity but also on the environmental factors (Vashpanov 1997). It is predicted that the change of photoinduced voltage in the process of physical adsorption of polar molecules is due to adsorption-induced electric effects. This may find application in sensor electronics for the development of gas sensing elements.

Another reason for the interest in investigating the photovoltaic properties of PS and PS-based structures is fundamental. Photoelectric phenomena are ideal processes for studying electronic processes in low-dimensional semiconductor systems: absorption and emission of electromagnetic waves, charge carriers transport, trapping and recombination, and establishing the energy structure of multilayered systems.

Usually, photovoltaic properties of experimental PS-based structures are studied in so-called "sandwich" configuration (Astrova et al. 1997; Balagurov et al. 1997; Smestad et al. 1992; Tsai et al. 1993; Yu and Wie 1993; Zheng et al. 1992). The thickness of the porous layer varies from one-tenth of a micron to several tens of microns. In some cases, PS layers are subjected to additional processing in order to passivate the surface, eliminate the oxide film, dry the material, etc. Front contact, which in different reports had areas ranging from one-hundredth of a square centimeter to several square centimeters, can be formed by depositing a metal (typically aluminum) layer on the PS surface at a sharp angle in order to avoid the penetration of metal into pores and, respectively, electrical shortening of silicon nanocrystals. Ohmic contact on the back side (nonworking) of the silicon substrate can be created by thermal vacuum deposition of an Al layer with further thermal annealing at 500–700°C before electrochemical etching. Generally, electrical properties of sandwich-like metal/PS/Si/metal structures are determined by a metal/PS Schottky barrier, PS layer, and PS/Si heterojunction. Respectively, photoinduced charge carriers in multilayered sandwich structures can be divided by different electrical barriers. Therefore, analyzing the properties of the structure it is necessary to technically realize the situations when the contribution of certain elements is prevailing.

Current–voltage characteristics (CVC) of PS-based structures with differently shaped front contacts made of different materials (Ca, Mg, Sb, Au, Ag) that formed Schottky barriers with porous layer show rectifier-like behavior. Current increases with increasing reverse bias voltage and light intensity. The spectral profile of photosensitivity is similar to that of the industrial silicon photodiodes. Since 500 nm visible light penetrates as deep as down to 10 μm into the PS layer, photons with $\lambda > 500$ nm are mainly absorbed in the silicon substrate. This explains the similarity in the spectral curve of photosensitivity between a PS-based structure and a silicon photodiode. In some cases, PS-based structures' photoresponse exhibits an additional band within 500–700 nm, which can be connected with energy states in PS nanocrystals as their electronic spectra are modified with respect to the bulk due to quantum confinement.

It is predicted that electronic levels responsible for radiative transitions are also revealed in photocurrent spectra. This is confirmed by the correlation between photoluminescence spectra and photoresponse spectra. High quantum efficiency can be due to a well-developed PS surface and passivation of the Si nanocrystals' surface that results in a low rate of surface recombination.

A photovoltaic signal due to the PS/Si heterojunction was observed in sandwich-like structures with thick (over few tens of microns) porous layer during the scanning of their cross section with a laser beam (Pulsford et al. 1993, 1994). Unlike photoluminescence, which was measurable everywhere along the cross section, photocurrent was only detected while scanning in the vicinity of the PS–Si heterojunction.

Photogeneration of charge carriers in different layers of a sandwich-like structure and the separation of nonequilibrium carriers by different electrical barriers can cause photoresponses of different polarity in a no-load regime (Laptev et al. 1997; Monastyrskii et al. 2001). In this case, the negative photovoltaic signal is related to the absorption of light by silicon substrate while the positive is related to the photoexcitation of trapping levels in the PS and on the PS surface.

Despite intensive research and new technical solutions, fundamental problems of PS-based photoelectric converters design and investigation remain not completely solved. Apart from the advantages of PS-based structures, discussed above, there exist certain difficulties in their application, connected with high electrical resistance, low mechanical strength, and thermal conductivity of the porous layer. At the current stage of development, effective charge transport realization is problematic. On top of that, some knowledge gaps exist when it comes to the complex carriers excitation processes and charge transport in silicon nanosystems, creating additional obstacles on the road to designing the optimal PT. One of the possible routes to overcome the difficulties is to create hybrid systems comprising nanosized PS and polymer components. In this respect, very little is known about the influence of organic components on the photovoltaic properties of hybrid nanosystems. Clarifying these aspects, however, would significantly help in solving the task of creating new functional nanomaterials for optoelectronics.

In the present work, we discuss the design of photosensitive structures based on PS and ways to improve their photoelectric conversion efficiency. In particular, the features of the electrochemical method of forming PS structures for photovoltaic applications are considered. Special attention is focused on studying the possibility of modifying PS layers by nanoclusters of metals and also "adsorption doping" of silicon nanosystems in order to improve the photovoltaic properties of PS nanostructures. In addition, the chapter shows that multilayer organic–inorganic nanosystems, such as hybrid conjugated polymer/PS structures, not only exhibit high sensitivity to electromagnetic radiation in a wide spectral range but can be also used to create selective and highly responsive photovoltaic gas sensors.

22.2 ELECTROCHEMICAL METHOD OF POROUS SILICON PREPARATION FOR PHOTOVOLTAIC APPLICATION

Formation of a porous layer in the surface of monocrystalline silicon by electrochemical etching in electrolytes based on hydrofluoric acid was first described in 1956 (Uhlir 1956). Compared to chemical etching and photostimulated chemical etching (Fathauer et al. 1992), which are also exploited in order to obtain PS, electrochemical anodizing allows more control and is a faster process, thus suitable for large-scale photoelectronic elements production technology.

In the process of electrochemical etching of silicon single crystals in HF solutions, the formation of narrow "etching channels" occurs. These channels are directed toward the bulk of the silicon plate.

Lasting anodizing not only the increases the length of channels but also causes pores to expand until dividing thin walls are partly etched. As a result, the remains of these walls end up on the surface of a silicon sample in the form of thin wires that are mainly perpendicular to the surface or directed along the crystallographic axes. The thickness of such wires usually ranges from a few nanometers to tens of nanometers.

The mechanism of the porous layer formation is closely connected with the peculiarities of multistage and concurrent chemical reactions that occur on the surface of the silicon plate at anodic polarization of the silicon electrode (Smith and Collins 1992; Zhang et al. 1989). The PS formation process is mainly determined by two factors: fluorine ions delivery to the reaction zone, and the presence of moving positive charge carriers at the near-surface layer of the silicon anode (Bisi et al. 2000; Lehmann and Gosele 1990; Lehmann et al. 1993). The first factor is connected with electrolyte and electrical parameters of the anodizing process, in particular, with anodic current density. Another factor of the electrochemical formation of the PS layer is related to the electrophysical properties of silicon substrate. Considering this, anodic processing of p-type and n-type silicon will be significantly different.

The scheme of setup used for obtaining PS in a galvanostatic regime is presented in Figure 22.1. A silicon plate (anode) and a platinum electrode (cathode) were put into a fluoropolymer cell filled with an electrolyte. Ethanol or water–ethanol solution was used with volume components ratio of $C_2H_5OH:HF$ = 1:1 or $C_2H_5OH:HF:H_2O$ = 1:1:1. Both n-Si and p-Si plates were used, doped with phosphorus and boron and having a specific resistivity of 4.5 and 10 Ohm·cm, respectively. Thick (400 μm) silicon plates with (100) and (111) crystallographic orientations having 100 mm diameter were used. Front and back surfaces of plates were polished to be reflective.

The necessary condition of the homogeneity of the PS layer was met by even current distribution across the silicon plate, ensured by the netlike shape of a platinum electrode and by thermal vacuum deposition of metallic (Au, Ag, or Al) ~0.5 μm film onto the back surface of the plate. Annealed for 20 min at 450 °C, metallic film also served as a contact for further measurements.

Anodic current density and etching time were 20–35 mA/cm² and 5–10 minutes, respectively. In order to have positive carriers in the near-surface layer of n-Si (which are needed for the anodic reaction to occur and for the formation of PS), the working surface of the plate was irradiated by a 500 W incandescent tungsten bulb during the entire process of electrochemical etching.

It has to be noted that in the case of anodic processing of p-type silicon, when equilibrium concentration of holes is large enough, illumination also plays a significant role in the formation of PS (this changes the PS formation rate, porosity level, etc.) and, importantly for applications, has a noticeable effect on the photoluminescent properties of PS (Suemune et al. 1992).

The wide spectrum of physical and chemical properties of PS is related to the peculiarities of its morphology. The microstructure of PS layers, in turn, is dependent not only on technological conditions (anodic current density, etching duration, type and temperature of the electrolyte used, irradiation parameters, etc.) but also on conductivity type, doping level and crystallographic orientation of initial monocrystalline silicon (Bisi et al. 2000).

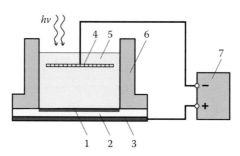

Figure 22.1 Scheme of the setup for obtaining PS layers: 1, PS layer; 2, silicon plate; 3, metallic contact; 4, platinum electrode; 5, electrolyte; 6, fluoropolymer cell; 7, stabilized current source.

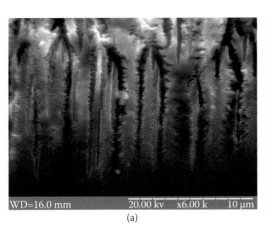
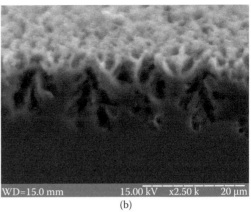

| WD=16.0 mm | 20.00 kv x6.00 k 10 μm | WD=15.0 mm | 15.00 kV x2.50 k 20 μm |
(a)   (b)

Figure 22.2 SEM images of the cross section of PS obtained on silicon substrates with n-type conductivity and different crystallographic orientations: (a) (100); (b) (111).

Conditions of electrochemical etching define not only the morphology but also the porosity of the porous layer: crucial structural characteristic of PS, which is determined as the ratio of voids volume to the total volume of the porous layer:

$$p = \frac{V - V_{Si}}{V} = \frac{V_{por}}{V}, \tag{22.1}$$

where V_{Si}, V_{por} and V represent the volumes of silicon walls between pores, volume of voids, and the total volume of the porous layer, respectively (Bisi et al. 2000, Galiy et al. 1998). With the increase of porosity, pores coagulate and etched walls form a disordered network of quantum wires and dispersed nanocrystals.

Investigations of the morphology of PS layers obtained on differently oriented n-Si show that long and narrow pores perpendicular to the surface and directed toward the bulk are formed on (100) single crystals (Figure 22.2). In the case of PS formation on single crystals with (111) crystallographic orientation, pores with sponge-like or coral-like structure are created.

Depending on pore diameter d_{por}, PS layers are classified as follows: nanoporous ($d_{por} < 5$ nm), mesoporous (5 nm $< d_{por} < 50$ nm) and macroporous ($d_{por} > 50$ nm). Nanoporous silicon is usually formed on silicon substrates of p-type conductivity at short durations of electrochemical etching (from a few seconds to several minutes) and low anodic current density. Mesoporous silicon is formed on silicon substrates of both electronic and hole conductivity type at the following anodizing conditions: process time from 2 to 20 minutes and current density 10 to 30 mA/cm². Macroporous silicon is usually formed on lightly doped n-Si at high anodic currents and longer anodizing times. Pore size can be increased by additional etching of PS in hydrofluoric acid (HF)-based electrolytes.

Thus, by adjusting technological regimes of electrochemical etching and by selecting appropriate initial silicon single crystals, structures can be obtained with the desired morphology and predicted functional properties, in particular, for the development of photodetectors that are highly sensitive to electromagnetic radiation in a wide spectral range.

22.3 RESEARCH TECHNIQUES OF PHOTOVOLTAIC PROPERTIES OF POROUS SILICON STRUCTURES

In order to study electrical and photoelectrical properties of PS-based sandwich structures, the samples with 0.5–1 cm² surface area were used and various types of electrical contacts attached to the porous layer were exploited (see Figure 22.3). Metallic contact that is typically used for connecting photosensitive structures and is 3–6 mm in diameter was formed by thermal deposition of semitransparent silver film. The angle between the flux of evaporated Ag atoms and normal to the PS surface was about 60°, which ensured

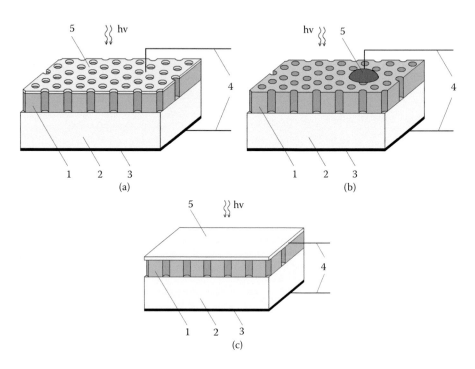

Figure 22.3 Schematic representation of PS-based photodetectors with semitransparent Ag contact (a), colloidal carbon contact (b) and optically transparent SnO₂ electrode (c): 1, PS layer; 2, silicon plate; 3, metallic (Au, Ag, or Al) film; 4, output photodetector; 5, contact with the PS.

electrical contact on the top of the porous layer preventing the penetration of Ag inside the pores. Another type of electrical contact that is about 3 mm in diameter was created using colloidal carbon (conductive paint could be used as well). Besides these, an optically transparent electrode (glass plate coated with SnO₂) can be used as a spring contact.

CVC were measured at room temperature with current flowing through the structure in the direction perpendicular to the surface. Photoelectric phenomena were investigated, by irradiating the structures from the side of the porous layer with 2 mW He–Ne laser (λ=630 nm) or 1 W white light-emitting diode (LED (FYLP–1W–UWB–A) that provided a light flux of 76 lumens. Spectral dependencies of the photoinduced signal were measured using standard optical equipment, including a grating monochromator and an incandescent bulb (2800 K). Photoresponse spectra were normalized with respect to a 2800 K black-body radiation curve (Planck curve) and were corrected taking into account the spectral sensitivity of the instrument. The spectral dependence of photoresponse on an industrial silicon photodiode was also measured for comparison.

To study the energy characteristics of the structures, a white LED was exploited as its emission intensity is directly proportional to the current. In the case of temperature dependency studies, experimental samples were put in the cryostat kept at 10⁻³ mmHg vacuum. Measurements were performed at linear heating of the samples from 80 K to 325 K with the rate of 0.1 K/sec. Kinetics of the photoresponse of the PS-based sandwich structures to rectangular light pulses was examined using a Hantek 1008B oscilloscope.

22.4 SANDWICH-LIKE STRUCTURES OF POROUS SILICON AND METAL NANOPARTICLES

Due to a well-developed pore system, PS is an ideal host for the deposit and infiltration of nanoobjects of different origin, thus allowing the creation of nanocomposites with predicted functional properties. In particular, it was established that the introduction of metallic impurities (gold, nickel, palladium, copper, and others) increases the efficiency of photoelectric, optical, and luminescent and catalytic processes in PS-based structures (Amran et al. 2013; Olenych 2014; Presting et al. 2004; Venger et al. 2004). Considering that photoelectric parameters of metal/PS/Si/metal sandwich-like structures are basically determined by the Schottky barrier, doping of porous layers by metals with a larger work function might enhance the efficiency of photovoltaic structures obtained on n-Si.

Blocking contact (Schottky barrier) between metal and degenerate semiconductor with electronic conductivity is formed when metal work function W_M is larger than semiconductor work function W_S (Figure 22.4). At the initial moment of time, the flow of electrons from the semiconductor exceeds the flow of electrons from metal. As a result, positive volume charge is accumulated in the near-surface area of the semiconductor and negative volume charge is accumulated in the near-surface area of the metal. This leads to the appearance of an electric field in the contact area and to the bending of energy bands, mostly in the near-contact region of the semiconductor. Contact potential φ_κ is given by the difference of both work functions:

$$\varphi_K = \frac{W_M - W_S}{e},\tag{22.2}$$

where e denotes the charge of an electron.

Bending of energy bands, in turn, leads to the decrease in concentration of primary carriers (electrons) and an increase in the concentration of secondary carriers (holes) in the near-surface layer of semiconductor. Under the influence of radiation, photogenerated carriers are separated by the electric field of the Schottky barrier, the penetration depth of which L_D depends on the concentration of doping impurity and contact potential φ_κ. Hence, by changing the height of the potential barrier, the photoelectric parameters of metal/PS/Si/metal sandwich-like structures can be controlled.

Thus, in order to study the influence of the metal modification of the porous layer on the photoelectric properties of Ag/PS/n-Si/Au structures, metals with different work functions were used: Ag (4.26–4.74 eV), Co (5.0 eV), Ni (5.05–5.35 eV), and Pd (5.22–5.6 eV).

The simplest way of doping PS layers by metals is by processing in solutions that contain ions of dopant metal. When the current flows through the electrolyte, metal ions transfer to cathode occurs and ions are deposited not only on surface but also in pores—this is the advantage compared to vacuum techniques typically used for depositing metallic nanoparticles (Karbovnyk et al. 2015). When deposited on the porous layer surface, metal ions that have larger positive electrochemical potential as compared to silicon are neutralized by capturing electrons from surface silicon atoms, and become the inception centers for the further growth of metal nanocrystals (Coulthard et al. 2000; Primachenko et al. 2005). Moreover, oxidation/reduction processes may be favorable for the formation of oxide film on the PS surface.

Doping of the PS layer with metals was done by an electrochemical technique with a constant current of 2–20 mA flow ing through solutions of salts of these metals in water, or in organic solvent (silver nitrate ($AgNO_3$), cobalt acetate ($Co(CH_3COO)_2$), nickel sulfate ($NiSO_4$) and palladium acetate ($Pd(CH_3COO)_2$). Introduction of metals into the porous layer was controlled by scanning electron microscopy (SEM) in the elastic electron scattering mode, and by energy dispersive X-ray microanalysis (EDXMA) (see Figure 22.5).

In the elastic electron scattering mode, brighter SEM image parts are identified as formed metal clusters. SEM image analysis of the cross section of experimental structures shows that a significant amount of

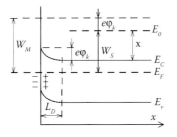

Figure 22.4 Energy band diagram of the blocking Schottky contact (W_M and W_S, metal and semiconductor work function, respectively; φ_κ, contact potential; χ, electron affinity; E_F, Fermi level energy; L_D, Debye screening length).

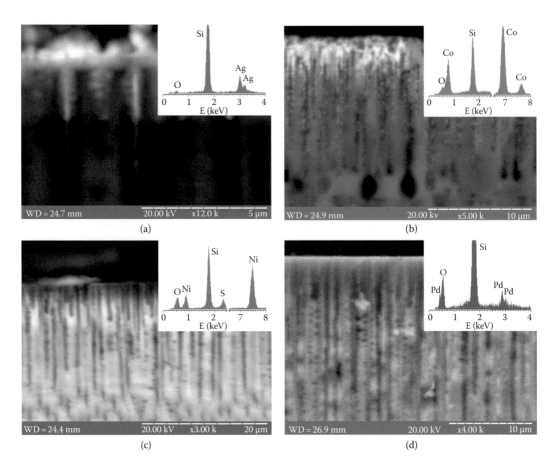

Figure 22.5 SEM images and EDXMA pattern of the cross sections of PS: modified with silver (a), cobalt (b), nickel (c), and palladium (d).

electrodeposited silver and cobalt resides near the surface of the PS. Pore surface in the upper part of the layer is well decorated by the particles of these metals. Deeper inside the porous layer, cobalt was located on the pore wall in the form of nanoparticles with sizes up to tens of nanometers. Since SEM images of the cross section of PS structures modified by nickel do not reveal any submicron particles, it can be assumed that nickel and palladium deposition in PS pores occur in the form of atoms or small nanoclusters.

Determination of the chemical composition of metal-modified PS layers was done based on the interpretation of energy spectra in the X-ray microanalysis mode. Along with the intensive maximum at 1.74 keV, which is characteristic for silicon, peaks corresponding to metal atoms were observed. In the case of modification by silver, peaks at 2.98 and 3.15 keV were detected and in the case of cobalt nanoparticles, doping maxima were observed at 0.78, 6.93, and 7.65 keV. Energies of 0.85 and 7.48 keV correspond to nickel atoms while energies of 0.45, 2.83, and 3.04 keV correspond to palladium atoms. The 0.52 keV peak that was observed in X-ray microanalysis of nickel- and palladium-modified PS structures corresponds to oxygen atoms and can be due to oxidation of the porous layer surface. In addition to nickel, sulfur was identified in the porous layer as confirmed by the peak at 2.31 keV. This proves the presence of residual electrolyte (used for deposition) in pores. It has to be noted that intensities of bands related to incorporated metal atoms decreased when increasing the distance between the PS surface and the analyzed region, which can serve as an additional argument in favor of prevailing deposition of metals near the surface.

Figure 22.6 shows dark-current CVC of Au/n-Si/PS/Ag sandwich-like structures that exhibit nonlinear behavior, possibly due to electrical barriers at metal/semiconductor contacts. Domination of one of the Schottky barriers determines the rectifier character of the CVC. Under the influence of surface radiation

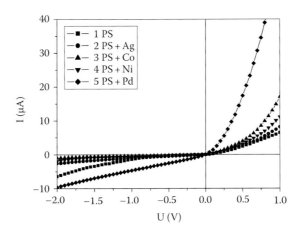

Figure 22.6 Dark-current CVC of the initial Au/n-Si/PS/Ag structure (1) and structures modified by silver (2), cobalt (3), nickel (4), and palladium (5).

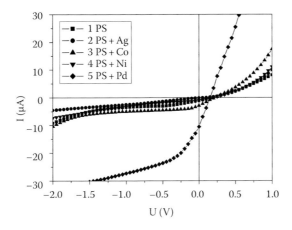

Figure 22.7 CVC of initial Au/n-Si/PS/Ag structure (1) and structures modified by silver (2), cobalt (3), nickel (4), and palladium (5) under the influence of irradiation by FYLP–1W–UWB–A LED.

with white light (76 lumen flux) photovoltage generation was observed and the reverse current was increased, confirming the photogeneration of free carriers in the structures under study (Figure 22.7).

It is worth mentioning that rectification coefficient and photovoltage/photocurrent values were larger for Au/n-Si/PS/Ag structures that had metals incorporated into the porous layer. This can be due to passivation of PS nanocrystal surfaces and due to modification of electronic parameters of these nano-crystals because of adsorption-induced electric effects (Chiesa et al. 2003; Olenych et al. 2013). Detected increase of electric conductivity of structures based on metal-doped PS can be connected with a larger area of metal–semiconductor contact and with the formation of additional paths for current flowing through the porous layer. Besides, electrodeposited dendrite-like metallic formations ensure efficiency in collecting photogenerated carriers and extracting them from the bulk of the PS layer.

Spectral dependencies of the photoresponse of Au/n-Si/PS/Ag barrier structures are presented in Figure 22.8. Photovoltage spectra in the no-load regime were similar to the spectrum of the photoresponse of silicon diode and PS–silicon heterojunctions (Olenych 2014; Olenych et al. 2013) and were characterized by a wide photosensitivity band within the range of 930–970 nm. However, the dependence of the spectral position of the photosensitivity maximum for Au/n-Si/PS/Ag structures on the type incorporated was observed. This may be caused by the shunting of silicon nanocrystals by deposited metals and by different levels of passivation of the porous layer.

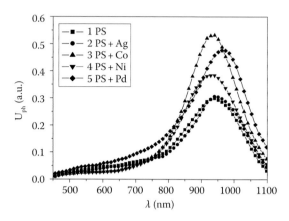

Figure 22.8 Spectral dependence of photovoltage of the initial Au/n-Si/PS/Ag structure (1) and structures modified by silver (2), cobalt (3), nickel (3), and palladium (4).

Since the passivation of silicon nanocrystals surface significantly influences relaxation processes in PS (Olenych et al. 2015a), the change of porous layer passivation by different metal clusters being reflected in temperature dependencies of the photoresponse of Au/n-Si/PS/Ag structures should be expected. In the 80–300 K range photovoltage temperature dependencies exhibit non-monotonic behavior as illustrated in Figure 22.9. For the initial Au/n-Si/PS/Ag structure and the structure modified by nickel, the photovoltaic signal maximum was observed at nearly 190 K. Sandwich structures with silver and cobalt incorporated into PS show two peaks of photoresponse at approximately 100 and 220 K. In the case of palladium-doped PS, the maximum photovoltage was detected at about 235 K. Moreover, the photovoltaic structure with the porous layer modified by palladium was characterized by photovoltage sign inversion in the 80–125 K temperature range. Such an effect was also observed for PS-based structures doped with gold (Primachenko et al. 2005; Venger et al. 2004).

The observed character of temperature dependencies of the photosignal may be explained by a number of reasons, in particular, by Fermi level temperature shift, existence of nonequilibrium carriers trapping levels both on the silicon nanocrystal surface and on the porous layer–surface boundary, etc. (Venger et al. 2002; Venger et al. 2004). For instance, trapping levels of a different nature and having different activation energies were revealed in the studies of thermally stimulated conductivity and depolarization of PS (Grigor'ev et al. 2006; Olenych et al. 2015a). Photosignal magnitude is dependent on how long carriers are being trapped

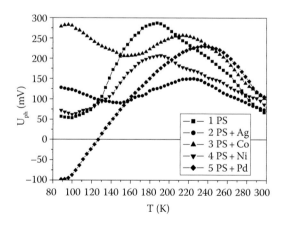

Figure 22.9 Temperature dependencies of the photovoltage for initial Au/n-Si/PS/Ag structure (1) and structures modified by silver (2), cobalt (3), nickel (4), and palladium (5) under the influence of PS surface irradiation with FYLP–1W–UWB–A LED.

Industrial nanosilicon

at these levels, and this trapping time increases as the temperature goes down. The inversion of the photo-voltage sign for palladium-modified Au/n-Si/PS/Ag structures in different temperature regions can be due to different trap depths for nonequilibrium electrons and holes.

In order to find additional information about photoelectric processes in Au/n-Si/PS/Ag structures, their energy characteristics were investigated. The character of photovoltage dependencies on the intensity of irradiation was similar to that of the silicon photodiode; however, some deviations from nonlinearity for photocurrent were observed (Figure 22.10). Sublinear dependence of photocurrent on the intensity of irradiation can also be related to the carrier trapping. Dependence of the extreme value of photovoltage for a given light intensity on the doping metal type can be due to contact potential and the passivation state of the PS nanocrystal surface.

In order to study temporal parameters of the photoresponse, pulsed infrared radiation with a wavelength of $\lambda = 940$ nm was exploited. This wavelength corresponds to the photosensitivity peak of Au/n-Si/PS/Ag structures. Results of the study of kinetics of the photoresponse to rectangular 0.5 ms 1 kHz light pulses are shown in Figure 22.11.

Analyzing the obtained dependencies, it can be concluded that metal-modified PS sandwich structures have slightly smaller photovoltaic rise times as compared to the initial sample. Reducing the photoresponse time due to doping with metals is probably caused by the change of passivation of PS nanocrystals. At the same time, better passivation decreases the coefficient of surface recombination, consequently increasing the efficiency of PS-based photodetectors.

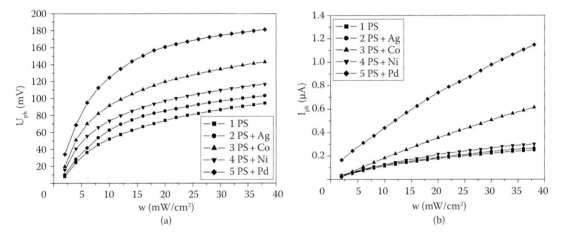

Figure 22.10 Photovoltage dependencies (a) and photocurrent dependencies (b) for initial Au/n-Si/PS/Ag structure (1) and structures modified by silver (2), cobalt (3), nickel (4), and palladium (5).

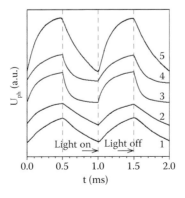

Figure 22.11 Kinetics of the photoresponse to rectangular 940 nm light pulses for initial Au/n-Si/PS/Ag structure (1) and structures modified by silver (2), cobalt (3), nickel (4), and palladium (5).

22.5 PHOTOVOLTAIC PROPERTIES OF POROUS SILICON/SILICON STRUCTURES MODIFIED BY MOLECULE ADSORPTION

Wide use of semiconductors in photoelectronics is based on the possibility to change charge carrier (holes or electrons) concentration by doping, that means by introducing impurity atoms. Doping of nanocrystals opens the way to creating new types of semiconductor nanostructures for a wide range of different applications (Bryan and Gamelin 2005; Norris et al. 2008). However, modifying properties of submicron-sized systems is often complicated. Semiconductor nanocrystals resist the introduction of doping impurities: on the one hand self-purification processes take place, on the other hand, even when doping atoms are introduced they considerably distort the nearby crystalline structure, leading to the degradation of the electrophysical parameters of nanostructures (Erwin et al. 2005; Mocatta et al. 2011). Problems with nanodoping are governed by fundamental differences in mechanisms that regulate impurity introduction in bulk materials and in nanoparticles. Unlike in macroscopic solids, where thermodynamics is the key factor, in nanostructures everything is determined by kinetics and, first of all, by surface kinetics (Erwin et al. 2005).

Taking into account dependencies of electrophysical properties of PS nanostructures on the environmental conditions and on the adsorption of polar molecules (Chiesa et al. 2003; Olenych et al. 2011; Osminkina et al. 2005; Ozdemir and Gole 2007; Vorontsov et al. 2007), in view of potential applications it is of interest to study the possibilities of controlling the conductivity type of PS nanocrystals by adsorption of molecules with pronounced acceptor of donor properties (e.g., bromine, iodine, or ammonia, respectively).

Atoms of the adsorbed substance, when delivered onto the surface of a semiconductor, tend to capture or lose an electron, depending on their physical and chemical properties. In particular, bromine and iodine atoms, due to their high electronegativity, localize a semiconductor electron near their position. On the contrary, ammonia and water, when on the semiconductor surface, lose the electron that goes into a conductivity band or to the surface center that is likely to form a chemical bond with adsorbed molecules. As a result, adsorbed molecules create acceptor E_{as} or donor E_{ds} local energy levels in the band gap on the semiconductor surface (Figure 22.12).

Free-charge carriers from a semiconductor can be trapped on a newly formed center consequently leading to the change in charge density and bending of energy bands of the semiconductor in the near-surface area (Sze 1969). As a result, the concentration of equilibrium carrier in the near-surface layer is determined not by bulk doping, but by the value and the sign of surface charge. Adsorption-induced electric effects have a significant effect on the physical processes in semiconductors, forming depletion regions or regions enriched with majority charge carriers leading to the change in electrical conductivity of the material. An inversion layer is formed when the bending of an energy band is so strong that the chemical potential level crosses

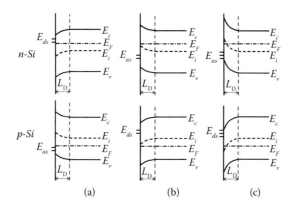

(a)   (b)   (c)

Figure 22.12 Energy band diagram of the n-type and p-type semiconductor surface in case of adsorption of acceptor and donor molecules: (a) the formation of an enriched layer; (b) the formation of a depleted layer; (c) the formation of conductivity type inversion layer (E_{as} and E_{ds}, acceptor and donor surface energy levels, respectively; E_F, Fermi level energy; E_i, mid-band gap level; L_D, Debye screening length).

the mid-band gap. Conductivity type of the inversion layer is governed by minority carriers and differs from that of the bulk.

Field-effect influence due to surface adsorption of polar molecules is especially significant, when the dimensions of the semiconductor crystal are comparable with Debye screening length L_D which is the case for mesoporous silicon nanocrystals. In this case, the modification of electrical parameters is expected across the entire volume of the porous layer, as illustrated in Figure 22.13.

Thus, electric conductivity of the porous layer obtained on n-Si has to increase due to the adsorption of NH_3 molecules with donor properties, and decrease due to the adsorption of acceptor-type I_2 or Br_2. Increased concentration of adsorbed halogen molecules can cause the inversion of the conductivity type of mesoporous silicon nanocrystals to p-type and the formation of electrical barriers on the PS/n-Si boundary that are similar to those in silicon diodes (see Figure 22.13). Carrier concentration due to adsorption doping of PS/p-Si structures should change in a similar fashion. However, in this case the rise of the PS electric conductivity is caused by adsorption of iodine and bromine molecules and the drop of conductivity of the porous layer (or even the inversion from p-type to n-type) is due to adsorbed ammonia molecules.

Eventually, sandwich structures that contain inversion PS layers resulting from adsorption doping can be interpreted as a number of p–n junctions connected in parallel. Influence of adsorption of polar molecules on the electronic structure of silicon nanocrystals is confirmed both by the change of conductivity behavior of PS/Si structures due to adsorption doping, and by the formation of photosensitive electrical barriers (Olenych et al. 2012b; Olenych et al. 2013; Olenych et al. 2014).

Doping of as-obtained mesoporous silicon grown on silicon substrates of electronic and hole conductivity type can be performed either from the liquid phase by adsorption of I_2 molecules from 1.5% and 10% ethanol solution of iodine, or from the gas phase by adsorption of Br_2 or NH_3 molecules in a specifically designed cell with a glass window and changeable gas environment. Investigations of electrical and photoelectrical properties of sandwich structures based on adsorption-doped PS were performed in an ammonia or bromine environment and also 1 hour after a short-time (1 to 3 sec) immersion of samples in the iodine solution.

Our studies reveal that the adsorption of polar molecules significantly affects the electrical conductivity of the PS/silicon substrate for the structures in both cases of direct and alternating currents. The character of conductivity changes occurring in the PS/p-Si and PS/n-Si structures turn out to be different.

It was found out that increasing NH_3 concentration in the experimental cell causes the increase of differential conductivity of the PS/n-Si structure and a decrease in conductivity of the PS/p-Si structure in AC mode (frequency and magnitude of the test signal were 1 MHz and 25 mV, respectively). On the contrary, increasing the Br_2 concentration leads to an increase in the differential conductivity and then reaches saturation in the case of the PS/p-Si structure (Figure 22.14). The same parameter for the PS/n-Si structure decreases according to the exponential law.

In the DC regime, initial PS/silicon substrate sandwich structures exhibit symmetrical though nonlinear CVC, confirming the presence of several potential barriers (Figures 22.15 and 22.16). As a result of adsorption doping, along with the conductivity change a more complex change of CVC of the PS/p-Si structure can be observed under the influence of ammonia adsorption, and the PS/n-Si structure under the influence of bromine adsorption. In these cases, a "rectifier effect" can be seen inherent to p–n junctions.

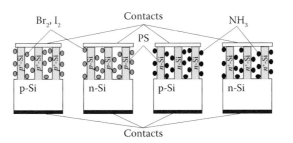

Figure 22.13 Schemes of experimental structures based on PS at the adsorption of acceptor (Br_2, I_2) and donor (NH_3) molecules.

Industrial nanosilicon

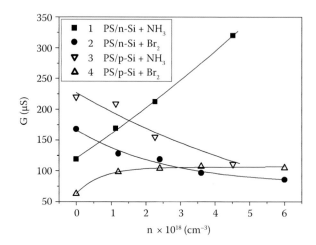

Figure 22.14 Dependencies of differential AC conductivity of the PS/n-Si (1,2) and PS/p-Si (3,4) structures on the concentration of NH_3 (curves 1 and 3) and Br_2 (curves 2 and 4).

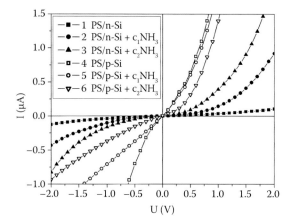

Figure 22.15 Dark CVC of the PS/n-Si (1,2,3) and PS/p-Si (4,5,6) structures at different concentrations of molecular ammonia: 0 (1,4), $c_1 = 1.13 \times 10^{18}$ cm^{-3} (2,5), $c_2 = 2.26 \times 10^{18}$ cm^{-3} (3,6).

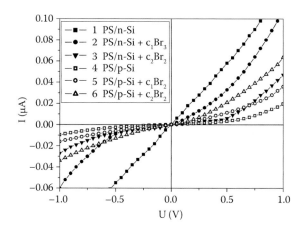

Figure 22.16 Dark CVC of the PS/n-Si (1,2,3) and PS/p-Si (4,5,6) structures at different concentrations of molecular bromine: 0 (1,4), $c_1 = 2.4 \times 10^{18}$ cm^{-3} (2,5), $c_2 = 3.6 \times 10^{18}$ cm^{-3} (3,6).

For an ammonia-doped PS/p-Si structure, the direct branch of the CVC corresponds to the negative potential at the porous layer, and to positive potential for a bromine-doped PS/n-Si structure. This indicates a predominance of one of the potential barriers.

It has to be noted that electrical parameters of PS/n-Si are changed in a similar fashion at adsorption doping of the porous layer from liquid phase. Adsorption of iodine molecules induces the inversion of conductivity type if PS is obtained on weakly doped n-Si. In addition, increasing the concentration of I_2 molecules in the solution used for doping (that means, increasing the number of adsorbed molecules) causes the rise of the rectifying coefficient according to the CVC of iodine-modified PS/n-Si structures (Figure 22.17).

Under the influence of He–Ne laser radiation with a wavelength of $\lambda = 0.63$ μm and an intensity of 60 mW/cm², the CVC of iodine-modified PS/n-Si structures were changed similarly to those of photodiode structures (see the inset in Figure 22.17). The increase of the amount of the adsorbed molecules caused the increase in photovoltage and photocurrent in the short-circuit mode. In the open circuit regime, the photogenerated electron–hole pairs are separated at the potential barrier. During this process the holes are accumulated in PS nanocrystals, creating a positive potential on the porous layer with respect to the silicon substrate.

In order to get additional information about the nature of the influence of acceptor and donor molecules adsorption on the photoelectrical properties of the structures based on mesoporous silicon, we investigated the spectral dependencies of sandwich structure photoresponse. The photovoltage spectra of modified samples in an open circuit regime were similar to the photoresponse spectrum of an industrial silicon photodiode and were characterized by the wide maximum in the 750–950 nm range (Figure 22.18). Apart from that, broadening of the photosensitivity spectral region toward higher energies for the structures based on adsorption-doped PS was observed.

A similarity in the spectral photoresponse dependence for discussed structures to that obtained for the silicon photodiode suggests that the photovoltage is associated with light absorption in the silicon substrate and separation of photocarriers at the silicon/PS boundary. This can serve as an additional argument in favor of the qualitative model, assuming that adsorption of chemically active or polar molecules imposes formation of the p–n junction due to the inversion of the conductivity type in PS nanocrystals.

It should be noted that desorption of molecules is accompanied by the decrease of the photovoltage. Hence, the defining condition for the photovoltaic signal to appear is still the adsorption doping of PS nanostructures.

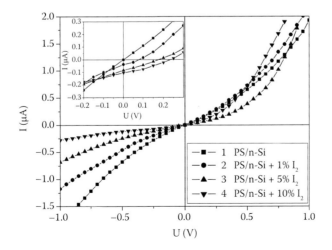

Figure 22.17 Dark-current CVC of initial PS/n-Si structure (1) and structures modified with iodine (2, 3, and 4). 2: I_2 adsorbed from 1% ethanol solution of iodine; 3: 5%; 4: 10%. Inset shows CVC of the same structures under the influence of He–Ne laser radiation with the intensity of 60 mW/cm².

Industrial nanosilicon

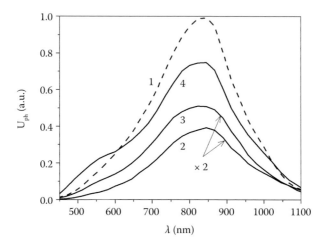

Figure 22.18 Spectral dependence of the photovoltage of a silicon photodiode (1), ammonia-doped PS/p-Si structure (2), bromine-doped (3), and iodine-doped (4) PS/n-Si structure.

22.6 PHOTODETECTORS BASED ON HYBRID STRUCTURES OF POROUS SILICON AND CONDUCTING POLYMERS

The possibility of using PS as a base material for growing hybrid structures broadens the spectrum of its applications, in particular in photodetectors. A specific niche is occupied by organic–inorganic semiconductor conjugated polymer/PS nanosystems (Nahor et al. 2011; Stakhira et al. 2005). An interaction between silicon polymer and nanoparticles in such hybrid structures may significantly alter their properties with respect to the properties of individual components—enhancing some of them, or creating new features (Aksimentyeva et al. 2007b; Aksimentyeva et al. 2015; Nguyen et al. 2000; Olenych et al. 2012a). Free films based on blends of colloidal dispersions of semiconductor nanocrystals and conjugated polymers have shown promise as composite materials for sensing, optoelectronics and photoelectric applications (Davidenko et al. 2011; Günes and Sariciftci 2008; Holder et al. 2008; Olenych et al. 2015b).

The molecular structure of polymers as well as their conformation and orientation can have a significant influence on the macroscopic properties of materials that are based on these polymers (Aksimentyeva et al. 2007a). A typical feature that makes electrically conductive polymers stand out among other high molecular compounds is the conjugated π-electron bonding system. Overlapping of electronic levels causes the formation of a single delocalized system, which serves as a source of charge carriers and paramagnetic centers in the polymer. The conducting polymers reveal electrical conductivity in the doped states, though remain insulators, when undoped (a neutral state). Due to doping–undoping processes, the electronic properties (e.g., the band gap) of the conjugated polymers can be varied substantially (Heeger 2002; Leclere et al. 2006; Pyshkina et al. 2010).

In recent years, extensive research has been devoted to conjugated polymer application in photoelectronics. As of today, there exist all-polymer photodiodes and solar cells with a decent efficiency (Admassie et al. 2006; Halls et al. 2002). It has to be noted that conjugated polymers have a capability to be doped not only by donor or acceptor elements but also by chemical and electrochemical methods (e.g. using protonic acids). A polymeric anion poly(styrenesulfonate) (PSS) often acts simultaneously as an acid dopant and an anionic surfactant, which stabilizes dispersion of the polymer (Ouyanga et al. 2004). A particularly interesting example of a conducting polymer is poly(3,4-ethylenedioxythiophene) usually abbreviated as PEDOT. It exhibits remarkable optical and electrical properties (Pyshkina et al. 2010). The use of PEDOT as an electron-blocking layer and as a collector of holes in photovoltaic structures was reported, but usually PEDOT doped by PSS (PEDOT:PSS) is used as a hole injecting layer. Promising power conversion efficiency has been reported for hybrid solar cells based on silicon nanowires and PEDOT:PSS (Wang et al. 2016; Yu et al. 2013). Due to PS structural peculiarities and the functional properties of a conjugated polymer, multilayer photodetectors based on hybrid

PEDOT/PS/Si structures are more efficient absorbers of electromagnetic radiation in a wide spectral range and demonstrate great potential for photoelectric applications.

Thin-film PEDOT coatings on PS surfaces were produced by electrochemical polymerization of 0.1 M solution of 3,4-ethylendioxythiophene in water–ethanol solvent (1:1), with 0.5 M H_2SO_4 used as an electrolyte. The PEDOT film was formed at a current density of 0.1 mA/cm^2 for 10 min. During the electrochemical polymerization, the monomer penetrates into the pores of the silicon and a conducting polymer is synthesized right on the surface of the electrode and inside the pores (Aksimentyeva et al. 2007b; Misra et al. 2001).

Measurements indicate that PEDOT/PS/n-Si hybrid structures have a rectifier-like CVC, with forward biasing corresponding to positive voltage on the polymer film (Figure 22.19). The rectifier-like CVC can be explained by contact phenomena, electric barriers in the porous layer and at the interfaces of PEDOT/PS, and the PS/silicon substrate. Since PS/n-Si heterostructures usually have varistor-like CVC, rectifier-like CVC of hybrid structures may be due to electric barrier at the PEDOT–PS boundary, or the modification of the porous layer by the polymer. Upon the illumination of the PS surface by He–Ne laser with an intensity of 60 mW/cm^2, the CVC of PEDOT/PS/n-Si hybrid structure changes similarly to a photodiode structure. In the photovoltaic mode, generated electron–hole pairs are divided by a potential barrier, with electrons being accumulated in the silicon substrate forming negative potential with respect to the polymer film.

Since PEDOT can be a catalyst in the oxidation–reduction reaction (Kumar et al. 2006), and efficiently localizes the free electrons due to electrostatic interactions (Ouyanga et al. 2004), we assume that inversion of n-type conductivity of silicon nanocrystals takes place with the appearance of a potential barrier at the border of the PS/silicon substrate, similar to adsorption doping of PS by chemically active or polar molecules.

When studying energy characteristics of PEDOT/PS/n-Si photovoltaic structures, we observed the decrease in the influence of light flux intensity on the magnitude of photovoltaic signal at increasing the excitation level and sublinear photocurrent dependence, which might be due to carriers trapping at traps (see the inset in Figure 22.19). Obtained results indicate that the structure of investigated systems is more complex than that of the diode.

Spectral and temperature dependencies of photovoltage of PEDOT/PS/n-Si sandwich structures are shown in Figure 22.20. The spectral sensitivity of a hybrid structure is comparable to the spectrum of the photoresponse of adsorption-doped PS-based heterostructures. Along with the photosensitivity maximum at 850 nm, the obtained photovoltaic structure demonstrated a slight increase of sensitivity in the 700–750 nm range, which may be because of light absorption by the PS layer.

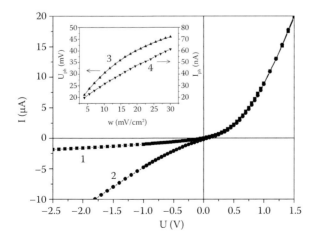

Figure 22.19 CVC of the PEDOT/PS/n-Si structure obtained in the dark (curve 1) and under irradiation by the He–Ne laser ($\lambda = 0.63$ μm) with the intensity of 60 mW/cm^2 (curve 2). Inset: dependences of photovoltage (3) and photocurrent (4) of PEDOT/PS/n-Si structure on the light intensity.

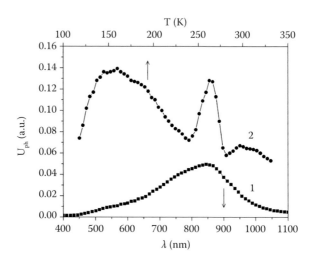

Figure 22.20 Spectral (1) and temperature (2) dependencies of photovoltage for hybrid PEDOT/PS/n-Si structure.

In measured temperature dependencies, peak values of photoresponse can be seen in the ranges of 150–190 K and 260–270 K, and a further decrease of this value upon heating up to 330 K (see Figure 22.20, curve 2). Like in photovoltaic sandwich structures based on metal-modified PS, the photosignal increase for hybrid organic–inorganic nanosystems in certain temperature ranges may be caused by the nonequilibrium carriers trapping level both in the porous layer and at the PEDOT/PS interface (Monastyrskii et al. 2014).

It has to be mentioned that PEDOT film not only modifies electronic parameters of PS nanostructures but also increases the efficiency of PEDOT/PS/n-Si photovoltaic structures by means of better passivation of silicon nanocrystals, decreasing the coefficient of the charge carriers surface recombination. Besides, thin-film polymer coating is transparent for visible radiation, which is an important requirement for charge photogeneration to be possible in PS structures.

22.7 APPLICATION OF PHOTOVOLTAIC POROUS SILICON STRUCTURES IN GAS SENSORS

Due to the large specific surface area, PS nanostructures are considered very promising materials for sensor applications (Baratto et al. 2002; Barillaro et al. 2003; Harraz 2014; Ozdemir and Gole 2007). A common feature of most gas sensors based on PS and conjugated polymeric/PS nanosystems is that most of them use adsorption-electric effects, resulting in changes of the electrical parameters of PS nanocrystals. On the other hand, surface modification of PS by polymer films causes the appearance of new properties or effects, depending on the gas environment, that may be easily registered. One among these effects is the emergence of a photovoltaic signal in PEDOT/PS/n-Si structures (Monastyrskii et al. 2014). In addition, a polymer film on the surface of the porous layer can serve as a catalytic material and, therefore, help to improve the sensitivity and selectivity of the gas adsorption sensors. This suggests some background for exploiting photovoltaic PEDOT/PS hybrid nanosystems in gas sensing devices.

The studies of the sensing capabilities of photoresponsive PEDOT/PS/n-Si structures were carried out in a sealed chamber, with the possibility to change the gas environment. Generation of photovoltage was induced by irradiation of hybrid structures on the polymer film side with white light LED. Tests were performed in different environments, including ammonia (NH_3) and nitrogen dioxide (NO_2). The choice of these gases is due to the importance of their control in the atmosphere, and their different adsorption properties: NH_3 molecules exhibit donor properties (Chiesa et al. 2003) and NO_2 shows acceptor behavior (Osminkina et al. 2005).

Appearance of the potential barrier and, therefore, the value of the photosignal is largely dependent on the surrounding environment. Increased concentration of ammonia molecules in air causes the reduction in the photovoltage between the top of PEDOT/PS/n-Si sensor structure and the silicon substrate (Figure 22.21). Sensitivity of photovoltage to adsorption of polar molecules in hybrid structures can be associated with adsorption-induced electric effects that cause a significant change in the electronic parameters of PS nanocrystals.

In contrast to the adsorption of NH_3 molecules, adsorption of NO_2 molecules, which have acceptor properties, results in a significant increase of photovoltaic signal (see Figure 22.21). For the same irradiation conditions, the change of photovoltage with increasing NO_2 concentration was greater than in the case of NH_3.

To evaluate the sensing properties of photovoltaic PEDOT/PS/n-Si structures, we calculated their "sensing ability" as the following ratio:

$$\gamma = \frac{1}{U} \frac{\Delta U}{\Delta C}, \tag{22.3}$$

where $\Delta U/U$ is the relative change of photovoltage and ΔC is the change in the concentration of analyzed gases. The estimated sensitivities of sensor structures depending on the concentration of molecules of ammonia and nitrogen dioxide are shown in Figure 22.22.

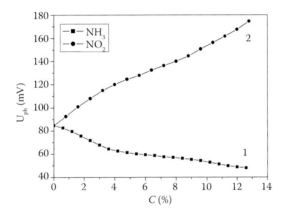

Figure 22.21 Dependence of photovoltage in PEDOT/PS/n-Si structure on the concentration of ammonia molecules (1) and nitrogen dioxide (2).

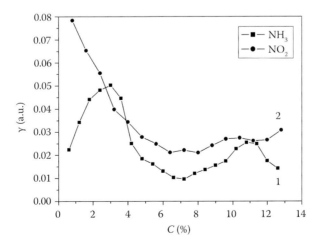

Figure 22.22 Dependence of sensing ability of PEDOT/PS/n-Si hybrid structure on the concentration of ammonia molecules (1) and nitrogen dioxide (2).

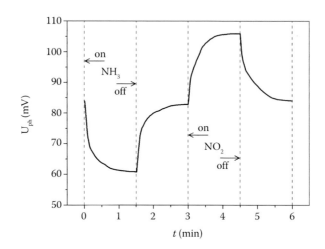

Figure 22.23 Response of the sensor PEDOT/PS/n-Si structure to the step-like change in the concentration of NH_3 and NO_2.

Analysis of the obtained dependencies have shown that the sensing ability of photovoltaic sensor based on hybrid PEDOT/PS/n-Si structures is largest for NO_2 molecules at low concentrations that can be used in the development of highly sensitive and selective gas analyzers.

Important parameters are the sensor response time and the recovery time. Dynamic dependencies shown in Figure 22.23 indicate that the response time of photovoltaic structures (responding to a sharp change in the concentration of the analyzed gases) is about 70–90 s and is small enough for microelectronic gas sensors.

It should be noted that the interaction of the working surface of the sensor with NH_3 and NO_2 molecules has the character of physical adsorption and is a reversible activation-free process, as evidenced by the almost complete recovery of the initial values of photovoltaic signal in the hybrid structure after purging and pumping gas from the chamber.

The dependence of photovoltage for PEDOT/PS/n-Si structures both from the type and the concentration of adsorbed molecules opens possibilities for designing new types of selective and highly sensitive gas sensors.

22.8 SUMMARY AND OUTLOOK

In this chapter, we have considered the main approaches to the design of photosensitive structures based on silicon nanosystems, specifically on PS. The interest in these materials is rapidly growing since the discovery of visible luminescence from PS. Outstanding electronic and optical properties of PS expand its applications to diverse fields such as chemical and biological sensors, light emitters and photodetectors, and energy conversion and storage devices. Here we focused on novel technical solutions and possible ways of enhancing the photovoltaic properties of PS-based structures. The porous structure of such a material favors the incorporation of electrically conductive fragments into its bulk that form additional pathways for current flow, and ensure efficient collecting of photogenerated carriers and their transport out of the porous layer.

Besides, PS can be looked at as a convenient model object to study electronic and photonic processes in disordered semiconductor systems, since it is relatively easily obtained technologically and possesses predicted structural (that means also electrical and optical) features. Based on comprehensive studies of CVC and spectral, thermal, and energy dependencies of the photoresponse nonequilibrium electronic processes in photovoltaic PS heterostructures were described.

We have shown that the adsorption of acceptor or donor molecules modifies the electrical and photoelectric properties of silicon nanosystems. Formation of electrical barriers in the structure of PS/silicon substrate in the case of polar molecules adsorption indicates a clear possibility for controlling the conductivity type in the PS nanocrystals by adsorption doping from gas or liquid phases. The obtained results can also be used for developing alternative energy sources, high-sensitivity sensors, and other electronic devices based on PS and some other systems involving semiconductor nanocrystals.

Particular attention should be paid to photosensitive hybrid organic–inorganic structures. One of the viable ways to increase the efficiency of PS-based photoelectric converters is to create nanosized silicon heterostructures with polymer components that exhibit a number of specific properties, such as intrinsic electronic conductivity, photosensitivity, and a capability of generating charge carriers. Moreover, better passivation of PS nanocrystals with polymer films decreases the surface recombination coefficient for charge carriers. Hybrid structures, due to nanosized components and peculiarities of their chemical and physical interaction, combine the best features of organic and inorganic materials. These structures show great promise for the development of new types of inexpensive electromagnetic radiation detectors for visible and near-infrared ranges, and can serve as a basis for optoelectronic or sensing systems with outstanding functional parameters.

REFERENCES

Admassie, S., Zhang, F., Manoj, A.G., Svensson, M., Andersson, M.R., and Inganäs, O. 2006. A polymer photodiode using vapour-phase polymerized PEDOT as an anode. *Solar Energy Materials and Solar Cells* 90: 133–141.

Ahmed, N.M., Al-Douri Y., Alwan, A.M., Jabbar, A.A., and Arif, G.E. 2013. Characteristics of nanostructure silicon photodiode using laser assisted etching. *Procedia Engineering* 53: 393–399.

Aksimentyeva, O., Konopelnyk, O., Bolesta, I., Karbovnyk. I., Poliovyi, D., and Popov, A.I. 2007a. Charge transport in electrically responsive polymer layers. *Journal of Physics: Conference Series* 93: 012042.

Aksimentyeva, O., Monastyrskyi, L., Savchyn, V., Stakhira, P., Vertsimakha, Ya., and Tsizh, B. 2007b. Electronic processes in the porous silicon-conducting polymer heterostructures. *Molecular Crystals and Liquid Crystals* 467: 73–83.

Aksimentyeva, O.I., Tsizh, B.R., Monastyrskii, L.S., Olenych, I.B., and Pavlyk, M.R. 2015. Luminescence in porous silicon—Poly(para–phenylene) hybrid nanostructures. *Physics Procedia* 76: 31–36.

Alferov, Z.I., Andreev, V.M., and Rumyantsev, V.D. 2004. Solar photovoltaics: Trends and prospects. *Semiconductors* 38: 899–908.

Amran, T.S., Hashim, M.R., Al-Obaidi, N.K., Yazid, H., and Adnan, R. 2013. Optical absorption and photoluminescence studies of gold nanoparticles deposited on porous silicon. *Nanoscale Research Letters* 8: 35.

Astrova, E.V., Lebedev, A.A., Remenyuk, A.D., Rud', Yu.V., and Rud', V.Yu. 1997. Photosensitivity of porous silicon-silicon heterostructures. *Semiconductors* 31: 121–123.

Bailey, S.G., and Flood, D.J. 1998. Space photovoltaics. *Progress in Photovoltaics: Research and Applications* 6: 1–14.

Balagurov, L.A., Yarkin, D.G., Petrovicheva, G.A., Petrova, E.A., Orlov, A.F., and Andryushin, Ya, S. 1997. Highly sensitive porous silicon based photodiode structures. *Journal of Applied Physics* 82: 4647–4650.

Baratto, C., Faglia, G., Sberveglieri, G., Gaburro, Z., Pancheri, L., Oton, C., and Pavesi, L. 2002. Multiparametric porous silicon sensors. *Sensors* 2: 121–126.

Barillaro, G., Nannini, A., and Pieri, F. 2003. APSFET: A new, porous silicon-based gas sensing device. *Sensors and Actuators B* 93: 263–270.

Bilyalov, R., Stalmans, L., Beaucarne, G., Loo R., Caymax, M., Poortmans, J., and Nijs, J. 2001. Porous silicon as an intermediate layer for thin-film solar cell. *Solar Energy Materials and Solar Cells* 65: 477–485.

Bisi, O., Ossicini, S., and Pavesi, L. 2000. Porous silicon: A quantum sponge structure for silicon based optoelectronics. *Surface Science Reports* 38: 1–126.

Bryan, J.D., and Gamelin, D.R. 2005. Doped semiconductor nanocrystals: Synthesis, characterization, physical properties, and applications. *Progress in Inorganic Chemistry* 54: 47–126.

Canham, L.T. 1990. Silicon quantum wire array fabrication by electrochemical and chemical dissolution of wafers. *Applied Physics Letters* 57: 1046–1048.

Chiesa, M., Amato, G., Boarino, L., Garrone, E., Geobaldo, F., and Giamello, E. 2003. Reversible insulator-to-metal transition in p+-type mesoporous silicon induced by the adsorption of ammonia. *Angewandte Chemie International Edition* 42: 5032–5035.

Coulthard, I., Sammyniaken, R., Naftel, S.J., Zhang, P., and Sham, T.K. 2000. Semiconductor growth and junction formation within nano-porous oxides. *Physica Status Solidi A* 182: 157–162.

Cullis, A.G., and Canham, L.T. 1991. Visible light emission due to quantum size effects in highly porous crystalline silicon. *Nature* 353: 335–338.

Cullis, A.G., Canham, L.T., and Calcott, P.D.J. 1997. The structural and luminescence properties of porous silicon. *Journal of Applied Physics* 82: 909–965.

Davidenko, N.A., Ishchenko, A.A., Skryshevsky, V.A., Studzinsky, S.L., and Mokrinskaya, E.V. 2011. Photoconductivity of polymeric composites doped with porous silicon nanoparticles and additions of ionic poly-methine dyes of different types. *Molecular Crystals and Liquid Crystals* 536: 93/[325]–98/[330].

Erwin, S.C., Lijun, Zu, Haftel, M.I., Efros, A.L., Kennedy, T.A., and Norris, D.J. 2005. Doping semiconductor nano-crystals. *Nature* 436: 91–94.

Fathauer, R.W., George, T., Ksendzov, A., and Vasquez, R.P. 1992. Visible luminescence from silicon wafers subjected to stain etches. *Applied Physics Letters* 60: 995–997.

Föll, H., Christophersen, M., Carstensen, J., and Hasse, G. 2002. Formation and application of porous silicon. *Materials Science and Engineering: R: Reports* 39: 93–141.

Galiy, P.V., Lesiv, T.I. Monastyrskii, L.S. Nenchuk, T.M., and Olenych, I.B. 1998. Surface investigations of nano-structured porous silicon. *Thin Solid Films* 318: 113–116.

Goetzberger, A., Hebling, C., and Schock, H.-W. 2003. Photovoltaic materials, history, status and outlook. *Materials Science and Engineering R* 40: 1–46.

Goswami, D.Y., Vijayaraghavan, S., Lu, S., and Tamm, G. 2004. New and emerging developments in solar energy. *Solar Energy* 76: 33–43.

Green, M.A. 2003a. Crystalline and thin-film silicon solar cells: State of the art and future potential. *Solar Energy* 74: 181–192.

Green, M.A. 2003b. *Third generation photovoltaics*. Berlin: Springer Verlag.

Grigor'ev, L.V., Grigor'ev, I.M., Zamoryanskaya, M.V., Sokolov, V.I., and Sorokin, L.M. 2006. Transport properties of thermally oxidized porous silicon. *Technical Physics Letters* 32: 750–753.

Günes, S., Sariciftci, N.S. 2008. Hybrid solar cells. *Inorganica Chimica Acta* 361: 581–588.

Halls, J.J.M., Walsh, C.A., Greenham, N.C., Marseglia, E.A., Friend, R.H., Moratti, S.C., and Holmes, A.B. 2002. Efficient photodiodes from interpenetrating polymer networks. *Nature* 376: 498–500.

Harraz, F.A. 2014. Porous silicon chemical sensors and biosensors: A review. *Sensors and Actuators B: Chemical* 202: 897–912.

Hasegawa, S., Maehashi, K., Nakashima H., Ito, T., and Hiraki, A. 1989. Growth and characterization of GaAs films on porous Si. *Journal of Crystal Growth* 95: 113–116.

Heeger, A.J. 2001. Semiconducting and metallic polymers: The fourth generation of polymeric materials. *Synthetic Metals* 125: 23–42.

Holder, E., Tessler, N., and Rogach, A.L. 2008. Hybrid nanocomposite materials with organic and inorganic components for optoelectronic devices. *Journal of Materials Chemistry* 18: 1064–1078.

Hsu, H.C., Cheng, C.S., Chang, C.C., Yang, S., Chang, C.S., and Hsieh, W.F. 2005. Orientation-enhanced growth and optical properties of ZnO nanowires grown on porous silicon substrates. *Nanotechnology* 16: 297–301.

Karbovnyk, I., Collins, J., Bolesta, I., Stelmashchuk, A., Kolkevych, A., Velupillai, S., Klym, H., Fedyshyn, O., Tymoshuk, S., and Kolych, I. 2015. Random nanostructured films for environmental monitoring and optical sensing: Experimental and computational studies. *Nanoscale Research Letters* 10: 151.

Kim, J., Moon, I.S., Lee, M.J., and Kim, D.W. 2007. Formation of a porous silicon layer by electrochemical etching and application to the silicon solar cell. *Journal of the Ceramic Society of Japan* 115: 333–337.

Kumar, S.S., Mathiyarasu, J., Phani, K.L.N., and Yegnaraman, V. 2006. Simultaneous determination of dopamine and ascorbic acid on poly(3,4-ethylenedioxythiophene) modified glassy carbon electrode. *Journal of Solid State Electrochemistry* 10: 905–913.

Laptev, A.N., Prokaznikov, A.V., and Rud', N.A. 1997. Hysteresis of the current-voltage characteristics of porous-silicon light-emitting structures. *Technical Physics Letters* 23: 440–442.

Leclere, P., Surin, M., Brocorens, P., Cavallini, M., Biscarini, F., and Lazzaroni, R. 2006. Supramolecular assembly of conjugated polymers: From molecular engineering to solid-state properties. *Materials Science and Engineering: R: Reports* 55: 1–56.

Lehmann, V., and Gosele, V. 1990. Porous silicon formation: A quantum wire effect. *Applied Physics Letters* 58: 856–858.

Lehmann, V., Jobst, B., Muschik, T., Kux, A., and Petrova-Koch, V. 1993. Correlation between optical properties and crystallite size in porous silicon. *Japanese Journal of Applied Physics* 32: 2095–2099.

Lin, T.L., Sadwick, L., Wang K.L., Kao, Y.C., Hull, R., Nieh, C.W., Jamieson, D.N., and Liu, J.K. 1987. Growth and characterization of molecular beam epitaxial GaAs layers on porous silicon. *Journal of Applied Physics* 51: 814–816.

Martin-Palma, R.J., Guerrero-Lemus, R., Moreno, J.D., Martinez-Duart, J.M., Gras, A., and Levy, D. 2000. Development and characterization of porous silicon based photodiodes. *Materials Science and Engineering: B* 69–70: 87–91.

Menna, P., Francia, G.D., and Ferrara, V.L. 1995. Porous silicon in solar cells: A review and a description of its application as an AR coating. *Solar Energy Materials and Solar Cells* 37: 13–24.

Misra, S., Bhattacharya, R., and Angelucci, R. 2001. Integrated polymer thin film macroporous silicon microsystems. *Journal of the Indian Institute of Science* 81: 563–567.

Mocatta, D., Cohen, G., Schattner, J., Millo, O., Rabani, E., and Banin, U. 2011. Heavily doped semiconductor nanocrystal quantum dots. *Science* 332: 77–81.

Monastyrskii, L.S., Aksimentyeva, O.I., Olenych, I.B., and Sokolovskii, B.S. 2014. Photosensitive structures of conjugated polymer—Porous silicon. *Molecular Crystals and Liquid Crystals* 589: 124–131.

Monastyrskii, L., Parandii, P., Panasiuk, M., and Olenych, I. 2001. Photovoltaic effect in porous silicon heterostructures. *Proceedings SPIE* 4425: 347–350.

Nahor, A., Berger, O., Bardavid, Y., Toker, G., Tamar, Y., Reiss, L., Asscher, M., Yitzchaik, S., and Sa'ar, A. 2011. Hybrid structures of porous silicon and conjugated polymers for photovoltaic applications. *Physica Status Solidi C* 8: 1908–1912.

Nguyen, T.P., Le Rendu, P., Tran, V.H., Parkhutik, V., and Esteve, R.F. 2000. Electrical and optical properties of conducting polymer/porous silicon structures. *Journal of Porous Materials* 7: 393–396.

Norris, D.J., Efros, A.L., and Erwin, S.C. 2008. Doped nanocrystals. *Science*, 319: 1776–1779.

Olenych, I.B. 2014. Electrical and photoelectrical properties of porous silicon modified by cobalt nanoparticles. *Journal of Nano- and Electronic Physics* 6: 04022.

Olenych, I.B., Aksimentyeva, O.I., Monastyrskii, L.S., Horbenko, Y.Y., and Yarytska, L.I. 2015b. Sensory properties of hybrid composites based on poly(3,4-ethylenedioxythiophene)-porous silicon–carbon nanotubes. *Nanoscale Research Letters* 10: 187.

Olenych, I.B., Aksimentyeva O.I., Monastyrskii, L.S., and Pavlyk, M.R. 2012a. Electrochromic effect in photoluminescent porous silicon—Polyaniline hybrid structures. *Journal of Applied Spectroscopy* 79: 495–498.

Olenych, I.B., Monastyrskii, L.S., Aksimentyeva, O.I., and Sokolovskii, B.S. 2011. Humidity sensitive structures on the basis of porous silicon. *Ukrainian Journal of Physics* 56: 1198–202.

Olenych, I.B., Monastyrskii, L.S., Aksimentyeva, O.I., and Sokolovskii, B.S. 2013. Effect of bromine adsorption on the charge transport in porous silicon—Silicon structures. *Electronic Materials Letter* 9: 257–260.

Olenych, I.B., Monastyrskii, L.S., Aksimentyeva, O.I., and Yarytska, L.I. 2014. Modification of the electrical properties of porous silicon by adsorption of ammonia molecules. *Universal Journal of Physics and Application* 2: 201–205.

Olenych, I.B., Monastyrskii, L.S., and Sokolovskii, B.S. 2012b. Photodetectors on the basis of porous silicon. *Journal of Nano- and Electronic Physics* 4: 03025.

Olenych, I., Tsizh, B., Monastyrskii, L., Aksimentyeva, O., and Sokolovskii, B. 2015a. Preparation and properties of nanocomposites of silicon oxide in porous silicon. *Solid State Phenomena* 230: 127–132.

Osminkina, L.A., Konstantinova, E.A., Sharov, K.S., Kashkarov, P.K., and Timoshenko, V.Yu. 2005. The role of boron impurity in the activation of free charge carriers in layers of porous silicon during the adsorption of acceptor molecules. *Semiconductors* 39: 347–350.

Osorio, E., Urteaga, R., Acquaroli, L.N., Garcia-Salgado, G., Juare, H., and Koropecki, R.R. 2011. Optimization of porous silicon multilayer as antireflection coatings for solar cells. *Solar Energy Materials and Solar Cells* 95: 3069–3073.

Ouyanga, J., Xua, Q., Chua, C.W., Yanga, Y., Lib, G., and Shinar, J. 2004. On the mechanism of conductivity enhancement in poly(3,4-ethylenedioxythiophene):poly(styrene sulfonate) film through solvent treatment. *Polymer* 45: 8443–8450.

Ozdemir, S., and Gole, J. 2007. The potential of porous silicon gas sensors. *Current Opinion in Solid State and Materials Science* 11: 92–100.

Peng, K.Q., and Lee, S.T. 2001. Silicon nanowires for photovoltaic solar energy conversion. *Advanced Materials* 23: 198–215.

Presting, H., Konle, J., Starkov, V., Vyatkin, A., and König, U. 2004. Porous silicon for micro-sized fuel cell reformer units. *Materials Science and Engineering: B* 108: 162–165.

Primachenko, V.E., Kononets, J.F., Bulakh, B.M., Venger, E.F., Kaganovich, E.B., Kizyak, I.M., Kirillova, S.I., Manoilov, E.G., and Tsyrkunov, Y.A. 2005. The electronic and emissive properties of Au-doped porous silicon. *Semiconductors* 39: 565–571 (2005).

Pulsford, N.J., Rikken, G.L.J.A., Kessener, Y.A.R.R., Lous, E.J., and Venhuizen, A.H.J. 1993. Carrier injection and transport in porous silicon Schottky diodes. *Journal of Luminescence* 57: 181–184.

Pulsford, N.J., Rikken, G.L.J.A., Kessener, Y.A.R.R., Lous, E.J., and Venhuizen, A.H.J. 1994. Behavior of a rectifying junction at the interface between porous silicon and its substrate. *Journal of Applied Physics* 75: 636–638.

Pyshkina, O., Kubarkov, A., and Sergeyev, V. 2010. Poly(3,4-ethylenedioxythiophene): Synthesis and properties. *Materials Science and Applied Chemistry* 21: 51–54.

Selj, J.H., Thogersen, A., Foss, S.E., and Marstein E.S. 2010. Optimization of multilayer porous silicon antireflection coatings for silicon solar cells. *Journal of Applied Physics* 107: 074904.

Smestad, G., Kunst, M., and Vial, C. 1992. Photovoltaic response in electrochemically prepared photoluminescent porous silicon. *Solar Energy Materials and Solar Cells* 26: 277–283.

Smith, R.L., and Collins, S.D. 1992. Porous silicon formation mechanisms. *Journal of Applied Physics* 71: R1–R12.

Stakhira, P., Aksimentyeva, O., Mykytyuk, Z., and Cherpak, V. 2005. Photovoltaic properties of heterostructure based on porous silicon and polyaniline. *Functional Materials* 12: 807–809.

Strehlke, S., Bastide, S., Guillet, J., and Levy-Clement, C. 2000. Design of porous silicon antireflection coatings for silicon solar cells. *Materials Science and Engineering B* 69–70: 81–86.

Suemune, I., Noguchi, N., and Yamanishi, M. 1992. Photoirradiation effect on photoluminescence from anodized porous silicons and luminescence mechanism. *Japanese Journal of Applied Physics* 31: L494–L497.

Sze, S.M. 1969. *Physics of semiconductor devices*. New York: Wiley.

Tsai, C., Li, K.-H., Campbell, J.C., and Tasch, A. 1993. Photodetectors fabricated from rapid-thermal-oxidized porous Si. *Applied Physics Letters* 62: 2818–2820.

Uhlir, A. 1956. Formation of porous silicon. *Bell System Technical Journal* 35: 333–339.

Ünal, B., Parbukov, A.N., and Bayliss, S.C. 2001. Photovoltaic properties of a novel stain etched porous silicon and its application in photosensitive devices. *Optical Materials* 17: 79–82.

Vashpanov, Y.A. 1997. Electronic properties of microporous silicon under illumination and with the adsorption of ammonia. *Technical Physics Letters* 23: 448–449.

Venger, E.F., Gorbach, T.Ya., Kirillova, S.I., Primachenko, V.E., and Chernobai, V.A. 2002. Changes in properties of a porous silicon/silicon system during gradual etching off of the porous silicon layer. *Semiconductors* 36: 330–335.

Venger, E.F., Kirillova, S.I., Kizyak, I.M., Manoilov, E.G., and Primachenko, V.E. 2004. The effect of a Au impurity on the photoluminescence of porous Si and photovoltage on porous-Si structures. *Semiconductors* 38: 113–119 (2004).

Vorontsov, A.S., Osminkina, L.A., Tkachenko, A.E., Konstantinova, E.A., Elenskii, V.G., Timoshenko, V.Y., and Kashkarov, P.K. 2007. Modification of the properties of porous silicon on adsorption of iodine molecules. *Semiconductors* 41: 953–957.

Wang, H., Wang, J., Hong, L., Tan, Y.H., Tan, C.S., and Rusli. 2016. Thin film silicon nanowire/PEDOT:PSS hybrid solar cells with surface treatment. *Nanoscale Research Letters* 11: 311.

Wei, G.-P., Zheng, Y.-M., Huang, Z.-J., Li Y., Feng, J.-W., and Mo Y.-W. 1994. Porous silicon and its application test for photovoltaic devices. *Solar Energy Materials and Solar Cells* 35: 319–324.

Wu, K.-H., and Li, C.-W. 2015. Light absorption enhancement of silicon-based photovoltaic devices with multiple bandgap structures of porous silicon. *Materials* 8: 5922–5932.

Yae, S. Kawamoto, Y., Tanaka, H., Fukumuro, N., and Matsuda, H. 2003. Formation of porous silicon by metal particle enhanced chemical etching in HF solution and its application for efficient solar cells. *Electrochemistry Communications* 5: 632–636.

Yamaguchi, M., and Luque, A. 1999. High efficiency and high concentration in photovoltaics. *IEEE Transactions on Electron Devices* 46: 2139–2144.

Yamaguchi, M., Ohshita, Y., Arafune, K., Sai, H., and Tachibana, M. 2006. Present status and future of crystalline silicon solar cells in Japan. *Solar Energy* 80: 104–110.

Yu, L.Z., and Wie, C.R. 1993. Study of MSM photodetector fabricated on porous silicon. *Sensors and Actuators A* 39: 253–257.

Yu, P., Tsai, C.-Y., Chang, J.-K., Lai, C.-C., Chen, P.-H., Lai, Y.-C., Tsai, P.-T., et al. 2013. 13% efficiency hybrid organic/silicon-nanowire heterojunction solar cell via interface engineering. *ACS Nano* 7: 10780–10787.

Zhang, X.G., Collins, S.D., and Smith, R.L. 1989. Porous silicon formation and electropolishing of silicon by anodic polarization in HF solution. *Journal of the Electrochemical Society* 136: 1561–1565.

Zheng, J.P., Jiao, K.L., Shen, W.P., Anderson, W.A., and Kwok, H.S. 1992. Highly sensitive photodetector using porous silicon. *Applied Physics Letters* 61: 459–461.

23 Silicon nanostalagmites for hybrid solar cells

Zingway Pei and Subramani Thiyagu

Contents

23.1 INTRODUCTION

In the past few years, there has been an enormous and growing interest in the development of nanostructure materials to improve the light harvesting efficiency for achieving high-efficiency Si solar cells while maintaining low cost. Feasible silicon nanostructures [1–4] such as silicon nanowires (SiNWs) and silicon nanoholes (SiNHs) have gained much attention due to their unique properties and possible applications in the fields of nanoelectronics [5–8], nanooptoelectronics [8,9], nanophotovolatics [9–11], and for sensor applications [12]. Various methods have been developed to prepare one-dimensional (1D) silicon nanostructures. They include laser ablation [13], physical vapor deposition [14], and chemical vapor deposition [15,16] which is a bottom-up approach [17]. However, most methods often result in randomly oriented SiNWs and their diameters and lengths have a wide distribution, which limit the applications to real optoelectronic devices. Nanoelectronics and nanooptoelectronics generally require vertically aligned, tunable length, and high-density nanowires to obtain processing compatibility.

Silicon thin-film solar cells are promising candidates for future generations of photovoltaic devices [18,19]. They offer cost effectiveness and the possibility of deposition on flexible substrates [20–22]. However, the efficiency of a thin-film Si solar cell is relatively low compared to a crystalline solar cell. The low absorption rate, relative poor material quality, and narrow absorption spectra are the major factors.

SiNWs are usually produced via a vapor–liquid–solid [VLS] growth mechanism [23], which introduces the 1D growing of a nanostructure by a metal nanocatalyst droplet containing gases such as silane, or grown from the gas phase by supplying Si vapor. However, the VLS growth mechanism generally requires high temperature, which is not a suitable method for an Si thin-film nanostructure. In particular, the microcrystalline silicon [μc-Si] solar cells grown on glass or plastic substrate cannot sustain high temperatures.

To obtain vertically oriented, length tunable, and high-density silicon nanostructures, some of VLS growth does not occur, wet chemical etching [24,25] through a predetermined template might be a possible method to achieve a light trapping structure at low cost. An efficient light management is essential to further improve the light confinement in the cells. Light trapping is the standard technique for improving thin-film silicon efficiencies and exploiting the sunlight. In particular, μc-Si solar cells have gained considerable attention in recent years. Wet chemical etched μc-Si surfaces using a catalytic etching method that forms the μc-Si nanostalagmites (μc-SiNSs) show an ultralow reflectance compared to Si thin-film layers. This ultralow reflectance is potentially fascinating for photovoltaic applications where enough absorption of solar light occurs in the thinnest Si layer possible. The μc-SiNSs structure can be used as a solar cell absorber.

23.2 FORMATION OF SILICON NANOSTALAGMITE

23.2.1 DEPOSITION OF MICROCRYSTALLINE SILICON

Detailed material studies combined with the preparation and characterization of μc-Si has brought success in the development of highly efficient μc-Si solar cells. As μc-Si is an indirect transition semiconductor, and has a low absorption coefficient in long wavelengths of 800 nm or more, the absorber thickness must be in the order of at least 1 μm. Initially, glass substrate is cleaned with organic solvents—acetone and isopropyl alcohol (IPA)—and in between, the glass substrate is rinsed with deionized (DI) water. Approximately 1 μm-thick μc-Si is deposited by plasma-enhanced chemical vapor deposition (PECVD) on a glass substrate. Silicon layers are deposited from a silane[SiH_4]-hydrogen[H_2] gas mixture. Deposition of μc-Si is controlled by the gas supply flow rate, substrate temperature, and deposition pressure.

23.2.2 PREPARATION OF POLYSTYRENE NANOSPHERE TEMPLATE

The use of self-assembled block copolymer structures to directly produce mesoscopic (1–100 nm) features requires potentially lower cost and can also be applied to large-scale methods as compared to conventional lithographic techniques, which may attract much research interest. Recently, nanopatterning by a template method through the use of block copolymer thin films received much attention [26–29]. The use of block copolymer could make a nanopattern in a simple and cost-effective way as compared to expensive deep ultraviolet (DUV) lithography and electron beam (E-beam) lithography. The structures of block copolymers are composed of two chemically distinct polymer chains through covalent bonding. Upon thermal annealing, block copolymers tend to self-assemble into nanometer-scale domains through nanophase separation into ordered morphologies such as sphere, cylinders, and lamellae, depending on the volume fraction of components of polymers [26,27,30]. Nanotemplates are achieved by removing one of the polymers with a specific solvent in this ordered structure.

The block copolymer used here was a poly(styrene-block-methyl methacrylate) (PS-b-PMMA) diblock copolymer with a molecular weight of 46.1k for poly(styrene) (PS) block and 21.0k for poly(methylmethacrylate (PMMA) block (Polymer Source Inc.). The ratio of polystyrene (PS) to PMMA is around 7:3. The incomplete nanophase separation in the conventional PS-b-PMMA is the origin for the failure of template formation. To prevent the sensitivity to temperature and time, an extra step to aid the nanophase separation was used. Tailoring the remaining PS to exposure the underlying layer could help the nanotemplate formation. The use of heated acetic acid to remove PMMA at the template formation step might help in tailoring the shape of the PS, thus enhancing the nanophase separation as shown in Figure 23.1. The PS will react with the heated acetic acid with corresponding modification in surface and shape. Therefore, a modified PS-b-PMMA block copolymer nanopatterning method was demonstrated.

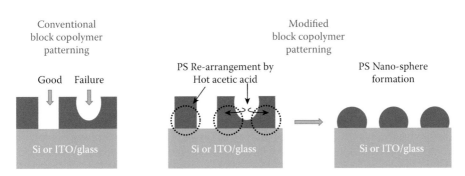

Figure 23.1 Schematic diagram for modified block copolymer nanopatterning to create PS nanosphere.

23.2.2.1 Sub-30 nm PS nanospheres fabrication

Silicon wafers were first cleaned by isopropanol and deionized water. After cleaning, a thin (~1.5 nm) native oxide layer was grown at the Si surface and was used as the substrate. The block copolymer solution was prepared by dissolving PS-b-PMMA in toluene. After mixing overnight, the PS-b-PMMA solution was spin coated on the Si substrate. The PS-b-PMMA thin-film was then annealed at 170°C under vacuum. This temperature was well above the glass transition temperature (Tg) of both PS and PMMA to allow the nanophase separation. Substrates were removed from the oven after 24 h in vacuum, followed by quenching to room temperature. The PS-b-PMMA film was then immersed in heated acetic acid, followed by a rinse in deionized water, resulting in the formation of PS nanospheres with a diameter of 30–50 nm as shown in the scanning electron microscopy (SEM) image in Figure 23.2 [31].

Evolution of the morphology of PS nanospheres on silicon substrates through a simple modification in immersion time in acetic acid highlights an easy way to create the nanospheres in the range of 30–40 nm. The density of PS nanospheres is as high as $10^{10}/cm^2$. The kinetics of PS nanosphere formation was assumed to be in correlation with the solubility of polystyrene in the acetic acid based on the time evolution. In general, the PS is relatively difficult to dissolve in acetic acid at room temperature whereas the PMMA was easily dissolved under the same conditions. The discrepancy in solubility was used to make a nanohole template in a conventional PS-b-PMMA nanopatterning method. However, at high temperature, the solubility of PS acetic acid was slightly improved. This helps the PS nanophase

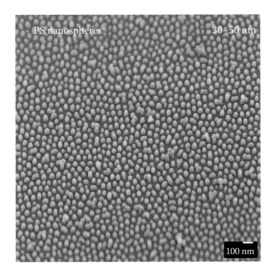

Figure 23.2 SEM image of PS nanospheres with diameter of 30–50 nm used as a template to form nanostalagmite.

to rearrange at high temperature during immersion in acetic acid, thereby the connecting PS chains are separated into several spheres with balanced force under rearrangement.

23.2.3 TEMPLATE-BASED METAL-ASSISTED CHEMICAL ETCHING METHOD

In recent years, a simple Ag catalytic etching technique has been used to prepare large-area, aligned silicon nanostructure arrays on single-crystal silicon wafers. The location, size, length, and orientation control of SiNW arrays have been established by catalytic etching through a prepatterned template. To avoid high cost photolithography, the templates were usually obtained by using self-assembly of PS spheres with a diameter of around half a micrometer. Further reducing the diameter and spacing of the PS spheres in the template could produce SiNWs with larger surface areas, which could enhance the photogenerated carrier collection efficiency in solar cells. In addition, small diameter SiNWs with higher aspect ratios could achieve ultralow reflectance over a broad range of wavelengths. As a consequence, the production of SiNWs with sub-100 nm diameter and spacing is strongly motivated by the desire to achieve an ultralow reflectance and excellent carrier collection efficiency along the radial direction. However, a high surface-to-volume ratio in the SiNWs may limit the open circuit voltage (V_{oc}) and fill factor (FF) of the resulting solar cells due to the increase in surface recombination. The length of the nanowires should, therefore, be limited. Figure 23.3 shows the schematic diagram for the fabrication of microcrystalline silicon nanostalagmite (μc-SiNS) arrays using a PS nanospheres template obtained by the modified block copolymer nanopatterning method. SiNSs were prepared on an indium tin oxide (ITO)/glass substrate thorough a template by chemical etching with a silver (Ag) catalyst. Approximately 1 μm thick microcrystalline silicon (μc-Si) is deposited by PECVD on a glass substrate. The template used polystyrene nanospheres with diameter 30–50 nm and density 10^{10}/cm^2 by nanophase separation of the PS containing block copolymers as shown in the SEM image of Figure 23.2. The formation of PS nanospheres is implemented through a modified block copolymer (PS-b-PMMA) process by removing PMMA and reshaping the remaining PS material. After the formation of PS nanospheres on top of μc-Si, the silver film was deposited using a thermal evaporator. The thickness of Ag film was 10 nm. Thickness of Ag film is determined by the template size; film thickness should be approximately one quarter of from the PS nanosphere size. The etching mixture consisting of hydrofluoric acid (HF) and H$_2$O$_2$, and deionized water, was used at room temperature. The concentrations of HF and H$_2$O$_2$ were 4.6 M and 0.44 M, respectively. The Ag layers adhering to the Si surfaces had a higher electronegativity than the Si, and electrons were therefore attracted to Ag from the Si, making the Ag layers negatively charged. The electrons from the negatively charged Ag layers were preferentially captured by the O$^-$ ions of the H$_2$O$_2$ and became O^{2-} ions. This charge transfer caused local oxidation of the Si underneath the Ag patterns.

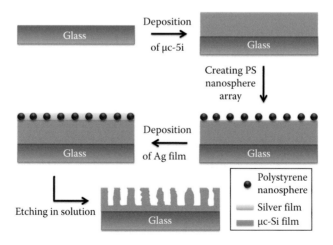

Figure 23.3 Schematic diagram for the fabrication of microcrystalline silicon nanostalagmites using template PS nanospheres based on modified block copolymer.

The resultant SiO_2 was then continuously etched away by HF, leading to the penetration of Ag into the Si substrates. The surfaces covered by the PS nanospheres did not possess this catalyzed chemical etching that left Si nanostalagmites with a diameter similar to the diameter of the PS nanospheres. The Si nanostalagmites are obtained when the Ag penetration reaches a certain depth. The etching duration was varied depending on the required length of the nanowires. After etching, the substrate was immersed in toluene to remove the PS nanospheres. The silver film was then removed by immersing it in boiling aqua regia. The chemical reaction equation is shown below:

$$H_2O_2 + 2H^+ + 2e^- \xrightarrow{\ Ag\ } 2H_2O \tag{23.1}$$

$$Si + 2H_2O \xrightarrow{\ Ag\ } SiO_2 + 4H^+ + 4e^- \tag{23.2}$$

$$SiO_2 + 6HF \rightarrow H_2SiF_6 + 2H_2O \tag{23.3}$$

Figure 23.4 shows the SEM images of the silicon nanostalagmite array (SiNS) on the glass substrate [32]. The length of SiNS can be controlled and varies linearly with the duration of the catalyst etching process. Figure 23.4a and b shows a high-magnification cross-sectional SEM image of µc-SiNSs fabricated with different catalyst chemical etching times of 30 and 90, producing nanostalagmites with a length of around 300 nm and 1 µm, respectively. The SiNSs are randomly distributed due to the PS nanosphere template being aperiodic, as shown in Figure 23.2. Although the SiNSs were not strictly vertical, this structure is still similar to that of antireflection feathers on the eyes of moths.

23.3 SILICON NANOSTALAGMITE PROPERTIES

23.3.1 LIGHT TRAPPING CONCEPT

Nanowires provide not only the advantage of more efficient charge transport over planar material but also present the potential for improved optical absorption characteristics. Nanostalagmites are similar in structure to nanowires; the Si nanostalagmites were expected to have an efficient light trapping effect. The light trapping effect is shown in schematic Figure 23.5. The incident light will have multiple internal reflections causing a long optical path for light absorption. In addition, to achieve optimal omnidirectional

30 sec

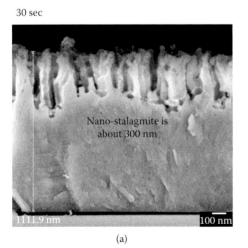

90 sec

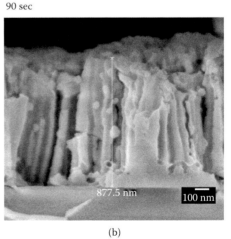

(a)   (b)

Figure 23.4 SEM image of microcrystalline silicon nanostalagmites arrays: high-magnification cross-sectional SEM images of sample after etching for (a) 30 s and (b) 90 s. (From Thiyagu, S., et al., *Nanoscale Res. Lett.*, 7, 171, 1–6, 2012.)

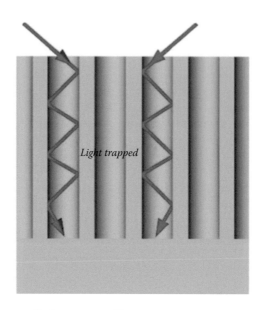

Figure 23.5 Schematic illustration of light trapping effect.

antireflection, the distance between the nanostructures is best when it is smaller than the wavelength of incident light. The prepared μc-SiNSs samples were black in appearance and highly nonreflective to the naked eye.

23.3.2 OPTICAL ABSORPTION

Optical measurements were performed on the microcrystalline samples before and after μc-SiNS fabrication. Transmittance and reflectance spectra of thin μc-Si film before (light gray) and after etching to form nanostalagmites (black) are shown in Figure 23.6a [32]. The transmittance for μc-SiNS over the entire spectral range from 300 to 800 nm is around 0.3%. In comparison, the planar control shows increased transmittance and strong interference patterns after 600 nm. The reflectance of μc-SiNSs and μc-Si film are also depicted in Figure 23.6a. The ultralow reflectance of the μc-SiNSs array of around 0.3% over 300–800 nm was observed. In comparison, the planar μc-Si film exhibits around 20–30% reflectance over a measured spectrum. The ultralow reflectance of SiNSs is caused by strong absorption due to the strong light trapping property of the dense SiNSs. The nanostalagmite simply acts as very effective antireflection layer. This might be due to the "moth-eye"-like biomimetic effect.

Acquisition of reflectance (R) and transmittance (T) spectra from μc-SiNSs allows further acquisition of their absorbance (A) spectra, expressed as A(%) = 100 – R(%) – T(%) as shown in Figure 23.6b. Photographs of bare μc-Si and a chemically etched substrate are shown in the inset to Figure 23.6b. The absorption of over 98% was obtained for SiNSs over the measured spectrum from 300 to 800 nm. This indicates that SiNSs could extend the absorption to the infrared regime, harvesting more light to increase photocurrent. Our nanostructure SiNS have low reflectance and high absorption, similarly silicon nanocones and nanodomes also had higher absorption. However, their process used reactive ion etching, which means a high vacuum was used. Here we demonstrated an alternative way by using simple and chemical processes to obtain a silicon nanostalagmite structure on glass/ITO substrate. This remarkable property suggests that SiNS arrays are an appropriate candidate for antireflective surfaces and absorption materials used in photovoltaic cells. The effect of light trapping can be understood by the absorption coefficient. By taking the reflectance (R) and transmittance (T_0) spectrum into account, the absorption coefficient can be calculated by the following equation:

$$T = \frac{T_0}{(1-R)} = e^{-\alpha \cdot d} \tag{23.4}$$

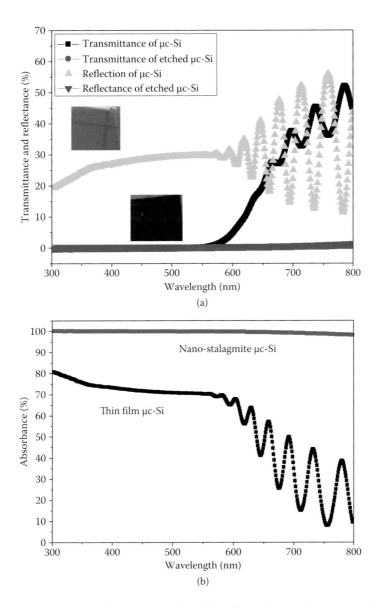

Figure 23.6 Optical measurement on thin microcrystalline silicon film with and without nanostalagmites arrays: (a) transmittance and reflectance; (b) absorption. (From Thiyagu, S., et al., *Nanoscale Res. Lett.*, 7, 171, 1–6, 2012.)

in which T is the transmission, α is the absorption coefficient, and d is the thickness of the µc-Si layer. Figure 23.7 depicts the absorption coefficient of the SiNSs layer [32]. The highest absorption coefficient for the SiNSs layer is also ~7 × 10^4/cm at 620 nm. The absorption coefficient at wavelengths shorter than 620 nm cannot be deduced because of measured "zero" transmittance, which is limited by the instrument. The absorption coefficient for the planar µc-Si is around 6 × 10^4/cm at 550 nm. For the planar µc-Si, the absorption decreased with the increase of the wavelength. However, after 600 nm, planar µc-Si shows clear interference in the transmittance and reflectance spectrum, and Equation 23.4 does not apply well. The absorption coefficient for µc-Si after 600 nm can only be used for estimating the advantage of the SiNSs layer. By this estimation, the average absorption coefficient for µc-Si at 750 nm is around 2 × 10^3/cm. In comparison, the α for SiNSs still retained around 6 × 10^4/cm. There is a difference of about 27 times. This indicates that the light in the SiNSs at this wavelength is multiply reflected 27 times by a rough estimation.

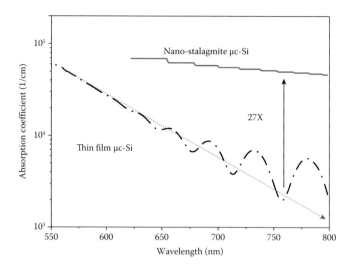

Figure 23.7 The absorption coefficient for thin microcrystalline silicon film with and without nanostalagmites arrays. (From Thiyagu, S., et al., *Nanoscale Res. Lett.*, 7, 171, 1–6, 2012.)

23.3.3 ANGLE-DEPENDENT SILICON NANOSTALAGMITE REFLECTION

A good antireflective coating should show low reflectance over a wide angle of incidence (AOI), which is important for applications in sunrise-to-sunset solar cells (Figure 23.8). Figure 23.9(a) shows that wide range of AOI in a wide range of wavelengths. We have measured all possible wavelengths in the range from 300 to 800 nm. One of the unique features of these catalytic chemical etching SiNS is that efficient light trapping occurs irrespective of the AOI. Angle-dependent reflectivity of SiNS arrays is shown in Figure 23.9b. The performance of SiNSs arrays showed a reduced dependence on the AOI and significantly higher absorption at any angle. At an AOI up to 60°, the total reflectance was maintained at approximately 0.3%.

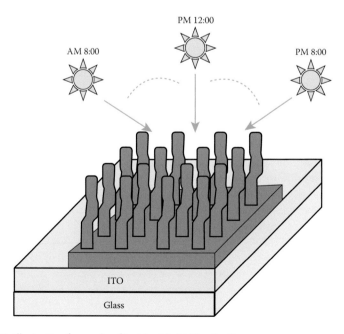

Figure 23.8 Schematic illustration for angle of incident light illumination on nanostructures.

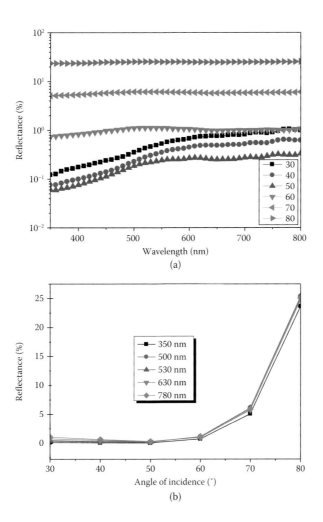

Figure 23.9 Optical measurement of (a) wide range of AOI in a wide range of wavelengths (b) angle-dependent reflectivity of SiNS.

23.4 SOLAR CELL USING LIGHT TRAPPING NANOSTRUCTURE

23.4.1 SOLAR CELL DESIGN

A solar cell is essential for a p–n junction with a large surface area. The n-type or p-type material is kept thin to allow light to pass through to the p–n junction. Light travels in packets of energy called photons. The generation of electric current happens inside the depletion zone of the p–n junction. The depletion region is the area around the p–n junction, where the electrons from the n-type silicon, have diffused into the holes of the p-type material. When a photon of light is absorbed by one of these atoms in the n-type silicon, it will dislodge an electron, creating a free electron and a hole. The free electron and hole have sufficient energy to jump out of the depletion zone. If a wire is connected from the cathode (n-type silicon) to the anode (p-type silicon), electrons will flow through the wire. The electron is attracted to the positive charge of the p-type material and travels through the external load (meter) creating a flow of electric current. The hole created by the dislodged electron is attracted to the negative charge of n-type material and migrates to the back electrical contact. As the electron enters the p-type silicon from the back electrical contact, it combines with the hole restoring the electrical neutrality. A schematic diagram and energy band diagram of a p-n junction solar cell is shown in Figure 23.10.

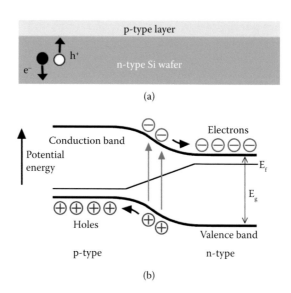

(a)

(b)

Figure 23.10 (a) Schematic diagram of planar p-n junction. (b) Energy band diagram of p-n junction solar cell.

23.4.2 RADIAL p-n JUNCTION

In a planar junction solar cell, generated charge carriers (electron or hole) are need to travel all the way from deep region to the junction without recombination. If the charge carriers to be the long diffusion length so that carriers can travel long distance but the material must have high crystallinity.

On the contrary, solar cell with radial junction and nanostructure arrays, the light will be trapped that most of the incident light is absorbed. In addition, radial junction provides a carrier transport direction transverse to the direction of light, that enable the photo-generated carrier collected in a very short distance. With the two specific advantages, the material quality could be tolerance from crystalline to polycrystalline or microcrystalline, for solar cells that sustain high efficiency. A schematic diagram of radial p-n junction is shown in Figure 23.11.

23.4.3 SOLAR CELL BASED ON ON P-TYPE POLYMER/N-TYPE Si NANOSTRUCTURE

To enhance the conversion efficiency, it is significant to implement a carrier collection structure on the SiNWs solar cells, such as a p-n junction diode as discussed in previous section. Unfortunately, the high temperature diffusion and surface passivation steps involved in silicon solar cell fabrication are not compatible with low-cost substrates. Also, high series and shunt resistances appear to limit the power conversion efficiency (PCE) of solar cells below 1% when using p-doped amorphous or nanocrystal silicon

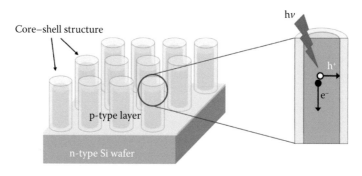

Figure 23.11 Schematic diagram of radial p-n junction.

Industrial nanosilicon

layers by chemical vapor deposition. To ameliorate this problem, conductive polymer is coated in the SiNW to form heterojunction solar cells that have several advantages, including large-area coverage, low-temperature process capability, easy preparation, low cost, and so on. The hole conducting (p-type) polymer poly(3,4-ethylenedioxythiophene): poly(styrenesulfonate) (PEDOT:PSS) was applied on the n-type SiNWs to form a heterojunction diode solar cell. The vertically aligned SiNW arrays have been fabricated by a catalytic etching process through a PS spheres template in a convenient method with controlled length, orientation, high density, and high aspect ratio. The interface between the PEDOT:PSS layer and SiNWs could possibly form good heterojunctions for electron-hole pairs (EHPs) separation because the highest occupied molecular orbital energy level of PEDOT:PSS is ~5.1 eV which is similar to the valence band energy of silicon. The SiNW/PEDOT:PSS heterojunction solar cell greatly shortens the carrier transport distance. Solar cell parameters such as open-circuit voltage (V_{oc}), FF, efficiency (η), and short circuit current (I_{sc}) have been determined as well as structural and morphological appearance in reflectance and SEM studies.

SiNWs were prepared on n-type silicon wafer thorough a template by chemical etching with a silver (Ag) catalyst. The template used polystyrene nanospheres with diameter 30–50 nm and density $10^{11}/cm^2$ by nanophase separation of the PS containing block copolymers.[18] The silver film was deposited using a thermal evaporator at the rate of 0.1 Å/s. The thickness of Ag film was 10 nm. A Teflon vessel was used as the container. For the solution etching process, an etching mixture consisting of HF, H_2O_2, and deionized water was used at room temperature. The concentrations of HF and H_2O_2 were 4.6 and 0.44 M, respectively. After etching, the substrate was immersed in toluene to remove PS nanospheres. The silver film was removed by immersion in boiling aqua regia (3:1 (v/v) HCl/HNO$_3$). Finally, the SiNWs were immersed in HF to remove the chemically contaminated oxide surface layer. Poly(3,4-ethylenedioxythiophene): poly(styrenesulfonate) (PEDOT:PSS) was used to form a SiNW/PEDOT:PSS heterojunction solar cell based on n-type SiNWs. Figure 23.12a shows a schematic diagram of SiNW/PEDOT:PSS heterojunction solar cell. For the device fabrication, Al electrical contacts were deposited on the back side of the Si surface. Then, the PEDOT:PSS was spin coated on the ITO-coated glass rather than being directly spin coated on the SiNW arrays. The top portion of the SiNWs was then immersed in the thin wet PEDOT:PSS film. The samples were subsequently annealed at 140°C for 10 min and each individual SiNW was expected to stick on an ITO electrode via PEDOT:PSS. The energy level of the heterojunction solar cells is sketched in Figure 23.12b.

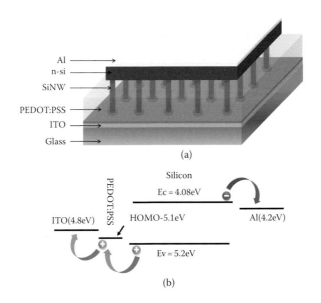

(a)

(b)

Figure 23.12 The (a) schematic and (b) energy band diagram of the SiNW/PEDOT:PSS heterojunction solar cells.

The morphologies of SiNWs were examined by SEM. Figure 23.13a and b shows the cross section SEM images of the as-prepared SiNWs arrays on n-type Si substrate. The SiNWs were fabricated with different etching times of 20 and 180 s; obtained nanowires had a length of around 250 nm and 2 µm, respectively. The length of SiNWs can be controlled and varied linearly with the etching duration of the catalyst etching process. SiNWs with controlled length and orientation with high density are about $10^{11}/cm^2$ with a high aspect ratio by the PS spheres template with diameter around 30–50 nm. Figure 23.13c and d shows the SEM images of the SiNWs secluded from the ITO-coated glass by a mechanical force. The PEDOT:PSS is covered almost the SiNWs entire surface. The polymer makes a continuous film on SiNWs arrays which penetrate far down into the gap between the nanowires. Figure 23.13c shows the cross-sectional SEM images of polymer filling SiNWs with an etching time of 180 s. Figure 23.13d shows the angle 45° SEM images of SiNWs surface-covered with PEDOT:PSS. PEDOT:PSS were covered just not only top of the Si surface also completely infiltrate into the gap of SiNWs arrays.

To investigate the characteristics of SiNW/PEDOT:PSS a heterojunction solar cell was fabricated with various etching times in SiNW formation from 10 to 180 s. The photovoltaic J-V characteristic of the SiNW/PEDOT:PSS heterojunction solar cell with various etching times was measured under an illumination intensity of 100 mA/cm² (AM 1.5G) as shown in Figure 23.14. The details of the photovoltaic characteristics of the device are listed in Table 23.1.

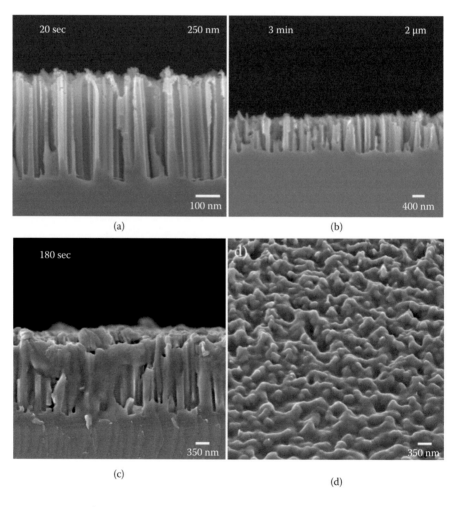

(a) (b)

(c) (d)

Figure 23.13 Cross-sectional SEM images of SiNW arrays after etching for (a) 20 and (b) 180 s and SiNWs covered with PEDOT with etching time of (c) 180 s and (d) angle of 45°.

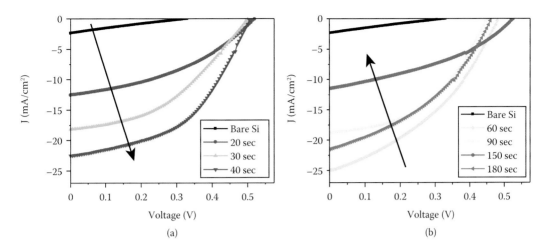

Figure 23.14 (a, b) Photocurrent–voltage characteristics of the SiNW/PEDOT:PSS heterojunction solar cell with varying etching time in SiNW formation.

Table 23.1 **The photovoltaic characteristics of the device**

ETCHING TIME, SECOND	NANOWIRE LENGTH,[a] nm	JSC, mA/cm²	Voc, V	FILL FACTOR	EFFICIENCY, %	RS, Ω-cm²	RSH, Ω-cm²
0	0	2.35	0.33	0.22	0.4	189	122
10	125	12.5	0.52	0.39	2.59	23.2	140
20	250	18.1	0.50	0.42	3.83	18.2	183
30	375	22.5	0.50	0.47	5.47	7.5	313
40	500	19.4	0.49	0.47	4.61	8.2	175
50	625	22.5	0.48	0.37	4.1	11.1	75
60	750	24.9	0.48	0.37	4.6	7.9	71
90	1125	21.5	0.46	0.38	3.87	8.9	100
120	1500	20.2	0.44	0.41	3.78	9.7	88
150	1875	18.7	0.45	0.45	3.89	16.1	100
180	2250	11.5	0.52	0.37	2.26	25	80

[a] The average etching rate is around 12.5 nm/s.

Figure 23.14 shows the photovoltaic effect of PEDOT:PSS on bare Si and on SiNWs. A strong increase in short-circuit current density (J_{sc}), open circuit voltage (V_{oc}) FF, and PCE was observed in the cells with SiNW structures. The J_{sc} improved from 2.35 to 22.5 mA/cm², the V_{oc} from 0.33 to 0.50 V, the *FF* from 0.22 to 0.47, and resulted in an enhancement of the PCE from 0.4% to 5.47%. The major reasons for increase in short-circuit current density were the larger contact area, carrier diffusion distance, ultralow reflectance, and light trapping effect in the SiNW arrays.

The electron–hole pair and collection efficiency are greatly improved in the SiNW/PEDOT:PSS solar cells because the diffusion distance from the core of the SiNW to the SiNW/PEDOT:PSS heterojunction is only several nanometers compared to the several micrometers in planar cells. The most photogenerated EHP in SiNWs, the hole in SiNWs will move toward PEDOT:PSS and electrons will move along SiNWs to the Al contact. The ultralow reflectance of SiNW arrays is about 0.1% over the entire spectral range from 300 to 800 nm. The ultralow reflectance of SiNW ultimately leads to strong absorption due to a strong light trapping effect inside the dense SiNWs. The spectral reflectance of the SiNWs was measured at different etching times, from 10 to 180 s as shown in Figure 23.15.

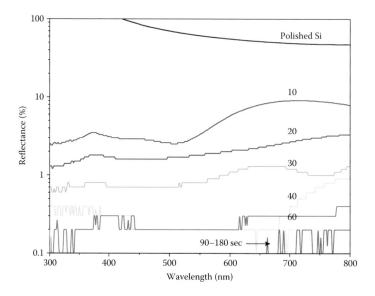

Figure 23.15 Reflectance spectra of SiNWs with various etching times from 10 to 180 s.

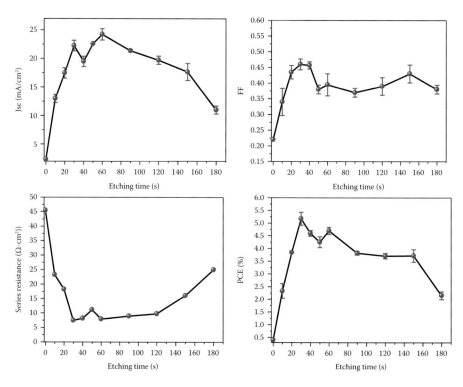

Figure 23.16 The dependence of photovoltaic (PV) parameters of SiNW/PEDOT:PSS heterojunction solar cells is plotted as a functions of etching time in SiNW formation.

For only 10 s of catalyst chemical etching, the reflectance drastically reduces from over 40% to less than 10% as compared to the surface polished Si wafer over the entire spectral range of 300 to 800 nm. For further increases in the etching time, the reflection decreases gradually. After 60 s, the reflection is approaching 0.1%. The length of SiNWs also greatly influences the properties of SiNW arrays and SiNW/PEDOT:PSS heterojunction solar cells. Lengthy catalyst etching time results in longer SiNWs.

In Figure 23.5, the J_{sc}, *FF*, Rs, and PCE parameters of the SiNW/PEDOT:PSS heterojunction cell are plotted as a function of the catalyst chemical etching time. For each second, several solar cells are fabricated and measured. The changes in efficiency mainly resulted from the changes in J$_{sc}$. It indicated that the J$_{sc}$ was the parameter that was affected most directly and significantly by the increase in etching time.

A similar dependence occurs in *FF*, and the increase in etching time of the SiNW caused a gradual decline in both of the parameters. The changes in Jsc were more attributed to the changes in series resistance (Rs) and shunt resistance (Rsh) (Table 23.1). In Figure 23.16, when the etching time increases, the current density (J_{sc}) curve was an appropriate mirror match with the series resistance (Rs) curve. The improvements in Jsc and Voc both correspond to the reduction in surface recombination, while lower FF stems from the significantly higher sheet resistance in the conductive polymer. The average values of the device performance parameters for each catalyst etching second are plotted. In Figure 23.16, the average J$_{sc}$ in the SiNW/PEDOT:PSS heterojunction cells increase from 12.5 to 24.9 mA/cm^2 as the catalyst etching time increases from 10 to 60 s; nevertheless, the average J$_{sc}$ decreases from 24.9 to 11.5 mA/cm^2 as the catalyst etching time increase from 60 to 180 s. The J$_{sc}$ decreases mainly due to the series and shunt resistance—but it might be due to a leakage of current. The carrier collection efficiency and overall efficiency decreased due to the long diffusion distance of carriers from the Si substrate.

23.5 SUMMARY

The SiNSs were prepared by catalytic etching process through a 30–50 nm PS nanosphere template. The length of nanostalagmite is defined by the duration of the etching process. The SiNSs arrays have low transmission, ultralow reflection of approximately 0.3%, and high absorption around 99% compared to planar material due to a strong light trapping effect. Reflection is also suppressed for a wide range of angles of incidence in a wide range of wavelengths. This indicates the extensive light trapping effect by the SiNS and could possibly harvest large amounts of solar energy in the infrared regime. To form a junction on SiNS for an efficient solar cell, the PEDOT:PSS was applied to form hybrid heterojunction solar cells. A short-circuit current of 21.1 mA/cm^2 and PCE of 5.7% was achieved in SiNS/PEDOT:PSS heterojunction solar cells, while the planar cells were only 0.4%. The performance of the solar cells was primary dependent on the etching time of SiNS. With a longer etching time, the performance of the solar cell begins to decrease. This is believed to be associated to the limited conductance of the conducting polymer, PEDOT:PSS.

REFERENCES

1. S. Thiyagu, H. J. Syu, C. T. Liu, C. C. Hsueh, S. T. Yang, and C. F. Lin, Low-pressure-assisted coating method to improve interface between PEDOT:PSS and silicon nanotips for high-efficiency organic/inorganic hybrid solar cells via solution process, *ACS Appl. Mater. Interfaces*, 2016, 8, 2406–2415.
2. S. Thiyagu, C. C. Hsueh, H. J. Syu, C. T. Liu, S. T. Yang, and C. F. Lin, Interface modification for efficiency enhancement in silicon nanohole hybrid solar cells, *RSC Adv.*, 2016, 6, 12374.
3. S. Thiyagu, H. J. Syu, C. C. Hsueh, C. T. Liu, T. C. Lin, and C. F. Lin, Optical trapping enhancement from high density silicon nanohole and nanowire arrays for efficient hybrid organic–inorganic solar cells, *RSC Adv.*, 2015, 5, 13224.
4. S. Thiyagu, C. C. Hsueh, C. T. Liu, H. J. Syu, T. C. Lin, and C. F. Lin, Hybrid organic-inorganic heterojunction solar cells with 12% efficiency by utilizing flexible film-silicon with hierarchical surface, *Nanoscale*, 2014, 6, 3361–3366.
5. W. Lu, and C. M. Lieber, Semiconductor nanowires, *J. Phys. D Appl. Phys.*, 2006, 39, R387–R407.
6. M. D. Kelzenberg, D. B. Turner-Evans, B. M. Kayes, M. A. Filler, M. C. Putnam, N. S. Lewis, and H. A. Atwater, Photovoltaic measurements in single-nanowire silicon solar, *Nano Lett.*, 2008, 8(2), 710–714.
7. T. H. Stelzner, M. Pietsch, G. Andrä, F. Falk, E. Ose, and S. H. Christiansen, Silicon nanowire-based solar cells, *Nanotechnology*, 2008, 19, 295203.
8. B. Tian, X. Zheng, T. J. Kempa1, Y. Fang, N. Yu, G. Yu, J. Huang, and C. M. Lieber, Coaxial silicon nanowires as solar cells and nanoelectronic power sources, *Nature*, 2007, 449, 885–889.
9. K. Peng, Z. Huang, and J. Zhu, Fabrication of large-area silicon nanowire p-n junction diode arrays, *Adv. Mater.*, 2004, 16, 73–76.

10. S. Thiyagu, B. P. Devi, and Z. Pei, Fabrication of large area high density, ultra-low reflection silicon nanowire arrays for efficient solar cell applications, *Nano Res.*, 2011, 4(11), 1136–1143.
11. Z. Pei, S. Thiyagu, M. S. Jhong, W. S. Hsieh, S. J. Cheng, M. W. Ho, Y. H. Chen, J. C. Liu, and C. M. Yeh, An amorphous silicon random nanocone/polymer hybrid solar cell, *Sol. Energ. Mater. Sol. Cells*, 2011, 95, 2431–2436.
12. O. H. Elibol, D. Morisette, D. Akin, J. P. Denton, and R. Bashir, Integrated nanoscale silicon sensors using top-down fabrication, *Appl. Phys. Lett.*, 2003, 83, 4613–4615.
13. A. M. Morales, and C. M. Lieber, A laser ablation method for the synthesis of crystalline semiconductor nanowires, *Science*, 1998, 279, 208–211.
14. D. P. Yu, Z. G. Bai, Y. Ding, Q. L. Hang, H. Z. Zhang, J. J. Wang, Y. H. Zou, W. Qian, G. C. Xiong, H. T. Zhou, and S. Q. Feng, Nanoscale silicon wires synthesized using simple physical evaporation, *Appl. Phys. Lett.*, 1998, 72, 3458–3460.
15. B. S. Kim, T. W. Koo, J. H. Lee, D. S. Kim, Y. C. Jung, S. W. Hwang, B. L. Choi, E. K. Lee, J. M. Kim, and D. Whang, Catalyst-free growth of single-crystal silicon and germanium Nanowires, *Nano Lett.*, 2009, 9, 864–869.
16. X. Liu, and D. Wang, Kinetically-induced hexagonality in chemically grown silicon nanowires, *Nano Res.*, 2009, 2, 575–582.
17. D. Whang, S. Jin, Y. Wu, and C. M. Lieber, Large-scale hierarchical organization of nanowire arrays for integrated nanosystems, *Nano Lett.*, 2003, 3, 1255–1259.
18. S. Hegedus, Thin film solar modules: The low cost, high throughput and versatile alternative to Si wafers, *Prog. Photovolt. Res. Appl.*, 2006, 14, 393–411.
19. M. A. Green, Consolidation of thin-film photovoltaic technology: The coming decade of opportunity, *Prog. Photovolt. Res. Appl.*, 2006, 14, 383–392.
20. B. Rech, and H. Wagner, Potential of amorphous silicon for solar cells, *Appl. Phys. A Mater. Sci. Process.*, 1999, 69, 155–167.
21. J. Meier, J. Spitznagel, U. Kroll, C. Bucher, S. Faÿ, T. Moriarty, and A. Shah, Potential of amorphous and microcrystalline silicon solar cells, *Thin Solid Films*, 2004, 518, 451–452.
22. A. Salleo, and W. S. Wong, Editors, *Flexible Electronics: Materials and Applications*, New York, Springer, 2009.
23. R. S. Wagner, and W. C. Ellis, Vapor-liquid-solid mechanism of single crystal growthapor-liquid-solid mechanism of single crystal growth, *Appl. Phys. Lett.*, 1964, 4, 89–90.
24. B. S. Kim, S. Sangwoo, S. J. Shin, K. M. Kim, and H. H. Cho, Micro-nano hybrid structures with manipulated wettability using a two-step silicon etching on a large area, *Nanoscale Res. Lett.*, 2011, 6, 333–342.
25. L. Linhan, G. Siping, S. Xianzhong, F. Jiayou, and W. Yan, Synthesis and photoluminescence properties of porous silicon nanowire arrays, *Nanoscale Res. Lett.*, 2010, 5, 1822–1828.
26. C. T. Black, R. Ruiz, G. Breyta, J. Y. Cheng, M. E. Colburn, K. W. Guarini, H. C. Kim, and Y. Zhang, Polymer self-assembly in semiconductor microelectronics, *IBM J. Res. Dev.*, 2007, 51, 605–608.
27. K. Asakawa, and T. Hiraoka, Nanopatterning with microdomains of block copolymers using reactive-ion etching selectivity, *Jpn. J. Appl. Phys.*, 2002, 41, 6112–6118.
28. W. A. Lopes, and H. M. Jaeger, Hierarchical self-assembly of metal nanostructures on diblock copolymer scaffolds, *Nature*, 2001, 414, 735–738.
29. T. Thurn-Albrecht, J. Schotter, G. A. Kästle, N. Emley, T. Shibauchi, L. Krusin-Elbaum, K. Guarini, C T. Black, M. T. Tuominen, and T. P. Russell, Ultrahigh-density nanowire arrays grown in self-assembled diblock copolymer templates, *Science*, 2000, 290, 2126–2129.
30. F. S. Bates, and G. H. Fredrickson, Block copolymers—Designer soft materials, *Phys. Today*, 1999, 52, 32–38.
31. S. Thiyagu, Z. Pei, M. W. Ho, S. J. Cheng, W. S. Hsieh, and Y. Y. Lin, A modified block copolymer nano-patterning method for high density sub-30 nm polystyrene nanosphere and gold nanomesh formation. *Nanosci. Nanotechnol. Lett.* 2011, 3(2), 215–221.
32. S. Thiyagu, B. P. Devi, Z. Pei, Y. H. Chen, and J. C. Liu, Ultra-low reflectance, high absorption microcrystalline silicon nanostalagmite, *Nanoscale Res. Lett.*, 2012, 7, 171, 1–6.

24

Bottom-up nanostructured silicon for thermoelectrics

Aikebaier Yusufu and Ken Kurosaki

Contents

24.1 INTRODUCTION

The energy demand has been continually increasing all over the world; however, over 60% of the energy obtained from primary sources is converted into waste heat across all fields, including industry, agriculture, and everyday life (Kajikawa 2006). Thus, various alternative energy saving/harvesting technologies, which would allow efficient reuse of waste heat (such as thermoelectric [TE] energy conversion) are attracting significant attention.

Generally, TE modules are fabricated by bonding p- and n-type TE materials into a π pattern, which is subsequently sandwiched between ceramic plates. In such a system, an electrical current is generated due to the Seebeck effect by creating a temperature difference between the upper and the lower sections of the module. Consequently, the applied electrical current generates a temperature gradient due to the Peltier effect (Figure 24.1). The described phenomena can be used for power generation and electronic refrigeration, respectively (Goldsmid 1964; Tritt and Subramanian 2006; Bell 2008; Yadav et al. 2011). According to Equation 24.1 below, the conversion efficiency (η) of a TE device depends on the figure of merit (ZT) of the utilized TE material and the temperature difference between its top and bottom parts:

$$\eta = \frac{T_{\mathrm{H}} - T_{\mathrm{C}}}{T_{\mathrm{H}}} \left[\frac{1 + ZT_{\mathrm{M}} - 1}{(1 + ZT_{\mathrm{M}})^{1/2} + (T_{\mathrm{C}} / T_{\mathrm{H}})} \right] \tag{24.1}$$

where T_{H} is the temperature of the hot side, T_{C} is the temperature of the cold side, and T_{M} is the average temperature (the relationships between η and ZT obtained at different values of T_{H} are plotted in Figure 24.2). Since the $(T_{\mathrm{H}}-T_{\mathrm{C}})/T_{\mathrm{H}}$ term in Equation 24.1 corresponds to Carnot efficiency, the magnitude of η monotonically increases with increasing ZT until it reaches the maximum Carnot efficiency value. The term ZT is defined by Equation 24.2, where S, σ, T, κ_{el}, and κ_{lat} are the Seebeck coefficient,

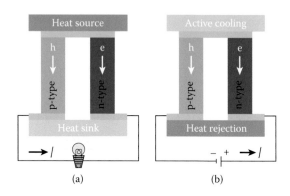

Figure 24.1 Thermoelectric modules made of n-type and p-type thermoelectric materials: (a) power generation and (b) refrigeration processes.

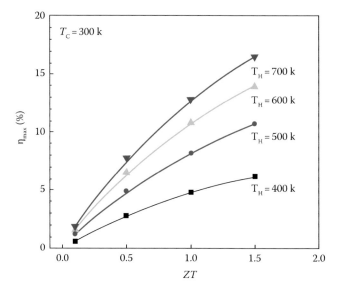

Figure 24.2 Maximum conversion efficiency η_{max} plotted as a function of material's figure of merit ZT at various temperatures. The lower temperature T_C was set to 300 K, and the upper temperature T_H was varied between 400, 500, 600, and 700 K.

electrical conductivity, absolute temperature, and electronic and lattice (phonon) components of the thermal conductivity (κ), respectively (Snyder and Toberer 2008):

$$ZT = \frac{S^2\sigma}{\left(\kappa_{el} + \kappa_{lat}\right)}T \tag{24.2}$$

As shown in Equation 24.2, the value of ZT can be increased by decreasing κ or by increasing either S or σ. However, σ is closely related to κ_{el} through the Wiedemann–Franz relationship described in Equation 24.3 below, where L is the Lorenz number (2.44×10^{-8} W·Ω·K^{-2} for metals):

$$\kappa_{el} = L\sigma T \tag{24.3}$$

The magnitudes of S, σ, and κ_{el} depend on the carrier concentration in a material, while the maximum value of $S^2\sigma$ corresponds to the optimized carrier concentration, indicating that ZT can be effectively increased by decreasing κ_{lat}.

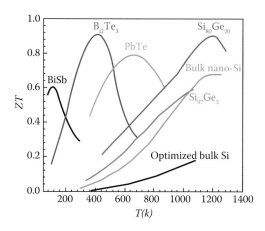

Figure 24.3 Temperature dependencies of the figure of merit ZT plotted for conventional TE materials: BiSb (Sootsman et al. 2009), Bi_2Te_3, PbTe (Sootsman et al. 2009), and $Si_{80}Ge_{20}$ (Sootsman et al. 2009). The ZT data for optimized bulk Si (Zhu et al. 2009), bulk nano-Si (Bux et al. 2009), and bulk nano-$Si_{97}Ge_3$ (Yusufu et al. 2014) are shown for comparison.

Table 24.1 Elemental characteristics of Bi and Te: reserves, distribution, and toxicity

	RESERVE (T)	DISTRIBUTION	LD_{50} (mg/kg)
Bi	320,000[a]	China (69.1%)[a]	–
Te	22,000[a]	USA (12.8%)[a]	83[b]

[a] Mineral Commodity Summaries 2010.
[b] Material Safety Data Sheet.

TE modules are silent, reliable, and scalable solid-state devices with no moving parts, which make them ideal candidates for small distributed power generation applications. However, the efficiency of the currently used TE devices is relatively low, and the utilized TE materials are composed of toxic or rare elements, which limit their power generation applications to specialized areas such as NASA spacecraft (Yang and Caillat 2006). More broad industrial applications (such as power generation from exhaust heat in automobiles) have not been widely considered. The temperature dependencies of ZT for well-known high-performance TE materials such as BiSb (Sootsman et al. 2009), Bi_2Te_3 (Sootsman et al. 2009), PbTe (Sootsman et al. 2009), and Si–Ge alloys (Sootsman et al. 2009) are shown in Figure 24.3. Among these compounds, the best bulk TE materials correspond to Bi_2Te_3-based ones, which can be used for refrigeration near room temperature (Poudel et al. 2008), and PbTe-based compounds utilized for power generation at middle temperatures (Yang and Caillat 2006); the ZT values obtained for these materials are close to unity, which is equivalent to the device efficiency of several percent. Table 24.1 summarizes the elemental characteristics of Bi and Te. Although the Bi reserves are much higher than those of Te, the former is mostly distributed in China. Moreover, Te is a toxic element, which makes the use of Bi_2Te_3 in consumer markets very problematic. Therefore, the development of high-efficiency TE materials from inexpensive, earth-abundant, and environmentally friendly elements is highly required. In this work, nanostructured Si is considered the best candidate material, which satisfies the above-mentioned criteria.

24.2 SILICON AS A THERMOELECTRIC MATERIAL

Table 24.2 compares various TE properties of Si and Bi_2Te_3 compounds. In particular, Si exhibits good electrical characteristics (such as high $S^2\sigma$), which are comparable to those of Bi_2Te_3, but its magnitude of κ_{lat} is very large (>100 W·m^{-1} K^{-1} at room temperature) (Weber and Gmelin 1991; Bux et al. 2009; Zhu et al. 2009) and corresponds to the maximum ZT of 0.01 at room temperature (Zhu et al. 2009). The κ_{lat} of an undoped Si single crystal is approximately 140 W·m^{-1} K^{-1} at room temperature,

Table 24.2 Thermoelectric properties of doped Si (Bux et al. 2009) and Bi$_x$Sb$_{2-x}$Te$_3$ (Lan et al. 2010)

	n_H	σ	S	$S^2\sigma$	K_{lat}	ZT_{MAX}
	10^{19} cm^{-3}	10^5 $\Omega^{-1}\cdot$m^{-1}	μV K^{-1}	mW$\cdot$m$^{-1}\cdot$K^{-2}	W$\cdot$m$^{-1}\cdot$K^{-1}	–
Doped Si	1.7	0.23	420	4.1	100	0.01
Bi$_x$Sb$_{2-x}$Te$_3$	1.8	0.95	220	4.6	1.4	1

Note: The data are obtained at room temperature.

which is almost 200 times larger than that of Bi$_2$Te$_3$ (Lan et al. 2010). If the κ_{lat} of Si could be lowered while maintaining high values of $S^2\sigma$, it would become the ideal TE material. Therefore, much effort has been spent on reducing the κ_{lat} of bulk Si. For instance, it can be significantly decreased by forming a solid solution with Ge, and its optimized composition (corresponding to the stoichiometric formula of Si$_{100-x}$Ge$_x$ [x = 20–30]) exhibits the highest ZT value of approximately 0.7–0.8 (Dismukes et al. 1964; Slack and Hussain 1991). However, various issues such as low ZT values and high costs of Ge limit possible industrial applications of these materials. Alloy scattering via Ge substitution is a very effective way of reducing the κ_{lat} of Si; on the other hand, it has been experimentally proved that nanostructuring also reduces the magnitude of κ_{lat}, leading to values that are similar or even lower than those obtained through Ge substitution, and, hence, higher values of ZT (Joshi et al. 2008; Wang et al. 2008; Bux et al. 2009; Zebarjadi et al. 2011; Bathula et al. 2012; Yu et al. 2012; Miura et al. 2015; Kurosaki et al. 2016).

24.3 ENHANCEMENT OF SILICON THERMOELECTRIC PROPERTIES VIA NANOSTRUCTURING

As described in the previous section, the significant reduction of κ_{lat} accompanied by a little decrease in σ is a key route for enhancing the ZT of Si. It is well known that κ_{lat} and σ are directly related to the mean free path (MFP) of phonons and electrons, respectively. For crystalline Si, the MFP of phonons corresponds to the micrometer scale and significantly exceeds that of electrons. Minnich et al. (2009) have shown that most of the heat is generated by the phonons with a MFP of more than 100 nm. At room temperature, the phonons with a MFP of more than 1 μm contribute to around 40% of κ (Regner et al. 2013), suggesting that the magnitude of κ can be potentially reduced without decreasing σ when the size of the material structure is smaller than the MFP of phonons and larger than that of electrons.

Using this strategy, Boukai et al. (2008) and Hochbaum et al. (2008) have fabricated Si nanowires with the diameters below the MFP of phonons. Hochbaum et al. (2008) produced p-type Si nanowires with diameters of 50–115 nm and evaluated their TE properties. The maximum ZT value of 0.6 was obtained at room temperature for the Si nanowires with a diameter of 52 nm. On the other hand, Boukai et al. (2008) fabricated p-type Si-nanowire arrays with widths of 10–20 nm and a thickness of 20 nm. An extremely small average value of κ (0.76 ± 0.15 W$\cdot$m$^{-1}\cdot$K^{-1}) was obtained for 10-nm-wide Si-nanowire arrays, which was lower than that of amorphous Si (1 W$\cdot$m$^{-1}\cdot$K^{-1}) and the minimum κ value for bulk Si (0.99 W$\cdot$m$^{-1}\cdot$K^{-1}). As a result, the high ZT value of 1 was achieved at 200 K because of the large reduction in κ.

High TE properties have also been reported for Si nanoribbons (Korotcenkov and Cho 2010; Tang et al. 2010; Yu et al. 2010; Boor et al. 2012). The ZT value of 0.4 was obtained at room temperature for the p-type Si ribbons fabricated via reactive ion etching of a Si-on-insulator substrate (width = 1–3 μm, length = 20–50 μm), which were composed of uniform arrays of nanoholes (55 nm pitch holey; Tang et al. 2010). In addition, Si nanofilms containing nanoholes with diameters of 11–16 nm, which were characterized by very low values of κ not exceeding 2 W$\cdot$m^{-1} K^{-1}, were produced via electron beam lithography (Yu et al. 2010). However, the drawback of such porous Si materials is that their magnitudes of σ decrease because of the reduction in μ, which mostly likely occurs due to the dispersion of charge carriers across SiOH and SiO$_x$H$_y$ surface layers (Korotcenkov and Cho 2010). Therefore, maintaining pristine surfaces is important for preventing the dispersion of charge carriers in nanohole-containing Si ribbons (Boor et al. 2012).

Although Si nanowires and nanoholes exhibit good TE properties, the practical applications of such finely structured Si materials are limited because they are not suitable for mass production. Therefore, materials with finely controlled nanostructures (even in the bulk state) are required. Bulk Si nanocrystals have recently received much attention as potential candidates for inexpensive, nontoxic, and highly efficient nanostructured TE materials. The TE properties of n-type bulk Si nanocrystals fabricated via ball milling (BM) combined with pressure sintering were studied by Bux et al. (2009). More specifically, Si powder was first ball-milled in a glove box under regulated atmosphere until it was converted into nanoscale particles. The obtained nanopowder was then sintered by using either spark plasma sintering (SPS) or hot pressing (HP) techniques to produce bulk Si nanocrystals with a finely controlled internal nanostructure. The material nanostructure corresponding to the best TE performance has been investigated by optimizing the utilized BM and pressure sintering conditions. The κ value of the bulk Si nanocrystals obtained at room temperature was approximately 12 $W^{-1}\cdot m^{-1}\cdot K^{-1}$, and the ZT magnitude of 0.7 was achieved at 1275 K. On the other hand, enhanced TE properties were reported for the bulk Si nanocrystals fabricated by sintering n-type Si nanoparticles, which were synthesized from $SiSl_4$ via a vapor-phase synthetic route (Miura et al. 2015). The obtained Si nanostructures exhibited the high ZT value of 0.6 at 1073 K, which was mainly due to the significant reduction in κ. The low κ values of Si nanocrystals can be explained by a model that takes into account nanoscale phonon scattering at crystal grain boundaries (Suzuki et al. 2012).

24.4 BOTTOM-UP AND TOP-DOWN SYNTHESIS OF NANOSTRUCTURED BULK SILICON FOR THERMOELECTRIC APPLICATIONS

Nanostructured bulk TE materials are typically manufactured via two routes. The first route corresponds to the bulk formation due to fabrication and consolidation of fine nanocrystals, while the second route occurs via natural precipitation of nanoscale particles (they can be called top-down and bottom-up processes, respectively). The top-down process is very efficient and has been applied to the synthesis of various TE materials including Si (Bux et al. 2009), BiSbTe (Poudel et al. 2008), and Si–Ge alloys (Zhu et al. 2009). For example, the BiSbTe bulk nanocrystals synthesized through the top-down process exhibit the maximum ZT value of 1.4 (Poudel et al. 2008), which is much higher than the ZT magnitude of 1 obtained for conventional bulk BiSbTe. As shown in Figure 24.4, the top-down structures are usually prepared by a method which combines the BM and HP, or SPS techniques. Unfortunately, this top-down process has some drawbacks. Since undesirable impurities can be introduced in the material structure during BM, the resulting nanocrystals may grow too much during HP/SPS and undergo oxidization during the BM and/or HP/SPS stages. On the other hand, bottom-up structures are usually prepared by melt solidification under appropriate cooling conditions (see Figure 24.5). The bottom-up process has been realized via a solid-state phase transformation such as second-phase precipitation (Pei et al. 2011; Biswas et al. 2012; Heinz et al. 2014; Yusufu et al. 2014), metastable phase decomposition (Heinz et al. 2014), or spinodal decomposition (Androulakis et al. 2007; Gelbstein et al. 2013). Using this technique, nanocrystals of various lead chalcogenides such as PbS (Zhao et al. 2014), PbSe (Lee et al. 2013; Zhao et al. 2014), and PbTe (Pei et al. 2011; Zhao et al. 2014) have been produced. For example, SrTe–PbTe nanocomposites (Biswas et al. 2012) prepared through the bottom-up process, in which nanoscale SrTe is naturally precipitated inside the PbTe matrix, exhibited a record high ZT of 2.2. Because Si nanopowders oxidize easily, it is difficult to apply

Industrial nanosilicon

(a)　　　　(b)　　　　(c)　　　　(d)

Figure 24.4 A schematic diagram of the top-down process: (a) chunks of starting materials, (b) nanopowders prepared by BM, (c) bulk formation via sintering, and (d) bulk materials composed of fine nanocrystals.

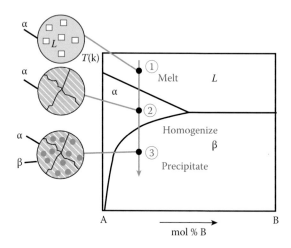

Figure 24.5 A schematic diagram of the bottom-up process. (1) The material is initially melted when the α phase nucleation occurs. (2) The α phase becomes homogenized. (3) The second (β) phase is formed inside the matrix phase (α phase) during the cooling stage. The size and distribution of the second phase can be controlled by varying cooling conditions (such as cooling speed).

the BM–HP/SPS method (top-down process) to manufacturing bulk nanostructured Si. Thus, the bottom-up technique (involving nanoscale second-phase precipitation) can be the best method for the fabrication of bulk Si nanostructures.

24.5 ENHANCED THERMOELECTRIC PROPERTIES OF BOTTOM-UP NANOSTRUCTURED SILICON–GERMANIUM ALLOYS

In this section, we demonstrate the enhancement of the TE properties of nanostructured bulk Si–Ge alloys synthesized by the bottom-up method. According to the Si–P phase diagram depicted in Figure 24.6 (Olesinski et al. 1985), the solubility of P in Si significantly depends on temperature: it reaches 2.4 at.% at 1453 K and is almost zero at room temperature. Thus, a P-rich phase will precipitate during solidification of super-heavily P-doped Si from a melt. In this study, we defined the nominal alloy composition as $Si_{100-x}Ge_xP_3$ ($x = 0, 1, 3, 5, 10,$ or 20), in which Ge species act as additional phonon scattering centers decreasing κ_{lat}.

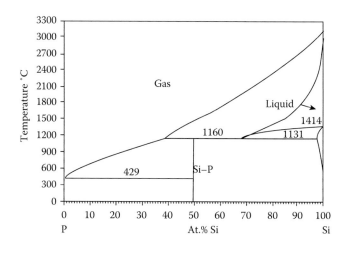

Figure 24.6 An Si–P binary phase diagram (Adapted from Olesinski, R.W., et al., *J. Phase Equil.* 6, 130–133, 1985.).

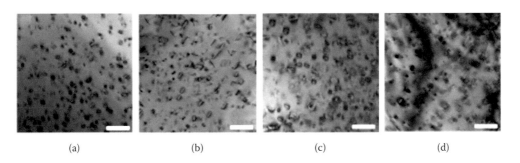

Figure 24.7 Bright-field TEM images of the samples with nominal compositions of $Si_{100-x}Ge_xP_3$. Panels (a), (b), (c), and (d) correspond to x = 0, 1, 3, and 20, respectively. All scale bars are 100 nm. The dark contrast areas around the precipitates are caused by strain. (Adapted from Yusufu, A., et al., *Nanoscale*, 6, 13921–13927, 2014.)

The correct amounts of the starting materials (Si, Ge, and P) were arc-melted, roughly crushed into microscale powders, and finally transformed into the bulk state via SPS under appropriate conditions.

The bright-field transmission electron microscopy (TEM) images of the representative Si–Ge alloy samples (depicted in Figure 24.7) contain nanoscale precipitates with various sizes below a few dozen nanometers. Two different types of precipitates exist: spherical precipitates with diameters of approximately 5 nm, and bar-shaped (plate-like) precipitates with lengths of approximately 20 nm (their high-resolution TEM images are shown in Figure 24.8). The energy dispersive X-ray spectrometry (EDS) point analysis revealed that the plate-like precipitate at point B contained more P species than the Si matrix at point A while no difference in Ge content was observed, indicating that the nanoscale precipitates were P-rich compounds. Furthermore, as shown in Figure 24.8b and c, the plate-like precipitates interacted with the

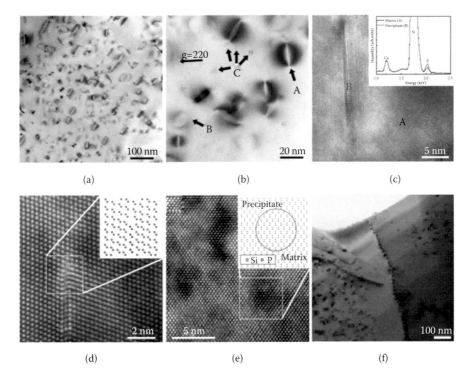

Figure 24.8 (a) An HRTEM image and EDS analysis of the plate-like precipitates. HRTEM lattice images and structural models of the (b) plate-like precipitates along the [110] direction and (c) nearly spherical precipitates, (d) a plate-like precipitate taken along the [110] direction and (e) nearly spherical precipitates; (f) a low-magnification bright-field image showing a grain boundary. (Adapted from Yusufu, A, et al., *Nanoscale*, 6, 13921–13927, 2014.)

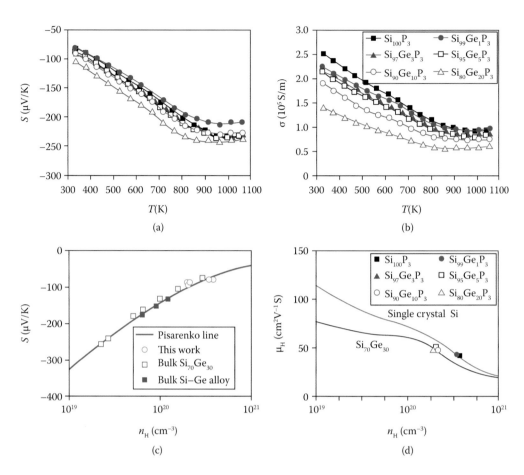

Figure 24.9 Temperature dependencies of the (a) Seebeck coefficient S and (b) electrical conductivity σ measured for the nanostructured bulk Si–Ge alloys synthesized via the bottom-up process. Relationships between (c) the Seebeck coefficient S and the Hall carrier concentration n_H and (d) the Hall carrier mobility μ_H and the Hall carrier concentration n_H for the same samples. The theoretical data for non-nanostructured Si–Ge alloys are shown for comparison. All measurements were performed at 300 K.

matrix phase in a semicoherent manner, while the spherical precipitates interacted with the matrix phase in a coherent manner (as will be discussed later, both coherent and semicoherent bonding types are very important for controlling the transport properties of phonons and electrons).

Figure 24.9a and b shows the temperature dependences of S and σ, respectively, for the nanostructured bulk Si–Ge alloys. Because the addition of P into Si–Ge alloys led to electron doping, all samples exhibited negative values of S over the entire temperature range. As the temperature increased, the absolute value of S increased as well, while the magnitude of σ decreased, demonstrating the typical behavior of degenerate semiconductors. According to the data depicted in Figure 24.9c, the relationship between S and the Hall carrier concentration n_H can be approximated for all samples with the Pisarenko line obtained by using a theoretical model, in which the nanostructure effect was not taken into account. Furthermore, the Hall mobility (μ_H) values calculated for all samples lie on the theoretical curves of μ_H versus n_H for Si–Ge alloys containing no nanostructures. Thus, the electrical properties of nanostructured bulk Si–Ge alloys are similar to those of traditional (non-nanostructured) Si–Ge alloys, indicating that the electron transport characteristics are hardly affected by the formation of nanoscale precipitates.

On the other hand, Figure 24.10a and b shows that the values of κ and κ_{lat} decreased throughout the entire measured temperature range. Our previous theoretical calculations (Yusufu et al. 2014) revealed that the observed significant reduction in κ_{lat} was due to the effective phonon scattering (caused not only by the Ge substitution of Si, but also by the formation of nanoscale precipitates). It has been

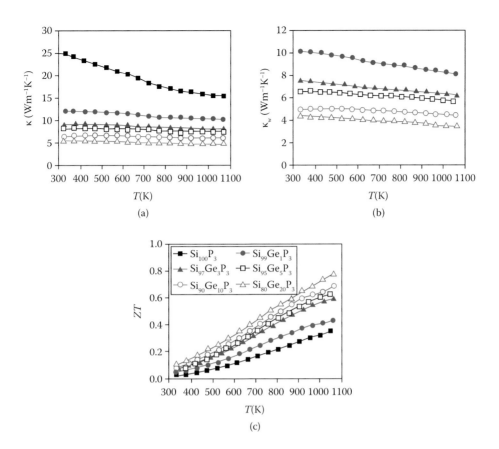

Figure 24.10 Temperature dependencies of the (a) total thermal conductivity κ, (b) lattice thermal conductivity κ_{lat}, and (c) dimensionless figure of merit ZT for nanostructured bulk Si–Ge alloys synthesized via the bottom-up process.

demonstrated for various materials that nanostructuring reduces κ_{lat}; however, in many cases, it also deteriorates carrier mobility to a certain degree (Hsu et al. 2004; Androulakis et al. 2006). In the present study, Si–Ge alloys containing nanoscale precipitates exhibited the values of carrier mobility similar to those obtained for non-nanostructured Si–Ge alloys, indicating that the nanoscale precipitates scattered phonons effectively, but almost did not scatter any electrons. It has been confirmed that the nanoscale precipitates bond to the matrix phases coherently or semicoherently (Yusufu et al. 2014), and the resulting types of matched interfaces have little effect on the electron transport properties (Lo et al. 2012; He et al. 2013; Tan et al. 2014). As a result, the enhanced ZT of 0.8 was obtained for the sample with $x = 20$ at 1073 K (Figure 24.10c), which was higher than that for non-nanostructured Si–Ge alloys with the same Ge content.

24.6 MELT SPINNING AS AN EFFECTIVE BOTTOM-UP METHOD FOR NANOSTRUCTURING BULK SILICON

Si reacts with transition metals to form a broad class of metal silicides (MSi_xs), some of which are metallic, while others are semiconducting. Recently, the nanocomposites composed of Si and MSi_x phases, in which MSi_x nanoparticles are uniformly dispersed across the Si matrix, have attracted much attention as materials with good TE properties (Mingo 2009; Kurosaki et al. 2016). In particular, the TE properties of several Si–MSi_x nanocomposites such as $Si_{80}Ge_{20}$–$CrSi_2$ (Zamanipour and Vashaee 2012), $Si_{92}Ge_8$–$MoSi_2$ (Favier et al. 2014), $Si_{100-x}Ge_x$ ($x = 10$–30)–WSi_2 (Mackey et al. 2015), and Si–VSi_2 (Yusufu et al. 2016) have been investigated, and their obtained ZT values were relatively high. However, these materials were synthesized via the top-down process, which consisted of the BM stage followed by HP or SPS, and all the related issues (such as incorporation of impurities and oxidization during synthesis) were observed. Therefore, our group has proposed melt spinning (MS) as an effective bottom-up method for fabricating Si–MSi_x nanocomposites.

In general, the MS process (which is characterized by extremely high cooling rates) is widely applied to the synthesis of amorphous alloys (Greer 1995). Recently, MS has been used to synthesize nanostructured TE materials—in particular, Bi_2Te_3-based materials (Xie et al. 2009). During MS, a molten alloy is injected into a high-speed rotating Cu roller at a high Ar gas pressure, which transforms it into an amorphous or nanocrystalline ribbon-like sample, owing to the very large magnitude of the cooling rate. Furthermore, the crystallinity and size of the sample microstructure can be mainly controlled by changing the rotation speed of the Cu roller. The schematic representation of the MS process and the appearance of the obtained ribbons are shown in Figure 24.11a and b, respectively.

On the other hand, according to the binary phase diagrams for Si and various transition metals, most of the MSi_x species have a common eutectic point with Si (see Figure 24.12). Among the large number of silicides, we selected $TiSi_2$, VSi_2, $CoSi_2$, and $NiSi_2$ as dispersed precipitates in bulk Si, which form Si–MSi_x nanocomposites. It was assumed that the position of the eutectic point and degree of lattice matching between Si and MSi_x were important for utilizing the obtained Si–MSi_x nanocomposites as TE materials. For example, in the Si–V system, the eutectic point between Si and VSi_2 exists at an Si-rich part, in which VSi_2 nanoparticles are distributed across the Si phase without interacting with each other when the melt with the eutectic composition rapidly cools down to the solid state. On the other hand, the lattice parameters of $NiSi_2$ (a = 0.54160 nm, JCPDS Card Nos. 00–043–0989) and $CoSi_2$ (a = 0.53640 nm, JCPDS Card Nos. 00–038–1449) are very similar to that of Si (a = 0.54301 nm, JCPDS Card Nos. 00–005–0565), which allows formation of coherent interfaces, in which a $NiSi_2$ or $CoSi_2$ phase coexists with an Si one.

Figure 24.12 shows the field-emission scanning electron microscopy (FE–SEM) images of the fabricated Si–MSi_2 nanocomposites (M = Ti, V, Ni, Co; all scale bars are equal to 200 nm, and the binary phase diagrams for the related Si–M system are shown in the right column for reference purposes). In the obtained FE–SEM images, the bright and dark areas represent the MSi_2 and Si phases, respectively. Normally, microscale structures are produced via traditional arc-melting, while nanoscale structures are obtained through the MS process, indicating that rapid cooling leads to a decrease in the structure size. For the Si–Ti, Si–V, and Si–Co systems, the MS process was performed at the eutectic points of Si–$TiSi_2$, Si–VSi_2, and Si–$CoSi_2$, respectively. As a result, distinct eutectic structures were confirmed for the Si–$TiSi_2$ and Si–$CoSi_2$ systems. On the other hand, a dot-type structure was obtained for the Si–$NiSi_2$ system, since the corresponding sample did not have a eutectic composition. Similar to the Si–$NiSi_2$ system, the dot-type structure was obtained for the Si–VSi_2 system as well despite its eutectic composition (this was probably due

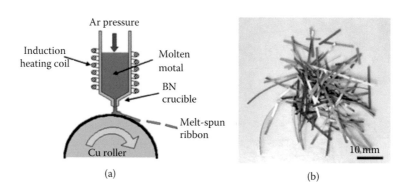

(a)

(b)

Figure 24.11 (a) A schematic representation of the MS process. (b) Nanocomposite MS ribbons composed of Si and metal silicides.

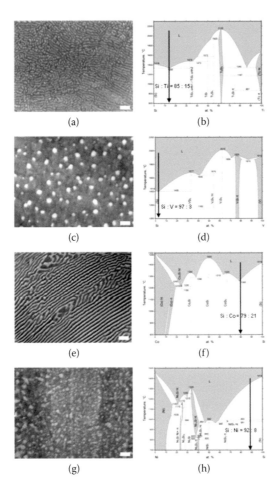

Figure 24.12 FE–SEM images of the MS ribbons composed of Si and various metal silicides (the corresponding phase diagrams are shown in the right-hand column). (a) and (b) Si–Ti system, ribbon composition: $Si_{85}Ti_{15}$. (c) and (d) Si–V system, ribbon composition: $Si_{97}V_3$. (e) and (f) Si–Co system, ribbon composition: $Si_{79}Co_{21}$. (g) and (h) Si–Ni system, ribbon composition: $Si_{92}Ni_8$. The bright and dark areas represent metal silicides and Si species, respectively. All scale bars are equal to 200 nm.

to the very small VSi_2 content (Si:V = 97:3, Si:VSi_2 = 91:3), which prevented bonding between species and the formation of an eutectic structure).

One advantage of the nanocomposite formation via the MS process is that its structure size can be easily controlled by changing the cooling rate (which depends on the rotation speed of the Cu roller). According to Figure 24.13, the structure size for the Si–VSi_2 nanocomposite clearly decreased with increasing cooling rate, while the traditional arc-melting led to the lowest value of the cooling rate (Tanusilp et al. 2016). However, the MS process is characterized by one drawback. In general, bulk TE materials are required for manufacturing TE generators or refrigerators. Thus, the ribbon samples synthesized via the MS process must be compacted and densified into bulks. However, during this procedure, the nanostructures are easily converted into microstructures. Figure 24.14 shows the microstructures obtained for the Si–$TiSi_2$ composite ([a] and [b] depict the MS ribbon sample and SPS sample, respectively), which indicate that the size of the resulting structure increases from the nanoscale to the microscale even during SPS characterized by relatively low sintering temperatures and short sintering times. Therefore, a special synthesis method for fabricating nanostructured bulk materials from MS ribbons has to be developed.

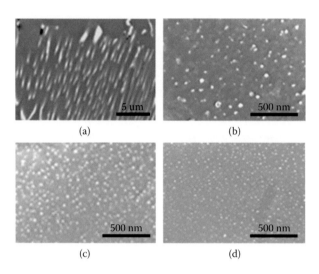

Figure 24.13 A decrease in the precipitate size caused by the increase in the cooling rate during the MS process (the sample composition was $Si_{97}V_3$). (a) An FE–SEM image of the polished surface of the arc-melted ingot. FE–SEM images of the MS ribbons obtained at the following rotation speeds of the copper roller: (b) 2000 rpm, (c) 4000 rpm, and (d) 6000 rpm (the increase in the rotation speed causes an increase in the cooling rate). The bright and dark areas represent VSi_2 and Si species, respectively.

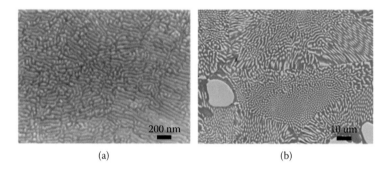

Figure 24.14 Transformation of the precipitate nanostructures into the microstructures observed during the forma-tion of bulk samples from the MS ribbons via SPS. FE–SEM images of the Si and $TiSi_2$ composites (Si : Ti = 85 : 15): (a) MS ribbons and (b) SPS bulk. The bright and dark areas represent $TiSi_2$ and Si species, respectively.

24.7 SUMMARY

The careful control of the Si structure on the nanoscale level leads to a reduction in its κ_{lat} value while main-taining superior electrical properties (such as high power factor), which results in a significant enhancement of its TE performance. The maximum ZT values for various nanostructured Si materials are summarized in Table 24.3, indicating that the ZT enhancement has been experimentally demonstrated not only for Si nano-wires or nanomesh structures but also for nanostructured bulk Si. In this work, two different methods for the synthesis of nanostructured bulk Si have been proposed: the top-down process (during which nanostructures are artificially formed) and the bottom-up process (during which nanostructures are naturally formed). In this section, we have demonstrated the enhanced TE properties of the nanostructured bulk Si synthesized by the bottom-up method developed in our group. Furthermore, we have introduced the advanced melt-spinning method, which allows fabrication of nanocomposites composed of Si and various metal silicides. It can be assumed that the bottom-up method for the synthesis of nanostructured Si will become prevalent due to its simplicity and high reproducibility. Hence, the nanostructured bulk Si synthesized via the bottom-up technique represents a high-performance, cost-effective, and environmentally friendly TE material.

Table 24.3 Maximum *ZT* values for various nanostructured Si and Si–Ge alloys

	ZT_{MAX}	T (K)	SIZE OF NANOSTRUCTURES	CONDUCTION TYPE	REMARKS	REF.
Si nanowires	~1	200	20 nm wide	p-type (7×10^{19} cm^{-3})	–	Boukai et al., 2008
	0.6	350	10 nm wide	p-type (2×10^{20} cm^{-3})	–	Boukai et al., 2008
	0.6	300	52 nm in diameter	p-type	–	Hochbaum et al., 2008
Si nanoholes	0.4	300	55 nm pitch hole	p-type (5×10^{19} cm^{-3})	#1	Tang et al., 2010
Nanostructured bulk Si (top-down process)	0.7	1275	Crystalline size: 5–20 nm	n-type (4.6×10^{20} cm^{-3})	–	Bux et al., 2009
	0.6	1073	Crystalline size: 32 nm	n-type (1.2×10^{20} cm^{-3})	#2, #3	Miura et al., 2015
	0.95	1173	Crystalline size: 10–30 nm	n-type	#4	Zhu et al., 2009
Nanostructured bulk Si (bottom-up process)	0.6	1050	Plate-like precipitates: 20 nm Spherical precipitates: 5 nm	n-type (2.1×10^{20} cm^{-3})	#5, #6	Yusufu et al., 2014
Nanostructured bulk Si–Ge alloys (top-down process)	1.3	1173	Crystalline size: 10–20 nm	n-type (2.2×10^{20} cm^{-3})	#7	Wang et al., 2008
	0.95	1073	Crystalline size: 5–50 nm	p-type	#7	Joshi et al., 2008

Note: #1, sample porosity is 35%; #2, prepared by plasma-enhanced chemical vapour deposition; #3, contains amorphous Si oxide; #4, contains 5 at.% of Ge; #5, P-rich phase bonded to Si matrix semi-coherently or coherently; #6, contains 3 at.% of Ge; #7, sample composition is $Si_{0.8}Ge_{0.2}$.

ACKNOWLEDGMENTS

A part of the experimental data presented in this section (Figures 24.7 through 24.10 and 24.12 through 24.14) was obtained by the authors' group due to the financial support through the Grant-in-Aid for Scientific Research program (Grant No. 25289220) funded by the Ministry of Education, Culture, Sports, Science, and Technology (Japan), the Precursory Research for Embryonic Science and Technology (PRESTO) program funded by the Japan Science and Technology Agency, and the Program for Creating Future Wisdom funded by Osaka University.

REFERENCES

Androulakis J, et al. (2007). Spinodal decomposition and nucleation and growth as a means to bulk nanostructured thermoelectrics: Enhanced performance in $Pb_{(1-x)}Sn_xTe$-PbS. *J. Am. Chem. Soc.* 129:9780–9788.

Androulakis JK, et al. (2006). Nanostructuring and high thermoelectric efficiency in p-type $Ag(Pb_{1-y}Sn_y)_mSbTe_{2+m}$. *Adv. Mater.* 18:1170–1173.

Bathula S, et al. (2012). Enhanced thermoelectric figure-of-merit in spark plasma sintered nanostructured n-type SiGe alloys. *Appl. Phys. Lett.* 101:213902.

Bell LE. (2008). Cooling, heating, generating power, and recovering waste heat with thermoelectric systems. *Science* 321:1457–1461.

Biswas K, et al. (2012). High-performance bulk thermoelectrics with all-scale hierarchical architectures. *Nature* 489:414–418.

Boor J, et al. (2012). Thermoelectric properties of porous silicon. *Appl. Phys. A* 107:789–794.

Boukai AI, Bunimovich Y, Tahir-Kheli J, Yu JK, Goddard WA, Heath JR. (2008). Silicon nanowires as efficient thermoelectric materials. *Nature* 451:168–171.

Bux SK, et al. (2009). Nanostructured bulk silicon as an effective thermoelectric material. *Adv. Funct. Mater.* 19:2445–2452.

Dismukes JP, Ekstrom L, Steigmeier EF, Kudman I, Beers DS. (1964). Thermal and electrical properties of heavily doped Ge-Si alloys up to 1300 K. *J. Appl. Phys.* 35:2899–2907.

Favier K, et al. (2014). Influence of in situ formed $MoSi_2$ inclusions on the thermoelectrical properties of an n-type silicon–germanium alloy. *Acta Mater.* 64:429–442.

Gelbstein Y, Davidow J, Girard SN, Chung DY, Kanatzidis M. (2013). Controlling metallurgical phase separation reactions of the $Ge_{0.87}Pb_{0.13}Te$ alloy for high thermoelectric performance. *Adv. Eng. Mater.* 3:815–820.

Goldsmid HJ. (1964). *Thermoelectric refrigeration.* New York, NY: Plenum Press.

Greer AL. (1995). Metallic glasses. *Science* 267:1947.

He J, Kanatzidis MG, Dravid VP. (2013). High performance bulk thermoelectrics via a panoscopic approach. *Mater. Today* 16:166–176.

Heinz NA, Ikeda T, Pei Y, Snyder GS. (2014). Applying quantitative microstructure control in advanced functional composites. *Adv. Funct. Mater.* 24:2135–2153.

Hochbaum AI, et al. (2008). Enhanced thermoelectric performance of rough silicon nanowires. *Nature* 451:163–167.

Hsu KF, et al. (2004). Cubic $AgPb_mSbTe_{2+m}$: Bulk thermoelectric materials with high figure of merit. *Science* 303:818–821.

Joshi G, et al. (2008). Enhanced thermoelectric figure-of-merit in nanostructured p-type silicon germanium bulk alloys. *Nano Lett.* 8:4670–4674.

Kajikawa T. (2006). Thermoelectric power generation system recovering industrial waste heat. In: *Thermoelectrics handbook: Macro to nano* (Rowe DM, ed.), pp. 50–51. Roca Baton, FL: CRC.

Korotcenkov G, Cho BK. (2010). Silicon porosification: State of the art. *Crit. Rev. Solid State Mater. Sci.* 35:153–260.

Kurosaki K, Yusufu A, Miyazaki Y, Ohishi Y, Muta H, Yamanaka S. (2016). Enhanced thermoelectric properties of silicon via nanostructuring. *Mater. Trans.* 57:1018–1021.

Lan Y, Minnich AJ, Chen G, Ren Z. (2010). Enhancement of thermoelectric figure-of-merit by a bulk nanostructuring approach. *Adv. Funct. Mater.* 20:357–376.

Lee Y, et al. (2013). High-performance tellurium-free thermoelectrics: All-scale hierarchical structuring of p-type PbSe-MSe systems (M = Ca, Sr, Ba). *J. Am. Chem. Soc.* 135:5152–5160.

Lo SH, He J, Biswas K, Kanatzidis MG, Dravid VP. (2012). Phonon scattering and thermal conductivity in p-type nanostructured PbTe-BaTe bulk thermoelectric materials. *Adv. Funct. Mater.* 22:5175–5184.

Mackey J, Dynys F, Sehirlioglu A. (2015). $Si/Ge–WSi_2$ composites: Processing and thermoelectric properties. *Acta Mater.* 98:263–274.

Mingo N, Hauser D, Kobayashi NP, Plissonnier M, Shakouri A. (2009). Nanoparticle-in-alloy approach to efficient thermoelectrics: Silicides in SiGe. *Nano Lett.* 9:711–715.

Minnich AJ, Dresselhaus MS, Ren ZF, Chen G. (2009). Bulk nanostructured thermoelectric materials: Current research and future prospects. *Energy Environ. Sci.* 2:466–479.

Miura A. Zhou S, Nozaki T, Shiomi J. (2015). Crystalline-amorphous silicon nanocomposites with reduced thermal conductivity for bulk thermoelectrics. *ACS Appl. Mater. Interface* 7:13484–13489.

Olesinski RW, Kanani N, Abbaschian GJ. (1985). The P-Si (phosphorus-silicon) system. *J. Phase Equil.* 6:130–133.

Pei Y, Heinz NA, LaLonde A, Snyder GJ. (2011). Combination of large nanostructures and complex band structure for high performance thermoelectric lead telluride. *Energy Environ. Sci.* 4:3640–3645.

Poudel B, et al. (2008). High-thermoelectric performance of nanostructured bismuth antimony telluride bulk alloys. *Science* 320:634–638.

Regner KT, Sellan DP, Su Z, Amon CH, McGaughey AJ, Malen JA. (2013). Broadband phonon mean free path contributions to thermal conductivity measured using frequency domain thermoreflectance. *Nat. Commun.* 4:1640.

Slack GA, Hussain MA (1991) The maximum possible conversion efficiency of silicon-germanium thermoelectric generators. *J. Appl. Phys.* 70:2694–2718.

Snyder GJ, Toberer ES. (2008). Complex thermoelectric materials. *Nat. Mater.* 7:105–114.

Sootsman JR, Chung DY, Kanatzidis MG. (2009). New and old concepts in thermoelectric materials. *Angew. Chem. Int. Ed.* 48:8616–8639.

Suzuki T, Ohishi Y, Kurosaki K, Muta H, Yamanaka S. (2012). Thermal conductivity of size-controlled bulk silicon nanocrystals using self-limiting oxidation and HF etching. *Appl. Phys. Express* 5:081302.

Tan G, et al. (2014). High thermoelectric performance of p-type SnTe via a synergistic band engineering and nanostructuring approach. *J. Am. Chem. Soc.* 136:7006–7017.

Tang J, et al. (2010). Holey silicon as an efficient thermoelectric material. *Nano Lett.* 10:4279–4283.

Tanusilp S, Kurosaki K, Yusufu A, Ohishi Y, Muta H, Yamanaka S. (2016). Enhancement of thermoelectric properties of bulk Si by dispersing size-controlled VSi_2. *J. Electron. Mater.* 46(5): p. 3249.

Tritt TM, Subramanian MA. (2006). Thermoelectric materials, phenomena, and applications: A Bird's eye view. *MRS Bull.* 31:188–198.

Wang XW, et al. (2008). Enhanced thermoelectric figure of merit in nanostructured n-type silicon germanium bulk alloy. *Appl. Phys. Lett.* 93:193121.

Weber L, Gmelin E. (1991). Transport properties of silicon. *Appl. Phys. A* 53:136–140.

Xie W, Tang X, Yan Y, Zhang Q, Tritt TM. (2009). Unique nanostructures and enhanced thermoelectric performance of melt-spun BiSbTe alloys. *Appl. Phys. Lett.* 94:102111.

Yadav GG, Susoreny JA, Zhang G, Yang H, Wu Y. (2011). Nanostructure-based thermoelectric conversion: An insight into the feasibility and sustainability for large-scale deployment. *Nanoscale* 3:3555–3562.

Yang J, Caillat T. (2006). Thermoelectric materials for space and automotive power generation. *MRS Bull.* 31:224–229.

Yu B, et al. (2012). Enhancement of thermoelectric properties by modulation-doping in silicon germanium alloy nanocomposites. *Nano Lett.* 12:2077–2082.

Yu JK, Mitrovic S, Tham D, Varghese J, Heath JR. (2010). Reduction of thermal conductivity in phononic nanomesh structures. *Nature Nanotechnol.* 5:718–721.

Yusufu A, et al. (2014). Bottom-up nanostructured bulk silicon: A practical high-efficiency thermoelectric material. *Nanoscale* 6:13921–13927.

Yusufu A, Kurosaki K, Ohishi Y, Muta H, Yamanaka S. (2016). Improving thermoelectric properties of bulk Si by dispersing VSi_2 nanoparticles. *Jpn. J. Appl. Phys.* 55:061301.

Zamanipour Z, Vashaee D. (2012). Comparison of thermoelectric properties of p-type nanostructured bulk $Si_{0.8}Ge_{0.2}$ alloy with $Si_{0.8}Ge_{0.2}$ composites embedded with $CrSi_2$ nano-inclusions. *J. Appl. Phys.* 112:093714.

Zebarjadi M, et al. (2011). Power factor enhancement by modulation doping in bulk nanocomposites. *Nano Lett.* 11:2225–2230.

Zhao LD, Dravid VP, Kanatzidis MG. (2014). The panoscopic approach to high performance thermoelectrics. *Energy Environ. Sci.* 7:251–268.

Zhu GH, et al. (2009). Increased phonon scattering by nanograins and point defects in nanostructured silicon with a low concentration of germanium. *Phys. Rev. Lett.* 102:196803.

25 Nanosilicon and thermoelectricity

Dario Narducci

Contents

25.1 INTRODUCTION

25.1.1 FUNDAMENTALS OF THERMOELECTRIC SCIENCE

Thermoelectricity is the onset of transport phenomena in which heat or charge current are cross–driven by temperature gradients or electric fields. The most common thermoelectric phenomena are heat currents flowing as a result of the application of an electric field, and electric fields setting up as a result of a temperature difference. Both phenomena, namely charge current driven by a temperature difference (Seebeck effect) and heat flow driven by an electric current (Peltier effect), were discovered in the first half of the 19th century and found almost immediate practical applications. The Peltier effect, discovered in 1834, found its first use in a solid-state cooler built by Lenz just 4 years later, while the usability of the Seebeck effect (observed for the first time by Volta in 1794 and rediscovered by Seebeck in 1821) to generate electric power was demonstrated by Altenkirch in 1909 [1].

Upon application of a temperature difference ΔT any material develops a voltage ΔV that, in the low ΔT limit, is proportional to ΔT. The proportionality factor

$$\alpha = -\frac{\Delta V}{\Delta T} \tag{25.1}$$

is named the *Seebeck coefficient*. Likewise, upon application of an electric field resulting in an electric current of density $\vec{j}_e$, a thermal current of density $\vec{j}_Q$ is observed flowing through the medium. The proportionality factor Π that links the two current densities as

$$\vec{j}_Q = \Pi \vec{j}_e \tag{25.2}$$

is known as the Peltier coefficient.

Both Peltier and Seebeck effects find proper theoretical description within the theory of linear nonequilibrium thermodynamics. The Onsager–Callen theory shows that in an electronic conductor (namely a metal, a semiconductor, or a dielectric) [2]

$$\begin{bmatrix} \vec{j}_e/e \\ \vec{j}_Q \end{bmatrix} = \begin{bmatrix} (T/e^2)\sigma & (T^2/e^2)\alpha\sigma \\ (T^2/e^2)\alpha\sigma & T^2\kappa \end{bmatrix} \begin{bmatrix} -(\vec{\nabla}\mu_e)/T \\ \vec{\nabla}(1/T) \end{bmatrix} \tag{25.3}$$

where κ is the open-circuit thermal conductivity, σ is the isothermal electrical conductivity, μ_e is the electron electrochemical potential, and $-e$ is the electron charge. Comparing Equation 25.3 to Equations 25.1 and 25.2, it follows immediately from the Onsager reciprocity relations that $\pi = T\alpha$. Note that both α and Π are temperature dependent.

It should be stressed that while the Seebeck effect is an open-circuit, zero-current effect, the Peltier effect is a short-circuit effect. Thus, while the application of the Peltier Equation (25.2) to Peltier coolers is immediate, the analysis of the physics of the Seebeck generator must account also for the fact that the application of a temperature difference in a closed circuit leads to a number of additional thermal and electric effects. The first analysis of the relationship between the thermodynamic efficiency η of a thermoelectric generator (TEG) seen as a thermal engine is due to Altenkirch [3] and was then refined and cast in its currently used form by Ioffe [4]. It accounts not only for the Seebeck effect, which sets the voltage difference between the hot and cold sides of the conductor, but also for the heat drifted by the Peltier effect upon electric current circulation, the Joule effect, and the amount of heat simply transferred through the conductor by thermal conduction. Typically, a TEG is made of p-type and n-type thermoelectric legs making a parallel thermal circuit and a series electric circuit (Figure 25.1). It could be shown [4,5] that when the resistive load matches the TEG electrical resistance, the thermoelectric efficiency η, namely the ratio between the output electric power w_e and the input thermal power w_Q, reads

$$\eta = \frac{T_H - T_C}{T_H} \frac{M-1}{M + T_C/T_H} \tag{25.4}$$

where T_C and T_H are respectively the temperatures of the cold and of the hot sides, while $M \equiv \sqrt{1 + Z_{pn}\bar{T}}$, with $\bar{T} \equiv (T_H + T_C)/2$ and

$$Z_{pn}\bar{T} \equiv \frac{(|\alpha_p| + |\alpha_n|)^2}{(\sqrt{\kappa_p/\sigma_p} + \sqrt{\kappa_n/\sigma_n})^2} \bar{T} \tag{25.5}$$

is the (dimensionless) thermoelectric figure of merit. Although Z_{pn} cannot be reduced to a combination of single-leg figures of merit, a thermoelectric figure of merit for the single leg (material) is usually defined:

$$Z\bar{T} \equiv \frac{\sigma\alpha^2}{\kappa} \bar{T} \tag{25.6}$$

Inspection of $Z\bar{T}$ clearly shows that the ideal material for a TEG should be an excellent electric conductor (high σ) with a large Seebeck coefficient and low thermal conductivity. It is clear that

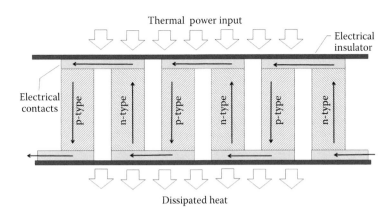

Figure 25.1 Layout of a typical thermoelectric generator. Alternated p- and n-type legs set up a series electrical circuit so that thermoelectric voltages sum up, while they are all in parallel with respect to the two heat sinks.

such a combination of properties is hard if not impossible to achieve. To better understand this point, Equation 25.3 shows that the entropy current density $\vec{j}_S \equiv \vec{j}_Q / T$ reads

$$\vec{j}_S = -(1/e^2)\alpha\sigma\vec{\nabla}\mu_e + T\kappa\vec{\nabla}(1/T) \tag{25.7}$$

Since however the same equation also shows that $\vec{\nabla}\mu_e = -\vec{j}_e e/\sigma$, then

$$\vec{j}_S = (1/e)\alpha\vec{j}_e + T\kappa\vec{\nabla}(1/T) \tag{25.8}$$

The first term in the right-hand side of the previous equation is the entropy transported by the carrier flux

$$\vec{j}_e S_j \equiv \vec{j}_{S,j} = (1/e)\alpha\vec{j}_e \tag{25.9}$$

while the second term accounts for the entropy flux of thermal origin. It can then be immediately concluded that α is directly proportional to the entropy per carrier S_j generated by the thermal engine. Thus, for any value of the total entropy generated by the engine, S_j will be proportional to the reciprocal carrier density—and so will be α. Therefore, since high electrical conductivities require large carrier densities, it can be concluded that no material may fulfill both requirements of displaying large α and σ as long as the carrier density remains the only material parameter that is operated on. Furthermore, simultaneous large σ and small κ require solids, where phonon and charge carrier mobilities are ruled by scattering centers that are highly effective at limiting phonon diffusion, while marginally affecting charge drift and diffusion. Although not being impossible in principle, this is not the case for most real-world situations, where grain boundaries, and extended and point defects are often comparably good at scattering both types of particles. As a conclusion, it should come as no surprise that for more than a century, no material whatsoever had displayed thermoelectric figures of merit larger than one. As a consequence, TEGs found almost no use in converting wasted heat into electric power up to the end of the last century.

25.1.2 HOW NANOTECHNOLOGY HAS AFFECTED THERMOELECTRICITY

Nanotechnology has given a great momentum to the conversion rate achievable by TEGs [6]. Thermoelectric figures of merit have grown by a factor of almost three over less than 15 years. This has been due both to novel classes of thermoelectric materials and to old materials that have been rejuvenated by nanostructuring. Nanowires and nanolayers, along with multilayered structures, have improved the thermoelectric characteristics of many age-old materials [7–9] either by decreasing their thermal conductivity or by increasing their power factor $P = \sigma\alpha^2$. Both approaches have led to an increase of the thermoelectric figure of merit, although the impact of P and κ on the power density that the device outputs is not equivalent [10]. As a result, TEGs are nowadays at the edge of industrial production, so requirements such as geoabundance and low raw material costs have entered the list of technological constraints that thermoelectric materials and devices have to comply with. Thus, interest toward silicon, silicon–germanium alloys, and silicides as thermoelectric materials has quite obviously renewed.

The intuition of the beneficial role that nanotechnology might have played on thermoelectrics dates back to 1993, when two famous papers by Hicks and Dresselhaus [11,12] predicted a potentially large increase of the power factor (and a lesser decrease of the thermal conductivity) in dimensionally constrained semiconductors. However, as of today nanotechnology has mostly succeeded at increasing only the σ/κ ratio of many materials. One of the most notable milestones in this matter was in 2008, when *Nature* published two papers in the same issue, by Hochbaum et al. [13] and Boukai et al. [14], on silicon nanowires (SiNWs). Authors independently demonstrated that phonon scattering at the outer roughened walls of SiNWs causes a drop of the thermal conductivity of single-crystalline silicon (c-Si) by almost two orders of magnitude, with only a marginal impact on its electrical conductivity. This piece of evidence has triggered a major effort to devise manufacturing techniques capable of industrial upscaling, quite naturally meeting silicon microelectronic technology that offers a rich portfolio of preparation methods [15–18].

Not only nanowires but also nanolayers have shown potential for the reduction of silicon κ. As in SiNWs, dimensional constraints by themselves are not enough to reduce thermal conductivity in two dimensions, and surface roughness is crucial [19,20].

In addition, a major side effect of the discovery of high ZT in dimensionally constrained systems has been the attempt to also use nanotechnology to enhance the thermoelectric figure of merit in bulk materials. It should be noted that, as a rule of thumb, the power output of a TEG scales with the size of its thermoelectric elements. Therefore, nanostructured bulk materials would be capable of much larger power outputs than nanowires or nanolayers. Nanocrystallinity and nanosized second phases (precipitates) have provided interesting opportunities. Precipitates of silicon dioxide [21] and nanovoids [22,23] have been found to increase the σ/κ ratio. Also, it should at least be mentioned that nanostructured bulk materials were possibly the only systems where enhancements of the thermoelectric power factor, the Holy Grail of nanostructured thermoelectrics, were reported [24,25].

This chapter will focus on the ways ZT may be improved in silicon by nanotechnology. It will focus on dimensionally constrained systems, namely silicon nanowires and nanolayers only. After providing an overview of the thermoelectric theory in nondimensionally constrained systems, that will be extended to one-dimensional (1D) and two-dimensional materials, the state of the art of silicon nanowires and nanolayers for thermoelectric applications. In all cases, the profound interplay between preparation methods and thermoelectric performances will be reviewed; both top-down and bottom-up methods will be described. The chapter will then end with some general evaluations about the practical usability of Si nanostructures in TEGs.

25.2 THERMOELECTRIC THEORY OF INFINITE SYSTEMS

25.2.1 ELECTRONIC TRANSPORT

In a crystalline system and within the linear response limit, both the electrical conductivity and the Seebeck coefficient may be derived by solving the pertinent Boltzmann transport equation (BTE) in the relaxation-time approximation. The time evolution of the one-particle distribution $f(\vec{r};t)$ in a system experiencing a perturbation (e.g., an applied electric field or a temperature gradient) may be written as [26]

$$\frac{\partial f}{\partial t} + \frac{d\vec{r}}{dt} \cdot \vec{\nabla}_{\vec{r}} f + \frac{d\vec{p}}{dt} \cdot \vec{\nabla}_{\vec{p}} f = \left(\frac{\partial f}{\partial t}\right)_c \tag{25.10}$$

where $\vec{p}$ is the particle momentum and the right-hand side of the previous equation adds the interaction (scattering) of the given particle with all other particles in the system. The relaxation-time approximation assumes that the effect of the external perturbation brings f out of equilibrium, while scattering restores the equilibrium distribution $f_0(\vec{r};t)$ with a time decay of the form

$$f(\vec{r};t) - f_0(\vec{r};t) \propto \exp(-t/\tau) \tag{25.11}$$

where τ is the relaxation time. Under the additional assumption that transient terms are negligible and that the deviation of f from f_0 and its gradient are much smaller than f and $\vec{\nabla} f$, Equation 25.10 may be rewritten as

$$f = f_0 - \tau\left(\vec{v} \cdot \vec{\nabla}_{\vec{r}} f_0 + \frac{\vec{F}}{m} \cdot \vec{\nabla}_{\vec{v}} f_0\right) \tag{25.12}$$

where $\vec{F} = \vec{p}/m$. Further analyses of the BTE are out of the scope of this chapter. The reader may refer to classic textbooks [26–28] for additional details. The BTE may be used to obtain all transport coefficients relevant to thermoelectricity.

Electrical conductivity follows, considering that an electric field $\vec{\varepsilon}$ causes a force on charge carriers that may be written as $\vec{F} = -e\vec{\varepsilon} = e\vec{\nabla}\varphi$. Thus, along a given direction (say x) $F = -dE/dx = ed\varphi/dx$. Since the applied electric field bends both the bands and the Fermi level, Equation 25.12 may be rewritten as

$$f = f_0 - \tau v_x\left(\frac{dE_f}{dx} + e\varepsilon\right)\frac{\partial f_0}{\partial E} \tag{25.13}$$

It may be shown [26] that the current density along x reads

$$j_x = -\frac{1}{4\pi^3}\int_{-\infty}^{+\infty}\int_{-\infty}^{+\infty}\int_{-\infty}^{+\infty} ev_x fd^3\vec{k}$$

$$= \frac{e^2}{3}\frac{d\Phi}{dx}\int_0^{+\infty}\tau v^2 g(E)\frac{\partial f_0}{\partial E}dE \tag{25.14}$$

where Φ is the electrochemical potential of the electron and $g(E)$ is the electronic density of states (DOS). Thus

$$\sigma = -\frac{e^2}{3}\int_0^{+\infty}\tau v^2 g(E)\frac{\partial f_0}{\partial E}dE \tag{25.15}$$

The electrical conductivity may be alternately expressed through the carrier density n. Since $n = \int_0^\infty g(E)f_0\,dE$, the following is immediately obtained

$$\sigma = e\mu\int_0^\infty g(E)f_0\,dE \tag{25.16}$$

Comparison of the last two equations further returns the drift mobility μ.

The Seebeck coefficient may be also obtained from Equations 25.12 and 25.13 by considering the effect of the simultaneous application of a temperature gradient and an electric field. In such a situation both f_0 and E_f depends on the position so that

$$\frac{df_0(E_f,T)}{dx} = \frac{\partial f_0}{\partial E}\frac{\partial E_f}{\partial x} - \frac{E - E_f}{T}\frac{\partial f_0}{\partial E}\frac{dT}{dx} \tag{25.17}$$

Thus the current density reads

$$j_x = -\frac{e}{3} \int_0^{+\infty} \tau v^2 \left(\frac{dE_f}{dx} + \frac{E - E_f}{T} \frac{dT}{dx} \right) g(E) \frac{\partial f_0}{\partial E} dE \tag{25.18}$$

Splitting j_x into the two terms depending on $-dT/dx$ and on dE_f/dx and casting them into Equation 25.3, under open circuit conditions (namely imposing $j_x = 0$) it becomes

$$\alpha = -\frac{1}{eT} \frac{\displaystyle\int_0^\infty \tau v^2 (E - E_f) \frac{\partial f_0}{\partial E} g(E) dE}{\displaystyle\int_0^\infty \tau v^2 \frac{\partial f_0}{\partial E} g(E) dE} \tag{25.19}$$

Equations 25.15 and 25.19 admit two rather popular rewritings. Limiting to semiconductors and assuming parabolic bands, the electronic dispersion relation in the momentum space k reads

$$E\left(k_x, k_y, k_z\right) = \frac{\hbar^2 k_x^2}{2m_x} + \frac{\hbar^2 k_y^2}{2m_y} + \frac{\hbar^2 k_z^2}{2m_z} \tag{25.20}$$

where $\hbar$ is the reduced Planck constant and m_i ($i = x, y, z$) is the effective mass along the three coordinate axes. Thus $g(E) = (2\pi)^{-2} \left(2m^*/\hbar^2\right)^{3/2} E^{1/2}$ so, since $E = mv^2/2$, Equation 25.16 may be restated as

$$\sigma = e\mu \frac{1}{(2\pi)^2} \left(\frac{2m^*}{\hbar^2} \right)^{3/2} \int_0^\infty \frac{E^{1/2}}{\exp\left(\dfrac{E - E_f}{k_B T} \right) + 1} dE \tag{25.21}$$

(where k_B is the Boltzmann constant) that immediately returns

$$\sigma = \frac{e\mu}{2\pi^2} \left(\frac{2k_B T}{\hbar^2} \right)^{3/2} \left(m_x m_y m_z \right)^{1/2} F_{1/2} \tag{25.22}$$

where F_i is the Fermi–Dirac integral

$$F_i(x, \eta) = \int_0^\infty \frac{x^i dx}{\exp(x - \eta) + 1} \tag{25.23}$$

and $\eta_f \equiv E_f/(k_B T)$. For the Seebeck coefficient, further assuming that $\tau = \tau_0 E^r$, Equation 25.19 may be rewritten as

$$\alpha = -\frac{1}{eT} \left(-E_f + \frac{\displaystyle\int_0^\infty E^{5/2+r} \frac{\partial f_0}{\partial E} dE}{\displaystyle\int_0^\infty E^{3/2+r} \frac{\partial f_0}{\partial E} dE} \right) \tag{25.24}$$

Integrating by parts, $\int_0^\infty E^s \frac{\partial f_0}{\partial E} dE = E^s f_0 \big|_0^\infty - \int_0^\infty s E^{s-1} f_0 \, dE$. However $E^s f_0 = 0$ in both integration limits so that

$$\alpha = -\frac{k_B}{e}\left(-\eta_f + \frac{(r+5/2)F_{r+3/2}}{(r+3/2)F_{r+1/2}}\right) \tag{25.25}$$

A final, commonly referenced expression for α in metals and degenerate semiconductors may be obtained by introducing the spectral electrical conductivity $\sigma(E)$. In view of Equation 25.15 it can be defined as

$$\sigma(E) \equiv -\frac{e^2}{3}\tau v^2 g(E) \tag{25.26}$$

Since $\partial f_0/\partial E$ is peaked around E_f, whenever $g(E)$ is approximately constant around the Fermi energy one may expand $\sigma(E)$ as $\sigma(E) = \sigma(E_f) + \sigma'(E_f)(E - E_f)$. Replacing it into Equation 25.19 and taking τ also to be almost constant around E_f leads to

$$\sigma = -\frac{1}{3eT}\frac{\sigma'(E_f)}{\sigma(E_f)}(k_B T)^2 \tag{25.27}$$

where $\partial f_0/\partial E$ was approximated by a boxcar function of half-width $k_B T$. Thus, the Mott formula is obtained:

$$\alpha = -\frac{k_B^2 T}{3e}\frac{\sigma'(E_f)}{\sigma(E_f)} = \frac{k_B^2 T}{3e}\frac{\ln \sigma'(E)}{dE}\bigg|_{E=E_f} \tag{25.28}$$

In ordinary three-dimensional (3D) systems, it is easy to verify that, as anticipated on the basis of purely thermodynamic arguments, α and n show opposite dependency on the reduced Fermi energy, so an increase of carrier density leads to an increase of σ and a decrease of $|\alpha|$. Thus it may be concluded that the only possible way to increase σ without decreasing $|\alpha|$ is either to increase τ_0 or to reduce m^*. Both approaches have been widely pursued, the former using energy filtering [29] or modulation doping [30], the latter through so-called valleytronics [31]. This apart, no other avenue is open to enhance the power factor.

25.2.2 THERMAL CONDUCTIVITY

The standard kinetic model relates the lattice thermal conductivity at temperature T to the phonon mean free path (MFP) $\Lambda(\omega, s, T)$ and to its group velocity (ω, s), both quantities depending on the phonon frequency ω and on its polarization s [26]:

$$\kappa(T) = \frac{1}{3}\sum_s \int_0^\infty c_V(\omega, s, T) v(\omega, s)\Lambda(\omega, s, T)\, d\omega \tag{25.29}$$

where $c_V(\omega, s, T)$ is the spectral specific heat at constant volume of the material and the sum runs over all polarizations. Further to phonon-phonon and phonon-electron scattering, the phonon MFP is limited by the presence of defects, that however have differential scattering capabilities depending on ω and on a typical length scale ℓ associated to the defect itself. In the copresence of more scattering effects, it is commonplace to write Λ according to Matthiensen's spectral rule as

$$\Lambda(\omega, s, T)^{-1} = \sum_i \Lambda_i(\omega, s, T)^{-1} \tag{25.30}$$

where Λ_i is the phonon MFP that would be observed if only the i-th scattering mechanism were active.

Equation 25.29 accounts for the vibrational density of states (VDOS) $\omega_s(\vec{k})$ through the specific heat and the phonon velocity. In the harmonic limit, the specific heat at constant volume reads [27]

$$c_V = \frac{1}{V} \sum_{\vec{k}_s} \frac{\partial}{\partial T} \frac{\hbar\omega_s(\vec{k})}{\exp(\hbar\omega_s(\vec{k})/(k_B T)) - 1} \tag{25.31}$$

where V is the volume of the crystal, and

$$\vec{v}(\omega, s) = \vec{\nabla}_{\vec{k}} \omega(\vec{k}, s) \tag{25.32}$$

As a result, minimization of κ calls for materials with a low specific heat, a condition met in lattices made of heavy atoms. In addition, it may be shown that strong anharmonicity, as observed, for example, in cubic I-V-VI$_2$ compounds or in tetrahedrites, help decrease κ [32,33]. Defect engineering has been also widely explored. Second phases or polycrystallinity quite obviously reduce Λ, and might be more effective at scattering phonons than electrons due to the large difference of (quasi) particle wavelengths [34]. This approach has been extensively applied in dimensionally constrained systems, and will be further discussed in the next sections.

25.3 THERMOELECTRIC THEORY OF DIMENSIONALLY CONSTRAINED SEMICONDUCTORS

25.3.1 ELECTRONIC TRANSPORT

In two seminal papers [11,12], Hicks and Dresselhaus analyzed how the reduced dimensionality of nano-structured systems might affect the thermoelectric figure of merit. In 2D systems (namely nanofilms or multilayered nanostructures) Equation 25.20 is replaced by

$$E(k_x, k_y) = \frac{\hbar^2 k_x^2}{2m_x} + \frac{\hbar^2 k_y^2}{2m_y} + \frac{\hbar^2 \pi^2}{2m_z a^2} \tag{25.33}$$

where a_{2D} is the thickness of the 2D structure. Thus, since $g(E)$ is independent of E, in view of Equations 25.22 and (25.25) it can easily be shown that along the dimensionally unconstrained directions (either x or y)

$$\sigma = \frac{e\mu}{2\pi a_{2D}} \left(\frac{2k_B T}{\hbar^2} \right) (m_x m_y)^{1/2} F_0 \tag{25.34}$$

and

$$\alpha = -\frac{k_B}{e} \left(-\eta_f + \frac{\hbar^2 \pi^2}{2m_z a_{2D}^2 k_B T} + \frac{(r+2)F_{r+1}}{(r+1)F_r} \right) \tag{25.35}$$

where anisotropicity of the effective mass is duly accounted for. Note that both the electrical conductivity and the Seebeck coefficient depend not only on the Fermi energy but also on a_{2D}. Thus, two independent leverages are available here to control the transport coefficients—while only the semiconductor doping level is of use in bulk materials. Comparing Equations 25.34 and 25.35, it can be observed how the adverse interdependency of α and σ on the carrier density breaks down through a_{2D}, enabling a remarkable enhancement of the power factor with respect to 3D systems. For Bi$_2$Te$_3$, a possible ZT value of 1.5 at 300 K was predicted [11].

In 1D systems (e.g., nanowires) the dispersion relation instead reads

$$E(k_x) = \frac{\hbar^2 k_x^2}{2m_x} + \frac{\hbar^2 \pi^2}{2m_y a_{1D}^2} + \frac{\hbar^2 \pi^2}{2m_z a_{1D}^2} \tag{25.36}$$

where a_{1D} is the diameter of the 1D structure. Thus, $g(E) \propto E^{-1/2}$ and Equations 25.22 and 25.25 lead to

$$\sigma = \frac{e\mu}{\pi a_{1D}^2}\left(\frac{2k_B T}{\hbar^2}\right)^{1/2} m_x^{1/2} F_{-1/2} \qquad (25.37)$$

and

$$\alpha = -\frac{k_B}{e}\left(-\eta_f + \frac{(r+3/2)F_{r+1/2}}{(r+1/2)F_{r-1/2}}\right) \qquad (25.38)$$

Here, different to the 2D case, the Seebeck coefficient does not depend on the reduced size while the electrical conductivity does. Nonetheless, a strategy to optimize both σ and α similar to that reported for 2D systems applies, leading to a prediction of a ZT value for $Bi_2 Te_3$ as large as 14 [12].

Figure 25.2 displays the dependency of the power factor on the Fermi energy and the pertinent reduced size in 1D and 2D systems.

25.3.2 HEAT TRANSPORT

Reduced dimensionality also favorably impacts the thermal conductivity. In view of Equation 25.29 and in the gray approximation $[\Lambda(\omega)=\text{const.}]$ it may be verified that any nanostructure with a characteristic size ℓ smaller than the phonon MFP will display a lower thermal conductivity than the bulk material. In silicon, the phonon MFP at room temperature is ≈ 200 nm [35] so submicrometric structures are expected to show reduced thermal conductivities. However, for this to be the case, outer structure surfaces must be capable of incoherently (diffusively) scattering phonons, namely roughened surfaces are needed with a root mean square larger than phonon wavelengths. Note that phonon MFP is always much larger than the charge carrier MFP. Thus, a wide range of a_{2D} and a_{1D} values exists to reduce κ without negatively affecting the carrier mobility.

As it will be shown in the next sections, most of the progress related to the introduction of nanotechnology in thermoelectricity has taken advantage of this possibility of reducing κ without any harm to the power factor. As of today, only very minor progress has been achieved on power factor enhancements.

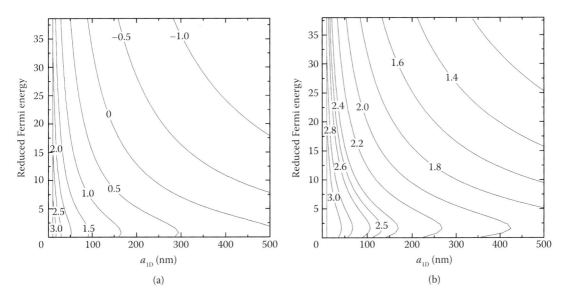

Figure 25.2 Predicted logarithm of the power factor (in W/mK)² for silicon in 1D (a) and 2D (b) systems at 300 K as a function of the Fermi energy and of the pertinent reduced size.

25.4 SILICON NANOWIRES

As mentioned, SiNWs have become extremely popular for thermoelectric applications since the discovery that the thermal conductivity of single-crystalline silicon might have been reduced by almost two orders of magnitude (with very minor impact on their electrical conductivity) by making nanowires with a controlled surface roughness [36]. The two back-to-back papers presenting this outstanding result anticipated the two lines of research that would have been followed to replicate the same ZTs using upscalable technologies. Boukai's collaboration [14] obtained nanowires using a typical top-down approach, making use of extreme lithography, while Hochbaum and coworkers [13] achieved the same result through a bottom-up approach, defining nanowires by electroless etching. In what follows, the development of both approaches will be presented and discussed.

25.4.1 TOP-DOWN APPROACH

The most common and natural approach to obtain SiNWs through top-down techniques relies on lithography, provided that resolutions around 30–100 nm may be achieved. In view of industrial applications, however, the use of deep ultraviolet (DUV) stepper systems, albeit common in nanoelectronics, is clearly unsuitable. In principle, electron beam lithography (EBL) might be considered. In EBL, a focused electron beam patterns an electron-sensitive resist (e.g., poly-methyl methacrylate—PMMA). Due to the high energy of the electron beam (10–100 keV), resolutions better than 50 nm may be easily attained. Unfortunately, as all scanning methods, EBL is also intrinsically slow, with manufacturing times ranging from a few hours to a few days. UV-assisted nanoimprint lithography might overcome such a limitation. The technique makes use of a stamp to imprint the desired geometry on a UV-sensitive polymer, which is then exposed to UV radiation before removing the stamp. Soft stamps (e.g., polydimethylsiloxane—PDMS) are widely preferred as they may adapt to the substrate, adding some tolerance toward planarity defects or particles contaminating the substrate. The stamp itself is often made by EBL followed by plasma etching. Of course, long fabrication times for the stamp itself are acceptable, and patterning costs are orders of magnitude lower than those typical of direct EBL.

Once the pattern is generated, it is transferred to the active material to either single-crystalline or polycrystalline silicon. This usually calls for some anisotropic etching. It is essential that the etching recipe has a high anisotropy to prevent mask underetching, that would obviously destroy the nanowire. To this purpose, capacitively-coupled etching is often preferred to more modern and efficient high-density plasmas for their slower etching rate. In addition, an etch-stopper (e.g., SiO_2) is needed [37].

Additional steps are then needed to integrate the nanowires in the TEG device. First, nanowires must be electrically passivated, a step usually carried out by SiO_2 deposition using low-pressure chemical vapor deposition (LPCVD). Silicon dioxide not only grants a pinhole-free conformal layer protecting nanowires (NWs) from the external environment, but it is also compatible with TEG operations as its very low thermal conductivity does not affect the thermoelectric efficiency of the SiNW array. Passivation is followed by the formation of metal contacts, serving to connect the wires to each other and to the external load through bonding pads. To this aim, the passivation is selectively removed from the contact area over the nanowires using conventional lithography. Then, a metal layer is deposited and patterned to create the metal connections. Metal contacts are extremely critical for thermoelectric applications, either in bulk or in nanoTEGs. Metal must ensure ohmic contacts and low electrical and thermal contact resistances [38]. A sensible candidate in this case is an Al–1%Si alloy. Further details on the process are reported elsewhere [37].

This ends the fabrication of the SiNW array. As a matter of fact, however, the planar geometry of the array (with SiNWs lying on the substrate) requires additional micromachining steps for the nanoTEG to provide acceptably large output current densities. The (partial) removal of the substrate prevents (or at least limits) thermal current shunts through the substrate, forcing most of the heat to flow through the nanowires and, as a result, keeping the temperature difference across the NWs as close as possible to the applied temperature drop.

It should be noted that the TEG that may be obtained through this procedure is basically a single layer of NWs covering the substrate surface. Thus, however large that NW efficiency might be, the electric power output they provide is limited by the input thermal power they admit [39]. This is a general, essential issue in all TEGs based on nanostructured material, and stands as the most important challenge that has to be overcome to make nanoTEGs of real technological interest. It is however quite intuitive that stacks of SiNW arrays may only lead to an enhancement in the output power. For this goal to be reached, however, alternate manufacturing top-down technologies are needed.

A possible way to attain stacked layers of nanowires makes use of so-called multisidewall patterning technology [40,41]. Nanowires are fabricated starting from a multilayer of SiO_2 and Si_3N_4, that are patterned to form blocks. An isotropic hydrofluoric acid (HF) etching creates niches on the patterned stack that are filled with polysilicon (deposited by LPCVD). Excess polysilicon is then removed by reactive ion etching (RIE) to obtain square-sectioned nanowires within the niches [42]. In principle, this method allows for an arbitrarily large number of NWs (grown parallel to the substrate) to be stacked on top of each other. The number of stackable layers is set by the verticality of the etching steps [42]. It should be noted that this procedure forcefully leads to polycrystalline NWs. If this further decreases their thermal conductivity, polycrystallinity also degrades the electrical conductivity. Nonetheless, thermoelectric figures of merit remain appreciably large, up to 0.09 at 300 K [20], with the obvious advantage that a larger filling factor could be achieved [39].

25.4.2 BOTTOM-UP APPROACH

In principle, bottom-up preparation of SiNWs may be either subtractive or additive. Nanowires may be obtained by selective etching of bulk silicon (e.g., by metal-assisted chemical etching—MACE) or by localized growth onto a suitable substrate.

Subtractive formation of SiNWs was largely inspired by the observation that suitable etching solutions containing strong oxidizers (e.g., H_2O_2) along with a Si complexing agent (e.g., F^-) could be used to obtain micro- and nanoporous silicon, either using electrochemical setups or by localized catalysis [43,44]. More generally, MACE enables selective localized etching of silicon by severely enhancing the etching rate through the use of metallic nanoparticles. When applied to silicon, it allows the formation of high-density SiNW bushes, rather conveniently aligned normal to the substrate [45–49]. While extremely convenient in applications ranging from photovoltaics to sensors, this technique found only few applications in thermoelectrics due to the hurdle of contacting the wire tips with scalable techniques [50–52], although evidence recently advanced about the possibility of growing metal pads on top of vertically oriented SiNWs [53] might revive the interest toward subtractive techniques.

Nanowires for thermoelectric applications are more commonly obtained by localized growth. A number of possible strategies may be considered. The most common approach makes use of vapor–liquid–solid (VLS) techniques [54,55]. The principle VLS implements is the formation of metal–silicon alloys upon exposure to a gaseous Si precursor of metal nanoparticles deposited on the substrate. If the alloy is liquid at the process temperature, then supersaturation in the liquid droplet leads to the precipitation of silicon at the droplet–substrate interface. This process ends up with the growth of crystalline silicon wires normal to the substrate. A likely process may also be promoted even if the metal–silicon alloy remains solid at the deposition temperature. In this case (referred to as vapor–solid–solid—VSS) it is mandatory to guarantee high surface and bulk Si mobility to enable Si atom migrations to the substrate interface. Among metals, gold is the most common choice as a growth promoter due to its chemical stability and to the low eutectic temperature of its alloy with silicon [56–58]. However, other metals have also been considered, including copper [59], aluminum [60,61], and silver [62].

As the metal particle size rules the diameter of the nanowire, their controlled deposition or formation is a substantial concern. If blank layers are needed, dewetting of a liquid metallic layer out of the substrate may be convenient. Partial melting or the formation of a liquid alloy leads to the formation of nanosized metal droplets, the size and density of which may be set by choosing a suitable layer thickness and annealing temperature [63]. Colloidal dispersions of nanoparticles are also often used [64]. In principle, this technique generates steep particle size distributions. When instead metal particles need to be deposited only on selected portions of the substrate (to grow nanowires only in those areas), galvanic displacement is the reference method. Coupled to more or less sophisticated patterning techniques (from lithography to screen printing, or shadow masking) it enables metal particle deposition only on selected areas, namely those where (electrically conducting) silicon is exposed [65–67].

Once the substrate has been seeded with metal nanoparticles, both VLS and VSS mechanisms may be implemented by supplying silicon in several ways. The most obvious choice is chemical vapor deposition (CVD), which provides easy scalability, doping methods, and enables the growth of NWs with high aspect ratios. Deposition temperature is set by the compromise between the high temperatures needed to decompose the gas precursor and the low temperature that makes the metal-catalyzed growth the only viable deposition process [67–70]. As an alternative, plasma-enhanced vapor deposition may replace CVD, where plasma predissociates the precursor, enabling a wider range of seeds to be used [71], although at the price of uncatalyzed concurrent deposition of amorphous silicon.

Atomic silicon may also be provided through physical vapor deposition (PVD) techniques. Laser ablation, with a laser beam irradiating a target made of a mixture of silicon and the catalyst, was studied. The ablated material is cooled down using an inert gas, enabling the growth of SiNWs even on nonsilicon substrates. The most relevant disadvantage of this approach is that since no metal nanoparticles are used as a growth seed, the geometry of the nanowires cannot be easily tuned. If this is not a concern, however, long SiNWs may be obtained with relatively short deposition times, as the growth rates are in the micrometer per minute range [72].

Silicon may also be brought to the catalytic metal particle through a liquid phase [73–75]. In the solution–liquid–solid (SLS) method, an organic solvent dissolves a silicon precursor. As an alternative, a supercritical organic fluid containing the silicon precursor at high pressure is put in contact with the metal catalyst. In both cases (and with close similarity to what happens with VLS), the substrate is then heated at a temperature high enough for the liquid metal–Si alloy to form, leading to the growth of the SiNW. As the method requires no vacuum, it clearly stands up for its low cost, although contamination may be here harder to avoid.

In summary, VLS, VSS, and SLS methods provide relatively easy, positively lower cost alternatives to top-down methods. Also, they enable the growth of single-crystalline nanowires with densities that are orders of magnitude larger than any known top-down method. All such advantages come, however, at a cost. Metallic particles may be a source of disabling contamination for SiNWs grown for thermoelectric applications. Gold (as most transition metals) is very well known to inject mid-gap localized states that severely affect carrier mobility. Thus, metal choice may be critical. Furthermore, SiNW doping suffers from the complex physical chemistry ruling Si alloying and dealloying, making a quantitative control of the doping level rather cumbersome. Finally, unless metal particles are deposited using a patterning technique, NW growth over the substrate surface occurs randomly, and wire diameter and length have finite-width distributions. This may or may not be a problem, depending on the availability of suitable strategies to ensure that both sides of the NW may be electrically and thermally contacted without introducing high contact resistances.

25.5 SILICON NANOLAYERS

Silicon nanolayers, namely silicon thin films with a thickness lower than about 200 nm, have been also investigated as an instance of dimensionally constrained silicon nanostructures. As long as their surfaces (both external and internal) have a convenient roughness so as to incoherently scatter phonons, they also show a reduced thermal conductivity.

Nanolayer growth requires no special techniques, as ultrathin films may be deposited using standard CVD or molecular beam epitaxy (MBE) techniques. However, use of silicon layers in TEGs runs into the same problems described for SiNWs, namely the need of removing or thinning the deposition substrate to prevent or limit the heat current shunt through the substrate itself.

Liu and Asheghi carried out a systematic investigation of the dependence of κ upon the layer thickness in single-crystalline silicon, showing that in undoped ultrathin silicon layers the lateral thermal conductivity for thicknesses between 20 and 100 nm displayed a large reduction resulting from phonon-boundary scattering (Figure 25.3). For 20-nm-thick silicon layers, κ at room temperature is ≈ 22 $Wm^{-1}K^{-1}$, which is about 1/7 of the bulk value for silicon (148 Wm^{-1} K^{-1}) [76]. Results were obtained in doped ultrathin films [77] (Figure 25.3). It is interesting to observe that thermal conductivity in nanocrystalline thin films was found to be basically the same as in single crystals ultrathin films, a finding that suggests that phonon scattering at the outer layer boundaries and at inner grain boundaries suppress the same part of the phonon spectrum.

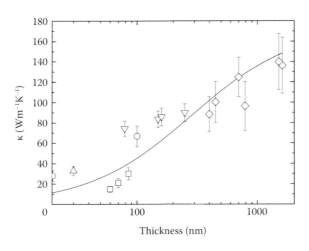

Figure 25.3 In-plane thermal conductivity of silicon films as a function of the film thickness. Data from Refs. [39] (□), [77] (○), [76] (△), [78] (▽), and [79] (◇).

Compared to SiNWs, the reduction in the thermal conductivity that can be achieved in nanolayers at the same nanoscale (film thickness for 2D structures, nanowire diameter for 1D structures) is lower, as possibly expected, so the achievable figures of merit are smaller. Nonetheless, nanolayers may be competitive with nanowires regarding the output power density. Geometric filling factors are larger here, as layer stacks are relatively easy to obtain and the fraction of passive, thermoelectrically inactive material embedding silicon is much smaller [39].

25.6 USE OF NANOSILICON IN THERMOELECTRIC DEVICES

In spite of the large research effort devoted to silicon nanowires and nanolayers, only very few are examples of TEGs based on them. Two of the main reasons for this seeming inconsistency are the thermomechanic stability of the Si nanostructures and the marginal power output that may be obtained even from relatively dense SiNW arrays.

As in all classes of TEGs, including in nanosilicon-based TEGs, the manufacturability of a module needs to guarantee the long-term mechanical stability of the whole device. Single-crystalline SiNWs are less stiff than bulk silicon but the need to suspend them across gaps to prevent thermal shunts makes the whole structure very fragile. In addition, unless very low temperature differences are applied, differential thermal expansion may be an additional cause of failure. For polycrystalline SiNWs, thermomechanical problems are quite relaxed since the wires are highly flexible [41,42] and capable of sustaining even large stresses without breakage. In this case though, electrical contacts may be critical. High doping of the contact areas may be very challenging, due to the easy outdiffusion of the dopant even at moderate temperatures [80].

However, none of these issues are impossible to overcome. The main factor hindering the exploitation of the enhanced thermoelectric efficiency of SiNWs is related to the power output they may yield. For SiNW arrays grown parallel to the substrate surface, the ratio between the total SiNW cross-section and the geometrical heat exchange surface is very unfavorable [39]. The supporting structure often leaves less than a few percent of thermoelectrically active area, so only a small fraction of the available heat current flows through the nanowires. This problem is only reduced when moving from embedded to suspended SiNWs, as heat is radiatively exchanged between facing supporting structures spaced by a few micrometers or less. Only vertically oriented SiNWs might escape this shortcoming, which makes effort in this direction especially worthwhile [81]. Power output also depends on the nanowire efficiency, which also in this case scales with the temperature difference through the standard Carnot factor, and applying a large temperature difference across nanostructures is not simple. Since the thermal resistance of a nanowire scales with its length, that hardly exceeds a few micrometers, the whole thermal chain from the hot to the cold sink

Industrial nanosilicon

unavoidably embeds larger thermal resistances [82]. Thus, the applied temperature difference mostly drops on supporting structures and at contacts, with only a minor fraction of it being experienced by the thermoelectric element.

As a factual display of these limitations, it is interesting to briefly review the most important attempts of embedding SiNWs into TEGs. In 2011, in what is possibly the first report of such a technology, a Singapore-based collaboration reported on a high-density SiNW–based TEG obtained through a top-down approach and using CMOS-compliant fabrication methods. Nanowires were oriented normal to the substrate and had a diameter around 80 nm and a length (height) of 1 μm. To reduce the electrical resistance, wires were bundled in groups of 540 × 540. This led to an open circuit voltage of 1.5 mV over a temperature difference of 0.12 K. The short-circuit current was reported to be of ≈4 μA, that is, an output power of 1.5 nW over a chip surface of 25 mm². No larger temperature difference could be tested due to the parasitic (series) thermal resistances [83]. In subsequent papers, power outputs could be raised to 4.6 nW (open circuit voltage of 2.7 mV) but the large discrepancy between the apparent and expected Seebeck coefficient is possibly an indication of the unfavorable partition of the applied temperature difference (95 K) and the actual temperature drop over the NWs [84].

Curtin et al. reported in 2012, the fabrication of a TEG using vertical SiNW arrays [85]. The device was fabricated using interference lithography. The resulting 80-nm SiNWs were 1 μm tall and a 15% packing density could be achieved. Mechanical stability issues were addressed by embedding the wires into spin-on glass. Authors reported a maximum power output of 29 μW over a temperature difference of 56 K over an active surface of a 2.5 × 10⁻⁵cm² out of a total surface of 4 cm².

A bottom-up approach was implemented by Davila, Fonseca, and coworkers. Nanowires were grown using a VLS technique across the gap between two Si walls, belonging to almost fully insulated supporting structures (Figure 25.4). This enabled larger temperature differences to be achieved, so that power outputs

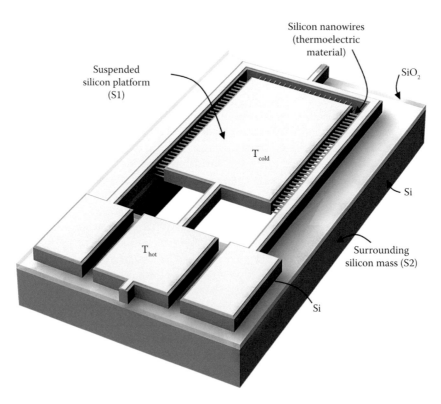

Figure 25.4 Schematics of a micro-nanoTEG [86] obtained by growing nanowires by VLS between facing sides of a microfabricated silicon support. Silicon nanowires connect a thermally insulated (suspended) silicon mass (S1) to the surrounding silicon bulk (S2). (Courtesy of Dr. Luis Fonseca, CSIC, Barcelona, Spain).

as large as 1.4 µW could be obtained over a chip surface of 1 mm^2 upon application of a temperature difference of 300 K [86,87].

A much larger literature is available concerning the use of silicon thin films. Most of them make use of films with thicknesses above or around the critical thickness relevant to enable a nanosize-related reduction of the thermal conductivity ($\approx$200 nm) and therefore will not be considered. The interested reader may refer to a recent review on this topic [37]. Limiting to nanofilms, recently Lopeandia and coworkers [88] announced the fabrication of a TEG using 100-nm ultrathin films of silicon suspended between facing micromachined silicon walls. Both p- and n-type layers were used. Authors reported a power output of 4.5 µW/cm^2 under a temperature difference of 5.5 K.

Apparently, a remarkable gap exists between the abundant literature on the Si nanostructure preparation and characterization and that concerning its use in real devices, witnessing again the currently marginal possibilities of deployment of nanowires and nanolayers in effective TEGs.

25.7 SUMMARY AND CONCLUSIONS

Nanostructured silicon has provided clear evidence of how nanotechnology may change the physical properties of an age-old material. In spite of the marginal thermoelectric figure of merit of bulk silicon ($ZT \approx 0.01$), silicon nanowires and nanolayers were shown to have the full potential to climb to high-efficiency thermoelectric conversion. The physical reasons for the improved ZT stand nowadays mostly on the capability of limiting phonon MFP without affecting hole and electron MFPs by using outer walls of dimensionally constrained structures as selective scatterers. Several approaches have been considered to achieve this goal, and a major research effort has been spent to attain techniques with good scalability. Nonetheless, very few examples are available of successful implementations of SiNWs within real TEGs. As seen, this is mostly due to the dramatic difference between high efficiency and high power output densities. Thermoelectric harvesters are solid-state thermal engines, and the electric power outputs they are capable of obviously depends on the thermal input they may accept. Nanostructures either oppose gigantic thermal resistances or dissipate most of the heat flow through thermal shunts. Furthermore, it is very difficult to have large temperature differences dropping onto micrometer-long thermoelectric legs. As a result, the current performances of NW-based TEGs are far from being competitive with standard bulk TEGs.

It might be tempting to conclude that SiNWs (and dimensionally constrained systems, more in general) were just a good exercise to learn the physics needed to master bulk materials. Although this conclusion is worth being considered and analyzed, if only for its coming from the very same scientist who introduced nanotechnology in thermoelectrics [6], a less negative outlook might be possible, even neglecting the possibility of further enhancements due to increased power factors. Output powers in the order of tens of nanowatt may appear negligible, but they are close to the operational range of ultralow power microelectronic components [89,90]. Thus, especially when no continuous operation is needed, nanowatt TEGs are already within the field of usability as integrated power sources. Also, the issue of the thermal chain is largely simplified when all parts of the TEG are comparable in size to the NW length. Therefore, it might be not unreasonable to foresee a role for SiNWs in integrated thermal microharvesters for (mobile) sensing networks. This apart, it is fair to conclude that no possibility is foreseeable for nanostructured silicon in bulk waste heat recovery unless major breakthroughs will overcome the many obstacles currently limiting power output densities.

ACKNOWLEDGMENTS

This work was supported by FP7-NMP-2013-SMALL-7, SiNERGY (Silicon Friendly Materials and Device Solutions for Microenergy Applications) Project, Contract n. 604169.

REFERENCES

1. H. Goldsmid. *Thermoelectric Refrigeration*. The International Cryogenics Monograph Series. Springer, Boston, MA, 2013.
2. C. Goupil. Thermodynamics of thermoelectricity. In M. Tadashi, editor, *Thermodynamics*. InTech Open Access Publisher, Rijeka, Croatia, 275–292, 2011.

3. E. Altenkirch. On the effectiveness of the thermopile. *Phys. Z.*, 10:560–580, 1909.

4. A. Ioffe. *Semiconductor Thermoelements and Thermoelectric Cooling*. Infosearch Ltd., London, 1957.

5. D. Narducci. Thermodynamic efficiency, power output and performance indices of classic and nanostructured thermoelectric materials. *J. Nanoeng. Nanomanuf.*, 1(1):63–70, 2011.

6. J. P. Heremans, M. S. Dresselhaus, L. E. Bell, and D. T. Morelli. When thermoelectrics reached the nanoscale. *Nat. Nanotechnol.*, 8(7):471–473, 2013.

7. G. J. Snyder and E. S. Toberer. Complex thermoelectric materials. *Nat. Mater.*, 7(2):105–114, 2008.

8. M. G. Kanatzidis. Nanostructured thermoelectrics: The new paradigm? *Chem. Mater.*, 22(3):648–659, 2010.

9. A. J. Minnich, M. S. Dresselhaus, Z. F. Ren, and G. Chen. Bulk nanostructured thermoelectric materials: Current research and future prospects. *Energy Environ. Sci.*, 2:466–479, 2009.

10. D. Narducci, E. Selezneva, A. Arcari, G. Cerofolini, E. Romano, R. Tonini, and G. Ottaviani. Enhanced thermoelectric properties of strongly degenerate polycrystalline silicon upon second phase segregation. *MRS Proc.*, vol. 1314, mrsf10-1314-ll05-16, 2011.

11. L. D. Hicks and M. S. Dresselhaus. Effect of quantum-well structures on the thermoelectric figure of merit. *Phys. Rev. B*, 47:12727–12731, 1993.

12. L. D. Hicks and M. S. Dresselhaus. Thermoelectric figure of merit of a one-dimensional conductor. *Phys. Rev. B*, 47:16631–16634, 1993.

13. A. I. Hochbaum, R. K. Chen, R. D. Delgado, W. J. Liang, E. C. Garnett, M. Najarian, A. Majumdar, and P. D. Yang. Enhanced thermoelectric performance of rough silicon nanowires. *Nature*, 451(7175):163–167, 2008.

14. A. I. Boukai, Y. Bunimovich, J. Tahir-Kheli, J. K. Yu, W. A. Goddard, and J. R. Heath. Silicon nanowires as efficient thermoelectric materials. *Nature*, 451(7175):168–171, 2008.

15. G. Pennelli. Top down fabrication of long silicon nanowire devices by means of lateral oxidation. *Microelectron. Eng.*, 86(11):2139–2143, 2009.

16. R. Juhasz, N. Elfstrom, J. Linnros, N. Elfström, and J. Linnros. Controlled fabrication of silicon nanowires by electron beam lithography and electrochemical size reduction. *Nano Lett.*, 5(2):275–280, 2004.

17. Z. Wang. *Nanowires and Nanobelts: Materials, Properties and Devices. Volume 1: Metal and Semiconductor Nanowires*. Springer, Boston, MA, 2013.

18. J. Colinge and J. Greer. *Nanowire Transistors: Physics of Devices and Materials in One Dimension*. Cambridge University Press, Cambridge, 2016.

19. S. Neogi and D. Donadio. Thermal transport in free-standing silicon membranes: Influence of dimensional reduction and surface nanostructures. *Eur. Phys. J. B*, 88(3):1–9, 2015.

20. F. Suriano, M. Ferri, F. Moscatelli, F. Mancarella, L. Belsito, S. Solmi, A. Roncaglia, S. Frabboni, G. Gazzadi, and D. Narducci. Influence of grain size on the thermoelectric properties of polycrystalline silicon nanowires. *J. Electron. Mater.*, 44(1):371–376, 2015.

21. N. Petermann, J. Stötzel, N. Stein, V. Kessler, H. Wiggers, R. Theissmann, G. Schierning, and R. Schmechel. Thermoelectrics from silicon nanoparticles: The influence of native oxide. *Eur. Phys. J. B*, 88(6):163, 2015.

22. B. Lorenzi, D. Narducci, R. Tonini, S. Frabboni, G. C. Gazzadi, G. Ottaviani, N. Neophytou, and X. Zianni. Paradoxical enhancement of the power factor of polycrystalline silicon as a result of the formation of nanovoids. *J. Electron. Mater.*, 43(10):3812–3816, 2014.

23. D. Narducci, B. Lorenzi, X. Zianni, N. Neophytou, S. Frabboni, G. C. Gazzadi, A. Roncaglia, and F. Suriano. Enhancement of the power factor in two-phase silicon-boron nanocrystalline alloys. *Phys. Status Solidi A*, 211(6):1255–1258, 2014.

24. N. Neophytou and H. Kosina. Optimizing thermoelectric power factor by means of a potential barrier. *J. Appl. Phys.*, 114(4):044315, 2013.

25. Y. Tian, M. R. Sakr, J. M. Kinder, D. Liang, M. J. MacDonald, R. L. J. Qiu, H. J. Gao, and X. P. Gao. One-dimensional quantum confinement effect modulated thermoelectric properties in InAs nanowires. *Nano Lett.*, 12:6492–6497, 2012.

26. G. Chen. *Nanoscale Energy Transport and Conversion*. Oxford University Press, Oxford, 2005.

27. J. Ziman. *Electrons and Phonons: The Theory of Transport Phenomena in Solids*. Oxford Classic Texts in the Physical Sciences. Oxford University Press, Oxford, 2001.

28. F. Reif. *Fundamentals of Statistical and Thermal Physics*. Waveland Press, Long Grove, IL, 2009.

29. D. Narducci, S. Frabboni, and X. Zianni. Silicon de novo: Energy filtering and enhanced thermoelectric performances of nanocrystalline silicon and silicon alloys. *J. Mater. Chem. C*, 3:12176–12185, 2015.

30. A. Mehdizadeh Dehkordi, M. Zebarjadi, J. He, and T. M. Tritt. Thermoelectric power factor: Enhancement mechanisms and strategies for higher performance thermoelectric materials. *Mater. Sci. Eng., R*, 97:1–22, 2015.

31. Y. Pei, X. Shi, A. LaLonde, H. Wang, L. Chen, and G. J. Snyder. Convergence of electronic bands for high performance bulk thermoelectrics. *Nature*, 473(7345):66–69, 2011.

32. E. Lara-Curzio, A. May, O. Delaire, M. McGuire, X. Lu, C.-Y. Liu, E. Case, and D. Morelli. Low-temperature heat capacity and localized vibrational modes in natural and synthetic tetrahedrites. *J. Appl. Phys.*, 115(19): 193515, 2014.

33. D. Morelli, V. Jovovic, and J. Heremans. Intrinsically minimal thermal conductivity in cubic I-V-VI semiconductors. *Phys. Rev. Lett.*, 101(3):035901, 2008.

34. L. Hu, T. Zhu, X. Liu, and X. Zhao. Point defect engineering of high-performance bismuth-telluride-based thermoelectric materials. *Adv. Funct. Mater.*, 24(33):5211–5218, 2014.

35. F. Yang and C. Dames. Mean free path spectra as a tool to understand thermal conductivity in bulk and nanostructures. *Phys. Rev. B*, 87(3):035437, 2013.

36. J. P. Feser, J. S. Sadhu, B. P. Azeredo, K. H. Hsu, J. Ma, J. Kim, M. Seong, et al. Thermal conductivity of silicon nanowire arrays with controlled roughness. *J. Appl. Phys.*, 112:114306, 2012.

37. D. Narducci, L. Belsito, and A. Morata. Silicon for thermoelectric energy harvesting applications. In D. Davila Pineda and A. Rezaniakolae, editors, *Thermoelectric Energy Conversion. Basic Concepts and Device Applications.* Wiley, Weinheim, pp. 53–91.

38. Y.-C. Lin and Y. Huang. Silicon and silicide nanowires: Applications, fabrication and properties. In Y. Huang and K.-N. Tu, editors, *Nanoscale Contact Engineering for Si Nanowire Devices*, Pages 413–451. Singapore: Pan Stanford, 2013.

39. D. Narducci, G. Cerofolini, M. Ferri, F. Suriano, F. Mancarella, L. Belsito, S. Solmi, and A. Roncaglia. Phonon scattering enhancement in silicon nanolayers. *J. Mater. Sci.*, 48(7):2779–2784, 2013.

40. G. Cerofolini, M. Ferri, E. Romano, F. Suriano, G. Veronese, S. Solmi, and D. Narducci. Terascale integration via a redesign of the crossbar based on a vertical arrangement of poly-Si nanowires. *Semicond. Sci. Technol.*, 25(9):095011, 2010.

41. G. F. Cerofolini, M. Ferri, E. Romano, F. Suriano, G. P. Veronese, S. Solmi, and D. Narducci. Crossbar architecture for tera-scale integration. *Semicond. Sci. Technol.*, 26(4):045005, 2011.

42. M. Ferri, F. Suriano, A. Roncaglia, S. Solmi, G. Cerofolini, E. Romano, and D. Narducci. Ultradense silicon nanowire arrays produced via top-down planar technology. *Microelectron. Eng.*, 88(6):877–881, 2011.

43. X. Li and P. Bonn. Metal-assisted chemical etching in Hf/H_2O_2 produces porous silicon. *Appl. Phys. Lett.*, 77(16):2572–2574, 2000.

44. S. Chattopadhyay, X. Li, and P. Bohn. In-plane control of morphology and tunable photoluminescence in porous silicon produced by metal-assisted electroless chemical etching. *J. Appl. Phys.*, 91(9):6134–6140, 2002.

45. M.-L. Zhang, K.-Q. Peng, X. Fan, J.-S. Jie, R.-Q. Zhang, S.-T. Lee, and N.-B. Wong. Preparation of large-area uniform silicon nanowires arrays through metal-assisted chemical etching. *J. Phys. Chem. C*, 112(12):4444–4450, 2008.

46. C. Chartier, S. Bastide, and C. Lévy-Clément. Metal-assisted chemical etching of silicon in $Hf-H_2O_2$. *Electrochim. Acta*, 53(17):5509–5516, 2008.

47. Z. Huang, X. Zhang, M. Reiche, L. Ltu, W. Lee, T. Shimizu, S. Senz, and U. Gsele. Extended arrays of vertically aligned sub-10 nm diameter [100] Si nanowires by metal-assisted chemical etching. *Nano Lett.*, 8(9):3046–3051, 2008.

48. Z. Huang, N. Geyer, P. Werner, J. De Boor, and U. Gsele. Metal-assisted chemical etching of silicon: A review. *Adv. Mater.*, 23(2):285–308, 2011.

49. M. Bollani, J. Osmond, G. Nicotra, C. Spinella, and D. Narducci. Strain-induced generation of silicon nanopillars. *Nanotechnology*, 24(33):335302, 2013.

50. B. Xu, C. Li, K. Thielemans, M. Myronov, and K. Fobelets. Thermoelectric performance of $Si_{0.8}Ge_{0.2}$ nanowire arrays. *IEEE T. Electron. Dev.*, 59(12):3193–3198, 2012.

51. M. Ghossoub, K. Valavala, M. Seong, B. Azeredo, K. Hsu, J. Sadhu, P. Singh, and S. Sinha. Spectral phonon scattering from sub-10 nm surface roughness wavelengths in metal-assisted chemically etched si nanowires. *Nano Lett.*, 13(4):1564–1571, 2013.

52. T. Zhang, S. Wu, J. Xu, R. Zheng, and G. Cheng. High thermoelectric figure-of-merits from large-area porous silicon nanowire arrays. *Nano Energy*, 13:433–441, 2015.

53. G. Pennelli and M. Macucci. High-power thermoelectric generators based on nanostructured silicon. *Semicond. Sci. Technol.*, 31(5):054001, 2016.

54. R. S. Wagner and W. C. Ellis. Vapor-liquid-solid mechanism of single crystal growth. *Appl. Phys. Lett.*, 4(5):89, 1964.

55. E. Givargizov and N. Sheftal'. Morphology of silicon whiskers grown by the VLS-technique. *J. Cryst. Growth*, 9:326–329, 1971.

56. K.-K. Lew and J. M. Redwing. Growth characteristics of silicon nanowires synthesized by vaporliquidsolid growth in nanoporous alumina templates. *J. Cryst. Growth*, 254(1–2):14–22, 2003.

57. T.-W. Ho and F. C.-N. Hong. A novel method to grow vertically aligned silicon nanowires on Si (111) and their optical absorption. *J. Nanomater.*, 2012:1–9, 2012.

Industrial nanosilicon

58. F. Thissandier, P. Gentile, N. Pauc, T. Brousse, G. Bidan, and S. Sadki. Tuning silicon nanowires doping level and morphology for highly efficient micro-supercapacitors. *Nano Energy*, 5:20–27, 2014.

59. Y. Yao and S. Fan. Si nanowires synthesized with Cu catalyst. *Mater. Lett.*, 61(1):177–181, 2007.

60. Y. Wang, V. Schmidt, S. Senz, and U. Gösele. Epitaxial growth of silicon nanowires using an aluminium catalyst. *Nat. Nanotechnol.*, 1(3):186–189, 2006.

61. S.-J. Whang, S. Lee, D.-Z. Chi, W.-F. Yang, B.-J. Cho, Y.-F. Liew, and D.-L. Kwong. B-doping of vapour-liquid-solid grown Au-catalysed and Al-catalysed Si nanowires: Effects of B_2H_6 gas during Si nanowire growth and B-doping by a post-synthesis in situ plasma process. *Nanotechnology*, 18(27):275302, 2007.

62. L. Weber. Equilibrium solid solubility of silicon in silver. *Metall. Mater. Trans. A*, 33(4):1145–1150, 2002.

63. A. Fasoli and W. Milne. Overview and status of bottom-up silicon nanowire electronics. *Mater. Sci. Semicond. Process.*, 15(6):601–614, 2012.

64. H.-Y. Tuan, D. C. Lee, T. Hanrath, and B. A. Korgel. Catalytic solid-phase seeding of silicon nanowires by nickel nanocrystals in organic solvents. *Nano Lett.*, 5(4):681–684, 2005.

65. L. Magagnin, V. Bertani, P. Cavallotti, R. Maboudian, and C. Carraro. Selective deposition of gold nanoclusters on silicon by a galvanic displacement process. *Microelectron. Eng.*, 64(1–4):479–485, 2002.

66. D. Gao, R. He, C. Carraro, R. T. Howe, P. Yang, and R. Maboudian. Selective growth of Si nanowire arrays via galvanic displacement processes in water-in-oil microemulsions. *J. Am. Chem. Soc.*, 127(13):4574–4575, 2005.

67. G. Gadea, A. Morata, J. D. Santos, D. Dávila, C. Calaza, M. Salleras, L. Fonseca, and A. Tarancón. Towards a full integration of vertically aligned silicon nanowires in MEMS using silane as a precursor. *Nanotechnology*, 26(19):195302, 2015.

68. S. Inasawa. In-situ observation of the growth of individual silicon wires in the zinc reduction reaction of $SiCl_4$. *J. Cryst. Growth*, 412:109–115, 2015.

69. A. I. Hochbaum, R. Fan, R. He, and P. Yang. Controlled growth of Si nanowire arrays for device integration. *Nano Lett.*, 5(3):457–460, 2005.

70. Y. Cui, L. J. Lauhon, M. S. Gudiksen, J. Wang, and C. M. Lieber. Diameter-controlled synthesis of single-crystal silicon nanowires. *Appl. Phys. Lett.*, 78(15):2214, 2001.

71. S. Hofmann, C. Ducati, R. J. Neill, S. Piscanec, A. C. Ferrari, J. Geng, R. E. Dunin-Borkowski, and J. Robertson. Gold catalyzed growth of silicon nanowires by plasma enhanced chemical vapor deposition. *J. Appl. Phys.*, 94(9):6005, 2003.

72. A. M. Morales. A Laser Ablation method for the synthesis of crystalline semiconductor nanowires. *Science*, 279(5348):208–211, 1998.

73. J. Holmes, K. Johnston, R. Doty, and B. Korgel. Control of thickness and orientation of solution-grown silicon nanowires. *Science*, 287(5457):1471–1473, 2000.

74. A. T. Heitsch, D. D. Fanfair, H. Y. Tuan, and B. A. Korgel. Solution-liquid-solid (SLS) growth of silicon nanowires. *J. Am. Chem. Soc.*, 130:5436–5437, 2008.

75. A. Dong, R. Tang, and W. E. Buhro. Solution-based growth and structural characterization of homo- and heterobranched semiconductor nanowires. *J. Am. Chem. Soc.*, 129(40):12254–12262, 2007.

76. W. Liu and M. Asheghi. Thermal conductivity measurements of ultra-thin single crystal silicon layers. *J. Heat Transfer*, 128(1):75–83, 2006.

77. W. Liu and M. Asheghi. Thermal conduction in ultrathin pure and doped single-crystal silicon layers at high temperatures. *J. Appl. Phys.*, 98(12):123523–123526, 2005.

78. N. Neophytou, X. Zianni, M. Ferri, A. Roncaglia, G. Cerofolini, and D. Narducci. Nanograin effects on the thermoelectric properties of poly-si nanowires. *J. Electron. Mater.*, 42(7):2393–2401, 2013.

79. E. Dimaggio and G. Pennelli. Reliable fabrication of metal contacts on silicon nanowire forests. *Nano Lett.*, 16(7):4348–4354, 2016.

80. D. Narducci. Explicitly accounting for the heat sink strengths in the thermal matching of thermoelectric devices. A unified practical approach. *Mater. Today Proc.*, 2(2):474–482, 2015.

81. Y. Li, K. Buddharaju, N. Singh, G. Q. Lo, and S. J. Lee. Chip-level thermoelectric power generators based on high-density silicon nanowire array prepared with top-down CMOS technology. *IEEE Electron Device Lett.*, 32(5):674–676, 2011.

82. Y. Li, K. Buddharaju, N. Singh, and S. Lee. Top-down silicon nanowire-based thermoelectric generator: Design and characterization. *J. Electron. Mater.*, 41(6):989–992, 2012.

83. B. M. Curtin, E. W. Fang, and J. E. Bowers. Highly ordered vertical silicon nanowire array composite thin films for thermoelectric devices. *J. Electron. Mater.*, 41(5):887–894, 2012.

84. D. Dávila, A. Tarancón, C. Calaza, M. Salleras, M. Fernández-Regúlez, A. San Paulo, and L. Fonseca. Monolithically integrated thermoelectric energy harvester based on silicon nanowire arrays for powering micro/nanodevices. *Nano Energy*, 1(6):812–819, 2012.

85. L. Fonseca, J.-D. Santos, A. Roncaglia, D. Narducci, C. Calaza, M. Salleras, I. Donmez, et al. Smart integration of silicon nanowire arrays in all-silicon thermoelectric micro-nanogenerators. *Semicond. Sci. Technol.*, 31(8):084001, 2016.

86. A. Perez-Marn, A. Lopeanda, L. Abad, P. Ferrando-Villaba, G. Garcia, A. Lopez, F. Muoz-Pascual, and J. Rodrguez-Viejo. Micropower thermoelectric generator from thin Si membranes. *Nano Energy*, 4:73–80, 2014.

87. B. Warneke, M. Last, B. Liebowitz, and K. S. J. Pister. Smart dust: Communicating with a cubic-millimeter computer. *Computer*, 34(1):44–51, 2001.

88. S. Luryi, J. Xu, and A. Zaslavsky. *Future Trends in Microelectronics: Frontiers and Innovations*. New York: Wiley, 2013.

89. Y. S. Ju and K. E. Goodson. Phonon scattering in silicon films with thickness of order 100 nm. *Appl. Phys. Lett.*, 74(20):3005–3007, 1999.

90. M. Asheghi, M. Touzelbaev, K. Goodson, Y. Leung, and S. Wong. Temperature-dependent thermal conductivity of single-crystal silicon layers in soi substrates. *J. Heat Transfer*, 120(1):30–36, 1998.

Industrial nanosilicon

Nanostructured silicon for thermoelectric applications

Giovanni Pennelli and Elisabetta Dimaggio

Contents

26.1 INTRODUCTION

26.1.1 THERMOELECTRICITY FUNDAMENTALS

Thermoelectricity is a well-known phenomenon, investigated since the beginning of the 19th century, when Thomas Johann Seebeck (Tallinn 1770; Berlin 1831) discovered that an electrical current flows if different metals, maintained at different temperatures, are connected together (Pennelli 2014). In simple words, if a difference of temperature $\Delta T = T_H - T_C$ (T_H stands for T_{Hot} and T_C stands for T_{Cold}; see Figure 26.1a) is maintained between the ends of a conductor, or of a semiconductor, a potential difference ΔV is generated. ΔV is proportional to ΔT: $\Delta V = S\Delta T$, where S is the Seebeck coefficient. The sign of the potential difference depends on the sign of the charge carriers in the conductor. Charge carriers tend to move from the hot end to the cold end of the material. In static conditions, when the electrical current I (current density J) is 0, the charges accumulate at the ends, so an electrical potential is generated and any further carrier movement is prevented. Therefore, in metals and in n-doped semiconductors, where charge carriers are electrons, the T_H end is positive with respect to the T_C end where electrons are accumulated, and S is negative as $V_C - V_H = S(T_H - T_C)$. Conversely, in p-doped semiconductors, holes are accumulated at the cold end, meanwhile an excess of negative charge is left at the hot end, so S is positive. In practical cases, the equipment (voltmeter) used for the measurement of the potential difference has a uniform temperature, which can be T_{Hot} or T_{Cold}. Figure 26.1b shows that if only one kind of material is used, it is not possible to measure any potential difference in a closed loop. Even in the presence of a temperature gradient, the total temperature difference in the loop is 0, therefore $\Delta V = S\Delta T = 0$, because the two ends, where the voltmeter is applied, are both at the cold temperature T_C. Using instead metals with different Seebeck coefficients, the measured voltage drop is proportional both to the temperature difference and to the difference of the Seebeck coefficients, as shown Figure 26.1c. The winning strategy for the achievement of a potential difference as high as possible is to combine semiconductors with complementary n and p doping, so that the difference of the Seebeck coefficients results in the sum of the absolute values of S for the p (positive S) and n (negative S) materials. The sketch shown in Figure 26.1d is a basic design of a thermoelectric module, made of n- and a p-doped pieces of a semiconductor, interconnected in a suitable configuration for the generation of electrical power.

As shown in Figure 26.1d, the two pieces of the semiconductor, which are named the "legs" of the thermo-electric module, are connected electrically in series and thermally in parallel.

In Figure 26.2a, it is shown how an electrical load (represented by the resistor $R_L = R_{Load}$) can be applied to a thermoelectric module, made of an n- doped semiconducting leg and a p- doped semiconducting leg. Holes diffuse from the hot part T_H to the cold part T_C, meanwhile electrons flow from the cold to the hot part. Interconnecting in series the legs, as shown in Figure 26.2, a net electrical current, due to the temperature difference $T_H - T_C$, is forced to flow into the electrical load R_L. Therefore, part of the thermal power taken from the heat source is converted into electrical power. A thermoelectric generator is made of many thermoelectric modules, connected both in series (for increasing the generated voltage) and in parallel (for increasing the delivered electrical current). As shown in Figure 26.2b, imposing an electrical current in the same elementary structure by means of a voltage generator, heat is transferred from the

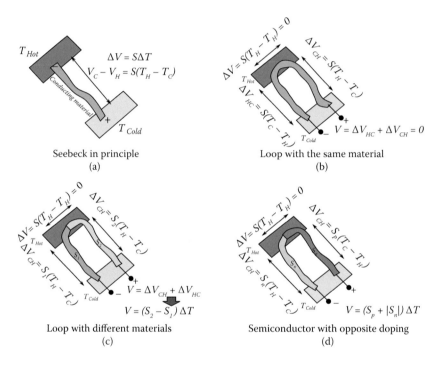

Figure 26.1 Sketches showing the principle of the Seebeck effect. (a) A potential difference is established between the ends of a conductor, which are maintained at a different temperature ($\Delta T = T_H - T_C$). (b) The total voltage drop in a loop with the same material is 0. (c) A voltage drop is generated if the loop is made by connecting different materials (with different Seebeck coefficients). (d) If in a loop, semiconductors with opposite doping are used; the total Seebeck coefficient is the sum of the absolute value of the two coefficients.

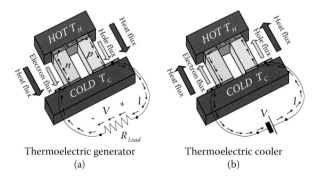

Figure 26.2 Sketch of (a) a thermoelectric generator and (b) a thermoelectric cooler. (From Pennelli, G., Beilstein J. Nanotechnol., 5, 1268, 2014. Open access license.)

cold to the hot source. This phenomenon is complementary with respect to the Seebeck effect, and it was investigated in the 19th century by Jean Charles Peltier (born at Ham in France in 1785 and died at Paris in 1845). Peltier found that if an electrical current is imposed on a piece of a conductor, a temperature difference is generated between its ends. Charge carriers bring heat, hence an electrical current J brings an heat flux ϕ (W/m^2) proportional to J: $\phi = \Pi J$, where Π is the Peltier coefficient. The Seebeck and the Peltier effects are strictly connected, and effectively they are two points of view of the same phenomenon: as the molecule of an ideal gas, charge carriers move from the hot to the cold part, and in doing that they bring heat. It can be demonstrated that the Peltier and the Seebeck coefficients are related by the relationship $\Pi = ST$ (where T is the absolute temperature). In principle, the same device can be used for both purposes: given a heat source and a heat sink, electrical power can be generated by the conversion of heat; conversely, if electrical power is given to the device by means of a power supply, heat can be driven from the cold part to the hot part. In the first case, the device works as thermoelectric generator, in the second case the device can be used for cooling down: this second configuration (Peltier cooler) has an important application in many portable fridges. Even if we talk about problems and solutions for thermoelectric generation in this chapter, the two aspects are strictly linked, and the same solutions are valid for both applications.

Given a conductor in electric equilibrium ($J = 0$), but not in thermal equilibrium ($\partial T/\partial x \neq 0$), the Seebeck coefficient S is formally defined as the ratio between the generated electric field ε and the temperature gradient $\partial T/\partial x$:

$$S(T) = \frac{\varepsilon}{\dfrac{\partial T}{\partial x}} \tag{26.1}$$

$S = S(T)$ is in general a function of the absolute temperature T. In a conductor, or in a semiconductor with uniform doping, an external electrical field ε (V/m) generates an electrical current $J = \sigma \varepsilon$ (J: current density, A/m^2), where σ is the electrical conductivity of the material, and it is expressed in Ω^{-1}m^{-1}. The thermal transport in a material is described by the heat diffusion equation, which relates the heat flux ϕ (expressed in W/m^2) and the temperature gradient $\partial T/\partial x$:

$$\phi = -k_t \frac{\partial T}{\partial x} \tag{26.2}$$

where k_t is the thermal conductivity, measured as W/mK. In a material where there is an electric field and a temperature gradient, both the Seebeck and the Peltier effects must be taken into account. In this case, the expression of the current is modified by the electric field due to the temperature gradient. From the thermal point of view, the heat driven by the electrical current must be taken into account in the heat transport equation. Therefore, the two equations for the electrical and thermal transport are modified as follows:

$$J = \sigma \varepsilon - S\sigma \frac{\partial T}{\partial x}$$
$$\phi = \Pi J - k_t \frac{\partial T}{\partial x}$$

These two equations of the thermal and electrical (thermoelectric) transport in a material can be formally derived from the Boltzmann's transport equation.

26.1.2 EFFICIENCY OF A THERMOELECTRIC GENERATOR

Figure 26.3 shows the electrical circuit, which represents the thermoelectric module sketched in Figure 26.2. The voltage V of the generator is proportional to the temperature difference $V = (S_p + |S_n|)\Delta T$). The Seebeck coefficient depends on the materials used for the fabrication of the legs, and it is of the order of

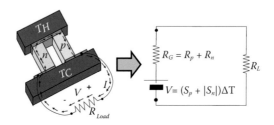

Figure 26.3 A thermoelectric generator is equivalent, from the electrical point of view, to a battery $V = (S_p + |S_n|)$ $(T_H - T_C)$ in series with a resistance $R_p + R_n$ which depends on the resistivity of the p and n legs.

hundreds of μV/K. Therefore, the voltage generated by a single module is in general quite small (in the order of tens of millivolts—it depends on the temperature drop). For this reason, a thermoelectric generator is in general made of several modules electrically connected in series. In this way, the generated voltage can reach values of the order of several volts. Conversely, if the device must be used for cooling, it is more convenient to arrange the modules in parallel so that higher currents can be achieved, and the heat flux due to the charged carriers is larger. The series resistance R_G ($R_{Generator}$) depends on the electrical conductivity of both legs (σ_n and σ_p) and on their geometrical factors (length and cross-section surfaces).

Heat is taken from the hot source and flows to the cold source through the legs. Only part of this heat is converted into electrical power. The efficiency η of the thermoelectric generator is defined in the usual way: it is the generated electrical power, which is the useful power, divided by the heat taken from the hot source. The electrical power can be written as the product of the load resistance R_L by the square of the generated electrical current: $P_{electrical} = R_L I^2$. Let us indicate with Φ_H (W) the total thermal power taken from T_H: it is given by the heat flux ϕ (W/m²) on the hot end of the leg, multiplied by the surface of the leg's cross section. The efficiency can be written as:

$$\eta = \frac{R_L I^2}{\Phi_H} \tag{26.3}$$

The value of η can be derived from the thermoelectric equations, and it depends on the electrical load R_L. With some calculations, the following relationship is achieved:

$$\eta = \frac{R_L I^2}{S T_H I - \dfrac{1}{2} R_G I^2 + K(T_H - T_C)} \tag{26.4}$$

where R_G is the total resistance of the module, determined by the material resistivity $\rho = 1/\sigma$ and by the geometrical parameters (length L and cross-sectional area A) of the legs: $R = \rho L/A$. K is the total thermal conductance of the legs, which depends on the material thermal conductivity k_t and on the geometrical parameters L and A: $K = k_t A/L$. Following this equation, the efficiency depends at first on the load resistance R_L, which gives the current I, and moreover it depends on the thermoelectric parameters of the material: the Seebeck coefficient S, the electrical conductivity σ (or resistivity ρ) and the thermal conductivity k_t. It must be noted that all these three parameters depend on the temperature. Therefore, the formula for the efficiency η is approximated, because the parameters S, σ, and k_t are evaluated at the average temperature of the legs: $\bar{T} = \dfrac{T_H + T_C}{2}$. The load resistance R_L appears explicitly in the expression of the efficiency, and moreover it determines the current I, which depends on R_L, V and R_G. Given the temperature difference $\Delta T = T_H - T_C$ and the material of the thermoelectric module, the generator potential $V = S\Delta T$ and internal resistance R_G are fixed. At this point, the efficiency is a function only of the load resistance R_L. The maximum efficiency achievable with a ΔT and with a material characterized by S, σ and k_t is given by a particular value of the load resistance R_L. By simple mathematical passages (solving for R_L the equation $\partial\eta/\partial R_L = 0$) it is possible to find the optimum electrical load R_L that gives the maximum efficiency:

$$R_{L\ optimum} = R_G \sqrt{Z\overline{T} + 1} \qquad (26.5)$$

The factor Z (K^{-1}) is the fundamental parameter which qualifies a material for thermoelectric applications:

$$Z = \frac{S^2 \sigma}{k_t} \qquad (26.6)$$

It must be noted that the electrical load that maximizes the efficiency is different from the condition which maximizes the output power, which is $R_L = R_G$. This means that if the maximum output power is required, the efficiency is reduced with respect to that achievable with the same temperature difference and device. The maximum achievable efficiency with $R_L = R_{optimum}$ is:

$$\eta_{max} = \frac{\Delta T}{T_H} \frac{\sqrt{Z\overline{T} + 1} - 1}{\sqrt{Z\overline{T} + 1} + \dfrac{T_C}{T_H}} \qquad (26.7)$$

This equation shows that the maximum efficiency is determined by two main factors. The first one is $\dfrac{\Delta T}{T_H}$, which is the Carnot efficiency. This factor indicates that, given an heat source with a temperature T_H and a heat sink with a temperature T_C, the maximum achievable efficiency is limited by the second principle of thermodynamics. The second factor depends on the parameter Z, which is always smaller than 1. It tends to 1 if the parameter Z, or better so-called figure of merit $Z\overline{T}$, tends to ∞: in this ideal case, the maximum efficiency coincides with the thermodynamic limit (Carnot efficiency). For this reason, the parameter Z, or the figure of merit ZT which is adimensional, must be as large as possible. The adimensional parameter ZT is used for indicating the thermoelectric potentialities of a material:

$$ZT = S^2 \frac{\sigma}{k_t} T \qquad (26.8)$$

A good thermoelectric material should have a Seebeck coefficient S and an electrical conductivity σ as large as possible, and a thermal conductivity k_t as small as possible. Even if the formula for the maximum efficiency can be formally derived by mathematical passages, simple considerations can enlighten the physical meaning of $Z = S^2 \dfrac{\sigma}{k_t}$. Given a temperature difference, a large value of S produces a higher voltage and current, and hence a larger electrical power. It must be noted that in the factor Z, it appears as the square of the Seebeck coefficient S. Part of the generated electrical power is converted back into heat for the Joule effect, driven by the material's resistivity (conductivity): therefore, the electrical resistivity ρ should be as small as possible, hence the electrical conductivity σ should be as large as possible. The last parameter to be considered is the thermal conductivity k_t. The thermal conduction of the legs causes the diffusion of heat from the hot source to the cold sink: this heat passes through the legs without giving any contribution to the generation of the electrical power, hence it is wasted. Therefore, k_t should be as low as possible.

26.1.3 WHY NANOSTRUCTURED SILICON FOR THERMOELECTRICITY?

A material largely used for thermoelectric applications, studied since 1954, is Bi_2Te_3, which is a narrow gap semiconductor with a Seebeck coefficient of the order of 0.1–0.3 mV/K and with a thermal conductivity of few W/mK (Goldsmid and Douglass 1954; Satterthwaite and Ure 1957). The ZT factor of Bi_2Te_3 reaches

0.7–0.8 (it depends on the doping concentration) at room temperature. Several other tellurium compounds show good thermoelectric potentialities (Hyun et al. 1998; Yamashita and Tomiyoshi 2004; Yan et al. 2010), and lead telluride compounds (Dughaish 2002; Gelbstein et al. 2005) can reach high Z factors on a large temperature range (in excess of 700 K). However, there are several drawbacks in the use of tellurium, because first it is a very poisonous element, hence the disposal of devices based on tellurium would raise serious environmental problems. Moreover, tellurium is a very rare element: its abundance on the earth's crust is only slightly higher than that of platinum. For this reason, in the last years several experimental studies have been devoted to materials and tellurium-free compounds which can exhibit a good figure of merit ZT on a temperature range as large as possible. To be used in thermoelectric applications, a material should also have a good mechanical strength, a good stability at high temperatures, and its characteristics should be stable for long periods, without showing any aging effect. In all these aspects, silicon is an excellent material: it is biocompatible, it has a very high mechanical strength, and it is very stable up to temperatures in excess of 900 K. Moreover, it is the second element in abundance (after oxygen) on the earth's crust. Furthermore, for its pervasiveness in the electronic market, silicon is probably the best-known material from the technological point of view, and it could be exploited in a worldwide network of fabrication facilities and commercial markets. From the thermoelectric point of view, silicon has a good Seebeck coefficient, which is several hundreds of μV/K (it depends on the doping), and an electrical conductivity which can be increased up to 10^5 $\Omega^{-1}m^{-1}$ by using a suitable doping concentration, either n or p type. Unfortunately, bulk silicon has a high thermal conductivity of 148 W/mK: this value prevents the use of bulk silicon for thermoelectric applications. However, in the last few years it has been found that thermal conductivity is significantly reduced in nanostructured silicon. Values of k_t smaller than 10 W/mK have been reported by several experimental works (Li et al. 2003; Boukay et al. 2008; Hochbaum et al. 2008). The thermal conductivity k_t in silicon, as in other semiconducting materials, is given by the contributions of both the charge carriers and of the phonon diffusion: $k_t = k_e + k_{ph}$, where k_e denotes heat conduction due to the electron (or hole) diffusion and k_{ph} denotes heat conduction due to the propagation of lattice vibrations (phonons). Hence, the Z factor can be written as:

$$Z = S^2 \frac{\sigma}{k_e + k_{ph}} \qquad (26.9)$$

The first contribution to the thermal conductivity, k_e, depends on electron (or holes) concentration and mobility, and it is related with the electrical conductivity σ through the well-known Wiedemann–Franz law:

$$\frac{k_e}{\sigma T} = \left(\frac{\pi}{3}\right)^2 \left(\frac{k}{q}\right)^2 = (156\,\mu VK^{-1})^2 \qquad (26.10)$$

Therefore, k_e increases with σ. However, for silicon (as for other semiconductors) its value is smaller than 1 W/mK even for high doping concentrations (i.e. even for high values of σ). This is not true for metals, which have a very high concentration of charge carriers. Metals are in general good heat conductors, because the k_e term is very large. Instead, for silicon, and in general for semiconducting materials, the concentration of the charge carrier is not as large as in metals, even for high doping concentration values. In semiconductors, the predominant contribution to the heat transport is given by the phonon propagation, which is taken into account by the k_{ph} term. As the mean free path of phonons is of the order of several tens of nanometers, their diffusion can be limited in structures with nanometric dimensions, where surfaces are nearer than the phonon mean free path. Several experimental works have demonstrated that the reduction of the thermal conductivity is stronger in nanowires with rough surfaces (Kim et al. 2011; Kraemer et al. 2011; Park et al. 2011; Feser et al. 2012; Lim et al. 2012). In particular, a strong reduction of the thermal conductivity has been observed in rough nanowires with a diameter between 80 and 120 nm (Feser et al. 2012; Pennelli et al. 2014). All these experimental results can be interpreted assuming that the thermal conductivity reduction is due to the phonon scattering on the nanostructures surfaces, which limits the phonon propagation. Conversely, the mean free path of electrons (and of holes) is very small with respect

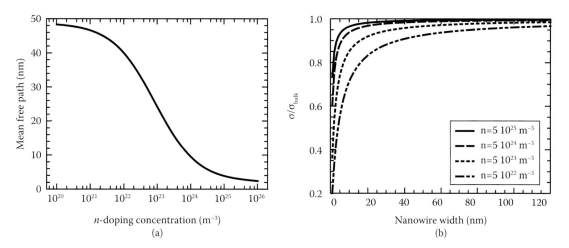

Figure 26.4 (a) The mean free path of electrons in doped silicon is reported as a function of doping concentration. (b) The electrical conductivity of a nanowire is reported as a function of the nanowire width, for different *n* doping concentration. (Reproduced from Pennelli, G., Top-down fabrication of silicon nanowire devices for thermoelectric applications: properties and perspectives, *Eur. Phys. J. B*, 88, 121, 2015. With kind permission of *The European Physical Journal [EPJ]*.)

to that of phonons (it is of the order of few nanometers for doped silicon). Therefore, electrons (and holes) propagation, which determines the electrical conductivity, is only slightly affected by the surface roughness: the nanowire diameter must be squeezed under 20 nm to have a strong effect on the electrical conductivity. Figure 26.4a (Pennelli 2015) shows the mean free path of the electrons as a function of doping concentration: it is smaller than 10 nm for a doping concentration greater than $10^{23}\,\text{m}^{-3}$. In Figure 26.4b, the electron conductivity in a rectangular nanowire is shown as a function of the nanowire width. A simple diffusive model for the nanowire scattering on the lateral surfaces has been considered. It can be seen that the electrical conductivity is very close to that of bulk silicon in nanowires wider than 50 nm.

An approximate value of the Seebeck coefficient for *n*- and *p*- doped semiconductors can be calculated by means of the following expressions:

$$S_n = \frac{-k}{q}\left(\frac{5}{2} - ln\,\frac{n}{N_C}\right) \tag{26.11}$$

$$S_p = \frac{k}{q}\left(\frac{5}{2} - ln\,\frac{p}{N_V}\right) \tag{26.12}$$

where k is the Boltzmann's constant and N_C and N_V are the equivalent density of states for, respectively, electrons in the conduction band and holes in the valence band. These expressions can be derived from the Boltzmann's transport equations, with some approximations which are valid for moderately doped semiconductors. Following these formulas, it is evident that the absolute value of S decreases with the increasing doping concentration. Even if more accurate models for the evaluation of S (Pennelli and Macucci 2013), valid also for heavily doped semiconductors, are used, the result is that S decreases with doping increasing, from roughly 1 mV/K for $n = 10^{15}\,\text{cm}^{-3}$ to about 200 μV/K for $n = 10^{19}\,\text{cm}^{-3}$. In general, all the semiconductors show a decreasing of S with doping increasing. Therefore, at first sight, it seems convenient to use low-doped semiconductors for thermoelectric applications. However, in bulk semiconductors, the thermal conductivity is dominated by the high value of k_{ph}, which is only slightly dependent on the doping. Therefore, the factor $Z = S^2\frac{\sigma}{k_t} \simeq S^2\frac{\sigma}{k_{ph}}$ can be improved by increasing the product $S^2\sigma$, which is called the power factor. As the electrical conductivity increases with doping in several orders of magnitude, S decreases by a factor of 5 or 6; Z is larger with high doping concentrations. Hence, bulk semiconductors must be heavily doped to have good thermoelectric properties. In nanostructured silicon, as in silicon nanowires, k_{ph} can be reduced to a few W/mK, and it becomes comparable with k_e. Therefore, a correct doping value must be accurately determined on the basis of the temperature difference (Pennelli et al. 2014).

26.2 LITHOGRAPHIC PROCESSES FOR Si-NANOSTRUCTURED THERMOELECTRIC GENERATORS

The reduction of the thermal conductivity in silicon nanowires and nanostructures has been demonstrated on devices made of one, or very few, nanostructures. These devices are of paramount importance, because they allow the investigation of the fundamental properties of the electrical and heat transport at nanoscale. However, for their nanoscopic nature, these devices can handle a reduced amount of current, as their cross section is very small. Therefore, their applications to practical purposes is very limited because they deliver a very small amount of power. Furthermore, these devices have a reduced mechanical strength, and they cannot be assembled for the exploitation of macroscopic hot and cold heat sources. For these reasons, the research is focusing on the development of processes and techniques for the fabrication of a collection of a large number of nanostructures, which need to be interconnected and provided with contacts for electrical and thermal conduction.

In this section, processes for the fabrication of well-organized planar arrays of silicon nanostructures will be presented. As mentioned at the end of the previous section, Si nanostructures, useful for thermoelectric applications, should have widths in the range 50–100 nm. These dimensions are not particularly critical, and can be achieved by electron beam lithography or even by advanced optical lithography. Electron beam lithography allows a rapid prototyping, which is useful in research laboratories. Conversely, optical lithography allows a massive, large scale, production. At the present state of the art, thermoelectric generators based on nanostructured silicon are still at the level of prototypes, developed by advanced e-beam lithography. The main point of lithography is to reproduce, on the surface of a silicon chip, a geometry with dimensions in the nanometric range. For this purpose, a very focused and narrow electron beam is moved onto the sample so that the geometry is drawn. Even if dedicated and very expensive equipment is available for doing this operation, a scanning electron beam (SEM) microscope, which is available in many experimental facilities dealing with solid-state physics, can be easily modified and used for e-beam lithography (Pennelli et al. 2003; Pennelli 2008). The basic principles of a lithographic step, based on e-beam, is shown in Figure 26.5. The beam is moved onto the sample for drawing the geometry, which is a line in the simple sketches of Figure 26.5. The sample is covered by an e-beam resist, which is a polymeric film sensitive to the irradiation of electrons. Many resists have been developed for e-beam lithography (Pennelli et al. 2013). One of the most effective and low-cost e-beam resists is polymethyl methacrylate (PMMA), which is made available by chemical suppliers in a large variety of molecular weights. PMMA is dissolved in an organic solvent, such as anisole, and then deposited by spinning onto the sample. The final thickness of the resist film is given by the PMMA molecular weight and concentration in the solvent and, in particular, by the spinning velocity and time. A right choice of all these parameters gives a suitable thickness in the range between a few tens of, and several hundred, nanometers. Under the effect of the e-beam exposure, the resist becomes soluble in a particular solvent such as methyl isobutyl ketone. After exposure and development, the sample appears as shown in the sketch in the middle of Figure 26.5. At this point, the nanometric geometry to be fabricated has been transferred onto the PMMA layer. In theory, the minimum geometry width achievable with the lithographic process is comparable with the diameter of the e-beam used for its writing, which is of the order of few nanometers. Practically, the resolution is limited by phenomena of forward and back electron scattering, and it is very difficult to achieve features smaller than 10 times the beam diameter. Once defined in the PMMA layer, the geometry must be transferred to the substrate. A first possibility for the exploitation of the shaped PMMA layer is to use it as a mask for the wet or dry etch of the substrate, as shown on the right-hand sketches of Figure 26.5. As an example, let us consider a silicon substrate with a silicon dioxide layer on the top. Buffered HF (BHF) allows the removal of the SiO_2 in the exposed areas.

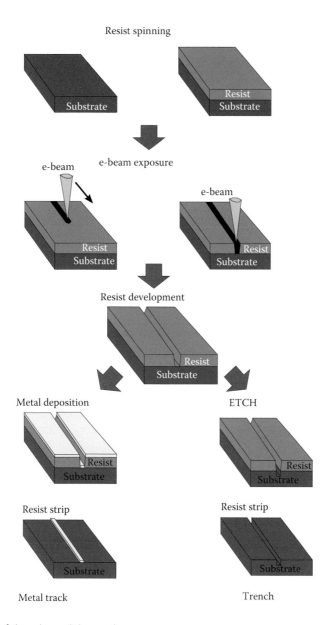

Resist spinning

e-beam exposure

Resist development

Metal deposition

ETCH

Resist strip

Resist strip

Metal track

Trench

Figure 26.5 Sketches of the e-beam lithography process.

BHF is selective on SiO_2 with respect to both the PMMA layer and the silicon underneath: this means that BHF etches very fast silica but leaves unaltered both silicon and PMMA. After the etch step, the geometry has been transferred to the SiO_2 layer, and the PMMA can be removed by cleaning the sample in a strong organic solvent, such as acetone. Another way of exploiting the shaped PMMA layer is to deposit a metal film, for example, by means of thermal evaporation. The metal will cover the top of the PMMA in unexposed areas, meanwhile it will be deposited onto the substrate in exposed areas where the resist has been removed (see the left-hand sketches of Figure 26.5). After the metal deposition, the sample is cleaned in acetone, which removes the resist and the metal over it. Therefore, the metal remains only in the exposed regions of the sample. This is so-called lift-off technique, which allows the definition of metal strips and metal pads to be used for interconnections and contacts.

Lithography, etch and metal track definition can be used for the fabrication of silicon nanodevices on a top layer of a silicon-on-insulator (SOI) substrate (Ciucci et al. 2005; Pennelli et al. 2006).

The SOI substrate is made of a monocrystalline thin silicon layer which is on the top of an amorphous, insulating, silicon dioxide layer. Underneath, the two layers are supported by a silicon substrate. SOI substrates are becoming a standard in integrated circuit (IC) fabrication, and there are several commercial substrates with a large variety of thicknesses both of the top Si layer and of the buried SiO_2 layer. Most of the experimental work presented in this chapter has been made on an SOI layer with an Si top layer 260 nm thick and a buried oxide layer 2 μm thick. The total thickness of the substrate (Si–SiO_2 and Si substrate) is 0.6 mm. Devices based on silicon nanowires can be fabricated following the steps shown in Figure 26.6. At first, a top SiO_2 layer is grown by thermal oxidation on the Si top layer of the SOI substrate. Silicon oxidation is performed in an oven under oxide flux, with temperatures in excess of 1000°C. For example, 10 minutes of oxidation of a silicon wafer at 1150°C in oxygen flux gives an oxide layer with a thickness of about 50 nm. This layer is shaped by e-beam lithography (following the sketches of Figure 26.5: PMMA spinning, e-beam exposure, development, wet etch by BHF, and cleaning in acetone). The use of e-beam lithography allows, in this same process step, both geometries with nanometric dimensions to be defined such as nanowires, and big structures to be used as interconnections and contacts. The SiO_2 layer is used as a mask for the etching of silicon, which can be performed either by wet etching or by plasma etching. Anisotropic etching in alkaline solutions, such as potassium hydroxide (KOH), results in nanowires with a very uniform trapezoidal cross section. Even if e-beam lithography allows the definition of very narrow nanowires, a further reduction can be obtained by silicon oxidation. As the oxide has a bigger volume with respect to silicon, a mechanical stress is generated during the growth in the trapezoidal cross section. Under stress, the oxide growth rate decreases, so the reduction of the silicon area is very well controllable. Figure 26.7 shows SEM photos of devices based on a single nanowire fabricated by means of this technique. It must be noted that, at the end of the fabrication process, the silicon nanowire is already positioned between contacts which are fabricated together (simultaneously) with the nanowire. This is the key point in the use of the top-down processes, which means to achieve the nanostructure starting from a silicon wafer and carving it with lithography, etches, and so on. Silicon nanowires can also be achieved by means of chemical deposition starting from raw materials (typically a gaseous silicon compound, such as silane, SiH_4). Even if a massive production of nanostructures is possible by means

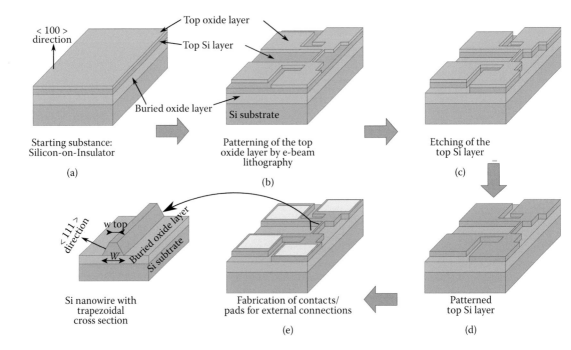

Figure 26.6 Process steps for the fabrication of a device based on a silicon nanowire with triangular cross section. (Reproduced from Pennelli, G., Top-down fabrication of silicon nanowire devices for thermoelectric applications: properties and perspectives, *Eur. Phys. J. B*, 88, 121, 2015. With kind permission of *The European Physical Journal [EPJ]*.)

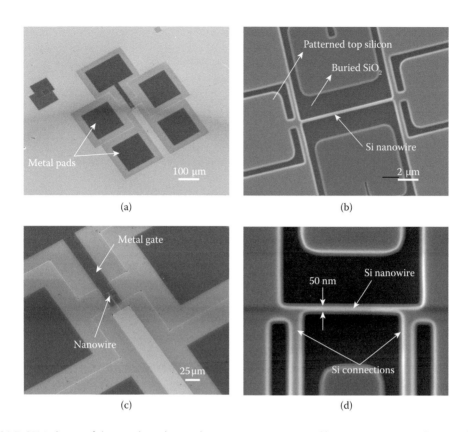

Figure 26.7 SEM photos of devices, based on a silicon nanowire positioned between contacts to be used for electrical characterization. (a) An overall illustration of a device, with metal pads for external connections. (b) Detail of a nanowire between contacts (without metal gate); a four contact configuration is used for a more precise characterization of the electrical transport. (c) A silicon nanowire, embedded in silicon dioxide, covered by a metal gate which wraps the device almost all around. (d) Detail of another silicon nanowire, embedded in silicon dioxide.

of these bottom-up techniques, there remains the problem of positioning the nanostructures between contacts and control electrodes, necessary to obtain a device. Figure 26.7 shows several top-down devices, at different magnifications. The nanowire is connected through silicon leads to the metal tracks and pads for electrical characterization. Metal tracks are fabricated with another lithographic step, which is precisely aligned to the structures achieved by patterning the silicon top layer.

The top-down process previously illustrated, can be used for the fabrication of large arrays made of a huge number of interconnected nanowires (Totaro et al. 2012; Pennelli et al. 2013). Figure 26.8 shows SEM photos of an array of silicon nanowires. Each nanowire of the array has a diameter of about 60 nm, and a length of 3 μm: nanowires with this length can be fabricated with a good reliability and reproducibility. The whole silicon nanowire network is 1 mm long between the contacts, and several hundreds of micrometers wide. Each nanowire of the array is equivalent to a resistor, both from the thermal and from the electrical point of view. Therefore, this array is equivalent to an electrical (or thermal) resistor network, as shown in the sketches of Figure 26.8. In the case of a defectless array, the nanowires parallel to the contacts are unessential, and the array is equivalent to very long nanowires placed between the contacts: each series of many nanowires, which are 3 μm long, is equivalent to a nanowire 1 mm long. Even the best fabrication process does not allow the reproducible and reliable fabrication of a single nanowire narrower than 80 nm and with a length in the order of millimeters. Instead, in the first few tens of micrometers, long nanowires are very reliable. Moreover, the network is very robust with respect to nanowire breakage, which could happen both during the fabrication process and during the practical use of the device. To this end, the nanowires which are placed in parallel to the contacts play an important role because they generate the nanowire network, increasing the number of interconnections. It is intuitive that a network remains interconnected even if

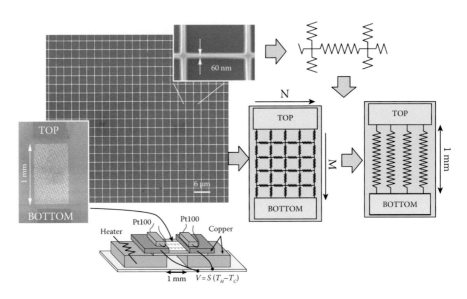

Figure 26.8 On the left: SEM images of a silicon nanowire network for thermoelectric generation. On the right: electrical (and thermal) sketches of the network, which is equivalent to many silicon nanowires in parallel, each with a millimetric length. On the bottom: sketch of the mechanical assembly for the measurement of the Seebeck coefficient. (Adapted with permission from Pennelli, G., et al., Seebeck coefficient of nanowires interconnected into large area networks, *Nano Letters*, 13, 2592 (2013). Copyright 2013 American Chemical Society.)

a large percentage of branches are removed (broken). Following the theory of percolation, it can be demonstrated (Pennelli et al. 2013) that the resistance of the whole network only doubles if the percentage of nanowire failure reaches 40 %, which means that almost one nanowire in every two is broken. It must be noted that, in principle, there are no limits to the network length and width, which can be in the range of several millimeters/centimeters. Therefore, the arrangement of the nanowires in a complex network allows the fabrication of macroscopic devices, whose properties of thermal and electrical transport are those of nanometric structures. In other words, the network arrangement allows, with macroscopic devices, the exploitation of the nanometric properties of matter. For example, the sketch at the bottom of Figure 26.8 shows a macroscopic assembly for the measurement of the Seebeck coefficient of the network.

The substrate, which has a high thermal conductivity, can strongly affect the thermoelectric potentialities of these devices. As they are fabricated on SOI substrates, a possibility is to suspend the nanostructures by underetching the buried silicon dioxide. Figure 26.9 shows a sketch of a device made of a central silicon mass, suspended by means of silicon nanostructures (Pennelli and Macucci 2016) which are anchored to the substrate at their ends. Instead of nanowires (one-dimensional structures), this device is based on silicon nanomembranes with a nanometric thickness t_h and a large side W, which is perpendicular to the silicon substrate. The thickness t_h is determined by the lithographic step (see the sketches on the right in Figure 26.9). The lithography is used for the definition of a suitable mask, which could be of SiO_2 or of metal. Then, an anisotropic vertical plasma etching allows the definition of the nanomembrane: the depth achievable with this etch step determines the large side width W perpendicular to the substrate. The hot source is applied to the central mass, and the heat flows through the nanomembranes toward the substrate, where the heat sink is applied. The reduction of the thermal conductivity is still determined by the phonon scattering on the nanomembrane walls. However, the effect of surface roughness on phonon propagation in two-dimensional structures has not yet been investigated as well as it has for silicon nanowires. Nanomembranes have a larger cross section with respect to nanowires, therefore, at first, they can deliver higher current (higher power), and moreover they have a stronger mechanical strength. Nanomembranes are placed with the large width W perpendicular to the silicon substrate, so that a large number of them can be packed in parallel in a small amount of space. If, for example, t_h = 100 nm and the distance between the nanomembranes is $2t_h$ = 200 nm, then 5000 parallel nanomembranes can be arranged in every millimeter of the device. In this way, devices with a total large cross section, which implies large currents, power,

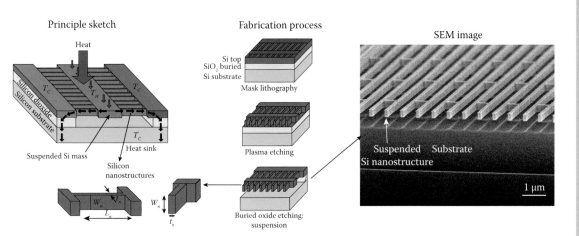

Figure 26.9 On the left: sketches of a device, based on suspended silicon nanomembranes. The large side *W* is perpendicular to the silicon substrate. In the middle: principle sketches of the fabrication process. On the right: SEM photo of a prototype. (Adapted from Pennelli, G. and Macucci, M., *Semicond. Sci. Technol.*, 31, 054001, 2016. With permission.)

and high mechanical strength, can be obtained. The length of the nanomembranes must be designed for a compromise between good thermal insulation, needed between the central hot mass and the cold support, and a good mechanical strength. An accurate modeling, by means of the finite-element technique (Pennelli and Macucci 2016), has demonstrated that lengths in the range 20–40 μm could be optimal for a device which could deliver power of several tens of Watts/mm^2 and with a mechanical strength of several N/mm^2.

26.3 NANOWIRE FORESTS FOR SILICON-BASED THERMOELECTRIC GENERATORS

In the previous section, the fabrication processes of silicon nanostructures, which are placed in a planar configuration parallel to the substrate, have been illustrated. These processes can exploit planar lithographic steps for the definition of small and long features. In this section, processes for the fabrication of nanowires, which are perpendicular to the silicon substrate, will be discussed. The aim is to fabricate many nanowires with a high aspect ratio (length: diameter width), as shown in the sketch of Figure 26.10. In this figure is shown a thermoelectric generator based on forests of silicon nanowires with different doping, interconnected in a suitable way for making a thermoelectric module. In principle, high resolution lithography can be used for the definition of a mask made of an array of dots with a small diameter (see Figure 26.10): these dots will determine the diameter of the nanowires. Nanowires can then be achieved by means of a highly selective vertical etch, based, for example, on plasma ion etching (reactive ion etching [RIE]). The length of the nanowires relies on the selectivity of this etch step, and in particular on the capability of obtaining a high aspect ratio. With the modern RIE, or even better deep-RIE (D-RIE) (Morton et al. 2008; Zeniou et al. 2014) techniques, an aspect ratio as high as 100 can be obtained. Therefore, nanowires with a diameter of the order of 100 nm and with a length of several micrometers can be fabricated by lithography and highly directional vertical etch (Stranz et al. 2012). However, several recent experimental works have been focused on a very simple and low-cost technique for the fabrication of silicon nanowires, which are perpendicular to the silicon substrate. This technique is based on the metal-assisted chemical etching (MACE) of silicon (see the sketch of Figure 26.11). At first, gold or silver nanoparticles are deposited on the surface of a silicon wafer. Then, the substrate is soaked in an aqueous solution containing both hydrofluoric acid (HF), which etches silicon dioxide but it has no effect on silicon, and an oxidizing agent, such as typically hydrogen peroxide H_2O_2. The oxidizing agent tends to withdraw electrons from the silicon substrate, leaving a hole. However, this process is prevented by the potential barrier between the silicon surface and the solution, due to the difference between the electrochemical potential of the solution and that of the silicon substrate. Therefore, bare silicon is not

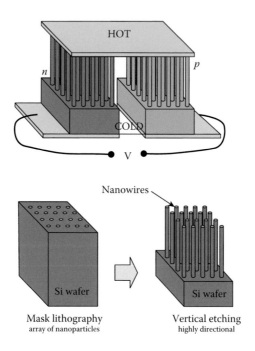

Figure 26.10 At the top: sketch of a thermoelectric module, based on an *n*-doped and a *p*-doped silicon nanowire forest. At the bottom: sketch of the fabrication process, based on lithography and highly directional vertical etching.

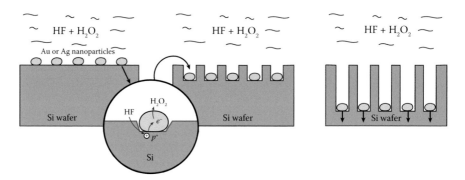

Figure 26.11 Sketches of the MACE of silicon. Gold or silver nanoparticles are deposited on the silicon surface, and then the etch is performed in an aqueous solution of HF and H_2O_2.

affected by the HF/H_2O_2 solution. At the silicon–metal interface, instead, the barrier is lower, electrons can be withdrawn from the silicon, and hence holes are generated very close to the metal surface: this means that silicon is oxidized under the metal nanoparticles for the catalytic effect of the metal, and then it is removed by HF etching. Therefore, as shown in Figure 26.11, silicon is etched under the nanoparticle which, consequently, buries into the silicon. Hence, the silicon–metal interface is maintained and the etch can further proceed, so the metal nanoparticle drills deep holes into the silicon. If the density of metal nanoparticles is high, vertical pillar structures, which are tall and very narrow (nanowires), can be achieved. The etch is catalyzed by an array of metal nanoparticles, which can be fabricated by lithography, metal evaporation, and lift-off, as described in the previous section. However, even if lithography gives very precise and well-organized arrays it is a very expensive process, in particular, if large surfaces must be considered as in the case of a thermoelectric generator. Therefore, different and less expensive techniques are in general used for the deposition of nanoparticles. Gold or silver nanoparticles can be fabricated by depositing an uniform, but sufficiently thin, metal layer. Then, a rapid thermal annealing process results in nanoparticles which are randomly placed on the silicon substrate, but which have an average diameter and distance that are controllable by adjusting the film thickness and the annealing parameters (temperature and time) (Pennelli and Nannini 2009). An even easier technique consists of the use of a solution based on HF and a metal salt, such as silver nitrate $AgNO_3$. By soaking a silicon substrate for a short time in a solution of $HF/AgNO_3$, silver is reduced in a process similar to the one already described for H_2O_2, and it is deposited on the silicon surface. Hence, the silicon surface becomes covered by silver nanoparticles whose average diameter depends on the $HF/AgNO_3$ concentration and on the soaking time. Typical process parameters can be: concentrations of 5M HF/0.05M $AgNO_3$; short soaking time between 30 seconds and 1 minute. After the deposition of the nanoparticles, the vertical etch can be performed in HF/H_2O_2 aqueous solution. The nanowire width is determined by the average diameter of the nanoparticles and it is only slightly affected by the HF/H_2O_2 concentration; meanwhile, the nanowire length is determined by the HF/H_2O_2 concentration and by the etch time. As alternative, a one-pot etch can be made by exploiting a metal salt (such as $AgNO_3$) as an oxidizing agent: instead of using the $HF/AgNO_3$ solution only for the deposition of Ag nanoparticles, the whole etching process can be performed with long soaking times in the same $HF/AgNO_3$ solution (Peng et al. 2010). In the same one-pot process, Ag nanoparticles are formed at the beginning, and then successive Ag reduction produces the vertical etch. The SEM photo of Figure 26.12 shows a cross section of a silicon nanowire forest, produced by means of the one-pot technique. The n-type silicon substrate, which has a resistivity of about $5\ \Omega \times cm$, has been etched in a solution of HF 4.5 M and $AgNO_3$ 0.01 M for 2 hours. At the end of the etching process, the substrate is covered with a thick dendritic silver layer, which is then removed by HNO_3 etching: typical etching parameters are 1 minute of etch in a HNO_3:H_2O solution 1:1 in volume. The SEM image of Figure 26.12 has been taken after the removal of the silver. A correct choice of the process parameters can result in nanowires with a length of several tens of micrometers and with an average diameter between 50 and 80 nm.

The main issue for the fabrication of thermoelectric generators based on vertical nanowires is to provide the nanowire forests with contacts necessary for both the electrical and the thermal transport. At the bottom, the nanowires are anchored to the silicon substrate, which is a good thermal and electrical conductor (if it has a suitable doping). The key point is to fabricate a common contact onto the top of the silicon nanowires. To this end, it has been demonstrated (Dimaggio and Pennelli 2016) that copper can be selectively grown on the silicon nanowire apexes by a simple electrodeposition process. At first, a thin layer of copper, which provides the seed for the successive electrodeposition, is deposited onto the vertexes of the silicon nanowires by means of thermal evaporation. Practically, after the MACE etching of the silicon nanowire forests and the removal of silver nanoparticles by HNO_3 etching, the sample is cleaned in HF: this HF cleaning, before metal evaporation, is essential for removing the thin silicon dioxide layer, which is generated around the nanowires during the HNO_3 etch. Then, a thin layer of chrome (thickness of 10–20 nm) followed by a Cu layer 50–80 nm thick are thermally evaporated in vacuum. The first Cr layer provides a good adhesion onto the nanowire apexes. Thermal evaporation is a directional process, therefore most of the metal is deposited on the top ends of the nanowires and

Figure 26.12 SEM photo of a cross section of a silicon nanowire forest. The silicon nanowires are perpendicular to the silicon substrate (at the bottom), and they are achieved by means of a MACE etch in $HF/AgNO_3$ solution.

Industrial nanosilicon

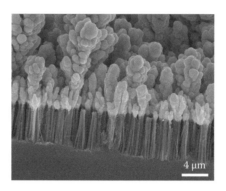

Figure 26.13 SEM photo of a cross section of a silicon nanowire forest, where copper has been selectively grown on the nanowire vertices, providing a top common electrical contact. The bottom contact is provided by the silicon substrate.

not on their vertical sides. Some metal is also deposited at the bottom between the nanowires, but this only improves the bottom contact. After the deposition of the Cu seed (anchored to the silicon by the Cr layer), the growth of copper is performed in an electrolytic cell filled with an aqueous solution of sulfuric acid (H_2SO_4) and copper sulfate ($CuSO_4$). The substrate is the negative electrode, and the cell is made in such a way that only the nanowire vertices are in contact with the solution. The positive counter electrode is made of a copper sheet. When a current is forced through the electrodes, copper grows selectively onto the silicon nanowires apexes, and forms a porous layer which connects all the tops of the forest. Figure 26.13 is a SEM image of a silicon nanowire forest anchored to the silicon substrate at the bottom end and with a common copper layer at the top. The copper has been grown with an electrode-position current of 220 mA/cm^2 for a time of 5 minutes. The selective growth of copper on top of the nanowires is a very strong process, reliable within a large range of electrodeposition parameters (current density and deposition time), and independent from the silicon nanowire doping. Moreover, it has been demonstrated that the copper top layer has a high electrical conductivity, of the order of 10^6 Ω^{-1} m^{-1}. Electrical characterization of the copper–silicon nanowire forest structure showed that, if the nanowires are sufficiently doped, a good ohmic contact is formed between the top contact and the nanowires (Dimaggio and Pennelli 2016).

26.4 FINAL CONSIDERATIONS

If materials are squeezed at nanometric dimensions, the phonon propagation becomes reduced by the surface scattering well before the electron (or hole) propagation can become seriously compromised. This opens important perspectives for thermoelectric conversion, as the conversion efficiency increases if materials with high electrical conductivity and low thermal conductivity are used. In the presence of a temperature difference, many conducting and semiconducting materials can generate an electrical signal for the thermoelectric effect. Devices for the measurement of the temperature (thermocouples) are based on this principle. However, even if the generated electrical power is enough for the measurement of the temperature, it is very small if compared with the thermal dissipation through the materials. For its low conversion efficiency, at the current state of the art, the direct generation of electrical power from thermal sources has only a limited range of practical applications. Therefore, there is a major research effort for the production of materials with low thermal conductivity and high electrical conductivity. Nanostructuring can offer a possible road for a strong reduction of thermal conductivity in semiconducting materials, and in particular in silicon which is a very abundant, biocompatible, and a technologically feasible material.

In this chapter, we reported and illustrated the main research streams which are focused on the exploitation of silicon nanostructures for the fabrication of devices for high-efficiency thermoelectric conversion. A lot of experimental work and process development is still necessary for the fabrication of a device, based on nanostructures, which is capable of handling a large amount of current. The main issue is essentially to be able to fabricate a high density of interconnected nanostructures on large areas. A possibility is to improve and perfect processes and techniques, such as lithography and etching, currently used for IC fabrication. In this way, on the same chip could be integrated both the circuit and the thermoelectric generator which provides its autonomous supply. However, this philosophy is limited by the planarity of the process, which allows the fabrication of nanostructures only in a thin layer on the surface of the chip. Another possibility is to exploit nanowire forests, which can be fabricated by highly directional vertical etching performed with low-cost techniques such as metal-assisted etching. This second philosophy is very promising, in particular for its low cost, but a lot of development is still necessary in order to obtain a good uniformity and repeatability on large areas. Moreover, the metal-assisted etching shows noticeable problems on heavily doped substrates. Nanowire forests can easily be fabricated but only on low-doped ($<10^{17}$ cm^{-3}) substrates. The electrical conductivity of the nanowires can then be enhanced by doping them with a process of thermal diffusion. However, the substrate, which is electrically in series with the nanowire forest, remains resistive thus the performances of the devices are degraded.

Another important point to be pursued in future research works is the improvement of techniques for the electrical, and in particular, the thermal characterization of nanostructures. The electrical conductivity can be easily measured, once provided contacts to the nanostructures. The most reliable thermal conductivity measurements of silicon nanowires have been performed on devices where heater and temperature sensors have been microfabricated very close to the nanostructure. However, standards and techniques for the reliable measurement of thermal conductivity of large collections of silicon nanowires still need to be developed and investigated.

REFERENCES

Boukay, A. *et al.* Silicon nanowires as efficient thermoelectric materials. *Nature Letters* **451**, 168–171 (2008).

Ciucci, S. *et al.* Silicon nanowires fabricated by means of an underetching technique. *Microelectronic Engineering* **78–79**, 338 (2005).

Dimaggio, E. & Pennelli, G. Reliable fabrication of metal contacts on silicon nanowire forests. *Nano Letters* **31**, 4348 (2016).

Dughaish, Z. Lead telluride as a thermoelectric material for thermoelectric power generation. *Physica B* **322**, 205 (2002).

Feser, J. *et al.* Thermal conductivity of silicon nanowire arrays with controlled roughness. *Journal of Applied Physics* **112**, 114306 (2012).

Gelbstein, Y., Dashevsky, Z. & Dariel, M. High performance n-type pbte-based materials for thermoelectric applications. *Physica B* **363**, 196 (2005).

Goldsmid, H. & Douglass, R. The use of semiconductors in thermoelectric refrigeration. *British Journal of Applied Physics* **5**, 386 (1954).

Hochbaum, A. I. *et al.* Enhanced thermoelectric performance of rough silicon nanowires. *Nature Letters* **451**, 163–167 (2008).

Hyun, D., Hwang, J., Oh, T., Shim, J. & Kolomoets, N. Electrical properties of the 85% bi2te3 -15% bi2se3 thermoelectric material doped with sbi3 and cubr. *The Journal of Physics and Chemistry of Solids* **59**, 1039 (1998).

Kim, H., Park, Y.-H., Kim, I., Kim, J., Choi, H.-J. & Kim, W. Effect of surface roughness on thermal conductivity of vls-grown rough si 1–x ge x nanowires. *Applied Physics A* **104**, 23 (2011).

Kraemer, D. *et al.* High-performance flat-panel solar thermoelectric generators with high thermal concentration. *Nature Materials* **10**, 532–538 (2011).

Li, D. *et al.* Thermal conductivity of individual silicon nanowires. *Applied Physics Letters* **83**, 2934–2936 (2003).

Lim, J., Hippalgaonkar, K., Andrews, S.C., Majumdar, A. & Yang, P. Quantifying surface roughness effects on phonon transport in silicon nanowires. *Nano Letters* **12**, 2475–2482 (2012).

Morton, K., Nieberg, G., Bai, S. & Chou, S. Wafer-scale patterning of sub- 40 nm diameter and high aspect ratio (50:1) silicon pillar arrays by nanoimprinting and etching. *Nanotechnology* **19**, 345301 (2008).

Park, Y.-H. *et al.* Thermal conductivity of vls-grown rough si nanowires with various surface roughnesses and diameters. *Applied Physics A* **104**, 7–14 (2011).

Peng, K. *et al.* Metal-particle- induced, highly localized site-specific etching of Si and formation of single-crystalline Si nanowires in aqueous fluoride solution. *Chemistry: A European Journal* **12**, 7942 (2010).

Pennelli, G. Fast, high bit number pattern generator for electron and ion beam lithographies. *Review of Scientific Instruments* **79**, 033902 (2008).

Pennelli, G. Review of nanostructured devices for thermoelectric applications. *Beilstein Journal of Nanotechnology* **5**, 1268 (2014).

Pennelli, G. Top-down fabrication of silicon nanowire devices for thermoelectric applications: properties and perspectives. *European Physics Journal B* **88**, 121 (2015).

Pennelli, G., D'Angelo, F., Piotto, M., Barillaro, G. & Pellegrini, B. A low cost high resolution pattern generator for electron-beam lithography. *Review of Scientific Instruments* **74**, 3579 (2003).

Pennelli, G. & Macucci, M. Optimization of the thermoelectric properties of nanostructured silicon. *Journal of Applied Physics.* **114**, 214507 (2013).

Pennelli, G. & Macucci, M. High-power thermoelectric generators based on nanostructured silicon. *Semiconductor Science and Technology* **31**, 054001 (2016).

Pennelli, G. & Nannini, A. Nanostructured multimetal granular thin films: How to control chaos. *e-Journal of Surface Science and Nanotechnology* **7**, 503 (2009).

Pennelli, G., Nannini, A. & Macucci, M. Indirect measurement of thermal conductivity in silicon nanowires. *Journal of Applied Physics* **115**, 084507 (2014).

Pennelli, G., Piotto, M. & Barillaro, G. Silicon single-electron transistor fabricated by anisotropic etch and oxidation. *Microelectronic Engineering* **83**, 1710 (2006).

Pennelli, G., Totaro, M., Piotto, M. & Bruschi, P. Seebeck coefficient of nanowires interconnected into large area networks. *Nano Letters* **13**, 2592 (2013).

Satterthwaite, C. & Ure, W. Electrical and thermal properties of bi2te3. *Physical Review* **105**, 1164 (1957).

Stranz, A., Kahler, J., Merzsch, A. & Peiner, E. Nanowire silicon as a material for thermoelectric energy conversion. *Microsystem Technologies* **18**, 857 (2012).

Totaro, M., Bruschi, P. & Pennelli, G. Top down fabricated silicon nanowire networks for thermoelectric applications. *Microelectronic Engineering* **97**, 157 (2012).

Yamashita, O. & Tomiyoshi, S. High performance n-type bismuth telluride with highly stable thermoelectric figure of merit. *Journal of Applied Physics* **95**, 6277 (2004).

Yan, X. *et al.* Experimental studies of anisotropic thermoelectric properties and structures of n-type bi2te2.7se0.3. *Nano Letters* **10**, 3373 (2010).

Zeniou, A., Ellinas, K., Olziersky, A. & Gololides, E. Ultra-high aspect ratio Si nanowires fabricated with plasma etching: Plasma processing, mechanical stability analysis against adhesion and capillary forces and oleophobicity. *Nanotechnology* **25**, 035302 (2014).

27 Nanoscale silicon in photonics and photovoltaics

*Fabio Iacona, Alessia Irrera, Salvatore Mirabella, Simona Boninelli,
Barbara Fazio, Maria Miritello, Antonio Terrasi, Giorgia Franzò, and
Francesco Priolo*

Contents

27.1 INTRODUCTION

Crystalline silicon (c-Si) is currently the most important semiconductor material for the electronic and photovoltaic industries. This leading position is due to a unique combination of advantageous properties: the availability of large and highly pure single crystals, effective conductivity engineering, a matching insulator, and natural abundance. These properties have driven the development of Si technology to its present maturity. Together with its 1.1-eV band gap, optimal for capturing the solar spectrum using a single-junction device, this maturity makes silicon almost ideally suited for photovoltaics applications. As a result, around 90% of solar panels in use today are based on silicon.

The optical properties of c-Si are relatively poor, owing to its indirect band gap that precludes the efficient emission and absorption of light. High optical performances are required to enable true optoelectronic integration and pave the way to faster, highly integrated, and low-cost devices. For photovoltaics applications, higher optoelectronic performance would enable a new generation of high-efficiency Si solar cells. Si nanostructures may offer a range of new opportunities for all of these applications.

Optical signals are now a well-established mean for efficient data transfer from long to short distances. On-chip data processing is still being performed electronically, however, even though processor clock speeds are limited by the interconnect problem (Miller 2009). The parasitic capacities generating at the several metal–insulator–metal capacitors present in complex multilevel metallization schemes constitute the major contribution to the delay in the signal propagation. The intrinsic resistivity of the metal lines, as well as the contact resistance at the various metal–metal interfaces constitute other relevant delay sources.

An important reduction of the delay times has been achieved by replacing the traditional metallization schemes based on Al and SiO$_2$ with new materials, such as Cu-based metal films and low dielectric constant insulating layers, but, as fast as the minimum feature size of the devices will further reduce, the delay due to metal interconnections will constitute again an unacceptable bottleneck for device performance. The next step-change is to introduce photonics into on-chip communications, which requires nanoscale optical sources, circuits, and detectors to encode and transmit data around the chip. Silicon is poised to deliver this functionality, as silicon photonic devices are complementary metal oxide semiconductor (CMOS)-compatible, and many highly performing devices have already been demonstrated (Liu et al. 2004; Fujita et al. 2005; Notomi et al. 2005; Rong et al. 2005; Reed et al. 2010; Matsuo et al. 2011). Photonic crystals (PhCs) add unique capabilities to this toolkit, as they offer extremely tight light confinement, thus providing strongly enhanced nonlinear effects (Notomi et al. 2005), modulators with very low switching energy (Matsuo et al. 2011), and the opportunity for enhancing and suppressing spontaneous emission (Fujita et al. 2005).

The most successful way to manipulate the energy structure of Si is to use quantum confinement in nanostructures (Ossicini et al. 2004). Semiconductor nanostructures are widely investigated in view of their exciting physical properties and with an eye to possible optoelectronic, photonic, and photovoltaic applications. Light emission in Si nanostructures was first demonstrated with porous Si in the 1990s (Canham 1990). Efficient devices were fabricated (Bisi et al. 2000), but the stability concerns were never fully solved, and the research focus shifted toward the more stable Si nanocrystals (NCs) and Si nanowires (NWs).

SiNCs have been studied intensively over the past decade (Pavesi and Turan 2010). The excitonic emission from SiNCs has two characteristic features: with decreasing crystal size, the spectrum shifts to the blue and its intensity increases. This observation indicates the quantization-related increase of the band gap and the enhancement of the radiative recombination rate, as momentum conservation is gradually relaxed with reducing grain size owing to the Heisenberg principle (Sykora et al. 2008), while the transition itself remains indirect. The electrical driving problem, related to the presence of the insulating SiO$_2$ matrix in which SiNCs are usually embedded, can be solved by using NWs to provide the quantum confinement. Indeed, NWs as small as 5 nm in diameter have been produced, luminescing at room temperature and exhibiting the characteristic blueshift indicative of quantum confinement (Artoni et al. 2012; Irrera et al. 2012; Pecora et al. 2012). NWs are ideal for light-emitting devices, as the continuity of conduction along the wire length affords easy current flow.

The ability to control the flow of light with PhCs (Joannopoulos et al. 2007) has added tremendous functionality to the Si photonic toolkit. The full benefit of PhC light confinement is derived from three-dimensional (3D) structures, but these tend to be very difficult to make and control. Instead, researchers have made impressive progress with two-dimensional (2D) structures that rely on total internal reflection for confinement in the third dimension (Krauss et al. 1996). Key examples for such 2D structures are silicon line-defect PhC waveguides with engineered dispersion (O'Faolain et al. 2010), and PhC nanocavities with ultrahigh quality factors that can achieve extreme light confinement in very small spaces (Song et al. 2005).

One of the greatest challenges facing mankind is the provision of clean energy; solar cells will make an important contribution to this challenge. Among the many materials proposed for making photovoltaic solar cells, silicon is the only one that combines suitable optoelectronic properties with abundance and technological availability. Currently, the highest demonstrated photovoltaic conversion efficiency of a Si solar cell is near 25% (Zhao et al. 1998). This is very close to the Shockley–Queisser limit of about 30% (Shockley and Queisser 1961), and therefore only limited further progress can be expected. Si cells currently dominate the market, as they can be produced cheaply on large substrates. Si nanostructures offer interesting features that could be used to modify absorption and extraction, in order to boost the efficiency of photovoltaic energy conversion (Polman and Atwater 2012; Priolo et al. 2014).

SiNCs present several properties of importance for photovoltaics: (i) the possibility of tuning the band gap and the recombination rate through quantum confinement and dedicated surface termination, (ii) the reduction of the density of states and discretization of energy levels that affect hot-carrier cooling processes, and (iii) the enhancement of the Coulomb interaction between carriers enclosed in small volumes, promoting collective effects such as multiple carrier generation (MEG) (Nozik 2002). However, the potential barrier of about 3 eV that aids quantum confinement also impedes carrier extraction. Several different approaches are under investigation, such as the fabrication of SiNCs/organic hybrid solar cells (Liu et al. 2010), but demonstrated power conversion efficiencies are limited to a few percent, because of the low carrier mobility. The ultimate solution would be a highly conductive medium with only a minimal conduction-band offset; a potential barrier as small as 0.1 eV would be sufficient to preserve advantages of quantum confinement while minimizing extraction loss. Nevertheless, the efficient extraction of free carriers from a confined environment remains a challenge.

An interesting solution to this problem may again be offered by SiNWs. In analogy with the light emission discussion, quantum confinement takes place in two dimensions, whereas carriers can be extracted along the third. SiNW solar cells have been fabricated both on a single wire (Kelzenberg et al. 2008) and on NW arrays (Kelzenberg et al. 2010). Two main schemes are used: axial junctions (Mohite et al. 2012), in which the p-i-n diode is fabricated along the length of the NW by varying the doping density during growth, and radial junctions (Tian et al. 2007), in which the diode is fabricated coaxially by a core–shell method.

The reduced absorption of thin-film solar cells requires the use of light trapping schemes to increase the effective absorption length of the material. Nanophotonic techniques are particularly promising for the purpose of light trapping, as they allow control of the flow of light on a wide length scale. The key figure of merit for any light trapping scheme is the Lambertian limit (Yablonovitch 1982), which has a maximum value of $4n^2$ and is a measure for the achievable pathlength compared with a single pass through the material; a photon may travel up to $4n^2$ times further than without light trapping, and hence increase its chances of being absorbed. The type of structure to be used in order to achieve light trapping in a thin-film solar cell ranges from random surface roughness, which excites many diffraction orders, to ordered gratings that only excite a few, although the optimum light trapping structure lies somewhere in-between. Accordingly, a number of structures that have been developed using numerical optimization techniques appear quasi random (Kroll et al. 2008) or periodic with a more complex unit cell (Martins et al. 2013).

This chapter will present an overview on some opportunities offered by Si nanostructures for photonics and photovoltaics. Several approaches, mainly based on nanoscale engineering, have been developed to extend Si capabilities. These methods either manipulate the intrinsic electronic properties of silicon using quantum size and surface effects, as in NCs and in NWs, or they add photonic functionality through wavelength-scale nanostructuring such as in PhCs.

27.2 SILICON NANOWIRES

27.2.1 SYNTHESIS: VAPOR–LIQUID–SOLID PROCESSES; METAL-ASSISTED WET ETCHING PROCESSES

The vapor–liquid–solid (VLS) mechanism allows the growth, catalyzed by metallic particles, of SiNWs from the vapor phase (Wagner and Ellis 1964). As schematically illustrated in Figure 27.1a, the process exploits the formation of a binary alloy between Si and some metals (Au, Cu, Pt). Above the eutectic temperature (which is lower than the melting temperature of both Si and the metal), an eutectic liquid phase is formed at the interface between a metal droplet and the Si substrate. Si atoms from a vapor phase can diffuse into the liquid phase which is formed at the Si–metal interface, and can definitely segregate at the eutectic–substrate interface, leading to the axial and epitaxial growth of SiNWs, with a mean diameter which can be made in the order of tens of nanometers. Au is the most commonly used catalyst for VLS growth, due to the very low eutectic temperature of the Au–Si alloy (363°C). It is used either as colloidal Au nanoparticles, or as evaporated or sputtered thin films followed by an annealing process to form droplets. The most used gaseous Si precursors are the typical molecules employed for the growth by chemical vapor deposition of Si and its compounds (SiH_4, $SiCl_4$); alternatively, a gaseous Si flux can be obtained by evaporation or sputtering of a solid source.

An important advantage of the bottom-up NW growth by the VLS mechanism is the possibility of in situ doping, by adding suitable precursors to the vapor phase (Schmid et al. 2009), since this avoids the use of damaging techniques, such as ion implantation. The control of dopant concentration, distribution,

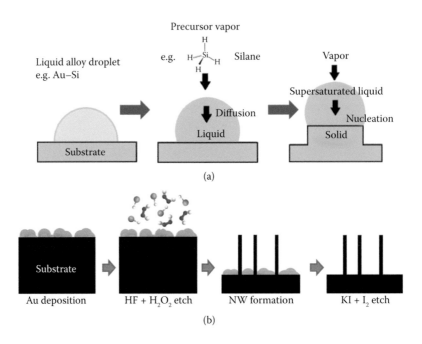

Figure 27.1 Schematic view of the processes used for SiNW synthesis. (a) VLS mechanism. From left to right, formation of the eutectic liquid phase, diffusion of Si atoms from the vapor to the liquid phase and their segregation at the eutectic/substrate interface are shown. (b) Metal-assisted wet etching process. From left to right, deposition of a thin metal layer, etching by a HF + H_2O_2 solution, SiNW formation and removal of the metal layer by a KI + I_2 solution are shown. All steps are performed at room temperature.

and activation is a key issue for the fabrication of high-quality devices for any future applications of SiNWs. A drawback related with the VLS growth of NWs is the incorporation of the metal catalyst due to the high temperature (den Hertog et al. 2008); indeed, Au acts as a deep nonradiative recombination center, thus negatively altering both electronic and optical properties of the wires. Finally, VLS-based techniques are hardly suitable for the fabrication of NWs with an extremely small diameter, such as to lead to the observation of quantum confinement effects, since the Gibbs–Thomson effect limits the minimum obtainable size (Dayeh and Picraux 2010).

The main alternative to VLS-based processes is represented by the top-down approach (Walavalkar et al. 2010). However, this way may require expensive and time-consuming lithographic processes (such as electron-beam lithography) to produce NW size close to that required for an efficient carrier confinement. Metal-assisted wet etching of Si substrates constitutes a widespread alternative, since it is a low cost, direct etching process, working at room temperature and usually employing an HF + H_2O_2 etching solution and $AgNO_3$ as a catalyst (Sivakov et al. 2010). However, this technique is also not able to control NW diameter in the few nm range; furthermore, the process, although not involving metal inclusion, leads to the formation of dendrites, whose removal could damage the NWs.

In agreement with the above considerations, photoluminescence (PL) from SiNWs due to quantum confinement has only been obtained by reducing, through thermal oxidation processes, the diameter of NWs obtained by plasma etching of a Si wafer (Walavalkar et al. 2010; Valenta et al. 2011) or by a $TiSi_2$-catalyzed VLS growth (Guichard et al. 2006). Other mechanisms invoked to explain PL from SiNWs are the presence of N-containing complexes (Shao et al. 2009) or the phonon-assisted low-temperature recombination of photogenerated carriers (Demichel et al. 2009).

A further process for the direct synthesis of ultrathin SiNWs exhibiting room-temperature light emission has been also proposed (Irrera et al. 2012; Priolo et al, 2014). The process, schematically illustrated in Figure 27.1b, is a modified metal-assisted wet etching where the catalyst, instead of being a salt, consists of an ultrathin metal film. An Au or Ag layer, having a thickness ranging from 2 to 10 nm, is deposited on cleaned Si wafers at room temperature by electron-beam evaporation. Under proper experimental conditions, the films are not continuous, but exhibit a peculiar mesh-like morphology, which leaves a relevant fraction of the Si surface uncovered. The metal-covered Si samples are etched at room temperature in an aqueous solution of HF and H_2O_2. Depending on the doping level of the substrate, etch rate values range from 250 to 500 nm min^{-1}. The thin metallic mesh injects holes into the underlying Si. Due to the presence of H_2O_2, in these regions Si oxidation occurs, while HF removes the formed SiO_2, producing the sinking of the metal into Si, and hence NW formation. The metal particles remain trapped at the bottom of the etched regions, and they can be effectively removed by dipping the sample in a $KI + I_2$ aqueous solution, without contaminating the NWs. The whole process works at room temperature and is maskless, cheap, fast, and compatible with Si technology. With respect to the use of salt, this strategy does not imply dendrite formation and, more importantly, allows a better control over the SiNW structural properties. It is indeed possible to synthesize SiNWs having a diameter compatible with the occurrence of quantum confinement phenomena. The key point which allows ultrathin NWs to be obtained is the correlation between the thickness of the metal films and the size of the Si regions which remain uncovered due to the nanoscale morphology of the films (Irrera et al. 2012).

Figure 27.2 reports a scanning electron microscopy (SEM) image in cross section of SiNWs synthesized by the metal-assisted wet etching technique, which displays a dense and uniform array of NWs. The NW length can be tuned within a wide range by changing the etching time (Irrera et al. 2011). Particularly relevant is the NW density (in the order of 1×10^{12} cm^{-2}) which is significantly higher than the typical values found for nanostructures grown by techniques exploiting the VLS mechanism. Extremely small NW diameters can be achieved (Irrera et al. 2012); in fact, values ranging from 5 to 9 nm have been reported and can be obtained in a controllable and reproducible way by properly tuning the nature and the thickness of the metal film which catalyzes the etching process.

27.2.2 OPTICAL PROPERTIES: PHOTO- AND ELECTROLUMINESCENCE; LIGHT ABSORPTION

Figure 27.3 reports typical room temperature PL spectra of SiNW arrays synthesized by metal-assisted etching having different diameters (excitation at 488 nm). The spectra consist of a broad band (full width at half maximum of about 150 nm) with the wavelength corresponding to the maximum of the PL emission, exhibiting a shift as a function of the NW mean size. In particular, the PL peak is centered at about 750 nm for NWs having a mean diameter of 9 nm while it is blueshifted at about 690 nm for NWs having a mean diameter of 7 nm, and further shifted at 640 nm for a size of 5 nm. This behavior strongly resembles that found in SiNCs (Iacona et al. 2000), indicating that quantum confinement effects are responsible for the light emission. Indeed emission is very intense and visible with the naked eye. The external quantum efficiency is higher than 0.5%, by taking into account the spatial emission profile of the SiNWs and under the conservative assumption that the exciting laser beam is totally absorbed by the material.

Figure 27.2 Cross section SEM image of a SiNW array obtained by the metal-assisted wet etching technique. NW length is about 2.5 μm.

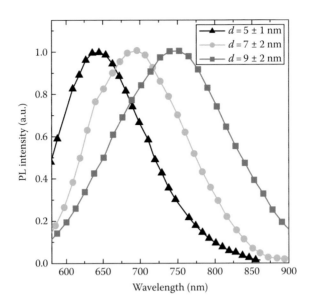

Figure 27.3 Normalized PL spectra obtained by exciting at room temperature SiNW samples having different sizes. (The source of the material Irrera, A., et al., Quantum confinement and electroluminescence in ultrathin silicon nanowires fabricated by a maskless etching technique, *Nanotechnology*, 23, 075204, 2012 is acknowledged.)

This value is comparable with the best efficiencies reported for porous Si (Cullis et al. 1997) and SiNCs (Dal Negro et al. 2006). The intensity of the PL signal exhibits a maximum at about 270 K; above this temperature a slight decrease is observed. The overall dependence on temperature is quite weak, since the intensity change between 11 and 270 K accounts for about one order of magnitude (Artoni et al. 2012). Typical PL lifetime values are of the order of a few tens of μs and increase by increasing the detection wavelength; these values are two orders of magnitude above those measured in pillars produced by e-beam lithography followed by oxidation (Walavalkar et al. 2010), demonstrating the superior structural quality of NWs synthesized by metal-assisted etching. While the PL intensity increases almost linearly by increasing the pump power, no noticeable variation of the lifetime with increasing the pump power is detected, suggesting that nonradiative processes do not play a relevant role in the luminescence properties of the system. The excitation cross section of SiNWs, calculated by measuring the rise time of the PL signal as a function of the pump power, is about 5×10^{-17} cm^2. All of the above characteristics of the PL emission from SiNWs closely resemble those found for SiNCs (Ossicini et al. 2004; Pavesi and Turan 2010), further supporting the view that quantum confinement effects are responsible for the photon emission process in SiNWs.

SiNWs synthesized by metal-assisted wet etching exhibit a peculiar fractal behavior (Fazio et al. 2016). The fractal parameter that takes into account both the gaps and the heterogeneity of the structure is the lacunarity, defined as the measure of how the space is filled, being related to its gappiness (alternation of full and empty space) (Plotnick et al. 1996). Light-scattering properties of SiNWs are quite important. Figure 27.4 shows the diffuse (hemispherical) optical reflectance of a SiNW sample characterized by a lacunarity with a maximum on the scale of 150–200 nm along the structure planar section, compared with that of an optically flat Si sample. The diffuse reflectance from SiNWs shows a sharp drop to near zero (below 1%) across the entire visible–near infrared (IR) range for wavelengths just below the Si band gap at 1.1 μm. This observation points to a strong absorption of the reflected light due to multiple scattering within the NW layer.

According to these observations, it can be concluded that light is strongly diffused within the thin NW layer. In particular, in the strongly absorbing visible region, light is neither reflected nor scattered out of the SiNW forest but remains mostly trapped in the NW layer by multiple scattering processes until it is eventually absorbed (by supposing that the scattering processes are preferentially in-plane because the fractal shape and the refractive index fluctuations are across the 2D planar structure). A key role in this behavior is played by the 2D random fractal texture of the NW forest. Indeed, even if the ultrasmall average diameter

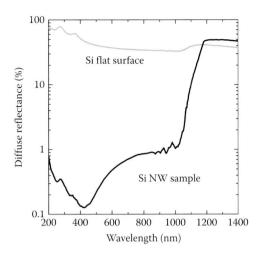

Figure 27.4 Diffuse (hemispherical) reflectance of a SiNW random fractal array and of a bulk c-Si flat surface.

could hardly lead to efficient light scattering by a single NW, the peculiar random fractal arrangement made of tightly spaced NWs separated by air voids introduces strong heterogeneities, and consequently highly efficient in-plane scattering pathways on a length scale between 10 and 1000 nm. Thus, one can think of light as scattered by "regions" of NWs with varying densities rather than by single NWs. When multiple scattering occurs, the light path length increases substantially, thus increasing the likelihood of absorption at the end (Holmberg et al. 2012). Furthermore, observing in detail the reflectance spectrum of SiNWs, a large and broad minimum peak at about 428 nm and approaching 0.1% can be observed, which indicates light over-trapping by the structure. This very low reflectance value places SiNWs synthesized by metal-assisted wet etching among the most promising Si nanostructures for applications in solar cells, being also very inexpensive and easily implementable over a large area.

27.2.3 APPLICATIONS: LEDs AND SOLAR CELLS

The capability of SiNWs to emit photons if electrically excited, and therefore to constitute the active region in Si-based light-emitting devices (LEDs) operating at room temperature to be employed in Si photonics, has been demonstrated through the fabrication of prototype devices (Irrera et al. 2012). These devices consisted of NWs obtained by metal-assisted etching of highly doped p-type single-crystal Si wafers, with a thick Au layer in the back side and a transparent conductive layer of aluminum zinc oxide in the front side.

Intense electroluminescence (EL) spectra have been reported at room temperature by forward biasing the devices; the emission starts at 2 V and its intensity increases roughly linearly by increasing the applied voltage (Irrera et al. 2012). Spectra consist of a broad band, centered at about 700 nm; its shape and position are similar to those of the PL spectra, indicating that both mechanisms of excitation involve the same emitting centers—quantum confined SiNWs.

One of the most fascinating applications of SiNWs in photovoltaics is the realization of p-type–intrinsic–n-type (p-i-n) coaxial SiNW solar cells (Tian et al. 2007). Single-crystalline SiNW p-cores were synthesized by means of a nanocluster-catalyzed VLS growth. Polycrystalline Si (poly-Si) shells were deposited at a higher temperature and lower pressure than for p-core growth to inhibit axial elongation of the SiNW core during the shell deposition, where PH_3 was used as the n-type dopant in the outer shell. The core/shell SiNWs have uniform diameters of about 300 nm. An advantage of this core/shell architecture is that carrier separation takes place in the radial versus the longer axial direction, with a carrier collection distance smaller or comparable to the minority carrier diffusion length (Kayes et al. 2005). Hence, photo-generated carriers can reach the p-i-n junction with high efficiency without substantial bulk recombination.

To characterize electrical transport through the p-i-n coaxial SiNWs, metal contacts were fabricated to the inner p-core and outer n-shell, as shown in Figure 27.5a and b. Dark current–voltage (*I-V*) curves from core–core and shell–shell configurations indicate that ohmic contacts are made to both core and

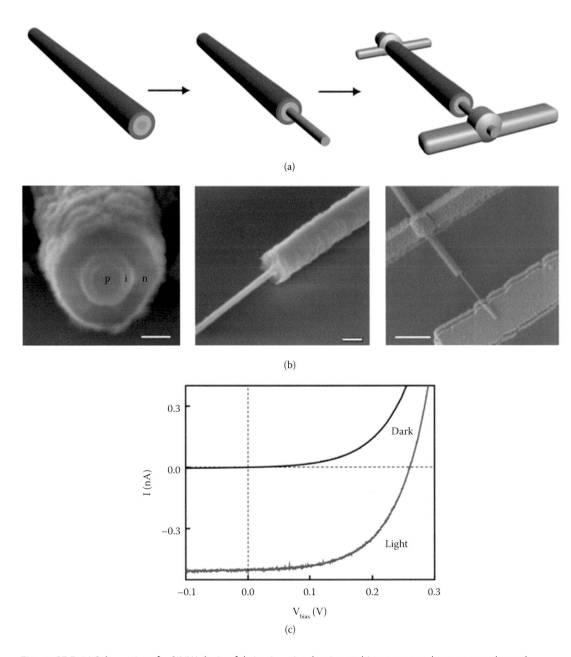

(a)

(b)

(c)

Figure 27.5 (a) Schematics of a SiNW device fabrication. A selective etching exposes the p-core and metal contacts are deposited on the p-core and n-shell. (b) SEM images corresponding to schematics in (a) scale bars are 100 nm (left), 200 nm (middle) and 1.5 mm (right) (c) dark and light *I–V* curves of the p-i-n SiNW photovoltaic device. (Reprinted by permission from Macmillan Publishers Ltd. *Nature*, Tian, B., et al., (2007), Coaxial silicon nanowires as solar cells and nanoelectronic power sources, *Nature, 449*, 885–890, copyright 2007.)

shell portions of the NWs. Furthermore, *I-V* curves recorded from different core–shell contact geometries show rectifying behavior, and demonstrate that the p-i-n coaxial SiNWs behave as well-defined diodes. The photovoltaic properties of the SiNW diodes were characterized by using a standard solar simulator. *I–V* data (shown in Figure 27.5c) yield an open-circuit voltage V_{oc} of 0.260 V, a short circuit current I_{sc} of 0.503 nA and a fill factor F_{fill} of 55.0%. The maximum power output for the SiNW device at 1-sun is about 72 pW. These values are constant for measurements made over a 7-month period, thus demonstrating excellent stability of the NW photovoltaic elements. The apparent photovoltaic efficiency of the device at the moment is 3.4% (upper bound), but might be improved through increased understanding of absorption and better coupling of light into the devices by vertical integration (Kayes et al. 2005) or multilayer stacking (Javey et al. 2007).

27.3 SILICON NANOCRYSTALS

27.3.1 SYNTHESIS AND STRUCTURE

SiNCs embedded in SiO_2 can be prepared by using several techniques, including plasma-enhanced chemical vapor deposition (PECVD) (Iacona et al. 2000), sputter deposition (Gourbilleau et al. 2001), ion implantation (Pavesi et al. 2000), and reactive evaporation (Zacharias et al. 2002). The process usually implies the synthesis of a SiO_x film (i.e., a SiO_2 film containing a Si excess) followed by a thermal annealing in inert atmosphere, at temperatures ranging between 1000 and 1250°C, to induce a thermodynamically-driven phase separation in the substoichiometric layer, leading to the formation of SiNCs embedded in an almost stoichiometric SiO_2 matrix.

SiNC formation can be very effectively followed by transmission electron microscopy (TEM). Figure 27.6a reports a dark field plan view TEM image of an SiO_x film synthesized by PECVD and annealed at 1250°C, showing the presence of a high density of small clusters, uniformly embedded in an amorphous SiO_2 matrix. On the basis of the electron diffraction pattern, reported in the inset and showing well distinct rings corresponding to the (111), (220), and (311) planes of Si, the bright spots in the TEM image can be unambiguously identified as SiNCs. Very detailed information about the SiNC structure can be obtained by collecting high-resolution TEM images. The image shown in Figure 27.6b has been obtained by using the energy filtered TEM (EFTEM) technique (Iacona et al. 2004). The crystalline core of the NC is surrounded by an amorphous Si shell, about 1 nm thick, demonstrating that, contrary to what happens with flat bulk Si/SiO_2 interfaces, a sharp $SiNC/SiO_2$ interface does not exist (Daldosso et al. 2003).

27.3.2 OPTICAL PROPERTIES: PHOTONIC AND PHOTOVOLTAIC APPLICATIONS

SiNCs exhibit a bright light emission in the range 700–1100 nm. As an example, Figure 27.7 reports the normalized PL spectra of SiNCs obtained by annealing SiO_x films with different Si concentration at 1250°C, synthesized by PECVD by properly changing the ratio of the gaseous precursors (SiH_4 and N_2O). PL measurements are performed at room temperature, by pumping with the 488 nm line of an Ar^+ laser. The PL signal shows a marked blueshift with decreasing Si content as a result of the smaller size of the SiNCs (Iacona et al. 2000; Priolo et al. 2001). Indeed, the average NC radius determined by TEM increases from 1.1 to 2.1 nm by increasing the Si content from 37 at.% to 44 at.%. A similar tuning of the wavelength of the emitted light can be also obtained by changing the annealing temperature for a fixed

(a)   (b)

Figure 27.6 (a) Dark field plan view TEM image of a SiO_x film annealed at 1250°C, showing the presence of a high density of SiNCs, uniformly embedded in an amorphous SiO_2 matrix. The electron diffraction pattern reported in the inset shows distinct rings corresponding to the (111), (220), and (311) lattice planes of Si. (b) High-resolution EFTEM image of a SiNC. The presence of the Si(111) lattice planes is clearly visible in the NC core.

50 nm

Industrial nanosilicon

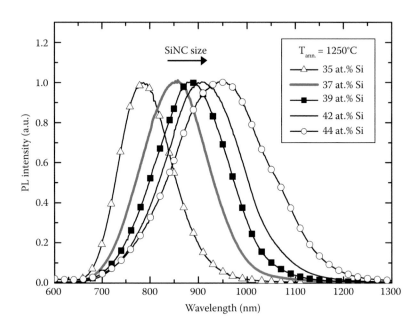

Figure 27.7 Normalized room-temperature PL spectra of SiNCs obtained by annealing SiO$_x$ films at 1250 °C with different Si concentrations. The PL signal shows a marked blueshift with decreasing Si content as a result of the smaller size of the SiNCs. (Reprinted with permission from Priolo, F., et al., (2001), Role of the energy transfer in the optical properties of undoped and Er-doped interacting Si nanocrystals, *J. Appl. Phys.*, 89, 264–272. Copyright 2001 American Institute of Physics.)

Si concentration in the SiO$_x$ films (Iacona et al. 2000). The data shown in Figure 27.7 are in qualitative agreement with the carrier quantum confinement theory, which predicts the progressive blueshift of the PL peak with decreasing the crystal size, due to the enlargement of the band gap of the SiNCs with respect to bulk c-Si (Delerue et al. 1993). However, the experimental PL data are remarkably shifted toward lower energies with respect to theoretical calculations of the SiNC band gap (Delerue et al. 1993). Indeed, most of the experimental results are consistent with a model in which light emission is not due to band-to-band recombination processes within the NCs, but the recombination occurs on intermediate states introducing energy levels inside the band gap and corresponding to SiO$_2$/Si interfacial states (Dinh et al.1996; Shimizu-Iwayama et al. 1998). The nature of these states is such that their energy level is not fixed (otherwise no dependence of the emitted wavelength on SiNC size could exist), but it depends on the NC size, so smaller grains, having a larger band gap, can also have energy levels at higher energy with respect to larger grains, characterized by a narrower band gap. This effect has been quantitatively explained with the trapping of an electron (or even an exciton) by Si=O bonds producing localized states in the band gap of SiNCs with diameter smaller than 3 nm (Wolkin et al. 1999).

The discovery that SiNCs can exhibit optical gain (Pavesi et al. 2000) opened the route toward the development of a Si laser. However, in order to reach the ambitious goal of an injection laser, a suitable strategy for the electrical excitation of SiNCs has to be developed. In this framework, SiNCs embedded in SiO$_2$ have been successfully employed to fabricate light-emitting MOS devices operating at room temperature (Franzò et al. 2002). The active region of a typical device consists of a highly doped n-type poly-Si film, a 80-nm-thick SiO$_x$ film (where SiNCs are formed due to a thermal annealing) and a highly doped p-type Si substrate. SiO$_x$ films are conductive due to the excess Si they contain and allow an electrical current to flow once the devices are biased. The main conduction mechanisms are direct or Fowler–Nordheim tunneling, or Poole–Frenkel, while SiNC excitation mainly occurs through impact of hot electrons (Franzò et al. 2002; Irrera et al. 2006). The emission microscopy (EMMI) image of a device based on a SiO$_x$ film is reported in Figure 27.8. Ring-shaped, metal contacts to the poly-Si film and to the Si substrate and a metal-free round area which allows the exit of the light are visible. The image reveals an intense emission, very stable and homogeneous over the whole active area; the emission starts at a voltage

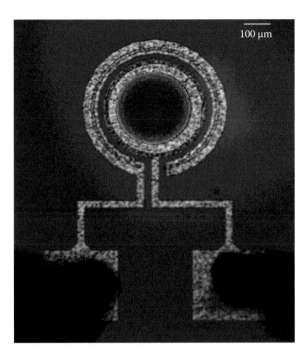

Figure 27.8 EMMI image of a device based on SiNCs operating at room temperature. The colors indicate different intensities of the emitted light, red being the highest. (Reprinted with kind permission from Springer Science+Business Media: *Appl. Phys. A: Mater. Sci. Process.*, Electroluminescence of silicon nanocrystals in MOS structures, 74, 2002, 1–5, Franzò, G., et al.)

of −15 V, with a current density of about 90 μA cm⁻². The EL spectrum of the device presents a main peak at 890 nm, which is very similar both in position and shape to the PL peak measured in the same sample. It is therefore straightforward to attribute the EL signal to electron–hole pair recombination in the SiNCs dispersed in the oxide layer. Significant improvements of the performances of devices based on SiNCs have been obtained by properly tuning the composition, the annealing temperature, and the thickness of the SiO_x active layer (Franzò et al. 2002; Irrera et al. 2002; Irrera et al. 2006). Also, more complex device structures have been proposed to optimize light emission from SiNCs (Walters et al. 2005). To improve the carrier injection efficiency, a field-effect EL mechanism has been employed (see Figure 27.9). In this excitation process, electrons and holes are sequentially injected in arrays of SiNCs embedded in the gate oxide of a transistor. The formed excitons recombine radiatively giving origin to the EL signal. This approach reduces nonradiative processes limiting EL efficiency, such as Auger recombination, and improves the device reliability since it involves a less extended oxide wear-out with respect to the impact excitation by hot carriers.

The chance of modulating the light absorption spectrum by exploiting the quantum confinement effect in semiconductor nanostructures is currently receiving a wide attention by the material science community for boosting novel technologies with higher efficiency in the sunlight–electricity conversion (Green 2001). The confinement electron wave function is well accomplished in NCs with sizes below the exciton Bohr radius, thus modifying the light absorption process through several interesting phenomena such as MEG (Schaller and Klimov 2004), intermediate band formation (allowing the absorption of lower energy photons) (Luque and Martì 1997), and optical band gap (E_g^{OPT}) widening (Conibeer et al. 2006). This last effect, related to the reduction of the NC size, produces a blueshift of the energy onset in the absorption spectrum, giving the chance to find the best agreement between the light absorption spectrum and the solar one, in terms of sunlight–electricity energy conversion.

In particular, the investigation and exploitation of quantum confinement in SiNCs is appealing because of possible application in photovoltaic cells whose technology is widely based on Si. More, given the non-toxicity and abundance of the material, the interest in SiNCs for photovoltaics is rising day by day,

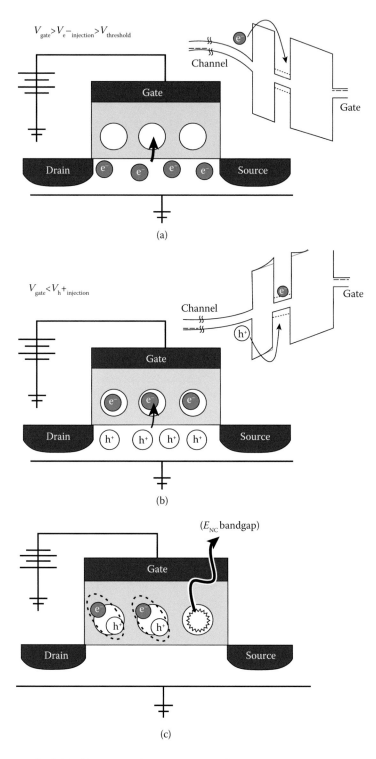

Figure 27.9 Schematic of a field-effect EL mechanism in a SiNC floating-gate transistor structure. Inset band diagrams depict the relevant tunneling processes. The array of SiNCs embedded in the gate oxide of the transistor can be sequentially charged with electrons (a) by Fowler–Nordheim tunneling, and holes (b) via Coulomb field enhanced Fowler–Nordheim tunneling to prepare excitons that radiatively recombine (c). (Reprinted by permission from Macmillan Publishers Ltd. *Nat. Mater.*, Walters, R.J., et al., (2005), Field-effect electroluminescence in silicon nanocrystals, 4, 143–146, copyright 2005.)

Industrial nanosilicon

also supported by the consolidated understanding of the optoelectronic properties and by the availability of technologically advanced synthesis techniques. It should be noted that, contrary to photonic applications, for photovoltaic purposes the indirect nature of the energy band gap in SiNCs is advantageous, since the photogenerated electron–hole pair has a longer lifetime with respect to direct band gap materials (Timmerman et al. 2008).

While it is well assessed, at least from a theoretical point of view, that a reduction of NC size induces an increase of E_g^{OPT}, an open question still remains on whether the size of NCs is the only parameter driving the modification of the light absorption spectrum in quantum structures. Figure 27.10 reports the thermal variation of E_g^{OPT} of SiNCs obtained by annealing SiO_x samples grown by PECVD (Mirabella et al. 2009). The increase in the Si content results in the formation of larger SiNCs, with lower E_g^{OPT}. Still, the U-shaped trend for E_g^{OPT}, with a minimum at 900–1000°C and an overall variation of about 1 eV, causes an unexpected red- and blueshift of the light absorption spectrum. On the other hand, the energy of the PL peak for the 43% Si sample decreases with annealing temperature as expected, given the Ostwald ripening mechanism which leads to larger SiNCs by increasing temperature (with lower PL energy). Thus, the SiNC size is not the only parameter determining the E_g^{OPT}. Actually, it was shown that the univocal increase of E_g^{OPT} above 900°C is caused by the amorphous-to-crystalline transition of NCs (Mirabella et al. 2009). Indeed, the light absorption of bulk c-Si is known to be shifted toward higher energy in comparison to amorphous Si (Tauc 1974). This evidence shows that the exact nature of the NC phase is crucial in the light absorption process, while the effect of the NC ripening can be somehow negligible. In other words, the size tuning of the NCs affects the photon emission process more effectively than the light absorption.

Si nanostructures not only allow to make traditional "photon-in/electron-out" photovoltaic devices; they can also form the basis of an autonomous device—a "solar shaper." Such a photon-in/photon-out scheme does not involve carrier extraction, but it would be used to extract energy from the parts of the solar spectrum that are otherwise not efficiently converted. It therefore rectifies the fundamental reason for the limited efficiency of photovoltaic conversion: the mismatch between the broadband character of solar radiation and the narrow window of optimal performance of a semiconductor photodiode. A solar shaper could form an add-on that supplements the traditional cell and converts the incoming broadband spectrum into a new spectrum that is narrower and better optimized for absorption by a single-junction solar cell. This would require the "cutting" of high-energy ultraviolet (UV) photons into multiple lower-energy quanta that can be converted more efficiently, as well as the "pasting" together of low-energy photons in the near IR that are otherwise not converted at all. SiNCs offer interesting possibilities that could be used for both processes.

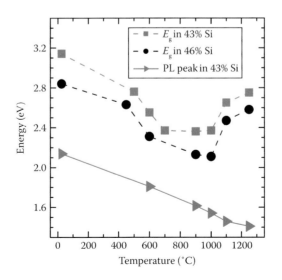

Figure 27.10 Optical band gap (symbols plus dashed lines) versus annealing temperature for SiNCs obtained by annealing two PECVD-grown SiO_x films (43 and 46 Si at.%, respectively). The energy of the PL peak is also reported for the 43% Si sample (triangle plus continuous line).

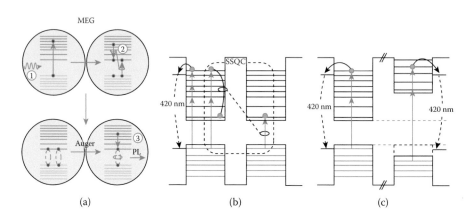

Figure 27.11 Schematic views of the photon cutting (a) and photon pasting (b, c) processes by SiNCs. (a) In MEG, a hot carrier is created on absorption of a high-energy photon (1), and then cools down by generating a second exciton (2). Efficient Auger interaction allows for PL from only one exciton (3). (b) Absorption of low-energy photons by free carriers. The resulting hot carrier can be trapped at a surface-related defect level (420 nm) or undergo spatially separated multiplication. (c) NCs with different band gap sizes can be used to absorb and trap IR photons of different energy. By tuning the NC diameter, absorption for specific IR photons can be optimized. (Reprinted by permission from Macmillan Publishers Ltd. *Nat. Nanotechnol.*, Priolo, F., et al., (2014), Silicon nanostructures for photonics and photovoltaics, *Nat. Nanotechnol.*, 9, 19–32, copyright 2014.)

For photon cutting, the MEG process could be used (Figure 27.11a). In this process, a highly energetic hot carrier relaxes by generating additional electron–hole pairs. Dedicated investigations confirmed that the efficiency of this process in NCs is considerably enhanced when compared with bulk materials of the same band gap (Trinh et al. 2008; Beard et al. 2010). Applied to photovoltaics, MEG would create more than one electron–hole pair per single absorbed high-energy photon, thereby creating additional photocurrent from the green and blue part of the solar spectrum. Indeed, theoretical evaluations indicate that total conversion efficiencies of up to about 50%, well above the Shockley–Queisser limit, are feasible (Hanna and Nozik 2006).

SiNCs may also be suitable for photon pasting. The most important properties in this respect are (1) the identification of phononless recombination, (2) a significant reduction of the cooling rate for hot carriers, (3) a large cross section of free carriers for the absorption of IR photons, and (4) defect-related hot-carrier PL. These properties open perspectives for photovoltaic conversion of IR photons that are otherwise lost. A specific approach that makes use of such combined effects is the absorption of low-energy photons by free carriers for the generation of hot excitons, followed by carrier multiplication, shown in Figure 27.11b, or trapping at the Si–O related defect and subsequent blue emission at about 420 nm (Tsybeskov et al. 1994; Brewer and Von Haeften 2009). Because the 420 nm wavelength of the hot defect-related emission is fixed, the absorption of the IR photons can be tailored by changing the NC size, as highlighted in Figure 27.11c.

27.3.3 ERBIUM DOPING OF Si NANOCRYSTALS

Another approach that has been used to efficiently produce photons from Si nanostructures exploits the introduction of light-emitting impurities, such as rare earth (RE) ions. The sharp emission lines exhibited by several RE ions embedded in solid matrices make available many additional emission wavelengths, spanning from the visible to the IR region. Among the various RE ions, particularly relevant is the emission at 1.54 µm of Er^{3+} ions, since this wavelength is strategic for telecommunication, corresponding to a minimum in the loss spectrum of silica optical fibers (Polman 1997). SiO_x films represent an ideal precursor for the synthesis of Er-doped SiNCs. Er incorporation can be accomplished during the SiO_x growth, by adding a proper organometallic Er compound (St. John et al. 1999), or by employing an Er source (typically Er_2O_3) in a sputter deposition process (Galli et al. 2006). Alternatively, Er ions can also be introduced by ion implantation (Franzò et al. 1999), either in as deposited SiO_x films or in films where crystalline or amorphous Si clusters are already formed. After the implantation step an annealing process at 900°C is needed, in order to recover the damage, but the same thermal process is also strongly advisable in samples doped during the growth in order to fully optically activate Er ions. The obtained material is usually described as Er-doped SiNCs, even if it has been demonstrated that, under the above described experimental conditions, amorphous Si nanoclusters strongly predominate over their crystalline counterpart (Franzò et al. 2003).

In this system, SiNCs act as efficient sensitizers for the RE which is excited much more efficiently than in pure SiO_2. Indeed, room temperature PL yields two orders of magnitude higher are observed for Er-doped SiO_2 in the presence of NCs than in pure SiO_2 (Fujii et al. 1997). Moreover, since Er is embedded within a larger gap matrix, the nonradiative decay channels typically limiting Er luminescence in bulk Si, back-transfer and Auger de-excitations (Polman 1997), might be absent in this case. This further improves the luminescence yield.

The sensitizing action of SiNCs is clearly demonstrated in Figure 27.12, where the room temperature PL spectra of SiO_x films in which Er has been implanted at different doses after SiNC formation are reported. Spectra were taken by exciting the samples with a 488-nm laser beam. As soon as Er is introduced, the SiNC signal at 0.85 μm decreases, while a new peak at around 1.54 μm, coming from the $^4I_{13/2} \rightarrow {}^4I_{15/2}$ intra 4f-shell Er transition, appears. With increasing Er dose, this phenomenon becomes particularly evident with a quenching of the visible PL due to NCs corresponding to a simultaneous strong enhancement in the Er-related signal.

The SiNC-mediated excitation of the Er ions is further demonstrated by the photoluminescence excitation (PLE) spectra reported in Figure 27.13, referring to the 1.54 μm line of Er in presence of SiNCs (line) and in SiO_2 (line and filled squares). Both samples have the same Er concentration. It is worth noticing that for Er in SiO_2 (whose PLE spectrum has been multiplied by a factor of four) photon emission can occur only at the resonant wavelengths of 378, 488, and 520 nm, corresponding to higher lying excited levels related to Er^{3+}. On the contrary, in the presence of SiNCs, Er can be excited within a much broader wavelength range. By comparing the PLE spectrum of Er in the presence of NCs with that obtained for SiNCs emitting at 0.98 μm (line and open circles in Figure 27.13), a strong similarity in the trend over a wide range is found. This suggests that the Er excitation is occurring via the SiNCs.

The Er excitation mechanism in presence of SiNCs is schematized in Figure 27.14 (Franzò et al. 2000). When Er-doped SiNCs are pumped with a laser beam, photons are absorbed by the SiNCs and promote

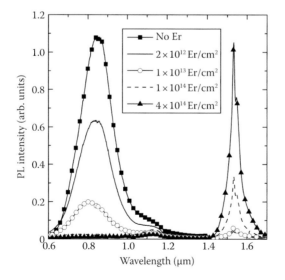

Figure 27.12 Room-temperature PL spectra of Er-implanted SiNCs at different Er doses. Er introduction induces the decrease of the SiNC signal at 0.85 μm, while an intense peak at around 1.54 μm, coming from the $^4I_{13/2} \rightarrow {}^4I_{15/2}$ intra 4f-shell Er transition, appears. (Reprinted with kind permission from Springer Science+Business Media: *Appl. Phys. A: Mater. Sci. Process.*, The excitation mechanism of rare-earth ions in silicon nanocrystals, 69, 1999, 3–12, Franzò, G., et al.)

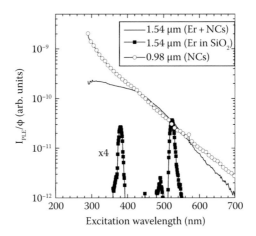

Figure 27.13 PLE spectra of Er in presence of SiNCs (line) and in SiO_2 (line and filled squares). The PLE spectrum of Er in SiO_2 has been multiplied by a factor of four. For comparison, the PLE spectrum of SiNCs emitting at 0.98 μm (line and open circles) is also shown.

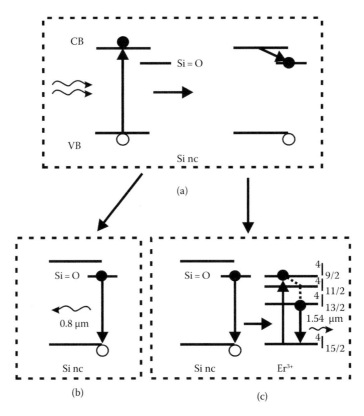

Figure 27.14 Schematic view of the Er excitation mechanism in the presence of SiNCs. (Reprinted with permission from Franzò, G., et al., (2000), Er^{3+} ions-Si nanocrystals interactions and their effects on the luminescence properties, *Appl. Phys. Lett.*, 76, 2167–2169. Copyright 2000 American Institute of Physics.)

an electron from the conduction band to the valence band. Theoretical calculations demonstrated that the electron in a conduction band is then trapped by an a Si=O interfacial state (a) (Wolkin et al. 1999). The recombination of the electron in the interfacial state with a hole in the valence band gives the typical SiNC light emission at about 0.8 μm (b). Alternatively, in presence of Er, the energy can be transferred to the Er ion to excite it (c). Since the 0.8 μm NC wavelength couples well with the $^4I_{9/2}$ level of the Er manifold, it is reasonable to believe that this is indeed the Er level first excited by the NCs. From this level a rapid relaxation occurs to the metastable $^4I_{13/2}$ level, with the final emission of photons at 1.54 μm (c). Processes (b) and (c) are in competition with one another. When an Er ion is close to a NC, that NC will become "dark" in the sense that the energy transfer to Er will be much more probable than the NC radiative emission.

The possibility of observing positive gain at 1.54 μm from Er-doped SiNCs has been extensively discussed, with a particular attention to gain-limiting effects, such as cooperative up-conversion and confined carrier absorption induced by excited NCs. SiNCs can have positive effects, such as the increase of the effective excitation cross section of Er and the lowering of the pump threshold for population inversion to occur, as well as negative aspects such as the confined carrier absorption of 1.54 μm signal photons (Pacifici et al. 2003). Nevertheless, the main limiting step for this system remains the fact that only a fraction of the Er^{3+} ions are indeed coupled to the NC sensitizers (Wojdak et al. 2004).

27.4 SILICON-BASED PHOTONIC CRYSTALS

27.4.1 PHOTONIC CRYSTAL FABRICATION AND DEFECT ENGINEERING

Planar PhCs are usually made by etching a periodic pattern (typically a triangular or square lattice of air holes) into a slab waveguide with a high refractive index. In the case of Si, a silicon-on-insulator planar waveguide is used as the starting material, and the patterned PhC membrane is released by removing the lower SiO$_2$ cladding with a HF-based wet etch (Figure 27.15a). Typical parameters for Si PhC slabs operating at λ = 1.55 μm are a lattice constant a of about 400 nm, a slab thickness of about 200 nm and a hole radius of about 100 nm.

Planar PhCs are able to provide efficient optical confinement in all three dimensions by combining the 2D photonic band gap confinement in the plane with total internal reflection (TIR) confinement in the vertical direction (Figure 27.15b). In order to represent the 3D nature of the structure, the TIR confinement is superimposed on the 2D band structure by means of the light line: only modes whose k-vectors lie below this line are guided by TIR and can propagate without losses. All other modes are subject to radiation losses and are therefore referred to as quasi-guided modes.

PhC nanocavities with ultrahigh quality factors can achieve extreme light confinement in a very small space and provide the ultimate interaction between light and matter (Akahane et al. 2004; Song et al. 2005). Because the electromagnetic field intensity and the light–matter interaction in an optical cavity scale as its Q/V ratio (with Q the quality factor and V its modal volume), the quest for PhC nanocavities with a high Q/V ratio is intense. State-of-the-art Si PhC cavities feature Q values in excess of 4×10^6 while keeping modal volumes below $(\lambda/n)^3$, where n is the effective index of the cavity mode (Taguchi et al. 2011), and performance is increasing constantly with improvements in Si nanofabrication.

Although Si photonics has almost reached maturity with respect to optical circuit elements, the light source continues to challenge researchers. Nanocavities may help to address this challenge, as they offer luminescence enhancement through the Purcell effect. The Purcell effect increases the radiative emission rate of an emitter interacting with an optical cavity by a factor proportional to the Q/V ratio.

PhC nanocavities can be understood as intentional "defects" introduced into the otherwise perfect photonic lattice. In analogy with solids, such defects create allowed energy states that correspond to the resonant frequencies of the cavities. The best-known examples for such point-defect cavities include the L3 nanocavity, obtained by removing three holes from the Γ-K direction of a triangular lattice (Corcoran et al. 2009). The inventive step in this case was the creation of a cavity mode with almost no k-vector components inside the light cone, which is achieved by a Gaussian-mode envelope. In Figure 27.15c, a SEM micrograph of an L3 Si PhC nanocavity, characterized by a period a = 420 nm, a normalized radius r/a = 0.28 and a slab thickness of 220 nm, is shown. These parameters produce an L3 nanocavity with a fundamental mode around 1.55 μm wavelength. The two holes adjacent to the cavity (indicated by arrows) are reduced in size and displaced laterally to increase the Q-value of the cavity. To obtain the maximum out-coupling efficiency in the vertical direction, a far-field optimization technique has been applied whereby

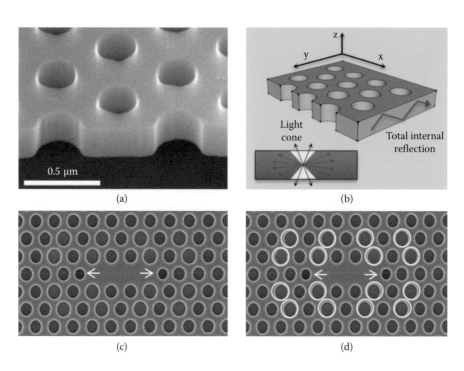

Figure 27.15 (a) SEM micrograph of a Si PhC operating at λ = 1.55 μm. (b) Schematic view of the optical properties of a PhC. (c, d) SEM micrographs of L3 nanocavities; arrows indicate holes whose radius has been reduced to increase the Q-factor of the cavity, while circled holes in (d) have been slightly enlarged for far-field optimization. (Reprinted by permission from Macmillan Publishers Ltd. *Nat. Nanotechnol.*, Priolo, F., et al., (2014), Silicon nanostructures for photonics and photovoltaics, 9, 19–32, copyright 2014.)

alternating holes around the cavity are enlarged (circled holes in Figure 27.15d) to form a "second-order grating," which significantly increases the vertical out-coupling efficiency.

An approach which allows the fabrication of an Si light source that can be electrically pumped, operates at sub-band gap wavelengths, works at room temperature, has a small size, and a narrow emission line-width, is the coupling of optically active defects in Si with a high Q PhC cavity. In fact, optically active defects offer an alternative and particularly powerful approach to improve Si luminescence in the important telecommunication windows (Canham et al. 1989; Henry et al. 1991). Hydrogen plasma treatment can be used to generate surface defects at which hydrogen may get trapped or to introduce hydrogen that can attach to preexisting defects created during the processing needed to fabricate the PhC. TEM analysis is a useful technique to better understand the nature and the location of the defects created by the hydrogen plasma treatment.

Figure 27.16a shows a plan view TEM image of the region containing the PhC using bright-field imaging. The black spots indicate the presence of extended defects induced by the hydrogen plasma treatment during reactive ion etching. Even though the defects are present throughout the entire Si surface, it is evident that their concentration increases toward the sidewalls of the holes. This effect is a consequence of the plasma treatment, during which hydrogen impacts on all exposed surfaces and produces extended damage. To better illustrate the structure of the resulting defect population, a defocused off-Bragg cross section view of the plasma-treated Czochralski silicon (Cz-Si) substrate is shown in Figure 27.16b. There are different types of defects as a function of penetration depth. Near the surface, concentrated within the first 10 nm, the defect population is dominated by nanobubbles, whose size is a few nm. Their exact nature is unknown, but they most likely consist of agglomerates of vacancies. Going further down, a preponderance of platelets (indicated by arrows in Figure 27.16b), whose mean diameter is about 10–15 nm, occupying the (100) plane (parallel to the surface) and the (111) planes is found. A high-resolution image of a few platelets is shown in the top left inset in Figure 27.16b. Moreover, some of the dark traces located between the two previously described regions exhibit the typical "coffee bean" shape indicative of dislocation loops.

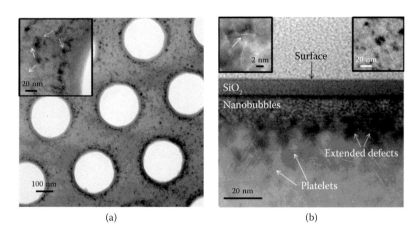

(a)   (b)

Figure 27.16 (a) Plan view TEM image of a Si PhC showing the defects created by hydrogen plasma treatment. The inset shows a zoom-in region close to a single hole, where some defects are indicated by arrows. (b) Cross-sectional view of a plasma-treated Cz-Si sample. Platelets and extended defects are indicated by arrows. In the top left inset, a high-resolution image of the platelets is given, while in the top right inset a typical coffee bean shape of a dislocation loop is shown. (Reprinted from Shakoor A, et al.: Room temperature all-silicon photonic crystal nanocavity light emitting diode at sub-bandgap wavelengths. *Laser Photon. Rev.* 2013. 7, 114–121. Copyright Wiley-VCH Verlag GmbH & Co. KGaA. With permission.)

One of them is shown in the top right inset of Figure 27.16b. The existence of these hydrogen plasma-induced defects has been widely described in the literature (Johnson et al. 1987; Nordmark et al. 2009).

27.4.2 OPTICAL PROPERTIES AND LED APPLICATIONS

Figure 27.17 shows the comparison of room temperature PL from hydrogen plasma-treated PhC nanocavities modes, with different lattice constants, with that of bulk Cz-Si. The PL signal from bulk Si is very weak, while in the treated PhC cavity a broad background signal with strong, sharp peaks corresponding to the fundamental mode of each PhC cavity is observed. The enhancement of background PL is due to incorporation of optically active defects by hydrogen plasma treatment while the peak reflects the enhancement due to a cavity resonance. Combining the enhancement of PL arising from hydrogen plasma treatment with the Purcell enhancement and improved extraction due to the nanocavity, a 40,000-fold enhancement of the PL signal at room temperature relative to the PL of bulk Si is obtained. As demonstrated in Figure 27.17, the emission line is easily tunable through a 1300–1600 nm wavelength range and also demonstrates the robustness and repeatability of this method. A Purcell enhancement factor of ~10, which increases the radiative recombination rate and suppresses thermal quenching, can be estimated (Hauke et al. 2010; Lo Savio et al. 2011; Shakoor et al. 2012).

Based on this result, an electrically driven nano-LED source can also be realized (Shakoor et al. 2013). Since PhC fabrication is fully compatible with ultralarge-scale integration processes (Settle et al. 2006), p-i-n junctions are created by using multiple lithography and ion implantation steps providing a monolithic source of electron–hole pairs to feed optically active defects. Here, fingers of doped regions, as marked in Figure 27.18a, are created, that extend into the PhC. This forces carriers to recombine in the cavity by virtue of the low resistivity of the two highly doped fingers. The conditions were carefully optimized to provide the best possible current injection.

Room temperature EL can be generated by applying a forward bias across the junction. The comparison between EL and PL is shown in Figure 27.18b, where the maximum power spectral density expressed in pW/nm for the fundamental cavity mode for both PL and EL is reported. PL is recorded for a pump power of 0.8 mW and at an excitation wavelength of 640 nm. EL is recorded at an applied voltage of 3.5 V, with a current of 156.5 µA, thus consuming an electrical power of 0.55 mW. Remarkably, the EL signal is more intense than the PL signal across the entire spectral range, in contrast to that usually observed, and is a testament to the potential of this system as an electrically driven source.

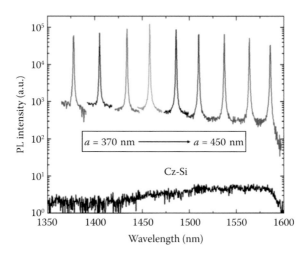

Figure 27.17 Room-temperature PL signal of hydrogen-treated PhC Si nanocavities. Peaks correspond to the fundamental mode of cavities with different lattice period *a*. The emission line of the plasma-treated cavity is over four orders of magnitude higher than bulk Cz-Si and tunable between 1300 and 1600 nm. (Reprinted from Shakoor, A., et al.: Room temperature all-silicon photonic crystal nanocavity light emitting diode at sub-bandgap wavelengths. *Laser Photon. Rev.* 2013. 7, 114–121. Copyright Wiley-VCH Verlag GmbH & Co. KGaA. With permission.)

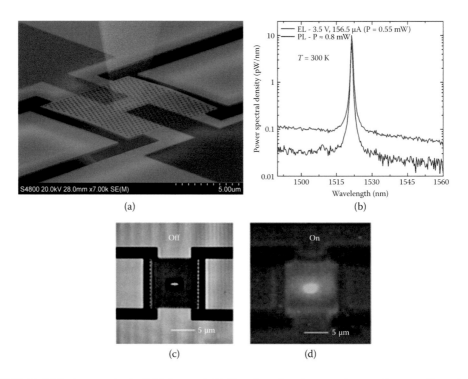

Figure 27.18 (a) SEM micrograph of a Si PhC nano-LED. The doped regions are shown schematically and extend into the PhC. (b) Comparison of EL and PL signals from a PhC nanocavity treated with hydrogen plasma. (c) Micrograph and (d) filtered IR picture of the device showing strong electrically driven emission from the PhC nanocavity at room temperature. (Reprinted from Shakoor, A., et al.: Room temperature all-silicon photonic crystal nanocavity light emitting diode at sub-bandgap wavelengths. *Laser Photon. Rev.* 2013. 7, 114–121. Copyright Wiley-VCH Verlag GmbH & Co. KGaA. With permission.)

Industrial nanosilicon

Figure 27.18 also shows an optical image of the device under zero voltage (Figure 27.18c) and with a 3.5 V bias applied (Figure 27.18d). The latter image was captured with an IR camera and the bright emission spot is clearly visible as soon as the voltage is turned on. The Q-factor of the cavity for the electrical devices was 4000, while for bare PhC cavities (before device fabrication) higher Q values are observed. This reduction is a consequence of imperfections introduced during the p-i-n junction fabrication process as well as free carrier absorption. In the fundamental cavity mode, the device gives a spectral density of 10 pW/nm (by considering the active area, 800 µW/nm/cm^2), which is by far the highest reported value for any Si-based electrically driven nanoemitter, thus encouraging further efforts for the realization of the first Si-based electrically pumped laser.

27.5 CONCLUSIONS

In this chapter, some interesting opportunities offered by Si nanostructures for the development of photonics and photovoltaics have been presented and discussed. All approaches are mainly based on nanoscale engineering and seem to be able to hugely extend the capabilities of bulk Si; NCs and NWs offer the possibility to manipulate the intrinsic electronic properties of silicon using quantum size and surface effects, while new photonic functionality can be added through wavelength-scale nanostructures such as PhCs. Recent advances on silicon nanostructures demonstrate that they may be able to combine quantum dot-like optical properties with suitable electrical conduction, and therefore to exhibit performances comparable to those of III–V semiconductors. In addition, PhCs may offer a very high Purcell enhancement and are therefore essential for any application involving radiative transitions. Silicon nanostructures have mainly been developed with light emission and propagation in mind, but many concepts can be applied to the complementary problem of light absorption in solar cells. Silicon is the only viable material for the large-scale production of solar cells, owing to its abundance and technological maturity, but it is necessary to improve its light trapping properties. The main challenge is to implement light trapping designs in conveniently conductive nanomaterials and to ensure that the designs can be realized on the required large areas in a cost-effective manner. Given the large market demand for new photonic technologies and the many innovative solutions provided by industrial and academic research groups, there is no doubt that Si technology will mature further and will become the dominant technology also in photonics and photovoltaics.

REFERENCES

Akahane Y, Asano T, Song BS, Noda S. (2004). High-Q photonic nanocavity in a two-dimensional photonic crystal. *Nature.* 425:944–947.

Artoni P, Irrera A, Iacona F, Pecora EF, Franzò G, Priolo F. (2012). Temperature dependence and aging effects on silicon nanowires photoluminescence. *Opt. Express.* 20:1483–1490.

Beard MC, Midgett AG, Hanna MC, Luther JM, Hughes BK, Nozik AJ. (2010). Comparing multiple exciton generation in quantum dots to impact ionization in bulk semiconductors: Implications for enhancement of solar energy conversion. *Nano Lett.* 10:3019–3027.

Bisi O, Ossicini S, Pavesi L. (2000). Porous silicon: A quantum sponge structure for silicon based optoelectronics. *Surf. Sci. Rep.* 38:1–126.

Brewer A, Von Haeften K. (2009). *In situ* passivation and blue luminescence of silicon clusters using a cluster beam/H$_2$O codeposition production method. *Appl. Phys. Lett.* 94:261102.

Canham LT. (1990). Silicon quantum wire array fabrication by electrochemical and chemical dissolution of wafers. *Appl. Phys. Lett.* 57:1046–1048.

Canham LT, Dyball MR, Leong WY, Houlton MR, Cullis AG, Smith P. (1989). Radiative recombination channels due to hydrogen in crystalline silicon. *Mater. Sci. Eng. B.* 4:41–45.

Conibeer G, et al. (2006). Silicon nanostructures for third generation photovoltaic solar cells. *Thin Solid Films.* 511–512:654–662.

Corcoran B, et al. (2009). Green light emission in silicon through slow-light enhanced third-harmonic generation in photonic-crystal waveguides. *Nat. Photonics.* 3:206–210.

Cullis AG, Canham LT, Calcott PDJ. (1997). The structural and luminescence properties of porous silicon. *J. Appl. Phys.* 82:909–965.

Daldosso N, et al. (2003). Role of the interface region on the optoelectronic properties of silicon nanocrystals embedded in SiO$_2$. *Phys. Rev. B.* 68:085327.

Dal Negro L, et al. (2006). Light emission efficiency and dynamics in silicon-rich silicon nitride films. *Appl. Phys. Lett.* 88:233109.

Dayeh SA, Picraux ST. (2010). Direct observation of nanoscale size effects in Ge semiconductor nanowire growth. *Nano Lett.* 10:4032–4039.

Delerue C, Allan G, Lannoo M. (1993). Theoretical aspects of the luminescence of porous silicon. *Phys. Rev. B.* 48:11024–11036.

Demichel O, et al. (2009). Recombination dynamics of spatially confined electron-hole system in luminescent gold catalyzed silicon nanowires. *Nano Lett.* 9:2575–2578.

den Hertog MI, et al. (2008). Control of gold surface diffusion on Si nanowires. *Nano Lett.* 8:1544–1550.

Dinh LN, Chase LL, Balooch M, Siekhaus WJ, Wooten F. (1996). Optical properties of passivated Si nanocrystals and SiO_x nanostructures. *Phys. Rev. B.* 54:5029–5037.

Fazio B, et al. (2016). Strongly enhanced light trapping in a two-dimensional silicon nanowire random fractal array. *Light Sci. Appl.* 5:e16062.

Franzò G, Boninelli S, Pacifici D, Priolo F, Iacona F, Bongiorno C. (2003). Sensitizing properties of amorphous Si clusters on the 1.54-μm luminescence of Er in Si-rich SiO_2. *Appl. Phys. Lett.* 82:3871–3873.

Franzò G, et al. (2002). Electroluminescence of silicon nanocrystals in MOS structures. *Appl. Phys. A Mater. Sci. Process.* 74:1–5.

Franzò G, Pacifici D, Vinciguerra V, Priolo F, Iacona F. (2000). Er^{3+} ions-Si nanocrystals interactions and their effects on the luminescence properties. *Appl. Phys. Lett.* 76:2167–2169.

Franzò G, Vinciguerra V, Priolo F. (1999). The excitation mechanism of rare-earth ions in silicon nanocrystals. *Appl. Phys. A Mater. Sci. Process.* 69:3–12.

Fujii M, Yoshida M, Kanzawa Y, Hayashi S, Yamamoto K. (1997). 1.54 μm photoluminescence of Er^{3+} doped into SiO_2 films containing Si nanocrystals: Evidence for energy transfer from Si nanocrystals to Er^{3+}. *Appl. Phys. Lett.* 71:1198–1200.

Fujita M, Takahashi S, Tanaka Y, Asano T, Noda S. (2005). Simultaneous inhibition and redistribution of spontaneous light emission in photonic crystals. *Science.* 308:1296–1298.

Galli M, et al. (2006). Strong enhancement of Er^{3+} emission at room temperature in silicon-on-insulator photonic crystal waveguides. *Appl. Phys. Lett.* 88:251114.

Gourbilleau F, Portier X, Ternon C, Voivenel P, Madelon R, Rizk R. (2001). Si-rich/SiO_2 nanostructured multilayers by reactive magnetron sputtering. *Appl. Phys. Lett.* 78:3058–3060.

Green MA. (2001). Third generation p hotovoltaics: Ultra-high conversion efficiency at low cost. *Prog. Photovolt. Res. Appl.* 9:123–125.

Guichard R, Barsic DN, Sharma S, Kamins TI, Brongersma ML. (2006). Tunable light emission from quantum-confined excitons in $TiSi_2$-catalyzed silicon nanowires. *Nano Lett.* 6:2140–2144.

Hanna MC, Nozik AJ. (2006). Solar conversion efficiency of photovoltaic and photoelectrolysis cells with carrier multiplication absorbers. *J. Appl. Phys.* 100:074510.

Hauke N, et al. (2010). Enhanced photoluminescence emission from two-dimensional silicon photonic crystal nanocavities. *New J. Phys.* 12:053005.

Henry A, Monemar B, Lindstrom J, Bestwick TD, Oehrlein GS. (1991). Photoluminescence characterization of plasma exposed silicon surfaces. *J. Appl. Phys.* 70:5597–5603.

Holmberg VC, Bogart TD, Chockla AM, Hessel CM, Korgel BA. (2012). Optical properties of silicon and germanium nanowire fabric. *J. Phys. Chem. C.* 116:22486–22491.

Iacona F, Bongiorno C, Spinella C, Boninelli S, Priolo F. (2004). Formation and evolution of luminescent Si nanoclusters produced by thermal annealing of SiO_x films. *J. Appl. Phys.* 95:3723–3732.

Iacona F, Franzò G, Spinella C. (2000). Correlation between luminescence and structural properties of Si nanocrystals. *J. Appl. Phys.* 87:1295–1303.

Irrera A, et al. (2002). Excitation and de-excitation properties of silicon quantum dots under electrical pumping. *Appl. Phys. Lett.* 81:1866–1868.

Irrera A, et al. (2006). Electroluminescence and transport properties in amorphous silicon nanostructures. *Nanotechnology.* 17:1428–1436.

Irrera A, et al. (2011). Size-scaling in optical trapping of silicon nanowires. *Nano Lett.* 11:4879–4884.

Irrera A, et al. (2012). Quantum confinement and electroluminescence in ultrathin silicon nanowires fabricated by a maskless etching technique. *Nanotechnology.* 23:075204.

Javey A, Nam S, Friedman RS, Yan H, Lieber CM. (2007). Layer-by-layer assembly of nanowires for three-dimensional, multifunctional electronics. *Nano Lett.* 7:773–777.

Joannopoulos J, Johnson SG, Meade R, Winn J. (2007). *Photonic crystals: Molding the flow of light*, 2nd edition. Princeton, NJ: Princeton University Press.

Johnson NM, Ponce FA, Street RA, Nemanich RJ. (1987). Defects in single-crystal silicon induced by hydrogenation. *Phys. Rev. B.* 35:4166–4169.

Kayes BM, Atwater HA, Lewis NS. (2005). Comparison of the device physics principles of planar and radial p-n junction nanorod solar cells. *J. Appl. Phys.* 97:114302.

Kelzenberg MD, et al. (2010). Enhanced absorption and carrier collection in Si wire arrays for photovoltaic applications. *Nat. Mater.* 9:239–244.

Kelzenberg MD, et al. (2008). Photovoltaic measurements in single-nanowire silicon solar cells. *Nano Lett.* 8:710–714.

Krauss TF, De La Rue RM, Brand S. (1996). Two-dimensional photonic-bandgap structures operating at near-infrared wavelengths. *Nature.* 383:699–702.

Kroll M, et al. (2008). Employing dielectric diffractive structures in solar cells—A numerical study. *Phys. Status Solidi A.* 205:2777–2795.

Liu A, et al. (2004). A high-speed silicon optical modulator based on a metal–oxide–semiconductor capacitor. *Nature.* 427:615–618.

Liu C-Y, Holman ZC, Kortshagen URL. (2010). Optimization of Si NC/P3HT hybrid solar cells. *Adv. Funct. Mater.* 20:2157–2164.

Lo Savio R, et al. (2011). Room-temperature emission at telecom wavelengths from silicon photonic crystal nanocavities. *Appl. Phys. Lett.* 98:201106.

Luque A, Martì A. (1997). Increasing the efficiency of ideal solar cells by photon induced transitions at intermediate levels. *Phys. Rev. Lett.* 78:5014–5017.

Martins ER, et al. (2013). Deterministic quasi-random nanostructures for photon control. *Nat. Commun.* 4:2665.

Matsuo S, et al. (2011). 20-Gbit/s directly modulated photonic crystal nanocavity laser with ultra-low power consumption. *Opt. Express.* 19:2242–2250.

Miller DAB. (2009). Device requirements for optical interconnects to silicon chips. *Proc. IEEE.* 97:1166–1185.

Mirabella S, et al. (2009). Light absorption in silicon quantum dots embedded in silica. *J. Appl. Phys.* 106:103505.

Mohite AD, et al. (2012). Highly efficient charge separation and collection across *in situ* doped axial VLS-grown Si nanowire p-n junctions. *Nano Lett.* 12:1965–1971.

Nordmark H, Holmestad R, Walmsley JC, Ulyashin A. (2009). Transmission electron microscopy study of hydrogen defect formation at extended defects in hydrogen plasma treated multicrystalline silicon. *J. Appl. Phys.* 105:033506.

Notomi M, Shinya A, Mitsugi S, Kira G, Kuramochi E, Tanabe T. (2005). Optical bistable switching action of Si high-Q photonic-crystal nanocavities. *Opt. Express.* 13:2678–2687.

Nozik AJ. (2002). Quantum dot solar cells. *Physica E.* 14:115–120.

O'Faolain L, et al. (2010). Loss engineered slow light waveguides. *Opt. Express.* 18:27627–27638.

Ossicini S, Pavesi L, Priolo F. (2004). *Light emitting silicon for microphotonics.* Berlin: Springer.

Pacifici D, Franzò G, Priolo F, Iacona F, Dal Negro L. (2003). Modeling and perspectives of the Si nanocrystals–Er interaction for optical amplification. *Phys. Rev. B.* 67:245301.

Pavesi L, Dal Negro L, Mazzoleni C, Franzò G, Priolo F. (2000). Optical gain in silicon nanocrystals. *Nature.* 408:440–444.

Pavesi L, Turan R. (2010*). Silicon nanocrystals; fundamentals, synthesis, and applications.* Weinheim: Wiley-VCH.

Pecora F, et al. (2012). Nanopatterning of silicon nanowires for enhancing visible photoluminescence. *Nanoscale.* 4:2863–2866.

Plotnick RE, Gardner RH, Hargrove WW, Prestegaard K, Perlmutter M. (1996). Lacunarity analysis: A general technique for the analysis of spatial patterns. *Phys. Rev. E.* 53:5461–5468.

Polman A. (1997). Erbium implanted thin film photonic materials. *J. Appl. Phys.* 82:1–39.

Polman A, Atwater HA. (2012). Photonic design principles for ultra-high efficiency photovoltaics. *Nat. Mater.* 11:174–177.

Priolo F, Franzò G, Pacifici D, Vinciguerra V, Iacona F, Irrera A. (2001). Role of the energy transfer in the optical properties of undoped and Er-doped interacting Si nanocrystals. *J. Appl. Phys.* 89:264–272.

Priolo F, Gregorkiewicz T, Galli M, Krauss TF. (2014). Silicon nanostructures for photonics and photovoltaics. *Nat. Nanotechnol.* 9:19–32.

Reed, GT, Mashanovich GZ, Gardes FY, Thomson DJ. (2010). Silicon optical modulators. *Nat. Photonics.* 4:518–526.

Rong H, et al. (2005). An all-silicon Raman laser. *Nature.* 433:725–728.

Schaller RD, Klimov VI. (2004). High efficiency carrier multiplication in PbSe nanocrystals: Implications for solar energy conversion. *Phys. Rev. Lett.* 92:186601.

Schmid H, Björk MT, Knoch J, Karg S, Riel H, Riess W. (2009). Doping limits of grown in situ doped silicon nanowires using phosphine. *Nano Lett.* 9:173–177

Settle M, Salib M, Michaeli A, Krauss TF. (2006). Low loss silicon on insulator photonic crystal waveguides made by 193 nm optical lithography. *Opt. Express.* 14:2440–2445.

Shakoor A, et al. (2013). Room temperature all-silicon photonic crystal nanocavity light emitting diode at sub-bandgap wavelengths. *Laser Photon. Rev.* 7:114–121.

Shakoor A, et al. (2012). Enhancement of room temperature sub-bandgap light emission from silicon photonic crystal nanocavity by Purcell effect. *Physica B.* 407:4027–4031.

Shao M, et al. (2009). Nitrogen-doped silicon nanowires: Synthesis and their blue cathodoluminescence and photoluminescence. *Appl. Phys. Lett.* 95:143110.

Shimizu-Iwayama T, Kurumado N, Hole DE, Townsend PD. (1998). Optical properties of silicon nanoclusters fabricated by ion implantation. *J. Appl. Phys.* 83:6018–6022.

Shockley W, Queisser HJ. (1961). Detailed balance limit of efficiency of p–n junction solar cells. *J. Appl. Phys.* 32:510–519.

Sivakov VA, Voigt F, Berger A, Bauer G, Christiansen SH. (2010). Roughness of silicon nanowire sidewalls and room temperature photoluminescence. *Phys. Rev. B.* 82:125446.

Song BS, Noda S, Asano T, Akahane Y. (2005). Ultra-high-Q photonic double-heterostructure nanocavity. *Nat. Mater.* 4:207–210.

St. John J, Coffer JL, Chen Y, Pinizzotto RF. (1999). Synthesis and characterization of discrete luminescent erbium-doped silicon nanocrystals. *J. Am. Chem. Soc.* 121:1888–1892.

Sykora M, Mangolini L, Schaller RD, Kortshagen U, Jurbergs D, Klimov VI. (2008). Size-dependent intrinsic radiative decay rates of silicon nanocrystals at large confinement energies. *Phys. Rev. Lett.* 100:067401.

Taguchi Y, Takahashi Y, Sato Y, Asano T, Noda S. (2011). Statistical studies of photonic heterostructure nanocavities with an average Q factor of three million. *Opt. Express.* 19:11916–11921.

Tauc J. (1974). *Amorphous and liquid semiconductors.* London: Plenum Press.

Tian B, et al. (2007). Coaxial silicon nanowires as solar cells and nanoelectronic power sources. *Nature.* 449:885–890.

Timmerman D, Izeddin I, Stallinga P, Yassievich IN, Gregorkiewicz T. (2008). Space-separated quantum cutting with silicon nanocrystals for photovoltaic applications. *Nat. Photonics.* 2:105–109.

Trinh MT, et al. (2008). In spite of recent doubts carrier multiplication does occur in PbSe nanocrystals. *Nano Lett.* 8:1713–1718.

Tsybeskov L, Vandyshev JV, Fauchet PM. (1994). Blue emission in porous silicon—Oxygen-related photoluminescence. *Phys. Rev. B.* 49:7821–7824.

Valenta J, Bruhn B, Linnros J. (2011). Coexistence of 1D and quasi-0D photoluminescence from single silicon nanowires. *Nano Lett.* 11:3003–3009.

Wagner RS, Ellis WC. (1964). Vapor-liquid-solid mechanism of single crystal growth. *Appl. Phys. Lett.* 4:89–90.

Walavalkar SS, Hofmann CE, Homyk AP, Henry MD, Atwater HA, Scherer A. (2010). Tunable visible and near-IR emission from sub-10 nm etched single-crystal Si nanopillars. *Nano Lett.* 10:4423–4428.

Walters RJ, Bourianoff GI, Atwater HA. (2005). Field-effect electroluminescence in silicon nanocrystals. *Nat. Mater.* 4:143–146.

Wojdak M, et al. (2004). Sensitization of Er luminescence by Si nanoclusters. *Phys. Rev. B.* 69:233315.

Wolkin MV, Jorne J, Fauchet PM, Allan G, Delerue C. (1999). Electronic states and luminescence in porous silicon quantum dots: The role of oxygen. *Phys. Rev. Lett.* 82:197–200.

Yablonovitch E. (1982). Statistical ray optics. *J. Opt. Soc. Am.* 72:899–907.

Zacharias M, Heitmann J, Scholz R, Kahler U, Schmidt M, Blasing J. (2002). Size-controlled highly luminescent silicon nanocrystals: A SiO/SiO_2 superlattice approach. *Appl. Phys. Lett.* 80:661–663.

Zhao J, Wang A, Green MA, Ferrazza F. (1998). Novel 19.8% efficient "honeycomb" textured multicrystalline and 24.4% monocrystalline silicon solar cells. *Appl. Phys. Lett.* 73:1991–1993.

Industrial nanosilicon

28 Silicon–carbon yolk–shell structures for energy storage application

Xuefeng Song, Zhuang Sun, Cheng Yang, and Lisong Xiao

Contents

28.1 INTRODUCTION

It is now universally accepted that the combustion of oil fuel is the major reason for modern cities' atmospheric contamination.[1,2] And, the decreasing amount of petroleum resources impels people to seek for other clean energy substitutes. Along with the development of renewable energy technology, such as solar, wind, and tidal energy, how to effectively store these kinds of intermittent energy becomes more important in this century than it was in the past. Lithium-ion battery (LIB) is considered to be one of the most important energy storage and conversion technologies[3] with prominent advantages of high energy and power densities,[4–8] which have witnessed large-scale application in portable electronic devices, communication facilities, stationary energy storage, and ever-enlarging markets of electrically powered vehicles.[9]

Graphite, the traditional anode material utilized in LIBs, does not meet the high-energy needs due to its limited theoretical specific capacity of ~370 mAh g^{-1}.[10] Thus, there has been a growing interest in developing alternative anode materials with low cost, enhanced safety, high-energy density, and long cycle life. A great deal of anode materials with enhanced storage capacity, high-energy density, and improved cycle characteristics have been proposed for LIBs in the last decade.[11] Table 28.1 compares the properties of different anode materials[12]; among them, silicon (Si) has attracted substantial attention as an alternative anode material for LIBs, primarily due to its (1) highest theoretical capacity (~3600 mAh g^{-1}) and high volume capacity (9786 mAh cm^{-3}); (2) relatively low working potential, making it suitable as an anode (0.5 V vs. Li/Li$^+$); and (3) the natural abundance of elemental Si and its environmental benignity.[13]

However, the practical application of Si anodes is still blocked and severely restricted due to the following three major problems: First is the pulverization-induced poor cycle life of Si caused by the huge volumetric fluctuations (>300%) during the lithiation/delithiation process, which results in drastic irreversible capacity loss and low Coulombic efficiency (CE) of the electrodes.[13] Second is the low electrical conductivity (1 × 10^{-3} S cm^{-1}) and low diffusion coefficients of lithium ions (10–13 to 10–12 cm^{-2} S^{-1}). The last is the formation of the solid–electrolyte interface (SEI) in the lithiated expanded state of Si, which can be broken

Table 28.1 Comparison of various anode materials

ANODE MATERIALS	Li	C	Li$_4$Ti$_5$O$_{12}$	Si	Sn	Sb	Al	Mg	Bi
Density (g cm^{-3})	0.53	2.25	3.5	2.33	7.29	6.7	2.7	1.3	9.78
Lithiated phase	Li	LiC$_6$	Li$_{12}$Ti$_5$O$_{12}$	Li$_{4.4}$Si	Li$_{4.4}$Sn	Li$_3$Sb	LiAl	Li$_3$Mg	Li$_3$Bi
Theoretical specific capacity (mAh g^{-1})	3862	372	175	4200	994	660	993	3350	385
Theoretical specific capacity (mAh cm^{-3})	2047	837	613	9786	7246	4422	2681	4355	3765
Volume change (%)	100	12	1	320	260	200	96	100	215
Potential vs. Li (~v)	0	0.05	1.6	0.4	0.6	0.9	0.3	0.1	0.8

Source: Zhang, W.-J., J. Power Sources, 196(1), 13–24, 2011. With permission.

during the delithiation as the structure shrinks. This leads to the re-exposure of the fresh Si surface to the electrolyte, resulting in the repetitive formation of the SEI.[14,15]

Hence, great efforts have been made to address these issues, such as the development of nanostructured Si,[16–18] Si alloy nanohybrids,[19] core–shell-structured Si composites,[20,21] the addition of the electrolyte additives,[22] and the use of novel binders.[22,23] Among them, significant efforts have been made on the core–shell-structured Si composites, in which Si is coated by active/inactive shells, accommodating the strain and maintaining the structural integrity. For example, core–shell Si/C,[24] Si-Cu/C,[25] Si/SiO$_x$/C,[26] and Si/SiO$_2$/C[27] composites have been successfully prepared, which exhibit better electrochemical performances than the pristine Si electrodes. However, the success of this approach is still limited because large Si volume change can only be tolerated to a limited degree, especially during deep charge–discharge processes.

Apart from the structural stability of the anode, designing a stable SEI may be critically dependent on the outer surface features of active materials. Yolk–shell-structured carbon/void/silicon (YSCVS) composite[28,29] is quite promising for practical applications (Figure 28.1a), because the void space between the outer carbon shell and the inside Si core reserves the room for volume changes of Si core without deforming the carbon shell and the SEI layer, which in turn allows for the formation of a stable SEI layer on the surface of the outer carbon shell.[29] Besides, the homogeneous carbon-coated shell can prevent the permeation of the electrolyte and thus avoid its direct contact with the Si cores. Therefore, the SEI will only be formed on the outer surface of the carbon shell, leading to the high CE and improved cycling stability.

The number of scientific publications during 2011–2015 on the topic of yolk–shell structures for LIBs rapidly increased, as shown in Figure 28.1b. The total number of publications is 57, and the citation of

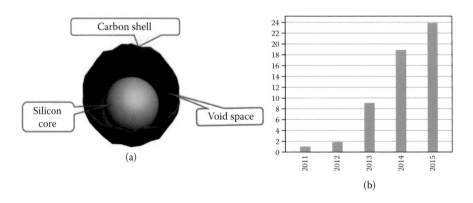

Figure 28.1 (a) Schematic representation of a typical Si@C yolk–shell structure. (b) The number of publications on the yolk–shell structures for Li-ion batteries application was sorted by year. Data were collected from the "Web of Science." The words "yolk–shell" and "Li-ion battery" were keyed into the "topic" search box.

these publications is 1089, which means an average citation of 19.11. A review article on this topic is timely, as a large number of synthesis approaches are becoming available for the preparation of the yolk–shell structures, and the yolk–shell-structured materials are already receiving increased commercial attention. In general, the morphologies of the YSCVS can be divided into zero-dimensional (0D), one-dimensional (1D), and three-dimensional (3D), which have been developed to provide better electrical conductivity, more contact points, larger tap densities, and so on. The focus here is placed on the recent research progresses in the systematic synthesis approaches to the YSCVS and its application of LIBs by reviewing published works on the YSCVS.

28.2 0D Si/C YOLK–SHELL-STRUCTURED ANODE MATERIALS

Among all the YSCVS structures used for LIBs, the most frequently used ones are the 0D Si/C yolk–shell structures, due to the cost-effectiveness, scalability, and commercial availability of the mechanically generated Si nanoparticles from the bulk polycrystalline Si. Besides, the methods for the synthesis of the Si/C yolk–shell nanocomposites are facile and controllable without special equipment.[28,30–35] Among all the strategies used for the synthesis of the 0D YSCVS, the selective etching methods have received a lot of attention and became the common methodology for the production of YSCVS. For the selective etching process, the Si cores are first coated with one or two layers of different materials. The inner layer of the two shells or the part of the core (shell) is selectively removed by dissolution using a solvent, that is, hydrofluoric acid (HF) or strong alkali solution.

28.2.1 SELECTIVE ETCHING OF INTERMEDIATE LAYER TO FORM THE 0D YSCVS

Selective etching or dissolution methods have received a lot of attention, and they are the common methodology for the production of the YSCVS. In this method, a core–shell particle consisting of a core and one layer of shell or two layers of shells with two different materials is first synthesized, and then parts of the core, the shell, or the inner shell are selectively removed, respectively. For example, a three-layer "sandwich" structure of silicon–silica–carbon was prepared by hydrolysis and condensation of tetraethyl orthosilicate (TEOS) and carbon precursor.[28,30–37] The silicon–silica–carbon was then converted into YSCVS by selectively etching the middle layer (silica layer) with an appropriate amount of aqueous HF.

In 2012, Li et al. have fabricated hollow core–shell-structured porous Si/C nanocomposites with void space up to tens of nanometers by using sacrificial silica layer (Figure 28.2a). The starting material was the powder of Si nanoparticles (~50 nm). After silica coating, carbon coating, and silica etching, the Si/void/C yolk–shell structure has been finally obtained. The schematic diagram of the synthesis process and the transmission electron microscope (TEM) images of the composites are shown in Figure 28.2a through e. Furthermore, they have also calculated the appropriate ratio of the diameter of the carbon sphere to the silicon core; when the ratio is ~1.5, the structure can maintain the integrity without breaking/degrading carbon shell (Figure 28.2f and g). As anode materials for LIBs, the hollow core–shell-structured porous Si/C nanocomposites exhibited superior capacity retention (86%) with a specific capacity of 650 mAh g^{-1} at a current density of 1 A g^{-1} after 100 cycles (Figure 28.2h).[30]

An outstanding study in 0D Si–C yolk–shell structures was conducted by Liu et al.[28] They fabricated this unique nanostructure from the commercial silicon nanoparticles, and the carbon shell formed in this study is very uniform by using polydopamines as carbon source. They also used the in situ TEM characterization to reveal the important details related to the volume changes in this configuration, as shown in Figure 28.3c. The successful lithiation of the Si particles indicates a good contact between the carbon shell and the Si particles. It is clear that the carbon shell remains intact after Si expansion even though the expansion causes the Si particles to impinge upon the carbon shell. The Si core freely expands/contracts in the void space without impacting carbon shell, indicating that there is enough void space to accommodate the full expansion of each Si particle. Because the overall shape of the yolk–shell structure does not change appreciably upon lithiation, it is expected that a battery electrode made of this Si/void/C structure would undergo minimal microstructural damage upon cycling (Figure 28.3a and b). In situ TEM images series captured from the device are shown in Figure 28.3c. In this series of images, the silicon particles are observed to expand within the space of the outer carbon shell. The entire volume expansion is accommodated within the available

Industrial nanosilicon

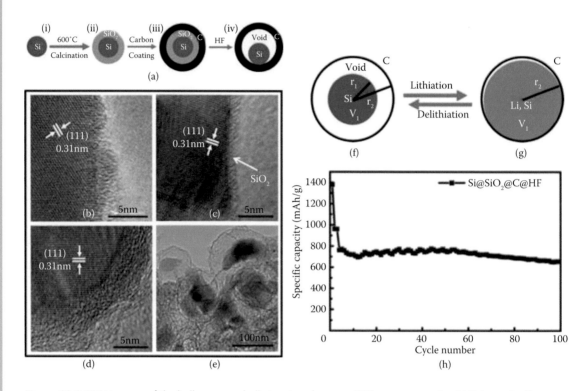

Figure 28.2 TEM images of the hollow core–shell-structured porous Si/C nanocomposite. (a) Schematic diagram of the synthesis process (from I to IV) of the hollow core–shell-structured composite. (b) TEM image of Si. (c) Si coated with a thin layer of SiO_2. (d) Si/SiO_2 composite formed in (b) was coated with a layer of carbon. (e) Si/SiO_2/C composite formed in (c) was etched with HF to remove the SiO_2 layer in-between, forming the desired hollow core–shell structure. Schematic figure showing the core–shell-structured porous Si/C nanocomposite before and after the lithiation–delithiation of Si core. (f) Before lithiation, a void space is present between the Si core and the carbon shell. (g) After lithiation, Li_xSi almost fully fills the void space. (h) Long-term cycling stability of the hollow core–shell-structured porous Si/C nanocomposite. (From Li, X., et al., *J. Mater. Chem.*, 22(22), 11014–11017, 2012. With permission.)

void space, and the shell does not rupture. As a result, outstanding cycle stability and high specific capacity are achieved; the capacity attained at the 1C rate is 1400 mAh g^{-1} after 1000 cycles, retaining 74% of the initial capacity with a CE of 99.8%.[28] These results are much better than that of Li et al.[30] and Zhou et al.[36] Furthermore, Chen et al.[35] and Iwamura et al.[37] have systematically investigated the effect of the buffer size to the silicon core by adjusting the thickness of the silica shell. According to their studies, the best capacity retention can be achieved with a void/Si volume ratio of approximately 3 due to its appropriate volume change tolerance and maintenance of good electrical contacts.

The Si core in the YSCVS can also be obtained without special equipment necessary for other approaches (such as laser irradiation and chemical vapor deposition) by using the magnesiothermic reduction method. For example, Ru et al. have reported a yolk–shell-structured porous Si/C microsphere by magnesiothermic reduction of silica spheres inside carbon shells (Figure 28.4). A carbon shell was first coated on the silica, followed by selective etching of silica to create the space required for expansion of Si during cycling. The carbon shells in this composite were designed to be porous by using dopamine as carbon source and triblock copolymer poly(ethylene oxide)-block-poly(propylene oxide)-block-poly(ethylene oxide) [PEO–PPO–PEO] as a pore-forming surfactant, thus facilitating diffusion of magnesium through the pore channels and following magnesiothermic reduction of the inner silica "yolk." Then, the magnesium oxide and the residual SiO_2 in the composites can be etched by HCl/HF solution to form the YSCVS.[38] In comparison with the yolk–shell structures reported previously,[29,31,36–38] the mesoporous-silicon /void/mesoporous-carbon has the advantage that both the core and shell are porous, which should make it much easier for the infiltration of the electrolyte, thus shortening the diffusion lengths of Li$^+$. The reversible

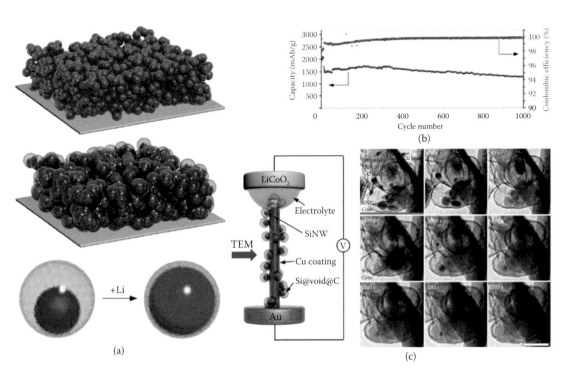

Figure 28.3 (a) Schematic of yolk–shell silicon structure, (b) cycling performance and Coulombic efficiency of Si/C yolk–shell-designed anodes, and (c) in situ TEM observations of the lithiation process. (From Liu, N., et al., *Nano Lett.*, 12(6), 3315–3321, 2012. With permission.)

specific capacity reached 790 mAh g^{-1} for the second cycle at a current density of 200 mA g^{-1} and then stabilized at ~530 mAh g^{-1} with a constant CE as high as 98%.[38] However, in this designed structure, large irreversible capacity loss was inevitable because the electrolyte can permeate the outer carbon shell and react with the surface of the Si core.

To prevent the electrolyte in the outer shell from reaching the surface of the silicon cores and to protect the silicon core from subsequent irreversible reaction with the electrolyte, the yolk–shell concept has been explored with carbon and SiO$_2$ as the dual shells (which can be described as Si/void/SiO$_2$/void/C). To form such a structure, Si nanoparticles are coated with a SiO$_2$ layer via the Stöber method and then the Si/SiO$_2$ particles are coated with a carbon layer (Figure 28.5a). Taking advantage of the inhomogeneous nature of the silica shell prepared by the Stöber method, proper etching via HF treatment can selectively etch a small portion of the outer layer and a large portion of the interior layer of the SiO$_2$ shell, leading to the formation of the Si/void/SiO$_2$/void/C structure. In this dual yolk–shell structure, carbon improves the electrical conductivity, whereas silica increases the mechanical stability and the void provides enough space for the volume expansion of the silicon core. Furthermore, the SiO$_2$ and C dual shells can offer a double barrier to prevent the electrolyte from reaching the surface of the Si nanoparticles and protect the anode from the subsequent irreversible reaction with the electrolyte. The dual yolk–shell structure has exhibited superior capacity retention (83%) with a specific capacity of 956 mAh g^{-1} at a current density of 460 mA g^{-1} after 430 cycles. In contrast, the capacity of the Si/C core–shell structures decreases rapidly in the first 10 cycles under the same experimental conditions (Figure 28.5f). Figure 28.5g shows the charge/discharge capacity at different current densities of the dual yolk–shell structure. Capacities above 950, 830, 610, 410, and 260 mAh g^{-1} are retained at current densities of 0.46, 0.9, 1.8, 3.7, and 7.5 A g^{-1}, respectively.[32]

Nevertheless, the silica shell is inactive for lithium-ion storage, and the irreversible electrochemical reactions between lithium ions and SiO$_2$ (forming lithium silicates) shells are inevitable during the initial cycles. Our group have obtained a silicon/double-shelled carbon (Si/DC) yolk-like nanostructure by designing a special template (Si/void/SiO$_2$ yolk–shell structure).[39] Firstly, the template was synthesized by

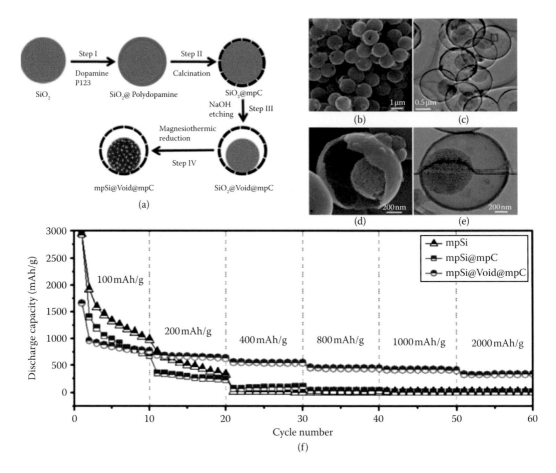

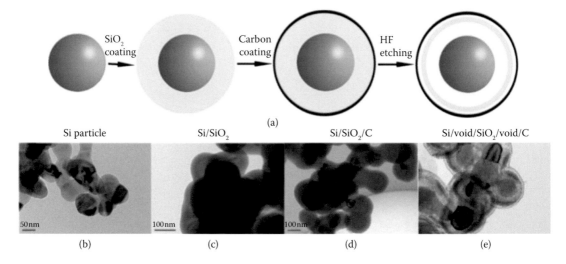

Figure 28.4 (a) Schematic illustration of the synthesis of mpSi/Void/mpC microspheres. SEM images of mpSi/Void/mpC (b and c) at different magnifications; TEM images of mpSi/Void/mpC (d and e) at different magnifications. The inset in (e) shows Si and C intensity line scans. Cycling stabilities of mpSi/Void/mpC, mpSi/mpC, and mpSi. Schematic illustration (f) of the Li+ insertion process for the yolk–shell-structured porous Si/C nanoparticles. (From Ru, Y., et al., *RSC Adv.*, 4(1), 71–75, 2014. With permission.)

Figure 28.5 (a) Schematic illustration of the fabrication process for the dual yolk–shell structure. (b) through (e) Corresponding TEM images of Si, Si/SiO₂, Si/SiO₂/C, and Si/void/SiO₂/void/C spheres. *(Continued)*

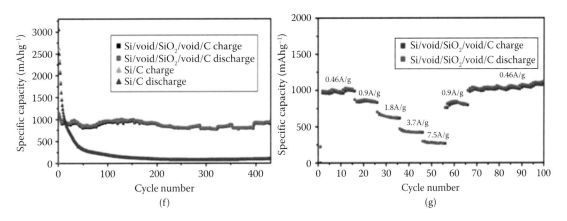

Figure 28.5 (Continued) (f) Cycling behavior of Si/void/SiO₂ /void/C and Si/C composites at a current density of 0.46 A g⁻¹ and (g) rate capability of Si/void/SiO₂/void/C composite at different current densities. (From Yang, L.Y., et al., *Sci. Rep.*, 5, 10908, 2015. With permission.)

using a vesicle template method. The Si/void/SiO₂ template was then coated with a glucose-derived polysaccharide on both the interior and the exterior surfaces of the porous SiO₂ shell. A subsequent annealing process at 800°C for 4 h under an argon atmosphere resulted in the carbonization of the polysaccharide component, leading to the formation of a Si/void@C/SiO₂/C structure. Finally, the SiO₂ shell was dissolved in HF to generate Si/DC powders with the Si/void/C/void/C configuration. The ultrathin double carbon shells in this designed structure not only prevent the electrolyte from reaching the inner core but also afford larger active specific area to trap lithium ions and improve the electrical conductivity of the composite. We have also synthesized Si/single-shelled (Si/SC) and Si/three-shelled (Si/TC) nanopowders, but the Si/DC exhibited the best results. As shown in Figure 28.6f, the specific discharge capacity of Si/DC remains as high as 943.8 mAh g⁻¹ at 50 mA g⁻¹ after 80 cycles, which is higher than that of the Si/SC (719.8 mAh g⁻¹) and the pristine Si electrode (115.3 mAh g⁻¹).[40]

Apart from the silica template, Li et al. also used nickel as a sacrificial layer to obtain graphene-encapsulated Si microparticles (Figure 28.7a). The Ni layer serves as both the catalyst for graphene growth and the sacrificial layer for providing void space. After etching away the Ni using FeCl₃ aqueous solution, the void space is opened up for Si microparticle expansion within the graphene cage. The graphene cage (~10 nm) possesses multilayered structure and can be observed from the TEM image. The mechanically robust graphene cage remains continuously throughout the curved regions, which acts as a buffer to accommodate the severe interior stresses during particle fracture (Figure 28.7b through d). The half-cell data in Figure 28.7e show that the reversible capacity of the graphene-encapsulated Si microparticle (Si MP) reached ~3300 mAh g⁻¹ at a current density of C/20 (1C = 4.2 A g⁻¹). It is worth noting that the electrochemical performances of this composite are the best as yet reported for the micro-scaled Si, and far surpass that of the pristine or amorphous-carbon-coated Si.[41] Furthermore, Zhang et al.[34] have chosen calcium carbonate as a template to synthesize granadilla-like outer carbon coating encapsulated silicon/carbon microspheres, which is discussed thoroughly in the part of 28.4.

28.2.2 PARTIAL REMOVAL OF THE SILICON CORE TO FABRICATE THE 0D YSCVS

Although selective etching sacrificial layers are very useful for the fabrication of the YSCVS with controllable structures, these methods have some intrinsic limitations as they usually involve multiple tedious steps, which cause difficulties for the large-scale production of the YSCVS. To overcome this problem, recently partial removal of the silicon core from the core/shell to form the YSCVS has been reported.[31,33,42,43] For example, Pan et al.[31] obtained the YSCVS by etching Si core/C shell composite with the NaOH solution (Figure 28.8a). In their method, the diameter of the Si nanoparticles and the void space between the Si nanoparticles and the carbon shell can be controlled by adjusting the NaOH etching time. Compared with the Si and Si/C electrodes, the YSCVS electrode shows higher specific capacity and better cycling stability (Figure 28.8f and g). The specific capacity of the YSCVS is about 804 mAh g⁻¹, which corresponds

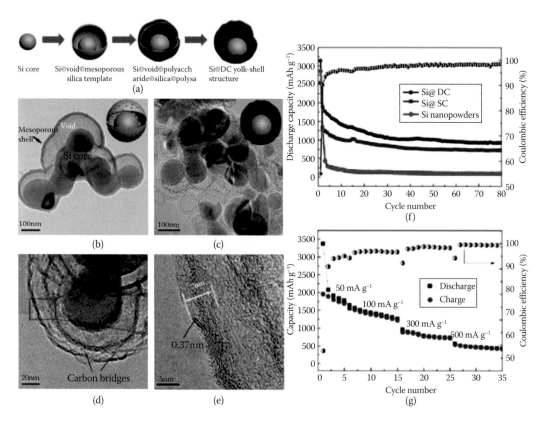

Figure 28.6 (a) Schematic procedures for producing Si/DC yolk-like powders. TEM images of (b) Si/void/meso-porous silica template and (c, d) Si/DC yolk-like structures. (e) HR-TEM image of the carbon shells in the Si/DC. (f) Discharge capacity and Coulombic efficiency of Si/DC nanoparticles in contrast with Si/SC and Si nanopowder. (g) Capacity retention curve at different current densities. (From Sun, Z., et al., *J. Electrochem. Soc.*, 162(8), A1530–A1536, 2015. With permission.)

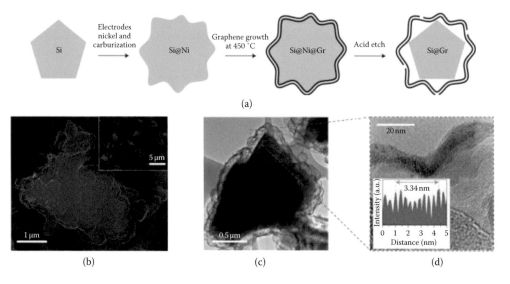

Figure 28.7 (a) Synthesis and characterization of graphene cage structure, and (b) SEM image of a graphene-encapsulated Si microparticles (SiMP/Gr). The inset gives a broader view, showing many Si microparticles encapsulated by the graphene cage. (c) TEM image of an individual particle of SiMP/Gr. (d) High-resolution TEM image of the graphene cage's layered structure. The intensity plot shows that 10 layers span a distance of 3.34 nm (average interlayer distance: 0.334 nm), a clear indication of graphene layers. *(Continued)*

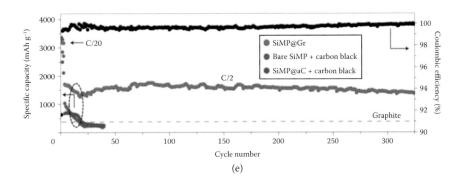

Figure 28.7 (Continued) (e) Half-cell delithiation capacity of SiMP/Gr with no electrically conductive additives. Bare and amorphous-carbon-coated Si MP are control samples with carbon black as conductive additives. The theoretical capacity (370 mAh g⁻¹) of a graphite electrode is shown as a grey dashed line. The Coulombic efficiency of the SiMP/Gr is plotted on the secondary y-axis. (From Li, Y., et al., *Nat. Energy*, 1, 16017, 2016. With permission.)

to 83% capacity retention.[31] Apart from the NaOH etching, the Si core in the core–shell Si/C composites can be also etched by the LiOH and HF solutions, reported by Su et al.[33] and Pang et al.,[43] respectively. It is easy to control the diameter of the Si core and the void space by using this method, and more importantly, different to the previous methods,[28,30,32,36–38] the route in this work avoided the introduction of SiO_2 intermediate layers, which would decrease the production cost-effectively.

However, the powders obtained from the above chemical etching methods have a critical disadvantage in terms of point contact between hollow carbon and Si, which results in large electrical resistance between them. Furthermore, the electrical conductivity becomes poorer after charging and discharging because the electrolyte is decomposed on the surface of carbon and Si to form the SEI. To solve this problem, Park et al. obtained the YSCVS by electroless etching (Figure 28.9a). They chose the Fe-phthalocyanine as the source of carbon and Fe^{3+}; after HF etching, Si core was partially dissolved to obtain void spaces between Si and carbon shell (Figure 28.9b and c). Si-encapsulating hollow carbon spheres have improved area contact between carbon and Si and result in excellent electrochemical performance including little capacity fading over 50 cycles and good rate capability (1100 mAh g⁻¹) at a 2C rate (4 A g⁻¹). Therefore, partial removal of the Si core by this method can improve the electrical contact between carbon and Si effectively.[42]

28.2.3 OTHER METHODS TO OBTAIN THE 0D YSCVS

In comparison with the methods to prepare yolk–shell structures reported previously, Li et al. have developed a novel approach to prepare the YSCVS (Figure 28.10a). When the mixture of mesoporous Si (MSi) and citric acid was heated up, the volume of air adsorbed by the MSi expanded and the viscoelastic citric acid layers inflated just like balloons, directly leading to the formation of the yolk–shell-structured MSi/C nanocomposites during the carbonization. The MSi/C nanocomposites possessed an MSi core with a diameter of 150 nm and a carbon shell with a diameter of 230 nm. Such nano- and mesoporous structures combined with voids between the MSi core and the carbon shell not only provides enough space for the volume expansion of the MSi during the lithiation but also accommodates the mechanical stresses/strains caused by the volume variation. Moreover, partial graphitization of the carbon contributed to the improved electrical conductivity and rate performance of the MSi/C. As a result, the prepared MSi/C exhibited an initial reversible capacity of 2599 mAh g⁻¹ and maintained 1265 mAh g¹ even after 150 cycles at 100 mA g⁻¹, with high CE above 99% (based on the weight of the MSi in the electrode).[44]

28.3 1D Si/C YOLK–SHELL STRUCTURE

The development of 1D nanostructured anode materials for high-performance LIBs, such as nanotubes, nanowires, and nanofibers, is stimulated by their high surface-to-volume ratios. Apart from the YSCVS with the 0D morphology, 0D Si nanoparticles encapsulated with 1D carbon materials have several advantages: Firstly, the 1D morphology features continuous 1D electronic pathways, allowing

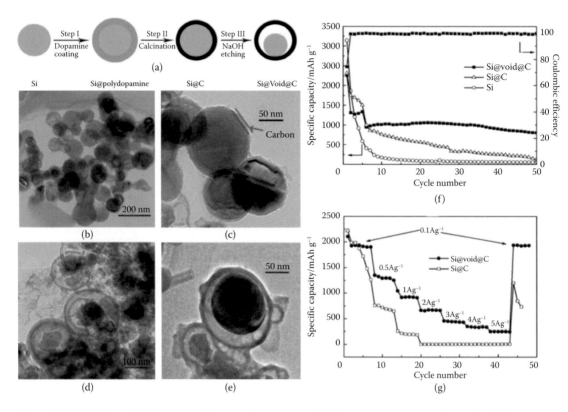

Figure 28.8 (a) Schematic illustration of the synthesis process of Si/C nanocomposites with a yolk–shell structure. TEM images of (b, c) Si/C and (d, e) Si/Void/C nanocomposites. (f) Cycling stability of the Si, Si/C, and Si/Void/C nanocomposites. All electrodes were cycled at 0.1 A g^{-1} for the first cycle, 0.5 A g^{-1} for four cycles, and 1 A g^{-1} for the later cycles. (g) Rate performance of the Si/C and Si/Void/C nanocomposites. (From Pan, L., et al., *Chem. Commun.*, 50(44), 5878–5880, 2014. With permission.)

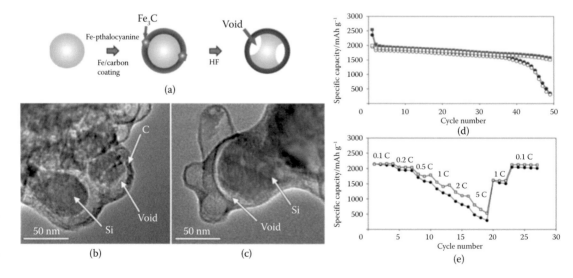

Figure 28.9 (a) Schematic diagram for the synthesis of Si-encapsulating hollow carbon by electroless etching. TEM images of the Si-encapsulating hollow carbon obtained via (b, c) electroless etching. (d) Cycle performance of the Si-encapsulating hollow carbon via electroless etching (square) and chemical etching (circle). (e) Rate performances of the Si-encapsulating hollow carbon via electroless etching (square) and chemical etching (circle). (From Park, Y., et al., *Adv. Energy Mater.*, 3(2), 206–212, 2013. With permission.)

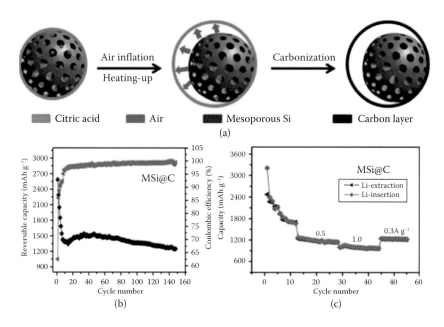

Figure 28.10 The formation process of the MSi/C nanocomposites with a yolk–shell structure (gray arrows represent the expansion direction of air). (b) Cycle performance of the MSi/C electrode under 100 mA g^{-1}. (c) Rate performance of the MSi/C electrode. (From Li, H.-H., et al. *RSC Adv.*, 4(68), 36218–36225, 2014. With permission.)

for efficient charge transport. Secondly, the ample empty space inside the hollow tubes allowed for Si expansion during electrochemical cycling. Lastly, the continuous 1D carbon matrix, which can directly connect to the current collector, provides fast channels for electron transfer and, therefore, enable outstanding high power and excellent rate capability.

28.3.1 0D Si NANOPARTICLES ENCAPSULATED IN 1D HOLLOW CARBON TUBE

Kong et al. have synthesized a 1D composite, in which Si nanoparticles were encapsulated in hollow carbon tubes. The encapsulation was assisted by the collection of loose lumps of electrospun polyacrylonitrile–Si hybrid nanofibers using water as the collector, and subsequently in situ coating the nanofibers with a thin layer of polydopamine (PDA), which is a good source for N-doped graphitized carbon. The hollow nature of the long nanofibers and the small size of the Si nanoparticles can be seen clearly in Figure 28.11b and c. Almost all the SEI can be considered to be only on the outside of the carbon tubes. Since the empty space inside the tubes provides adequate space for Si expansion, there are no changes of the interface between the electrode and electrolyte. As a result, the stable SEI can be retained during cycling. Therefore, encapsulating of Si nanoparticle in a carbon tube can address problems of mechanical instability, current collector contact, and unstable SEI challenges for the Si electrode. The electrochemical test demonstrated that a high capacity of 500 mAh g^{-1} can be delivered at a current density of 5 A g^{-1}. The capacity of Si/C nanofiber is as high as 72.6% of the theoretical capacity after 50 cycles (Figure 28.11d). At higher charge–discharge current density, the capacity of Si/C nanofiber is also much higher than that of counterparts, verifying the relatively good structural stability that ensures repeated alloying/de-alloying of Si with lithium ions at high rates without structural collapse (Figure 28.11e).[45]

Wu et al. also reported a 0D Si nanoparticle encapsulated in hollow graphitized carbon nanofibers. The process flow is described in Figure 28.12a. Firstly, Si nanoparticles were mixed with silicon dioxide precursor solution (TEOS). Secondly, Si nanoparticles were embedded in Si nanofibers by electrospinning process. Finally, a thin layer of carbon was coated onto the composite nanofiber by thermal carbonization of the polystyrene. The carbon-coated fibers were etched in HF aqueous solution to remove the SiO$_2$. Thus, a structure with Si nanoparticles encapsulated inside continuous hollow carbon tubes with adequate empty space for volume expansion was synthesized. The hollow nature of the nanofibers and the small size of

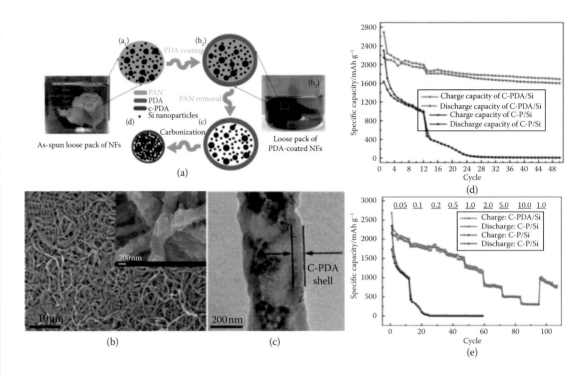

Figure 28.11 (a) An illustration showing the preparation process and the cross-sectional view of the morphology of a typical single-hybrid nanofiber produced in each step. Morphologies of the C-PDA–Si NFs by FESEM (b) and TEM (c). (d) Cycling performance at 50 (first to twelfth cycle) followed by 100 (thereafter) mA g⁻¹, and (e) rate charging capacity of the C-P–Si NFs and C-PDA–Si NFs (current density: A g⁻¹). (From Kong, J., et al. *Nanoscale*, 5(7), 2967–2973, 2013. With permission.)

the Si nanoparticles can be seen clearly in Figure 28.12b and c. As shown in Figure 28.12e, the first-cycle reversible lithium extraction capacity of the SiNP/CT structure including all the Si and carbon mass is 969 mAh g⁻¹ at a charge/discharge current density of 1 Ag⁻¹. At higher charge/discharge current densities ranging from 0.8 to 8 Ag⁻¹, high and stable capacities of 1000 to 700 mAhg⁻¹ of the electrodes are demonstrated (Figure 28.12f).[46]

28.3.2 1D Si NANOWIRE ENCAPSULATED IN 1D HOLLOW CARBON TUBE

Although the continuous 1D carbon matrix can directly connect to the current collector for those 0D silicon/1D carbon tube composites, the electrochemical performance is drastically plagued by the point-to-point interface contact mode between Si and C. This contact mode may, firstly, retard the fast transport of both electrons and lithium ions from/to the Si and, secondly, make the contact frequently ineffective due to the accidental change of the Si/C interface during cycling, thus forming a barrier for the electron and charge transport. To avoid this problem, Wang et al. have developed a 1D Si/1D carbon hybrid, in which the Si nanowires are encapsulated in the graphitic tubes (Figure 28.13a–d). This unique 1D/1D hybrid structure possesses built-in void space in between the wire and the tube, and at the meantime holds line-to-line contact between Si and C components, which not only allows for the free expansion of Si without rupturing the graphitic tubes but also enable the formation of highly efficient channels for the fast transport of both electrons and lithium ions when being used as the electrodes in LIBs. As a result, this structure exhibits good rate capability (1200 mAh g⁻¹ at 12.6 A g⁻¹) and remarkable cycling stability (1100 mAh g⁻¹ at 4.2 A g⁻¹ over 1000 cycles) (Figure 28.13e–g).[47]

Table 28.2 summarizes the preparation methods and electrochemical performances of the YSCVS composites. Firstly, it is worth to note that the Si cores in YSCVS are mostly nanoscale. However, the large volume change of the crystalline Si during the electrochemical reaction can cause GPa levels of stress to build up,[48] especially at high charge rates. Nanostructures of the Si cores are, therefore, necessary to

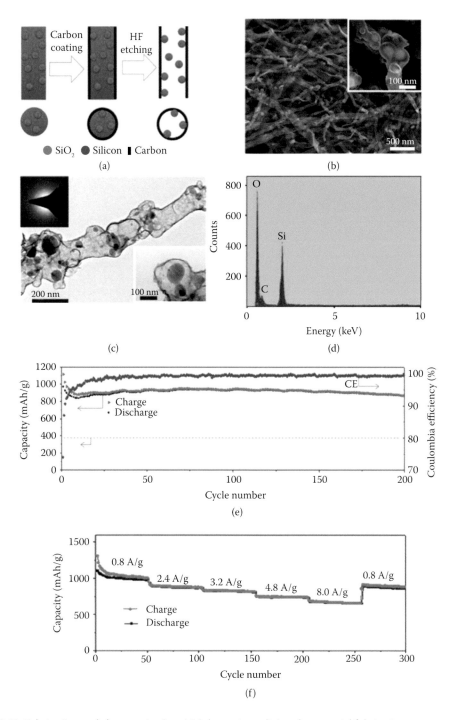

Figure 28.12 Fabrication and characterization. (a) Schematic outlining the material fabrication process. Si nanoparticles in SiO₂ nanofibers were first prepared by electrospinning. After carbon coating and removal of SiO₂ core, the Si nanoparticles encapsulated in carbon tubes (SiNP/CT) structure were obtained. (b) SEM images of the synthesized SiNP/CT. (c) TEM images of the synthesized samples. Lower inset shows TEM image with higher magnification; and upper inset shows SAED pattern of the sample. (d) EDS spectrum of the synthesized SiNP/CT samples. Electrochemical characterizations of the as-synthesized SiNP/CT. (e) Charge–discharge cycling test of the SiNP/CT electrodes at a current density of 1 A g⁻¹, showing 10% loss after 200 cycles. Dashed line indicates the theoretical capacity of traditional graphite anode. (f) Capacity retention of the SiNP/CT electrode cycled at various current densities ranging from 0.8 to 8 A g⁻¹. (From Wu, H., et al. *Nano Lett.*, 12(2), 904–909, 2012. With permission.)

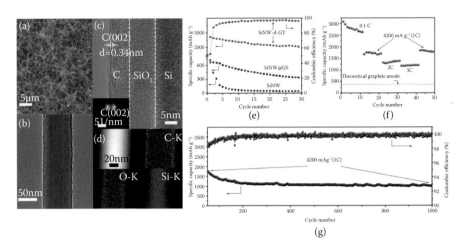

Figure 28.13 The morphology and structure of the SiNW/SiO₂/GS nanocables. (a) SEM image and (b, c) TEM images of the individual SiNW/SiO₂/GS nanocable. The fast Fourier-transformed electron diffraction (FFTED) pattern inserted in (c) is related to the graphitic character of the carbon sheath. (d) Scanning transmission electron microscopy (STEM) and elemental mapping images of a fraction of individual nanocable. Electrochemical characteristics of the SiNW-d-GT electrode. (e) Comparison of capacity retention of different electrodes. The current rate is 0.5C for both SiNW-d-GT and SiNW/GS, and 0.2C for SiNW. The Coulombic efficiency (CE) for SiNW-d-GT is exhibited as well. (f) Charge (delithiation) capacities during cycling at various rates as marked. (g) Capacity and CE of SiNW-d-GT continually cycled at a constant current rate of 1C (4200 mA g⁻¹). (From Wang, B., et al. *Adv. Mater.*, 25(26), 3560–3565, 2013. With permission.)

be adopted to relieve the stress at the surfaces and provide necessary void space for expansion.[49] Ex situ experiments have shown that cracks become less frequent for C–Si pillars in the diameter range of 240–360 nm,[50] and in situ experiments have shown that no cracks for particles with diameters below 150 nm in spite of the very high (minutes or less) lithiation rates.[51] Secondly, the synthesis of the YSCVS could be mainly classified into the following three strategies: (1) removing the intermediate SiO₂ layers of the Si/SiO₂/C precursors,[28,30,32,35–39] (2) directly etching Si cores in core–shell Si/C composites using HF, NaOH, or LiOH solutions,[31,33,43] and (3) etching the intermediate Ni or calcium carbonate templates to obtain the YSCVS.[41] Thirdly, in these papers,[28,30,31,35,43] note that the binders and the additives in the electrolytes also played important roles in improving the lithium-storage capability. Otherwise, the performances of Si-based anodes were greatly limited.[36–38] Therefore, the diameter of the silicon core, the synthesis methods, carbon source, Si content, the kinds of binders, and the additives in the electrolytes all can influence the overall performances of the YSCVS electrodes. Despite all the successes in this field, many synthesis routes and new applications are emerging. It is clear from the many examples discussed in this chapter that the synthesis of the YSCVS still presents significant conceptual challenges. Most of the used methods involved multistep approaches, and work succeeded only on specific cases. One of the most significant challenges is the lack of general high-throughput methods for the preparation of the high-quality YSCVS with controllable sizes, compositions, shapes, and architectures at a low cost.

28.4 3D Si/C YOLK–SHELL STRUCTURE

Although the Si-based anodes with 0D or 1D Si/C yolk–shell structure have exhibited good electrochemical performances for LIBs, there are still some challenges to the widespread application of Si batteries. Firstly, the low-dimensional Si/C yolk–shell composites generally have low tap density, resulting in severe encumbrance for industrial application. Secondly, the silicon–carbon hybrids with the 1D yolk–shell structure usually provide insufficient protection against the electrolyte ingress and the direct contact of Si nanoparticles with the electrolyte, leading to the adverse formation of SEI on the surface of Si nanoparticles. Finally, most syntheses of 1D yolk–shell Si/C hybrids require particular equipment as well as expensive and complex preparation technology, which seriously hinders the extensive use and development of Si-based anodes with 1D Si/C yolk–shell structure.

Table 28.2 Comparison of the preparation methods and performances of the YSCVS composites

REF.	Si SIZED (nm)	CARBON SOURCE	METHODS	Si (%)	VC OR FEC INVOLVED	BINDERS	CUR-RENT (mA g⁻¹)	REVERSIBLE CAPACITY (mAh g⁻¹)
[30]	50	Acetylene	HF–SiO$_2$	30	Yes	CMC	1000	654 (100)
[28]	100	Dopamine	HF–SiO$_2$	71	Yes	Sodium alginate	4000	1500 (1000)
[35]	50	Dopamine	HF–SiO$_2$	48	Yes	CMC	2000	528 (300)
[36]	100	Sucrose	HF–SiO$_2$	28	No	PVDF	50	618 (20)
[37]	100	Polyvinyl-chloride	HF–SiO$_2$	21	No	PVDF	200	350 (20)
[38]	500	Dopamine	NaOH–SiO$_2$	44	No	PVDF	200	530 (100)
[32]	100	Polyvinylidene fluoride	HF–SiO$_2$	64	No	PAA	460	956 (430)
[39]	100	Glucose	HF–SiO$_2$	56	No	PVDF	50	944 (80)
[31]	100	Dopamine	NaOH–Si	75	Yes	CMC	1000	804 (50)
[43]	100	Polyacrylonitrile	HF–Si	65	No	Sodium alginate	250	700 (1000)
[33]	50–100	RF	LiOH–Si	67	No	PVDF	50	628 (100)
[41]	1–3 µm	Polydopamine	FeCl$_3$–Ni	91	Yes	PVDF	2100	1400 (300)
[44]	150	Citric acid (CA)	Inflations of CA	64	No	PVDF	100	1265 (150)
[45]	50	Dopamine hydrochloride	DMF–PAN	47.9	No	No	100	1601 (50)
[46]	100	Polystyrene	HF–SiO$_2$	47	No	None	1000	872 (200)
[47]	50	Methane	HF–SiO$_2$	70	No	PVDF	4200	1100 (1000)

To improve the tap density and limit the SEI formation of the yolk–shell-structured silicon/carbon nanocomposites, significant progress of Si-based anodes with 3D Si/C yolk–shell structure has been achieved. For instance, the pomegranate-like Si structures[52] and granadilla-like silicon-based anode materials[53] have been reported to substantially improve the electrochemical performances of Si-based anodes.

Cui and coworkers[52] reported a bottom-up microemulsion approach for the synthesis of highly spherical silicon pomegranate microbeads with the 3D Si/C yolk–shell structure, as shown in Figure 28.14. Inspired by the structure of a pomegranate, single silicon nanoparticles are encapsulated by a conductive carbon layer, leaving enough room for volume expansion and contraction in the following lithiation and delithiation processes. An ensemble of these hybrid nanoparticles was then encapsulated by a thicker carbon layer in micrometer-size pouches as an electrolyte barrier. Firstly, the carbon component functions not only as a conducting framework but also as an electrolyte-blocking layer, so SEI forms mostly outside the secondary particle. Secondly, the void spaces inside the secondary particle are well defined and evenly distributed around each nanoparticle, effectively accommodating the volume expansion of the silicon without rupturing the carbon shell or changing the secondary particle size.

Benefiting from this hierarchical arrangement, the pomegranate design exhibits outstanding battery performance. Its reversible capacity reached 2350 mAh g^1 at a rate of C/20 (1C=charge/discharge in 1 h). Since silicon is only 77% of the mass of the pomegranate structure, the capacity with respect to silicon is as high as 3050 mAh g^{-1}. From the 2nd to 1000th cycle at a rate of C/2, the capacity retention was more

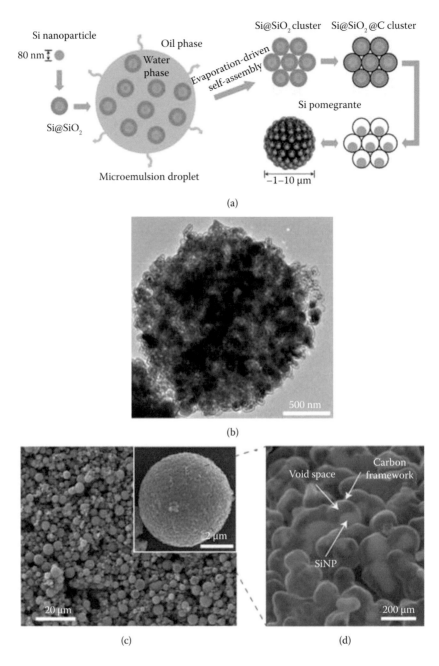

Figure 28.14 (a) Schematic of the fabrication process for Si/C pomegranates. (b) TEM image of one Si-C pomegranate particle. (c) SEM images of Si/C pomegranates showing the micrometer-sized and spherical morphology. (d) Magnified SEM image showing the local structure of silicon nanoparticles and the conductive carbon framework with well-defined void space between.

than 97%. After 1000 cycles, over 1160 mAh g^{-1} capacity remained, more than three times the theoretical capacity of graphite. The cycle stability (0.003% decay per cycle) is among the best cycling performances of silicon anodes reported to date.

Until now, all the previous yolk–shell composites were produced through the complex templating method, in which a SiO$_2$ sacrificial layer was first produced by hydrolysis of the expensive TEOS.[28,52] Then, a carbon coating was formed on the surface of the SiO$_2$ sacrificial layer. At the end, void space could be created by selectively etching SiO$_2$ using HF. Unfortunately, HF solution is not environmental friendly.

In consideration of this fact, inexpensive and environmentally benign methods are highly desirable for the synthesis of yolk–shell-structured silicon anodes.

To settle this issue, Zhang et al.[34] reported a green and facile method to synthesize granadilla-like YSCVS, in which a void space is created between the inside silicon nanoparticle and the outer carbon shell. The Si/C granadillas were produced through a modified templating method, wherein calcium carbonate functioned as a sacrificial layer, acetylene as a carbon precursor, and a dilute hydrochloric acid was employed to etch the calcium carbonate. The as-prepared Si/C granadillas coated with the integrated carbon layers on the outer surface (outer shell) are composed of the interconnected porous carbon network supported yolk–shell carbon@void@silicon anode (CVS) nanobeads with the Si nanoparticle core and the conductive amorphous carbon shell (inside shell), resulting in a unique double carbon-shell structure. Magnified image clearly reveals that all the translucent CVS nanobeads were well coated by the thin and homogeneous outer carbon coatings and supported by a well-connected 3D carbon network. Rationally designed robust 3D carbon framework, interconnected yolk–shell carbon/ silicon nanobeads, and the unique double carbon-shell endow these granadilla-like Si/C composites with excellent electrochemical performance. The granadillas with 30% silicon content deliver a reversible capacity of around 1100 mAh g^{-1} at a current density of 250 mA g^{-1} after 200 cycles. Besides, this composite exhibits an excellent rate performance of about 830 and 700 mAh g^{-1} at the current densities of 1 and 2 A g^{-1}, respectively.

Remarkable progress has been accomplished to coat a stable protection layer to maintain the SEI by combining a conformal coating with internal void space.[53] For example, Si/C yolk–shell nanostructures,[28,47,48,54] pomegranate-like Si structures,[52] and granadilla-like CVS nanobeads[34] have been reported to substantially improve the cycling life of Si-based anodes. However, these structures with void space were all based on Si nanostructures, which introduced new challenges to the practical application of Si-based anode materials.[14,55–60] Firstly, the nanostructured materials are barely packed, which makes it difficult to achieve robust electronic and ionic connections between neighboring nanoparticles. Therefore, there are enormous challenges to achieve high mass loading of active materials and high volumetric capacity. Secondly, the synthesis of most nanostructures require expensive and multistep procedures. For example, silicon nanoparticles are produced by pyrolysis of silane gas, whereas the well-designed void space is introduced using a sacrificial template, which makes these Si nanostructures still too expensive for large-scale practical application.

Considering these concerns, Cui and coworkers designed a nonfilling carbon-coated porous Si microparticle (nC-pSi MP) core–shell structure as the anode material.[61] In this structure, the core is a porous Si microparticle (pSi MP), which is composed of interconnected Si primary nanoparticles, and the shell is a confining carbon layer. This kind of confining carbon layer can allow Li$^+$ to pass through. Moreover, only the outer surface of pSi MP was coated with carbon. The interior pore structures were left unfilled, different from all the demonstrations in the previous literature, where the carbon coating penetrates into the structures. Such a design provides various attractive characteristics for large-volume-change anode materials as follows: (1) a commercially available SiO microparticle source and a simple preparation procedure make the process highly cost-effective and scalable. (2) The interconnected Si primary nanoparticles formed by thermal disproportionation of SiO microparticles ensure the size of the primary Si building blocks is <10 nm, below the critical fracture size. Thermal disproportionation also results in densely packed primary Si nanoparticles, allowing for good electronic conductivity among neighboring particles. (3) Carbon has multiple functions, such as coating the exterior surface of the Si microparticles, preventing electrolyte diffusion into the interior pore space and restricting SEI formation to the outer surface. (4) The nonfilling coating retains enough internal void space to accommodate the volume expansion of Si nanoparticles as well as keeps the carbon shell intact during the electrochemical cycling (Figure 28.15). The nonfilling coating introduces less carbon to the composites not only increasing specific capacity but also increasing the initial CE as a result of less Li$^+$ trapping in amorphous carbon.

In the case of nC-pSi MP structure, the composite structure exhibits excellent cycling stability with high reversible specific capacity (~1500 mAh g^{-1} and 1000 cycles) at the rate of C/4. The nC-pSiMPs involve accurate void space to adopt Si expansion while not losing packing density, which allows for a

Industrial nanosilicon

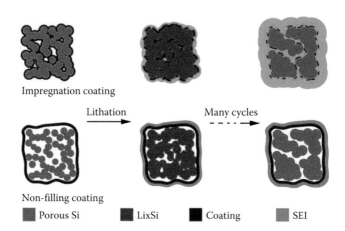

Figure 28.15 Schematic of coating design on pSi MPs and their structural evolution during cycling. For impregnation coating, a carbon layer is coated on each of the Si nanoparticle domains. Upon first cycling, the tremendous volume expansion of Si domains breaks the coating, exposing the silicon surface to the electrolyte, and resulting in excessive SEI formation. For a nonfilling coating, however, carbon only coats the outside of the microparticle, leaving the internal void space for Si expansion. Upon (de)lithiation, the outer carbon layer remains intact. As a result, the SEI outside the microstructure is not ruptured during cycling and remains thin.

high volumetric capacity ($\sim$1000 mAh cm^{-3}). As a result, the anodes can be deeply cycled up to 1000 times with capacity remaining around 1500 mAh g^{-1}. The real capacity can reach higher than 3 mAh cm^{-2} without obvious capacity decay after 100 cycles. In addition, the material synthesis and electrode fabrication processes are simple, scalable, highly reproducible, and compatible with slurry coating manufacturing technology. As a result, the nC-pSi MPs presented here display great promise for future mass production as a high-performance composite anode.

28.5 CONCLUSION

The controllable synthetic strategies of silicon–carbon yolk–shell structures via various routes have been introduced in this chapter. YSCVS composite holds potential for practical applications, because the void space between the outer carbon shell and the Si core reserves the room for volume changes of Si core without deforming the carbon shell and the SEI layer, which in turn allows for the formation of a stable SEI layer on the surface of the outer carbon shell. Undeniably, there are a number of challenges that need to be addressed in the controllable synthesis of these YSCVS. For example: (1) although a substantial series of YSCVS with different directions have been accessed via various templating routes associated with etching, most of them suffer from the complex multistep and tedious process; (2) the low-dimensional Si/C yolk–shell composites generally have low tap density, resulting in severe encumbrance for industrial application; and (3) the design and development of 3D Si/C yolk–shell structures with high mass loading, superior performance, and large tap density has so far proven to be a difficult task via conventional template routes, because of the difficult accessibility of appropriate templates. Up to date, there are few works to solve these issues with promise. Except for selective etching based on templating methods, further development of advanced synthesis methods for 3D Si/C yolk–shell structures would be significantly helpful to push this exploration forward.

REFERENCES

1. Wang Y, He P, Zhou H. Li-Redox Flow batteries based on hybrid electrolytes: At the cross road between Li-ion and Redox flow batteries. *Advanced Energy Materials*, 2012, 2(7): 770–779.
2. Jeong G, Kim Y-U, Kim H, et al. Prospective materials and applications for Li secondary batteries. *Energy & Environmental Science*, 2011, 4(6): 1986–2002.
3. Tarascon JM, Armand M. Issues and challenges facing rechargeable lithium batteries. *Nature*, 2001, 414(6861): 359–367.

4. Balogun M-S, Li C, Zeng Y, et al. Titanium dioxide@titanium nitride nanowires on carbon cloth with remarkable rate capability for flexible lithium-ion batteries. *Journal of Power Sources*, 2014, 272: 946–953.

5. Balogun M-S, Qiu W, Wang W, et al. Recent advances in metal nitrides as high-performance electrode materials for energy storage devices. *Journal of Materials Chemistry A*, 2015, 3(4): 1364–1387.

6. Balogun M-S, Yu M, Huang Y, et al. Binder-free Fe_2N nanoparticles on carbon textile with high power density as novel anode for high-performance flexible lithium ion batteries. *Nano Energy*, 2015, 11: 348–355.

7. Balogun M-S, Yu M, Li C, et al. Facile synthesis of titanium nitride nanowires on carbon fabric for flexible and high-rate lithium ion batteries. *Journal of Materials Chemistry A*, 2014, 2(28): 10825–10829.

8. Xiao LS, Sehlleier YH, Dobrowolny S, et al. Si-CNT/rGO Nanoheterostructures as high-performance lithium-ion-battery anodes. *ChemElectroChem*, 2015, 2: 1983–1990.

9. Zhu G-N, Wang Y-G, Xia Y-Y. Ti-based compounds as anode materials for Li-ion batteries. *Energy & Environmental Science*, 2012, 5(5): 6652–6667.

10. Wu H, Yu G, Pan L, et al. Stable Li-ion battery anodes by in-situ polymerization of conducting hydrogel to conformally coat silicon nanoparticles. *Nature Communications*, 2013, 4: 1–6.

11. Chan CK, Peng H, Liu G, et al. High-performance lithium battery anodes using silicon nanowires. *Nat Nano*, 2008, 3(1): 31–35.

12. Zhang W-J. A review of the electrochemical performance of alloy anodes for lithium-ion batteries. *Journal of Power Sources*, 2011, 196(1): 13–24.

13. Szczech JR, Jin S. Nanostructured silicon for high capacity lithium battery anodes. *Energy & Environmental Science*, 2011, 4(1): 56–72.

14. Wu H, Cui Y. Designing nanostructured Si anodes for high energy lithium ion batteries. *Nano Today*, 2012, 7(5): 414–429.

15. Wu H, Chan G, Choi JW, et al. Stable cycling of double-walled silicon nanotube battery anodes through solid-electrolyte interphase control. *Nature Nanotechnology*, 2012, 7(5): 310–315.

16. Kim H, Seo M, Park M-H, et al. A Critical size of silicon nano-anodes for lithium rechargeable batteries. *Angewandte Chemie*, 2010, 49(12): 2146–2149.

17. Yao Y, Mcdowell MT, Ryu I, et al. Interconnected silicon hollow nanospheres for lithium-ion battery anodes with long cycle life. *Nano Letters*, 2011, 11(7): 2949–2954.

18. Kim H, Han B, Choo J, et al. Three-dimensional porous silicon particles for use in high-performance lithium secondary batteries. *Angewandte Chemie*, 2008, 47(52): 10151–10154.

19. Iwamura S, Nishihara H, Ono Y, et al. Li-Rich Li-Si alloy as a lithium-containing negative electrode material towards high energy lithium-ion batteries. *Scientific Reports*, 2015, 5: 8085.

20. Gao P, Fu J, Yang J, et al. Microporous carbon coated silicon core/shell nanocomposite via in situpolymerization for advanced Li-ion battery anode material. *Physical Chemistry Chemical Physics*, 2009, 11(47): 11101–11105.

21. Cui L-F, Ruffo R, Chan CK, et al. Crystalline-amorphous core–shell silicon nanowires for high capacity and high current battery electrodes. *Nano Letters*, 2009, 9(1): 491–495.

22. Choi N-S, Yew KH, Lee KY, et al. Effect of fluoroethylene carbonate additive on interfacial properties of silicon thin-film electrode. *Journal of Power Sources*, 2006, 161(2): 1254–1259.

23. Kovalenko I, Zdyrko B, Magasinski A, et al. A major constituent of brown algae for use in high-capacity li-ion batteries. *Science*, 2011, 334(6052): 75–79.

24. Ng SH, Wang J, Wexler D, et al. Highly reversible lithium storage in spheroidal carbon-coated silicon nanocomposites as anodes for lithium-ion batteries. *Angewandte Chemie*, 2006, 45(41): 6896–6899.

25. Kang Y-M, Park M-S, Lee J-Y, et al. Si–Cu/carbon composites with a core–shell structure for Li-ion secondary battery. *Carbon*, 2007, 45(10): 1928–1933.

26. Hu Y-S, Demir-Cakan R, Titirici M-M, et al. Superior storage performance of a Si@SiOx/C nanocomposite as anode material for lithium-ion batteries. *Angewandte Chemie*, 2008, 47(9): 1645–1649.

27. Su L, Zhou Z, Ren M. Core double-shell Si@SiO2@C nanocomposites as anode materials for Li-ion batteries. *Chemical Communications*, 2010, 46(15): 2590–2592.

28. Liu N, Wu H, Mcdowell MT, et al. A yolk-shell design for stabilized and scalable li-ion battery alloy anodes. *Nano Letters*, 2012, 12(6): 3315–3321.

29. Liu N, Lu Z, Zhao J, et al. A pomegranate-inspired nanoscale design for large-volume-change lithium battery anodes. *Nature Nanotechnology*, 2014, 9(3): 187–192.

30. Li X, Meduri P, Chen X, et al. Hollow core-shell structured porous Si-C nanocomposites for Li-ion battery anodes. *Journal of Materials Chemistry*, 2012, 22(22): 11014–11017.

31. Pan L, Wang H, Gao D, et al. Facile synthesis of yolk-shell structured Si-C nanocomposites as anodes for lithium-ion batteries. *Chemical Communications*, 2014, 50(44): 5878–5880.

32. Yang LY, Li HZ, Liu J, et al. Dual yolk-shell structure of carbon and silica-coated silicon for high-performance lithium-ion batteries. *Scientific Reports*, 2015, 5: 10908.

33. Su L, Xie J, Xu Y, et al. Preparation and lithium storage performance of yolk-shell Si@void@C nanocomposites. *Physical Chemistry Chemical Physics*, 2015, 17(27): 17562–17565.

34. Zhang L, Rajagopalan R, Guo H, et al. A green and facile way to prepare granadilla-like silicon-based anode materials for Li-ion batteries. *Advanced Functional Materials*, 2016, 26(3): 440–446.

35. Chen S, Gordin ML, Yi R, et al. Silicon core-hollow carbon shell nanocomposites with tunable buffer voids for high capacity anodes of lithium-ion batteries. *Physical Chemistry Chemical Physics*, 2012, 14(37): 12741–12745.

36. Zhou X-Y, Tang J-J, Yang J, et al. Silicon@carbon hollow core–shell heterostructures novel anode materials for lithium ion batteries. *Electrochimica Acta*, 2013, 87: 663–668.

37. Iwamura S, Nishihara H, Kyotani T. Effect of buffer size around nanosilicon anode particles for lithium-ion batteries. *The Journal of Physical Chemistry C*, 2012, 116(10): 6004–6011.

38. Ru Y, Evans DG, Zhu H, et al. Facile fabrication of yolk-shell structured porous Si-C microspheres as effective anode materials for Li-ion batteries. *RSC Advances*, 2014, 4(1): 71–75.

39. Sun Z, Song X, Zhang P, et al. Controlled synthesis of yolk-mesoporous shell Si@SiO$_2$ nanohybrid designed for high performance Li ion battery. *RSC Advances*, 2014, 4(40): 20814–20820.

40. Sun Z, Tao SY, Song X, et al. A silicon/double-shelled carbon yolk-like nanostructure as high-performance anode materials for lithium-ion battery. *Journal of The Electrochemical Society*, 2015, 162(8): A1530–A1536.

41. Li Y, Yan K, Lee H-W, et al. Erratum: Growth of conformal graphene cages on micrometre-sized silicon particles as stable battery anodes. *Nature Energy*, 2016, 1: 16017.

42. Park Y, Choi N-S, Park S, et al. Si-encapsulating hollow carbon electrodes via electroless etching for lithium-ion batteries. *Advanced Energy Materials*, 2013, 3(2): 206–212.

43. Pang C, Song H, Li N, et al. A strategy for suitable mass production of a hollow Si@C nanostructured anode for lithium ion batteries. *RSC Advances*, 2015, 5(9): 6782–6789.

44. Li H-H, Wang J-W, Wu X-L, et al. A novel approach to prepare Si/C nanocomposites with yolk-shell structures for lithium ion batteries. *RSC Advances*, 2014, 4(68): 36218–36225.

45. Kong J, Yee WA, Wei Y, et al. Silicon nanoparticles encapsulated in hollow graphitized carbon nanofibers for lithium ion battery anodes. *Nanoscale*, 2013, 5(7): 2967–2973.

46 Wu H, Zheng G, Liu N, et al. Engineering empty space between Si nanoparticles for lithium-ion battery anodes. *Nano Letters*, 2012, 12(2): 904–909.

47. Wang B, Li X, Zhang X, et al. Contact-engineered and void-involved silicon/carbon nanohybrids as lithium-ion-battery anodes. *Advanced Materials*, 2013, 25(26): 3560–3565.

48. Sethuraman VA, Chon MJ, Shimshak M, et al. In situ measurements of stress evolution in silicon thin films during electrochemical lithiation and delithiation. *Journal of Power Sources*, 2010, 195(15): 5062–5066.

49. Goldman JL, Long BR, Gewirth AA, et al. Strain anisotropies and self-limiting capacities in single-crystalline 3D silicon microstructures: Models for high energy density lithium-ion battery anodes. *Advanced Functional Materials*, 2011, 21(13): 2412–2422.

50. Lee SW, Berla LA, Mcdowell MT, et al. Reaction front evolution during electrochemical lithiation of crystalline silicon nanopillars. *Israel Journal of Chemistry*, 2012, 52(11–12): 1118–1123.

51. Liu XH, Zhong L, Huang S, et al. Size-dependent fracture of silicon nanoparticles during lithiation. *ACS Nano*, 2012, 6(2): 1522–1531.

52. Liu N, Lu Z, Zhao J, et al. A pomegranate-inspired nanoscale design for large-volume-change lithium battery anodes. *Nature Nanotechnology*, 2014, 9(3): 187–192.

53. Obrovac MN, Chevrier VL. Alloy negative electrodes for Li-ion batteries. *Chemical Reviews*, 2014, 114(23): 11444–11502.

54. Li X, Gu M, Hu S, et al. Mesoporous silicon sponge as an anti-pulverization structure for high-performance lithium-ion battery anodes. *Nature Communications*, 2014, 5: 4105.

55. Chan CK, Peng H, Liu G, et al. High-performance lithium battery anodes using silicon nanowires. *Nature Nanotechnology*, 2008, 3(1): 31–35.

56. Liu G, Xun S, Vukmirovic N, et al. Polymers with tailored electronic structure for high capacity lithium battery electrodes. *Advanced Materials*, 2011, 23(40): 4679.

57. Magasinski A, Dixon P, Hertzberg B, et al. High-performance lithium-ion anodes using a hierarchical bottom-up approach. *Nature Materials*, 2010, 9(5): 461–461.

58. Park M-H, Kim MG, Joo J, et al. Silicon nanotube battery anodes. *Nano Letters*, 2009, 9(11): 3844–3847.

59. Evanoff K, Magasinski A, Yang J, et al. Nanosilicon-coated graphene granules as anodes for Li-ion batteries. *Advanced Energy Materials*, 2011, 1(4): 495–498.

60. Xu Y, Liu Q, Zhu Y, et al. Uniform nano-Sn/C composite anodes for lithium ion batteries. *Nano Letters*, 2013, 13(2): 470–474.

61. Lu Z, Liu N, Lee H-W, et al. Nonfilling carbon coating of porous silicon micrometer-sized particles for high-performance lithium battery anodes. *ACS Nano*, 2015, 9(3): 2540–2547.

Nanosilicon for quantum information

Davide Rotta and Enrico Prati

Contents

29.1 INTRODUCTION

Silicon has raised an increasing interest in the development of future devices for quantum information processing. Quantum computing consists of the implementation of quantum algorithms on a physical substrate made of interacting qubits, the quantum version of digital bits. Quantum algorithms are those algorithms conceived to solve hard computational problems, such as prime factorization of large integers and minimization of multidimensional functionals based on the encoding of the information by quantum states. In the simplest version, the quantum states arise from two-level systems and they provide the basis of the physical qubits. Such a physical basis is provided by quantum objects, like individual particles or superconducting currents, enabling the existence of a ground and an excited state, which can live simultaneously in a quantum superposition weighted by complex numbers. The possibility to encode the information in a superposition of states and to exponentially increase the number of possible states by using many qubits leads to the intrinsic parallelism that can be exploited by quantum algorithms (DiVincenzo 1995).

Therefore, if on one hand quantum computing is empowered by such quantum states, at the same time it is affected by their fragility as quantum coherence is undermined by unavoidable noise and interferences. Depending on the physical system employed to encode the physical qubits, the typical decoherence time of the qubit in the best working conditions may range from picoseconds to milliseconds, which would make quantum computation based on such physical qubits unfeasible. The quantum state cannot on average survive long enough to be carried across space and during time to complete its task. Part of the correlated and systematic errors are cancelled by a virtual layer by implementing methods such as the Hahn spin echo sequence (Hahn 1950), dynamical decoupling (Viola et al. 1999) and decoherence-free subspaces (Grace et al. 2006). Even if such virtual qubits, constituted by one or more physical qubits equipped with control electronics, are useful for extending the lifetime of the quantum states, the latter is still insufficient to develop fault-tolerant quantum information processing. To overcome such an issue, not longer after the proposal of the first quantum algorithms, practical quantum information processing was made possible by the discovery of quantum error correction codes (Preskill 1998). Such correction codes employ several virtual qubits to double check that the original state has been preserved without destroying the information, and by recovering it in the case where an error has occurred. There is a variety of different quantum error correction codes, but in general at least a factor of 10 more physical qubits are needed to ensure that quantum information survives for sufficient time to carry out the computation. The circuit of virtual qubits employed to manage the quantum information from the input time is called a logical qubit. Even simple algorithms and quantum simulation of molecules require millions to billions of physical qubits (Jones et al. 2012), so it appears that only scalable and serial fabrication manufacturing can assess such need. This is the point where silicon technology comes in. Even if currently the superconductor architecture has led to the first successful implementation of quantum information processing in systems of hundreds of qubits (Barends et al. 2016), it is likely that in the future only a silicon platform will address such a high-demanding number of qubits. Indeed, while superconducting qubits are constituted by micron-scale rings, making the control of coherence across space potentially difficult when billions of qubits will be required, silicon qubits can be packed at a length scale of tens of nanometers thanks to the most advanced technology nodes. Silicon carries an additional benefit in addition to its capability for large-scale fabrication and control by multimillion gate lines, which is the relatively long coherence time of ^{28}Si, where the disturbance of nuclear spins of natural silicon is removed extending the decoherence time up to 10 seconds for isotopically purified silicon at 1.8 K (Morton et al. 2011; Veldhorst et al. 2014; Tyryshkin et al. 2012). Silicon offers at least three different methods to encode qubits, each carrying advantages and disadvantages. On a physical basis, the multiple spin states of pairs or triplets of electrons, of holes, and also nuclear spins of impurity atoms such as individual P atoms can be employed. The successful encoding of physical qubits in silicon is making it more than just a promising material for quantum computing.

29.1.1 QUANTUM INFORMATION PROCESSING: BASICS

Quantum information processing involves all the techniques to process quantum information stored in a physical system from the physical to the virtual layer, the quantum error correction, the logical, and finally the application layer (Jones et al. 2012). In other words, it embraces all the methodologies to exploit quantum systems for solving computational problems. Although significantly different from classical computation, as it is applied to quantum systems, quantum information processing has some analogies with classical computation.

Classical computation was envisaged as a digital computation paradigm, meaning that the basic unit of classical information is a binary unit, namely a bit, which is generally defined as a memory register with a state variable assuming only two possible values: either 0 or 1.

Multiple bit registers can be used to encode numbers, letters, or symbols, according to the binary numerical system or the ASCII code for example, or more complex information such as geographic coordinates or multimedia files.

In this framework, information is processed by computers by applying a sequence of logic gates (namely the Boolean operations AND, OR, XOR, NAND) to specific bit states.

Such a sequence builds up an algorithm, with the aim to obtain the desired output solving specific computational problems, spanning from simple arithmetic functions to simulations and more complex data processing.

Typically, a bit is physically defined in electronic devices by two distinct operating voltages (e.g., dynamic random-access memory [DRAM]) or resistance (e.g., resistive random-access memory [ReRAM]) values. Other technologies, such as photonics and spintronics, employ alternative state variables, namely light intensity or magnetic polarization. Such state variables are mesoscopic variables, arising from the collective behavior of a large number of particles, which can be described to a great extent by means of classical

Industrial nanosilicon

mechanics laws. Differently, quantum systems involving for example an isolated conduction electron bound to a semiconductor quantum dot, obey quantum mechanics.

Let us consider a two-level quantum system defined by the Hamiltonian $\hat{H}$. The system eigenstates $|0\rangle$ and $|1\rangle$ are the solutions of the Schrödinger equation with eigenvalues E_0 and E_1, respectively:

$$\hat{H}|0\rangle = E_0|0\rangle$$
$$\hat{H}|1\rangle = E_1|1\rangle \tag{29.1}$$

Two only eigenstates constitute the background for a quantum bit (or qubit), which has a fundamental difference from classical bits; besides the two eigenstates $|0\rangle$ and $|1\rangle$ its state can be every possible linear superposition of such two states:

$$|\psi\rangle = \alpha|0\rangle + \beta|1\rangle \tag{29.2}$$

where α and β are complex coefficients bound to the normalization condition: $|\alpha|^2 + |\beta|^2 = 1$. As a consequence, the classical domain defined by only two states is replaced in quantum information processing by a Hilbert vector space generated by such two states.

The qubit state can be equivalently mapped in polar coordinates:

$$|\psi\rangle = \cos(\theta)|0\rangle + e^{i\phi}\sin(\theta)|1\rangle \tag{29.3}$$

so that a graphical representation is possible in terms of the Bloch sphere, depicted in Figure 29.1.

It is evident here how the states of a classical bit reduce to just the two poles on the sphere surface representing the qubit operating space. From another point of view, two complex numbers are required to describe a qubit state instead of just a single integer number, namely 0 or 1, needed for a bit.

This consideration suggests that a larger content of information is potentially available in a quantum system, which has important consequences when considering many-qubit systems. In fact, the state space of 2 qubits ζ is the vector product of all the qubit state spaces ζ_i, so $\zeta = \zeta_1 \otimes \zeta_2$ and the system is described by such a state:

$$|\psi\rangle_{12} = \left(\alpha_1|0\rangle_1 + \beta_1|1\rangle_1\right) \otimes \left(\alpha_2|0\rangle_2 + \beta_2|1\rangle_2\right) \tag{29.4}$$

or equivalently:

$$|\psi\rangle_{12} = c_1|00\rangle + c_2|10\rangle + c_3|10\rangle + c_4|11\rangle \tag{29.5}$$

where the two numbers inside kets synthetically denote the states of the first and second qubit respectively, and c_i are complex coefficients with the condition: $\Sigma|c_i|^2 = 1$.

This means that a two-qubit state is generally a superposition of all the four two-qubit eigenstate: the information stored by the two-qubit system includes all the possible results due to the quantum superposition principle.

The quantum mixture of eigenstates can be manipulated to perform some algorithms and it can be maintained unless such state is measured, since measurement makes the system collapse irreversibly in one of the four eigenstates with corresponding probability $|c_i|^2$

Generalizing to N qubits, a typical state will be of the form:

$$|\psi\rangle_{123\ldots N} = c_1|000\ldots0\rangle + c_2|000\ldots1\rangle + \ldots + c_{2^N}|111\ldots1\rangle \tag{29.6}$$

Industrial nanosilicon

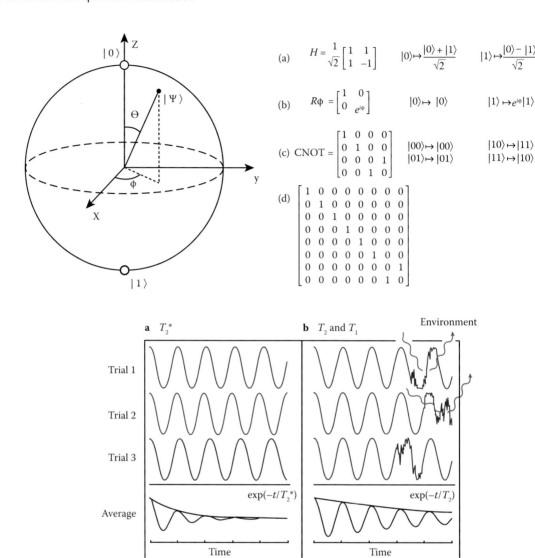

Figure 29.1 Top left: Representation of a general qubit state on the Bloch sphere. Top right: Example of qubit logic gates and corresponding effect on the logical basis states: one-qubit Hadamard (a) and phase gate (b), two-qubit controlled-NOT gate (c), and three-qubit Toffoli gate. Bottom: Schematic depiction of the physical phenomena underlying dephasing (a) and decoherence (b). (a) An oscillator with frequency varying by trial averages to an oscillation decaying with apparent dephasing timescale T_2^*. (b) A quantum oscillator interacting with the environment may have phase-kicks in a single trial that harm coherence in quantum computation, and lead to an average decay process of timescale T_2 typically longer than T_2^*. Relaxation processes are similar, and cause decay on the timescale $T_1 \geq T_2/2$. (Bottom picture taken from Ladd, T.D., et al., *Nature*, 464(7285), 45–53, 2010. With permission.)

Remarkably, if we were able to apply a logic gate (a NOT for example) on such a register as typically done in classical algorithms, the result would be an N-qubit state which is the superposition of the same gate applied to each of the 2^N classical states.

Quantum algorithms exploit this kind of parallelism, which turns out to be extremely useful in particular computational problems, such as the factorization of large numbers (computational resources for the Shor's quantum algorithm follow polynomial scaling with respect to the number digits instead of exponential) or minimization problems including quantum simulation of multiatomic molecules or a research in a database through Grover's algorithm. Thus, the task of a good quantum algorithm is to select the desired result among all the possible ones through a proper sequence of logic gates between the qubits, which ends in destructive interference for the wrong results and constructive for the sought-after ones.

To induce oscillations between the two states of a qubit and transform one into the other, one may induce evolution by the coupling with an external applied field. The simplest example is given by an individual spin coupled with an oscillating magnetic field

$$\boldsymbol{B} = B_0\boldsymbol{z} + B_1\left(\cos(\omega t)\boldsymbol{x} + \sin(\omega t)\boldsymbol{y}\right) \tag{29.7}$$

It is possible to demonstrate that the transition probability between the two states is:

$$P(0 \to 1) = \left(\frac{\omega_1}{\Omega}\right)^2 \sin^2\left(\frac{\Omega t}{2}\right) \tag{29.8}$$

where $\Omega = [(\omega - \omega_0)^2 + \omega_1^2)]^{1/2}$ and $\omega_i = \gamma B_i$ where γ is the gyromagnetic ratio. Such time-varying probability of finding the qubit in the two states is called Rabi oscillation.

29.1.2 UNIVERSAL QUANTUM COMPUTING

The power of quantum information processing currently relies mainly on two different quantum computing models. On the one hand, there is universal digital quantum computing, also called circuital quantum computing, based on a set of time-reversible quantum logic ports required to cover the spectrum of all possible unitary operations on the qubits. It is possible to show that there are discrete sets of quantum gates which can be used to construct, to arbitrary accuracy, any unitary quantum circuit on n qubits, which actually makes this set of gates universal. According to the Kitaev–Solovay theorem, under some technical conditions, every universal set of quantum gates is equivalent to every other set of quantum gates. Three common universal sets of quantum logic ports are: a triplet of gates consisting of the Hadamard gate and a phase rotation gate, both acting on a single qubit, and the controlled-NOT (CNOT) gate acting on two qubits; alternatively, the Deutsch gates, acting on three qubits; finally, reversible Toffoli gates which can be reduced to Deutsch gates. The matrix representation of such gates is shown in Figure 29.1.

On the other hand, there are analog quantum computers. Quantum annealing has been proposed as the quantum analogue of simulated annealing to solve optimization problems that can be reduced to finding the ground state of an Ising spin system. Such problems are of the quadratic unconstrained binary optimization (QUBO) type (Wang et al. 2016). In simulated annealing, thermal fluctuations are simulated on a traditional CPU to let the system hop from state to state over intermediate energy barriers to search for the desired lowest-energy state. In quantum annealing, quantum-mechanical tunneling through the energy barriers replaces and is supposed to outperform thermal fluctuations. Similar to simulated annealing, quantum annealing is a generic algorithm, applicable, in principle, to any QUBO problem. It provides a method to reach a solution of a specified optimality level within a given finite number of annealing runs. By tuning the local fields and coupling strengths of the Ising system and by running the evolution to the solution sufficiently slowly (adiabatically, as quantitatively described by the adiabatic theorem [Morita and Nishimori 2008]), the system moves from the ground state of the transverse field to the ground state of the coupling strength system, which represents the problem to be solved. The existing example of D-Wave machines manages eight superconducting qubit registers but at the same time their multi-micrometric size limits the coherence across many registers (Ladd et al. 2010).

Adiabatic quantum computers are a subclass of quantum annealing processors, where a two-dimensional (2D) array of interacting qubits described by generic Hamiltonian dominated by the kinetic terms is slowly evolved from its ground state to the ground state of the final interacting Hamiltonian encoding the computational problem. Also adiabatic quantum computing is universal for a class of problems, namely for quantum Merlin Arthur (QMA) complexity class constituted by the quantum analogue of the nonprobabilistic complexity class NP. The Hamiltonian of a QMA-complete problem can be expressed as a 2D grid of qubits (Oliveira and Terhal 2005).

A digital quantum computer is more suitable to run codes, while adiabatic quantum computing is suited to running optimization problems. Hybrid examples employing both, realized by using superconducting qubits have also been reported (Barends et al. 2016). It is worth noting that, for its formal analogy with

Hopfield networks employed in machine learning, quantum annealing is a promising method for the implementation of quantum deep-learning algorithms.

29.1.3 DIVINCENZO CRITERIA

After a brief introduction of the main theoretical concepts underlining quantum information processing, we now focus on its physical implementation.

Although several systems can be considered as possible candidates to store and manipulate quantum information, there is not apparently a unique "perfect" system for the implementation of qubits. In fact, on the one hand the need to preserve the coherence of quantum states points to systems that are well isolated from noisy environments. On the other hand, good isolation against unwanted noise from the environment normally also means that a controlled manipulation of the qubit is more difficult and generally slower.

In order to better address the requirement for physical implementation of quantum information processing, D. P. DiVincenzo systematically reported the following five criteria (DiVincenzo 2000):

1. A scalable and well-characterized qubit—the physical system that defines the qubit must be a two-level physical system obeying quantum mechanics. Its behavior (relevant parameters) and time evolution (Hamiltonian) must be known and under control. Furthermore, scalability requires that an N-qubit generates a 2^N-dimensional vector space, which is the core of the potential of quantum computation against classical.

2. Initialization of the qubits to a fiducial initial state—the qubit states must be set in a precise quantum state at the beginning of the computation with high fidelity. Basically, even error-free computation would lead to wrong results if starting from an incorrect initialization. Proper initialization is generally accomplished by letting the system relax to its ground state or by driving the system in a default state through specific selection rules such as Coulomb or Pauli blockade (Kouwenhoven and Marcus 1997; Kouwenhoven et al. 2001; van der Wiel et al. 2002; Hanson et al. 2007).

3. Quantum state coherence much longer than gate operation time—unitary evolution of the qubits must be preserved during computation, so interaction with the environment must be limited in order to avoid bit-flip errors as happens in classical computers. Furthermore, the introduction of unwanted coupling terms in the Hamiltonian of the systems (e.g., electrostatic or magnetic field fluctuations, and thermal fluctuations) or their time-dependent fluctuations lead to a drifted evolution of the qubit states and consequently to the accumulation of phase errors as shown in Figure 29.1 In detail, interaction with the environment eventually induces relaxation from the excited states to the ground state on a time scale T_1, and random phase shifts to the ideal system time evolution on a time scale $T_2 < 2\ T_1$ where the mentioned T_1 and T_2 are called relaxation time and decoherence time, respectively, from spin resonance terminology. Moreover, since qubit operations are benchmarked after being averaged over many experiments, variability of the qubit parameters among different trials similarly causes loss of quantum information on a time scale T_2^* called dephasing time, which is generally faster than T_2. Consequently, logic gates must be performed much faster than decoherence mechanisms.

4. A universal set of gates—quantum information processing with N qubits requires a set of logic gates spanning the entire Hilbert space defined by such qubits. It can be demonstrated that the capability to perform single qubit rotations on each qubit and a two-qubit gate between each couple of qubits suffices to provide a complete set. In the case of quantum annealing, this criterion may be replaced by the request of accurate adiabatic operations and controlled interaction between qubits.

5. Qubit-specific readout of the final state—the final state of each qubit must be correctly collected in the end of the algorithm. For example, if the final qubit state is $|\psi\rangle = \alpha|0\rangle + \beta|1\rangle$ then the measurement outcome will be either eigenvalue 0 with probability $|\alpha|^2$ or eigenvalue 1 with probability $|\beta|^2$.

From a technological point of view, the first criterion about scalability also involves the capability to build up a large number of identical qubits (at least those required for a specific algorithm implementation) with limited variability of the qubit figures of merit and fair tolerance against process variability.

Moreover, the inherent quantum character of qubits makes them more fragile against noise and error prone compared to classical bits so that Criterion 3 is normally unfulfilled and quantum error correction techniques are actually mandatory to detect and correct errors that cannot be completely avoided during computation

(Preskill 1998; Gottesman 2009; Devitt et al. 2013). Such techniques require a large number of ancilla qubits and are the most demanding part of quantum algorithms in terms of space as well as time resources. As a result, realistic quantum computers will be composed of millions to few billions of physical qubits depending on the problem size, including data qubits and ancillae as well as a number of communication qubits for the coherent transfer of quantum information (Copsey and Oskin 2003; Taylor et al. 2005; Rotta et al. 2014). Each qubit must be individually controlled by dedicated classical control circuitry that will have to meet stringent demands of high density and operating speed with limited noise and power consumption.

Several architectures were proposed and to some extent demonstrated for quantum information processing (Ladd et al. 2010), including qubits based on photons (O'Brien 2007), superconductors (Devoret and Schoelkopf 2013), neutral atoms or ion traps (Monroe and Kim 2013), and solid-state implementation such as atoms and quantum dots (Buluta et al. 2011; Awschalom et al. 2013). Nevertheless, after the previous considerations about large-scale circuits, semiconductor qubits turn out to be natural hosts for quantum computation and in particular Si qubits could take advantage of better coherence properties due to a reduced density of nuclear spins on the one hand, and the strong technological background developed by classical nanoelectronics (Morton et al. 2011).

29.2 FABRICATION

Silicon qubits have been realized in research labs and more recently by adopting a preindustrial process compatible with an industrial complementary metal-oxide semiconductor (CMOS) process. This section describes the different physical implementations of silicon qubits based on donors and quantum dots, and the corresponding fabrication schemes.

29.2.1 Si/SiGe HETEROSTRUCTURES QUANTUM DOTS

One of the most popular methods to create Si quantum dots is inherited from previous research on III–V semiconductors (Koppens et al. 2006) and makes use of depletion gates to design and confine submicron semiconducting regions from a 2D electron gas.

A high mobility 2D quantum well is built by epitaxial growth of a thin stressed Si film sandwiched between two SiGe barriers (Ge content is normally 20%–30%). Epitaxial growth of Si/SiGe heterostructures results in defect-free and sharp interfaces, characterized by extremely high mobility in the range of $10^4 \, cm^2/Vs$, which is an almost ideally clean environment for Si quantum dots. A 2D electron gas is then accumulated within the well, supplied by a buried delta-doping layer, by electrostatic accumulation via a global top gate or by photogeneration of carriers at low temperature (Simmons et al. 2009; Maune et al. 2012).

The quantum dots that define the qubit are finally defined by applying negative voltage to some depletion top gate electrodes defined by lithography as in the example of Figure 29.2. Such gates drain carriers from specific regions of the 2D gas, building up the in-plane confinement required to form a few-electron quantum dot.

29.2.2 PLANAR Si-SiO$_2$ METAL-OXIDE SEMICONDUCTOR QUANTUM DOTS

Although Si/SiGe heterostructures yield high-quality Si quantum dots, their fabrication process involves slow and expensive thin-film deposition processes like molecular beam epitaxy (MBE). To this extent, the use of CMOS-compatible processes on bare Si wafers is preferable. Several works demonstrated the feasibility of Si quantum dots within a metal-oxide semiconductor (MOS) stack device with similar geometry to that utilized in Si/SiGe heterostructures (Angus et al. 2007; E. Nordberg et al. 2009a; Nguyen et al. 2013; Hao et al. 2014). In this case, the gate oxide is grown on the Si substrate by deposition of an insulating thin film (high-k oxides e.g., Al_2O_3 or HfO_2 in widespread use in microelectronics are normally deposited by chemical vapor deposition (CVD) or atomic layer deposition) or by thermal oxidation (resulting in high-quality SiO_2). Finally, lithographically defined depletion metal gates are deposited by ordinary physical vapor deposition techniques, namely e-beam evaporation.

The final device has similar behavior as the aforementioned Si/SiGe quantum dot, but no quantum well is formed and the 2D electron gas is accumulated at the oxide interface by field-effect through a global top gate. Since SiO_2 is amorphous, dedicated processing is needed to reduce the density of interface defects down to acceptable values (for example a density of $10^{-11} \, cm^{-2}$ corresponds to an average interdefect distance of about 30 nm) (E. P. Nordberg et al. 2009b). Notably, the use of Al for metal gates also enables even more complex device layouts (see for example Figure 29.3) featuring multiple levels of metal gates (which are isolated by partial oxidation of the gate themselves) to improve the gate density and the consequent electrostatic control of the quantum dots (Lim et al. 2009; Lai et al. 2011; Lim et al. 2011; Yang et al. 2012; Yang et al. 2013; Veldhorst et al. 2014).

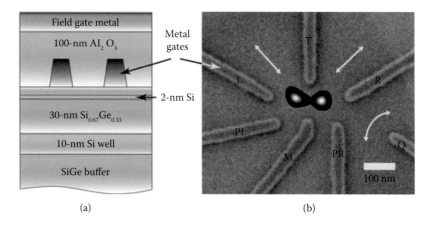

(a)　　　　　　　　　　　(b)

Figure 29.2 (a) Cross-section of a double quantum dot device based on Si/SiGe technology showing undoped heterostructure, dielectric, and gate stack. (b) Top view scanning electron micrograph of the corresponding device before dielectric isolation and field gate deposition. A numerical simulation of the electron density for the two-electron (1,1) state is superimposed on the micrograph. (Picture taken from Maune, B.M., et al., *Nature*, 481(7381), 344–347, 2012. With permission.)

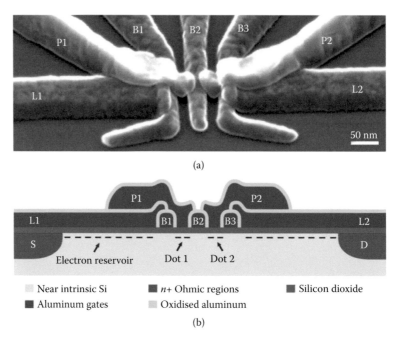

Figure 29.3 (a) Scanning electron micrograph of a CMOS quantum dot device with three superimposed metal gate layers. (b) Schematic cross-section view of the Si MOS double quantum dot. The architecture is defined by B1, B2, and B3 (barrier gates), L1 and L2 (lead gates), and P1 and P2 (plunger gates). The gates are separated by an Al_2O_3 layer (light gray). Positive voltages applied to the lead and plunger gates induce an electron layer (black dashes) underneath the SiO_2. By tuning the barrier gates, Dot 1 and Dot 2 are formed. The coupling of the dots is adjusted using the middle barrier (B2). The regions colored with red are the n+ source (S) and drain (D) contacts formed via diffused phosphorus. (Picture taken from Lai, N.S., et al., *Sci. Rep.*, 1, 110, 2011. With permission.)

29.2.3 Si-NANOWIRE QUANTUM DOTS AND PRE-INDUSTRIAL CMOS-COMPATIBLE PROCESSES

Besides in-plane confinement of a 2D electron gas, an alternative method to provide 0D confinement in Si is by limiting the length of a nanowire through electrostatic gating.

High-quality Si and Si–Ge core-shell nanowires can be grown by CVD, then harvested and deposited on a different substrate where electrical connections are defined (Hu et al. 2007).

A top-down process based on silicon-on-insulator (SOI) technology is a viable alternative to overcome the difficulty related to the nanowire manipulation. To some extent, this approach represents the extreme forefront of thin-body, silicon-on-insulator field-effect transistor (SOI-FET) technology scaled down to a few nm. Selective etching of a few-nm-wide line on an SOI substrate ends up in the creation of Si nanowires that, after deposition of gate oxide and metal lines, lead to ultrascaled silicon-on-insulator nanowire field-effect transistors (SOI-NW-FET) featuring a quantum dot or a single impurity atom as a channel (Voisin et al. 2014).

A hole spin qubit has been observed in a silicon p-transistor fabricated by a microelectronics technology based on 300 mm SOI wafers as shown in Figure 29.4 (Voisin et al. 2016; Maurand et al. 2016). The channel of the transistor is derived from silicon nanowire field-effect transistors. It consists of a 10-nm-thick and 20-nm-wide undoped silicon channel with p-doped source and drain contact regions, and two 30-nm-wide parallel top gates, covered by insulating silicon nitride spacers at the sides. The holes are accumulated below the two gates and a spin-blockade condition is achieved by exploiting an unpaired hole spin in

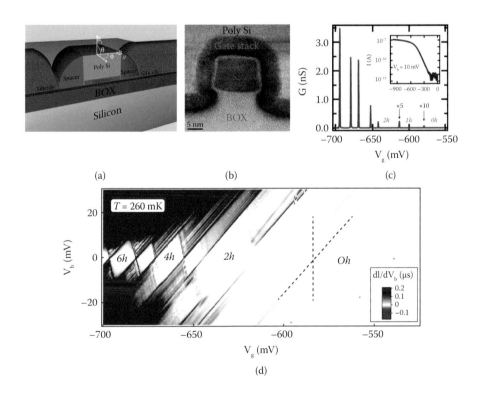

(a)   (b)   (c)

(d)

Figure 29.4 (a) Simplified schematics of a CMOS nanowire transistor. A quantum dot is formed below the gate inside a thin SOI channel (red). The dot is isolated from the silicided source and drain (green) by the presence of spacers. (b) False colored transmission electron microscope (TEM) micrograph of a nanowire transistor section. The silicon channel (in red) at the top of a SOI wafer with a 145nm buried oxide (BOX) layer (blue) is surrounded by a gate stack of SiO_2 and high-k dielectric (in yellow). The gate (in gray) is realized in polysilicon. (c) Measured differential conductance taken at zero bias as a function of gate voltage (V_g) at a temperature of $T = 4$ K showing Coulomb oscillations. Inset shows the transistor characteristic, namely current versus V_g at room temperature. (d) Stability diagram dI/dV vs. (V_b, V_g) performed at 260 mK showing the addition of the first holes in the quantum dot. (Picture taken from Voisin, B., et al., *Nano Lett.*, 16(1), 88–92, 2016. With permission.)

one of the two quantum dots, so Rabi oscillations have been observed. A set of modules for controlling physical qubits, carrying qubits across a microchip and including T-shaped connection has been proposed (Rotta et al. 2014).

29.2.4 DONORS

Shallow donor impurities such as phosphorus and arsenic in silicon at cryogenic temperatures behave as hydrogen-like atoms embedded in the Si crystal lattice that can be modeled in effective mass approximation. Besides four valence electrons that construct bonds with first-neighbor Si atoms, one excess electron is weakly bound to the impurity, with ionization energy of about 45 meV, and sets a favorable background for the implementation of qubits based on electron or nuclear spin qubits (Kane 1998; Pla et al. 2012). Indeed, bulk donors are ideally isolated from material interfaces and crystal defects that may introduce charge noise. Moreover, natural Si can be purified up to an almost spin-less ^{28}Si crystal with about 100 ppm of ^{29}Si. Hyperfine interaction, which is one of the main sources of spin decoherence, are thus significantly reduced, leading to coherence times of the order of seconds (Tyryshkin et al. 2003; Tyryshkin et al. 2012; Muhonen et al. 2014).

Given the nanometer scale of the impurity orbitals, nanometer scale precision for the impurity placement is needed to achieve full control of the qubit, which also becomes a serious scalability issue in view of multiqubit devices (Hollenberg et al. 2006; Copsey and Oskin 2003). Figure 29.5 shows the deterministic doping technique, which allows the controlled placement of single atoms with precision of the order of 10 nm through low current and low energy implantation processes monitored by dedicated electronics capable of detecting single implantation events (van Donkelaar et al. 2015; Prati et al. 2012). A more precise, though slower, technique exploits high vacuum scanning tunneling microscopy (STM) lithography to manipulate an atomically smooth surface. Using an STM tip, it is thus possible to move individual atoms and construct single atom devices or very small quantum dots with atomic resolution as shown in Figure 29.6 (Ruess et al. 2004; Mahapatra et al. 2011).

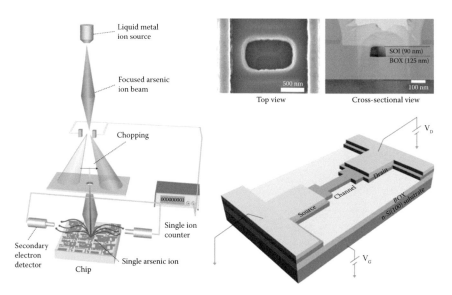

Figure 29.5 Left: Schematic illustration of the deterministic doping setup, which includes the metal ion beam source and focusing system, a beam shutter, and the control setup for the detection of single ion implantation inside nanodevices. Top right: A top view of the channel of a device similar to those measured. The channel is visible through the aperture required for single ion implantation. The 90 nm Si channel on the 125 nm BOX of the SOI substrate is shown in the cross-sectional view. Bottom right: A schematic of the measurement setup for quantum transport electrical characterization. (Picture adapted from Prati, E., et al., *Nat. Nanotechnol.*, 7(7), 443–447, 2012. With permission.)

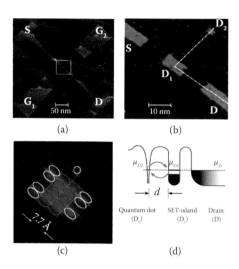

Figure 29.6 (a) Filled state STM image of a single electron device fabricated via atomic precision STM lithography process on Si. (b) High resolution detail of (a) with highlight of source S, drain D and two quantum dots D_1 and D_2 which are defined by roughly 4 and 120 P atoms, respectively. The D_2 area overlaid by a grid with Si (001) dimer row spacing. (c) The potential sites for P incorporation are chains of at least three adjacent Si dimers, shown by the green ellipses. (d) Schematic representation of the potential landscape along the white dotted line in (b), showing the quantum dot D_2, the single electron transistor (SET) island D_1 used for charge sensing, and the drain terminal D. (Picture taken from Mahapatra, S., et al., *Nano Lett.*, 11(10), 4376–4381, 2011. With permission.)

29.3 PROPERTIES

This section presents the different architectures for quantum computation in silicon.

The different systems are compared in terms of the relevant physical parameters, such as the definition of the logical basis and the interactions with the environment that impact their coherence properties and operation.

29.3.1 SPIN QUBITS

The qubit logical basis states correspond to a single electron spin being oriented either up or down along the Z direction (Kane 1998; Loss and DiVincenzo 1998; Kloeffel and Loss 2013; Morello et al. 2009). The two-spin energy eigenstates are separated by the presence of an external magnetic field applied parallel to the Z direction as well (Zeeman splitting), which also provides Z rotations through the bare time evolution of such spin system. X rotation—spin-flip gate—is performed by means of electromagnetic coupling with microwave pulses at the resonance frequency, analogously to electron spin resonance experiments. A microwave field is usually introduced by placing the sample in a macroscopic resonating cavity (Kane 1998; Laucht et al. 2015) or through the use of qubit-specific resonating microstrips adjacent to the qubit (Morello et al. 2009; Pla et al. 2012). Initialization and readout are carried out by means of spin-charge conversion (Kouwenhoven and Marcus 1997; Kouwenhoven et al. 2001; van der Wiel et al. 2002; Hanson et al. 2007), which is based on spin-dependent tunneling time of the electron between the dot and a reservoir, as explained with more detail in Figure 29.7 (Simmons et al. 2009; Huebl et al. 2010; Morello et al. 2010; Mahapatra et al. 2011). It is one of the most studied solid-state qubits, encompassing the different implementations based on an electron spin bound to a quantum dot (Koppens et al. 2006; Kawakami et al. 2014; Veldhorst et al. 2014) or to an impurity atom (Morello et al. 2010; Pla et al. 2012) and a similar architecture based on single nuclear spin featuring coherence times of seconds (Muhonen et al. 2014; Pla et al. 2014).

29.3.2 SINGLET–TRIPLET QUBITS

Singlet–triplet qubits are based on the ground state (the singlet) and an excited state (a triplet) formed by a pair of either two excess electrons or holes confined in a double quantum dot (see Figure 29.8)

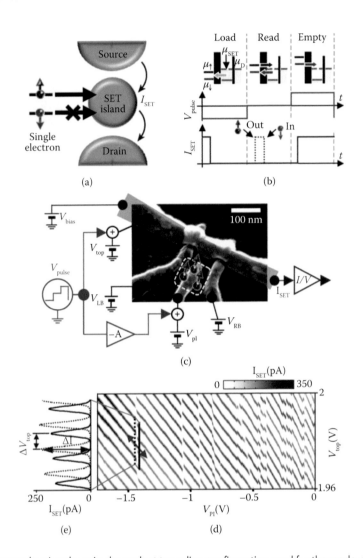

Figure 29.7 (a) Diagram showing the spin-dependent tunneling configuration used for the readout of donor spin qubits, where a single electron can tunnel from the donor impurity onto the island of an SET only when in a spin-up state. (b) Pulsing sequence for single-shot spin readout, and corresponding SET response current I_{SET}. The dashed peak in I_{SET} is the expected signal from a spin-up electron tunneling out the donor and immediately substituted by a spin-down electron tunneling in. The diagrams at the top depict the electrochemical potentials of the electron site ($\mu_{\uparrow\downarrow}$), of the SET island (μ_{SET}) and of the drain contact (μ_D). (c) Scanning electron micrograph of a donor-based spin qubit. The area where the P donors are implanted is marked by the dashed square. Both DC voltages and pulses are applied to the gates as indicated. The red shaded area represents the electron layer induced by the top gate and confined beneath the SiO_2 gate oxide layer. (d) SET current I_{SET} as a function of the voltages on the top and the plunger gates, V_{top} and V_{pl}, at B = 0. The lines of SET Coulomb peaks are broken by charge transfer events. (e) Line traces of I_{SET} along the solid and dashed lines in (d). Ionizing the donor shifts the sequence of SET current peaks by an amount $\Delta V_{top} = \Delta q/C_{top}$, causing a change ΔI in the SET current. (Reprinted by permission from Macmillan Publishers Ltd. *Nature*, Morello, A., et al., Single-shot readout of an electron spin in silicon, 467, 687–691, copyright 2010.)

(Petta et al. 2005; Maune et al. 2012; Shulman et al. 2012). From the four two-spin states arising, the $S_z = 0$ subspace is selected to define the qubit logic basis:

$$|0\rangle = \frac{1}{\sqrt{2}}\left(|\uparrow\rangle - |\downarrow\rangle\right) \text{ and } |1\rangle = \frac{1}{\sqrt{2}}\left(|\uparrow\rangle + |\downarrow\rangle\right) \qquad (29.9)$$

which gives the qubit name, as they correspond to the spin singlet (S = 0) state and to one of the three triplet (S = 1) states. The degeneration with the remaining triplet states with $S_z = \pm 1$ is removed by an

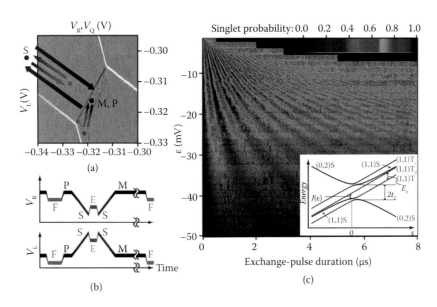

Figure 29.8 (a) Charge stability diagram of a double quantum dot, reporting the differential transconductance of the charge-sensing quantum point contact as a function of the control gate voltages V_L, V_R, and V_Q. The red and green arrows superimposed indicate the pulse sequence to obtain Rabi oscillations. (b) Schematic representation of the same pulse sequence for control gate voltages V_R and V_L as functions of time. Gate levels F, P, S, E, and M denote the corresponding positions in the stability diagram in (a). (c) Rabi oscillations of the singlet probability as a function of detuning ε and exchange-pulse duration (step E in the pulse sequence). Inset: energy diagram at the (0,2)–(1,1) anticrossing, showing energies of the qubit states (1,1)S and (1,1)T_0 and the Zeeman split (1,1)$T_\pm$ states as functions of detuning, ε. In the diagram $J(\varepsilon)$ is the exchange-energy splitting between qubit states, E_Z is the Zeeman splitting between triplet states, and tc is the tunnel coupling. (Picture adapted from Maune, B.M., et al., *Nature*, 481(7381), 344–347, 2012. With permission.)

external magnetic field. Since the electrons are arranged in two quantum dots, electrical control of the chemical potential of the dots and of the interdot tunnel coupling, respectively, allows a tuning of the exchange interaction and to control the qubit state time evolution (Shi et al. 2011; Stepanenko et al. 2012). The exchange interaction drives Z rotations, whereas a magnetic field gradient between the two dots is usually introduced by a neighboring permanent magnet to prompt X rotations. Finally, qubit readout is based on Pauli blockade and obtained by means of charge-sensing technique when the system is configured with either both electrons bound to one quantum dot for the singlet, or one electron in each quantum dot for the triplet (Johnson et al. 2005; Lai et al. 2011; Nguyen et al. 2013; Weber et al. 2014).

29.3.3 CHARGE QUBITS

In a charge qubit the two-qubit states are defined in a double quantum dot by a single electron charge occupying either the first or the second dot (Hayashi et al. 2003; Petersson et al. 2010; Kim et al. 2015; Cao et al. 2013). Quantum dot potential wells and the interdot tunneling barrier are electrostatically defined by means of metal gates. Coherent qubit rotations can be achieved by controlling the bias between the dots and interdot tunnel coupling as described in Figure 29.9. Qubit dynamics is extremely fast, leading to the potential for 10 GHz frequency operation and to the concurrent drawback of very fast dephasing (ns timescale) induced by electrical noise.

29.3.4 EXCHANGE-ONLY QUBITS

Exchange-only qubits are also known as hybrid qubits for their hybrid features between spin and charge qubits. Such qubits employ three electrons bound to either a double or a triple quantum dot. The qubit is defined within the spin subspace with total spin $S = \frac{1}{2}$ and $S_z = -\frac{1}{2}$ and corresponding eigenstates:

$$|0\rangle = |S\rangle|\downarrow\rangle \text{ and } |1\rangle = \frac{1}{\sqrt{3}}|T_0\rangle|\downarrow\rangle - \frac{1}{\sqrt{3}}|T_-\rangle|\uparrow\rangle \tag{29.10}$$

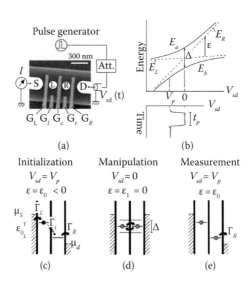

Figure 29.9 (a) Schematic measurement circuit combined with scanning electron microscope image of a double quantum dot employed as charge qubit. Etching (upper and lower dark regions) and negatively biased gate electrodes (G_L, G_l, G_C, G_r, and G_R) define a double quantum dot (L and R) between the source (S) and drain (D). (b) Energy levels at small detuning ε of the bonding (E_b) and antibonding (E_a) states, which are the logical basis eigenstates during the manipulation, and localized states (E_L and E_R) during initialization. A typical pulsed voltage $V_{sd}(t)$ is shown at the bottom. (c–e) Energy diagrams of the double quantum dot for $\varepsilon_1 = 0$ during (c) initialization, (d) coherent oscillation, and (e) measurement process. (From Hayashi, T., et al., *Phys. Rev. Lett.*, 91(22), 226804, 2003. With permission.)

Here the qubit logic states are conveniently defined by means of double kets, representing the vector product of the spin state of the first two electrons

$$| S \rangle = \frac{1}{\sqrt{2}}\left(| \uparrow \rangle - | \downarrow \rangle\right) \quad | T_0 \rangle = \frac{1}{\sqrt{2}}\left(| \uparrow \rangle + | \downarrow \rangle\right) \quad | T_- \rangle = | \downarrow\downarrow \rangle \tag{29.11}$$

and of the third one, respectively.

In fact, although in the earliest implementation the three spins are arranged each in a distinct quantum dot (DiVincenzo et al. 2000), a more compact version, shown in Figure 29.10, was conceived later where two of them are bound to the same quantum dot (Shi et al. 2012) for which effective parametrization has been obtained (Ferraro et al. 2014; Ferraro et al. 2015b) and universal set of gates exists (De Michielis et al. 2015) as well as coherent adiabatic passage (Ferraro et al. 2015a).

In both the configurations, the total spin is conserved and manipulation is driven by exchange interaction of the electron spins, which is in turn finely controlled from the external by modulating the interdot tunnel coupling via a pulsed gate technique (Laird et al. 2010; Medford et al. 2013; Shi et al. 2014; Kim et al. 2014). This method is faster (rotation frequency of the order of GHz) than single spin manipulation (which has a μs scale dynamics) and all-electrical (no magnetic field is required to rotate the qubit) due to its charge-like behavior. On the other way around, such qubit has better prospects in terms of coherence compared to bare charge qubits (Koh et al. 2013) due to the inherent spin character that is fundamental to the qubit logic basis definition.

29.4 PROSPECTS

Silicon qubits have the additional advantage of a good compatibility with its own classical control electronics, which is mainly based on CMOS technology as well. Cryogenic control electronics appears to be required to manipulate arrays of qubits for both minimizing noise and increasing bandwidth. Dedicated silicon cryogenic amplifiers (Prati and Shinada 2013; Tagliaferri et al. 2016) and commercial

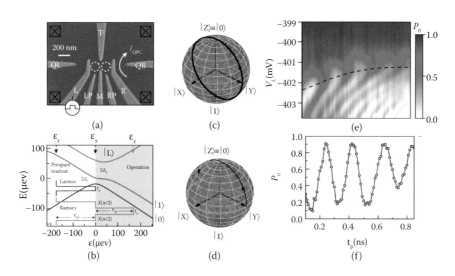

Figure 29.10 (a) Scanning electron microscope image of an exchange-only qubit device electrically defined by metal top gates in a Si/SiGe heterostructure. Operating point is set by applying voltages to the metal gates. Readout is achieved by monitoring the current through the quantum point contact I_{QPC}. (b) Diagram of the calculated energy levels E versus detuning ε. The ground state and first excited state correspond to the logical |0> and |1> state, respectively. (c, d) Evolution of the qubit state during X (e) and Z (d) rotations. (e, f) Rabi oscillations at variable V_L voltages. (Picture adapted from Kim, D., et al., *Nature*, 511(7507), 70–74, 2014. With permission.)

field-programmable gate array systems (Conway Lamb et al. 2016) have been reported to operate at temperatures between 1.6 K and 4.2 K. Such additional degree of integration with control electronics needed to assess fault-tolerant error correction makes silicon increasingly interesting for carrying quantum information processing.

REFERENCES

Angus, S. J. et al. 2007. Gate-defined quantum dots in intrinsic silicon. *Nano Letters*, 7(7), 2051–2055.

Awschalom, D. D. et al. 2013. Quantum spintronics: Engineering and manipulating atom-like spins in semiconductors. *Science (New York, NY)*, 339(6124), 1174–1179.

Barends, R. et al. 2016. Digitized adiabatic quantum computing with a superconducting circuit. *Nature*, 534(7606), 222–226. doi: 10.1038/nature17658.

Buluta, I., Ashhab, S. & Nori, F. 2011. Natural and artificial atoms for quantum computation. *Reports on Progress in Physics*, 74(10), 104401.

Cao, G. et al. 2013. Ultrafast universal quantum control of a quantum-dot charge qubit using Landau-Zener-Stückelberg interference. *Nature Communications*, 4, 1401.

Conway Lamb, I. D. et al. 2016. An FPGA-based instrumentation platform for use at deep cryogenic temperatures. *Review of Scientific Instruments*, 87(1), 14701. doi: 10.1063/1.4939094.

Copsey, D. & Oskin, M. 2003. Toward a scalable, silicon-based quantum computing architecture. *IEEE Journal of Selected Topics in Quantum Electronics*, 9(6), 1552–1569.

De Michielis, M. et al. 2015. Universal set of quantum gates for double-dot exchange-only spin qubits with intradot coupling. *Journal of Physics A: Mathematical and Theoretical*, 48(6), 65304.

Devitt, S. J., Munro, W. J. & Nemoto, K. 2013. Quantum error correction for beginners. *Reports on progress in physics. Physical Society (Great Britain)*, 76(7), 76001.

Devoret, M. H. & Schoelkopf, R. J. 2013. Superconducting circuits for quantum information: An outlook. *Science (New York, NY)*, 339(6124), 1169–1174.

DiVincenzo, D. P. 1995. Quantum computation. *Science*, 270(5234), 255–261. doi: 10.1126/science.270.5234.255.

DiVincenzo, D. P. 2000. The physical implementation of quantum computation. Available at: http://arxiv.org/abs/quant-ph/0002077 [Accessed July 29, 2014].

DiVincenzo, D. P. et al. 2000. Universal quantum computation with the exchange interaction. *Nature*, 408(6810), 339–342.

Ferraro, E. et al. 2014. Effective Hamiltonian for the hybrid double quantum dot qubit. *Quantum Information Processing*, 13(5), 1155–1173. doi: 10.1007/s11128-013-0718-2.

Ferraro, E. et al. 2015a. Coherent tunneling by adiabatic passage of an exchange-only spin qubit in a double quantum dot chain. *Physical Review B*, 91(7), 75435. doi: 10.1103/PhysRevB.91.075435.

Ferraro, E. et al. 2015b. Effective Hamiltonian for two interacting double-dot exchange-only qubits and their controlled-NOT operations. *Quantum Information Processing*, 14(1), 47–65. doi: 10.1007/s11128-014-0864-1.

Gottesman, D. 2009. An introduction to quantum error correction and fault-tolerant quantum computation, 0, 1–46. Available at: http://arxiv.org/abs/0904.2557 (accessed April 8, 2017).

Grace, M. et al. 2006. Encoding a qubit into multilevel subspaces. *New Journal of Physics*, 8(3), 35–35

Hahn, E. L. 1950. Spin echoes. *Physical Review*, 80(4), 580–594.doi: 10.1103/PhysRev.80.580.

Hanson, R. et al. 2007. Spins in few-electron quantum dots. *Reviews of Modern Physics*, 79(4), 1217–1265. doi: 10.1103/RevModPhys.79.1217.

Hao, X. et al. 2014. Electron spin resonance and spin-valley physics in a silicon double quantum dot. *Nature Communications*, 5, 3860.

Hayashi, T. et al. 2003. Coherent manipulation of electronic states in a double quantum dot. *Physical Review Letters*, 91(22), 226804. doi: 10.1103/PhysRevLett.91.226804.

Hollenberg, L. et al. 2006. Two-dimensional architectures for donor-based quantum computing. *Physical Review B*, 74(4), 45311. doi: 10.1103/PhysRevB.74.045311.

Hu, Y. et al. 2007. A Ge/Si heterostructure nanowire-based double quantum dot with integrated charge sensor. *Nature Nanotechnology*, 2(10), 622–625.

Huebl, H. et al. 2010. Electron tunnel rates in a donor-silicon single electron transistor hybrid. *Physical Review B*, 81(23), 235318. doi: 10.1103/PhysRevB.81.235318.

Johnson, A. C., Petta, J. R. & Marcus, C. M. 2005. Singlet-triplet spin blockade and charge sensing in a few-electron double quantum dot. *Physical Review B*, 72(16), 165308. doi: 10.1103/PhysRevB.72.165308

Jones, N. C. et al. 2012. Layered architecture for quantum computing. *Physical Review X*, 2(3), 31007. doi: 10.1103/PhysRevX.2.031007.

Kane, B. 1998. A silicon-based nuclear spin quantum computer. *Nature*, 133–137. Available at: http://www.nature.com/nature/journal/v393/n6681/abs/393133a0.html [Accessed June 3, 2014].

Kawakami, E. et al. 2014. Electrical control of a long-lived spin qubit in a Si/SiGe quantum dot. *Nature Nanotechnology*, 9(9), 666–670.

Kim, D. et al. 2014. Quantum control and process tomography of a semiconductor quantum dot hybrid qubit. *Nature*, 511(7507), 70–74.

Kim, D. et al. 2015. Microwave-driven coherent operation of a semiconductor quantum dot charge qubit. *Nature Nanotechnology*, 10(3), 243–247. doi: 10.1038/nnano.2014.336.

Kloeffel, C. & Loss, D. 2013. Prospects for spin-based quantum computing in quantum dots. *Annual Review of Condensed Matter Physics*, 4(1), 51–81. doi: 10.1146/annurev-conmatphys-030212-184248.

Koh, T. S., Coppersmith, S. N. & Friesen, M. 2013. High-fidelity gates in quantum dot spin qubits. *Proceedings of the National Academy of Sciences of the United States of America*, 110(49), 19695–19700.

Koppens, F. H. L. et al. 2006. Driven coherent oscillations of a single electron spin in a quantum dot. *Nature*, 442(7104), 766–771.

Kouwenhoven, L. P., Austing, D. G. & Tarucha, S. 2001. Few-electron quantum dots. *Reports on Progress in Physics*, 64(6), 701–736.

Kouwenhoven, L. P. & Marcus, C. 1997. Electron transport in quantum dots. *Mesoscopic Electron Transport*, 345(Kluwer), 105–214. doi: 10.1007/978-94-015-8839-3_4.

Ladd, T. D. et al. 2010. Quantum computers. *Nature*, 464(7285), 45–53.

Lai, N. S. et al. 2011. Pauli spin blockade in a highly tunable silicon double quantum dot. *Scientific Reports*, 1, 110.

Laird, E. A. et al. 2010. Coherent spin manipulation in an exchange-only qubit. *Physical Review B*, 82(7), 75403. doi: 10.1103/PhysRevB.82.075403.

Laucht, A. et al. 2015. Electrically controlling single-spin qubits in a continuous microwave field. *Science Advances*, 1(3), e1500022.

Lim, W. H. et al. 2009. Observation of the single-electron regime in a highly tunable silicon quantum dot. *Applied Physics Letters*, 95(24), 242102.

Lim, W. H. et al. 2011. Spin filling of valley-orbit states in a silicon quantum dot. *Nanotechnology*, 22(33), 335704.

Loss, D. & DiVincenzo, D. P. 1998. Quantum computation with quantum dots. *Physical Review A*, 57(1), 120–126. doi: 10.1103/PhysRevA.57.120.

Mahapatra, S., Büch, H. & Simmons, M. Y. 2011. Charge sensing of precisely positioned p donors in Si. *Nano Letters*, 11(10), 4376–4381.

Maune, B. M. et al. 2012. Coherent singlet-triplet oscillations in a silicon-based double quantum dot. *Nature*, 481(7381), 344–347.

Maurand, R., Jehl, X., Kotekar-Patil, D., Corna, A., Bohuslavskyi, H., Laviéville, R., De Franceschi, S. 2016. A CMOS silicon spin qubit. *Nature Communications*, 7, 13575. http://doi.org/10.1038/ncomms13575.

Medford, J. et al. 2013. Self-consistent measurement and state tomography of an exchange-only spin qubit. *Nature Nanotechnology*, 8(9), 654–659.

Monroe, C. & Kim, J. 2013. Scaling the ion trap quantum processor. *Science (New York, NY)*, 339(6124), 1164–1169.

Morello, A. et al. 2009. Architecture for high-sensitivity single-shot readout and control of the electron spin of individual donors in silicon. *Physical Review B*, 80(8), 81307. doi: 10.1103/PhysRevB.80.081307

Morello, A. et al. 2010. Single-shot readout of an electron spin in silicon. *Nature*, 467(7316), 687–691.

Morita, S. & Nishimori, H. 2008. Mathematical foundation of quantum annealing. *Journal of Mathematical Physics*, 49(12), 125210. doi: 10.1063/1.2995837.

Morton, J. J. L. et al. 2011. Embracing the quantum limit in silicon computing. *Nature*, 479(7373), 345–353.

Muhonen, J. T. et al. 2014. Storing quantum information for 30 seconds in a nanoelectronic device. *Nature Nanotechnology*, 9(12), 986–991.

Nguyen, K. T. et al. 2013. Charge sensed Pauli blockade in a metal-oxide-semiconductor lateral double quantum dot. *Nano Letters*, 13(12), 5785–5790.

Nordberg, E. P. et al. 2009b. Charge sensing in enhancement mode double-top-gated metal-oxide-semiconductor quantum dots. *Applied Physics Letters*, 95(20), 202102. doi: 10.1063/1.3259416.

Nordberg, E. P. et al. 2009a. Enhancement-mode double-top-gated metal-oxide-semiconductor nanostructures with tunable lateral geometry. *Physical Review B*, 80(11), 115331. doi: 10.1103/PhysRevB.80.115331

O'Brien, J. L. 2007. Optical quantum computing. *Science (New York, NY)*, 318(5856), 1567–1570.

Oliveira, R. & Terhal, B.M. 2005. The complexity of quantum spin systems on a two-dimensional square lattice. *Quantum Information and Computation*, 8(10), 900–924.

Petersson, K. D. et al. 2010. Quantum coherence in a one-electron semiconductor charge qubit. *Physical Review Letters*, 105(24), 246804. doi: 10.1103/PhysRevLett.105.246804.

Petta, J. R. et al. 2005. Coherent manipulation of coupled electron spins in semiconductor quantum dots. *Science (New York, NY)*, 309(5744), 2180–2184.

Pla, J. J. et al. 2014. Coherent control of a single Si 29 nuclear spin qubit. *Physical Review Letters*, 113(24), 1–5.

Pla, J. J. et al. 2012. A single-atom electron spin qubit in silicon. *Nature*, 489(7417), 541–545.

Prati, E. & Shinada, T. 2013. *Single-atom nanoelectronics*. Pan Stanford Publishing, Singapore. p. 200.

Prati, E. et al. 2012. Anderson-Mott transition in arrays of a few dopant atoms in a silicon transistor. *Nature Nanotechnology*, 7(7), 443–447.

Preskill, J. 1998. Reliable quantum computers. *Proceedings of the Royal Society A: Mathematical, Physical and Engineering Sciences*, 454(1969), 385–410.

Rotta, D., De Michielis, M., Ferraro, E., Fanciulli, M., and Prati, E. 2016. Maximum density of quantum information in a scalable CMOS implementation of the hybrid qubit architecture. *Quantum Information Processing*, 15(6), 2253–2274.

Ruess, F. J. et al. 2004. Toward atomic-scale device fabrication in silicon using scanning probe microscopy. *Nano Letters*, 4(10), 1969–1973. doi: 10.1021/nl048808v.

Shi, Z. et al. 2011. Tunable singlet-triplet splitting in a few-electron Si/SiGe quantum dot. *Applied Physics Letters*, 99(23), 1–4.

Shi, Z. et al. 2012. Fast hybrid silicon double-quantum-dot qubit. *Physical Review Letters*, 108(14), 140503. doi: 10.1103/PhysRevLett.108.140503.

Shi, Z. et al. 2014. Fast coherent manipulation of three-electron states in a double quantum dot. *Nature Communications*, 5, 3020.

Shulman, M. D. et al. 2012. Demonstration of entanglement of electrostatically coupled singlet-triplet qubits. *Science (New York, NY)*, 336(6078), 202–205.

Simmons, C. B. et al. 2009. Charge sensing and controllable tunnel coupling in a Si/SiGe double quantum dot. *Nano Letters*, 9(9), 3234–3238.

Stepanenko, D. et al. 2012. Singlet-triplet splitting in double quantum dots due to spin-orbit and hyperfine interactions. *Physical Review B*, 85(7), 75416. doi: 10.1103/PhysRevB.85.075416.

Tagliaferri, M. L. V. et al. 2016. Modular printed circuit boards for broadband characterization of nanoelectronic quantum devices. *IEEE Transactions on Instrumentation and Measurement*, 65(8), 1827–1835.

Taylor, J. M. et al. 2005. Fault-tolerant architecture for quantum computation using electrically controlled semiconductor spins. *Nature Physics*, 1(3), 177–183. doi: 10.1038/nphys174

Tyryshkin, A. et al. 2003. Electron spin relaxation times of phosphorus donors in silicon. *Physical Review B*, 68(19), 193207. doi: 10.1103/PhysRevB.68.193207.

Tyryshkin, A. M. et al. 2012. Electron spin coherence exceeding seconds in high-purity silicon. *Nature Materials*, 11(2), 143–147.

van der Wiel, W. et al. 2002. Electron transport through double quantum dots. *Reviews of Modern Physics*, 75(1), 1–22.

van Donkelaar, J. et al. 2015. Single atom devices by ion implantation. *Journal of Physics: Condensed Matter*, 27(15), 154204. doi: 10.1088/0953-8984/27/15/154204.

Veldhorst, M. et al. 2014. An addressable quantum dot qubit with fault-tolerant control-fidelity. *Nature Nanotechnology*, 9(12), 981–985.

Viola, L., Knill, E. & Lloyd, S. 1999. Dynamical decoupling of open quantum systems. *Physical Review Letters*, 82(12), 2417–2421. doi: 10.1103/PhysRevLett.82.2417.

Voisin, B. et al. 2014. Few-electron edge-state quantum dots in a silicon nanowire field-effect transistor. *Nano Letters*, 14(4), 2094–2098.

Voisin, B. et al. 2016. Electrical control of g-factor in a few-hole silicon nanowire MOSFET. *Nano Letters*, 16(1), 88–92. doi: 10.1021/acs.nanolett.5b02920.

Wang, C., Chen, H. & Jonckheere, E. 2016. Quantum versus simulated annealing in wireless interference network optimization. *Scientific Reports*, 6, 25797.

Weber, B. et al. 2014. Spin blockade and exchange in Coulomb-confined silicon double quantum dots. *Nature Nanotechnology*, (April), 2–7. Available at: http://www.ncbi.nlm.nih.gov/pubmed/24727686 [Accessed May 27, 2014].

Yang, C. H. et al. 2012. Orbital and valley state spectra of a few-electron silicon quantum dot. *Physical Review B*, 86(11), 115319. doi: 10.1103/PhysRevB.86.115319.

Yang, C. H. et al. 2013. Spin-valley lifetimes in a silicon quantum dot with tunable valley splitting. *Nature Communications*, 4(c), 2069.

Index